Advances in Plant Biology

Volume 5

Series Editor

John J. Harada
Davis, USA

Advances in Plant Biology provides summaries and updates of topical areas of plant biology. This series focuses largely on mechanisms that underlie the growth, development, and response of plants to their environment. Each volume contains primarily on information at the molecular, cellular, biochemical, genetic and genomic level, although they will focused on information obtained using other approaches.

More information about this series at http://www.springer.com/series/8047

Steven M. Theg • Francis-André Wollman
Editors

Plastid Biology

Springer

Editors
Steven M. Theg
Department of Plant Biology
Univeristy of California-Davis
Davis
California
USA

Francis-André Wollman
Physiologie Membranaire et Moléculaire du Chloroplaste
Institut de Biologie Physico-Chimique
Paris
France

ISBN 978-1-4939-1135-6 ISBN 978-1-4939-1136-3 (eBook)
DOI 10.1007/978-1-4939-1136-3
Springer New York Heidelberg Dordrecht London

Library of Congress Control Number: 2014947238

Printed on acid-free paper

Springer is part of Springer Science+Business Media (www.springer.com)

Preface

Photosynthesis is the process through which the energy inherent in sunlight is captured in the chemical bonds of reduced carbon compounds, thereby providing the food upon which almost all life depends. In addition, the production of oxygen as a result of the utilization of water as the ultimate electron donor to the photosynthetic electron transport chain has transformed our atmosphere, allowing for the emergence of oxygenic respiration, without which there would be no human life on Earth.

Photosynthesis is carried out in plants and algae in chloroplasts. Given their central role in energy transduction in the biosphere, chloroplasts have been the focus of attention for generations of scientists. This volume brings together many aspects of modern research into plastids relating to their biogenesis, functioning in photosynthesis and utility for biotechnology.

Plastids had their origins in free living photosynthetic bacteria and took up residence in the primitive eukaryotic cells through endosymbiosis. While they have lost most of their DNA to the nucleus, they retain a functioning genome and are capable of a limited but critical amount of semi-autonomous protein synthesis. Accordingly, we start this volume with a series of three chapters devoted to the handling of the genetic information contained within the plastid genome and crosstalk between the chloroplast and nucleus as the information encoded in both locations is decoded. Following this are five chapters that examine the biogenesis and differentiation of the plastid itself and the sub-structures found at the plastid surface and within the internal thylakoid system. Also included here is a treatment of the unusual non-photosynthetic plastids found within the Apicoplexa, a group of parasitic protists responsible for a number of important human diseases.

Despite having their own genomes, the vast majority of plastid proteins are synthesized in the cytosol and taken up into and subsequently distributed within the organelle. The next six chapters of the volume describe these processes, as well as the roles of molecular chaperones and proteases in protein homeostasis. This is followed by three chapters dedicated to critical aspects of chloroplast physiology relating to dissipation of excess light energy, control of electron transport and ion homeostasis. Finally, the book ends with two chapters discussing the emerging roles of plastids in biotechnology, one as a platform for synthesis of useful proteins, made desirable because of the superior containment of transgenes within this organelle than when inserted in nuclear genomes, and the other as a source of hydrogen production to be used as biofuel.

Each of the chapters has been written by leading authorities in their respective research areas. Many chapters are the result of collaborations between experts in different laboratories, giving a broader than usual perspective on a given topic. In each case, readers will find well-crafted chapters containing information and insights for both novices and experts alike.

We are grateful to our many friends and scholars who contributed these outstanding chapters. The breadth of their knowledge and clarity of their writing have made for a unique and readable volume bringing together many disparate but interconnected topics relating to plastid biology. We are also indebted to those at Springer, especially Kenneth Teng and Brian Halm, who oversaw this project in its final stages of production.

Davis, CA, USA Steven M. Theg
Paris, France Francis-André Wollman

Contents

Part III Biogenesis of Chloroplast Proteins

Part IV Chloroplast Photophysiology

Part V Chloroplast Biotechnology

Contributors

Zach Adam The Robert H. Smith Institute of Plant Sciences and Genetics in Agriculture, The Hebrew University, Rehovot, Israel

Veronica Albrecht-Borth Australian National University, Canberra, Australia

Roberto Bassi Dipartimento di Biotecnologie, Università di Verona, Verona, Italy

María Bernal Plant Nutrition Department, Estación Experimental De Aula Dei, Consejo Superior de Investigaciones Científicas (CSIC), Zaragoza, Spain

Department of Plant Physiology, Ruhr University Bochum, Bochum, Germany

Alexandra-Viola Bohne Molekulare Pflanzenwissenschaften, Biozentrum LMU München, Planegg-Martinsried, Germany

Kathrin Bolte Laboratory for Cell Biology, Philipps University of Marburg, Marburg, Germany

Thomas Börner Institute of Biology, Humboldt University Berlin, Berlin, Germany

Jeffrey L. Caplan Department of Plant and Soil Sciences, Delaware Biotechnology Institute, University of Delaware, Newark, DE, USA

Pierre Cardol Laboratoire de Génétique des Microorganismes, Institut de Botanique, Université de Liège, Liège, Belgium

Carole Dabney-Smith Department of Chemistry and Biochemistry, Miami University, Oxford, OH, USA

Luca Dall'Osto Dipartimento di Biotecnologie, Università di Verona, Verona, Italy

Savithramma P. Dinesh-Kumar Department of Plant Biology and The Genome Center, College of Biological Sciences, University of California, Davis, CA, USA

Manuel Dubald Bayer CropScience, Morrisville, NC, USA

Giovanni Finazzi Centre National Recherche Scientifique, Unité Mixte Recherche 5168, Laboratoire Physiologie Cellulaire et Végétale, Grenoble, France

Commissariat à l'Energie Atomique et Energies Alternatives, l'Institut de Recherches en Technologies et Sciences pour le Vivant, Grenoble, France

Université Grenoble Alpes, Grenoble, France

Institut National Recherche Agronomique, Grenoble, France

Elzbieta Glaser Department of Biochemistry and Biophysics, Stockholm University, Stockholm, Sweden

Michel Goldschmidt-Clermont University of Geneva, Geneva 4, Switzerland

Marc Hanikenne Functional Genomics and Plant Molecular Imaging, Center for Protein Engineering (CIP), PhytoSystems, B22, Department of Life Sciences, University of Liège, Liège, Belgium

Thomas Happe Fakultät für Biologie und Biotechnologie, AG Photobiotechnologie, Ruhr-Universität Bochum, Bochum, Germany

Franziska Hempel Laboratory for Cell Biology, Philipps University of Marburg, Marburg, Germany

LOEWE-Zentrum für Synthetische Mikrobiologie (Synmikro), Marburg, Germany

Anja Hemschemeier Fakultät für Biologie und Biotechnologie, AG Photobiotechnologie, Ruhr-Universität Bochum, Bochum, Germany

Kentaro Inoue Department of Plant Sciences, University of California, Davis, CA, USA

Paul Jarvis Department of Plant Sciences, University of Oxford, Oxford, UK

Giles N. Johnson Faculty of Life Sciences, University of Manchester, Manchester, UK

Felix Kessler Laboratoire de Physiologie Végétale, Université de Neuchâtel, Neuchâtel, Switzerland

Helmut Kirchhoff Institute of Biological Chemistry, Washington State University, Pullman, WA, USA

Andreas Klingl LOEWE-Zentrum für Synthetische Mikrobiologie (Synmikro), Marburg, Germany

Philipp Knörzer Fakultät für Biologie und Biotechnologie, AG Photobiotechnologie, Ruhr-Universität Bochum, Bochum, Germany

Amutha Sampath Kumar Department of Plant and Soil Sciences, Delaware Biotechnology Institute, University of Delaware, Newark, DE, USA

Julia Legen Institute of Biology, Humboldt University Berlin, Berlin, Germany

Yonghua Li-Beisson Institut de Biologie Environnementale et Biotechnologie, CEA/CNRS/Aix Marseille Université, Saint-Paul-lez-Durance, France

Uwe-G. Maier Laboratory for Cell Biology, Philipps University of Marburg, Marburg, Germany

LOEWE-Zentrum für Synthetische Mikrobiologie (Synmikro), Marburg, Germany

Jodi Maple-Grødem Centre for Organelle Research, University of Stavanger, Stavanger, Norway

Centre for Movement Disorders, Stavanger University Hospital, Stavanger, Norway

Jun Minagawa National Institute for Basic Biology (NIBB), Myodaiji, Okazaki, Japan

Jörg Nickelsen Molekulare Pflanzenwissenschaften, Biozentrum LMU München, Planegg-Martinsried, Germany

Gilles Peltier Institut de Biologie Environnementale et Biotechnologie, CEA/CNRS/Aix Marseille Université, Saint-Paul-lez-Durance, France

Barry J. Pogson Australian National University, Canberra, Australia

Saul Purton Institute of Structural and Molecular Biology, University College London, London, UK

Silvia Ramundo Departments of Molecular Biology and Plant Biology, University of Geneva, Geneva, Switzerland

Cécile Raynaud Institut de Biologie des Plantes, Paris-Sud University, Orsay, France

Jean-David Rochaix Departments of Molecular Biology and Plant Biology, University of Geneva, Geneva, Switzerland

Alexander Ruban School of Biological and Chemical Sciences, Queen Mary University of London, London, UK

Wataru Sakamoto Institute of Plant Science and Resources, Okayama University, Kurashiki, Okayama, Japan

Christian Schmitz-Linneweber Institute of Biology, Humboldt University Berlin, Berlin, Germany

Michael Schroda Department of Molecular Biotechnology & Systems Biology, TU Kaiserslautern, Kaiserslautern, Germany

Lilach Sheiner Center for Tropical and Emerging Global Diseases & Department of Cellular Biology, University of Georgia, Athens, GA, USA

Amanda Storm Department of Chemistry and Biochemistry, Miami University, Oxford, OH, USA

Boris Striepen Center for Tropical and Emerging Global Diseases & Department of Cellular Biology, University of Georgia, Athens, GA, USA

Ghislaine Tissot-Lecuelle Alganelle, La Motte-Servolex, France

Raphael Trösch Institute of Biology, Humboldt University of Berlin, Berlin, Germany

Eugen-Ioan Urzica Department of Chemistry and Biochemistry, UCLA, Los Angeles, CA, USA

Peter Westhoff Institut für Entwicklungs- und Molekularbiologie der Pflanzen, Heinrich-Heine-Universität, Düsseldorf, Germany

Felix Willmund Department of Molecular Biotechnology & Systems Biology, TU Kaiserslautern, Kaiserslautern, Germany

Stefan Zauner Laboratory for Cell Biology, Philipps University of Marburg, Marburg, Germany

Petya Zhelyazkova Institute of Biology, Humboldt University Berlin, Berlin, Germany

Part I
Genetic Material and its Expression

Chapter 1
Chloroplast Gene Expression—RNA Synthesis and Processing

Thomas Börner, Petya Zhelyazkova, Julia Legen and Christian Schmitz-Linneweber

Abstract Both transcription and transcript processing are more complex in chloroplasts than in bacteria. Plastid genes are transcribed by a plastid-encoded RNA polymerase (PEP) and one (monocots) or two (dicots) nuclear-encoded RNA polymerase(s) (NEP). PEP is a bacterial-type multisubunit enzyme composed of core subunits (coded for by the plastid rpoA, B, C1 and C2 genes) and additional protein factors encoded in the nuclear genome. The nuclear genome of Arabidopsis contains six genes for sigma factors required by PEP for promoter recognition. NEP activity is represented by phage-type RNA polymerases. Factors supporting NEP activity have not been identified yet. NEP and PEP use different promoters. Both types of RNA polymerase are active in proplastids and all stages of chloroplast development. PEP is the dominating transcriptase in chloroplasts.

Chloroplast RNA processing consists of hundreds of mostly independent events. In recent years, much progress has been made in identifying factors behind RNA splicing and RNA editing. Namely, pentatricopeptide repeat (PPR) proteins have come into focus as RNA binding proteins conferring specificity to individual processing events. Also, studies on chloroplast RNases have helped considerably to understand chloroplast RNA turnover. Such mechanistic insights are set in contrast to how little we know about the regulatory role of RNA processing in chloroplasts.

Keywords Chloroplast transcription · Chloroplast RNA polymerase · Chloroplast promoter · Chloroplast RNA processing · Chloroplast RNA-binding proteins · PPR proteins · Chloroplast splicing · Chloroplast editing · Chloroplast RNA degradation · Chloroplast nucleases

Abbreviations

CRS2	Chloroplast RNA splicing 2 protein
IR	Inverted repeat
NEP	Nuclear-encoded plastid RNA polymerase

C. Schmitz-Linneweber (✉) · T. Börner · P. Zhelyazkova · J. Legen
Institute of Biology, Humboldt University Berlin, Chausseestr. 117,
10115 Berlin, Germany
e-mail: smitzlic@rz.hu-berlin.de

S.M. Theg, F.-A. Wollman (eds.), *Plastid Biology,* Advances in Plant Biology 5,
DOI 10.1007/978-1-4939-1136-3_1,

Nt	Nucleotides
PEP	Plastid-encoded plastid RNA polymerase
PPR	Pentatricopeptide repeat
TAC	Transcriptionally active chromosome
TFs	Transcription factors
TPR	Tetratricopeptide repeat
TSSs	Transcription start sites

1.1 Introduction

Chloroplasts, which have their own genomes (plastomes) and specific machineries for gene expression, evolved from a bacterium that was related to the extant cyanobacteria. During evolution, the majority of the cyanobacterial genes were lost or transferred to the nucleus; only a few genes, mainly those required for photosynthesis and gene expression, are currently retained in the plastome ([84, 321]; see Chap. 3). Despite the lower gene content, however, the transcriptional apparatus of higher-plant chloroplasts is more complex than that of bacteria. For example, bacteria use a multisubunit RNA polymerase to transcribe all of their genes. Chloroplasts in angiosperms and possibly in the moss, *Physcomitrella,* possess a homologous enzyme, but additionally require one or more single-subunit phage-type RNA polymerases for transcription. In contrast, the chloroplasts of algae and the lycophyte, *Selaginella,* have a simpler, more archaic apparatus that seems to rely solely on the bacteria-type multisubunit enzyme for transcription [320]. RNA processing is also more complex in chloroplasts than in bacteria, as virtually all chloroplast mRNAs, rRNAs and tRNAs are subjected to maturation, which involves trimming of the 5′ and/or 3′ ends. To become functional, many transcripts require additional *cis*- and/or *trans*-splicing, and (in the case of most land plants) editing of their nucleotide sequences [14]. Transcription and RNA processing seem to take place in close proximity, since components of both processes are found together with DNA in the nucleoids of chloroplasts [176]. In addition to tRNAs and rRNAs, many other non-coding RNAs (including a large number of antisense RNAs) have recently been found in plastids, partly through deep-sequencing strategies [58, 81, 109, 169, 188, 316, 338, 340]. Many of the detected non-coding RNAs are the products of transcription from own promoters [306, 340]; these non-coding RNAs could play a role in regulating gene expression, thus further increasing the complexity of plastid RNA metabolism [77, 108, 267, 316, 337]. A number of the recently described small plastid RNAs, however, are identical to the 3′ and 5′ end regions of mature mRNAs protected from degradation by RNA-binding proteins or stem-loop structures, and are therefore thought to represent by-products of RNA degradation and processing with questionable potential for regulatory functions [239, 340]. A well-investigated example of a plastid non-coding RNA is the *Chlamydomonas tscA* RNA which functions in trans-splicing [233].

This chapter focuses on recent studies dealing with the function of RNA polymerases in plastid gene expression and the role of RNA-binding proteins in the processing of chloroplast transcripts. For more information, a number of recent reviews provide more details on the evolution and regulation of chloroplast transcription, the function of plastid sigma factors, and on plastid RNA processing [14, 155, 160, 262, 320].

1.2 RNA Synthesis

1.2.1 The Plastid-Encoded Plastid RNA Polymerase (PEP) is a Bacteria-Type Multisubunit RNA Polymerase

Homologs of the cyanobacterial RNA polymerase subunits α, β, β′ and β″ are encoded by the plastid *rpoA, B, C* and *C1* genes; together, these form the core of the *p*lastid-*e*ncoded plastid RNA *p*olymerase (PEP; [111, 198, 269, 272]). Similar to the gene organization in bacteria, *rpoA*, which encodes the α subunit of PEP, is found in a gene cluster with several genes encoding ribosomal proteins [223], while *rpoB*, *rpoC* and *rpoC1,* encoding the β, β′ and β″ subunits, respectively, together form an operon [127, 269]. The PEP β and β′ subunits can serve as functional substitutes for the homologous subunits of the *E. coli* RNA polymerase [265]. PEP is sensitive to tagetitoxin, an inhibitor of bacterial transcription [178], further demonstrating the high degree of conservation between the plastid-encoded and eubacterial RNA polymerases. However, the PEP α subunit does not substitute for the *E. coli* homolog in transplastomic tobacco plants [285]. As the bacterial polymerase, the chloroplast core enzyme requires a sigma (σ) factor for promoter recognition and initiation of transcription [162]. While *Chlamydomonas reinhardtii* has only one nuclear gene encoding a sigma factor [26], land plants and the red algae, *Cyanidioschyzon merolae* and *Cyanidium caldarium*, possess several sigma factor genes ([154, 165, 180], for reviews on higher plant sigma factors see [262, 290, 291]). It is not yet known whether the less complex organization of the transcriptional apparatus in algae (PEP alone and fewer sigma factors) is causally related to the lower degree of transcriptional regulation in algal chloroplasts versus those of higher plants [62, 76].

PEP can be isolated from plastids as a soluble enzyme or an insoluble form, also known as *t*ranscriptionally *a*ctive *c*hromosome (TAC), which contains DNA, RNA, the PEP subunits, and a large number of other proteins [37, 89, 144, 164, 215, 230]. Similar to isolated nucleoids [241], TAC exhibits *in vitro* transcriptional activity. The soluble PEP fraction isolated from mustard (*Sinapis alba*) etioplasts, referred to as PEP-B, consists of only the core subunits (Fig. 1.1a; [217, 276]. However, the existence of transcription factors in very low amounts and/or only loosely associated with PEP-B cannot be completely ruled out. Soluble PEP preparations from photosynthetically active plastids, called PEP-A, contain the

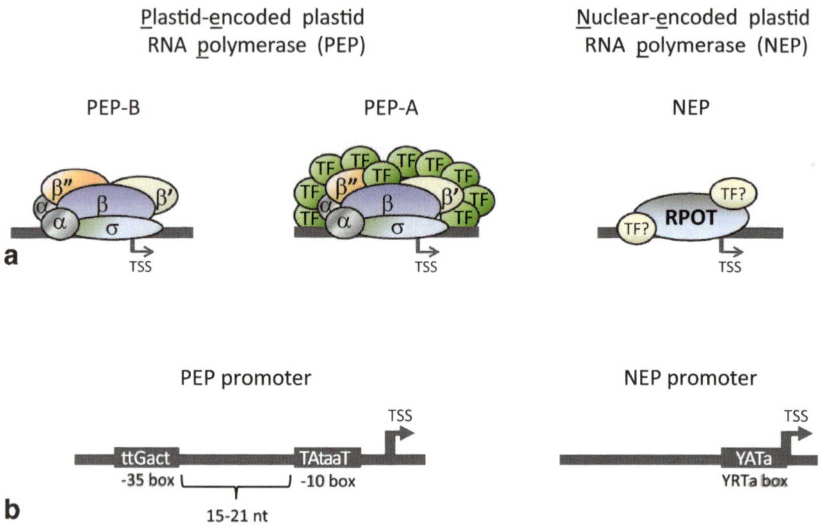

Fig. 1.1 Plastid RNA polymerases and their promoters. **a** PEP-A and PEP-B represent the soluble forms of PEP isolated from chloroplasts and etioplasts, respectively. PEP-B comprises the core subunits 2 α, 1 β, 1 β′ and 1 β″. For promoter recognition and transcription initiation, a σ factor is needed. PEP-A has a more complex structure and consists of the core subunits, the σ factor, and auxiliary factors such as transcription factors (TFs) like the PAPs (see text). For RNA synthesis, the nuclear-encoded plastid RNA polymerase (NEP) requires only the catalytic subunit, RPOT. Unknown TFs support promoter recognition and regulation. **b** Structures of the PEP and NEP promoters, with consensus sequences as found in the barley plastome. Typical PEP promoters resemble bacterial promoters with −10 and −35 consensus sequences, while typical NEP promoters have a YRT core motif. Note, however, that many PEP and NEP promoters do not conform to the depicted structures. The transcription start sites (TSSs) are indicated by *arrows*

PEP core subunits associated with ~10 nuclear-encoded proteins (Fig. 1.1a). PEP complexes have been assessed in etioplasts and chloroplasts; other plastid types have not yet been analyzed in terms of their protein compositions. The proteins associated with the core subunits of PEP (the *P*EP-*a*ssociated *p*roteins, or PAPs) in PEP-A preparations [276] are also observed as components of TAC (the pTACs). Experimental data support the view that the PAPs/pTACs are required for transcription and its regulation under light conditions [122, 197, 215, 217, 218]. Additional factors involved in transcription and the regulation of gene expression can be found in nucleoid preparations [138, 176, 228]. The combination of PEP with its accessory proteins may help establish nuclear control over plastid transcription and adapt transcription to endogenous and exogenous cues [276]. This is also true for the sigma factors, which confer promoter recognition to PEP. The PEP sigma factors of higher plants belong to the eubacterial σ70 family [173]. *Arabidopsis* has six different sigma factors [74, 154, 260, 262]. Sigma factors do not co-purify with PEP, perhaps because they are not needed for the elongation phase of RNA synthesis [276]. In addition, highly purified PEP complexes do not contain the plastid transcription kinase, cpCK2, or the *c*hloroplast *s*ensor *k*inase, CSK [276], even though these enzymes are believed to regulate transcription by phosphorylating PEP subunits and sigma factors in a photosynthesis/redox-dependent manner [10, 11, 36, 126, 163, 197, 224, 225, 302]. Experimental data support the involvement of sigma factors in the regulation of plastid transcription during development and in response to changing environmental conditions (reviewed in [154, 155, 260, 262]). Transcription of plastid genes is also controlled by hormones, but future studies will be needed to identify the factors responsible for mediating the effects of hormones on plastid transcription [160, 344, 345].

1.2.2 PEP Promoters

Given the bacterial origin of PEP, it is unsurprising that many of the promoters utilized by PEP resemble the *E. coli* σ70 promoter architecture, which harbors both −35 and −10 consensus sequence elements [75, 85, 282]. The *E. coli* RNA polymerase can accurately transcribe from such PEP promoters [34, 35]. In *Chlamydomonas* chloroplasts, however, most promoters lack a conserved −35 sequence element; instead, extended −10 boxes and/or more remote sequences confer full promoter strength [24, 116, 133, 140, 141]. Furthermore, neither the −10 nor the −35 box seem to be essential for a functional PEP promoter in higher plants. According to a plastome-wide search for conserved PEP promoter motifs, the −10 element "TAtaaT" (upper-case letters indicate overrepresented nucleotides >1 bit) is located 3–9 nucleotides (nt) upstream of the transcription start site of 89% of all primary (unprocessed) transcripts in the chloroplasts of mature barley leaves, and the −35 element "ttGact" can be found 15–21 nt upstream of 70% of the PEP promoters harboring this −10 motif (Fig. 1.1b; [340]). Comparable whole-genome analyses are not yet available for algae and dicots. The −10 and −35 boxes can be complemented or replaced by other sequences, most of which have not yet been identified. For instance, the mustard *psbA* promoter harbors a regulatory element (TATATA) between the −10 and −35 promoter elements; *in vitro*, this regulatory element promotes a basal level of transcription in the absence of the −35 region in plastid extracts from dark- and light-grown plants. However, the −35 element is essential for the full promoter activity required during active photosynthesis [64, 161], and it is needed for *in vitro* transcription in barley chloroplasts [137]. In the case of the wheat *psbA* promoter, an extended −10 sequence (TGnTATAAT) is utilized as the sole *psbA* promoter element by PEP in mature chloroplasts. PEP obtained from developing chloroplasts in the leaf base, however, requires both the −10 and −35 boxes, suggesting that different transcription factors may participate during chloroplast development [248]. Several *cis*-elements required for the binding of regulatory proteins in the context of PEP promoters have been described. A 22-bp sequence, known as the AAG box, plays an important role in regulating the blue light-responsive promoter of *psbD* (which encodes the photosystem II reaction center chlorophyll protein, D2) by providing a binding site for the AAG-binding factor, PTF1, which acts as a positive regulator [7, 137]. The blue-light dependent activation of the *psbA* and *psbD* promoters in *Arabidopsis* chloroplasts depends on the sigma factor, SIG5, whose expression is stimulated by blue light [204]. SIG5 is also responsible for the enhanced transcription of *psbD* and several other genes under various stress conditions ([193]; Yamburenko et al., unpubl. data). Similarly, a transcription factor binds to a sequence −3 to −32 nt upstream of the *rbcL* transcription start site and enhances transcription [136]. *In silico* analyses suggest that there are many more, yet-uncharacterized nuclear-encoded plastid transcription factors [258, 312].

Similar to most protein-encoding genes/operons and the rRNA gene cluster, the majority of tRNA genes are transcribed by PEP from typical σ^{70}-like promoters upstream of the transcription start site [155]. In addition, some reports suggest that several tRNAs are transcribed from gene-internal promoters; these include the spinach *trnS*, *trnR* and *trnT* [53, 86, 323], the mustard *trnS*, *trnH* and *trnR* [156, 195, 196], and the *Chlamydomonas trnE* [119]. However, the exact tRNA-related internal promoter elements and the polymerase(s) capable of recognizing them have not yet been elucidated.

1.2.3 The Nuclear-Encoded Plastid RNA Polymerase (NEP) is Represented by Phage-Type RNA Polymerases

In stark contrast to the bacterial RNA polymerase, PEP is not sufficient to transcribe all plastid genes in higher plants. Instead, a *n*uclear-*e*ncoded plastid RNA *p*olymerase (NEP) activity participates in and is essential for plastid transcription [1, 102, 271]. The first evidence for the existence of one or more NEP enzymes came from studies on the effect of translation inhibitors on cytoplasmic and plastid ribosomes [65]. Active RNA synthesis occurs in ribosome-deficient plastids, suggesting a nuclear location for the gene(s) responsible for this activity [39, 95, 102, 271]. Moreover, transcription takes place in plastids of the parasitic plant, *Epifagus virginiana,* even though its plastome lacks genes encoding the core subunits of PEP [68, 189]. Similarly, plastid genes are transcribed in PEP-knockout transplastomic tobacco plants, but these plants have an albino phenotype, suggesting that NEP alone cannot provide for photosynthetically active chloroplasts [1, 88, 151].

NEP is represented by one or more phage-type RNA polymerases in higher plants [97, 98, 153], encoded by the *RpoT* (*R*NA *po*lymerase of the phage *T3*/*T7* type) genes [97]. In contrast to the multi-subunit PEP, these phage-type enzymes are composed of only a single catalytic subunit, possibly associated with only one or a few auxiliary factor(s) (see below; Fig. 1.1a; [146]). While monocots and the basal angiosperm, *Nuphar*, contain only one plastid phage-type RNA polymerase (RPOTp; [46, 66, 148, 332]), eudicots have two of these enzymes, RPOTp and RPOTmp, the latter of which is targeted to both plastids and mitochondria [98, 99, 142, 147]. Knocking out the *RpoTp* or *RpoTmp* genes in *Arabidopsis* yields plants with delayed chloroplast biogenesis and slightly altered leaf morphogenesis, while *RpoTp*/*RpoTmp* double mutants exhibit a more severe phenotype characterized by extreme growth retardation [110]. Transgenic tobacco and *Arabidopsis* plants overexpressing RPOTp show increased transcription from a set of NEP promoters [159], and RPOTp recognizes distinct NEP promoters *in vitro* [146]. Even though RPOTmp fails to drive transcription from NEP promoters *in vitro* [146], the enzyme plays a distinct role in plastid transcription during the early developmental stages of *Arabidopsis* [54].

Specific antibodies detect both RPOTp and RPOTmp in the stroma and membrane fractions of plastids (J. Sobanski et al., unpublished data, [5, 46]) and the two phage-type polymerases can be prepared from plastids in both soluble and membrane-bound forms (J. Sobanski et al., unpublished data, [5, 6]). The RING H2-protein mediates the binding of RPOTmp to the stromal side of the thylakoid membrane in spinach [6]. RPOTp and RPOTmp are not detected in purified PEP fractions, PEP-containing TAC preparations, or the proteome of plastid nucleoids [176, 199, 215, 276], most likely because the phage-type polymerases are much less abundant than the PEP subunits in chloroplasts.

The phage T7 RNA polymerase is a genuine single-subunit enzyme; the complete process of transcription (including promoter recognition, initiation, elongation and termination) is performed by a single protein, regardless of whether the DNA template is linear, circular or supercoiled [277]. Similarly, the *Arabidopsis* RPOTp polymerase is able to correctly recognize promoters, transcribe the gene, and stop at a (bacterial) terminator without additional factors in *in vitro* assays, provided that the DNA templates are in the supercoiled conformation [146]. However, *Arabidopsis* RPOT polymerases are also capable of correctly initiating transcription *in vitro* on linear double-stranded DNA templates if the base sequence of the promoter is altered to prevent base pairing (i.e., if the promoter region is already in a partially open state; A. Bohne and T. Börner, unpublished data). This finding suggests that, similar to the related phage-type RNA polymerases in yeast and human mitochondria [59, 179, 232, 284], RPOT polymerases need additional factors to melt the DNA duplex at promoter regions *in organello*. However, such factors have not yet been identified in plants [231]. As shown for PEP (see above), transcription by NEP is also affected by developmental and environmental cues (reviewed in [155, 160]). In the case of the Type II Pc promoter of spinach chloroplasts, a specific transcription factor, CDF2, is involved in the development-dependent decision on whether to use the NEP promoter or the PEP promoter for transcription of the *rrn* genes [23]. Future work is warranted to identify additional NEP-interacting factors and the signaling pathways responsible for regulating NEP activity.

1.2.4 NEP Promoters

In green chloroplasts, PEP transcripts are overrepresented, while most of the transcripts generated by NEP are of low abundance and not easily detectable [101, 158]. Therefore, the NEP transcription start sites have been identified in plants lacking PEP activity [1, 112, 264, 273, 287, 340]. Based on their architectures, the NEP promoters can be grouped into three types: Type-Ia, Type-Ib, and Type-II [158, 319]. The majority of the analyzed NEP promoters belong to the Type-I NEP promoters, which are characterized by a conserved YRTa core motif located a few nucleotides upstream of the transcription start site (Fig. 1.1b; [340]). The plastid promoters share the YRTa motif with many plant mitochondrial promoters [112]. The similarity of the NEP and mitochondrial promoters is not surprising, since the NEP-encoding genes originated from duplication(s) of the gene encoding the mitochondrial RNA polymerase [320]. NEP accurately initiates transcription at the *Oenothera berteriana* mitochondrial *atpA* promoter when integrated into the tobacco plastome, suggesting that there are relationships not only between the promoters and RNA polymerases of plant mitochondria and chloroplasts, but also among the factor(s) involved in promoter recognition [27]. The Type-I promoters are further divided into two subclasses, Type-Ia and -Ib. Type-Ia promoters have only the YRTa box as a conserved sequence motif. No sequence elements outside of this core motif have significant influence on *in vitro* transcription from the tobacco *rpoB* Type-Ia promoter [157]. However, deletion analysis of the 5′-flanking region of the *Arabidopsis rpoB* fused to GUS and transiently expressed in the chloroplasts of cultured tobacco cells suggests the existence of additional regulatory elements upstream of the YRTa sequence [113]. The Type-Ib NEP promoters carry an additional conserved sequence motif ($ATAN_{0-1}GAA$), called the "GAA box", located approximately 18–20 nt upstream of the YRTa motif [319]. Deletion analysis of the tobacco Type-Ib P*atpB*-289 promoter reveals that the GAA box plays a functional role in promoter recognition both *in vivo* and *in vitro* [129, 325]. There is no Type-Ib promoter in the barley chloroplast genome, suggesting that this promoter type may not be used by NEP in the plastids of Poaceae and perhaps other monocots [340].

Transcription from Type-II NEP promoters is YRTa-independent, and is instead controlled by "non-consensus" promoter elements [160]. The best investigated example is the tobacco *clpP* NEP promoter, whose core sequence comprises the region −5 to +25 with respect to the transcription initiation site [275]. Interestingly, the *clpP* NEP promoter sequence is conserved among monocots, dicots and *C. reinhardtii*, but is not required to drive transcription in rice and *Chlamydomonas.* However, when introduced into tobacco, the rice sequence is efficiently utilized as a promoter. This promoter sequence might therefore be recognized by a distinct transcription factor or NEP enzyme that is present in dicots but not monocots, such as PROTmp [159, 275]. The Pc promoter of the *rrn* operon in spinach chloroplasts represents another non-YRTa NEP promoter [155]. The promoter region of the *rrn* operon is highly conserved in plants and contains the −10 and −35 PEP promoter elements, which drive PEP-mediated transcription of the operon in barley, tobacco, maize, and later in the development of *Arabidopsis* chloroplasts [1, 54, 112, 282, 307]. However, in spinach, as well as during the early developmental stages of *Arabidopsis* chloroplasts, NEP initiates at the Pc promoter located between the conserved PEP promoter elements [9, 54, 114, 115, 287]. Approximately 70% of the more than 200 NEP promoters used in the PEP-deficient plastids of *albostrians* barley have a YRTa box as the only conserved promoter element, and thus belong to Type-Ia. The remaining 30% of the NEP promoters lack YRTa, as well as any other consensus motif in the region −50 downstream to +25 upstream of the transcription start sites [340]. Thus, the Type-II promoters may be regarded as a group of apparently unrelated promoters defined by the lack of YRTa.

1.2.5 Division of Labor among Different Plastid RNA Polymerases

The algae investigated to date and the lycophyte, *Selaginella moellendorffii*, do not show NEP activity; instead, PEP transcribes all of their chloroplast genes (reviewed in [320]). Angiosperms and most likely also the moss, *Physcomitrella patens*, rely on NEP in addition to PEP for plastid transcription, although the advantage of this is a matter of some debate. The establishment of NEP activity is believed to have evolved in land plants to offset elevated levels of point mutations in PEP promoters, which may have occurred due to enhanced UV irradiation after the water-to-land transition [175]. This view is supported by two observations: in the absence of PEP, numerous NEP promoters are activated in barley plastids [340]; and a NEP promoter that is inactive in wild-type *Arabidopsis*, compensates when transcription is abolished from the *atpB* PEP promoter in a sigma factor-6 knockout line [261]. An additional or alternative advantage of a second RNA polymerase activity in plastids might be stronger control of organellar transcription by the nuclear genome.

A division of labor between PEP- and NEP- mediated transcription was first proposed by Hess et al. [102] and further elaborated by Mullet [192] and Hajdukiewicz et al. [88]. Initial studies suggested that NEP plays a role in transcribing housekeeping genes, while PEP is responsible for transcribing the photosynthetic genes [1, 88, 102, 112, 130, 308]. However, later studies showed that there is no strict division of labor between the two polymerases with respect to the functional classes of plastid genes they transcribe (housekeeping/non-photosynthetic vs. photosynthetic). Many housekeeping genes have both PEP and NEP promoters, and certain non-photosynthetic genes are transcribed only by PEP in green leaves (e.g., [88, 307, 340]). A few potential NEP promoters may exist upstream of photosynthetic genes in normal green chloroplasts (Fig. 1.2; [340]), and more than 200 new NEP promoters are activated in the leaf plastids of a barley mutant lacking PEP activity, resulting in the NEP-mediated transcription of virtually all plastid genes ([339]; see also [151]).

The transcriptional activity of plastid genes massively increases with the onset of chloroplast development (reviewed in [155]). In addition, the transcription of the *rpoB-C1-C2* genes is NEP-dependent [102] and precedes the strong transcription of photosynthetic genes during chloroplast development in barley [18] and pea leaves [61]. These data, together with the detection of NEP promoters upstream of housekeeping genes (see above), led researchers to suggest that NEP might be responsible for the basal transcriptional activity in the plastids of non-green cells. With the onset of chloroplast development from non-green proplastids, increased NEP activity would transcribe the genes encoding the core subunits of PEP. Then,

PEP would take over transcription and provide the high transcriptional activity needed for further chloroplast development, including the assembly of the photosynthetic apparatus [88, 192]. Indeed, NEP promoters are more active in early leaf development, while the transcriptional activity of PEP increases during chloroplast maturation [18, 54, 58, 66, 130, 288, 342]. However, these roles of NEP and PEP in chloroplast development have not yet been directly demonstrated. More recent data show that both PEP and NEP are present and active in all investigated green and non-green tissues during all developmental stages of the leaf [38, 42, 57, 58, 125, 288, 305, 342]. Nevertheless, PEP is clearly the predominating RNA polymerase in photosynthetically active chloroplasts (Fig. 1.2; [340]). PEP transcribes the vast majority of plastid genes, including all photosynthetic genes. In mature barley chloroplasts, active NEP promoters (but no PEP promoters) were mapped within 750 nt upstream of the *rpl23* and *rpoB* coding sequences. However, *rpl23* is part of a PEP-controlled gene cluster [128, 174], leaving *rpoB-C-C1* as the only known example of an exclusively NEP-dependent transcript in monocots [340]. Although chloroplast genes can be transcribed from promoters located even further upstream of the coding region [308], no PEP-dependent transcription start sites is seen in the 2 kb region upstream of the annotated *rpoB* gene in the barley plastome (Fig. 1.2). Given that multiple promoters are very common in plastids and a large percentage of genes/operons have both NEP and PEP promoters [155, 340], it is remarkable that the expression of the genes encoding the ß, ß′ and ß″ PEP subunits is entirely dependent on NEP in both monocots and dicots [157, 287, 340].

The nuclear genomes of the eudicots harbor two genes for NEP activity, *RPOTp* and *RPOTmp* [98], suggesting that there is also a division of labor between the two NEP polymerases. Indeed, several studies suggest that RPOTp and RPOTmp display their major activities in different tissues and developmental stages. In *Arabidopsis*, *RPOTmp* promoter activity is detected in young, non-green cells of different organs, whereas *RPOTp* expression is mainly observed in green, photosynthetically active tissues [67]. In agreement with this observation, Courtois et al. [54] found that RPOTmp is needed for the synthesis of rRNAs from the Pc promoter in *Arabidopsis* seeds during imbibition, while at later stages, PEP becomes the principle polymerase responsible for *rrn* transcription [54]. Furthermore, lack of RPOTmp activity resulted in lower accumulation of several chloroplast transcripts in young *Arabidopsis* seedlings upon illumination [8, cf. 147]. However, several lines of evidence suggest that RPOTp is also present and required early in development, and that RPOTmp may also play a role in mature chloroplasts. The activity of RPOTmp in mature chloroplasts can be deduced from the use of NEP promoters in *Arabidopsis* mutants lacking RPOTp. However, the strong NEP promoter that drives transcription of the essential *ycf1* gene in wild-type dicot chloroplasts is not used in very young RPOTp mutant seedlings, hinting that RPOTp may play a role at this early stage of development [288]. In addition, knocking out or knocking down RPOTp decreases the levels of transcripts originating from NEP promoters in both mature and developing *Arabidopsis* chloroplasts (the effect is more pronounced in the latter; [288]). RPOTp appears to prefer Type-I promoters, while RPOTmp

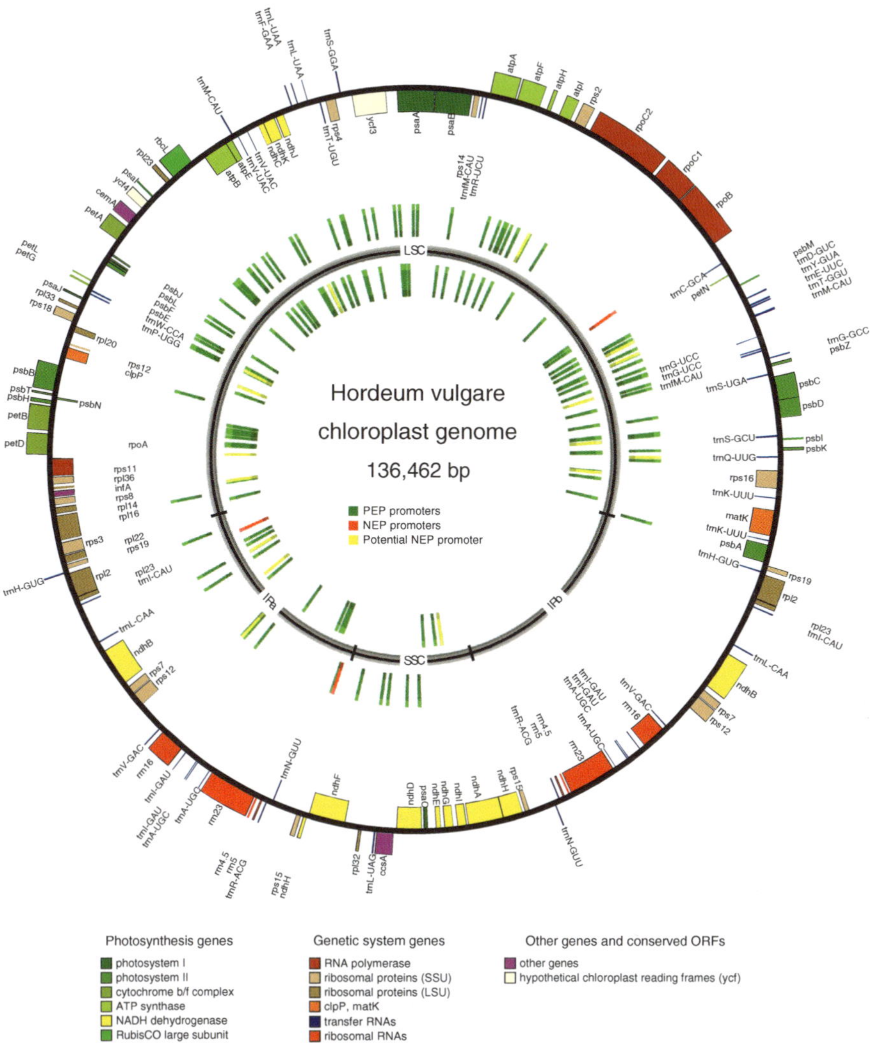

Fig. 1.2 Distribution of PEP- and NEP-dependent transcription start sites (TSSs) in mature barley chloroplasts. The *outer circle* depicts the gene organization of the barley chloroplast genome (NC_008590). The graphical representation was created using OGDraw (OrganellarGenomeDRAW; http://ogdraw.mpimp-golm.mpg.de/; [166]) and further modified. Genes at the inside and outside of the circle are transcribed clockwise and counterclockwise, respectively. Genes are color coded based on the function of the proteins they encode (see the legend *below* the circle). The *inner circle* depicts the genomic distribution of the TSSs mapped in mature barley chloroplasts as follows: *green*—PEP-dependent TSSs; *red*—NEP-dependent TSSs; *yellow*—potential NEP-dependent TSSs. TSSs mapped to the inverted repeat (IR) are shown only within IRa. The image was generated using CGView (Circular Genome Viewer; http://wishart.biology.ualberta.ca/cgview/; [281])

prefers Type-II promoters. Overexpression of RPOTp enhances the usage of Type-I promoters [159]. Similarly, usage of the non-consensus Type-II promoters of the *clpP* gene and the *rrn* operon is unaffected and enhanced, respectively, by the lack of RPOTp activity. However, most of the Type-I NEP promoters are still active in the absence of RPOTp, suggesting that RPOTmp can recognize Type-I promoters [288].

1.3 RNA Processing

Early on, transcription was recognized as a major point of gene regulation in bacteria, epitomized by the operon model of Jacob and Monod [118]. In addition to the core transcriptional machinery, a number of factors (repressors or activators of transcription) are known to determine the usage of bacterial promoters. Such modulators of transcription initiation are DNA-binding proteins, and include the famous *trp* repressor [250]. Bacterial RNAs are translated as they are transcribed, so there is very little posttranscriptional RNA processing. Splicing, RNA editing and intercistronic processing are rare events in bacteria; thus, transcription initiation and RNA degradation largely determine mRNA expression and eventual protein production [83]. Although non-coding RNAs have lately come into focus as regulators of gene expression in bacteria, prokaryotes undergo relatively little regulated RNA processing.

In chloroplasts, however, every primary RNA is subject to some form of modification after transcription [278]. As in bacteria, chloroplasts express the majority of their genes as polycistronic RNAs. However, the bacterial concept of the operon as a cluster of co-regulated genes does not fully apply to plastids. Instead of being directly translated, numerous polycistronic transcripts function as precursors that are cleaved into smaller polycistronic or monocistronic RNAs, many of which still require splicing and/or RNA editing to become functional [14, 278]. Thus, RNA maturation further increases the complexity of RNA populations arising from most genes. Major events in plastid RNA maturation (*e.g.,* 5′- and 3′-end processing and intercistronic processing) involve the action of ribonucleases that have low sequence specificity, and the extent of processing is often determined by barriers such as RNA-binding protein and the presence of secondary structures [14, 278].

This part of our review focuses on the poorly understood complexity of posttranscriptional processes in chloroplasts. We will summarize the most important findings on the central processes of RNA splicing, editing and end maturation, and then focus on studies that point to the potential regulatory functions of these RNA processing steps. In contrast to translational regulation, which is discussed in the accompanying article by Nickelsen et al. (Chap. 3), only a few studies demonstrate that RNA processing has a true rate-limiting effect on chloroplast gene expression. We will not attempt a detailed discussion of the large body of work on the mechanistic aspects of RNA processing. For this, we direct the reader to recent reviews on the individual RNA processing steps of splicing, editing and RNA degradation [50, 117, 279].

1.3.1 Chloroplast RNA Splicing

1.3.1.1 Chloroplast Introns and Factors

The two dominant classes of introns found in the chloroplast genes are the group I and group II introns, which are archaic introns believed to be the precursors of the eukaryotic spliceosomal introns [45, 104, 247, 270, 310]. Group I and group II introns are structurally different, and harbor subdomains that have specific functions in the splicing reaction [242]. For example, the group II introns share six secondary domains that fold into a structure that is held together by tertiary interactions within the intron and with exonic sequences [185]. This structure brings together the splice sites, intron-internal guiding sequences, and the branch point. The number of introns and their positions within the genome are relatively stable; the chloroplast genes of land plants usually contain around 20 introns, all but one of which fall into group II (for example: 17 intron in maize chloroplasts, 21 in Arabidopsis thaliana chloroplasts, [252]). These introns disrupt protein-encoding genes as well as those for tRNAs. In chlorophyte algae, group I introns are far more dominant, and the overall intron number per genome is more variable than that in land plants (e.g. 7 introns in *C. reinhardtii*, 27 in *Pseudendoclonium*; [181, 219]). In addition, some chlorophytes also have introns in their rRNA-encoding genes [235]. These introns are all ribozymes by definition, and bacterial group I and group II introns can be made to self-splice *in vitro* [242]. However the chloroplast introns require *trans*-acting factors for excision [252]. A large and growing set of nuclear-encoded proteins important for chloroplast splicing have been identified over the past 15 years. These factors are not related to the nuclear spliceosomal machinery, but instead have been evolutionarily recruited from very different sources. For example, the maize chloroplast RNA splicing 2 protein (CRS2) is a modified peptidyl-tRNA-hydrolase [120], while the *Chlamydomonas Raa2* is derived from pseudouridine synthase [213]. Other known splicing factors contain various RNA binding domains, including the CRM domain found in ribosome-assembly factors [16], the abundant RRM domain [257], the mTERF domain [92], and the organelle-specific PPR domain [19, 52, 55, 135]. In accordance with their diverse origins, the target ranges of these factors differ somewhat, although they overlap. The known factors and their target introns are listed in Table 1.1.

In terms of molecular functions, these factors are believed to help mold the intron into a structure that allows splicing to occur. Intron folding could, for example, be promoted by high-affinity, sequence-specific interactions that stabilize otherwise transient RNA-internal interactions [208]. Proteins could also block competing non-productive folding pathways, or act as helicases to actively resolve misfolded RNA structures [90, 100]. Finally, the proteins may help juxtapose the 5′-splice site

Table 1.1 Chloroplast RNA Splicing Factors

Name	Type[a]	Spec[b]	Loc[c]	Target site(s)	Evidence	Mutant phenotype	Reference
APO1	Zinc-finger-like	A.t	cp	*psaA, psaB, ycf3 int.2, clpP int1, petD, ndhA, ndhB*	Genetic	Albino, pale	Watkins et al. [318]
CAF1	CRM	Z.m; A.t	cp	*pet Dint., trnG int., rps16 int., rpl16 int., ycf3 int.1, clpP int.1, rpoC1 int., ndhA int.*	Genetic	Albino	Asakura and Barkan [2], Ostheimer et al. [207]
CAF2	CRM	Z.m; A.t	cp	*rps12 int.1; petB int., ndhB int., ndhA int., ycf3 int.1*	Genetic	Albino	Asakura and Barkan [2], Ostheimer et al. [206]
CFM2	CRM	A.t	cp	*trnL int., ndhA int., ycf3 int.1, clpP int.2*	Genetic	Pale green	Asakura and Barkan 2007
CFM3a	CRM	A.t	cp/mt	*ndhB int., rpl16 int., rps16 int.,trnG int., petB int., petD int.*	Genetic	Pale green, stunted growth	Asakura et al. [3]
cNAPL	Nucleosome assembly-like	C.r	cp	*tscA*	Biochemical	–	Glanz et al. [79]
CPN60	Heat shock protein	C.r	cp	*psaA*	Biochemical	–	Balczun et al. [12]
CRS1	CRM	Z.m; A.t	cp	*atpF*	Genetic	Pale	Asakura and Barkan [2], Jenkins et al. [120, 121], Ostersetzer et al. [208]
CRS2	CRM	Z.m	cp	*rps12 int.1, ycf3 int.1, clpP int.1*	Genetic	Ivory	Asakura and Barkan [2], Ostheimer et al. [207]
HCF107	HAT		cp	*psbH*	Genetic	hcf	Hammani et al. [94]
HCF152	PPR	A.t	cp	*petB int.*	Genetic	hcf	Meierhoff et al. [165]
MatK		N.t	cp	*atpF int., trnK int., trnA int., trnI int., trnV int., rpl2 int., rps12 int.2*	Biochemical	Lethal	Zoschke et al. [306]
OTP51	PPR-LAGLIDADG		cp	*ycf3 int.2*	Genetic	Albino	de Longevialle et al. [55]
PPR4	PPR	Z.m	cp	*rps12 int.1*	Genetic	Albino	Schmitz-Linneweber et al. [257]
PPR5	PPR	Z.m	cp	*trnG-UCC int.*	Genetic	Albino, pale	Beick et al. [19]
Raa1	OPR	C.r	cp	*psaA int1 and int2*	Genetic	No phototrophic growth	Merendino et al. [184]

Table 1.1 (continued)

Name	Type[a]	Spec[b]	Loc[c]	Target site(s)	Evidence	Mutant phenotype	Reference
Raa2	Pseudouridine synthase	C.r	cp (membrane of cps)	*psaA int. 1*	Genetic	Light sensitive (ML and HL), no phototrophic growth	Perron et al. [213]
Raa3		C.r	cp (stroma of cps)	*psaA int. 2*	Genetic	Light sensitive (ML and HL), no phototrophic growth	Rivier et al. [233]
RAA4	Pseudo uridine synthase	C.r	cp	*tscA, psaA int.,*	Genetic	Pale, lethal	Glanz et al. [80]
Rat1 and 2		C.r	cp	*psaA int.1 (tscA)*	Genetic	PSI deficiency	Balczun et al. 2005
RH3	RRM	Z.m, A.t	cp	*trnI, trnA, rps12 int.1, rps12 int.2, rpl2, 23Srrn, ycf3 int.1 and 2*	Genetic	Pale	Asakura et al. [4]
RNC1	Nuclease III	Z.m	cp	*trnV int., trnK int., trnI int., trnA int., rps12 int.2, ndhBint.,petB int., pet Dint., trnG int.*	Genetic	Albino, pale	Watkins et al. [317]
THA8	PPR	Z.m	cp	*Ycf3 int.2, trnA*	Genetic	Embryo lethal,	Khrouchtchova et al. [135]
WHY	whirly	Z.m	cp	*atpF int.*	Genetic	Albino, pale	Prikryl et al. [220]
WSL	PPR	O.s.	cp	*rpl2 int.*	Genetic	Albino, striped	Tan et al. [293]
WTF1	PORR	Z.m	cp	*petB int., ndhB int., petD int., trnG int., rpl16, rps12 int.2, atpF int., trnA int., trnI int., trnK int., trnV int.*	Genetic		Kroeger et al. 2009
Zm-mTERF4	mTERF	Z.m.	cp	*trnI int., trnA int., rpl2 int., atpF int., ycf3-intron 2*	Genetic	Albino	Hammani and Barkan [92]

[a] Type of domains encoded by the respective gene; *CRM* chloroplast RNA maturation; *PPR* pentatricopeptide repeat; *HAT* Half A TPR repeat; *PORR* Plant Organellar RNA Recognition domain; OPR Octotricopeptide repeat; mTERF mitochondrial transcription terminator domain

[b] Species: *At Arabidopsis thaliana*; *Cr Chlamydomonas reinhardtii*; *Os Oryza sativa*; So *Spinacia oleracea*; *Zm Zea amys*

[c] Localization of the protein: *cp* chloroplast; *mt* mitochondria

with the internal branch point, allowing an intron-internal phosphodiester bridge to form and freeing the 3′-OH group of the 5′-exon. The latter is brought into proximity with the 3′-splice site, the two exons are fused, and the intron is released as a circular structure known as the lariat. It is not yet clear how chloroplast factors fulfill this role at an atomic level; few biochemical or structural studies have addressed the exact binding sites of splice factors on their target introns and how these factors change the conformation of their intron ligands. For the maize factor, CRS1, we know that binding to its single target, the *atpF* intron, triggers structural changes in a particular intronic domain [208]. Footprinting analyses have demonstrated that CRS1 facilitates the internalization of intronic elements required for the core of the functional ribozyme [208]. In the future, it will be important to understand how chloroplast splicing factors act on and affect the structures of their target introns.

In addition to the nuclear-encoded splicing factors, there is also one chloroplast-encoded protein essential for splicing a set of introns: MatK. Canonical bacterial group II introns harbor reading frames for maturase proteins that specifically support the splicing of their own introns and are required for the mobility of group II introns (bacterial introns can reverse-splice into novel genomic locations, a process not happening in chloroplasts and thus not further discussed here, [149]). With one exception, the introns of the land plant chloroplasts have lost their maturase reading frames. The sole maturase left in the chloroplast, MatK, resides in the *trnK* gene and has been implicated in splicing a subset of introns characterized by specific structural elements [103, 311]. MatK was recently demonstrated to associate *in vivo* with these introns [343], but we need further structural insights into how, where and why MatK attaches to its target introns in chloroplasts.

1.3.1.2 Regulation of Chloroplast RNA Splicing

RNA splicing is an essential process, making it an ideal step for switching on or off the gene expression of intron-containing reading frames. Unspliced chloroplast RNAs accumulate to high levels, and changes in the ratio of spliced to unspliced mRNAs in different tissues have been described in maize (for the *atpF*, *petD*, *petB*, *rpl16*, and *ycf3* introns, [13, 182]), potato (*atpF*, *ndhB*, [305]), for the mustard *trn*G intron , and the tomato *ndhB* intron [125]. The latter is believed to involve inhibition of the first splicing reaction [125], but we do not yet fully understand how these shifts in splicing efficiency occur. The existing studies largely agree, however, that splicing is most effective in chloroplasts, whereas non-photosynthetic tissues show relative over-accumulations of unspliced precursor RNAs. Unexpectedly, light does not seem to generally activate splicing in land plants [13, 156]. However, it does appear to have a positive effect on the splicing of the *psbA* group I introns in *C. reinhardtii* chloroplasts [60]. At present, it is unclear if these findings reflect an active change in splicing efficiency, or if there are changes in the stability of spliced versus unspliced transcripts. It is even less clear whether the observed changes impact the amount of proteins produced from these mRNAs, *i.e.*, whether splicing can indeed be rate-limiting for gene expression. In *Chlamydomonas*, a mutation in a group

I intron of the *psbA* mRNA reduces the levels of both mature mRNA and PsbA protein by two-fold [150]. In this case, splicing could be a true rate-limiting step; however, it seems doubtful that such correlations between splicing rate and protein production exist for many spliced RNAs. There is evidence in *C. reinhardtii* that the amount of chloroplast mRNAs exceeds the capacity of the translational apparatus [62], suggesting that smaller changes in splicing might not affect the eventual protein levels. Also, there is growing evidence that many chloroplast mRNAs are regulated at the level of translation, *i.e.*, after splicing (see Chap. 3). Nevertheless, for selected introns or under selected conditions, splicing could become rate limiting for gene expression.

We can only speculate on how this could be accomplished. Most simply, nucleus-encoded chloroplast splicing factors could become rate-limiting for splicing. A correlation of splice factor abundance and the splicing rate of a target mRNA has been demonstrated for CRS1 and its target, *atpF* [294], but few other splicing factors have been measured in a comprehensive fashion (under different conditions, in plants of different ages, etc.).

Next to such direct effects by varying amounts of splicing factors, we can imagine indirect effects from the transcriptional and translational machineries. Splicing efficiencies and transcription rates have not yet been formally correlated, but the speed of an intron's production could impact its folding status and thus its splicing efficiency. The different chloroplast RNA polymerases can be expected to have different transcription elongation rates, and each polymerase could be tuned to different velocities depending on external and internal cues (*e.g.*, changes in phosphorylation) [295]. This could affect the folding and subsequent splicing of all chloroplast introns [210]. In addition, it is well known that translation in bacterial systems can impact transcription rates, and recent data show that transcription and translation are physically linked in prokaryotes [40, 222]. To date, no evidence suggests that translation would be uncoupled from transcription in chloroplasts. Thus, if emerging transcripts are rapidly associating with ribosomes, the latter could drive into the intronic structures, almost certainly decreasing splicing. To prevent this, splicing would have to be finished before the start codons emerge from the polymerase. Detailed studies on the kinetics of transcription, translation and splicing of selected messages are needed to answer such questions.

Alterations of the Mg2+ concentrations in chloroplasts may offer a regulatory mechanism that is completely independent of protein co-factor activity. Group II introns fold into catalytically active conformations only in the presence of Mg2+ ions [226], and the concentration of free Mg2+ is dependent on chloroplast biogenesis and the activity of Mg2+ transporters in the chloroplast envelope [107]. Thus, regulation of Mg2+ availability could also limit splicing.

In summary, there is currently no direct evidence that introns benefit chloroplasts by regulating gene expression. However, the evolutionary stability of introns in land plants suggests that other benefits may exist. The ultimate test of the putative advantage of having an intron is, of course, to remove it. This was recently done for the two group II introns in the tobacco *ycf3* gene [214], an assembly factor for photosystem I [238]. While the loss of *ycf3* intron 2 has no phenotypic consequence, deletion of intron 1 decreases photosynthetic activity [214]. This is because intron 2 remains unspliced in the intron 1 deletion strains, disrupting *ycf3* expression. Importantly, this demonstrates that an intron can have a *cis*-acting effect on the expression of its own gene. It is unclear how this *cis*-interaction occurs on a molecular level, but it may be related to a physical interaction of the introns necessary for the splicing of intron 2.

While the interaction of intron 1 and intron 2 in *ycf3* is positive, an intron in the *ndhA* mRNA in spinach has a negative effect. In the latter case, an RNA editing event downstream of the intron takes place only in the absence of the intron, *i.e.,* after splicing [254]. It will be interesting to explore whether splicing can also affect other gene expression events, particularly translation and transcription (see above for possible kinetic interactions between these processes).

1.3.2 Chloroplast RNA Editing

1.3.2.1 Chloroplast RNA Editing Sites and Factors

The term "RNA editing" describes a variety of base conversion, deletion and insertion processes in various organisms [82]. In chloroplasts, RNA editing is restricted to nucleotide conversions from C to U or, less frequently, from U to C, and is achieved by amination and deamination reactions [51]. Most editing sites are located in coding regions and affect the coding potential of the mRNA. *Cis*-acting sequences adjacent to editing sites determine the specificity of these events. In recent years, it has been demonstrated that the PPR proteins are responsible for recognizing these sequence elements [253]. For the majority of editing sites, only one responsible PPR protein has been identified. For a few sites in mitochondria, however, the knockout of one PPR protein reduces but does not abolish editing, suggesting that the remaining editing is carried out by another PPR protein or other factor [335]. Similar observations have not yet been made in chloroplasts. Notably, while some PPR proteins seem to be responsible for only a single editing site, most PPR proteins recognize multiple sites (and in most cases show sequence similarities in their *cis*-sequences) [93].

PPR proteins have been identified for almost all of the 34 sites in *Arabidopsis* chloroplasts (Table 1.2). The PPR proteins were identified as the long-sought recognition factors based on mutant analyses and their ability to bind to the *cis*-elements *in vitro* (for recent reviews see [44, 51]). The PPR proteins share similarities with other helix-loop-helix proteins, particularly the pumilio proteins, which also bind RNA [56]. Recently, atomic structures of PPR proteins have been solved [134, 263, 333]. Together with previous data, they support the idea that amino acids from two consecutive. PPR repeats bind one nucleotide [see also 73, 221]. These structural models will certainly support the current efforts to predict bindings sites of PPR proteins computationally [17].

Intriguingly, not all PPR proteins can serve as editing factors; this is the function of a specific subclass of this large family, called the "E/DYW" PPR proteins. DYW

Table 1.2 Chloroplast RNA Editing Factors

Name	Type[a]	Spec[b]	Loc[c]	Target site(s)[d]	Evidence[e]	Mutant phenotype[f]	Reference
CRR4	E	*At*	cp	*ndhD*eU2TM	Genetic; biochemical	wt, NDH defective	Kotera et al. [143], Okuda et al. [200]
CRR21	E	*At*	cp	*ndhD*eU383SL	Genetic	wt, NDH defective	Okuda et al. [201]
OTP80	E	*At*	cp	*rpl23*eU89SL	Genetic	wt	Hammani et al. [81]
OTP81	DYW	*At*	cp	*rps12*i114eU58	Genetic	wt	Hammani et al. [81]
OTP85	DYW	*At*	cp	*ndhD*eU674SL	Genetic	wt	Hammani et al. [81]
OTP86	DYW	*At*	cp	*rps14*eU80SL	Genetic	wt	Hammani et al. [81]
RARE1	DYW	*At*	cp	*accD*eU794SL	Genetic	wt	Robbins et al. [234]
LPA66	DYW	*At*	cp	*psbF*eU77SF	Genetic	Pale-green; reduced PSII	Cai et al. [41]
YS1	DYW	*At*	cp	*rpoB*eU338SF	Genetic	virescent	Zhou et al. [341]
AtECB2	DYW	*At*	cp	*accD*eU794SL	Genetic	Albino, seedling lethal	Yu et al. [334]
CLB19	E	*At*	cp	*rpoA*eU200SF *clpP*eU559HY	Genetic	pyg, seedling lethal	Chateigner-Boutin et al. [49]
CRR22	DYW	*At*	cp	*ndhD*eU887PL *ndhB*eU746SF *rpoB*eU551SL	Genetic	wt, NDH defective	Okuda et al. [202]
CRR28	DYW	*At*	cp	*ndhB*eU467PL *ndhD*eU878SL	Genetic	wt, NDH defective	Okuda et al. [202]
OTP82	DYW	*At*	cp	*ndhB*eU836SL *ndhG*eU50SF	Genetic	wt	Okuda et al. [203]
OTP84	DYW	*At*	cp	*psbZ*eU50SL *ndhB*eU1481PL *ndhF*eU290SL	Genetic	wt, partially NDH defective	Hammani et al. [81]

Table 1.2 (continued)

Name	Type[a]	Spec[b]	Loc[c]	Target site(s)[d]	Evidence[e]	Mutant phenotype[f]	Reference
MORF2 (RIP2)	MORF	*At*	cp	ndhDeU2TM ndh-BeU1255HYps-bZeU50SL	Genetic	Albino	Takenaka et al. [258]
MORF9 (RIP9)	MORF	*At*	cp	*ndhD*eU2TM *ndh-B*eU1255HY-*petL*eU5PL	Genetic	Variegated, pale	Takenaka et al. [258]
LPA66	DYW	*At*	cp	*psbF*eU77SF	Genetic	hcf, pale green	Cai et al. [41]
RIP1 (MORF8)		*At*	cp/mp	*rps12(*i1)58 *petL*eU-5PL*ndhD*eU2TM*ndh-B*eU467PL*rpoC1eU*-488PL*ndhB*eU586HY	Genetic	Dwarf, pale green	Bentolila et al. [21]
VAC1	DYW	*At*	cp	*ndhF*eU290SL *accD*eU794SL	Genetic	Albino	Tseng et al. [301]
P25		*Nt*	cp	*psbL*eU1TM	Biochemical		Hirose and Sugiura [105]
P70		*Nt/ So*	cp	*psbE*eU72PS	Biochemical		Hirose and Sugiura [105], Miyamoto et al. [186]
P56		*Nt*	cp	*petB*eU204PL	Biochemical		Hirose and Sugiura [105], Miyamoto et al. [186]
CP31A/CP31B	RRM	*At / Nt*	cp	13 sites partially affected	Genetic	wt	Hirose and Sugiura [105], Tillich et al. [298]
ORRM1	MORF and RRM	*At / Zm*	cp	12 sites in Arabidopsis /9 sites in maize	genetic	wt	Sun et al. [283]

a Type of domains encoded by the respective gene; *E* E-type PPR protein domain; *DYW* DYW-type PPR protein; *RRM* protein containing RNA recognition motif(s); *MORF* multiple organellar RNA editing factor

b Species: *At Arabidopsis thaliana; Nt Nicotiana tabacum; So Spinacia oleracea; Zm Zea mays*

c Localization of the protein: *cp* chloroplast; *mt* mitochondria

d Editing site of respective factor with nucleotide position and amino acid conversion

e Type of experimental evidence;

stands for a C-terminal extension that includes the name-giving trio of amino acids [170]. The DYW domain has weak similarities to the cytidine deaminases, and is thus believed to harbor the enzymatic activity that carries out base deamination [243]. In this model, the DYW domain provides the enzymatic activity in *cis* when present in the PPR protein, and also in *trans* through heterodimer formation [236, 243]. Future studies are needed to provide enzymatic proof for this hypothesis. In an alternative model, additional proteins carry the necessary enzymatic activity and are recruited via PPR proteins. Indeed, PPR proteins are part of large, RNA-associated protein complexes [19, 71, 257], where editing PPR proteins interact with each other and with other factors [289]. These interacting proteins are believed to form the core of a larger structure that we call the "organellar editosome."

Recently, a novel class of proteins was identified as part of this editosome, the so-called MORF/RIP proteins [22, 289]. The MORF proteins form a small family in land plants, but are absent from chlorophytes. Most of the members of this family are imported into mitochondria, but at least two are also found in chloroplasts (MORF2 and MORF9), and another one, MORF8, is dually targeted to mitochondria and chloroplasts [289] for a comparison of the MORF and RIP nomenclature, please see [22]. Most of the analyzed organellar editing sites show reduced editing in the absence of either factor, demonstrating that the two MORF proteins act together at the same sites, which was substantiated by yeast-two-hybrid (Y2H) and pull-down experiments showing that the MORF proteins interact with each other and with PPR proteins [289]. The specificity of this interaction is low, however, because interactions occur also between plastid MORF proteins and mitochondrial PPR proteins, which presumably do not occur *in vivo*. It can be expected that the nature of the interactions between MORFs and PPR proteins will be scrutinized in the near future.

Another group of proteins that have been implicated in RNA editing are the chloroplast ribonucleoprotein (cpRNPs; [300]). They are required for the editing of specific sites in a tobacco *in vitro* RNA editing system [105], and null mutants of the *Arabidopsis* cpRNP, CP31A, display reduced editing at multiple sites [298]. The cpRNPs, which are very abundant RNA-binding proteins, are believed to help govern the conformation and/or stability of transcripts [194, 298] and thus play indirect roles in RNA editing. Their direct binding to *cis*-elements seems unlikely, as the PPR proteins perform this essential job, and PPR knockout phenotypes are much more severe than those observed for cpRNP mutants [253]. Finally, the recent discovery of yet another RNA binding protein involved in editing suggests that the complexity of the editing apparatus has been greatly underetimated in the past [283]. How the many newly identified factors (and further proteins) constitute the chloroplast editosome on individual editing sites is certainly one of the challenges lying ahead.

1.3.2.2 Regulation of Chloroplast RNA Editing

Only few chloroplast editing sites are conserved over longer evolutionary distances, within the embryophytes [72, 297]. Usually, editing sites evolve rapidly, at rates similar to those of synonymous codon positions [268, 299]. Differences in editing sites are observed between closely related taxa, and even between species of a single genus (e.g. [70, 72, 246, 255]). This strongly suggests that editing events *per se* are meaningless to the chloroplast; for most editing sites, chloroplast function is not affected by whether C-to-T editing occurs or a T is already encoded. This notion is supported by mutational and cell-biological analyses of an editing site in the essential *atpA* gene of tobacco chloroplasts. In this case, editing must occur to provide the proper amino acid at this position [256], but replacement of the edited C with a T on the DNA level did not result in any phenotypic alteration [256]. Thus, it does not seem to matter whether the T(U) is provided by RNA editing or by a DNA mutation. Finally, for a number of sites in chloroplasts (more so in mitochondria), the loss of a responsible pentatricopetide repeat protein abolishes editing but does not trigger any phenotypic change, indicating that the editing event itself is unimportant [93, 309, 336].

Of course, these findings strongly suggest that RNA editing does not play any regulatory function. In fact, it has been suggested that RNA editing is an evolutionary answer to genomic stress rather than an effort to increase the complexity and regulatory power of gene expression [175]. Organelles are obligate endosymbionts that persist asexually in their host cells and go through frequent bottlenecks during host reproduction. This lifestyle is known to lead to the accumulation of deleterious mutations that cannot be removed by means of recombination [171]. However, chloroplast genomes evolve much more slowly than the nuclear genomes of plants [172, 209, 322], suggesting that nuclear genes may suppress negative mutations within the organellar genome by providing repair factors that can reverse point mutations on the RNA level. In the case of editing, the involved repair factors are the PPR proteins. This model is supported by the finding that plant genomes use PPR proteins to suppress deleterious mitochondrial mutations that, if left unchecked, lead to cytoplasmic male sterility (CMS, [47]). In fact, plant breeders have selected successfully for these suppressors (which are called "restorers of fertility") multiple times in recent agricultural history [47].

In sum, there is reason to doubt that RNA editing evolved because of the need to regulate gene expression in chloroplasts. However, individual editing events may have been hijacked for regulatory purposes. Below, we summarize the few putative points of regulation that have been identified to date.

Most editing sites in chloroplasts appear to be fully edited (e.g. [48, 91, 240, 296]), leaving little room for regulation by the resulting protein products. However, a few sites show fluctuations in the ratio between edited and unedited messages over time and space, or in response to environmental clues [25, 106, 131, 132, 187, 237]. Notably, however, such quantitative editing changes are likely to be superseded by much larger variations in the abundances of the respective transcripts [211]. Thus, processes other than editing (*e.g.*, transcription and RNA degradation) have a much larger bearing on the eventual output of gene expression. An interesting exception might be the editing of the *rpoB* mRNA, which encodes an essential subunit of PEP. The PPR protein, YS1, is required for *rpoB* editing; it is believed to potentially limit PEP activity, thereby regulating the expression of tRNAs, particularly *trnE* [341]. *trnE* is required for plastid translation and additionally serves as a starting point for tetrapyrrole synthesis, which is crucial to the development of chloroplasts in the light [341]. Thus, an editing event in a chlorophyll synthesis-related gene might impact chloroplast biogenesis. Future studies on manipulating YS1 levels and correlating the expression of YS1 with chloroplast biogenesis under different conditions will be needed to further support this model. Consistent with the above findings, the *rpoA* mRNA encoding the α-subunit of PEP is also only partially edited, forming another potential link with RNA polymerase activity and chloroplast biogenesis [106]. In general, detailed investigations into the regulation of editing factors, particularly the PPR proteins and the recently identified MORF proteins, will be needed to clarify the role of RNA editing in the rate-limiting of chloroplast gene expression.

1.3.3 RNA Cleavage and Degradation

The half-lives of mRNAs are in the range of minutes in prokaryotes, but mRNAs can remain stable for up to hours in chloroplasts [139]. This reflects the fact that chloroplasts "live" in a very stable environment (the plant cell) where it is less crucial to rapidly adjust gene expression to changing external conditions (compared to the situation in a free-living bacterium). Nevertheless, the chloroplast harbors an extensive set of nucleolytic enzymes whose regulatory functions are just beginning to be understood [279].

1.3.3.1 Chloroplast RNases

Both endo- and exonuclease activities, which are mediated by nuclear-encoded ribonucleases (RNases), have been reported to participate in rRNA maturation, tRNA maturation, intercistronic mRNA processing, and RNA decay in plastids [14, 31, 278, 279]. Some plastid RNases are homologous to bacterial ribonucleases, and are likely to fulfill homologous functions. In many other cases, however, the enzymes and their precise functions have not yet been elucidated. In fact, there are a number of nucleases that are predicted to reside in the chloroplast, but still lack experimental verification or molecular characterization [279]. Among the best characterized plastid ribonucleases are the RNases that participate in 5′ and 3′ RNA maturation.

Processed 5′ RNA ends are thought to be generated *via* either a 5′-to-3′ exonuclease pathway or endonucleolytic cleavage [244, 278]. Homologs of the *E. coli* RNase E and the *B. subtilis* RNase J may act as major plastid endonucleases [279]. The *Arabidopsis* RNase E has a function comparable to its *E. coli* counterpart: it prefers 5′ monophosphorylated (processed) substrates; it is inhibited by structured RNA; and it preferentially cleaves AU-rich sequences [191, 249]. Recent analyses found that RNase E null mutants in *Arabidopsis* show multiple defects in the processing of polycistronic precursor transcripts [314]. The processing of the mRNA for the ribosomal protein, L22, is most severely affected, perhaps explaining the ribosome deficiency observed in RNase E mutants [314].

Another endonuclease, RNase J also exhibits endonucleolytic activity; however, unlike RNase E, RNase J is insensitive to the number of phosphates at the 5′ end [266]. Moreover, similar to its *B. subtilis* homolog, plastid RNase J acts as a 5′-to-3′ exonuclease and prefers 5′-monophosphorylated RNAs [266]. RNase E and -J endonucleolytic activities are thought to initiate intercistronic mRNA processing, which is followed by exonucleolytic trimming of the novel transcript ends [14]. In fact, RNase J may take part in the otherwise poorly understood 5′-to-3′ trimming of RNAs, and it could act as surveillance enzyme that eliminates long antisense RNAs, such as those arising from read-through transcription [266].

A further endonuclease found in chloroplasts is CSP41 (chloroplast stem–loop-binding protein of 41 kDa). This protein has been demonstrated to bind chloroplast RNAs [329, 330] and cleave them *in vitro* with a preference for stem-loop RNA segments [28, 328]. In *Arabidopsis,* two genes encode CSP41 proteins, which are involved in a dazzling and not yet fully understood variety of chloroplast tasks. The loss of CSP41 proteins leads to pleiotropic molecular phenotypes; these include decreased steady-state levels of multiple chloroplast RNAs, and decreased plastid transcription and translation rates [20, 32, 227]. The underlying molecular function(s) of CSP41 are not yet fully understood, however, in part because the proteins associate with various chloroplast structures and machineries. For example, CSP41 proteins are components of the PEP in mustard [218], and CSP41 mutants show decreased transcriptional activity [32], suggesting that the proteins play a role in transcription. In contrast, however, other proteomic studies failed to find CSP41 proteins in PEP preparations [215, 286], and transcriptional activity can be secondarily influenced by defects in chloroplast translation since PEP expression requires plastid ribosomes. Thus, additional approaches will be needed to verify the proposed role of CSP41 in transcription.

CSP41 from *C. reinhardtii* is also found in preparations of chloroplast 70S ribosomes [326], in preparations of the 30S ribosomal subunit [199], and in complexes containing the ribosomal proteins, L5 and L31 [212]. This could indicate that CSP41 plays a role in translation. However, CSP41 proteins are found together with pre-ribosomal particles [20] and bind *in vivo* to chloroplast rRNA in *Arabidopsis* [227]. Thus, a role in ribosome biogenesis seems more likely.

The case is further complicated by the finding that CSP41b interacts in the cytosol with heteroglycans [69], pointing to potential functions outside of nucleic acid metabolism. Further genetic analyses will be required to identify the primary molecular lesion(s) in CPS41 mutants.

The best characterized plastid exonuclease is the bacterial homolog of polynucleotide phosphorylase, or short PNPase, which participates in the processing, polyadenylation and degradation of chloroplast RNAs [31, 78, 278]. PNPase catalyzes both processive 3′-to-5′ degradation and RNA polymerization [331], and appears to act as a major 3′-to-5′ exonuclease for processing the 3′ termini of mRNAs [313].

Recently, a thorough mutational study of PNPase *in vivo* and *in vitro* demonstrated that PNPase promotes rRNA and mRNA 3′-end maturation and RNA degradation [78]. The ability of PNPase to degrade RNA is blocked by either stable secondary structures (*e.g.,* the stem-loops frequently found in chloroplast 3′-UTRs) or by proteins tightly bound to 3′-UTRs [14, 278, 339]. PNPase also seems to be required for the removal of excised introns, although it is unclear whether this ability impacts splicing efficiency or (more likely) is just a scavenging function [78]. In any case, an enzyme that can function in both degrading and stabilizing chloroplast RNAs would obviously be a natural target for regulating gene expression.

Finally, for the sake of completeness, we will mention three additional chloroplast RNases, all of which are involved in the processing of rRNAs and tRNAs. The maturation of rRNAs is believed to involve the 3′-to-5′ exonuclease, RNase R [30], while tRNA maturation involves the endonucleases, RNaseP and RNase Z, which produce the 5′ and 3′ ends of tRNAs, respectively [43, 251, 292, 315]. While these enzymes are essential, we do not yet know whether their activities regulate chloroplast translation by limiting the amounts of tRNA or rRNA.

1.3.3.2 Intercistronic mRNA Processing

Plastid RNA metabolism is characterized by excessive intercistronic mRNA processing (*i.e.,* increased processing of polycistronic transcripts between the coding regions). Initially, it was thought that intercistronic processing is mediated by site-specific endonucleases that generate processed 5′ and 3′ ends mapping to adjacent nucleotides [31]. However, it was later observed that the 5′ processed end of *petD* and the 3′ end of the upstream gene (*petB*) overlapped by approximately 30 nt, and thus could not have been generated by a single cleavage event [15]. A similar phenomenon exists for other adjacent processed RNAs in maize. A detailed analysis of the processed termini mapping to the *atpI-atpH* and *psaJ-rpl33* intergenic regions led to the emergence of a model in which the maize PPR10 binds to these intergenic regions and blocks 5′-to-3′ and 3′-to-5′ exonuclease activity, and thus defines the corresponding 5′- and 3′-processed plastid ends [216]. Indeed, recombinant PPR10 is sufficient to block 5′-to-3′ and 3′-to-5′ exonuclease activity *in vitro* [221]. Moreover, when PPR10 is supplemented *in vitro* with a generic 5′-to-3′ exonuclease, a 5′ end is generated that precisely matches the PPR10-dependent terminus generated *in vivo* [221]. In addition, other PPR and PPR-like proteins (RNA-binding proteins with helical repeat architectures, including CRP1, HCF152, PGR3, PPR38, MRL1, MCA1, Mbb1, NAC2 and HCF107) mediate the accumulation of RNAs with processed 5′/3′ termini mapping to intergenic regions [15, 29, 96, 124, 167, 183, 245, 304, 327]. Other non-PPR-like RNA-binding proteins are also likely to be capable of protecting RNAs against exonucleolytic attack, as recently shown for PrfB3 [280]. Taken together, these observations indicate that the PPRs (and other RNA binding proteins) make major contributions to 5′- and 3′-processed end formation by binding to target RNAs and protecting adjacent regions by blocking exonucleases [14].

Such an event should logically be accompanied by the presence of short RNA fragments *in vivo*; these would represent the PPR footprints (minimal PPR binding sites) that are protected from complete elimination by nucleases [216]. Indeed, small RNAs (sRNAs) corresponding to the PPR10 binding sites are found in the transcriptomes of several angiosperms [123, 190, 216, 239]. More than 80 sRNAs exist in the chloroplast transcriptomes of *Arabidopsis* and barley; some of them can be correlated with PPR binding sites, while most of the others co-localize with known mRNA ends [239, 339]. If, as predicted by this model, all sRNAs are linked to stabilizing proteins, then there should be at least one stabilizing protein (on average) for each chloroplast mRNA ([239, 339], and own unpublished results). Given that transcript termini are widely stabilized in angiosperm chloroplasts and green algae (well-studied examples are NAC2 for the *petD* mRNA and MCA1 for the *petA* mRNA in *Chlamydomonas*, [32, 205, 229, 259]), we can conclude that this is an evolutionarily conserved mechanism by which transcripts are defined in chloroplasts.

1.3.3.3 Regulation of RNA Degradation

Similar to the situations with RNA editing and splicing, there are various options for regulating gene expression by RNA stability, yet relatively few studies actually show situations in which RNA degradation becomes rate-limiting. However, intriguing examples come from work on RNA stabilizing factors in *Chlamydomonas* and on the regulation of the chloroplast PNPase in *Arabidopsis*.

As noted above, one of the many tasks of PPR proteins in chloroplast RNA metabolism is protecting transcript ends against the action of exonucleases, thereby increasing the half-lives of chloroplast messages. It is undisputed that this job is essential for chloroplast gene expression, but is it a point of regulation? In an elegant and laborious genetic approach, Raynaud et al. prepared a series of transgenic *Chlamydomonas* lines with decreasing amounts of the PPR protein, MCA1 [229], thereby incrementally stabilizing the *Chlamydomonas petA* mRNA [87]. They found a correlation between the amount of MCA1, the amount of *petA* mRNA and the translation rate of *petA* leading to the product, cytochrome f [229]. In line with its regulatory importance, MCA1 is a short-lived protein that responds rapidly to changing physiological conditions (*e.g.*, nitrogen starvation or culture age), triggering changes in the mRNA levels of *petA* [229]. It was recently shown that the unassembled cytochrome f induces the degradation of MCA1, thus constituting a negative feed-back loop for the regulation of cytochrome b6f biogenesis [33]. MCA1 forms complexes with TCA1, which aids it in stabilizing the petA transcripts. Both proteins support *petA* translation and (as a complex) connect and regulate RNA stability and translation [33]. Regulatory links between RNA stabilization and RNA translation also exist for the PPR protein, PPR10 [216], the HAT protein, HCF107 [94], and the *Chlamydomonas* protein, NAC2 [259], suggesting that this may be an emerging general feature of chloroplast gene regulation. PPR10 and HCF107 capture short sequence stretches in a single-stranded state, which would otherwise be found in secondary structural elements that inhibit translation ([94, 221], see also Chap. 3 for further details). These are among the rare examples wherein chloroplast proteins are known to induce changes in RNA secondary structure.

As we have seen, chloroplast RNA degradation depends on the action of protecting proteins and the antagonism of enzymatic activities. Regulation could occur on both sides of this balancing act. Indeed, there is evidence that at least one of the chloroplast RNases, PNPase, is a regulated protein. As noted above, PNPase can both degrade RNAs and add A-rich tails to chloroplast RNAs. This bidirectional activity is reversible and governed by the nucleotide diphosphate-inorganic phosphate ratio [331]. In *Chlamydomonas*, artificial reduction of PNPase levels renders cells unable to acclimate to phosphorus scarcity, leading to the mis-regulation of numerous nuclear genes [177, 331]. The transcript and protein levels of PNPase itself are negatively regulated under phosphorus deprivation, and are under the control of general factors responsible for governing the response to phosphorus deficiency in *Chlamydomonas* [324]. Chloroplast transcript levels are modulated by PNPase activity under phosphorus deprivation, demonstrating that this protein links phosphate availability to chloroplast RNA levels. We do not yet understand how this impacts chloroplast development, but it is the best known example for the regulation of a chloroplast RNase. In the future, it will be interesting to understand to what extent other abiotic stimuli (aside from phosphorus limitation) elicit responses in chloroplast RNA processing and stability.

1.3.4 RNA Processing: Outlook

Chloroplast RNA processing has been a challenging puzzle for the past two decades. Efforts to unravel its complexity have necessarily focused on identifying the underlying machineries, and were thus mehanistic by nature. Today, we know that a large and growing number of factors from various protein families are involved in chloroplast gene expression; some of them are essential, while others have more subtle effects. However, we do not yet know how the individual factors interact with target RNAs, or how they act to change RNA conformations and thus prepare/carry out discrete RNA processing steps. Likewise, while it is clear that these factors act in larger-order structures, little is known about interacting partners and the possibility for chloroplast spliceosomes, editosomes, degradosomes, and so on. Therefore, one future goal will certainly be to understand the action of individual, exemplary factors at the molecular and atomic levels, in conjunction with their immediate molecular environment. Structure-function studies are needed for many of the identified factors (*e.g.*, the PPR proteins). Another major challenge lying ahead is to find ways to identify regulatory steps amidst the post-transcriptional complexity. It is likely that holistic approaches (such as the measurement of many factors at the same time) will be necessary. Whole-genome, transcriptomic and proteomic studies are available for chloroplasts, but have not yet been brought together and scrutinized for the various regulatory steps in the chloroplast gene expression network. Modeling of such large datasets and other approaches from systems biology will be needed to help us predict the regulatory steps in chloroplast gene expression.

Finally, we need to address the spatial organization of chloroplast gene expression. Recent publications show that there are different RNA pools in the chloroplast; some of them are translated, while others are translation-inactive [63]. In *Chlamydomonas*, translated RNAs are found in specialized regions of the chloroplast, called "T-zones" [303]. In the future, genetic approaches will help us identify factors involved in localizing RNAs, while novel high-resolution microscopic techniques will help us understand the distribution of unprocessed, processed, and translated RNAs in the chloroplast. Furthermore, a three-dimensional analysis of chloroplast gene expression can be expected to open up novel inroads into understanding the remaining mysteries of RNA processing in the chloroplast.

Acknowledgements The experimental work of the authors was supported by the Deutsche Forschungsgemeinschaft (SFB 429 to TB and CSL; Emmy-Noether-Program to CSL). PZ was a fellow of the Helmholtz Graduate School "Molecular Cell Biology" at the Max Delbrück Center for Molecular Medicine and Humboldt University, Berlin, Germany. We are very grateful to Konrad Förstner (Institute for Molecular Infection Biology, University of Würzburg, Germany) for support in the preparation of Fig. 1.2.

References

1. Allison LA, Simon LD, Maliga P (1996) Deletion of rpoB reveals a second distinct transcription system in plastids of higher plants. EMBO J 15:2802–2809
2. Asakura Y, Barkan A (2006) Arabidopsis orthologs of maize chloroplast splicing factors promote splicing of orthologous and species-specific group II introns. Plant Physiol 142(4):1656–1563
3. Asakura Y, Bayraktar OA, Barkan A (2008) Two CRM protein subfamilies cooperate in the splicing of group IIB introns in chloroplasts. RNA 14(11):2319–2332
4. Asakura Y, Galarneau E, Watkins KP, Barkan A, van Wijk KJ (2012) Chloroplast RH3 DEAD box RNA helicases in maize and Arabidopsis function in splicing of specific group II introns and affect chloroplast ribosome biogenesis. Plant Physiol 159(3):961–974
5. Azevedo J, Courtois F, Lerbs-Mache S (2006) Sub-plastidial localization of two different phage-type RNA polymerases in spinach chloroplasts. Nucleic Acids Res 34:436–444
6. Azevedo J, Courtois F, Hakimi MA, Demarsy E, Lagrange T, Alcaraz JP, Jaiswal P, Marechal-Drouard L, Lerbs-Mache S (2008) Intraplastidial trafficking of a phage-type RNA polymerase is mediated by a thylakoid RING-H2 protein. Proc Natl Acad Sci U S A 105:9123–9128
7. Baba K, Nakano T, Yamagishi K, Yoshida S (2001) Involvement of a nuclear-encoded basic helix-loop-helix protein in transcription of the light-responsive promoter of psbD. Plant Physiol 125:595–603
8. Baba K, Schmidt J, Espinosa-Ruiz A, Villarejo A, Shiina T, Gardestrom P, Sane AP, Bhalerao RP (2004) Organellar gene transcription and early seedling development are affected in the rpoT;2 mutant of Arabidopsis. Plant J 38:38–48
9. Baeza L, Bertrand A, Mache R, Lerbs-Mache S (1991) Characterization of a protein binding sequence in the promoter region of the 16S rRNA gene of the spinach chloroplast genome. Nucleic Acids Res 19:3577–3581

10. Baginsky S, Tiller K, Link G (1997) Transcription factor phosphorylation by a protein kinase associated with chloroplast RNA polymerase from mustard (Sinapis alba). Plant Mol Biol 34:181–189
11. Baginsky S, Tiller K, Pfannschmidt T, Link G (1999) PTK, the chloroplast RNA polymerase-associated protein kinase from mustard (Sinapis alba), mediates redox control of plastid in vitro transcription. Plant Mol Biol 39:1013–1023
12. Balczun C, Bunse A, Schwarz C, Piotrowski M, Kück U (2006) Chloroplast heat shock protein Cpn60 from *Chlamydomonas reinhardtii* exhibits a novel function as a group II intron-specific RNA-binding protein. FEBS Lett 580(18):4527–4532
13. Barkan A (1989) Tissue-dependent plastid RNA splicing in maize: transcripts from four plastid genes are predominantly unspliced in leaf meristems and roots. Plant Cell 1:437–445
14. Barkan A (2011) Expression of plastid genes: organelle-specific elaborations on a prokaryotic scaffold. Plant Physiol 155:1520–1532
15. Barkan A, Walker M, Nolasco M, Johnson D (1994) A nuclear mutation in maize blocks the processing and translation of several chloroplast mRNAs and provides evidence for the differential translation of alternative mRNA forms. Embo J 13:3170–3181
16. Barkan A, Klipcan L, Ostersetzer O, Kawamura T, Asakura Y, Watkins KP (2007) The CRM domain: an RNA binding module derived from an ancient ribosome-associated protein. RNA 13:55–64
17. Barkan A, Rojas M, Fujii S, Yap A, Chong YS, Bond CS, Small I (2012) A combinatorial amino acid code for RNA recognition by pentatricopeptide repeat proteins. PLoS Genet 8(8):e1002910
18. Baumgartner BJ, Rapp JC, Mullet JE (1993) Plastid genes encoding the transcription/translation apparatus are differentially transcribed early in barley (*Hordeum vulgare*) chloroplast development (evidence for selective stabilization of *psbA* mRNA). Plant Physiol 101:781–791
19. Beick S, Schmitz-Linneweber C, Williams-Carrier R, Jensen B, Barkan A (2008) The pentatricopeptide repeat protein PPR5 stabilizes a specific tRNA precursor in maize chloroplasts. Mol Cell Biol 28:5337–5347
20. Beligni MV, Mayfield SP (2008) *Arabidopsis thaliana* mutants reveal a role for CSP41a and CSP41b, two ribosome-associated endonucleases, in chloroplast ribosomal RNA metabolism. Plant Mol Biol 67:389–401
21. Bentolila S, Heller WP, Sun T, Babina AM, Friso G, van Wijk KJ, Hanson MR (2012) RIP1, a member of an Arabidopsis protein family, interacts with the protein RARE1 and broadly affects RNA editing. Proc Natl Acad Sci USA 109(22):E1453–1461
22. Bentolila S, Oh J, Hanson MR, Bukowski R (2013) Comprehensive high-resolution analysis of the role of an Arabidopsis gene family in RNA editing. PLoS Genet 9(6):e1003584
23. Bligny M, Courtois F, Thaminy S, Chang CC, Lagrange T, Baruah-Wolff J, Stern D, Lerbs-Mache S (2000) Regulation of plastid rDNA transcription by interaction of CDF2 with two different RNA polymerases. EMBO J 19:1851–1860
24. Blowers AD, Ellmore GS, Klein U, Bogorad L (1990) Transcriptional analysis of endogenous and foreign genes in chloroplast transformants of *Chlamydomonas*. Plant Cell 2:1059–1070
25. Bock R, Hagemann R, Kossel H, Kudla J (1993) Tissue- and stage-specific modulation of RNA editing of the *psbF* and *psbL* transcript from spinach plastids-a new regulatory mechanism? Mol Gen Genet 240:238–244
26. Bohne AV, Irihimovitch V, Weihe A, Stern DB (2006) *Chlamydomonas reinhardtii* encodes a single sigma^{70}-like factor which likely functions in chloroplast transcription. Curr Genet 49:333–340
27. Bohne AV, Ruf S, Borner T, Bock R (2007) Faithful transcription initiation from a mitochondrial promoter in transgenic plastids. Nucleic Acids Res 35:7256–7266
28. Bollenbach TJ, Stern DB (2003) Secondary structures common to chloroplast mRNA 3′-untranslated regions direct cleavage by CSP41, an endoribonuclease belonging to the short chain dehydrogenase/reductase superfamily. J Biol Chem 278:25832–25838

29. Boudreau E, Nickelsen J, Lemaire S, Ossenbuhl F, Rochaix J (2000) The *Nac2* gene of *Chlamydomonas* encodes a chloroplast TPR-like protein involved in *psbD* RNA stability. EMBO J 19:3366–3376
30. Bollenbach TJ, Lange H, Gutierrez R, Erhardt M, Stern DB, Gagliardi D (2005) RNR1, a 3′-5′ exoribonuclease belonging to the RNR superfamily, catalyzes 3′ maturation of chloroplast ribosomal RNAs in *Arabidopsis thaliana*. Nucleic Acids Res 33:2751–2763
31. Bollenbach TJ, Schuster G, Portnoy V, Stern D (2007) Processing, degradation, and polyadenylation of chloroplast transcripts. In: Bock R (ed) Cell and molecular biology of plastids. Springer, Berlin, pp 175–211
32. Bollenbach TJ, Sharwood RE, Gutierrez R, Lerbs-Mache S, Stern DB (2009) The RNA-binding proteins CSP41a and CSP41b may regulate transcription and translation of chloroplast-encoded RNAs in *Arabidopsis*. Plant Mol Biol 69:541–552
33. Boulouis A, Raynaud C, Bujaldon S, Aznar A, Wollman FA, Choquet Y (2011) The nucleus-encoded trans-acting factor MCA1 plays a critical role in the regulation of cytochrome f synthesis in *Chlamydomonas* chloroplasts. Plant Cell 23:333–349
34. Boyer SK, Mullet JE (1986) Characterization of *P. sativum* chloroplast *psbA* transcripts produced *in vivo*, *in vitro* and in *E. coli*. Plant Mol Biol 6:229–243
35. Bradley D, Gatenby AA (1985) Mutational analysis of the maize chloroplast ATPase-beta subunit gene promoter: the isolation of promoter mutants in *E. coli* and their characterization in a chloroplast *in vitro* transcription system. EMBO J 4:3641–3648
36. Bräutigam K et al (2009) Dynamic plastid redox signals integrate gene expression and metabolism to induce distinct metabolic states in photosynthetic acclimation in *Arabidopsis*. Plant Cell 21:2715–2732
37. Briat JF, Laulhere JP, Mache R (1979) Transcription activity of a DNA-protein complex isolated from spinach plastids. Eur J Biochem 98:285–292
38. Brosch M, Krause K, Falk J, Krupinska K (2007) Analysis of gene expression in amyloplasts of potato tubers. Planta 227:91–99
39. Bünger W, Feierabend J (1980) Capacity for RNA synthesis in 70S ribosome-deficient plastids of heat-bleached rye leaves. Planta 149:163–169
40. Burmann BM, Schweimer K, Luo X, Wahl MC, Stitt BL, Gottesman ME, Rosch P (2010) A NusE: NusG complex links transcription and translation. Science 328:501–504
41. Cai W, Ji D, Peng L, Guo J, Ma J, Zou M, Lu C, Zhang L (2009) LPA66 is required for editing *psbF* chloroplast transcripts in arabidopsis. Plant Physiol 150:1260–1271
42. Cahoon AB, Harris FM, Stern DB (2004) Analysis of developing maize plastids reveals two mRNA stability classes correlating with RNA polymerase type. EMBO J 5:801–806
43. Canino G, Bocian E, Barbezier N, Echeverria M, Forner J, Binder S, Marchfelder A (2009) Arabidopsis encodes four tRNase Z enzymes. Plant Physiol 150:1494–1502
44. Castandet B, Araya A (2011) RNA editing in plant organelles. Why make it easy? Biochemistry (Mosc) 76:924–931
45. Cech TR (1986) The generality of self-splicing RNA: relationship to nuclear mRNA splicing. Cell 44:207–210
46. Chang CC, Sheen J, Bligny M, Niwa Y, Lerbs-Mache S, Stern DB (1999) Functional analysis of two maize cDNAs encoding T7-like RNA polymerases. Plant Cell 11:911–926
47. Chase CD (2007) Cytoplasmic male sterility: a window to the world of plant mitochondrial-nuclear interactions. Trends Genet 23:81–90
48. Chateigner-Boutin AL, Hanson MR (2003) Developmental co-variation of RNA editing extent of plastid editing sites exhibiting similar cis-elements. Nucleic Acids Res 31:2586–2594
49. Chateigner-Boutin AL, Ramos-Vega M, Guevara-Garcia A, Andres C, de la Luz Gutierrez-Nava M, Cantero A, Delannoy E, Jimenez LF, Lurin C, Small I, Leon P (2008) CLB19, a pentatricopeptide repeat protein required for editing of rpoA and clpP chloroplast transcripts. Plant J 56:590–602
50. Chateigner-Boutin A, Small I (2011) Organellar RNA editing. In: Wilusz J (ed) Wiley interdisciplinary reviews: RNA. Wiley, London, pp 493–506

51. Chateigner-Boutin AL, Small I (2011) Organellar RNA editing. Wiley Interdiscip Rev RNA 2:493–506
52. Chateigner-Boutin AL, des Francs-Small CC, Delannoy E, Kahlau S, Tanz SK, de Longevialle AF, Fujii S, Small I (2011) OTP70 is a pentatricopeptide repeat protein of the E subgroup involved in splicing of the plastid transcript rpoC1. Plant J 65:532–542
53. Cheng YS, Lin CH, Chen LJ (1997) Transcription and processing of the gene for spinach chloroplast threonine tRNA in a homologous *in vitro* system. Biochem Biophys Res Commun 233:380–385
54. Courtois F, Merendino L, Demarsy E, Mache R, Lerbs-Mache S (2007) Phage-type RNA polymerase RPOTmp transcribes the *rrn* operon from the PC promoter at early developmental stages in *Arabidopsis*. Plant Physiol 145:712–721
55. de Longevialle AF, Hendrickson L, Taylor NL, Delannoy E, Lurin C, Badger M, Millar AH, Small I (2008) The pentatricopeptide repeat gene OTP51 with two LAGLIDADG motifs is required for the cis-splicing of plastid ycf3 intron 2 in *Arabidopsis thaliana*. Plant J 56:157–168
56. Delannoy E, Stanley WA, Bond CS, Small ID (2007) Pentatricopeptide repeat (PPR) proteins as sequence-specificity factors in post-transcriptional processes in organelles. Biochem Soc Trans 35:1643–1647
57. Demarsy E, Courtois F, Azevedo J, Buhot L, Lerbs-Mache S (2006) Building up of the plastid transcriptional machinery during germination and early plant development. Plant Physiol 142:993–1003
58. Demarsy E, Buhr F, Lambert E, Lerbs-Mache S (2012) Characterization of the plastid-specific germination and seedling establishment transcriptional programme. J Exp Bot 63:925–939
59. Deshpande AP, Patel SS (2012) Mechanism of transcription initiation by the yeast mitochondrial RNA polymerase. Biochim Biophys Acta 1819:930–938
60. Deshpande NN, Bao Y, Herrin DL (1997) Evidence for light/redox-regulated splicing of *psbA* pre-RNAs in *Chlamydomonas* chloroplasts. RNA 3:37–48
61. DuBell AN, Mullet JE (1995) Differential transcription of pea chloroplast genes during light-induced leaf development (continuous far-red light activates chloroplast transcription). Plant Physiol 109:105–112
62. Eberhard S, Drapier D, Wollman FA (2002) Searching limiting steps in the expression of chloroplast-encoded proteins: relations between gene copy number, transcription, transcript abundance and translation rate in the chloroplast of *Chlamydomonas reinhardtii*. Plant J 31:149–160
63. Eberhard S, Loiselay C, Drapier D, Bujaldon S, Girard-Bascou J, Kuras R, Choquet Y, Wollman FA (2011) Dual functions of the nucleus-encoded factor TDA1 in trapping and translation activation of atpA transcripts in *Chlamydomonas reinhardtii* chloroplasts. Plant J 67:1055–1066
64. Eisermann A, Tiller K, Link G (1990) *In vitro* transcription and DNA binding characteristics of chloroplast and etioplast extracts from mustard (*Sinapis alba*) indicate differential usage of the *psbA* promoter. EMBO J 9:3981–3987
65. Ellis RJ, Hartley MR (1971) Sites of synthesis of chloroplast proteins. Nature 233:193–196
66. Emanuel C, Weihe A, Graner A, Hess WR, Börner T (2004) Chloroplast development affects expression of phage-type RNA polymerases in barley leaves. Plant J 38:460–472
67. Emanuel C, von Groll U, Müller M, Börner T, Weihe A (2006) Development- and tissue-specific expression of the *RpoT* gene family of *Arabidopsis* encoding mitochondrial and plastid RNA polymerases. Planta 223:998–1009
68. Ems SC, Morden CW, Dixon CK, Wolfe KH, dePamphilis CW, Palmer JD (1995) Transcription, splicing and editing of plastid RNAs in the nonphotosynthetic plant *Epifagus virginiana*. Plant Mol Biol 29:721–733
69. Fettke J, Nunes-Nesi A, Fernie AR, Steup M (2011) Identification of a novel heteroglycan-interacting protein, HIP 1.3, from *Arabidopsis thaliana*. J Plant Physiol 168:1415–1425
70. Fiebig A, Stegemann S, Bock R (2004) Rapid evolution of editing sites in a small non-essential plastid gene. Nucl Acids Res 7:3615–3622

71. Fisk DG, Walker MB, Barkan A (1999) Molecular cloning of the maize gene *crp1* reveals similarity between regulators of mitochondrial and chloroplast gene expression. Embo J 18:2621–2630
72. Freyer R, Kiefer-Meyer MC, Kossel H (1997) Occurrence of plastid RNA editing in all major lineages of land plants. Proc Natl Acad Sci U S A 94:6285–6290
73. Fujii S, Bond CS, Small ID (2011) Selection patterns on restorer-like genes reveal a conflict between nuclear and mitochondrial genomes throughout angiosperm evolution. Proc Natl Acad Sci U S A 108:1723–1728
74. Fujiwara M, Nagashima A, Kanamaru K, Tanaka K, Takahashi H (2000) Three new nuclear genes, sigD, sigE and sigF, encoding putative plastid RNA polymerase sigma factors in *Aarabidopsis thaliana*. FEBS Lett 481:47–52
75. Gatenby AA, Castleton JA, Saul MW (1981) Expression in *E. coli* of maize and wheat chloroplast genes for large subunit of ribulose bisphosphate carboxylase. Nature 291:117–121
76. Geimer S, Belicova A, Legen J, Slavikova S, Herrmann RG, Krajcovic J (2009) Transcriptome analysis of the *Euglena gracilis* plastid chromosome. Curr Genet 55:425–438
77. Georg J, Honsel A, Voss B, Rennenberg H, Hess WR (2010) A long antisense RNA in plant chloroplasts. New Phytol 186:615–622
78. Germain A, Herlich S, Larom S, Kim SH, Schuster G, Stern DB (2011) Mutational analysis of *Arabidopsis* chloroplast polynucleotide phosphorylase reveals roles for both RNase PH core domains in polyadenylation, RNA 3′-end maturation and intron degradation. Plant J 67:381–394
79. Glanz S, Bunse A, Wimbert A, Balczun C, Kück U (2006) A nucleosome assembly protein-like polypeptide binds to chloroplast group II intron RNA in *Chlamydomonas reinhardtii*. Nucleic Acids Res 34(18):5337–5351
80. Glanz S, Jacobs J, Kock V, Mishra A, Kück U (2012) Raa4 is a trans-splicing factor that specifically binds chloroplast tscA intron RNA. Plant J 69(3):421–431
81. Gonzalez-Ibeas D, Blanca J, Donaire L, Saladie M, Mascarell-Creus A, Cano-Delgado A, Garcia-Mas J, Llave C, Aranda MA (2011) Analysis of the melon (*Cucumis melo*) small RNAome by high-throughput pyrosequencing. BMC Genomics 12:393
82. Gott JM, Emeson RB (2000) Functions and mechanisms of RNA editing. Annu Rev Genet 34:499–531
83. Gottesman S, Storz G (2012) Bacterial small RNA regulators: versatile roles and rapidly evolving variations. Cold Spring Harb Perspect Biol 3
84. Green BR (2011) Chloroplast genomes of photosynthetic eukaryotes. Plant J 66:34–44
85. Gruissem W, Zurawski G (1985) Analysis of promoter regions for the spinach chloroplast *rbcL*, *atpB* and *psbA* genes. EMBO J 4:3375–3383
86. Gruissem W, Elsner-Menzel C, Latshaw S, Narita JO, Schaffer MA, Zurawski G (1986) A subpopulation of spinach chloroplast tRNA genes does not require upstream promoter elements for transcription. Nucleic Acids Res 14:7541–7556
87. Gumpel NJ, Ralley L, Girard-Bascou J, Wollman FA, Nugent JH, Purton S (1995) Nuclear mutants of *Chlamydomonas reinhardtii* defective in the biogenesis of the cytochrome b6f complex. Plant Mol Biol 29:921–932
88. Hajdukiewicz PT, Allison LA, Maliga P (1997) The two RNA polymerases encoded by the nuclear and the plastid compartments transcribe distinct groups of genes in tobacco plastids. EMBO J 16:4041–4048
89. Hallick RB, Lipper C, Richards OC, Rutter WJ (1976) Isolation of a transcriptionally active chromosome from chloroplasts of *Euglena gracilis*. Biochemistry 15:3039–3045
90. Halls C, Mohr S, Del Campo M, Yang Q, Jankowsky E, Lambowitz AM (2007) Involvement of DEAD-box proteins in group I and group II intron splicing. Biochemical characterization of Mss116p, ATP hydrolysis-dependent and -independent mechanisms, and general RNA chaperone activity. J Mol Biol 365:835–855
91. Halter CP, Peeters NM, Hanson MR (2004) RNA editing in ribosome-less plastids of iojap maize. Curr Genet 45:331–337

92. Hammani K, Barkan A (2014) An mTERF domain protein functions in group II intron splicing in maize chloroplasts. Nucleic Acids Res 42(8):5033–5042
93. Hammani K, Okuda K, Tanz SK, Chateigner-Boutin AL, Shikanai T, Small I (2009) A study of new *Arabidopsis* chloroplast RNA editing mutants reveals general features of editing factors and their target sites. Plant Cell 21:3686–3699
94. Hammani K, Cook WB, Barkan A (2012) RNA binding and RNA remodeling activities of the half-a-tetratricopeptide (HAT) protein HCF107 underlie its effects on gene expression. Proc Natl Acad Sci U S A 109:5651–5656
95. Han C-D, Patrie W, Polacco M, Coe EH (1993) Aberrations in plastid transcripts and deficiency of plastid DNA in striped and albino mutants in maize. Planta 191:552–563
96. Hattori M, Sugita M (2009) A moss pentatricopeptide repeat protein binds to the 3′ end of plastid clpP pre-mRNA and assists with mRNA maturation. FEBS J 276:5860–5869
97. Hedtke B, Börner T, Weihe A (1997) Mitochondrial and chloroplast phage-type RNA polymerases in *Arabidopsis*. Science 277:809–811
98. Hedtke B, Börner T, Weihe A (2000) One RNA polymerase serving two genomes. EMBO Rep 1:435–440
99. Hedtke B, Legen J, Weihe A, Herrmann RG, Börner T (2002) Six active phage-type RNA polymerase genes in *Nicotiana tabacum*. Plant J 30:625–637
100. Herschlag D (1995) RNA chaperones and the RNA folding problem. J Biol Chem 270:20871–20874
101. Hess WR, Börner T (1999) Organellar RNA polymerases of higher plants. Int Rev Cytol 190:1–59
102. Hess WR, Prombona A, Fieder B, Subramanian AR, Börner T (1993) Chloroplast rps15 and the rpoB/C1/C2 gene cluster are strongly transcribed in ribosome-deficient plastids: evidence for a functioning non-chloroplast-encoded RNA polymerase. EMBO J 12:563–571
103. Hess WR, Hoch B, Zeltz P, Hübschmann T, Kössel H, Börner T (1994) Inefficient rpl2 splicing in barley mutants with ribosome-deficient plastids. Plant Cell 6:1455–1465
104. Hetzer M, Wurzer G, Schweyen R, Mueller M (1997) Trans-activation of group II intron splicing by nuclear U5 snRNA. Nature 386:417–420
105. Hirose T, Sugiura M (2001) Involvement of a site-specific trans-acting factor and a common RNA-binding protein in the editing of chloroplast mRNAs: development of a chloroplast in vitro RNA editing system. Embo J 20:1144–1152
106. Hirose T, Kusumegi T, Tsudzuki T, Sugiura M (1999) RNA editing sites in tobacco chloroplast transcripts: editing as a possible regulator of chloroplast RNA polymerase activity. Mol Gen Genet 262:462–467
107. Horlitz M, Klaff P (2000) Gene-specific *trans*-regulatory functions of magnesium for chloroplast mRNA stability in higher plants. J Biol Chem 275:35638–35645
108. Hotto AM, Huston ZE, Stern DB (2010) Overexpression of a natural chloroplast-encoded antisense RNA in tobacco destabilizes 5S rRNA and retards plant growth. BMC Plant Biol 10:213
109. Hotto AM, Schmitz RJ, Fei Z, Ecker JR, Stern DB (2011) Unexpected diversity of chloroplast noncoding RNAs as revealed by deep sequencing of the *Arabidopsis* transcriptome. G3 1:559–570
110. Hricova A, Quesada V, Micol JL (2006) The SCABRA3 nuclear gene encodes the plastid RpoTp RNA polymerase, which is required for chloroplast biogenesis and mesophyll cell proliferation in *Arabidopsis*. Plant Physiol 141:942–956
111. Hu J, Bogorad L (1990) Maize chloroplast RNA polymerase: the 180-, 120-, and 38-kilodalton polypeptides are encoded in chloroplast genes. Proc Natl Acad Sci U S A 87:1531–1535
112. Hübschmann T, Börner T (1998) Characterisation of transcript initiation sites in ribosome-deficient barley plastids. Plant Mol Biol 36:493–496
113. Inada H, Seki M, Morikawa H, Nishimura M, Iba K (1997) Existence of three regulatory regions each containing a highly conserved motif in the promoter of plastid-encoded RNA polymerase gene (*rpoB*). Plant J 11:883–890

114. Iratni R, Baeza L, Andreeva A, Mache R, Lerbs-Mache S (1994) Regulation of rDNA transcription in chloroplasts: promoter exclusion by constitutive repression. Genes Dev 8:2928–2938
115. Iratni R, Diederich L, Harrak H, Bligny M, Lerbs-Mache S (1997) Organ-specific transcription of the *rrn* operon in spinach plastids. J Biol Chem 272:13676–13682
116. Ishikura K, Takaoka Y, Kato K, Sekine M, Yoshida K, Shinmyo A (1999) Expression of a foreign gene in *Chlamydomonas reinhardtii* chloroplast. J Biosci Bioeng 87:307–314
117. Jacobs J, Kuck U (2011) Function of chloroplast RNA-binding proteins. Cell Mol Life Sci 68:735–748
118. Jacob F, Perrin D, Sanchez C, Monod J (1960) [Operon: a group of genes with the expression coordinated by an operator.]. C R Hebd Seances Acad Sci 250:1727–1729
119. Jahn D (1992) Expression of the *Chlamydomonas reinhardtii* chloroplast $tRNA^{Glu}$ gene in a homologous *in vitro* transcription system is independent of upstream promoter elements. Arch Biochem Biophys 298:505–513
120. Jenkins BD, Barkan A (2001) Recruitment of a peptidyl-tRNA hydrolase as a facilitator of group II intron splicing in chloroplasts. Embo J 20:872–879
121. Jenkins BD, Kulhanek DJ, Barkan A (1997) Nuclear mutations that block group II RNA splicing in maize chloroplasts reveal several intron classes with distinct requirements for splicing factors. Plant Cell 9(3):283–296
122. Jeon Y, Jung HJ, Kang H, Park YI, Lee SH, Pai HS (2012) S1 domain-containing STF modulates plastid transcription and chloroplast biogenesis in *Nicotiana benthamiana*. New Phytol 193:349–363
123. Johnson C, Bowman L, Adai AT, Vance V, Sundaresan V (2007) CSRDB: a small RNA integrated database and browser resource for cereals. Nucleic Acids Res 35:D829–D833
124. Johnson X et al (2010) MRL1, a conserved pentatricopeptide repeat protein, is required for stabilization of rbcL mRNA in *Chlamydomonas* and *Arabidopsis*. Plant Cell 22:234–248
125. Kahlau S, Bock R (2008) Plastid transcriptomics and translatomics of tomato fruit development and chloroplast-to-chromoplast differentiation: chromoplast gene expression largely serves the production of a single protein. Plant Cell 20:856–874
126. Kanekatsu M, Munakata H, Furuzono K, Ohtsuki K (1993) Biochemical characterization of a 34 kDa ribonucleoprotein (p34) purified from the spinach chloroplast fraction as an effective phosphate acceptor for casein kinase II. FEBS Lett 335:176–180
127. Kaneko T et al (1996) Sequence analysis of the genome of the unicellular cyanobacterium *Synechocystis sp.* strain PCC6803. II. Sequence determination of the entire genome and assignment of potential protein-coding regions (supplement). DNA Res 3:185–209
128. Kanno A, Hirai A (1993) A transcription map of the chloroplast genome from rice (*Oryza sativa*). Curr Genet 23:166–174
129. Kapoor S, Sugiura M (1999) Identification of two essential sequence elements in the non-consensus type II P*atpB*-290 plastid promoter by using plastid transcription extracts from cultured tobacco BY-2 cells. Plant Cell 11:1799–1810
130. Kapoor S, Suzuki JY, Sugiura M (1997) Identification and functional significance of a new class of non-consensus-type plastid promoters. Plant J 11:327–337
131. Karcher D, Bock R (2002) The amino acid sequence of a plastid protein is developmentally regulated by RNA editing. J Biol Chem 277:5570–5574
132. Karcher D, Bock R (2002) Temperature sensitivity of RNA editing and intron splicing reactions in the plastid ndhB transcript. Curr Genet 41:48–52
133. Kasai S, Yoshimura S, Ishikura K, Takaoka Y, Kobayashi K, Kato K, Shinmyo A (2003) Effect of coding regions on chloroplast gene expression in *Chlamydomonas reinhardtii*. J Biosci Bioeng 95:276–282
134. Ke J, Chen RZ, Ban T, Zhou XE, Gu X, Tan MH, Chen C, Kang Y, Brunzelle JS, Zhu JK, Melcher K, Xu HE (2013) Structural basis for RNA recognition by a dimeric PPR-protein complex. Nat Struct Mol Biol 20(12):1377–1382. doi:10.1038/nsmb.2710. [Epub 2013 Nov 3]

135. Khrouchtchova A, Monde RA, Barkan A (2012) A short PPR protein required for the splicing of specific group II introns in angiosperm chloroplasts. RNA 18:1197–1209
136. Kim JW, Park JK, Kim BH, Lee JS, Sim WS (2002) Molecular analysis of the accumulation of the transcripts of the large subunit gene of ribulose-1,5-bisphosphate carboxylase/oxygenase by light. Mol Cells 14:281–287
137. Kim M, Thum KE, Morishige DT, Mullet JE (1999) Detailed architecture of the barley chloroplast *psbD-psbC* blue light-responsive promoter. J Biol Chem 274:4684–4692
138. Kindgren P, Kremnev D, Blanco NE, de Dios Barajas Lopez J, Fernandez AP, Tellgren-Roth C, Small I, Strand A (2012) The plastid redox insensitive 2 mutant of *Arabidopsis* is impaired in PEP activity and high light-dependent plastid redox signalling to the nucleus. Plant J 70:279–291
139. Klaff P, Gruissem W (1991) Changes in chloroplast mRNA stability during leaf development. Plant Cell 3:517–529
140. Klein U, De Camp JD, Bogorad L (1992) Two types of chloroplast gene promoters in *Chlamydomonas reinhardtii*. Proc Natl Acad Sci U S A 89:3453–3457
141. Klein U, Salvador ML, Bogorad L (1994) Activity of the *Chlamydomonas* chloroplast *rbcL* gene promoter is enhanced by a remote sequence element. Proc Natl Acad Sci U S A 91:10819–10823
142. Kobayashi Y, Dokiya Y, Sugita M (2001) Dual targeting of phage-type RNA polymerase to both mitochondria and plastids is due to alternative translation initiation in single transcripts. Biochem Biophys Res Commun 289:1106–1113
143. Kotera E1, Tasaka M, Shikanai T (2005) A pentatricopeptide repeat protein is essential for RNA editing in chloroplasts. Nature 433(7023):326–330
144. Krause K, Krupinska K (2000) Molecular and functional properties of highly purified transcriptionally active chromosomes from spinach chloroplasts. Physiol Plant 109:188–195
145. Kroeger TS, Watkins KP, Friso G, van Wijk KJ, Barkan A (2009) A plant-specific RNA-binding domain revealed through analysis of chloroplast group II intron splicing. Proc Natl Acad Sci USA 106(11):4537–4542
146. Kühn K, Bohne AV, Liere K, Weihe A, Börner T (2007) *Arabidopsis* phage-type RNA polymerases: accurate in vitro transcription of organellar genes. Plant Cell 19:959–971
147. Kühn K et al (2009) Phage-type RNA polymerase RPOTmp performs gene-specific transcription in mitochondria of *Arabidopsis thaliana*. Plant Cell 21:2762–2779
148. Kusumi K, Yara A, Mitsui N, Tozawa Y, Iba K (2004) Characterization of a rice nuclear-encoded plastid RNA polymerase gene *OsRpoTp*. Plant Cell Physiol 45:1194–1201
149. Lambowitz AM, Zimmerly S (2004) Mobile group II introns. Annu Rev Genet 38:1–35
150. Lee J, Herrin DL (2003) Mutagenesis of a light-regulated psbA intron reveals the importance of efficient splicing for photosynthetic growth. Nucleic Acids Res 31:4361–4372
151. Legen J, Kemp S, Krause K, Profanter B, Herrmann RG, Maier RM (2002) Comparative analysis of plastid transcription profiles of entire plastid chromosomes from tobacco attributed to wild-type and PEP-deficient transcription machineries. Plant J 31:171–188
152. Lee K, Lee HJ, Kim DH, Jeon Y, Pai HS, Kang H (2014) A nuclear-encoded chloroplast protein harboring a single CRM domain plays an important role in the Arabidopsis growth and stress response. BMC Plant Biol 14:98
153. Lerbs-Mache S (1993) The 110-kDa polypeptide of spinach plastid DNA-dependent RNA polymerase: single-subunit enzyme or catalytic core of multimeric enzyme complexes? Proc Natl Acad Sci U S A 90:5509–5513
154. Lerbs-Mache S (2011) Function of plastid sigma factors in higher plants: regulation of gene expression or just preservation of constitutive transcription? Plant Mol Biol 76:235–249
155. Liere K, Börner T (2007) Transcription and transcriptional regulation in plastids. In: Bock R (ed) Cell and molecular biology of plastids. Springer, Berlin, pp 121–174
156. Liere K, Link G (1994) Structure and expression characteristics of the chloroplast DNA region containing the split gene for tRNA$^{(Gly)}$ (UCC) from mustard (*Sinapis alba* L.). Curr Genet 26:557–563

157. Liere K, Maliga P (1999) *In vitro* characterization of the tobacco *rpoB* promoter reveals a core sequence motif conserved between phage-type plastid and plant mitochondrial promoters. EMBO J 18:249–257
158. Liere K, Maliga P (2001) Plastid RNA polymerases in higher plants. In: Aro E-M, Andersson B (eds) Regulation of photosynthesis. Springer, Netherlands, pp 29–49
159. Liere K, Kaden D, Maliga P, Börner T (2004) Overexpression of phage-type RNA polymerase RpoTp in tobacco demonstrates its role in chloroplast transcription by recognizing a distinct promoter type. Nucleic Acids Res 32:1159–1165
160. Liere K, Weihe A, Börner T (2011) The transcription machineries of plant mitochondria and chloroplasts: composition, function, and regulation. J Plant Physiol 168:1345–1360
161. Link G (1984) DNA sequence requirements for the accurate transcription of a protein-coding plastid gene in a plastid *in vitro* system from mustard (*Sinapis alba L.*). EMBO J 3:1697–1704
162. Link G (1996) Green life: control of chloroplast gene transcription. BioEssays 18:465–471
163. Lisitsky I, Schuster G (1995) Phosphorylation of a chloroplast RNA-binding protein changes its affinity to RNA. Nucleic Acids Res 23:2506–2511
164. Little MC, Hallick RB (1988) Chloroplast *rpoA*, *rpoB*, and *rpoC* genes specify at least three components of a chloroplast DNA-dependent RNA polymerase active in tRNA and mRNA transcription. J Biol Chem 263:14302–14307
165. Liu B, Troxler RF (1996) Molecular characterization of a positively photoregulated nuclear gene for a chloroplast RNA polymerase sigma factor in *Cyanidium caldarium*. Proc Natl Acad Sci U S A 93:3313–3318
166. Lohse M, Drechsel O, Bock R (2007) OrganellarGenomeDRAW (OGDRAW): a tool for the easy generation of high-quality custom graphical maps of plastid and mitochondrial genomes. Curr Genet 52:267–274
167. Loiselay C, Gumpel NJ, Girard-Bascou J, Watson AT, Purton S, Wollman FA, Choquet Y (2008) Molecular identification and function of cis- and trans-acting determinants for petA transcript stability in *Chlamydomonas reinhardtii* chloroplasts. Mol Cell Biol 28:5529–5542
168. de Longevialle AF, Hendrickson L, Taylor NL, Delannoy E, Lurin C, Badger M, Millar AH, Small I (2008) The pentatricopeptide repeat gene OTP51 with two LAGLIDADG motifs is required for the cis-splicing of plastid ycf3 intron 2 in *Arabidopsis thaliana*. Plant J 56(1):157–168
169. Lung B, Zemann A, Madej MJ, Schuelke M, Techritz S, Ruf S, Bock R, Huttenhofer A (2006) Identification of small non-coding RNAs from mitochondria and chloroplasts. Nucleic Acids Res 34:3842–3852
170. Lurin C et al (2004) Genome-wide analysis of *Arabidopsis* pentatricopeptide repeat proteins reveals their essential role in organelle biogenesis. Plant Cell 16:2089–2103
171. Lynch M, Blanchard JL (1998) Deleterious mutation accumulation in organelle genomes. Genetica 102-103:29–39
172. Lynch M, Koskella B, Schaack S (2006) Mutation pressure and the evolution of organelle genomic architecture. Science 311:1727–1730
173. Lysenko E (2006) Analysis of the evolution of the family of the *Sig* genes encoding plant sigma factors. Russ J Plant Physiol 53:605–614
174. Lyubetsky VA, Zverkov OA, Rubanov LI, Seliverstov AV (2011) Modeling RNA polymerase competition: the effect of sigma-subunit knockout and heat shock on gene transcription level. Biol Direct 6:3
175. Maier UG, Bozarth A, Funk HT, Zauner S, Rensing SA, Schmitz-Linneweber C, Borner T, Tillich M (2008) Complex chloroplast RNA metabolism: just debugging the genetic programme? BMC Biol 6:36
176. Majeran W, Friso G, Asakura Y, Qu X, Huang M, Ponnala L, Watkins KP, Barkan A, van Wijk KJ (2011) Nucleoid-enriched proteomes in developing plastids and chloroplasts from maize leaves: a new conceptual framework for nucleoid functions. Plant Physiol 158:156–89

177. Marchive C et al (2009) Abnormal physiological and molecular mutant phenotypes link chloroplast polynucleotide phosphorylase to the phosphorus deprivation response in *Arabidopsis*. Plant Physiol 151:905–924
178. Mathews DE, Durbin RD (1990) Tagetitoxin inhibits RNA synthesis directed by RNA polymerases from chloroplasts and *Escherichia coli*. J Biol Chem 265:493–498
179. Matsunaga M, Jaehning JA (2004) Intrinsic promoter recognition by a "core" RNA polymerase. J Biol Chem 279:44239–44242
180. Matsuzaki M et al (2004) Genome sequence of the ultrasmall unicellular red alga *Cyanidioschyzon merolae* 10D. Nature 428:653–657
181. Maul JE, Lilly JW, Cui L, dePamphilis CW, Miller W, Harris EH, Stern DB (2002) The *Chlamydomonas reinhardtii* plastid chromosome: islands of genes in a sea of repeats. Plant Cell 14:2659–2679
182. McCullough AJ, Kangasjarvi J, Gengenback BG, Jones RJ (1992) Plastid DNA in developing maize endosperm. Plant Physiol 100:958–964
183. Meierhoff K, Felder S, Nakamura T, Bechtold N, Schuster G (2003) HCF152, an *Arabidopsis* RNA binding pentatricopeptide repeat protein involved in the processing of chloroplast *psbB-psbT-psbH-petB-petD* RNAs. Plant Cell 15:1480–1495
184. Merendino L, Perron K, Rahire M, Howald I, Rochaix JD, Goldschmidt-Clermont M (2006) A novel multifunctional factor involved in trans-splicing of chloroplast introns in Chlamydom*onas. Nucleic Acids Res* 34(1):262–274
185. Michel F, Umesono K, Ozeki H (1989) Comparative and functional anatomy of group II catalytic introns—a review. Gene 82:5–30
186. Miyamoto T, Obokata J, Sugiura M (2002) Recognition of RNA editing sites is directed by unique proteins in chloroplasts: biochemical identification of cis-acting elements and trans-acting factors involved in RNA editing in tobacco and pea chloroplasts. Mol Cell Biol 22(19):6726–6734
187. Miyata Y, Sugita M (2004) Tissue- and stage-specific RNA editing of rps 14 transcripts in moss (*Physcomitrella patens*) chloroplasts. J Plant Physiol 161:113–115
188. Mohorianu I, Schwach F, Jing R, Lopez-Gomollon S, Moxon S, Szittya G, Sorefan K, Moulton V, Dalmay T (2011) Profiling of short RNAs during fleshy fruit development reveals stage-specific sRNAome expression patterns. Plant J 67:232–246
189. Morden CW, Delwiche CF, Kuhsel M, Palmer JD (1992) Gene phylogenies and the endosymbiotic origin of plastids. Biosystems 28:75–90
190. Morin RD, Aksay G, Dolgosheina E, Ebhardt HA, Magrini V, Mardis ER, Sahinalp SC, Unrau PJ (2008) Comparative analysis of the small RNA transcriptomes of *Pinus contorta* and *Oryza sativa*. Genome Res 18:571–584
191. Mudd EA, Sullivan S, Gisby MF, Mironov A, Kwon CS, Chung WI, Day A (2008) A 125 kDa RNase E/G-like protein is present in plastids and is essential for chloroplast development and autotrophic growth in *Arabidopsis*. J Exp Bot 59:2597–2610
192. Mullet JE (1993) Dynamic regulation of chloroplast transcription. Plant Physiol 103:309–313
193. Nagashima A, Hanaoka M, Shikanai T, Fujiwara M, Kanamaru K, Takahashi H, Tanaka K (2004) The multiple-stress responsive plastid sigma factor, SIG5, directs activation of the *psbD* blue light-responsive promoter (BLRP) in *Arabidopsis thaliana*. Plant Cell Physiology 45:357–368
194. Nakamura T, Ohta M, Sugiura M, Sugita M (2001) Chloroplast ribonucleoproteins function as a stabilizing factor of ribosome-free mRNAs in the stroma. J Biol Chem 276:147–152
195. Neuhaus H, Link G (1990) The chloroplast *psbK* operon from mustard (*Sinapis alba* L.): multiple transcripts during seedling development and evidence for divergent overlapping transcription. Curr Genet 18:377–383
196. Nickelsen J, Link G (1990) Nucleotide sequence of the mustard chloroplast genes *trnH* and *rps19*. Nucleic Acids Res 18:1051

197. Ogrzewalla K, Piotrowski M, Reinbothe S, Link G (2002) The plastid transcription kinase from mustard (*Sinapis alba* L.). A nuclear-encoded CK2-type chloroplast enzyme with redox-sensitive function. Eur J Biochem 269:3329–3337
198. Ohyama K et al (1986) Chloroplast gene organization deduced from complete sequence of liverwort *Marchantia polymorpha* chloroplast DNA. Nature 322:572–574
199. Olinares PD, Ponnala L, van Wijk KJ (2010) Megadalton complexes in the chloroplast stroma of *Arabidopsis thaliana* characterized by size exclusion chromatography, mass spectrometry, and hierarchical clustering. Mol Cell Proteomics 9:1594–1615
200. Okuda K, Nakamura T, Sugita M, Shimizu T, Shikanai T (2006) A pentatricopeptide repeat protein is a site recognition factor in chloroplast RNA editing. J Biol Chem 281: 37661–37667
201. Okuda K, Myouga F, Motohashi R, Shinozaki K, Shikanai T (2007) Conserved domain structure of pentatricopeptide repeat proteins involved in chloroplast RNA editing. Proc Natl Acad Sci USA 104:8178–8183
202. Okuda K, Chateigner-Boutin AL, Nakamura T, Delannoy E, Sugita M, Myouga F, Motohashi R, Shinozaki K, Small I, Shikanai T (2009) Pentatricopeptide repeat proteins with the DYW motif have distinct molecular functions in RNA editing and RNA cleavage in Arabidopsis chloroplasts. Plant Cell 21:146–156
203. Okuda K, Hammani K, Tanz SK, Peng L, Fukao Y, Myouga F, Motohashi R, Shinozaki K, Small I, Shikanai T (2009) The pentatricopeptide repeat protein OTP82 is required for RNA editing of plastid ndhB and ndhG transcripts. Plant J 61:339–349
204. Onda Y, Yagi Y, Saito Y, Takenaka N, Toyoshima Y (2008) Light induction of *Arabidopsis SIG1* and *SIG5* transcripts in mature leaves: differential roles of cryptochrome 1 and cryptochrome 2 and dual function of SIG5 in the recognition of plastid promoters. Plant J 55:968–978
205. Ossenbuhl F, Nickelsen J (2000) cis- and trans-acting determinants for translation of psbD mRNA in *Chlamydomonas reinhardtii*. Mol Cell Biol 20:8134–8142
206. Ostheimer GJ, Williams-Carrier R, Belcher S, Osborne E, Gierke J, Barkan A (2003) Group II intron splicing factors derived by diversification of an ancient RNA-binding domain. EMBO J 22(15):3919–3929
207. Ostheimer GJ, Hadjivassiliou H, Kloer DP, Barkan A, Matthews BW (2005) Structural analysis of the group II intron splicing factor CRS2 yields insights into its protein and RNA interaction surfaces. J Mol Biol 345(1):51–68
208. Ostersetzer O, Cooke AM, Watkins KP, Barkan A (2005) CRS1, a chloroplast group II intron splicing factor, promotes intron folding through specific interactions with two intron domains. Plant Cell 17:241–255
209. Palmer JD (1990) Contrasting modes and tempos of genome evolution in land plant organelles. Trends Genet 6:115–120
210. Pan T, Sosnick T (2006) RNA folding during transcription. Annu Rev Biophys Biomol Struct 35:161–175
211. Peeters NM, Hanson MR (2002) Transcript abundance supercedes editing efficiency as a factor in developmental variation of chloroplast gene expression. Rna 8:497–511
212. Peltier JB, Cai Y, Sun Q, Zabrouskov V, Giacomelli L, Rudella A, Ytterberg AJ, Rutschow H, van Wijk KJ (2006) The oligomeric stromal proteome of *Arabidopsis thaliana* chloroplasts. Mol Cell Proteomics 5:114–133
213. Perron K, Goldschmidt-Clermont M, Rochaix JD (1999) A factor related to pseudouridine synthases is required for chloroplast group II intron trans-splicing in *Chlamydomonas reinhardtii*. Embo J 18:6481–6490
214. Petersen K, Schottler MA, Karcher D, Thiele W, Bock R (2011) Elimination of a group II intron from a plastid gene causes a mutant phenotype. Nucleic Acids Res 39:5181–5192
215. Pfalz J, Liere K, Kandlbinder A, Dietz KJ, Oelmuller R (2006) pTAC2, −6, and −12 are components of the transcriptionally active plastid chromosome that are required for plastid gene expression. Plant Cell 18:176–197

216. Pfalz J, Bayraktar OA, Prikryl J, Barkan A (2009) Site-specific binding of a PPR protein defines and stabilizes 5′ and 3′ mRNA termini in chloroplasts. EMBO J 28:2042–2052
217. Pfannschmidt T, Link G (1997) The A and B forms of plastid DNA-dependent RNA polymerase from mustard (*Sinapis alba L.*) transcribe the same genes in a different developmental context. Mol Gen Genet 257:35–44
218. Pfannschmidt T, Ogrzewalla K, Baginsky S, Sickmann A, Meyer HE, Link G (2000) The multisubunit chloroplast RNA polymerase A from mustard (*Sinapis alba L.*). Integration of a prokaryotic core into a larger complex with organelle-specific functions. Eur J Biochem 267:253–261
219. Pombert JF, Otis C, Lemieux C, Turmel M (2005) The chloroplast genome sequence of the green alga *Pseudendoclonium akinetum* (Ulvophyceae) reveals unusual structural features and new insights into the branching order of chlorophyte lineages. Mol Biol Evol 22:1903–1918
220. Prikryl J, Watkins KP, Friso G, van Wijk KJ, Barkan A (2008) A member of the Whirly family is a multifunctional RNA- and DNA-binding protein that is essential for chloroplast biogenesis. Nucleic Acids Res 36(16):5152–5165
221. Prikryl J, Rojas M, Schuster G, Barkan A (2011) Mechanism of RNA stabilization and translational activation by a pentatricopeptide repeat protein. Proc Natl Acad Sci USA 108:415–420
222. Proshkin S, Rahmouni AR, Mironov A, Nudler E (2010) Cooperation between translating ribosomes and RNA polymerase in transcription elongation. Science 328:504–508
223. Purton S, Gray JC (1989) The plastid *rpoA* gene encoding a protein homologous to the bacterial RNA polymerase alpha subunit is expressed in pea chloroplasts. Mol Gen Genet 217:77–84
224. Puthiyaveetil S et al (2008) The ancestral symbiont sensor kinase CSK links photosynthesis with gene expression in chloroplasts. Proc Natl Acad Sci U S A 105:10061–10066
225. Puthiyaveetil S, Ibrahim IM, Allen JF (2012) Oxidation-reduction signalling components in regulatory pathways of state transitions and photosystem stoichiometry adjustment in chloroplasts. Plant Cell Environ 35:347–359
226. Pyle A (2002) Metal ions in the structure and function of RNA. J Biol Inorg Chem 7:679–690
227. Qi Y, Armbruster U, Schmitz-Linneweber C, Delannoy E, de Longevialle AF, Ruhle T, Small I, Jahns P, Leister D (2011) *Arabidopsis* CSP41 proteins form multimeric complexes that bind and stabilize distinct plastid transcripts. J Exp Bot 63:1251–1270
228. Qiao J, Ma C, Wimmelbacher M, Bornke F, Luo M (2011) Two novel proteins, MRL7 and its paralog MRL7-L, have essential but functionally distinct roles in chloroplast development and are involved in plastid gene expression regulation in *Arabidopsis*. Plant Cell Physiol 52:1017–1030
229. Raynaud C, Loiselay C, Wostrikoff K, Kuras R, Girard-Bascou J, Wollman FA, Choquet Y (2007) Evidence for regulatory function of nucleus-encoded factors on mRNA stabilization and translation in the chloroplast. Proc Natl Acad Sci U S A 104:9093–9098
230. Reiss T, Link G (1985) Characterization of transcriptionally active DNA-protein complexes from chloroplasts and etioplasts of mustard (*Sinapis alba* L.). Eur J Biochem 148:207–212
231. Richter CV, Bals T, Schunemann D (2010) Component interactions, regulation and mechanisms of chloroplast signal recognition particle-dependent protein transport. Eur J Cell Biol 89:965–973
232. Ringel R, Sologub M, Morozov YI, Litonin D, Cramer P, Temiakov D (2011) Structure of human mitochondrial RNA polymerase. Nature 478:269–273
233. Rivier C, Goldschmidt-Clermont M, Rochaix JD (2001) Identification of an RNA-protein complex involved in chloroplast group II intron trans-splicing in *Chlamydomonas reinhardtii*. EMBO J 20:1765–1773
234. Robbins JC, Heller WP, Hanson MR (2009) A comparative genomics approach identifies a PPR-DYW protein that is essential for C-to-U editing of the *Arabidopsis* chloroplast accD transcript. RNA 15:1142–1153

235. Rochaix JD, Rahire M, Michel F (1985) The chloroplast ribosomal intron of *Chlamydomonas reinhardii* codes for a polypeptide related to mitochondrial maturases. Nucleic Acids Res 13:975–984
236. Rudinger M, Szovenyi P, Rensing SA, Knoop V (2011) Assigning DYW-type PPR proteins to RNA editing sites in the funariid mosses *Physcomitrella patens* and *Funaria hygrometrica*. Plant J 67:370–380
237. Ruf S, Kössel H (1997) Tissue-specific and differential editing of the two *ycf3* editing sites in maize plastids. Curr Genet 32:19–23
238. Ruf S, Kossel H, Bock R (1997) Targeted inactivation of a tobacco intron-containing open reading frame reveals a novel chloroplast-encoded photosystem I-related gene. J Cell Biol 139:95–102
239. Ruwe H, Schmitz-Linneweber C (2012) Short non-coding RNA fragments accumulating in chloroplasts: footprints of RNA binding proteins? Nucleic Acids Res 40:3106–3116
240. Ruwe H, Castandet B, Schmitz-Linneweber C, Stern DB (2013) *Arabidopsis* chloroplast quantitative editotype. FEBS Lett 587(9):1429–1433
241. Sakai A, Takano H, Kuroiwa T (2004) Organelle nuclei in higher plants: structure, composition, function, and evolution. Int Rev Cytol 238:59–118
242. Saldanha R, Mohr G, Belfort M, Lambowitz AM (1993) Group I and group II introns. FASEB J 7:15–24
243. Salone V, Rudinger M, Polsakiewicz M, Hoffmann B, Groth-Malonek M, Szurek B, Small I, Knoop V, Lurin C (2007) A hypothesis on the identification of the editing enzyme in plant organelles. FEBS Lett 581:4132–4138
244. Salvador ML, Suay L, Klein U (2011) Messenger RNA degradation is initiated at the 5′ end and follows sequence- and condition-dependent modes in chloroplasts. Nucleic Acids Res 39:6213–6222
245. Sane AP, Stein B, Westhoff P (2005) The nuclear gene HCF107 encodes a membrane-associated R-TPR (RNA tetratricopeptide repeat)-containing protein involved in expression of the plastidial *psbH* gene in *Arabidopsis*. Plant J 42:720–730
246. Sasaki T, Yukawa Y, Miyamoto T, Obokata J, Sugiura M (2003) Identification of RNA editing sites in chloroplast transcripts from the maternal and paternal progenitors of tobacco (*Nicotiana tabacum*): comparative analysis shows the involvement of distinct *trans*-factors for *ndhB* editing. Mol Biol Evol 20:1028–1035
247. Sashital DG, Cornilescu G, McManus CJ, Brow DA, Butcher SE (2004) U2-U6 RNA folding reveals a group II intron-like domain and a four-helix junction. Nat Struct Mol Biol 11:1237–1242
248. Satoh J, Baba K, Nakahira Y, Tsunoyama Y, Shiina T, Toyoshima Y (1999) Developmental stage-specific multi-subunit plastid RNA polymerases (PEP) in wheat. Plant J 18:407–415
249. Schein A, Sheffy-Levin S, Glaser F, Schuster G (2008) The RNase E/G-type endoribonuclease of higher plants is located in the chloroplast and cleaves RNA similarly to the *E. coli* enzyme. RNA 14:1057–1068
250. Schevitz RW, Otwinowski Z, Joachimiak A, Lawson CL, Sigler PB (1985) The three-dimensional structure of trp repressor. Nature 317:782–786
251. Schiffer S, Rosch S, Marchfelder A (2002) Assigning a function to a conserved group of proteins: the tRNA 3′-processing enzymes. EMBO J 21:2769–2777
252. Schmitz-Linneweber C, Barkan A (2007) RNA splicing and RNA editing in chloroplasts. In: Bock R (ed) Cell and molecular biology of plastids. Springer, Berlin, pp 213–248
253. Schmitz-Linneweber C, Small I (2008) Pentatricopeptide repeat proteins: a socket set for organelle gene expression. Trends Plant Sci 13:663–670
254. Schmitz-Linneweber C, Tillich M, Herrmann RG, Maier RM (2001) Heterologous, splicing-dependent RNA editing in chloroplasts: allotetraploidy provides *trans*-factors. Embo J 20:4874–4883
255. Schmitz-Linneweber C, Regel R, Du TG, Hupfer H, Herrmann RG, Maier RM (2002) The plastid chromosome of *Atropa belladonna* and its comparison with that of *Nicotiana taba-*

cum: the role of RNA editing in generating divergence in the process of plant speciation. Mol Biol Evol 19:1602–1612

256. Schmitz-Linneweber C, Kushnir S, Babiychuk E, Poltnigg P, Herrmann RG, Maier RM (2005) Pigment deficiency in nightshade/tobacco cybrids is caused by the failure to edit the plastid ATPase alpha-subunit mRNA. Plant Cell 17:1815–1828
257. Schmitz-Linneweber C, Williams-Carrier R, Williams P, Kroeger T, Vichas A, Barkan A (2006) A pentatricopeptide repeat protein binds to and facilitates the *trans*-splicing of the maize chloroplast *rps12* pre-mRNA. Plant Cell 18:2650–2663
258. Schwacke R, Fischer K, Ketelsen B, Krupinska K, Krause K (2007) Comparative survey of plastid and mitochondrial targeting properties of transcription factors in *Arabidopsis* and rice. Mol Genet Genomics 277:631–646
259. Schwarz C, Elles I, Kortmann J, Piotrowski M, Nickelsen J (2007) Synthesis of the D2 protein of photosystem II in *Chlamydomonas* is controlled by a high molecular mass complex containing the RNA stabilization factor Nac2 and the translational activator RBP40. Plant Cell 19:3627–3639
260. Schweer J (2010) Plant sigma factors come of age: flexible transcription factor network for regulated plastid gene expression. Endocytobio Cell Res 20:1–20
261. Schweer J, Loschelder H, Link G (2006) A promoter switch that can rescue a plant sigma factor mutant. FEBS Lett 580:6617–6622
262. Schweer J, Turkeri H, Kolpack A, Link G (2010) Role and regulation of plastid sigma factors and their functional interactors during chloroplast transcription-recent lessons from *Arabidopsis thaliana*. Eur J Cell Biol 89:940–946
263. Schwinghammer K, Cheung AC, Morozov YI, Agaronyan K, Temiakov D, Cramer P (2013) Structure of human mitochondrial RNA polymerase elongation complex. Nat Struct Mol Biol 20(11):1298–1303. doi:10.1038/nsmb.2683. [Epub 2013 Oct 6]
264. Serino G, Maliga P (1998) RNA polymerase subunits encoded by the plastid *rpo* genes are not shared with the nucleus-encoded plastid enzyme. Plant Physiol 117:1165–1170
265. Severinov K, Mustaev A, Kukarin A, Muzzin O, Bass I, Darst SA, Goldfarb A (1996) Structural modules of the large subunits of RNA polymerase. Introducing archaebacterial and chloroplast split sites in the beta and beta′ subunits of *Escherichia coli* RNA polymerase. J Biol Chem 271:27969–27974
266. Sharwood RE, Halpert M, Luro S, Schuster G, Stern DB (2011) Chloroplast RNase J compensates for inefficient transcription termination by removal of antisense RNA. RNA 17:2165–2176
267. Sharwood RE, Hotto AM, Bollenbach TJ, Stern DB (2011) Overaccumulation of the chloroplast antisense RNA AS5 is correlated with decreased abundance of 5S rRNA *in vivo* and inefficient 5S rRNA maturation *in vitro*. RNA 17:230–243
268. Shields DC, Wolfe KH (1997) Accelerated evolution of sites undergoing mRNA editing in plant mitochondria and chloroplasts. Mol Biol Evol 14:344–349
269. Shinozaki K et al (1986) The complete nucleotide sequence of the tobacco chloroplast genome: its gene organization and expression. EMBO J 5:2043–2049
270. Shukla GC, Padgett RA (2002) A catalytically active group II intron domain 5 can function in the U12-dependent spliceosome. Mol Cell 9:1145–1150
271. Siemenroth A, Wollgiehn R, Neumann D, Börner T (1981) Synthesis of ribosomal RNA in ribosome-deficient plastids of the mutant "albostrians" of *Hordeum vulgare* L. Planta 153:547–555
272. Sijben-Muller G, Hallick RB, Alt J, Westhoff P, Herrmann RG (1986) Spinach plastid genes coding for initiation factor IF-1, ribosomal protein S11 and RNA polymerase alpha-subunit. Nucleic Acids Res 14:1029–1044
273. Silhavy D, Maliga P (1998) Mapping of promoters for the nucleus-encoded plastid RNA polymerase (NEP) in the *iojap* maize mutant. Curr Genet 33:340–344
274. Small I, Peeters N (2000) The PPR motif-a TPR-related motif prevalent in plant organellar proteins. Trends Biochem Sci 25:46–47

275. Sriraman P, Silhavy D, Maliga P (1998) The phage-type P*clpP*-53 plastid promoter comprises sequences downstream of the transcription initiation site. Nucleic Acids Res 26:4874–4879
276. Steiner S, Schröter Y, Pfalz J, Pfannschmidt T (2011) Identification of essential subunits in the plastid-encoded RNA polymerase complex reveals building blocks for proper plastid development. Plant Physiol 157:1043–1055
277. Steitz TA (2009) The structural changes of T7 RNA polymerase from transcription initiation to elongation. Curr Opin Struct Biol 19:683–690
278. Stern DB, Goldschmidt-Clermont M, Hanson MR (2010) Chloroplast RNA metabolism. Annu Rev Plant Biol 61:125–155
279. Stoppel R, Meurer J (2011) The cutting crew-ribonucleases are key players in the control of plastid gene expression. J Exp Bot 63:1663–1673
280. Stoppel R, Lezhneva L, Schwenkert S, Torabi S, Felder S, Meierhoff K, Westhoff P, Meurer J (2011) Recruitment of a ribosomal release factor for light- and stress-dependent regulation of petB transcript stability in *Arabidopsis* chloroplasts. Plant Cell 23:2680–2695
281. Stothard P, Wishart DS (2005) Circular genome visualization and exploration using CGView. Bioinformatics 21:537–539
282. Strittmatter G, Gozdzicka-Jozefiak A, Kössel H (1985) Identification of an rRNA operon promoter from *Zea mays* chloroplasts which excludes the proximal *tRNAVal* from the primary transcript. EMBO J 4:599–604
283. Sun T, Germain A, Giloteaux L, Hammani K, Barkan A, Hanson MR, Bentolila S (2013) An RNA recognition motif-containing protein is required for plastid RNA editing in Arabidopsis and maize. Proc Natl Acad Sci USA 110(12):E1169–1178. doi:10.1073/pnas.1220162110. [Epub 2013 Mar 4]
284. Surovtseva YV, Shutt TE, Cotney J, Cimen H, Chen SY, Koc EC, Shadel GS (2011) Mitochondrial ribosomal protein L12 selectively associates with human mitochondrial RNA polymerase to activate transcription. Proc Natl Acad Sci U S A 108:17921–17926
285. Suzuki JY, Maliga P (2000) Engineering of the *rpl23* gene cluster to replace the plastid RNA polymerase alpha subunit with the *Escherichia coli* homologue. Curr Genet 38:218–225
286. Suzuki JY, Jimmy Ytterberg A, Beardslee TA, Allison LA, Wijk KJ, Maliga P (2004) Affinity purification of the tobacco plastid RNA polymerase and in vitro reconstitution of the holoenzyme. Plant J 40:164–172
287. Swiatecka-Hagenbruch M, Liere K, Börner T (2007) High diversity of plastidial promoters in *Arabidopsis thaliana*. Mol Genet Genomics 277:725–734
288. Swiatecka-Hagenbruch M, Emanuel C, Hedtke B, Liere K, Börner T (2008) Impaired function of the phage-type RNA polymerase RpoTp in transcription of chloroplast genes is compensated by a second phage-type RNA polymerase. Nucleic Acids Res 36:785–792
289. Takenaka M, Zehrmann A, Verbitskiy D, Kugelmann M, Hartel B, Brennicke A (2012) Multiple organellar RNA editing factor (MORF) family proteins are required for RNA editing in mitochondria and plastids of plants. Proc Natl Acad Sci U S A 109:5104–5109
290. Tanaka K, Oikawa K, Ohta N, Kuroiwa H, Kuroiwa T, Takahashi H (1996) Nuclear encoding of a chloroplast RNA polymerase sigma subunit in a red alga. Science 272:1932–1935
291. Tanaka K, Tozawa Y, Mochizuki N, Shinozaki K, Nagatani A, Wakasa K, Takahashi H (1997) Characterization of three cDNA species encoding plastid RNA polymerase sigma factors in *Arabidopsis thaliana*: evidence for the sigma factor heterogeneity in higher plant plastids. FEBS Lett 413:309–313
292. Thomas BC, Li X, Gegenheimer P (2000) Chloroplast ribonuclease P does not utilize the ribozyme-type pre-tRNA cleavage mechanism. RNA 6:545–553
293. Tan J, Tan Z, Wu F, Sheng P, Heng Y, Wang X, Ren Y, Wang J, Guo X, Zhang X, Cheng Z, Jiang L, Liu X, Wang H, Wan J (2014) A novel chloroplast-localized pentatricopeptide repeat protein involved in splicing affects chloroplast development and abiotic stress response in rice. Mol Plant pii: ssu054. [Epub ahead of print]

294. Till B, Schmitz-Linneweber C, Williams-Carrier R, Barkan A (2001) CRS1 is a novel group II intron splicing factor that was derived from a domain of ancient origin. RNA 7:1227–1238
295. Tiller K, Link G (1993) Phosphorylation and dephosphorylation affect functional characteristics of chloroplast and etioplast transcription systems from mustard. EMBO J 12:1745–1753
296. Tillich M, Funk HT, Schmitz-Linneweber C, Poltnigg P, Sabater B, Martin M, Maier RM (2005) Editing of plastid RNA in *Arabidopsis thaliana* ecotypes. Plant J 43:708–715
297. Tillich M, Poltnigg P, Kushnir S, Schmitz-Linneweber C (2006) Maintenance of plastid RNA editing activities independently of their target sites. EMBO Rep 7:308–313
298. Tillich M, Hardel SL, Kupsch C, Armbruster U, Delannoy E, Gualberto JM, Lehwark P, Leister D, Small ID, Schmitz-Linneweber C (2009) Chloroplast ribonucleoprotein CP31A is required for editing and stability of specific chloroplast mRNAs. Proc Natl Acad Sci U S A 106:6002–6007
299. Tillich M, Le Sy V, Schulerowitz K, von Haeseler A, Maier UG, Schmitz-Linneweber C (2009) Loss of matK RNA editing in seed plant chloroplasts. BMC Evol Biol 9:201
300. Tillich M, Beick S, Schmitz-Linneweber C (2010) Chloroplast RNA-binding proteins: repair and regulation of chloroplast transcripts. RNA Biol 7:172–178
301. Tseng CC, Sung TY, Li YC, Hsu SJ, Lin CL, Hsieh MH (2010) Editing of accD and ndhF chloroplast transcripts is partially affected in the *Arabidopsis vanilla cream1* mutant. Plant Mol Biol 73:309–323
302. Türkeri H, Schweer J, Link G (2012) Phylogenetic and functional features of the plastid transcription kinase cpCK2 from *Arabidopsis* signify a role of cysteinyl SH-groups in regulatory phosphorylation of plastid sigma factors. FEBS J 279:395–409
303. Uniacke J, Zerges W (2009) Chloroplast protein targeting involves localized translation in *Chlamydomonas*. Proc Natl Acad Sci U S A 106:1439–1444
304. Vaistij F, Boudreau E, Lemaire S, Goldschmidt-Clermont M, Rochaix J (2000) Characterization of Mbb1, a nucleus-encoded tetratricopeptide repeat protein required for expression of the chloroplast *psbB/psbT/psbH* gene cluster in *Chlamydomonas reinhardtii*. Proc Natl Acad Sci U S A 97:14813–14818
305. Valkov VT, Scotti N, Kahlau S, Maclean D, Grillo S, Gray JC, Bock R, Cardi T (2009) Genome-wide analysis of plastid gene expression in potato leaf chloroplasts and tuber amyloplasts: transcriptional and posttranscriptional control. Plant Physiol 150:2030–2044
306. Vera A, Sugiura M (1994) A novel RNA gene in the tobacco plastid genome: its possible role in the maturation of 16S rRNA. EMBO J 13:2211–2217
307. Vera A, Sugiura M (1995) Chloroplast rRNA transcription from structurally different tandem promoters: an additional novel-type promoter. Curr Genet 27:280–284
308. Vera A, Hirose T, Sugiura M (1996) A ribosomal protein gene (*rpl32*) from tobacco chloroplast DNA is transcribed from alternative promoters: similarities in promoter region organization in plastid housekeeping genes. Mol Gen Genet 251:518–525
309. Verbitskiy D, Hartel B, Zehrmann A, Brennicke A, Takenaka M (2011) The DYW-E-PPR protein MEF14 is required for RNA editing at site matR-1895 in mitochondria of *Arabidopsis thaliana*. FEBS Lett 585:700–704
310. Villa T, Pleiss J, Guthrie C (2002) Spliceosomal snRNAs: Mg^{2+}-dependent chemistry at the catalytic core? Cell 109:149–152
311. Vogel J, Hübschmann T, Börner T, Hess WR (1997) Splicing and intron-internal RNA editing of *trnK-matK* transcripts in barley plastids: support for MatK as an essential splice factor. J Mol Biol 270:179–187
312. Wagner R, Pfannschmidt T (2006) Eukaryotic transcription factors in plastids-bioinformatic assessment and implications for the evolution of gene expression machineries in plants. Gene 381:62–70
313. Walter M, Kilian J, Kudla J (2002) PNPase activity determines the efficiency of mRNA 3′-end processing, the degradation of tRNA and the extent of polyadenylation in chloroplasts. EMBO J 21:6905–6914

314. Walter M, Piepenburg K, Schöttler MA, Petersen K, Kahlau S, Tiller N, Drechsel O, Weingartner M, Kudla J, Bock R. (2011) Knockout of the plastid RNase E leads to defective RNA processing and chloroplast ribosome deficiency. Plant J 64:851–863
315. Wang MJ, Davis NW, Gegenheimer P (1988) Novel mechanisms for maturation of chloroplast transfer RNA precursors. EMBO J 7:1567–1574
316. Wang L, Yu X, Wang H, Lu YZ, de Ruiter M, Prins M, He YK (2011) A novel class of heat-responsive small RNAs derived from the chloroplast genome of Chinese cabbage (*Brassica rapa*). BMC Genomics 12:289
317. Watkins KP, Kroeger TS, Cooke AM, Williams-Carrier RE, Friso G, Belcher SE, van Wijk KJ, Barkan A (2007) A ribonuclease III domain protein functions in group II intron splicing in maize chloroplasts. Plant Cell 19(8):2606–2623
318. Watkins KP, Rojas M, Friso G, van Wijk KJ, Meurer J, Barkan A (2011) APO1 promotes the splicing of chloroplast group II introns and harbors a plant-specific zinc-dependent RNA binding domain. Plant Cell 23(3):1082–1092
319. Weihe A, Börner T (1999) Transcription and the architecture of promoters in chloroplasts. Trends Plant Sci 4:169–170
320. Weihe A, Liere K, Börner T. (2012) Transcription and transcription regulation in chloroplasts and mitochondria of higher plants. In: Bullerwell CE (ed) Organelle genetics. Springer, Berlin, pp 297–325
321. Wicke S, Schneeweiss GM, dePamphilis CW, Muller KF, Quandt D (2011) The evolution of the plastid chromosome in land plants: gene content, gene order, gene function. Plant Mol Biol 76:273–297
322. Wolfe KH, Li WH, Sharp PM (1987) Rates of nucleotide substitution vary greatly among plant mitochondrial, chloroplast, and nuclear DNAs. Proc Natl Acad Sci U S A 84:9054–9058
323. Wu C-Y, Lin C-H, Chen L-J (1997) Identification of the transcription start site for the spinach chloroplast serine tRNA gene. FEBS Lett 418:157–161
324. Wykoff DD, Grossman AR, Weeks DP, Usuda H, Shimogawara K (1999) Psr1, a nuclear localized protein that regulates phosphorus metabolism in *Chlamydomonas*. Proc Natl Acad Sci USA 96:15336–15341
325. Xie G, Allison LA (2002) Sequences upstream of the YRTA core region are essential for transcription of the tobacco *atpB* NEP promoter in chloroplasts *in vivo*. Curr Genet 41:176–182
326. Yamaguchi K, Beligni MV, Prieto S, Haynes PA, McDonald WH, Yates JR 3rd, Mayfield SP (2003) Proteomic characterization of the *Chlamydomonas reinhardtii* chloroplast ribosome. Identification of proteins unique to th e70 S ribosome. J Biol Chem 278:33774–33785
327. Yamazaki H, Tasaka M, Shikanai T (2004) PPR motifs of the nucleus-encoded factor, PGR3, function in the selective and distinct steps of chloroplast gene expression in *Arabidopsis*. Plant J 38:152–163
328. Yang J, Stern DB (1997) The spinach chloroplast endoribonuclease CSP41 cleaves the 3′-untranslated region of petD mRNA primarily within its terminal stem-loop structure. J Biol Chem 272:12874–12880
329. Yang J, Usack L, Monde RA, Stern DB (1995) The 41 kDa protein component of the spinach chloroplast petD mRNA 3′ stem-loop: protein complex is a nuclear encoded chloroplast RNA-binding protein. Nucleic Acids Symp Ser 237–239
330. Yang J, Schuster G, Stern DB (1996) CSP41, a sequence-specific chloroplast mRNA binding protein, is an endoribonuclease. Plant Cell 8:1409–1420
331. Yehudai-Resheff S, Zimmer SL, Komine Y, Stern DB (2007) Integration of chloroplast nucleic acid metabolism into the phosphate deprivation response in *Chlamydomonas reinhardtii*. Plant Cell 19:1023–1038
332. Yin C, Richter U, Börner T, Weihe A (2010) Evolution of plant phage-type RNA polymerases: the genome of the basal angiosperm *Nuphar advena* encodes two mitochondrial and one plastid phage-type RNA polymerases. BMC Evol Biol 10:379
333. Yin P, Li Q, Yan C, Liu Y, Liu J, Yu F, Wang Z, Long J, He J, Wang HW, Wang J, Zhu JK, Shi Y, Yan N (2013) Structural basis for the modular recognition of single-stranded RNA by PPR proteins. Nature 504(7478):168–171. doi:10.1038/nature12651. [Epub 2013 Oct 27]

334. Yu QB, Jiang Y, Chong K, Yang ZN (2009) AtECB2, a pentatricopeptide repeat protein, is required for chloroplast transcript accD RNA editing and early chloroplast biogenesis in *Arabidopsis thaliana*. Plant J 59:1011–1023
335. Zehrmann A, Verbitskiy D, Hartel B, Brennicke A, Takenaka M (2011) PPR proteins network as site-specific RNA editing factors in plant organelles. RNA Biol 8:67–70
336. Zehrmann A, van der Merwe J, Verbitskiy D, Hartel B, Brennicke A, Takenaka M (2012) The DYW-class PPR protein MEF7 is required for RNA editing at four sites in mitochondria of *Arabidopsis thaliana*. RNA Biol 9:155–161
337. Zghidi-Abouzid O, Merendino L, Buhr F, Malik Ghulam M, Lerbs-Mache S (2011) Characterization of plastid *psbT* sense and antisense RNAs. Nucleic Acids Res 39:5379–5387
338. Zhao T, Li G, Mi S, Li S, Hannon GJ, Wang XJ, Qi Y (2007) A complex system of small RNAs in the unicellular green alga *Chlamydomonas reinhardtii*. Genes Dev 21:1190–1203
339. Zhelyazkova P, Hammani K, Rojas M, Voelker R, Vargas-Suarez M, Börner T, Barkan A (2012) Protein-mediated protection as the predominant mechanism for defining processed mRNA termini in land plant chloroplasts. Nucleic Acids Res 40:3092–3105
340. Zhelyazkova P, Sharma CM, Förstner KU, Liere K, Vogel J, Börner T (2012) The primary transcriptome of barley chloroplasts: numerous noncoding RNAs and the dominating role of the plastid-encoded RNA polymerase. Plant Cell 24:123–136
341. Zhou W, Cheng Y, Yap A, Chateigner-Boutin AL, Delannoy E, Hammani K, Small I, Huang J (2009) The *Arabidopsis* gene YS1 encoding a DYW protein is required for editing of rpoB transcripts and the rapid development of chloroplasts during early growth. Plant J 58:82–96
342. Zoschke R, Liere K, Börner T (2007) From seedling to mature plant: *Arabidopsis* plastidial genome copy number, RNA accumulation and transcription are differentially regulated during leaf development. Plant J 50:710–722
343. Zoschke R, Nakamura M, Liere K, Sugiura M, Börner T, Schmitz-Linneweber C (2010) An organellar maturase associates with multiple group II introns. Proc Natl Acad Sci U S A 107:3245–3250
344. Zubo YO et al (2008) Cytokinin stimulates chloroplast transcription in detached barley leaves. Plant Physiol 148:1082–1093
345. Zubo YO, Yamburenko MV, Kusnetsov VV, Börner T (2011) Methyl jasmonate, gibberellic acid, and auxin affect transcription and transcript accumulation of chloroplast genes in barley. J Plant Physiol 168:1335–1344

Chapter 2
Chloroplast Gene Expression—Translation

Jörg Nickelsen, Alexandra-Viola Bohne and Peter Westhoff

Abstract Translation has often been shown to represent the rate-limiting step of chloroplast gene expression. Genetic and biochemical analyses indicate that numerous nucleus-encoded protein factors in concert with their cognate target sites on chloroplast mRNAs are involved in determining protein-specific synthesis rates. In this chapter, we summarize the constituents of the chloroplast translational apparatus as well as the molecular principles underlying its spatiotemporal regulation.

Keywords Protein synthesis · Chloroplast translation · Chloroplast translational apparatus · Translation factors · T-zones · Helical repeat proteins

Abbreviations

CES	Control by epistasy of synthesis
CPSGs	Chloroplast stress granules
HCF	High chlorophyll fluorescence
LDMs	Low density membranes
NTRC	NADPH-dependent thioredoxin reductase C
OPR	octotricopeptide repeat
PABP	Poly(A) binding protein
PET	Photosynthetic electron transport
PPR	Pentatricopeptide repeat
PSI	Photosystem I
PSII	Photosystem II
PSRPs	Plastid-specific ribosomal proteins
RRF	Ribosome recycling factor
SD	Shine-Dalgarno

J. Nickelsen (✉) · A.-V. Bohne
Molekulare Pflanzenwissenschaften, Biozentrum LMU München,
Großhaderner Str. 2-4, 82152 Planegg-Martinsried, Germany
e-mail: joerg.nickelsen@lrz.uni-muenchen.de

P. Westhoff
Institut für Entwicklungs- und Molekularbiologie der Pflanzen, Heinrich-Heine-Universität,
Universitätsstrasse 1, 40225 Düsseldorf, Germany
e-mail: west@uni-duesseldorf.de

S.M. Theg, F.-A. Wollman (eds.), *Plastid Biology,* Advances in Plant Biology 5,
DOI 10.1007/978-1-4939-1136-3_2,

TMs	Thylakoid membranes
TPR	Tetratricopeptide repeat
UTRs	Untranslated regions

2.1 Introduction

Due to their endosymbiotic origin, chloroplasts contain their own gene expression machinery which is basically of prokaryotic origin. Although this system serves the synthesis of only less than ca. 100 proteins, it is crucial for mainly maintaining photosynthetic functions but also others like, e.g., fatty acid synthesis (see Chap. 1). One intrinsic complication for the control of plastid gene expression is the necessity of coordination with nuclear gene expression because most of the chloroplast protein complexes are assembled from subunits which are encoded either in the nucleus or in the plastid genome. As we know today, this coordination is mainly mediated by nucleus-encoded factors which control almost all steps of chloroplast gene expression ranging from transcription and RNA metabolism to translation and protein complex assembly [for recent overviews see 4 and Chap. 2 of this issue; 101, 115].

Despite the fact that all plastid gene expression steps appear to be dependent/affected by nucleus-encoded trans-acting factors, especially chloroplast translation has been considered to play a key role for determining the levels of photosynthetic proteins. This idea is based on findings that often chloroplast mRNA levels are not limiting for protein synthesis and, moreover, protein levels can vary considerably for instance in response to light, whereas the corresponding mRNA levels remain mostly constant [22, 50]. Furthermore, it should be taken into account that translation per se represents a highly energy consuming process which demands tight regulation to maintain cellular energy economy [123].

In line with this, several aspects of translational regulation in chloroplasts have been addressed in recent reviews [53, 71, 114, 123]. Here, we summarize recent developments and outcomes to provide a state-of-the-art view on the various facets of chloroplast protein synthesis. Aspects of chloroplast translation that are related to the connection between protein synthesis and the biogenesis and assembly of photosynthetic complexes will not be addressed here but are covered in Chap. 13 of this issue. The same holds for the biotechnological aspects of optimized chloroplast translation which will be reviewed in Chap. 20.

2.2 Technical Considerations for Measuring Translational Activities in Chloroplasts

The combined use of both genetic and biochemical approaches has dramatically improved our current knowledge on the molecular details of chloroplast translation. This includes the identification of the basic translational machinery, gene-specific

Table 2.1 Cloned nucleus-encoded factors for chloroplast translation from higher plants and *C. reinhardtii*

Factor	Homologies	Target gene(s)	Organism	Reference
cPDI	Protein disulfide isomerase	*psbA*	*C. reinhardtii*	Kim and Mayfield [41]
cPAB1	Poly(A)-binding protein	*psbA*	*C. reinhardtii*	Yohn et al. [117]
TBA1	Oxidoreductase	*psbA*	*C. reinhardtii*	Somanchi et al. [98]
HCF173	Short Chain Dehydrogenase	*psbA*	*A. thaliana*	Schult et al. [90]
DLA2	E2 subunit pyruvate dehydrogenase	*psbA*	*C. reinhardtii*	Bohne et al. [9]
TBC2	OPR protein	*psbC*	*C. reinhardtii*	Auchincloss et al. [2]
NAC2	TPR protein	*psbD*	*C. reinhardtii*	Boudreau et al. [11]
RBP40	–	*psbD*	*C. reinhardtii*	Schwarz et al. [91]
AC115	–	*psbD*	*C. reinhardtii*	Rattanachaikunsopon et al. [79]
TAB2	ATAB2	*psaB*	*C. reinhardtii*	Dauvillée et al. [19]
ATAB2	Tab2	PSI and PSII subunits	*A. thaliana*	Barneche et al. [5]
CRP1	PPR protein	*petA/petD*	*Z. mays*	Fisk et al. [26]
TCA1	–	*petA*	*C. reinhardtii*	Raynaud et al. [80]
PGR3	PPR protein	*petL*	*A. thaliana*	Cai et al. [15]
PR10	PPR protein	*atpH*	*Z. mays*	Pfalz et al. [74]
TDA1	OPR protein	*atpA*	*C. reinhardtii*	Eberhard et al. [23]
CRR2	PPR protein	*ndhB*	*A. thaliana*	Hashimoto et al. [30]

regulatory protein factors as well as—at least for some of these—their cognate target RNA elements on chloroplast transcripts. Furthermore, the relatively broad spectrum of model systems which have been utilized for analysing chloroplast gene expression has enabled researchers to compare translational processes/principles from single-celled microalgae like *Chlamydomonas reinhardtii* to vascular plants like *Arabidopsis thaliana* or *Zea mays*.

2.2.1 Genetic Approaches

2.2.1.1 Model Organisms

The identification of nucleus-encoded factors involved in chloroplast translation started with systematic forward genetic screens of photosynthetic mutants from *C. reinhardtii*, *A. thaliana* and maize. Defects in photosynthetic activities are characterized by, e.g., increased chlorophyll fluorescence and, thus, give rise to the so-called *hcf* (high chlorophyll fluorescence) phenotype. Some of these *hcf* mutants were subsequently found to be affected in the synthesis of distinct chloroplast proteins (Table 2.1). The half-a-tetratricopeptide repeat protein HCF107 was the first factor identified by this approach in *A. thaliana* [86]. In *C. reinhardtii*, however, photosynthesis is dispensable provided that cells are fed by a reduced carbon-source, namely acetate. As a consequence, photosynthetic mutants are fertile and accessible to further genetic manipulations like crossings and/or easy selection of suppressor strains which paved the way for genetic dissection of more complex regulatory networks [123].

With the comprehensive determination of the entire genome sequences of all abovementioned model systems together with dramatically improved mass spectrometrical detection of low abundant regulatory factors of chloroplast gene expression, nowadays, reverse genetic approaches more-and-more come to the fore to decipher the function of distinct protein factors for chloroplast translation. These include RNAi-based methods as well as the screening of mutant libraries for the identification of knock-out lines [76, 89].

2.2.1.2 Chloroplast Reporter Genes

With the development of efficient chloroplast transformation techniques for *C. reinhardtii* as well as tobacco more than 20 years ago, the targeted manipulation of chloroplast genomes at least for these two species became easily feasible [13, 105]. This methodological breakthrough resulted in the identification of numerous cis-acting elements—mainly located within the 5′ untranslated regions (UTRs) of plastid mRNAs—which affect chloroplast translation based on chimeric reporter gene assays. Reporters for measuring plastid translation rates include classical heterologous genes like *aadA*, *uidA*, GFP or *lucCt* (luciferase) as well as endogenous coding regions from *petA* and *atpA* [for an overview see 45, 123].

Unfortunately, no efficient chloroplast transformation systems for either *A. thaliana* or maize are available to date, hence, a genetically combined analysis of both nuclear mutants as well as transgenic chloroplasts harbouring chimeric reporter gene constructs is still hampered. Currently, transplastomic lines in different nuclear mutant backgrounds can only be generated in *C. reinhardtii* and, indeed, this has accelerated the identification of the interplay and fine mapping of regulatory cis-acting elements and the recognizing translational factors.

2.2.2 *Measurement of in vivo Protein Synthesis*

2.2.2.1 Pulse Labelling of Proteins

To quantitate translational activities more directly, labelling of proteins via the metabolic incorporation of ^{35}S or ^{14}C radioisotopes is routinely used for both algae and higher plants [e.g. 22, 67, 72]. Selective inhibition of cytoplasmic translation by cycloheximide treatment prior to labelling ensures detection of only chloroplast-synthesized polypeptides. However, as discussed by Zerges and Hauser [123] several limitations might affect pulse labelling assays. These include analyses of rapidly degraded proteins which might escape detection despite the fact that they are produced at normal rates. Furthermore, secondary effects of starvation phases prior to addition of radioisotopes or varying uptake kinetics of radiolabel under different culture conditions or between different strains might interfere with the interpretation of data. In addition, efficient incorporation of label during the relatively short pulse phase of ca. 5–30 min. is restricted to those chloroplast proteins with a high synthesis rate such as major subunits of PSII or the Rubisco large subunit RbcL. However, weakly synthesized or very small proteins often are not labelled with sufficient efficiency to allow their unambiguous visualization on autoradiographs. Nevertheless, pulse labelling experiments have been widely used and allow a rapid initial evaluation of molecular phenotypes in a given set of photosynthetic mutants.

2.2.2.2 Polysomal Loading

An alternative approach to estimate the translational status of a certain chloroplast mRNA is represented by so-called polysomal loading experiments [e.g. 90, 91]. These include the separation of high molecular weight polysome-fractions on appropriate sucrose gradients and the subsequent detection of particular transcripts within these fractions by either Northern or RT-PCR analyses. This technique enables solid conclusions on whether the translation of an mRNA is affected at the level of initiation (in this case no polysomal distribution is detected). However, blocks in elongation or termination phases or precise quantification of protein synthesis rates cannot be deduced from such data sets. Therefore, pulse labelling results are often required to complement polysomal analyses.

2.2.3 Reconstituted Systems

2.2.3.1 In Vitro Translation Systems

While the abovementioned methods are applicable to a variety of evolutionary distant model organisms, the analysis of chloroplast translation by in vitro systems is strictly restricted to tobacco chloroplasts. In 1996, Hirose and Sugiura for the first time reported on the successful preparation of translationally active chloroplast extracts which were capable of mediating accurate translation from exogenously added mRNA molecules [32]. In 2007, the same group introduced a non-radioactive improved version of this system which possesses a more than 100-fold higher protein synthesis activity [119]. By using these systems, several burning questions regarding translational processes have been answered as specified below. This underlines the power of a biochemical tool like an in vitro translation system for in-depth analysis of chloroplast translation and strongly argues for further efforts in establishing related systems for the abovementioned model systems.

2.2.3.2 RNA Binding Assays

Besides the thorough analysis of cis-acting determinants on plastid mRNAs, in vitro reconstituted systems can also be utilized to identify the regulative trans-acting protein factors by biochemical means [32]. Indeed, the analysis of RNA/protein interactions between chloroplast proteins and in vitro synthesized RNA probes from a huge variety of plastid genes by UV-crosslinking or gel shift assays has revealed the existence of numerous polypeptides interacting especially with the untranslated regions of plastidial mRNAs [for a review see 65]. Some of these proteins have been purified and identified by mass spectrometrical analysis creating the entry point of the abovementioned reverse genetic analysis of the plastid protein synthesis machinery (see Sect. 2.2.1.1).

2.3 Constituents of the Chloroplast Translational Apparatus

As outlined above the chloroplast translation machinery is of prokaryotic origin and, thus, it is basically assumed that the principle of its working mode and its regulation resembles those of bacteria. Due to evolutionary constraints imposed by the endosymbiotic lifestyle of chloroplasts, then, this machinery has been considerably modified by additional factors and principles which allowed the integration of the former cyanobacterium into an intracellular environment [6, 71]. This view is supported by the fact that many general constituents of this apparatus share strong homologies with their bacterial counterparts but additional—often gene-specific—regulators with novel molecular features entered the scene during endosymbiosis (Fig. 2.1).

2.3.1 Basic Elements and Factors

2.3.1.1 Chloroplast tRNAs and Codon Usage

Chloroplasts like mitochondria do not encode an entire minimal set of 32 tRNAs which would be required to read all codons according to Crick's wobble rules. For instance the tobacco chloroplast genome contains 30 tRNA genes while the one of *C. reinhardtii* harbours only 29 [54, 96]. Recently, it has been demonstrated that superwobbling mechanisms compensate for this apparent deficiency, i.e., tRNAs with a non-modified U residue in the wobble position are capable of recognizing all four nucleotides in the third codon position [83]. However, superwobbling is accompanied by reduced translational efficiencies as indicated by targeted manipulations of tobacco plastid glycine tRNA genes [83].

Hence, tRNA levels and concomitant codon usage appear to represent crucial determinants for manifesting synthesis levels of distinct chloroplast proteins. Indeed,

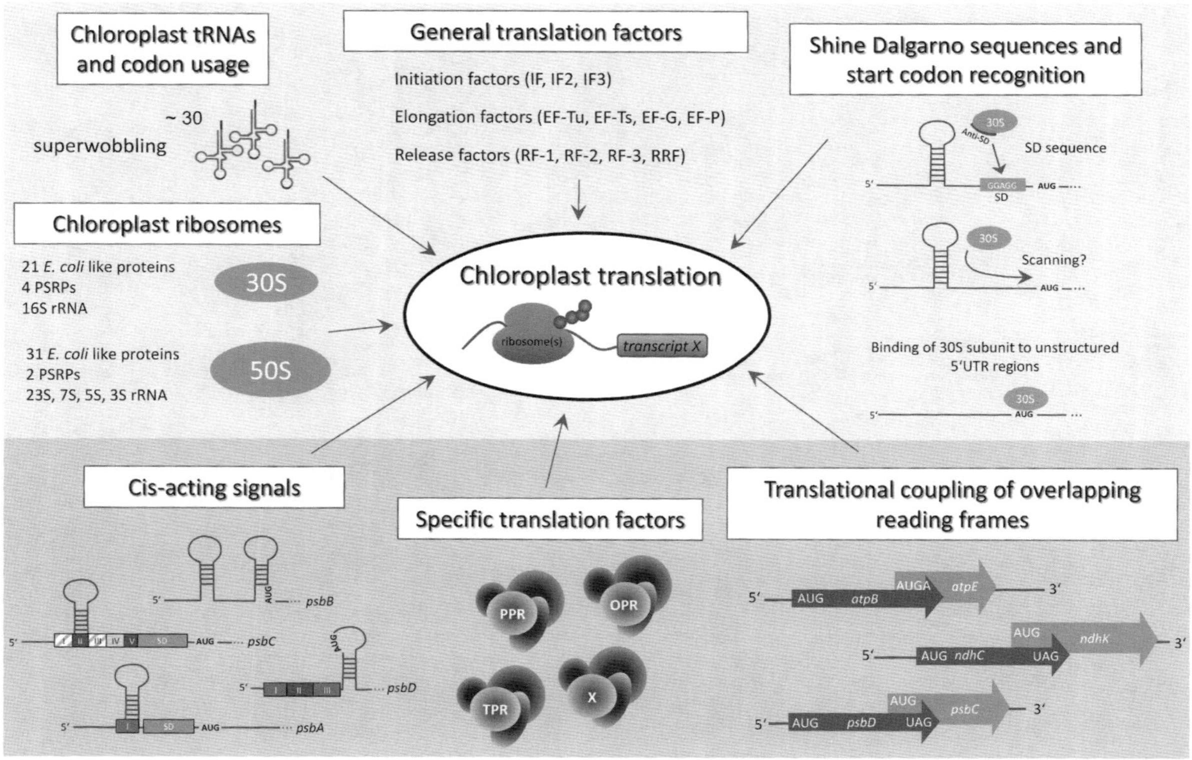

Fig. 2.1 Schematic illustration of factors and elements constituting the chloroplast translation machinery. Basic elements and general factors are delineated in the *upper* part. Gene-specific elements and factors involved in the translation of particular chloroplast transcripts are exemplified in the *lower* part. Illustration of PSII-specific cis-acting signals was adapted from Marin-Navarro et al. [53]. For further explanation see text

the codon usage in chloroplasts is different from that in prokaryotic and eukaryotic nuclear genomes. The question on a connection between codon usage and translational efficiencies has been further addressed by utilizing the above mentioned in vitro translation system from tobacco (2.2.3.1). In some cases, a clear correlation between translation rate and the usage of synonymous codons was observed, but surprisingly, frequent exceptions were also dismantled [63, 64]. This led to speculations that the frequent use of weakly translated codons like the chloroplast arginine codon CGA represent a mechanism to regulate translation, possibly in concert with other yet to be identified factors [64].

2.3.1.2 Chloroplast Ribosomes and General Translation Factors

As mentioned above chloroplast ribosomes—like mitochondrial ones—are of the bacterial-like 70S type and besides the ribosomal RNAs contain 58 proteins, 33 of these form part of the large 50S subunit and 25 build the small 30S subunit [8, 123]. 52 of them are orthologues to *E. coli* ribosomal proteins while six are chloroplast-specific and thus were named PSRPs (plastid-specific ribosomal proteins) [116]. Another plastid-specific feature of the otherwise prokaryotic ribosomes are extensions at the N- and C- termini of some of the ribosomal proteins, and consequently, plastid ribosomes are larger than their bacterial counterparts.

The very limited set of ca. 100 mRNAs being translated in chloroplasts together with the structural specialities of the ribosomal machinery led to speculations whether these apparent differences as compared to the bacterial translation systems reflect plastid-specific functions of ribosomes, e.g., light-regulated protein synthesis or thylakoid membrane associated translation of integral membrane proteins. Detailed cryo-electron microscopic analysis of ribosomes from spinach and *C. reinhardtii* has then revealed that PSRPs appear to fulfil more structural roles, e.g., by compensating for missing 16S rRNA elements [51, 94]. A recent reverse-genetic study with *A. thaliana* supports this notion in part and shows that PSRP3, 4 and 5 behave like typical classical ribosomal proteins. In contrast, a knock-down of PSPR2 and 5 does not result in a detectable phenotype in ribosome synthesis and translation [106].

On the other hand, plastid-specific extensions of the conserved ribosomal S21 protein from spinach or the S2 protein from *C. reinhardtii* were shown to be localized at the mRNA exit channel of the 30S subunit [51, 94]. Here, the initial interactions—like the recognition of the Shine-Dalgarno sequence that is located in the 5'UTR of mRNAs—take place and, thus, this region would be ideally suited to mediate chloroplast-specific translation initiation mechanisms. Supportive of that idea, UV-crosslinking experiments using purified ribosomes revealed that the plastid S2 protein directly interacts with RNA whereas the bacterial one does not [51]. Another ribosomal protein directly binding to 5′ UTR sequences in both bacteria and plastids, namely S1, is closely localized to S2 at the mRNA exit channel. However, the plastid S1 is much smaller than the bacterial one and exhibits different RNA binding affinities thereby potentially enabling a plastid-style positioning of the ribosome on the mRNA [56, 97].

Additional outcomes of structural analyses of chloroplast ribosomes are factors that form transient interactions with ribosomes. These include a ribosome recycling factor (RRF) or the plastid-specific PSRP1 protein which shares homologies with the *E. coli* cold-shock protein pY [94]. Both inhibit translation by binding within the intersubunit space of 70S ribosomes where they block the mRNA and tRNA binding sites and are involved in storage and stabilization of ribosomes under stress conditions [95]. It has been proposed that PSRP1 might be involved in overall chloroplast translational regulation by balancing the pools of stored and actively translating ribosomes in a light-dependent manner [95].

In *C. reinhardtii*, two other proteins, named RAP38 and RAP41, have been identified which co-sediment with 70S particles [116]. These two proteins, exhibit homologies with the higher plant plastid RNA binding proteins CSP41a and CSP41b which were implicated with functions during diverse processes such as transcription, RNA stabilization and ribosome assembly [7, 10, 78]. However, neither CSP41a nor CSP41b from *A. thaliana* are associated with polysomal fractions leaving open the question on a direct function of these factors during translation [10].

In addition to 70S ribosomes, chloroplasts contain an entire set of 11 bacterial-type non-ribosomal translation factors which are involved in initiation of translation (IF, IF2 and IF3), elongation (EF-Tu, EF-Ts, EF-G and EF-P) of polypeptide chains and RF-1, RF-2 , RF-3 and RRF for release/recycling of ribosomes [for an overview see 8, 123]. Interestingly, both RRF and EF-G have been postulated to act in concert with the abovementioned PSRP1 factor during reactivation of ribosomes after stress/dark-phases at least in spinach [95]. Expression profiles of these factors in various organisms suggest that they might also be involved in the light/redox-dependent induction of general chloroplast translational activities [53].

Mutations in the genes encoding the plastidial release factor 2 (*hcf109/AtprfB2*) lead to a pleiotropic photosynthetic phenotype, since the synthesis of several plastid-encoded subunits belonging to different thylakoid membrane protein complexes is disturbed [57, 58]. In contrast, the knock-out of AtprfB3, a PrfB-like protein, affects specifically the biogenesis of the cytochrome b_6f complex. This protein/gene can only be found in the nuclear genomes of vascular plants and appears to be involved in the light, redox and stress control of this thylakoid membrane complex [102].

Plastid translational activity is essential for photosynthesis but other cellular functions can also depend on chloroplast-encoded proteins like, e.g., the D subunit of the acetyl-CoA carboxylase (AccD) which is required for fatty acid synthesis. Nevertheless, single ribosomal components/subunits might be dispensable as suggested by the absence of some of the ribosomal protein genes from non-photosynthetic parasitic plants. Recent systematic reverse genetic approaches using transplastomic tobacco lines now revealed that many but not all of these genes are essential. For instance, the ribosomal proteins from the large subunit, namely Rpl20, Rpl22, Rpl23, and Rpl32 as well as Rps2, Rps3, Rps4, Rps16, and Rps18 from the small subunit cannot be inactivated whereas deletion strains lacking Rpl33, Rpl36 and Rps15 can be generated [27, 82, 83]. The resulting phenotypes of knockout lines revealed a specialized function for Rpl32 during acclimation of tobacco plants to cold stress and for Rpl36 in control of leaf morphology [27, 83]. This suggests that chloroplast translation influences nuclear gene expression presumably via retrograde signalling pathways documenting a tight integration of chloroplast translation into the overall cellular gene expression activity [27, 73].

2.3.1.3 Shine-Dalgarno Sequences and Start Codon Recognition

As outlined in detail by Zerges and Peled-Zehavi and Danon translation initiation, especially, the recruitment of the small ribosomal subunit and the subsequent recognition of the AUG start codon fundamentally differ between prokaryotes and eukaryotes [71, 121]. In the cytosol of eukaryotic cells, translation initiation follows the so-called scanning model which predicts the 40S ribosomal subunit to be bound immediately downstream of the mRNA's 5′ cap structure via an interaction between the cap-binding eIF4F complex and other eIFs associated with the 40S subunit. This complex formation is assisted by a further interaction of the eIF4F complex with the poly(A) binding protein (PABP) bound to the mRNA's 3′poly A tail. Subsequently, the 40S subunit—with the help of an RNA helicase—scans the 5′ UTR for the first appropriate AUG start codon in a 5′ to 3′ direction.

In contrast to this scenario, the small 30S ribosomal subunit of bacteria directly recognizes the AUG start codon. This recognition is mediated by the Shine-Dalgarno (SD) ribosome binding site, a short purine-rich stretch (GGAGG) ca. 10 nt upstream of the start codon which base pairs with the 3′ end of the 16S rRNA moiety of the 30S ribosomal subunit. As a consequence, translation initiation does not require any scanning mechanisms and, thus, allows the simultaneous translation of multiple reading frames on polycistronic transcripts.

Chloroplasts with their cyanobacterial history were assumed to utilize the second, bacteria-like mechanism for translation initiation. In agreement with this, one third of chloroplast genes contain bona fide SD-sequences at an appropriate distance to the AUG start codon [103]. However, two thirds do not contain SD-elements what has provoked a long-standing discussion on the role of chloroplast SD-like sequences as well as of alternative cis-acting determinants during translation initiation [for recent overviews see 3; as well as references herein; 53, 71]. In brief, the targeted mutagenesis of chloroplast genes in *C. reinhardtii* and tobacco revealed transcript-specific effects of SD-mutations. For instance in case of the *psbA*, *psbD* and *psbC* mRNAs from *C. reinhardtii*, altered SD-sequences clearly affect protein synthesis rates while this does not hold for other transcripts like those from the *petD*, *atpB*, *rps14*, and *rps7* genes [24, 55, 67, 85, 126]. Similarly, in vitro translation assays with tobacco chloroplast extracts revealed a non-homogeneous picture with regard to the functionality of SD-sequences [34–36, 75].

In conclusion, the available data suggest multiple alternative pathways for translation initiation in chloroplasts apart from the "classical" SD-element-mediated one. Basically, these include three different mechanisms which might also occur in prokaryotes. First, it has been proposed that other regions of the 16S rRNA apart from those complementary to the SD-sequence base pair with elements in 5′ UTRs of chloroplast mRNAs [48]. For instance in tobacco, the *psbA* 5′ UTR contains two crucial elements for translation initiation (AAG and UGAUGAU) which are positioned close to the start codon and have the capacity to base pair with the 3′ end of the 16S rRNA [32]. Secondly, a scanning procedure related to the eukaryotic one described above was suggested by the preferential use of upstream start codons during the analysis of chloroplast reporter gene constructs [21, 34]. Finally, recent analyses revealed that the absence of secondary structure elements at translation initiation sites represents a critical determinant for enabling access of the translational apparatus to the start codon in the absence of SD-sequences in both bacteria and chloroplasts [87]. This is illustrated by the *hcf107* mutant of *A. thaliana* which is affected in psbH synthesis and hence deficient in photosystem II [25]. The defect in psbH translation is associated with a lack of all *psbH*-containing transcripts that are processed and have *psbH* as their leading cistron; only non-processed *psbH* transcripts accumulate in the *hcf107* mutant background. These non-processed *psbH* transcripts contain a stable stem and loop structure just in front of *psbH* that encompasses the translational start site. The translational start site becomes only available when the stem loop structure is cleaved and/or when the secondary structure is unfolded by the half-a-tetratricopeptide repeat protein HCF107 [25, 29]. Thus, unstructured single-stranded RNA regions in 5′ UTRs might serve as landing pads for the 30S ribosomal subunit, and thus circumvent the need of precise base pairing of the 16S rRNA pairing via SD-elements.

2.3.2 Gene Specific Elements and Factors for Translation Initiation

The abovementioned general principles of translational control in chloroplasts already open a wide area for regulatory control possibilities which might resemble those in cyanobacteria. However, one particular outcome of recent scientific work on chloroplast biology is that the synthesis of many if not all chloroplast-encoded proteins is dependent on gene-specific translational regulators. These appear to be new inventions having embellished the basic prokaryotic apparatus during the evolutionary development of an organelle as discussed by Barkan [4]. Thus, these factors and their molecular working modes are likely to provide new insights of how gene expression machineries can be modified by the creation of new molecular principles in evolutionary terms, and thus represent a driving force for the development of eukaryotic cells. The phylogenetic development of chloroplast-specific translational mechanisms is likely to have occurred by a co-evolution of the two basic components of every gene expression system, i.e., the chloroplast-encoded cis-acting elements on mRNAs and their cognate interactors which are normally represented by nucleus-encoded polypeptides.

2.3.2.1 Cis-acting Signals on Chloroplast RNAs

Translation initiation is generally considered as the rate-limiting step determining individual protein synthesis rates independent of the system under investigation

[46]. Therefore, most attention has been attracted by the 5′ UTRs of chloroplast mRNAs where initiation takes place. However as mentioned above, the 3′ UTRs of cytoplasmic mRNAs interact with their 5′ regions and thereby promote translation initiation. Interestingly in some cases, also chloroplast translation appears to depend—at least to some extent—on 3′ RNA regions suggesting that related principles of translational regulation occur also in the organelle [40, 84]. Whether this indeed represents a considerable general phenomenon remains to be shown by future scientific work.

The targeted manipulation of many 5′ UTR regions from chloroplast genes in both *C. reinhardtii* and tobacco, chloroplast in vitro translation assays as well as RNA binding studies have revealed a complex picture involving various different cis-regulating RNA elements for plastid protein synthesis [for comprehensive reviews see 53, 71]. Intriguingly, these elements are often found close to elements required for the stabilization and/or processing of chloroplast mRNAs suggesting a tight correlation of these processes via 5′ UTRs of plastid messages (see also III.B.2 and IV.C).

Basically, these regulatory elements fall into three categories, i.e., the above described prokaryotic elements like SD-sequences, elements which are recognized by specific trans-acting factors and elements that define secondary RNA structures which positively or negatively affect the entry of the translational apparatus, namely the ribosomal subunits. Often it cannot be clearly distinguished between these possibilities because the underlying molecular mechanisms involve a mixture of different determinants. Especially, stem-loop structures within 5′ UTRs have attracted much attention. The first evidence for the role of such a hairpin-like element was obtained for the *C. reinhardtii psbC* gene encoding the inner antenna protein CP43 of PSII. The stem region of the *psbC* translational element contains bulged residues due to imperfect complementarity which are critical for translational activity. Increased stability of the stem region by the introduction of mutations or a vice versa reduced stability in suppressor mutants led to reduced or rescued CP43 synthesis, respectively [81, 125].

Furthermore, conserved RNA stem-loop structures in the vicinity of the AUG start codon as well as an unstructured region located upstream have been shown to be critical for the translation of the *petD* mRNA in *C. reinhardtii* [31, 47]. Similar to a stem-loop close to the *psbA* SD-sequence in *C. reinhardtii*, these structures appear to serve as recognition signals for trans-acting regulatory factors. Other stem-loops appear to constitute negatively acting elements due to the fact that they incorporate crucial basic translational elements like the SD-sequence (if present) or the AUG start codon into their double stranded stem region thereby abolishing ribosomal access to the mRNA. These cases include, e.g., the *C. reinhardtii psbD* and *psbB* mRNAs with structured AUG start codons and the *C. reinhardtii psaB* and *Z. mays atpH* mRNA with affected SD-sequences [44, 77, 99, 113]. The stem-loop structure within the 5′ UTR of *psbH* genes of higher plantsis another example for a structured AUG codon [25, 29]. It is generally assumed that the conformation of these regions is released by directly or indirectly interacting protein factors. Hammani et al. provide convincing evidence that the binding of the he half-a-tetratricopeptide peptide repeat protein

HCF107 to its binding site within that stem-loop structure is required to open the secondary structure and thereby to enhance translation of *psbH* [64]. This view is further supported by genetic suppressor analyses affecting the *psbD* stem loop and by in vitro RNA binding data of the *atpH* 5′ UTR as specified below [44, 77]. In case of the *psbD* and *petD* mRNA also unstructured elements were experimentally mapped which are required for translational activities. The same holds for distinct elements in the tobacco *psbA* 5′ UTR which have been postulated to interact with the 16S rRNA thereby substituting for a lacking SD element [32].

Besides the signals in 5′ UTRs, other crucial cis-acting determinants have been identified which are involved in the translation of overlapping reading frames of polycistronic plastid RNA transcripts. This appears particularly interesting because this gene organization raises questions as to whether a translational coupling exists between the reading frames similar to some bacterial or viral cases [39]. This would imply that the downstream cistron is translated exclusively by those ribosomes that completed the translation of the upstream cistron. Such a translational coupling usually leads to a rather low translation efficiency of the downstream cistron because many ribosomes are released at the stop codon of the upstream reading frame [37]. Recently three out of four cases of overlapping protein coding regions from the tobacco chloroplast genome, namely *ndhC-ndhK*, *atpB-atpE* and *psbD-psbC*, have been investigated in detail by using the tobacco in vitro translation system [1, 104, 120]. Interestingly, in all three cases the gene pairs encode subunits of the same chloroplast multiprotein complex, i.e., the NAD(P)H dehydrogenase, the ATP synthase and PSII, respectively. Since these subunits should accumulate to stoichiometrically related amounts, a translational coupling mechanism alone would not guarantee proper synthesis rates of both polypeptides. Indeed, different molecular mechanisms for the translation of these cistrons were revealed including a translational coupling between the *ndhC-ndhK* and *psbD-psbC* units [1, 120]. However, additional—yet to be dismantled—translational pathways are acting on these overlapping genes which allow meeting a 1:1 stoichiometry of the respective subunit pairs. It has been proposed that a rigid stem-loop structure encompassing the *psbC* SD-sequence as well as the AUG start codon is unwinded by ribosomes translating the upstream *psbD* cistron thereby facilitating translation initiation from these sites [1]. In case of *atpB-atpE*, AtpE was shown to be synthesized independently of AtpB via cis elements located ca. 25 nt upstream of its start codon within the *atpB* coding region [104].

2.3.2.2 Nucleus-Encoded Factors for Chloroplast Translation

Besides the nature of the regulatory chloroplast-encoded elements, especially, the identification and characterization of often gene-specific translational regulator proteins have been in the focus of research work on plastid protein synthesis. Both, genetic and biochemical approaches have led to the identification of some of such factors which are listed in Table 2.1 [for an overview see also 53, 114]. One interesting outcome of this work is that similar to the situation for control of plastid

RNA metabolism many of the trans-acting factors belong to the family of so-called "helical repeat proteins" 4, 101; Table 2.1]. This family comprises proteins containing the TPR (tetratricopeptide repeat), PPR (pentatricopeptide repeat) or OPR (octotricopeptide repeat) motifs which show a different evolutionary distribution amongst photoautotrophic organisms. Whilst TPR proteins mediating mainly protein/protein interactions are ubiquitously found from cyanobacteria to higher plants, the PPR family of RNA-binding proteins is not present in prokaryotes and contains only few members in eukaryotic algae but is predominantly present in higher plant genomes. In contrast, RNA recognizing OPR motifs are mostly identified in algal genomes but are absent or almost absent in cyanobacteria and higher plant genomes, respectively [23]. Thus, it is tempting to speculate that OPR proteins substitute for PPR proteins with regard to their organellar functions in algal systems.

The helical repeat proteins include, e.g., CRP1 and PPR10 from *Z. mays* as well as PGR3 from *A. thaliana* and Tda1 from *C. reinhardtii* which are involved in the translation of *petA/petD*, *atpH*, *petL* and *atpA* mRNAs, respectively [15, 77, 86, 87]. Moreover, factors sharing homologies with redox-sensitive or C-metabolic enzymes like protein disulfide isomerases (cPDI), oxidoreductases (TBA1), short chain dehydrogenases (HCF173) and the E2 subunit of the plastid pyruvate dehydrogenase (DLA2) all being involved in *psbA* mRNA translation were discovered supporting the idea of a tight coupling of chloroplast translation and the redox and/or the C-metabolic status of the cell [12, 41, 90, 98]. Finally, so-called "pioneer proteins" with no obvious similarities to any other described protein motif suggest that novel molecular principles of plastid protein synthesis still await their discovery (Table 2.1). However, for some of these factors like RBP40 or TCA1 recent work provides some initial insights into their precise working modes as outlined below. Both genetic analyses as well as biochemical assays detecting a high number of chloroplast RNA binding activities suggest that the list of known trans-acting translational factors still is very short and awaits dramatic extensions in the near future [65] (Fig. 2.2).

2.4 Regulatory Principles of Chloroplast Translation

Chloroplast translation is known to be highly regulated especially in response to different environmental stimuli with light being the most recognized one. However, developmental and metabolic signals must also be integrated into the translational regulatory system for instance during gametic differentiation, circadian rhythms or changing nutrient availability [121]. Putative mediators for regulatory processes are the abovementioned gene-specific trans-acting factors which have the potential not only to enable but also to modulate and thereby fine-tune chloroplast gene expression in response to various stimuli. Since these factors are usually identified and characterized by analysing knock-out mutants it remained an open question whether they are merely constitutive or fulfil regulatory functions. However, in an elegant work Raynaud et al. [80] showed that the RNA stability factor MCA1

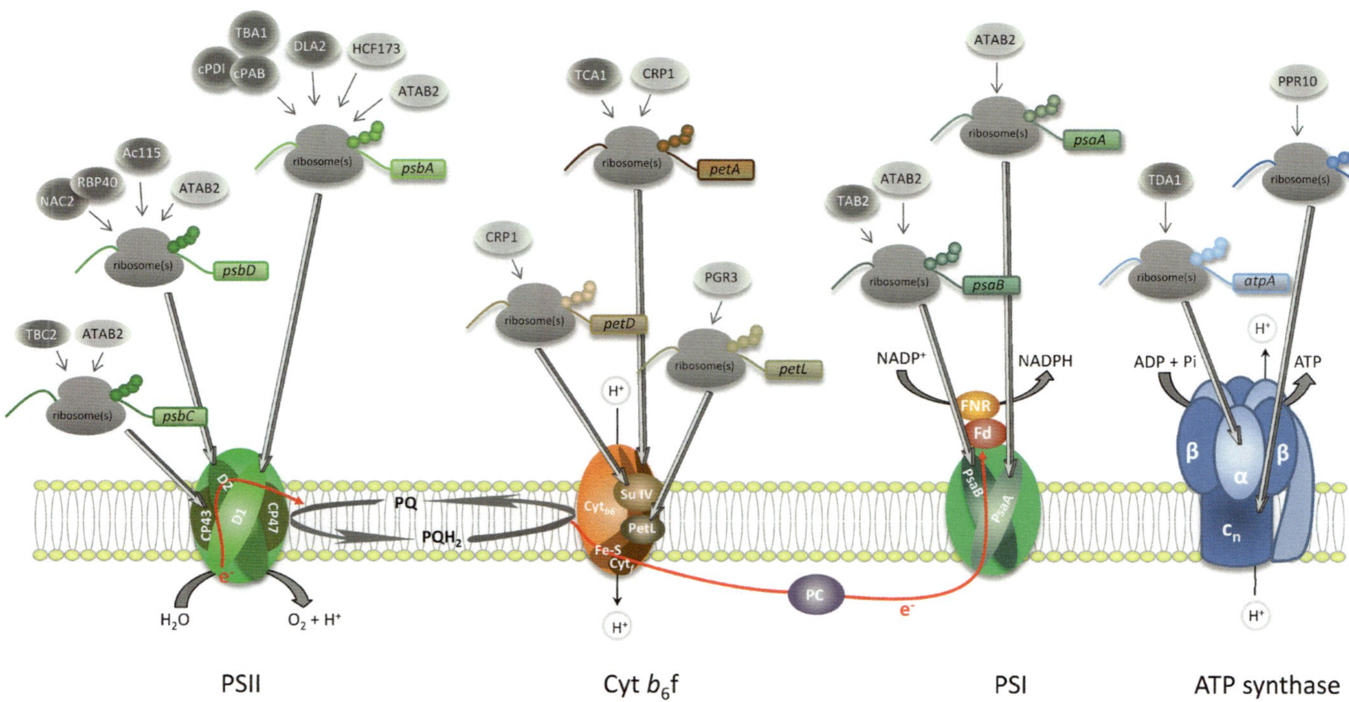

Fig. 2.2 Illustration of cloned nucleus-encoded factors involved in translation of subunits of thylakoid membrane complexes. Shown are photosystems I and II (PSI and PSII), the Cytochrome *b6*f complex (Cyt *b6*f), and the ATP synthase. Transcript specific factors involved in synthesis of respective proteins are depicted in *light grey* (higher plants) and *dark grey* (*C. reinhardtii) circles*. For further explanation see text and Table 2.1

and the translational activator TCA1 (Table 2.1) both acting on the *petA* 5′ UTR indeed regulate cytochrome f synthesis in *C. reinhardtii*. The authors constructed a set of strains with varying levels of these factors and found that both factors limit *petA* gene expression as would be expected for a regulatory factor. Moreover, both MCA1/TCA1 and Cytf levels decreased dramatically under different environmental conditions, i.e., stationary growth phase and nitrogen limitation, substantiating the idea that the levels of trans-acting nucleus-encoded factors determine and regulate chloroplast gene expression levels [80]. In line with this, also the RNAi-mediated silencing of the *NAC2* gene encoding a *psbD* mRNA stability factor from *C. reinhardtii* suggests that NAC2 is rate-limiting for D2 synthesis (C. Schwarz and J. Nickelsen, unpublished data). Thus, the emerging picture is that gene-specific translational activators do not only complement for the lack of genetic autonomy of chloroplasts like, e.g., nucleus-encoded ribosomal proteins, but alternatively, they mediate regulatory functions for adaptive cellular processes. Some of the better characterized examples are outlined in the following sections.

2.4.1 Regulation by Light and Redox/Metabolic Status

It has been known for a long time that light plays a crucial role for chloroplast protein synthesis levels based on findings that synthesis rates of chloroplast-encoded photosynthetic proteins like the large subunit of the Rubisco or the reaction center proteins D1 and D2 of PSII increase upon illumination despite the fact that their respective mRNA levels stay mostly constant [for an overview see 122]. When light regulation is addressed two principally different aspects have to be considered which are connected to the light intensities under investigation. Under normal light in growth phases of cells, the subunits of photosynthetic complexes are made in a concerted fashion to guarantee their stoichiometric synthesis for complex de novo assembly. This situation fundamentally differs from high light conditions (>700 μE) when the chloroplast translation machinery is mainly engaged in the repair synthesis of the PSII D1 protein which rapidly turns over in a light-dependent manner. Thus, especially for the D1 protein two different modes of protein synthesis must be considered.

Another important question concerns the nature of the chloroplast light signalling pathway influencing chloroplast translation. In the most favoured model, a redox signal generated by the photosynthetic electron transport chain is transferred to the translational apparatus via the redox status of the plastoquinone pool or via the chloroplast thioredoxin system [for a review see 6].

2.4.1.1 *psbA* mRNA Translation

Chloroplast D1 synthesis represents probably the most extensively studied case for light regulation and has been comprehensively reviewed [53, 62, 71, 123]. In brief, light regulated *psbA* mRNA translation has been shown to be mediated via its 5′ UTR in both *C. reinhardtii* and tobacco [55, 100], however, the involved cis-elements appear to be different. In *C. reinhardtii*, a stem-loop region preceding the putative SD-sequence was shown to be critical for translation whereas in tobacco mutation of the SD-sequence had no effect but instead three other elements within the 5′ UTR appeared to be critical for D1 synthesis [32, 100]. These include two sequences (RBS1 and RBS2) that were proposed to base-pair with the 3′ end of the 16S rRNA and an AU rich box located exactly between RBS1 and RBS2. The AU-box has been postulated to be looped out upon interaction with the small ribosomal subunit thereby serving as the recognition site for (a) yet to be cloned translational factor(s) [32].

However, the most elaborate—but also the most controversial—model has been proposed for *psbA* mRNA translation for D1 repair synthesis in *C. reinhardtii* [6, 123]. This model predicts that high-light induced regulation is mediated via redox reactions coupled to the photosynthetic electron transport (PET). Biochemical isolation of a multisubunit complex (the so-called RB complex) by affinity chromatography using the *psbA* 5′ UTR as affinity ligand revealed four associated proteins, i.e., a 60 kDa protein disufide isomerase (cPDI), a 47/70 kDa poly (A) binding protein (cPAB1), a yet to be identified 55 kDa protein (RB55) and RB38, an RNA binding protein of 38 kDa (Table 2.1). The in vitro binding of the RB complex to the *psbA* 5′ UTR via RB47 has been shown to be sensitive to redox reagents like DTT or reduced thioredoxin in vitro and parallels the level of *psbA* mRNA translation under different light/redox conditions or in different genetic backgrounds in vivo [19, 107, 117]. This led to a model for *psbA* mRNA translational regulation which involves a redox-controlled regulation of cPDI and subsequently RB47 via thioredoxin which becomes reduced by the PET chain. Thus, PET would function as the light sensor in this scenario.

An additional player in this redox control network has been identified by isolation of the *TBA1* locus from *C. reinhardtii* using a forward genetic approach (Table 2.1). *TBA1* encodes an oxidoreductase which has been hypothesized to play an opposing role to cPDI by mediating the re-reduction of RB47 upon illumination after its former oxidation by cPDI during dark phases [98]. Interestingly, a distinct mechanism coupled to the energy status of the chloroplast appears to act in addition to redox reactions on the RB complex, i.e., its inactivation via ADP-dependent phosphorylation of cPDI in the dark. This primary repression is then relieved via PET in the light and, as a consequence, makes the RB complex accessible to redox control via PET, a step termed "priming" [108]. While this model provides an attractive explanation of how light can be signalled to the translational apparatus via the PET, some of its details are at odds with other published data as discussed in detail by Zerges and Hauser [123]. One particular point concerns the specificity of the RB complex formation on the *psbA* mRNA which is questioned by the fact that its RB38-component has recently been shown to be involved in *psbD*—instead of *psbA*—mRNA translation [see below and 91]. Secondly, the model would predict an activation of the RB complex under reducing conditions in the chloroplast, however, high light stress resulting in preferential D1 repair synthesis results in inactivation of PSII and PSI activities and, thus, oxidizing conditions which would inactivate the RB complex rather than activating it [28, 68].

One reason for difficulties in developing a coherent model for *psbA* mRNA translation might be the fact that two distinct but co-existing translational systems involving different trans-acting factors mediate D1 synthesis, one operating during D1 synthesis for de novo synthesis of PSII and the other one during D1 repair synthesis. This idea is mainly supported by two lines of evidence mainly obtained from work on *C. reinhardtii*. First, the analysis of appropriate chloroplast reporter gene constructs revealed that D1 repair synthesis, in contrast to de novo synthesis, is not controlled by the *psbA* 5′ UTR suggesting that not initiation but elongation of translation represents the regulatory target [60, 61, 111]. Secondly, de novo and repair synthesis of D1 were demonstrated to be spatially separated in *C. reinhardtii* [109; see also V.]. Whereas de novo D1 synthesis takes place at centered regions around the pyrenoid of algal chloroplasts named T-(translation) zones, repair synthesis is distributed all over the thylakoid membrane lobes of *C. reinhardtii* constituting a clear separation between both processes [109]. Furthermore, based on FISH data, a directed light-dependent transport of mRNAs, ribosomes, and translational activators like RBP40 to T-zones was observed, raising questions on the molecular mechanisms determining mRNA and translation factor localization. In case of the *psbA* mRNA, a putative candidate for an RNA targeting factor has been identified by in vitro RNA binding analyses which was named RBP63 [70]. RBP63 recognizes a stretch of consecutive A residues in the 5′ UTR which is critical for translation of the *psbA* message [14, 70]. Interestingly, RBP63 is associated with stromal thylakoid membranes and, thus, was proposed to be involved in targeting of the *psbA* mRNA to distinct sites where PSII biogenesis takes place. Intriguingly, recent mass spectrometrical analyses identified RBP63 as the E2 subunit (DLA2) of the chloroplast pyruvate dehydrogenase (cpPDC) from *C. reinhardtii* [12]. Furthermore, molecular characterization of RBP63/DLA2 revealed that it is indeed involved in targeting of *psbA* mRNA to T-zones around the pyrenoid and thereby fine-tunes D1 synthesis in response to chloroplast C-metabolism and light conditions [12].

In line with this, a similar *psbA* mRNA tethering function has been proposed for the HCF173 protein of *A. thaliana* [90]. The HCF173 protein is attached to membranes and is associated with the *psbA* RNA. In its absence, D1 synthesis is drastically impaired and *hcf173* mutants are not able of photoautotrophic growth. HCF173 protein is part of a higher molecular weight complex suggesting that other factors are involved in controlling D1 synthesis [90]. However, their molecular identification has not been achieved yet.

2.4.1.2 *psbD* mRNA Translation

A second, well characterized example for translational control in *C. reinhardtii* is represented by the *psbD* mRNA encoding the immediate D1 assembly partner D2 in the reaction center of PSII. Similar to D1, D2 synthesis is regulated via the

5′ UTR of its message [67]. In contrast to the A-rich element within the *psbA* 5′ UTR recognized by DLA2, the *psbD* message reveals an U-rich element located 14–25 nt upstream of the AUG start codon which has been proven to be required for its translation [67]. This element is recognized by the abovementioned RBP40 factor both in vitro and in vivo. Interestingly, this interaction requires the mRNA stabilization factor NAC2 which together with RBP40 forms part of a high molecular weight complex suggesting a tight interrelationship between processes of RNA stabilization and translation [69, 91]. Based on the analysis of suppressor mutations and in vitro RNA mapping experiments it could be shown that binding of this NAC2/RBP40 complex induces a conformational change of the downstream RNA region at the AUG start codon. As a consequence, a stem-loop structure containing the start codon within its stem region is melted and the AUG codon is exposed thereby enabling access of the small ribosomal subunit and the initiator tRNA to the mRNA and subsequent initiation of D2 synthesis [44; Fig. 2.1]. It is well established that similar to D1 also D2 synthesis is induced by light [50]. Recent analyses now provided first evidence that the light signal is transmitted by the NAC2/RBP40 complex [92]. In the light, this complex forms in a redox-dependent manner, i.e., it is sensitive to treatment with reduced DTT or glutathione. In the dark, RBP40 detaches from the complex and consequently D2 synthesis is turned down. Intriguingly, a single Cys residue in RBP40 forms an intermolecular disulfide bridge with NAC2 which becomes reduced in the dark thereby coupling *psbD* gene expression to the redox status of the chloroplast [92]. However, in contrast to the above mentioned view that reducing power generated by the PET activates translation in the light, in case of *psbD* mRNA translation reduction of the NAC2/RBP40 disulfide bond causes inactivation. Even more surprising, reduction occurs in the dark raising questions on the source of the reducing electrons. In vitro studies now suggest that the disulfide bond is reductively released by the chloroplast NADPH-dependent thioredoxin reductase C (NTRC) which transfers electrons from NADPH generated by the oxidative pentose phosphate pathway [42]. This suggests a crosstalk between chloroplast gene expression and chloroplast carbon metabolism during dark adaption of algal cells [92]. Together with previous reports on the induction of chloroplast translation by acetate treatment of *C. reinhardtii* cells in the dark and the abovementioned function of the DLA2 subunit of the plastid PDC for D1 de novo synthesis, this adds another piece of evidence that plastid C-metabolism has severe impacts on the regulation of translation at least in *C. reinhardtii* [12, 59].

2.4.1.3 *rbcL* mRNA Translation

The synthesis of the chloroplast-encoded large subunit of the CO_2 fixing Rubisco enzyme (RbcL) is also highly regulated by light. Especially under oxidative stress, *rbcL* mRNA translation was found to be repressed [38, 93]. This repression appears to be induced by the redox state of the chloroplast glutathione pool which serves as a redox sensor under high light conditions [38]. Interestingly, the RbcL protein itself contains an intrinsic non-specific RNA binding activity located within its N-terminus [118]. It has therefore been hypothesized that the binding of RbcL to its own mRNA inhibits its translation when either its redox-controlled interaction with the chloroplast chaperone system or the assembly of the Rubisco subunit is disturbed [17]. The precise molecular mechanism of repression remains elusive but it has been shown that the early translational elongation phase appears to be affected under high light conditions because *rbcL* mRNA then shifts from polysomes to monosomes [17].

2.4.1.4 *psaB* mRNA Translation

Another interesting example of light controlled translational regulation is represented by the case of the *psaB* gene encoding the reaction center PsaB protein of PS I. PsaB synthesis has been shown to be controlled by the nucleus-encoded *Tab2* locus in *C. reinhardtii*. TAB2 represents a novel type of RNA binding protein with a characteristic WLL motif at its C-terminus which directly interacts with the *psaB* 5′ UTR [19]. Interestingly, this translation factor shares homologies with several orthologues in both eukaryotic and prokaryotic organisms performing oxygenic photosynthesis which range from *A. thaliana* to *Synechocystis* sp. PCC 6803 [19]. This structural evolutionary conservation is at least partially paralleled by a functional conservation of TBA2 since the *C. reinhardtii* TAB2 gene can partially complement a mutant of the *A. thaliana* homolog ATAB2. However, the *A. thaliana* factor has at least two targets—one for each PS—indicating a broader function for thylakoid membrane biogenesis in higher plants. A particularly interesting finding was that nuclear ATAB2 expression is tightly controlled by low fluence blue light via the photoreceptors CRY1 and CRY2 thereby providing an entry point to explain how blue light influences chloroplast—or more precisely thylakoid membrane—biogenesis during plant development [5].

2.4.2 Regulation of Translation via Interconnected Steps of Gene Expression

Besides external stimuli like light, internal regulatory principles underlie translational control in chloroplasts. The best described example is represented by the so-called CES (control by epistasy of synthesis) principle which links chloroplast translation to the assembly status of multisubunit photosynthetic complexes via feedback control mechanisms at least in *C. reinhardtii*. The underlying molecular details will be described in Chap. 13.

Another gene expression step connected to translation is the processing/stabilization of chloroplast transcripts (see Chap. 1). The long standing question whether RNA processing—especially of polycistronic transcripts—is required for translation or its regulation is still lively discussed, but current new insights into the dual role of some trans-acting factors in both RNA metabolism and translation start to shed some more light onto this issue [4]. It was principally believed that the reason for extensive RNA processing in chloroplasts is the acceleration of translational events via monocistronic mRNAs. Recent comparative analyses on the selection of SD-sequences suggested that at least in some cases internal SD-sequences are inefficiently recognized favouring a scanning model for translation initiation [21, 33]. This indeed would argue for a critical role of polycistronic transcript processing for translational control. However, other lines of evidence support the idea that polycistronic chloroplast transcripts are well translated [3, 119].

In the context of this question, one interesting observation during the analysis of several nuclear mutants affected in chloroplast gene expression was that many of the identified factors appeared to be required for both mRNA stabilization and translation. This holds, e.g., for CRP1 and PPR10 from maize or Hcf107, CRR2, PGR3 from *A. thaliana* [15, 26, 30, 74, 86]. Recently, a model for the connection between RNA stabilization/processing and translation initiation has been proposed based on a detailed in vitro analysis of recombinant PPR10 protein from maize [77]. PPR10 recognizes a defined single-stranded RNA region comprising 17 nt in the intergenic region between *atpI* and *atpH* [74, 77]. This interaction serves as a barrier against exonucleolytic attack from both the 5′ end as well as the 3′ end of transcripts thereby defining the transcripts termini. In addition, PPR10 binding remodels the downstream RNA conformation releasing the *atpH* SD-sequence from an RNA duplex and thus makes it accessible to the small ribosomal subunit [77]. A similar mechanism has been proposed for the half-a-tetratricopeptide repeat protein HCF107 in enhancing PsbH synthesis [29].

This scenario of a dual function for RNA metabolism and translation in higher plants is reminiscent of the situation for *psbD* gene expression in *C. reinhardtii* which depends on the RNA stability factor NAC2 and the translational activator RBP40 (see above). In this case, the artificial stabilization of *psbD* transcripts in a mutant *nac2* background via the transgenic introduction of poly(G) tracts into the *psbD* 5′ UTR led to accumulation of *psbD* mRNA but their translation was still hampered indicating that NAC2 is also required for D2 synthesis [67]. This function is—as mentioned above—exerted via its redox-controlled interaction with RBP40 which both together fulfil a PPR10-like concerted function for chloroplast gene expression. Similar insertions of poly(G) sequences into the 5′UTRs of *petD*, *psbB*, and *petA* transcripts all revealed the same picture, i.e., that chloroplast RNA stabilization factors are also required for translational control [20, 49, 112]. Furthermore apart from NAC2 and RBP40, complex formation between factors for RNA stabilization and translation has been described for MCA1 and TCA1 controlling *petA* gene expression in *C. reinhardtii* [12]. MCA1 represents the first algal PPR protein whose function has been to *petA* mRNA stabilization [49; see also Chap. 1]. MCA1 and TCA1 recognize adjacent targets on the first nucleotides of the *petA* message and act together to control *petA* gene expression with MCA1 serving also as a translational enhancer [12, 49, 80]. Consequently, the MCA1/TCA1 pair of trans-acting factors has been designated as single "*petA* gene expression system" underlining the fact that a single factor—or in the cases of NAC2/RBP40 or MCA1/TCA1 single protein complexes—mediate dual functions, i.e., RNA stabilization and translation initiation [12, 91]. Interestingly, MCA1 was recently shown to be involved in the CES regulatory pathway, too, and thus connects processes of protein synthesis and assembly [12; see Chap. 13]. Taken together, the emerging picture on the posttranscriptional control of chloroplast gene expression in both algae and plants is that of a tight connection of RNA metabolism and protein synthesis which is manifested by factors/complexes of factors which mediate their functions mainly via the 5′ UTRs of chloroplast transcripts.

2.5 Spatial Organization of Chloroplast Translation

One of the most fascinating new aspects on the regulation of chloroplast protein synthesis concerns the spatial organization of chloroplast translational events. Early electronmicroscopic studies as well as sucrose density gradient separations had demonstrated that polyribosomes associate with thylakoid membranes (TMs) [16, 43, 52]. This supported the idea that chloroplast-encoded integral membrane proteins are co-translationally inserted into TMs. This is also suggested by the fact that ca. 50 % of these polysomes could only be released from TMs by the addition of puromycin catalysing a release of the nascent polypeptide chain from ribosomes (for a review see [121]). At least three different chloroplast subcompartments have been identified which contain the abovementioned regulatory activators, i.e., the chloroplast stroma, the TM system—especially non-appressed thylakoids—and a membrane subfraction named low density membranes (LDMs) which resembles the inner envelope membrane with regard to its acyl lipid composition but associates with thylakoids in a Mg^{2+} dependent manner [124].

Intriguingly, the precise in situ localization of mRNAs as well as ribosomes and/or translational activators in the chloroplast of *C. reinhardtii* by FISH analyses using confocal microscopy revealed the existence of distinct so-called T-zones (for translation zones) which are located close to the periphery of the pyrenoid. Pyrenoids represent algae-specific structures which mainly consist of concentrated Rubisco enzyme and, thus, are centers where plastid CO_2 fixation takes place. At these suborganellar structures, especially the synthesis and assembly of PSII subunits is proceeding while *psaA* mRNAs did not co-localize to them indicating that PSI biogenesis is separated from T-zones [109]. A further specialization of T-zones is that they are dedicated only to the translation during de novo synthesis of PSII but D1 repair synthesis is distributed all along thylakoid membranes and thus spatially separated from T-zones [109]. This provides further compelling evidence for the idea that two completely different systems for *psbA* mRNA translation probably co-exist in chloroplasts (see above). How *psbA* mRNA is sorted between these two systems remains to be shown, nevertheless, one candidate for a membrane targeting factor for *psbA* mRNA is represented by the abovementioned DLA2 subunit of the pyruvate dehydrogenase which is required for proper localization of the *psbA* mRNA [9].

The translation of *rbcL* mRNA in *C. reinhardtii* is also highly localized, i.e., it is selectively enriched at the outer perimeter of the pyrenoid but not organized in T-zones [111]. Since *rbcL* mRNA depleted of ribosomes via lincomycin treatment did not localize to the pyrenoid a translation dependent targeting mechanism, presumably via the nascent RbcL polypeptide chain was suggested [111].

In addition to the observed light-dependent localizations of *psbA*/*psbC* and *rbcL* mRNAs, at two distinct regions at the periphery of the pyrenoid, another mRNA localization process was observed in cells which were exposed to high light or oxidative stress. Under these conditions, chloroplast stress granules (cpSGs) are formed in the internal perimeter of the pyrenoid where—embedded in pockets of stroma—plastid mRNAs are stored in a translationally repressed form [110]. One of the essential cpSGs-forming components has been hypothesized to be the RbcL subunit itself which has—as mentioned above—intrinsic RNA binding activity [111]. Furthermore, a flux of mRNAs between cpSGs and polysomes was observed suggesting a very dynamic mRNA trafficking in chloroplasts [111]. Additional genetic support for the existence of storage particles for chloroplast mRNA was obtained when the OPR protein Tda1 was analysed. Tda1 is specifically required for the translation of the *atpA* mRNA in *C. reinhardtii* and has been shown to act via its 5′ UTR [24]. Interestingly, the OPR domain of Tda1 is involved in translational activation while the distinct N-terminus recognizes the 5′ UTR and mediates a fine-tuning of *atpA* mRNA partition between polysomes and a high molecular weight, non-polysomic, storage particle. These storage particles might be identical or related to cpSGs as speculated by the authors [24]. Clearly, more work is required to obtain a more comprehensive picture on the spatial organization of chloroplast gene expression. Especially, in chloroplasts of higher plants a pyrenoid is lacking raising questions on the organizing structural element in these organelles.

2.6 Conclusions and Future Perspectives

Translation represents a key step during chloroplast gene expression and, accordingly, is highly regulated for instance by light. With the recent identification and molecular characterization of the involved regulatory factors, we are now starting to gain first insights on the regulatory principles which underlie a concerted synthesis of components for photosynthetic complexes. The overall picture emerging reveals helical repeat proteins playing an essential role for the organization of regulatory units which often physically and functionally connect RNA metabolism and translation initiation. These units also appear to represent the targets for signal transduction pathways affecting chloroplast protein synthesis and, as a consequence, chloroplast biogenesis. For the future, the interconnection between these pathways, which utilize redox signals or other metabolic semaphores, and chloroplast translation will represent an important issue of research efforts which might then dismantle novel unexpected crosstalks between different plastid biosynthetic pathways. A second focus of chloroplast cell biology will be the visualization of the three-dimensional spatial network which ensures chloroplast protein synthesis to take place at those sites where e.g. photosynthetic complexes are assembled. Maybe one of the most intriguing suggestions of recent RNA localization studies is that thylakoid membrane ultrastructure does not only determine photosynthetic performance but also affects gene expression.

Acknowledgments This work is supported by grants from the Deutsche Forschungsgemeinschaft to J.N. (Ni390/4-2) and P.W. (SFB TR1).

References

1. Adachi Y, Kuroda H, Yukawa Y, Sugiura M (2011) Translation of partially overlapping *psbD-psbC* mRNAs in chloroplasts: the role of 5′-processing and translational coupling. Nucleic Acids Res. doi:10.1093/nar/gkr1185
2. Auchincloss AH, Zerges W, Perron K, Girard-Bascou J, Rochaix J-D (2002) Characterization of Tbc2, a nucleus-encoded factor specifically required for translation of the chloroplast *psbC* mRNA in *Chlamydomonas reinhardtii*. J Cell Biol 157:953–962
3. Barkan A (1988) Proteins encoded by a complex chloroplast transcription unit are each translated from both monocistronic and polycistronic mRNAs. EMBO J 7:2637–2644
4. Barkan A (2011) Expression of plastid genes: organelle-specific elaborations on a prokaryotic scaffold. Plant Physiol 155:1520–1532
5. Barneche F, Winter V, Crevecoeur M, Rochaix JD (2006) ATAB2 is a novel factor in the signalling pathway of light-controlled synthesis of photosystem proteins. EMBO J 25:5907–5918
6. Barnes D, Mayfield SP (2003) Redox control of posttranscriptional processes in the chloroplast. Antioxid Redox Signal 5:89–94
7. Beligni MV, Mayfield SP (2008) *Arabidopsis thaliana* mutants reveal a role for CSP41a and CSP41b, two ribosome-associated endonucleases, in chloroplast ribosomal RNA metabolism. Plant Mol Biol 67:389–401
8. Beligni M, Yamaguchi K, Mayfield S (2004) The translational apparatus of *Chlamydomonas reinhardtii* chloroplast. Photosynth Res 82:315–325
9. Bohne A-V, Schwarz C, Schottkowski M, Lidschreiber M, Piotrowski M, Zerges W, Nickelsen J (2013) Reciprocal regulation of protein synthesis and carbon metabolism for thylakoid membrane biogenesis. PLoS Biol 11:e1001482
10. Bollenbach TJ, Sharwood RE, Gutierrez R, Lerbs-Mache S, Stern DB (2009) The RNA-binding proteins CSP41a and CSP41b may regulate transcription and translation of chloroplast-encoded RNAs in Arabidopsis. Plant Mol Biol 69:541–552
11. Boudreau E, Nickelsen J, Lemaire SD, Ossenbühl F, Rochaix J-D (2000) The Nac2 gene of Chlamydomonas encodes a chloroplast TPR-like protein involved in *psbD* mRNA stability. EMBO J 19:3366–3376
12. Boulouis A, Raynaud C, Bujaldon S, Aznar A, Wollman F-A et al (2011) The nucleus-encoded trans-acting factor MCA1 plays a critical role in the regulation of cytochrome f synthesis in *Chlamydomonas* chloroplasts. Plant Cell 23:333–349
13. Boynton JE, Gillham NW, Harris EH, Hosler JP, Johnson AM et al (1988) Chloroplast transformation in *Chlamydomonas* with high velocity microprojectiles. Science 240:1534–1538
14. Bruick RK, Mayfield SP (1998) Processing of the *psbA* 5′ untranslated region in *Chlamydomonas reinhardtii* depends upon factors mediating ribosome association. J Cell Biol 143:1145–1153

15. Cai W, Okuda K, Peng L, Shikanai T (2011) PROTON GRADIENT REGULATION 3 recognizes multiple targets with limited similarity and mediates translation and RNA stabilization in plastids. Plant J 67:318–327
16. Chua NH, Blobel G, Siekevitz P, Palade GE (1976) Periodic variations in the ratio of free to thylakoid-bound chloroplast ribosomes during the cell cycle of *Chlamydomonas reinhardtii*. J Cell Biol 71:497–514
17. Cohen I, Knopf JA, Irihimovitch V, Shapira M (2005) A proposed mechanism for the inhibitory effects of oxidative stress on Rubisco assembly and its subunit expression. Plant Physiol 137:738–746
18. Danon A, Mayfield SP (1994) Light-regulated translation of chloroplast messenger RNAs through redox potential. Science 266:1717–1719
19. Dauvillée D, Stampacchia O, Girard-Bascou J, Rochaix J-D (2003) Tab2 is a novel conserved RNA binding protein required for translation of the chloroplast *psaB* mRNA. EMBO J 22:6378–6388
20. Drager RG, Girard-Bascou J, Choquet Y, Kindle KL, Stern DB (1998) In vivo evidence for 5′->3′ exoribonuclease degradation of an unstable chloroplast mRNA. Plant J 13:85–96
21. Drechsel O, Bock R (2011) Selection of Shine-Dalgarno sequences in plastids. Nucleic Acids Res 39:1427–1438
22. Eberhard S, Drapier D, Wollman F-A (2002) Searching limiting steps in the expression of chloroplast-encoded proteins: relations between gene copy number, transcription, transcript abundance and translation rate in the chloroplast of *Chlamydomonas reinhardtii*. Plant J 31:149–160
23. Eberhard S, Loiselay C, Drapier D, Bujaldon S, Girard-Bascou J et al (2011) Dual functions of the nucleus-encoded factor TDA1 in trapping and translation activation of *atpA* transcripts in *Chlamydomonas reinhardtii* chloroplasts. Plant J 67:1055–1066
24. Fargo DC, Zhang M, Gillham NW, Boynton JE (1998) Shine-Dalgarno-like sequences are not required for translation of chloroplast mRNAs in *Chlamydomonas reinhardtii* chloroplasts or in *Escherichia coli*. Mol Gen Genet 257:271–282
25. Felder S, Meierhoff K, Sane AP, Meurer J, Driemel C et al (2001) The nucleus-encoded HCF107 gene of Arabidopsis provides a link between intercistronic RNA processing and the accumulation of translation-competent *psbH* transcripts in chloroplasts. Plant Cell 13:2127–2141
26. Fisk DG, Walker MB, Barkan A (1999) Molecular cloning of the maize gene *crp1* reveals similarity between regulators of mitochondrial and chloroplast gene expression. EMBO J 18:2621–2630
27. Fleischmann TT, Scharff LB, Alkatib S, Hasdorf S, Schöttler MA et al (2011). Nonessential plastid-encoded ribosomal proteins in tobacco: a developmental role for plastid translation and implications for reductive genome evolution. Plant Cell 23:3137–3155
28. Förster B, Mathesius U, Pogson BJ (2006) Comparative proteomics of high light stress in the model alga *Chlamydomonas reinhardtii*. Proteomics 6:4309–4320
29. Hammani K, Cook WB, Barkan A (2012) RNA binding and RNA remodeling activities of the half-a-tetratricopeptide (HAT) protein HCF107 underlie its effects on gene expression. Proc Natl Acad Sci U S A 109(15):5651–5656
30. Hashimoto M, Endo T, Peltier G, Tasaka M, Shikanai T (2003) A nucleus-encoded factor, CRR2, is essential for the expression of chloroplast *ndhB* in *Arabidopsis*. Plant J 36:541–549
31. Higgs DC, Shapiro RS, Kindle KL, Stern DB (1999) Small cis-acting sequences that specify secondary structures in a chloroplast mRNA are essential for RNA stability and translation. Mol Cell Biol 19:8479–8491
32. Hirose T, Sugiura M (1996) Cis-acting elements and trans-acting factors for accurate translation of chloroplast *psbA* mRNAs: development of an in vitro translation system from tobacco chloroplasts. EMBO J 15:1687–1695
33. Hirose T, Sugiura M (1997) Both RNA editing and RNA cleavage are required for translation of tobacco chloroplast *ndhD* mRNA: a possible regulatory mechanism for the expression of a chloroplast operon consisting of functionally unrelated genes. EMBO J 16:6804–6811

34. Hirose T, Sugiura M (2004) Functional Shine-Dalgarno-Like sequences for translational initiation of chloroplast mRNAs. Plant Cell Physiol 45:114–117
35. Hirose T, Sugiura M (2004) Multiple elements required for translation of plastid *atpB* mRNA lacking the Shine-Dalgarno sequence. Nucleic Acids Res 32:3503–3510
36. Hirose T, T. Kusumegi, M. Sugiura (1998) Translation of tobacco chloroplast *rps14* mRNA depends on a Shine-Dalgarno-like sequence in the 5′- untranslated region but not on internal RNA editing in the coding region. FEBS 430:257–260
37. Inokuchi Y, Hirashima A, Sekine Y, Janosi L, Kaji A (2000) Role of ribosome recycling factor (RRF) in translational coupling. EMBO J 19:3788–3798
38. Irihimovitch V, Shapira M (2000) Glutathione redox potential modulated by reactive oxygen species regulates translation of rubisco large subunit in the chloroplast. J Biol Chem 275:16289–16295
39. Jackson RJ, Kaminski A, Pöyry TAA (2007) Coupled termination-reinitiation events in mRNA translation. In: Mathews MB, Sonenberg N, Hershey JWB (eds) Translational control in biology and medicine. Cold Spring Harbor Laboratory Press, New York, pp 197–223
40. Katz YS, Danon A (2002) The 3′-untranslated region of chloroplast *psbA* mRNA stabilizes binding of regulatory proteins to the leader of the message. J Biol Chem 277:18665–18669
41. Kim J, Mayfield SP (1997) Protein disulfide isomerase as a regulator of chloroplast translational activation. Science 278:1954–1957
42. Kirchsteiger K, Pulido P, González M, Cejudo FJ (2009) NADPH thioredoxin reductase C controls the redox status of chloroplast 2-Cys peroxiredoxins in *Arabidopsis thaliana*. Mol Plant 2:298–307
43. Klein RR, Mason HS, Mullet JE (1988) Light-regulated translation of chloroplast proteins. Transcripts of *psaA-psaB*, *psbA*, and *rbcL* are associated with polysomes in dark-grown and illuminated barley seedlings. J Cell Biol 106:289–301
44. Klinkert B, Elles I, Nickelsen J (2006) Translation of chloroplast *psbD* mRNA in *Chlamydomonas* is controlled by a secondary RNA structure blocking the AUG start codon. Nucleic Acids Res 34:386–394
45. Koop H-U, Herz S, Golds T, Nickelsen J (2007) The genetic transformation of plastids. In: Bock R (ed) Cell and molecular biology of plastids, Vol 19. Springer, Berlin, pp 457–510
46. Kozak M (1999) Initiation of translation in prokaryotes and eukaryotes. Gene 234:187–208
47. Kramzar L, Mueller T, Erickson B, Higgs D (2006) Regulatory sequences of orthologous *petD* chloroplast mRNAs are highly specific among *Chlamydomonas* species. Plant Mol Biol 60:405–422
48. Kuroda H, Suzuki H, Kusumegi T, Hirose T, Yukawa Y et al (2007) Translation of *psbC* mRNAs starts from the downstream GUG, not the upstream AUG, and requires the extended Shine-Dalgarno sequence in tobacco chloroplasts. Plant Cell Physiol 48:1374–1378
49. Loiselay C, Gumpel NJ, Girard-Bascou J, Watson AT, Purton S et al (2008). Molecular identification and function of cis- and trans-acting determinants for *petA* transcript stability in *Chlamydomonas reinhardtii* chloroplasts. Mol Cell Biol 28:5529–5542
50. Malnoe P, Mayfield SP, Rochaix JD (1988) Comparative analysis of the biogenesis of photosystem II in the wild-type and Y-1 mutant of *Chlamydomonas reinhardtii*. J Cell Biol 106:609–616
51. Manuell AL, Quispe J, Mayfield S (2007) Structure of the chloroplast ribosome: novel domains for translation regulation. PLoS Biol 5:1785–1797
52. Margulies MM, Michaels A (1974) Ribosomes bound to chloroplast membranes in *Chlamydomonas reinhardtii*. J Cell Biol 60:65–77
53. Marín-Navarro J, Manuell A, Wu J, P Mayfield S (2007) Chloroplast translation regulation. Photosynth Res 94:359–374
54. Maul JE, Lilly JW, Cui L, dePamphilis CW, Miller W et al (2002) The *Chlamydomonas reinhardtii* plastid chromosome: islands of genes in a sea of repeats. Plant Cell 14:2659–2679
55. Mayfield SP, Cohen A, Danon A, Yohn CB (1994) Translation of the *psbA* mRNA of *Chlamydomonas reinhardtii* requires a structured RNA element contained within the 5′ untranslated region. J Cell Biol 127:1537–1545
56. Merendino L, Falciatore A, Rochaix J-D (2003) Expression and RNA binding properties of the chloroplast ribosomal protein S1 from *Chlamydomonas reinhardtii*. Plant Mol Biol 53:371–382

57. Meurer J, Berger A, Westhoff P (1996) A nuclear mutant of Arabidopsis with impaired stability on distinct transcripts of the plastid *psbB, psbD/C, ndhH,* and *ndhC* operons. Plant Cell 8:1193–1207
58. Meurer J, Lezhneva L, Amann K, Gödel M, Bezhani S et al (2002) A peptide chain release factor 2 affects the stability of UGA-containing transcripts in Arabidopsis chloroplasts. Plant Cell 14:3255–3269
59. Michaels A, Herrin DL (1990) Translational regulation of chloroplast gene expression during the light-dark cell cycle of *Chlamydomonas*: evidence for control by ATP/energy supply. Biochem Biophys Res Commun 170:1082–1088
60. Minai L, Wostrikoff K, Wollman F-A, Choquet Y (2006) Chloroplast biogenesis of photosystem II cores Involves a series of assembly-controlled steps that regulate translation. Plant Cell 18:159–175
61. Mühlbauer SK, Eichacker LA (1998) Light-dependent formation of the photosynthetic proton gradient regulates translation elongation in chloroplasts. J Biol Chem 273:20935–20940
62. Mulo P, Sakurai I, Aro E-M (2012) Strategies for *psbA* gene expression in cyanobacteria, green algae and higher plants: from transcription to PSII repair. Biochim Biophys Acta (BBA)—Bioenerg 1817:247–257
63. Nakamura M, Sugiura M (2007) Translation efficiencies of synonymous codons are not always correlated with codon usage in tobacco chloroplasts. Plant J 49:128–134
64. Nakamura M, Sugiura M (2011) Translation efficiencies of synonymous codons for arginine differ dramatically and are not correlated with codon usage in chloroplasts. Gene 472:50–54
65. Nickelsen J (2003) Chloroplast RNA-binding proteins. Curr Genet 43:392–399
66. Nickelsen J, van Dillewijn J, Rahire M, Rochaix JD (1994) Determinants for stability of the chloroplast *psbD* RNA are located within its short leader region in *Chlamydomonas reinhardtii*. EMBO J 13:3182–3191
67. Nickelsen J, Fleischmann M, Boudreau E, Rahire M, Rochaix JD (1999) Identification of *cis*-acting RNA leader elements required for chloroplast *psbD* gene expression in *Chlamydomonas*. Plant Cell 11:957–970
68. Nishiyama Y, Allakhverdiev SI, Murata N (2006) A new paradigm for the action of reactive oxygen species in the photoinhibition of photosystem II. Biochim Biophys Acta (BBA)—Bioenerg 1757:742–749
69. Ossenbühl F, Nickelsen J (2000) *Cis*- and *trans*-acting determinants for translation of *psbD* mRNA in *Chlamydomonas reinhardtii*. Mol Cell Biol 20:8134–8142
70. Ossenbühl F, Hartmann K, Nickelsen J (2002) A chloroplast RNA binding protein from stromal thylakoid membranes specifically binds to the 5′ untranslated region of the *psbA* mRNA. Eur J Biochem 269:3912–3919
71. Peled-Zehavi H, Danon A (2007) Translation and translational regulation in chloroplasts. In: Bock R (ed) Cell and molecular biology of plastids, Vol 19. Springer, Berlin, pp 249–281
72. Peng L, Maa J, Chi W, Guo J, Zhua S et al (2006) LOW PSII ACCUMULATION1 is involved in efficient assembly of photosystem II in *Arabidopsis thaliana*. Plant Cell 18:955–969
73. Pesaresi P, Masiero S, Eubel H, Braun HP, Bhushan S et al (2006) Nuclear photosynthetic gene expression is synergistically modulated by rates of protein synthesis in chloroplasts and mitochondria. Plant Cell 18:970–991
74. Pfalz J, Bayraktar OA, Prikryl J, Barkan A (2009) Site-specific binding of a PPR protein defines and stabilizes 5′ and 3′ mRNA termini in chloroplasts. EMBO J 28:2042–2052
75. Plader W, Sugiura M (2003) The Shine-Dalgarno-like sequence is a negative regulatory element for translation of tobacco chloroplast *rps2* mRNA: an additional mechanism for translational control in chloroplasts. Plant J 34:377–382
76. Pootakham W, Gonzalez-Ballester D, Grossman AR (2010) Identification and regulation of plasma membrane sulfate transporters in *Chlamydomonas*. Plant Physiol 153:1653–1668
77. Prikryl J, Rojas M, Schuster G, Barkan A (2011) Mechanism of RNA stabilization and translational activation by a pentatricopeptide repeat protein. Proc Natl Acad Sci U S A 108:415–420
78. Qi Y, Armbruster U, Schmitz-Linneweber C, Delannoy E, de Longevialle AF et al (2011) Arabidopsis CSP41 proteins form multimeric complexes that bind and stabilize distinct plastid transcripts. J Exp Bot 63:1251–1270

79. Rattanachaikunsopon P, Rosch C, Kuchka MR (1999) Cloning and characterization of the nuclear AC115 gene of *Chlamydomonas reinhardtii*. Plant Mol Biol 39:1–10
80. Raynaud C, Loiselay C, Wostrikoff K, Kuras R, Girard-Bascou J et al (2007) Evidence for regulatory function of nucleus-encoded factors on mRNA stabilization and translation in the chloroplast. Proc Natl Acad Sci U S A 104:9093–9098
81. Rochaix JD, Kuchka M, Mayfield S, Schirmer-Rahire M, Girard-Bascou J et al (1989) Nuclear and chloroplast mutations affect the synthesis or stability of the chloroplast *psbC* gene product in *Chlamydomonas reinhardtii*. EMBO J 8:1013–1021
82. Rogalski M, Ruf S, Bock R (2006) Tobacco plastid ribosomal protein S18 is essential for cell survival. Nucleic Acids Res 34:4537–4545
83. Rogalski M, Karcher D, Bock R (2008) Superwobbling facilitates translation with reduced tRNA sets. Nat Struct Mol Biol 15:192–198
84. Rott R, Levy H, Drager RG, Stern DB, Schuster G (1998) 3′-processed mRNA is preferentially translated in Chlamydomonas reinhardtii chloroplasts. Mol Cell Biol 18:4605–4611
85. Sakamoto W, Chen X, Kindle KL, Stern DB (1994) Function of the Chlamydomonas reinhardtiipetD 5′ untranslated region in regulating the accumulation of subunit IV of the cytochrome b6/f complex. Plant J 6:503–512
86. Sane AP, Stein B, Westhoff P (2005) The nuclear gene HCF107 encodes a membrane-associated R-TPR (RNA tetratricopeptide repeat)-containing protein involved in expression of the plastidial psbH gene in Arabidopsis. Plant J 42:720–730
87. Scharff LB, Childs L, Walther D, Bock R (2011) Local absence of secondary structure permits translation of mRNAs that lack ribosome-binding sites. PLoS Genet 7:e1002155.
88. Schmitz-Linneweber C, Williams-Carrier R, Barkan A (2005) RNA immunoprecipitation and microarray analysis show a chloroplast pentatricopeptide repeat protein to be associated with the 5′ region of mRNAs whose translation it activates. Plant Cell 17:2791–2804
89. Schroda M (2006) RNA silencing in Chlamydomonas: mechanisms and tools. Curr Genet 49:69–84
90. Schult K, Meierhoff K, Paradies S, Toller T, Wolff P et al (2007) The nuclear-encoded factor HCF173 is involved in the initiation of translation of the psbA mRNA in Arabidopsis thaliana. Plant Cell 19:1329–1346
91. Schwarz C, Elles I, Kortmann J, Piotrowski M, Nickelsen J (2007) Synthesis of the D2 protein of photosystem II in Chlamydomonas is controlled by a high molecular mass complex containing the RNA stabilization factor Nac2 and the translational activator RBP40. Plant Cell 19:3627–3639
92. Schwarz C, Bohne A-V, Wang F, Cejudo FJ, Nickelsen J (2012) An intermolecular disulfide-based light switch for chloroplast *psbD* gene expression in *Chlamydomonas reinhardtii*. Plant J 72(3):378–389
93. Shapira M, Lers A, Heifetz PB, Irihimovitz V, Osmond CB et al (1997) Differential regulation of chloroplast gene expression in Chlamydomonas reinhardtii during photoacclimation: light stress transiently suppresses synthesis of the Rubisco LSU protein while enhancing synthesis of the PS II D1 protein. Plant Mol Biol 33:1001–1011
94. Sharma MR, Wilson DN, Datta PP, Barat C, Schluenzen F et al (2007) Cryo-EM study of the spinach chloroplast ribosome reveals the structural and functional roles of plastid-specific ribosomal proteins. Proc Natl Acad Sci U S A 104:19315–19320
95. Sharma MR, Dönhöfer A, Barat C, Marquez V, Datta PP et al (2010) PSRP1 is not a ribosomal protein, but a ribosome-binding factor that is recycled by the ribosome-recycling factor (RRF) and elongation factor G (EF-G). J Biol Chem 285:4006–4014
96. Shinozaki K, Ohme M, Tanaka M, Wakasugi T, Hayashida N et al (1986) The complete nucleotide sequence of the tobacco chloroplast genome: its gene organization and expression. EMBO J 5:2043–2049
97. Shteiman-Kotler A, Schuster G (2000) RNA-binding characteristics of the chloroplast S1-like ribosomal protein CS1. Nucleic Acids Res 28:3310–3315
98. Somanchi A, Barnes D, Mayfield SP (2005) A nuclear gene of Chlamydomonas reinhardtii, Tba1, encodes a putative oxidoreductase required for translation of the chloroplast psbA mRNA. Plant J 42:341–352

99. Stampacchia O, Girard-Bascou J, Zanasco JL, Zerges W, Bennoun P et al (1997) A nuclear-encoded function essential for translation of the chloroplast *psaB* mRNA in *Chlamydomonas*. Plant Cell 9:773–782
100. Staub JM, Maliga P (1994) Translation of *psbA* mRNA is regulated by light via the 5′-untranslated region in tobacco plastids. Plant J 6:547–553
101. Stern DB, Goldschmidt-Clermont M, Hanson MR (2010) Chloroplast RNA metabolism. Ann Rev Plant Biol 61:125–155
102. Stoppel R, Lezhneva L, Schwenkert S, Torabi S, Felder S et al (2011) Recruitment of a ribosomal release factor for light- and stress-dependent regulation of *petB* transcript stability in Arabidopsis chloroplasts. Plant Cell 23:2680–2695
103. Sugiura M, Hirose T, Sugita M (1998) Evolution and mechanism of translation in chloroplasts. Annu Rev Genet 32:437–459
104. Suzuki H, Kuroda H, Yukawa Y, Sugiura M (2011) The downstream *atpE* cistron is efficiently translated via its own cis-element in partially overlapping *atpB-atpE* dicistronic mRNAs in chloroplasts. Nucleic Acids Res doi:10.1093/nar/gkr644
105. Svab Z, Hajdukiewicz PT, Maliga P (1990) Stable transformation of plastids in higher plants. Proc Natl Acad Sci U S A 87:8526–8530
106. Tiller N, Weingartner M, Thiele W, Maximova E, Schöttler MA et al (2012) The plastid-specific ribosomal proteins of *Arabidopsis thaliana* can be divided into non-essential proteins and genuine ribosomal proteins. Plant J 69:302–316
107. Trebitsh T, Danon A (2001) Translation of chloroplast *psbA* mRNA is regulated by signals initiated by both photosystems II and I. Proc Natl Acad Sci U S A 98:12289–12294
108. Trebitsh T, Levitan A, Sofer A, Danon A (2000) Translation of chloroplast *psbA* mRNA is modulated in the light by counteracting oxidizing and reducing activities. Mol Cell Biol 20:1116–1123
109. Uniacke J, Zerges W (2007) Photosystem II assembly and repair are differentially localized in *Chlamydomonas*. Plant Cell 19:3640–3654
110. Uniacke J, Zerges W (2008) Stress induces the assembly of RNA granules in the chloroplast of *Chlamydomonas reinhardtii*. J Cell Biol 182:641–646
111. Uniacke J, Zerges W (2009) Chloroplast protein targeting involves localized translation in *Chlamydomonas*. Proc Natl Acad Sci U S A 106:1430–1444
112. Vaistij FE, Boudreau E, Lemaire SD, Goldschmidt-Clermont M, Rochaix J-D (2000) Characterization of Mbb1, a nucleus-encoded tetratricopeptide-like repeat protein required for expression of the chloroplast *psbB/psbT/psbH* gene cluster in *Chlamydomonas reinhardtii*. Proc Natl Acad Sci U S A 97:14813–14818
113. Vaistij FE, Goldschmidt-Clermont M, Wostrikoff K, Rochaix JD (2000) Stability determinants in the chloroplast *psbB/T/H* mRNAs of *Chlamydomonas reinhardtii*. Plant J 21:469–482
114. Wobbe L, Schwarz C, Nickelsen J, Kruse O (2008) Translational control of photosynthetic gene expression in phototrophic eukaryotes. Physiol Plant 133:507–519
115. Woodson JD, Chory J (2008) Coordination of gene expression between organellar and nuclear genomes. Nat Rev Genet 9:383–395
116. Yamaguchi K, Beligni MV, Prieto S, Haynes PA, McDonald WH et al (2003) Proteomic characterization of the *Chlamydomonas reinhardtii* chloroplast ribosome. J Biol Chem 278:33774–33785.
117. Yohn CB, Cohen A, Rosch C, Kuchka MR, Mayfield SP (1998) Translation of the chloroplast *psbA* mRNA requires the nuclear-encoded poly(A)-binding protein, RB47. J Cell Biol 142:435–442
118. Yosef I, Irihimovitch V, Knopf JA, Cohen I, Orr-Dahan I et al (2004) RNA binding activity of the ribulose-1,5-bisphosphate carboxylase/oxygenase large subunit from *Chlamydomonas reinhardtii*. J Biol Chem 279:10148–10156
119. Yukawa M, Kuroda H, Sugiura M (2007) A new *in vitro* translation system for non-radioactive assay from tobacco chloroplasts: effect of pre-mRNA processing on translation *in vitro*. Plant J 49:367–376
120. Yukawa M, Sugiura M (2008) Termination codon-dependent translation of partially overlapping *ndhC-ndhK* transcripts in chloroplasts. Proc Natl Acad Sci U S A 105:19550–19554

121. Zerges W (2000) Translation in chloroplasts. Biochimie 82:583–601
122. Zerges W (2004) Regulation of translation in chloroplasts. In: Daniell H, Chase CD (eds) Molecular biology and biotechnology of plant organelles. Springer, Dordrecht, pp 347–383
123. Zerges W, Hauser C (2009) Protein synthesis in the chloroplast. In: Stern DB (ed) The *Chlamydomonas* sourcebook: organellar and metabolic processes, Vol 2. Elsevier, Oxford, pp 967–1025
124. Zerges W, Rochaix J-D (1998) Low density membranes are associated with RNA-binding proteins and thylakoids in the chloroplast of *Chlamydomonas reinhardtii*. J Cell Biol 140:101–110
125. Zerges W, Girard-Bascou J, Rochaix JD (1997) Translation of the chloroplast *psbC* mRNA is controlled by interactions between its 5′ leader and the nuclear lo*ci TBC1* and *TBC3* in *Chlamydomonas reinhardtii*. Mol Cell Biol 17:3440–3448
126. Zerges W, Auchincloss AH, Rochaix J-D (2003) Multiple translational control sequences in the 5′ leader of the chloroplast *psbC* mRNA interact with nuclear gene products in *Chlamydomonas reinhardtii*. Genetics 163:895–904

Chapter 3
The Chloroplast Genome and Nucleo-Cytosolic Crosstalk

Jean-David Rochaix and Silvia Ramundo

Abstract The biogenesis of the photosynthetic apparatus of eukaryotic organisms depends on the concerted action of the chloroplast and nucleo-cytosolic genetic systems. This coordination is mediated through a large number of nucleus-encoded plastid factors that act at all major steps of chloroplast gene expression. Moreover the nucleus is capable of sensing changes in the state of the chloroplast caused by environmental fluctuations through complex signaling chains which originate from the plastid compartment. These pathways are currently under intense investigation using a wide range of molecular-genetic, biochemical and system biology approaches.

Keywords Chloroplast · Transcription · Translation · Redox control · Retrograde signaling · Stress response · Repressible chloroplast gene expression

Abbreviations

ABA	Abscisic acid
CES	Control by epistasy of synthesis
Cytb_6f	Cytochrome b_6f complex
HAT	Half-a-tetratricopeptide
HON	Happy on norfluorazon
MEcPP	Methylerythritol cyclodiphosphate
MgProto	Mg-protoporphyrin IX
OPR	Octatricopeptide repeats
PAP	Phosphoadenosine 5'-phosphate
PChlide	Protochlorophyllide
PEP	Plastid encoded polymerase
PhANG	Photosynthesis-associated nuclear gene
PPR	Pentatricopeptide repeats
Prot IX	Protoporphyrin IX
PSI	Photosystem I
PSII	Photosystem II

J.-D. Rochaix (✉) · S. Ramundo
Departments of Molecular Biology and Plant Biology, University of Geneva, Geneva, Switzerland
e-mail: Jean-David.Rochaix@unige.ch

S.M. Theg, F.-A. Wollman (eds.), *Plastid Biology*, Advances in Plant Biology 5, DOI 10.1007/978-1-4939-1136-3_3, © Springer Science+Business Media New York 2014

PTK	Plastid transcription kinase
ROS	Reactive oxygen species
SMR	Small mutS-related
TPR	Tetratricopeptide repeats

3.1 Introduction

Chloroplasts are complex organelles which are involved in several important cellular metabolic pathways in photosynthetic eukaryotic organisms such as photosynthesis, carbon, nitrogen and sulfur metabolism, amino acid, tetrapyrrole and lipid biosynthesis. The origin of chloroplasts can be traced to an endosymbiotic event which occurred more than a billion years ago when a cyanobacterium invaded a primitive eukaryotic cell. This invasion was followed by a gradual transfer of genetic information from the invader to the host nucleus. However this transfer did not reach completion and a small portion of the cyanobacterial genome was maintained and exists today as chloroplast genome. Although the size of this genome is modest, ranging between 60 and 200 kb and containing between 60 and 120 genes, it has acquired important cellular functions because its complete loss has not been observed. Chloroplast genomes are part of an organellar protein synthesizing system which operates in an autonomous way although many of its components are encoded by nuclear genes and need to be imported into the chloroplast compartment. Moreover, the chloroplast protein synthesizing system cooperates closely with its counterpart in the nucleo-cytosol, as seen for example by the fact that the four major protein complexes of the photosynthetic apparatus, photosystem II (PSII), photosystem I (PSI), the cytochrome b_6f complex (Cytb_6f) and the ATP synthase, consists of subunits some of which are encoded by the nuclear genome while others are encoded by the chloroplast genome. These different subunits need to be synthesized and assembled in a coordinate way. This is not a trivial task as the ploidy levels between chloroplast and nuclear genomes differ dramatically. A typical mesophyll cell contains ca. 100 chloroplasts, each of which comprises close to 100 chloroplast genomes, whereas nuclear genes are present in only two copies per cell in diploid organisms. In recent years a great deal of progress has been achieved in understanding at least part of the molecular basis of this coordinate interplay between nucleus and chloroplast. From this analysis it has become apparent that a surprisingly large number of nucleus-encoded chloroplast proteins are required for this process and that most of them act at various steps of chloroplast gene expression, including transcription, RNA processing and splicing, RNA editing, RNA stability, translation and assembly of the photosynthetic complexes.

The nucleus does not only encode proteins required for chloroplast biosynthesis and functioning, it is also able to sense the state of the plastid and to respond to signals originating from this organelle. This process is called retrograde signaling. Although numerous studies have been devoted to this topic, the corresponding signaling chains are still largely unknown. The aim of this chapter is to provide an

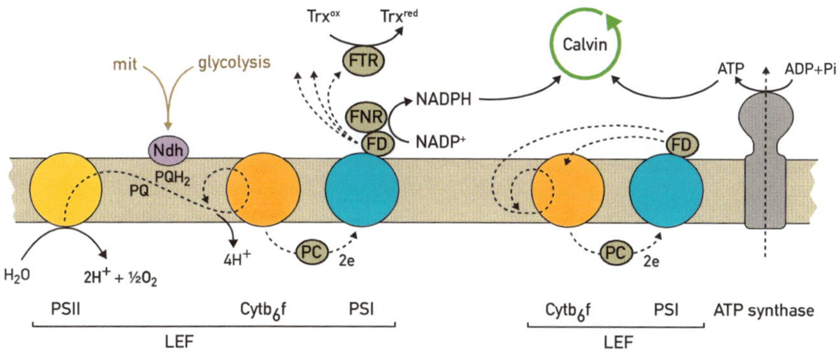

Fig. 3.1 Photosynthetic electron transport chain. The four major photosynthetic complexes PSII, Cytb_6f complex, PSI, ATP synthase in the thylakoid membrane are shown. Electron pathways are indicated by broken lines. The chlororespiratory chain which feeds reducing equivalents from mitochondria and glycolysis into the PQ pool is marked in *brown*. *LEF* linear electron transport chain, *CEF* cyclic electron transport chain, *FD* ferredoxin, *PC* plastocyanin, *FNR* ferredoxin-NADP reductase, *FTR* ferredoxin-thioredoxin reductase, *Ndh* NADH/NADPH dehydrogenase, *PQ* plastoquinone, *PQH_2* plastoquinol, *Calvin* Calvin-Benson cycle

updated view on where we stand in our understanding of the molecular basis of this intricate crosstalk between nucleo-cytosol and plastids. Earlier studies on this topic have been reviewed in several excellent reviews [36, 91, 105].

The reactions of photosynthesis involve light-induced charge separations across the thylakoid membrane within PSII and PSI which set into motion electron flow along the photosynthetic electron transfer chain (Fig. 3.1). Electrons are extracted from water by PSII and transferred subsequently to the plastoquinone pool, the Cytb_6f complex and plastocyanin which is able to re-reduce photo-oxidized PSI. The final electron acceptor of PSI, ferredoxin, reduces $NADP^+$ to NADPH. This electron flow is coupled with the pumping of protons into the thylakoid lumen and the proton gradient leads to the production of ATP through the ATP synthase. Both NADPH and ATP are then used for driving CO_2 assimilation by the Calvin-Benson cycle. Reduced ferredoxin is also able to transfer electrons to other metabolic pathways involved in nitrogen and sulfur metabolism. The redox state of some of these electron carriers, in particular that of the plastoquinone pool, depends on the relative excitation of PSII and PSI and also on the photorespiratory chain which feeds stromal reducing equivalents into the PQ pool. Both photosystems contain their own light-harvesting systems which have distinct pigment compositions and light absorption properties with PSII and PSI absorption peaks in the red and far-red regions, respectively. Hence any change in spectral light quality will lead to unequal excitation of the two photosystems and change the redox state of the plastoquinone pool. In turn this will elicit in the short term a compensatory response through a rebalancing of the antenna systems and in the long term it will impact both chloroplast and nuclear gene expression in such a way that the photosystem which is limiting is upregulated, thereby allowing for a readjustment of the stoichiometry of the two photosystems and improved photosynthetic performance [3]. Moreover if the upstream light reactions operate at a higher rate than the downstream Calvin-Benson cycle as a result of a sudden increase in light irradiance, the acceptor side of PSI will be over-reduced, leading both to the formation of reactive oxygen species and changes in nuclear gene expression. However the signaling chains between chloroplast and nucleus involved in these processes are still poorly understood.

3.2 Nucleus-Encoded Trans-Acting Factors Involved in Chloroplast Gene Expression

Nuclear genes encoding chloroplast proteins can be recognized by the fact that these proteins carry transit sequences at their N-terminal end which act as targeting signals. Based on conserved features of these transit peptides it has been estimated that between 3000 and 4000 nuclear genes encode chloroplast proteins [1]. This large set includes genes of enzymes involved in several metabolic pathways, of subunits of the photosynthetic complexes, and of proteins of the plastid genetic system and factors required for its expression. This genetic system comprises chloroplast DNA with a transcriptional apparatus consisting of at least two distinct RNA polymerases in land plants. The first, called PEP (plastid encoded polymerase) resembles the corresponding bacterial enzyme and contains the α, β, β' and β'' subunits encoded by the plastid genes *rpoA*, *rpoB, rpoC1* and *rpoC2*, respectively [53], and is associated with several σ factors in land plants. The other plastid RNA polymerase, called NEP, consists of a single nucleus-encoded subunit which is related to bacteriophage T7-type RNA polymerases [50]. In addition, plastids contain 70S-type ribosomes consisting of both chloroplast- and nucleus-encoded subunits, most of which are related to bacterial ribosomes, and the common translation initiation, elongation and termination factors (Fig. 3.2). Although the bacterial and chloroplast protein synthesizing systems are very similar in structure, there are some differences. Several plastid proteins of *C. reinhardtii* are considerably larger than their bacterial counterparts with N-terminal extensions and an insertion sequence which may play a role in mRNA recognition and translation initiation [156]. Moreover, a large number of ancillary nucleus-encoded factors are required for plastid gene expression, a feature which is not observed in bacterial systems.

3.2.1 Chloroplast Transcription Factors

Amongst chloroplast transcription factors six sigma factors have been identified, a surprisingly large number given the modest size of the plastid genomes. The activity of some of these factors depends on their phosphorylation status, which changes during plastid development and which is strongly influenced by the light environment [139], in particular when plants are shifted from the dark to the light or from light preferentially absorbed by PSII to light preferentially absorbed by PSI and

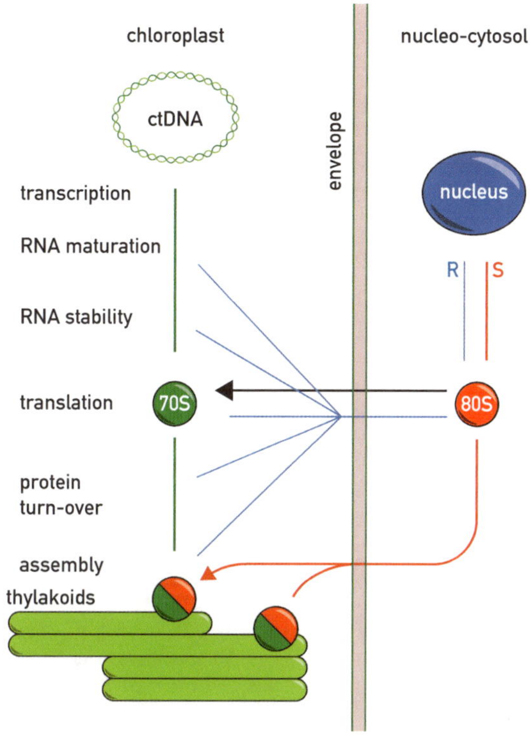

Fig. 3.2 Biogenesis of the photosynthetic apparatus. The chloroplast and nucleocytosol are separated by the chloroplast envelope. Nuclear genes coding for chloroplast structural proteins (*S*) and for factors required for chloroplast gene expression (R) are transcribed, the corresponding mRNAs are translated on cytoplasmic 80S ribosomes and their products are imported into the chloroplast. S-proteins involved in photosynthesis are assembled together with chloroplast-encoded subunits into functional photosynthetic complexes. R-factors are involved in the different steps of chloroplast gene expression as indicated

vice-versa. Amongst proteins associated with PEP, a Ser/Thr kinase of the casein kinase II family, called PTK (plastid transcription kinase), was identified [93]. *In vitro*, this kinase was shown to be activated by reduced glutathione and its activity was reduced by phosphorylation [8]. PTK acts as a global regulator of chloroplast transcription, a process which depends on its phosphorylation state and that of σ factors [72, 93] (Fig. 3.3). Globally, phosphorylation and dephosphorylation of the σ factors promote etioplast and chloroplast-specific promoter usage, respectively [139]. However transcription can be up- or down-regulated depending on which σ factor is phosphorylated and which promoter is used [123]. Also, some σ factors contain several phosphorylation sites, some of which are substrates for PTK [8, 139]. One well-studied example is SIG1 whose phosphorylation state changes when plants are shifted from light favouring PSI to light favouring PSII or vice-versa [124] (Fig. 3.3). In this case the PQ pool is reduced which leads to dephosphorylation of SIG1and in turn to an increase of transcription of the *psaA* and *psaB* genes encoding the two reaction center proteins of PSI. The reciprocal event occurs when the PQ pool is oxidized: SIG1 is phosphorylated and transcription of the *psaA* and *psaB* genes is reduced. Thus these changes lead to a readjustment of the stoichiometry of PSII and PSI and to a rebalancing of the absorbed light excitation energy between the two photosystems, thereby optimizing photosynthetic activity. One possible mechanism for this is that the phosphorylation of PTK is controlled through

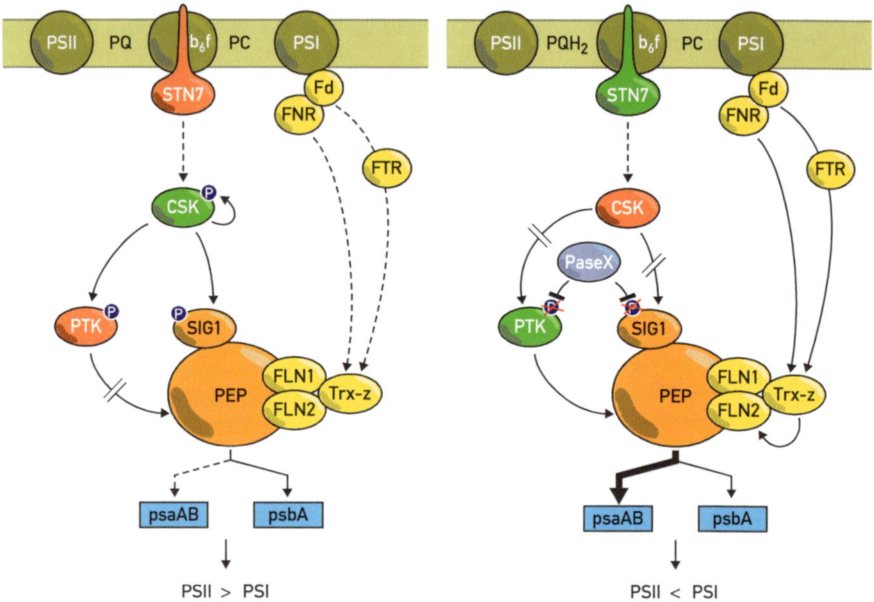

Fig. 3.3 Model for the link between the redox state of the photosynthetic electron transport chain and the chloroplast transcriptional apparatus. The *left* and *right* panels refer to an oxidized and reduced PQ pool, respectively. An oxidized PQ pool leads to a decrease in *psaAB* expression whereas a reduced PQ pool leads to an increase in *psaAB* expression; *psaAB* indicates that the *psaA* and *psaB* genes are part of the same operon in land plants. The kinases STN7, PTK, CSK and factors involved in these processes are indicated.(reproduced from ref. [115] with permission)

the STN7 kinase which is known to be activated when the plastoquinone pool is reduced (Fig. 3.4) [17]. In contrast to the PSI genes, expression of *psbA* encoding the reaction center protein of PSII is not altered significantly upon changes in the redox state of the PQ pool. Additional targets of PTK are small DNA binding proteins whose phosphorylation states change during these light shifts. These changes in phosphorylation appear to play an important role for mediating the redox signals from the photosynthetic electron transport chain to the chloroplast transcription machinery and appear to require in addition the action of thiol redox signals [128]. Although the underlying mechanisms are still unknown, one possibility is that these changes in thiol redox state could promote conformational protein changes, thereby granting access of masked sites to phosphorylation [122].

Another player in this chloroplast signaling network is the CSK kinase which shares structural features with cyanobacterial sensor histidine kinases [108]. This kinase, which is conserved in all major plant and algal lineages except *C. reinhardtii,* appears to be involved directly or indirectly in chloroplast gene expression based on the fact that the levels of several chloroplast transcripts are altered in CSK knock-out lines of Arabidopsis [108]. Moreover, CSK autophosphorylation occurs when the PQ pool is oxidized, and can be correlated with the phosphorylation of SIG1 and the decrease in *psaAB* gene expression. Based on these observations and

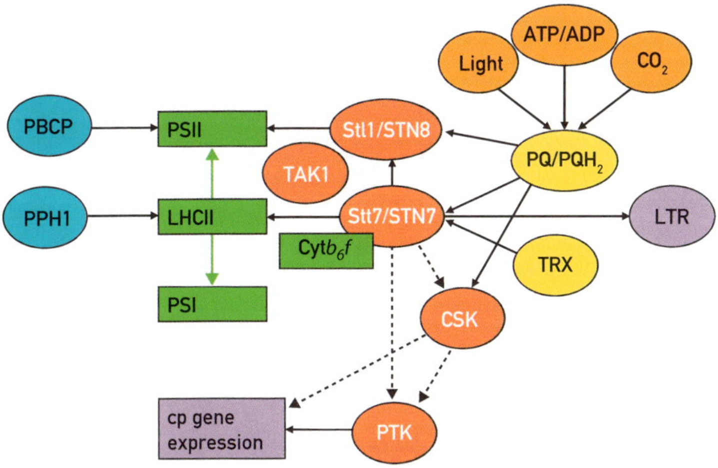

Fig. 3.4 Chloroplast signaling chains. The redox state of the PQ pool is modulated by the light irradiance, ATP/ADP ratio and ambient CO_2 level. The protein kinases Stl1/STN8, Stt7/STN7, CSK, PTK and TAK1 are shown with their targets indicated by arrows. *Broken arrows* indicate putative targets. *LTR* long term response involving retrograde signaling mediated through Stt7/STN7

the fact that CSK interacts with PTK and SIG1 in yeast two-hybrid assays, a model was proposed by Puthiyaveetil [107, 109] in which the activity of CSK is regulated by the redox state of the PQ pool through the STN7 kinase (Fig. 3.3). CSK phosphorylates and inactivates PTK when the PQ pool is oxidized. CSK also would phosphorylate SIG1, which causes the repression of the *psaAB* genes but *psbA* expression is not significantly affected. In contrast, reduction of the PQ pool would lead to dephosphorylation of SIG1 and derepression of the *psaAB* genes. This event is postulated to be mediated by an unidentified phosphatase which could also dephosphorylate PTK.

It is still unclear how the activation of CSK is linked to the redox state of the PQ pool and to what extent it is functionally linked with the STN7 kinase. Because *stn7* mutants are unable to acclimate properly [100], it is likely that these two kinases operate in the same pathway, although their activity is regulated in opposite ways by the redox state of the PQ pool. A reduced PQ pool activates STN7 and inactivates CSK and the opposite occurs when the plastoquinone pool is oxidized. Recent studies have revealed that a thioredoxin-like protein called Trx-z is also involved in this plastid signaling network [119]. An Arabidopsis knock-out mutant of Trx-z is strongly impaired in chloroplast development and displays an albino phenotype [5, 119], an unusual feature as Trx mutants usually show a mild phenotype because of functional redundancy within the Trx family. Loss of Trx-z led to the specific decrease of the expression of genes transcribed by PEP. Trx-z interacts with two fructose-like proteins, FLN1 and FLN2 based on yeast two hybrid assays. Although these two proteins are members of the pfbB-sugar kinase family, both have amino acid substitutions within the predicted active site and lack detectable sugar-phosphorylating activity. These features suggest that these proteins have acquired new regulatory functions, a view which is further supported by the fact that FLN2 undergoes redox changes during changes in the light environment which are mediated by Trx-z [119]. These observations open the possibility that Trx-z and the FLNs participate in a signaling chain which imparts redox regulation to PEP and could couple transcriptional control to light signals through the photosynthetic electron transport chain. A complex network of interacting proteins emerges from these studies in which signals are transmitted from the thylakoid membrane to the plastid gene expression system through multiple phosphorylation events and thiol-dependent protein redox changes (Fig. 3.3).

3.2.2 Factors Involved in Post-Transcriptional Steps of Chloroplast Gene Expression

Genetic studies with mutants of *C. reinhardtii*, maize and *Arabidopsis* affected in photosynthetic activity have revealed a large number of nucleus-encoded factors which are required for chloroplast gene expression and which act at different post-transcriptional steps [36, 129] (Fig. 3.2). Some of these proteins are involved in RNA stability and processing (M factors), splicing, translation (T-factors), and in the assembly of photosynthetic complexes. In *C. reinhardtii* most of these factors act in a gene-specific manner on the corresponding 5'UTRs, whereas in land plants the loss of these factors has a more pleiotropic effect with several chloroplast genes deficient in their expression. Many of the genes affected in these mutants have been identified and characterized. Some are well-conserved in oxygenic photosynthetic organisms while others appear to be only present in specific organisms. Overall they can be classified in several groups.

The first group comprises proteins derived from genes involved in RNA metabolism (Table 3.1). Examples include a pseudouridine synthase gene which is required for trans-splicing of the chloroplast *psaA* transcripts in *C. reinhardtii* [99]. Interestingly, this gene has nucleotide changes in the active sites which abolish pseudouridine synthase activity. Another example is the maize CRS2 gene required for the splicing of several chloroplast group II introns and which is related to peptidyl-tRNA hydrolase enzymes [54]. However this protein lacks several conserved amino acids that are important for the peptidyl-tRNA hydrolase activity and is unable to rescue a mutant of *E. coli* lacking this activity, strongly suggesting that it has lost its initial function. These two cases show that during evolution enzymes have been recruited for performing novel functions in RNA metabolism but at the cost of their original activity.

Some nucleus-encoded factors contain CRM domains which appear to be derived from an RNA binding module in an ancient ribosome-associated protein in archae and bacteria [12]. The notion that CRM domains bind RNA is based on the fact that a single CRM domain has RNA binding activity *in vitro* [12], that three characterized proteins from maize, CRS1, CAF1 and CAF2 containing multiple CRM do-

Table 3.1 Proteins involved in chloroplast post-transcriptional steps of chloroplast gene expression containing domains derived from catalytic domains

Chlamydomonas	Catalytic domain	Function	Ref
Raa2	Pseudouridine synthase	Splicing	[99]
Raa3	Pyridoxamine 5'-phosphate oxidase	Splicing	[114]
Maize			
CRS2	Peptidyl-tRNA hydrolase	Splicing	[54]
CRS1/Lp	CRM (ribosome assembly)	Splicing	[97]
CAF1/Lp	CRM	Splicing	[98]
CAF2/Lp	CRM	Splicing	[98]
CFM2	CRM	Splicing	[6]

Lp Ortholog exists in land plants, *CRM* Conserved domain involved in chloroplast RNA splicing and ribosome maturation [12]

mains are involved in plastid RNA metabolism [97, 138], and that structural studies of the CRM module reveal similarity to known RNA binding domains [98].

The second group comprises a large set of genes which encode proteins with degenerate repeats of 34–38 amino acids. Several families have been identified within this group such as TPR (tetratricopeptide repeats), PPR (pentatricopeptide repeats) and OPR proteins (octatricopeptide repeats) (Table 3.2). Proteins with similar repeats are widespread in nature and the atomic structure of several of them have been solved: β catenin with 12 ARM repeats of 42 amino acids [52], the A subunit of protein phosphatase 2A with 15 HEAT repeats of 39 amino acids [47], *Pumilio* with 8 Puf repeats of 36 amino acids [38] and protein phosphatase 5 with 3 TPR repeats [32]. These structural studies revealed that each repeat comprises two or three α helices and that the repeats are often tandemly organized forming a superhelix whose internal side is thought to be involved in protein-protein interactions or in RNA binding as shown for *Pumilio*, a translational repressor of the Drosophila *hunchback* mRNA [38]. A major technical hurdle is that chloroplast TPR proteins are highly insoluble making their biochemical and functional characterization difficult, although RNA binding activity could be demonstrated recently with HCF107, a protein which contains HAT motifs (half-a-tetratricopeptide) which are variants of TPR repeats [48]. Specific RNA binding was shown for a few PPR proteins which are structurally similar to TPR proteins [125]. They include HCF152 from Arabidopsis required for RNA processing [76, 89] and PPR10 from maize which binds RNA segments from two different transcription units [102] (see Table 3.2). The PPR10 protein binds two short 15 nt long RNA elements in two intercistronic regions where it blocks 5' to 3' as well as 3' to 5' exonucleolytic digestion. Exonucleolytic decay starts distal to the PPR10 binding site and may be initiated by an endonucleolytic event, e.g., by chloroplast RNaseE. In the end, RNA degradation leads to the complete elimination of the target RNAs with exception of the small 20 nt stretch protected by PPR10 [102]. Interestingly, such potential footprints have been detected in small RNA databases in cereals that include suggested binding sites of the PPR proteins CRR2 and PPR10 and HCF152 [86, 102]. Thus some PPR proteins appear to act as bidirectional roadblocks against exonucleases. In this sense

Table 3.2 Nucleus-encoded repeat-containing proteins involved in chloroplast gene expression

Protein	Organism	Process	Target	Ref
PPR proteins				
PPR2	Zm	Translation		[149]
PPR4	Zm	Splicing	*rps12*	[118]
PPR5	Zm/At	Translation	*trnG*	[16]
PPR10	Zm	RNA processing	*atpI-atpH ir*, *psaJ-rpl33 ir*	[106]
PPR8522/DGT1	Zm/At	Plastid transcription	*Emb*	[127]
CRP1	Zm	RNA processing	*petB-petD* ir, *psaC*	[11]
HCF152	At	RNA processing	*psbH-petB* ir	[76]
CRR4	At	RNA editing	*ndhD-1*	[20]
CRR22	At	RNA editing	*ndhB-7, ndhD-5, rpoB-3*	[94]
OTP82	At	RNA editing	*ndhG-1, ndhB-9*	[94]
Mrl1	Cr/Lp	RNA stability	*rbcL*	[56]
Mca1	Cr	RNA stability	*petA* 5'UTR	[73]
TRP proteins				
HCF107/Mbb1	At/Cr	RNA processing	*psbB* 5'UTR	[141]
Nac2	Cr	RNA processing	*psbD* 5'UTR	[90]
OPR proteins				
Tbc2	Cr	Translation	*psbC* 5'UTR	[7]
Tab1	Cr	Translation	*psaB* 5'UTR	[110]
Tda1	Cr	Translation	*atpA* 5'UTR	[37]
Raa1	Cr	Splicing/RNA processing	*psaA*	[78]
Rat2	Cr	Splicing/RNA processing	*psaA*	[9]

Zm Zea mays, *At Arabidopsis thaliana*, *Cr Chlamydomonas reinhardtii*, *Lp* indicates that an ortholog exists in land plants, *ir* intergenic region; target refers to the specific chloroplast RNA target sites of the corresponding factors, *emb* embryo lethal

they could provide an alternative to RNA hairpins for defining the termini of chloroplast mRNAs. Moreover PPR10 binding changes the local RNA structure [106] near the start codon of *atpH* RNA and unmasks a ribosome binding site. Although several other PPR proteins are known to be required for translation of specific chloroplast mRNAs, it is not yet known whether such PPR proteins also act by creating a local environment favourable for ribosome entry.

The PPR protein family contains 450 members in Arabidopsis, several of which are involved in RNA metabolism, in particular in RNA processing, translation and RNA editing [117]. This large set of proteins could provide a reservoir of trans-acting factors which act on specific chloroplast RNA targets at various post-transcriptional steps. These repeats appear to define a code which connects the succession and amino acid sequence of the repeats with the nucleotide sequence of the target RNA. It was indeed shown that PPR elements bind specific RNA nucleotides through the combinatorial action of two amino acids in each repeat [13]. Such a code would provide a molecular basis for the specific protein-RNA interactions which underlie the different post-transcriptional steps in chloroplast gene expression. The crystal structure of the related Puf repeats has given rise to models in which each repeat of the protein would bind a single or several nucleotides of its RNA target sequence [38, 147, 148].

Interestingly while only 11 PPR protein genes could be detected in the Chlamydomonas genome, a new family of proteins with OPR repeats has evolved in this alga. The first OPR protein identified was Tbc2, a factor required for the translation of the *psbC* mRNA [7] and subsequently several other factors involved in RNA processing, splicing and translation were identified (Table 3.2). In the case of Tab1, which is involved in the translation of the *psaB* mRNA of *C. reinhardtii,* a truncated portion of this protein displayed partial specificity of binding to the *psaB* 5'UTR and U-rich RNA *in vitro,* although competition experiments revealed that it also binds to other non-target chloroplast RNAs [110]. Other OPR proteins are Rat2 and Raa, involved in *psaA* RNA processing and trans-splicing [9, 78], and Tda1, specifically required for the translation of the *atpA* mRNA as well as in trapping of a subset of untranslated *atpA* transcripts into non-polysomic complexes [37]. Altogether the OPR family of *C. reinhardtii* includes more than a 100 proteins, several of which are likely to be implicated in chloroplast, and perhaps also mitochondrial, gene expression. These assigned functions of OPR proteins are clearly compatible with the large number of nucleus-encoded factors identified through genetic screens of photosynthetic mutants of *C. reinhardtii*. Whereas OPR proteins are very rare in plants, they are present in several bacteria and protozoans. In particular, OPR repeats appear to be structurally related to the RAP RNA binding domain which is abundant in apicomplexans [69]. Moreover some of the OPR proteins contain a region related to the FAST1 kinase-like domain [137], raising the possibility that the activity of some of these trans-acting factors may be modulated by reversible phosphorylation.

Another group of nucleus-encoded trans-acting factors involved in chloroplast gene expression includes proteins unrelated to any known protein. These "pioneer" proteins are usually specific to a given organism and not present in other plants and algae. The precise function of these proteins is still largely unknown.

3.2.3 CES Control

A major problem in the assembly of photosynthetic complexes is the coordinate synthesis of the different subunits. Subunits which accumulate in excess are usually degraded by chloroplast proteases [24]. Moreover, a fine control is exerted at the assembly stage through the CES (Control by Epistasy of Synthesis) in *C. reinhardtii* which was first identified and characterized for the Cytb_6f complex [25] and later extended for PSII [81], PSI [153], ATP synthase [33] and Rubisco [152]. The CES process involves an assembly-dependent translation of specific subunits of a complex. As an example, in the case of Cytb_6f the rate of synthesis of Cyt*f* at the level of translation initiation is substantially decreased in the absence of cytochrome b_6 or the PetD subunit. It results from an exposed regulatory motif within the C-terminal region of unassembled Cyt*f* that inhibits the translation of its *petA* mRNA through the 5'UTR. New insights into the mechanism of CES have come from the functional analysis of Mca1 and Tca1, which are required for the stable accumulation and translation, respectively, of the chloroplast *petA* mRNA coding for cytochrome *f* [113]. In transformed strains in which the amount of Mca1 and Tca1 were decreased, the level of *petA* mRNA and translation of cytochrome *f* also diminished, indicating that these factors are limiting for cytochrome *f* accumulation. Because Mca1 has a short half-life, its abundance varies rapidly under physiological conditions in which the demand for Cytb_6f changes. Such a case occurs under nitrogen starvation or in aging cultures in which a decrease of both factors can be correlated with a loss of *petA* mRNA and Cytb_6f [113]. Interestingly, the degradation of Mca1 is triggered through its interaction with unassembled Cyt*f* and involves the same residues of Cyt*f* that constitute the repressor motif during CES, i.e. mutations which abolish CES also prevent Mca1 turnover. Thus while some of the nucleus-encoded factors involved in chloroplast gene expression may be constitutively required, others such as Mca1 clearly have a regulatory role in this process.

3.2.4 Redox Control of Post-Transcriptional Steps of Chloroplast Gene Expression

A hallmark of chloroplast protein synthesis is the involvement of numerous nucleus-encoded translational activators, some of which are part of protein complexes. The activity of some of these activators is influenced by the light environment and can also be affected by the nutrients in the growth medium. Indeed a large increase in the rate of synthesis of the major thylakoid proteins occurs upon a transition from dark to light in plants and algae [74]. A particularly striking example is the D1 reaction center protein of PSII. Because of the highly oxidizing water splitting reaction catalyzed by PSII, photodamage to D1 is unavoidable and damaged D1 needs to be replaced continuously by newly synthesized protein. A protein complex consisting of RB60, RB55, RB47 and RB38 was characterized in *C. reinhardtii* which binds specifically to the 5'UTR of *psbA* mRNA encoding D1 [29]. This binding is strongly enhanced by light and mediated through RB47, a member of the polyA binding protein family [159]. Mutants lacking this protein are no longer able to synthesize D1 [159, 160]. This observation, together with the correlation between the light-stimulated RNA binding activity of the *psbA* mRNA complex and the association of *psbA* mRNA with polyribosomes and its translation, strongly suggest that the complex plays an important role in the initiation of translation of this mRNA. The levels of the proteins of the complex remain the same during the light shift, indicating a post-translational regulation. Indeed, *in vitro* the RNA binding activity of this complex appears to be regulated by the redox state of the complex which in turn is regulated by thioredoxin [31]. The disulfide isomerase RB60 which carries a redox-active regulatory site was identified as the redox sensor of the *psbA* RNA binding complex [58, 140]. It regulates, together with the oxidoreductase Tba1 [126], the RNA binding activity of RB 47 in response to the redox state of the chloroplast by reacting with a regulatory disulfide of RB47 indicating that these two redox partners are most likely coupled [2]. In its reduced but not in its oxidized state RB47 binds to the *psbA* 5'UTR, which correlates with an increase of translation. Furthermore RB60 appears to mediate the attenuation of *psbA* mRNA translation in the dark through phosphorylation of RB60 by ADP [30].

Another component of the *psbA* RNA complex, RB38, binds specifically to U-rich stretches of the *psbA* 5'UTR [14]. However RB38 was renamed RB40 and shown to be involved in *psbD,* encoding D2, the partner reaction center protein of D1, rather than in *psbA* translation [120]. Translation of *psbD* mRNA is also increased by light. By binding to the *psbD* 5'UTR, RB40 destabilizes a stem-loop-structure which provides access of the small ribosomal subunit to the *psbD* initiation codon [60, 96]. As in the case of *psbA*, D2 synthesis in the light appears to be mediated through the formation of a dynamic high MW complex in which Nac2 and RB40 form an interprotein disulfide bridge [120]. This bridge is opened in the dark by NTRC, the NADPH-dependent thioredoxin reductase, a process which leads to the dissociation of RB40 from the complex and correlates with the arrest of translation initiation [121]. This system might establish a direct link between chloroplast gene expression and carbon metabolism during cell adaptation in the dark.

3.3 Retrograde Signaling

Retrograde signaling is usually defined as the process through which changes in the developmental and metabolic state of plastids are transmitted to the nucleus where they elicit specific changes in nuclear gene expression. This communication between plastid and nucleus is essential for ensuring a coordination between nuclear and chloroplast gene expression for proper chloroplast biosynthesis and maintenance, as well as for adaptation under adverse environmental conditions. Although the identity of the plastid-derived signals is still not clear, at least four distinct sources of these signals have been considered. They include the tetrapyrrole biosynthetic pathway, plastid protein synthesis, the redox state of the plastid and in particular that of the photosynthetic electron transport chain, and reactive oxygen species (ROS) within the organelle.

3.3.1 Role of Tetrapyrroles in Retrograde Signaling

The first evidence for the involvement of tetrapyrroles in retrograde signaling came from studies of Chlamydomonas which revealed that the level of LHCII mRNA was decreased in mutants with defects in the tetrapyrrole pathway (Fig. 3.5) or in wild-type cells treated with inhibitors of chlorophyll synthesis [55]. The role of tetrapyrroles in retrograde signaling in this alga was further confirmed by the observation that feeding of Mg-protoporphyrin IX (MgProto) to Chlamydomonas cells upregulates the genes of HemA (glutamyl-tRNA reductase) and the chaperone

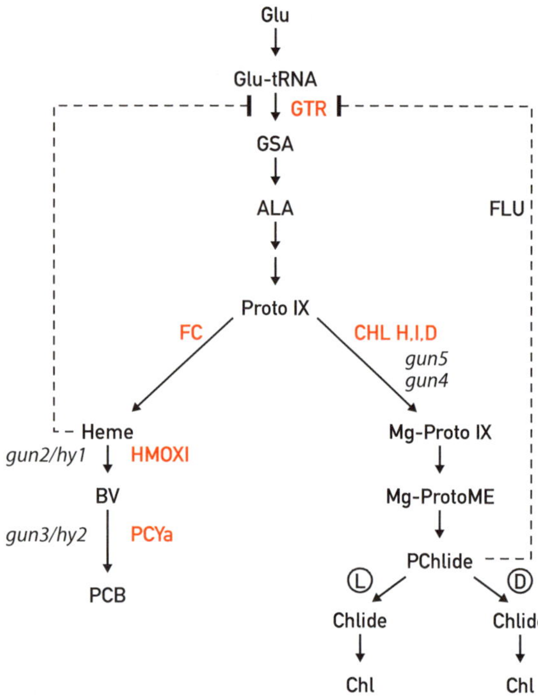

Fig. 3.5 Tetrapyrrole pathway. The heme and chlorophyll biosynthetic pathway branch at protoporphyrin IX (Prot IX). GTR, glutamine tRNA reductase, subjected to feedback inhibition by heme and FLU. In most land plants conversion of PChlide (protochlorophyllide) to Chlide is light-dependent (*L*) but in *Chlamydomonas* this conversion also occurs in the dark (*D*). Through its negative feedback on GTR, FLU prevents overaccumulation of PChlide in the dark. The steps affected by the *gun* and *hy* mutations are indicated. *GSA* glutamate 1-semialdehyde, *ALA* 5-aminolevulinic acid, *Chl* chlorophyll, *FC* ferrochelatase, *HMOX1* heme oxygenase, *BV* biliverdin, *PCY* bilin reductase, *PCB* phytocyanobilin

genes Hsp70A, Hsp70B and HSP70E [65]. To rule out the possible role of ROS in this response, these feeding experiments were performed in the dark. However light induction of HSP70A still occurs in mutants of *C. reinhardtii* affected in the Mg-chelatase which are deficient in chlorophyll and accumulate reduced amounts of Mg-tetrapyrroles. In these mutants the level of free heme is increased and the role of heme in this process is further supported by the observation that feeding of hemin to cells grown in the dark activates HSP70A expression [143]. A response element called PRE was identified in the HSP70A promoter which mediates induction by Mg Proto, heme and light and which is also present in other nuclear genes induced by hemin [142, 143]. Further support for the convergence of these signals in the same pathway arises from the observation that an extended treatment of cells with hemin abolishes the induction of HSP70A by Mg Proto or light [143]. Recently a genome-wide transcriptional profiling was performed to assess the global impact of these tetrapyrroles on nuclear gene expression in *C. reinhardtii*. Close to 1000 genes were identified whose expression changed transiently and which encode enzymes of the tricarboxylic acid cycle, heme-binding proteins, stress-response proteins and proteins involved in protein folding and degradation [144]. Because these tetrapyrroles are unlikely to be found in the natural environment of the alga, it is probable that they act as messengers in intracellular pathways that signal changes in the environment. However the drastic changes in mRNA levels were not paralleled by similar changes in protein amount [144].

Further evidence for the role of tetrapyrroles in retrograde signaling in Chlamydomonas is based on the recent identification of a functional bilin biosynthesis pathway in this alga although, as other chlorophytes, it does not produce phytochromes which are known to use bilins as chromophores [34]. In this alga a plastid-localized heme oxygenase HMOX1 and a ferredoxin-dependent bilin reductase (PCYA) yield bilin metabolites which appear to play an important role for phototrophic growth by detoxifying molecular oxygen produced by photosynthesis and for enhanced chlorophyll synthesis in the light. Thus bilins appear to be widespread as retrograde signals, at least in chlorophytes. This study also revealed a cytosolic heme oxygenase that enables the cells to scavenge extracellular heme as an iron source.

Additional compelling evidence of the involvement of tetrapyrroles as mediators for plastid-to-nucleus communication was provided by the recent identification of a tetrapyrrole-regulated ubiquitin ligase as cell cycle coordinator from organelle to nuclear DNA replication in the primitive red alga *Cyanidioschyzoan merolae* [61, 62, 135].

The situation is less clear in land plants. Genetic screens for mutants deficient in retrograde signaling have been performed which take advantage of the fact that treatment of plants with norflurazon, an inhibitor of carotenoid biosynthesis, leads to photooxidation of chloroplasts and to a decrease in the accumulation of LHCII mRNA [75]. By screening for derepression of a reporter gene driven by the LHCII promoter in a transgenic Arabidopsis line treated with norflurazon, several *gun* (*g*enome *un*coupled) mutants were isolated [134]. Loss of function of these GUN genes leads to an increased expression of LHCII genes, or more generally, of photosynthesis-associated nuclear genes when chloroplast development is inhibited. The *gun2* and *gun3* mutations affect the heme oxygenase and phytochromobilin synthase genes, respectively, which are responsible for the conversion of heme into phytochromobilin, the chromophore of phytochromes (Fig. 3.5). The *gun2* and *gun3* mutants are allelic to the long hypocotyl photomorphogenic mutants *hy1* and *hy2* which develop long hypocotyls when grown in the light, a phenotype which occurs in wild-type plants grown in the dark. The *gun5* mutation affects CHLH, one of the three subunits of Mg-chelatase which converts ProtoIX into Mg-ProtoIX [83], whereas *gun4* affects a regulatory protein which stimulates the activity of this enzyme by facilitating the binding of its substrate and/or the release of its product [68]. In contrast, although the *cs* and *ch42* mutants deficient in CHLI accumulate decreased amounts of Mg-ProtoIX and methylester of Mg-ProtoIX (Mg-Proto-IX-ME), they do not display a *gun* phenotype [83]. Similarly the *chlm* and *crd1* mutants affected in the steps following Mg-ProtoIX in the tetrapyrrole pathway are not *gun* mutants [84]. While the involvement of the tetrapyrrole pathway in retrograde signaling is undisputed, it is not yet clear whether the effect is direct or indirect and the identity of the signaling molecules is uncertain. It was first proposed that Mg-ProtoIX acts as a signal [130] and that it can cross the chloroplast envelope and accumulate in the cytosol [4]. However careful measurements of Mg-ProtoIX and Mg-ProtoIX-ME in the wild type and in several *gun* mutants led to the conclusion that Mg-ProtoIX does not act as a signaling molecule in Arabidopsis [84, 87].

Perturbation of the tetrapyrrole pathway may lead to photo-oxidative damage and to the formation of ROS and it may also change the redox state of the plastid which would in turn trigger retrograde signaling [85].

Whereas *gun1–5* were identified as loss-of-function mutants, *gun6* was isolated as a gain-of-function *gun* mutant which overexpresses the plastid ferrochelatase FC1 [151]. In this mutant the increased flux through the heme branch of the tetrapyrrole pathway leads to an increase in PhANG (photosynthesis-associated nuclear gene) expression, although the total heme level is not increased compared to wild-type plants. Therefore the inhibitory action of free heme on glutamate-tRNA reductase at an early step in the tetrapyrrole biosynthetic pathway and on chlorophyll synthesis would not occur. However, transient changes in heme level may play a role in this signaling pathway and heme itself may act as a positive signal, especially because blocking heme catabolism and phytochromobilin synthesis also enhances PhANG expression [151]. Surprisingly, overexpression of FC2, a second plastid ferrochelatase in Arabidopsis, does not lead to a *gun* phenotype, raising the possibility that the two ferrochelatases FC1 and FC2 contribute to distinct pools of heme [151]. Because heme can be exported from chloroplasts [136], it could then increase PhANG expression by interacting with transcription factors in the nucleo-cytosol as in the case of yeast [161] and humans [158], or it could act indirectly by binding to hemoproteins, chaperones or transporters. Proteins of the SOUL family which have been shown to bind heme could be involved in these processes of heme trafficking and signaling [26]. Thus as proposed for Chlamydomonas [143], heme could act as a positive retrograde signal for the expression of specific nuclear genes in land plants and play a role in the coordination with chloroplast development. However the underlying molecular mechanisms are still largely unknown.

3.3.2 Plastid Protein Synthesis and Retrograde Signaling

Because plastid and nuclear gene expression are tightly coordinated, it is not surprising that a deficiency in chloroplast protein synthesis has a strong impact on nuclear gene expression. The first evidence for such a process came from studies of the barley *albostrians* mutant [21]. In this variegated mutant the white tissue is deficient in plastid ribosomes and shows decreased activity of enzymes involved in photosynthesis, suggesting that cytoplasmic synthesis of plastid proteins may be controlled by plastid RNA [21]. These studies were later extended by showing that transcription of the genes encoding the light-harvesting system were down-regulated in the *albostrians* mutant [51].

It is now well established that treatment of wild-type plants with inhibitors of plastid translation such as chloramphenicol, lincomycin, erythromycin and streptomycin leads to decreased expression of nuclear genes of plastid proteins involved in photosynthesis in land plants [92, 131]. The inhibitory effect on nuclear gene expression was however only detected within the first 2–3 days of seedling development [92], and moreover, it was also seen in the dark in the pea *lip1* (*l*ight-*i*ndependent *p*hotomorphogenesis 1) and Arabidopsis *cop1* (*co*nstitutively *p*hotomorphogenic1) mutants which have a photomorphogenic development in the dark, thus indicating that light is not essential for this signaling pathway [46]. Similar effects were also observed by treating plants with tagetitoxin, a specific inhibitor of plastid RNA polymerase [112] or by nalidixic acid, an inhibitor of plastid DNA replication and transcription [46]. Taken together these studies indicate that a plastid-derived signal which is dependent on plastid transcription and/or translation is essential for the biogenesis of this organelle.

Amongst the isolated *gun* mutants only *gun1* retains the ability to express nuclear genes involved in photosynthesis when it is grown in the presence of an inhibitor of plastid protein synthesis early in seedling development before the photosynthetic machinery is fully functional [83]. The GUN1 signaling pathway is thus linked to disruption of chloroplast translation. The fact that *gun1* was isolated by selection on norflurazon suggests that chloroplast photo-oxidation induced by norflurazon interferes with signals originating from the plastid protein synthesis pathway. Moreover, seedling development in *gun1* is perturbed at the level of hypocotyl elongation and cotyledon expansion in response to sucrose, abscisic acid (ABA) and redox changes in the photosynthetic electron transport chain [27]. In the presence of these compounds developmental arrest is more severe compared to wild-type seedlings. GUN1 is a chloroplast PPR protein which contains in addition a SMR domain (small MutS-related) in its C-terminal part which is found in proteins involved in DNA repair and recombination. Based on immunofluorescence GUN1 colocalizes with pTAC2, a protein associated with transcriptionally active complexes in chloroplasts [63, 101], indicating that it is a plastid nucleoid protein.

Global gene expression responses of wild type, *gun1* and *gun5* to norflurazon revealed that they share a large number of genes which are derepressed as compared to wild type, strongly suggesting that GUN1 and GUN5 act in the same signaling pathway [63]. Interestingly, the promoters of these target genes contain the core of the light-responsive and the abscisic acid (ABA) response element. Amongst several ABA-deficient or ABA-sensitive mutants tested, only *abi4* displayed a *gun* phenotype, i.e., increased Lhcb mRNA accumulation in seedlings treated with lincomycin [63]. ABI4 appears to act downstream of GUN4 based on the observation that overexpression of ABI4 which encodes the transcription factor APETALA-2 suppressed the *gun1* phenotype [63]. The picture which emerges from this study is that three retrograde signaling chains which become apparent when perturbations occur in the tetrapyrrole pathway, in plastid gene expression or in the redox state of the photosynthetic electron transport chain, converge within the chloroplast upstream or at the level of GUN1. The GUN1-derived signal is then transmitted to ABI4, which binds to the Lhcb promoter where it interferes with the binding of G-box binding factors required for the light-dependent expression of photosynthetic genes. In this respect it is interesting to note that studies on the nuclear chloroplast transcriptome under 101 different conditions revealed 23 distinct regulons, i.e., sets of genes that are coexpressed under the different conditions [18]. Amongst the three types of transcriptional regulation which were apparent, one reveals a master switch which either induces or represses the same large set of genes [71]. In this respect

ABI4 may be the proposed master switch gene which controls nuclear genes in response to environmental signals and developmental cues in the chloroplast [63].

A key question is how the signals generated in the chloroplast are transmitted across the envelope membranes to the nucleus. Significant progress in this area has been achieved recently through the identification of PTM, a chloroplast envelope-bound homeodomain (PHD) transcription factor with transmembrane domains which mediates signals from undeveloped or damaged chloroplasts to the nucleus [132]. Upon release from the chloroplast envelope by an unknown protease, the N-terminal part of this factor moves to the nucleus where it transmits multiple retrograde plastid signals and activates transcription of *ABI4* which acts as a repressor of photosynthetic genes [132]. Although PTM proteins are conserved in land plants they appear to be absent from Chlamydomonas revealing major differences in retrograde signaling between plants and green algae. It should be noted that transcription factors tethered to the ER and Golgi membrane which are released from the membrane by proteolytic cleavage have been characterized and shown to be involved in both stress and developmental signaling pathways [23].

3.3.3 Chloroplast Redox Changes and Retrograde Signaling

Redox changes in the chloroplast can occur within the photosynthetic electron transport chain upon changes in environmental conditions such as light quality and quantity, CO_2 and nutrient availability and heat stress. These changes occur mostly at the level of the plastoquinone pool as the result of unequal excitation of the two photosystems or of limitations in the electron acceptor capacity in the Calvin–Benson cycle and in other downstream reactions. Changes in gene expression caused by changes in the redox state of the PQ pool occur in chloroplasts [103] and have been clearly demonstrated in some algae for nuclear genes [35, 39]. In contrast, in land plants the role of the redox poise of the PQ pool is less important for nuclear genes, although PQ redox control could be demonstrated for the genes of plastocyanin, cytosolic ascorbate reductase and ELIP2 of Arabidopsis [59, 104, 155]. Changes in PQ redox poise have been induced by shifting plants from light preferentially absorbed by PSII (PSII light) to light preferentially absorbed by PSI (PSI light) which favours reduction and oxidation of the PQ pool, respectively. Although the transcription of many nuclear genes depends on photosynthetic electron transport, regulation by the PQ redox state could only be demonstrated for a small portion of these genes. From a set of ca. 2000 genes of Arabidopsis responding to light changes corresponding to a shift from light preferentially absorbed by PSII to light preferentially absorbed by PSI, less than 15 % were shown to be strictly regulated at the RNA level by the redox state of the PQ pool [42, 43]. This low percentage may be due to the fact that expression profiles were determined at the end of the acclimation period when a new expression equilibrium had been reached. To identify genes whose expression is affected only transiently after the light switch and which might escape detection under steady state conditions, the kinetics of gene expression was determined during PSI to PSII or PSII to PSI light shift experiments [22]. These experiments revealed a larger set of genes with changed expression levels, although in most cases these changes were rather modest, less than two fold. Comparison of the expression profiles during PSI–PSII and PSII–PSI light shifts revealed opposite patterns with genes involved in metabolism down- and up regulated, respectively, whereas genes associated with photosynthesis were down regulated in a PSII–PSI shift and expression was more balanced in a PSI–PSII shift [22]. In such experiments genes which respond rapidly are likely to be directly regulated by the redox signals. A set of 54 and 29 genes corresponding to PSI–PSII and PSII–PSI shifts were identified in this way. These genes encode mostly proteins with regulatory functions such as transcription factors and protein interaction and modification factors. One of them is SIG1, a transcription factor of plastid encoded RNA polymerase. Very little overlap with expression profiles from ROS and other signaling pathways could be detected, suggesting that the responses to redox changes of the photosynthetic electron transport chain are distinct. One problem is that most of these changes in gene expression are modest and it remains to be seen whether the observed changes in transcript levels are accompanied by corresponding changes in their protein products [22].

One obvious target for sensing the redox state of the plastoquinone pool is the STN7 protein kinase which is known to be activated when the PQ is reduced and plastoquinol binds to the Qo site of the Cytb_6f complex. Amongst 937 genes found to be significantly regulated in the wild type during these light shift experiments 800 did no longer respond in the *stn7* mutant, indicating that most of these genes are under redox control [22]. However the downstream components of this retrograde signaling chain are still largely unknown.

Although changes in Lhcb gene expression caused by changes in the redox state of the PQ pool have been clearly demonstrated in some green algae [35], the situation is different in land plants. Here changes in the antenna size of PSII and PSI resulting from changes in PQ redox state are determined by variations in the Lhcb transcript levels which are controlled at the post-transcriptional rather than transcriptional level [45]. Another difficulty in these studies is that it is not easy to distinguish specific effects caused by changes in the PQ pool redox poise from signals mediated by photoreceptors [15]. Additional plastid redox sensors include ferredoxin and other redox-active thiol group-containing proteins, and antioxidants which are coupled to the photosynthetic electron transport. In all these cases a change in the redox state of a compound triggers signals of unknown nature which are transmitted across the chloroplast envelope to activate a cytosolic signaling pathway which in turn modulates the expression of a specific group of nuclear genes.

3.3.4 Reactive Oxygen Species Involved in Retrograde Signaling

Electron escape out of the photosynthetic electron transport chain can occur at different sites and is enhanced especially under adverse growth conditions such as elevated temperature, drought or high light when the capacity of the natural electron sinks of the electron transport chain and the detoxification capacity of the plastids is exceeded. At the level of PSII, charge recombination and conversion of singlet to triplet chlorophyll occurs, which then reacts with triplet oxygen to form singlet oxygen. At the acceptor side of PSI, electrons can react with oxygen to form superoxide, hydrogen peroxide and hydroxyl radicals. Because these different ROS are usually produced together it has been difficult to determine which ROS is responsible for a specific change in nuclear gene expression. However, in the case of the *flu* mutant of Arabidopsis, it has been possible to find conditions under which only singlet oxygen is produced [79, 80]. This mutant lacks the FLU protein required for the negative feedback control which occurs at an early step in the tetrapyrrole biosynthetic pathway (Fig. 3.5). In this way protochlorophyllide accumulation is prevented in the dark in wild-type plants but not in *flu* because the conversion of protochlorphyllide to chlorophyllide is light-dependent in land plants. Accumulation of this highly photodynamic chlorophyll precursor in dark-grown *flu* mutant plants leads to photo-oxidation and to the production of singlet oxygen upon exposure of the plants to light. A suppressor screen of the *flu* mutant identified the Executer 1 (Ex1) protein which is associated with the thylakoid membrane [146]. This protein qualifies as a singlet oxygen-signaling molecule based on the observation that the *flu* phenotype is suppressed in the *ex1 flu* double mutant. Interestingly, this suppressed strain still overaccumulates PChlide in the dark and releases singlet oxygen after exposure to light, but it no longer undergoes growth arrest and cell death. It is therefore likely that Ex1 is part of a singlet oxygen-sensitive programmed cell death response. Moreover, a paralog of Ex1 called Ex2 exists, and *ex1ex2 flu* plants are unable to induce most of the singlet oxygen-responsive genes [70]. Because singlet oxygen has a short half-life and limited diffusion [64], it is unlikely to act over long distances, especially beyond the chloroplast. In this case β-cyclocitral, which is produced through the oxidation of β-carotene by singlet oxygen, could transmit the signal across the chloroplast envelope to the nucleo-cytosol [95].

A genetic screen of *C. reinhardtii* cells for mutants more tolerant to singlet oxygen-producing chemicals identified the *Sor1* gene coding for a transcription factor which links reactive electrophile signaling to singlet oxygen acclimation. This response involves increased expression of *Sor1* itself and of a large number of oxidative stress response and detoxification genes [44].

Amongst the ROS H_2O_2 appears to be a promising candidate for intra- and intercellular signaling because of its longer half-life and lower toxicity [88, 145]. Its formation is increased under many stress conditions, especially high light, which lead to changes in nuclear gene expression. Some of the genes involved in this response are the cytosolic ascorbate peroxidases APX1 and APX2, the zinc finger transcription factors ZAT10 and ZAT12, and the chloroplast chlorophyll binding protein ELIP2. Most of these genes are upregulated upon treatment with H_2O_2. Moreover APX2 and ZAT10 are downregulated upon infiltration of leaves with catalase after high light treatment [57]. These two genes are upregulated by ABA, drought, and changes in glutathione metabolism and PQ redox poise [10, 82, 155]. Under stress conditions, a crosstalk between singlet oxygen and H_2O_2 signaling is apparent when the latter compound stimulates the oxidation of Q_A, the primary electron acceptor of PSII [67]. This enhances the photosynthetic yield and thereby attenuates the overexcitation of PSII during exposure to high light and consequently the formation of singlet oxygen. How H_2O_2 signaling occurs is still poorly understood. In this respect it is interesting to note that aquaporins can channel H_2O_2 across membranes [19] and that oxidative gating of aquaporin channels can be induced by H_2O_2 and hydroxyl radicals [157]. If H_2O_2 can traverse biological membranes, the question arises how the cells distinguish between H_2O_2 generated in the chloroplast from that produced in the cytosol, for example as a consequence of a pathogen attack at the plasma membrane, especially given the fact that the corresponding stress reactions are different.

3.3.5 Additional Metabolites Involved in Retrograde Signaling during Stress Responses

High light stress can induce other metabolites that act in retrograde signaling besides singlet oxygen and ROS. Recent genetic studies have revealed that plastids can sense external cues by accumulating specific metabolites which then move from the plastid to the nucleus to convey the information to the nucleus [41, 150]. This is the case for 3'-phosphoadenosine 5'-phosphate (PAP) and perhaps also for methylerythritol cyclodiphosphate (MEcPP) which accumulate under excess light and drought. PAP migrates from the chloroplast to the cytosol and subsequently to the nucleus where it interacts with XRNs, 5' to 3' exonucleases, and regulates stress-responsive genes [40]. PAP therefore qualifies as the first plastid retrograde signal that is generated in the plastid and moves to the nucleus to induce a response. The other metabolite MEcPP is a precursor of isoprenoids generated in the methylerythritol phosphate pathway in the chloroplast [154]. However it is not yet clear whether MEcPP migrates to the nucleus or additional downstream signals are involved in this retrograde signaling which ultimately regulates nuclear genes involved in stress responses.

3.3.6 Repressible Chloroplast Gene Expression in Chlamydomonas

While specific inhibitors of chloroplast gene expression have been used successfully to study retrograde signaling, a major disadvantage is that they affect all chloroplast genes and probably also most of the mitochondrial genes. This induces a major stress response which probably involves many signaling pathways and thereby makes the analysis more difficult. To elicit a more targeted response, a repressible chloroplast gene expression system was established in Chlamydomonas [111, 133]. This system is based on the use of nucleus-encoded chloroplast proteins that are required for chloroplast gene expression and that act at a post-transcriptional level in a gene-specific manner. The Nac2 protein was chosen because this protein is specifically required for the accumulation of the chloroplast *psbD* mRNA coding for the

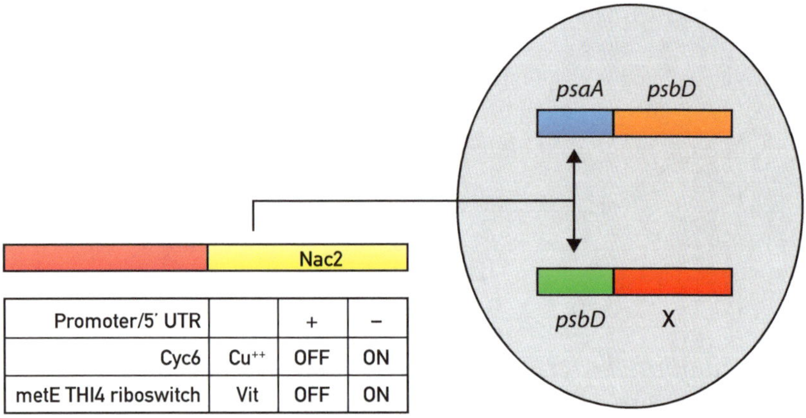

Promoter/5′ UTR		+	−
Cyc6	Cu++	OFF	ON
metE THI4 riboswitch	Vit	OFF	ON

Fig. 3.6 Repressible chloroplast gene expression system in *Chlamydomonas*. The coding sequence of *Nac2* is fused to either the nuclear *Cyc6* promoter or the combined *MetE* methionine synthase promoter and *Thi4* 5'UTR containg the thiamine pyrophosphate responsiveThi4 riboswitch. The Nac2 protein is targeted to the chloroplast where it interacts with the *psbD* 5'UTR for stabilization of the *psbD* mRNA. The *Nac2* dependence can be transferred to any chloroplast gene by fusing it to the *psbD* 5'UTR. The expression of *psbD* can be made independent of Nac2 by fusing the *psbD* coding sequence to another chloroplast 5'UTR. The Cyc6 promoter is active in copper-free medium and repressed in the presence of copper. The MetE promoter/Thi4 5'UTR allows expression of *Nac2* only in absence of vitamin B12 and thiamine pyrophosphate

PSII reaction center protein D2, and it acts on its 5'UTR [66, 90]. Thus *nac2* mutant cells lack *psbD* mRNA but accumulate other chloroplast mRNAs as in the wild type. Any chloroplast gene can be chosen as target by fusing the *psbD* 5'UTR to its coding sequence because the *psbD* 5'UTR is necessary and sufficient to convey Nac2 dependence on the expression of the downstream coding sequence [90]. The basic idea was to drive Nac2 with an inducible/repressible nuclear promoter/5'UTR in a *nac2* mutant background so that *psbD* is no longer expressed when Nac2 expression is repressed (Fig. 3.6). In one case the copper-repressible Cyc6 promoter of cytochrome *c6* was used which is repressed in cells grown in copper-containing, but not in copper-depleted medium [77]. Another possibility is to use the MetE promoter of methionine synthase which is repressed by vitamin B12, in combination with the thiamine pyrophosphate riboswitch contained in the Thi4 5'UTR which induces an unproductive alternative splicing event when thiamine is supplied to the growth medium [28]. Although both systems have been used for Nac2 repression, the vitamin repressible system proved to be more reliable and made it possible to engineer strains in which repression of essential chloroplast genes can be controlled in a tunable and reversible way [111, 133].

In particular, this system was used to conditionally downregulate expression of the essential plastid genes *rpoA* and *rps12* coding for the α subunit of the chloroplast RNA polymerase and the ribosomal protein Rps12, respectively [111]. Rps12 plays a key role in the decoding center of the ribosome, and is therefore indispensable for translation. Inhibition of chloroplast translation caused some unexpected

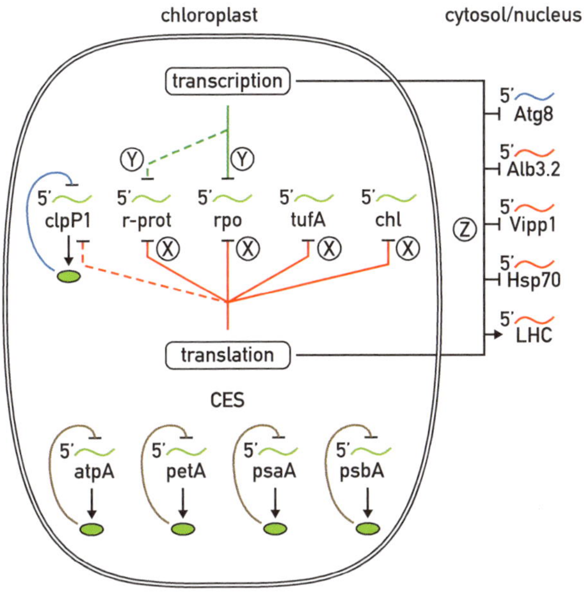

Fig. 3.7 Negative regulatory feedback loops in the chloroplast gene circuitry. *Left*: chloroplast compartment; *right*: nucleo-cytosol. *ClpP1* subunit of chloroplast ATP-dependent protease, *r-prot* chloroplast ribosomal proteins, *rpo* subunits of chloroplast RNA polymerase, *tufA* elongation factor, *chl* subunits of the light-independent protochlorophyllide reductase. Nucleus-encoded mRNAs coding for chloroplast and cytoplasmic proteins are shown in *red* and *blue*, respectively, on the *right*. Negative regulatory feedback loops are revealed through repression of transcription (*green lines*) or translation (*red lines*). Factors involved are still unknown (X, Y, Z) except for the ClpP1 protein which represses accumulation of its own mRNA directly or indirectly (*green circular line*). The feedback loops act mostly at the level of RNA accumulation in contrast to CES (Control of Epistasy of synthesis), an assembly-dependent feedback process in which unassembled CES subunits inhibit directly or indirectly their own translation (*green circular lines*) (for review see [24]) (reproduced from ref. [111] with permission)

effects. While the levels of the mRNAs of the components of the photosynthetic machinery changed only slightly upon arrest of chloroplast translation, a large increase in RNA occurred for several plastid mRNAs, including those of the RNA polymerase, several ribosomal proteins, elongation factor TufA, ClpP1, factors involved in light-independent chlorophyll synthesis and several plastid open reading frames of unknown function. Similar, although not identical, results were obtained after repression of chloroplast transcription. They could be explained by regulatory negative feedback loops for these chloroplast transcripts which might function as compensatory mechanisms to the decrease in gene expression (Fig. 3.7). It is also possible that ribosome-associated nucleolytic activities could degrade these mRNA after translation, but this process would no longer occur in the absence of translation and transcription. It is interesting to note that similar feedbacks have also been found in bacteria for ribosomal operons, raising the possibility that these feedback mechanisms have been conserved during evolution.

In Chlamydomonas, repression of chloroplast transcription and translation affects not only chloroplast RNA levels but also the accumulation of a specific set of nuclear RNAs, suggesting the existence of a retrograde signaling pathway similar to the one in land plants [111]. While mRNA levels of the LHCII proteins are down-regulated, those of VIPP1, Alb3.2 and several HSP proteins are increased (Fig. 3.7). It is possible that VIPP1 and Alb3.2 which, are both involved in some unknown way in lipid trafficking during thylakoid membrane biogenesis, are also implicated in membrane remodelling under conditions of limited chloroplast protein synthesis, a process which may also involve some HSP proteins. Besides mRNAs coding for chloroplast proteins, those of some cytosolic proteins are also upregulated [111]. The case of ubiquitin is particularly intriguing as it raises the possibility that under conditions which compromise chloroplast gene expression, cytosolic protein degradation is enhanced. Whether this process would act on proteins targeted to the chloroplast or whether it could also involve extraction of proteins from the plastid for degradation remains to be explored. In this respect, a mechanism of this sort has been studied in mitochondria where proteins from the inner mitochondrial membrane are extracted and subsequently presented to the cytoplasmic proteasome for degradation through the action of Cdc48 [49] whose mRNA is also increased when chloroplast gene expression is repressed in Chlamydomonas [111]. Finally, it is conceivable that compromised chloroplast gene expression will seriously affect plastid protein homeostasis and thereby induce several chloroplast chaperones in a similar way as described for the mitochondrial unfolded protein response [162]. In this regard, it is interesting to note that a recent genetic screen with Arabidopsis carried out in the presence of norflurazon, but under conditions that minimize photo-oxidative stress, led to the identification of a novel class of mutants named "*h*appy *o*n *n*orfluorazon" (*hon*) which are able to green in the presence of the herbicide [116]. These *hon* mutants are affected in plastid protein synthesis and degradation and they appear to activate an acclimatory response prior to the exposure to stress which attenuates their susceptibility to stress treatment. This example illustrates both the importance and the potential of future studies aiming to understand more deeply the mechanisms regulating chloroplast protein homeostasis.

3.4 Conclusions and Perspectives

Chloroplast biogenesis and function depends on the concerted action of two genetic systems located in the nucleo-cytosol and chloroplast. Coordination of these two systems requires an elaborate bidirectional cross-talk between them. A surprisingly large number of nuclear genes has been identified whose products are targeted to the chloroplast where they act at different levels of chloroplast gene expression including transcription, RNA processing, RNA stability, RNA editing, splicing, translation and assembly of the photosynthetic complexes. Several of these nuclear genes and their products have been identified and characterized. Some of these factors have short half-lives and their level varies rapidly upon changes in the environmental cues such as irradiance, nutrients and temperature, strongly suggesting that they play important regulatory roles.

Less is known about retrograde signaling, although several potential sources of retrograde signals have been uncovered involving the tetrapyrrole and isoprenoid pathway, plastid protein synthesis, the redox state of the photosynthetic electron transport chain, and a selected set of ROS which convey information about the state of plastids to the nuclear genome so that it can elicit a response for proper adjustment of the chloroplast. High throughput sequencing, transcriptomic, proteomic and metabolic studies are providing a wealth of data and some useful leads to uncover the mechanisms of plastid-to-nucleus communication. However, the major challenge remains the identification and characterization of the molecular players of these signaling cascades. The development of innovative genetic, biochemical and cell biology approaches will probably make this goal feasible.

Acknowledgements We thank Nicolas Roggli for preparing the figures. The work in the authors' laboratory was supported by grant 31003A_133089/1 from the Swiss National Foundation.

References

1. Abdallah F, Salamini F, Leister D (2000) A prediction of the size and evolutionary origin of the proteome of chloroplasts of *Arabidopsis*. Trends Plant Sci 5:141–142
2. Alergand T, Peled-Zehavi H, Katz Y, Danon A (2006) The chloroplast protein disulfide isomerase RB60 reacts with a regulatory disulfide of the RNA-binding protein RB47. Plant Cell Physiol 47:540–548
3. Allen JF (1993) Control of gene expression by redox potential and the requirement for chloroplast and mitochondrial genomes. J Theor Biol 165:609–631
4. Ankele E, Kindgren P, Pesquet E, Strand A (2007) *In vivo* visualization of Mg-protoporphyrin IX, a coordinator of photosynthetic gene expression in the nucleus and the chloroplast. Plant Cell 19:1964–1979
5. Arsova B, Hoja U, Wimmelbacher M, Greiner E, Ustun S, Melzer M, Petersen K, Lein W, Bornke F (2010) Plastidial thioredoxin z interacts with two fructokinase-like proteins in a thiol-dependent manner: evidence for an essential role in chloroplast development in *Arabidopsis* and *Nicotiana benthamiana*. Plant Cell 22:1498–1515
6. Asakura Y, Barkan A (2007) A CRM domain protein functions dually in group I and group II intron splicing in land plant chloroplasts. Plant Cell 19:3864–3875
7. Auchincloss AH, Zerges W, Perron K, Girard-Bascou J, Rochaix JD (2002) Characterization of Tbc2, a nucleus-encoded factor specifically required for translation of the chloroplast psbC mRNA in *Chlamydomonas reinhardtii*. J Cell Biol 157:953–962
8. Baginsky S, Tiller K, Pfannschmidt T, Link G (1999) PTK, the chloroplast RNA polymerase-associated protein kinase from mustard (*Sinapis alba*), mediates redox control of plastid *in vitro* transcription. Plant Mol Biol 39:1013–1023
9. Balczun C, Bunse A, Hahn D, Bennoun P, Nickelsen J, Kuck U (2005) Two adjacent nuclear genes are required for functional complementation of a chloroplast trans-splicing mutant from *Chlamydomonas reinhardtii*. Plant J 43:636–648
10. Ball L et al (2004) Evidence for a direct link between glutathione biosynthesis and stress defense gene expression in *Arabidopsis*. Plant Cell 16:2448–2462
11. Barkan A, Walker M, Nolasco M, Johnson D (1994) A nuclear mutation in maize blocks the processing and translation of several chloroplast mRNAs and provides evidence for the differential translation of alternative mRNA forms. EMBO J 13:3170–3181

12. Barkan A, Klipcan L, Ostersetzer O, Kawamura T, Asakura Y, Watkins KP (2007) The CRM domain: an RNA binding module derived from an ancient ribosome-associated protein. RNA 13:55–64 [Epub 2006 Nov 14]
13. Barkan A, Rojas M, Fujii S, Yap A, Chong YS, Bond CS, Small I (2012) A combinatorial amino acid code for RNA recognition by pentatricopeptide repeat proteins. PLoS Genet 8:e1002910
14. Barnes D, Cohen A, Bruick RK, Kantardjieff K, Fowler S, Efuet E, Mayfield SP (2004) Identification and characterization of a novel RNA binding protein that associates with the 5'-untranslated region of the chloroplast psbA mRNA. Biochemistry 43:8541–8550
15. Beck CF (2005) Signaling pathways from the chloroplast to the nucleus. Planta 222:743–756
16. Beick S, Schmitz-Linneweber C, Williams-Carrier R, Jensen B, Barkan A (2008) The pentatricopeptide repeat protein PPR5 stabilizes a specific tRNA precursor in maize chloroplasts. Mol Cell Biol 28:5337–5347
17. Bellafiore S, Barneche F, Peltier G, Rochaix JD (2005) State transitions and light adaptation require chloroplast thylakoid protein kinase STN7. Nature 433:892–895
18. Biehl A, Ritchly E, Noutsos C, Salamini F, Leister D (2004) Analysis of 101 nuclear transcriptomes reveals 23 distinct regulons and their relationship to metabolism, chromosomal gene distribution and coordination of nuclear and plastid gene expression. Gene 354:110–116
19. Bienert GP, Moller AL, Kristiansen KA, Schulz A, Moller IM, Schjoerring JK, Jahn TP (2007) Specific aquaporins facilitate the diffusion of hydrogen peroxide across membranes. J Biol Chem 282:1183–1192
20. Boussardon C, Salone V, Avon A, Berthome R, Hammani K, Okuda K, Shikanai T, Small I, Lurin C (2012) Two interacting proteins are necessary for the editing of the NdhD-1 site in *Arabidopsis* plastids. Plant Cell 24:3684–3694
21. Bradbeer JW, Atkinson YE, Börner T, Hagemann R (1979) Cytoplasmic synthesis of plastid polypeptides may be controlled by plastid-synthesized RNA. Nature 279:816–817
22. Brautigam K et al (2009) Dynamic plastid redox signals integrate gene expression and metabolism to induce distinct metabolic states in photosynthetic acclimation in *Arabidopsis*. Plant Cell 21:2715–2732
23. Chen YN, Slabaugh E, Brandizzi F (2008) Membrane-tethered transcription factors in *Arabidopsis thaliana*: novel regulators in stress response and development. Curr Opin Plant Biol 11:695–701
24. Choquet Y, Vallon O (2000) Synthesis, assembly and degradation of thylakoid membrane proteins. Biochimie 82:615–634
25. Choquet Y, Stern DB, Wostrikoff K, Kuras R, Girard-Bascou J, Wollman FA (1998) Translation of cytochrome f is autoregulated through the 5' untranslated region of petA mRNA in *Chlamydomonas* chloroplasts. Proc Natl Acad Sci U S A 95:4380–4385
26. Cornah JE, Roper JM, Pal Singh D, Smith AG (2002) Measurement of ferrochelatase activity using a novel assay suggests that plastids are the major site of haem biosynthesis in both photosynthetic and non-photosynthetic cells of pea (*Pisum sativum* L.). Biochem J 362:423–432
27. Cottage A, Mott EK, Kempster JA, Gray JC (2010) The *Arabidopsis* plastid-signalling mutant gun1 (genomes uncoupled1) shows altered sensitivity to sucrose and abscisic acid and alterations in early seedling development. J Exp Bot 61:3773–3786
28. Croft M, Moulin M, Webb ME, Smith A (2007) Thiamine biosynthesis in algae is regulated by riboswitches. Proc Natl Acad Sci U S A 104:20770–20775 [Epub 2007 Dec 18]
29. Danon A, Mayfield S (1991) Light-regulated translational activators: identification of chloroplast gene specific mRNA binding proteins. EMBO J 10:3993–4001
30. Danon A, Mayfield SP (1994) ADP-dependent phosphorylation regulates RNA-binding in vitro: implications in light-modulated translation. EMBO J 13:2227–2235
31. Danon A, Mayfield SP (1994) Light-regulated translation of chloroplast messenger RNAs through redox potential. Science 266:1717–1719
32. Das AK, Cohen PW, Barford D (1998) The structure of the tetratricopeptide repeats of protein phosphatase 5: implications for TPR-mediated protein-protein interactions. EMBO J 17:1192–1199

33. Drapier D, Rimbault B, Vallon O, Wollman FA, Choquet Y (2007) Intertwined translational regulations set uneven stoichiometry of chloroplast ATP synthase subunits. EMBO J 26:3581–3591
34. Duanmu D et al (2013) Retrograde bilin signaling enables *Chlamydomonas* greening and phototrophic survival. Proc Natl Acad Sci U S A 110(9):3621–3626
35. Durnford DG, Falkowski PG (1997) Chloroplast redox regulation of nuclear transcription during photoacclimation. Photosynth Res 53:229–241
36. Eberhard S, Finazzi G, Wollman FA (2008) The dynamics of photosynthesis. Annu Rev Genet 42:463–515
37. Eberhard S, Loiselay C, Drapier D, Bujaldon S, Girard-Bascou J, Kuras R, Choquet Y, Wollman FA (2011) Dual functions of the nucleus-encoded factor TDA1 in trapping and translation activation of atpA transcripts in *Chlamydomonas reinhardtii* chloroplasts. Plant J 67(6):1055–1066
38. Edwards TA, Pyle SE, Wharton RP, Aggarwal AK (2001) Structure of Pumilio reveals similarity between RNA and peptide binding motifs. Cell 105:281–289.
39. Escoubas JM, Lomas M, LaRoche J, Falkowski PG (1995) Light intensity regulation of cab gene transcription is signaled by the redox state of the plastoquinone pool. Proc Natl Acad Sci U S A 92:10237–10241
40. Estavillo GM et al (2011) Evidence for a SAL1-PAP chloroplast retrograde pathway that functions in drought and high light signaling in *Arabidopsis*. Plant Cell 23:3992–4012
41. Estavillo GM, Chan KX, Phua SY, Pogson BJ (2013) Reconsidering the nature and mode of action of metabolite retrograde signals from the chloroplast. Front Plant Sci 3:300
42. Fey V, Wagner R, Brautigam K, Pfannschmidt T (2005) Photosynthetic redox control of nuclear gene expression. J Exp Bot 56:1491–1498
43. Fey V, Wagner R, Brautigam K, Wirtz M, Hell R, Dietzmann A, Leister D, Oelmuller R, Pfannschmidt T (2005) Retrograde plastid redox signals in the expression of nuclear genes for chloroplast proteins of *Arabidopsis thaliana*. J Biol Chem 280:5318–5328
44. Fischer BB et al (2012) SINGLET OXYGEN RESISTANT 1 links reactive electrophile signaling to singlet oxygen acclimation in *Chlamydomonas reinhardtii*. Proc Natl Acad Sci U S A 109:E1302–E1311
45. Frigerio S et al (2007) Photosynthetic antenna size in higher plants is controlled by the plastoquinone redox state at the post-transcriptional rather than transcriptional level. J Biol Chem 282:29457–29469
46. Gray JC, Sullivan JA, Wang JH, Jerome CA, MacLean D (2003) Coordination of plastid and nuclear gene expression. Philos Trans R Soc Lond B Biol Sci 358:135–144; discussion 144–145
47. Groves MR, Hanlon N, Turowski P, Hemmings BA, Barford D (1999) The structure of the protein phosphatase 2A PR65/A subunit reveals the conformation of its 15 tandemly repeated HEAT motifs. Cell 96:99–110
48. Hammani K, Cook WB, Barkan A (2012) RNA binding and RNA remodeling activities of the half-a-tetratricopeptide (HAT) protein HCF107 underlie its effects on gene expression. Proc Natl Acad Sci U S A 109:5651–5656
49. Heo JM et al (2010) A stress-responsive system for mitochondrial protein degradation. Mol Cell 40:465–480
50. Hess WR, Borner T (1999) Organellar RNA polymerases of higher plants. Int Rev Cytol 190:1–59
51. Hess WR, Hübschmann T, Börner T (1994) Ribosome defcient plastids of albostrians barley: extreme representatives of non-photosynthetic plastids. Endocytobiosis Cell Res 10:65–80
52. Huber AH, Nelson WJ, Weis WI (1997) Three-dimensional structure of the armadillo repeat region of beta-catenin. Cell 90:871–882
53. Igloi GL, Kössel H (1992) The transcriptional apparatus of chloroplasts. Crit Rev Plant Sci 10:525–558
54. Jenkins BD, Barkan A (2001) Recruitment of a peptidyl-tRNA hydrolase as a facilitator of group II intron splicing in chloroplasts. EMBO J 20:872–879

55. Johanningmeier U, Howell SH (1984) Regulation of light-harvesting chlorophyll-binding protein mRNA accumulation in *Chlamydomonas reinhardi*. Possible involvement of chlorophyll synthesis precursors. J Biol Chem 259:13541–13549
56. Johnson X et al (2010) MRL1, a conserved pentatricopeptide repeat protein, is required for stabilization of rbcL mRNA in *Chlamydomonas* and *Arabidopsis*. Plant Cell 22:234–248
57. Karpinski S, Reynolds H, Karpinska B, Wingsle G, Creissen G, Mullineaux P (1999) Systemic signaling and acclimation in response to excess excitation energy in *Arabidopsis*. Science 284:654–657
58. Kim J, Mayfield SP (1997) Protein disulfide isomerase as a regulator of chloroplast translational activation. Science 278:1954–1957
59. Kimura M, Manabe K, Abe T, Yoshida S, Matsui M, Yamamoto YY (2003) Analysis of hydrogen peroxide-independent expression of the high-light-inducible ELIP2 gene with the aid of the ELIP2 promoter-luciferase fusions. Photochem Photobiol 77:668–674
60. Klinkert B, Elles I, Nickelsen J (2006) Translation of chloroplast psbD mRNA in *Chlamydomonas* is controlled by a secondary RNA structure blocking the AUG start codon. Nucleic Acids Res 34:386–394
61. Kobayashi Y, Kanesaki Y, Tanaka A, Kuroiwa H, Kuroiwa T, Tanaka K (2009) Tetrapyrrole signal as a cell-cycle coordinator from organelle to nuclear DNA replication in plant cells. Proc Natl Acad Sci U S A 106:803–807
62. Kobayashi Y, Imamura S, Hanaoka M, Tanaka K (2011) A tetrapyrrole-regulated ubiquitin ligase controls algal nuclear DNA replication. Nat Cell Biol 13:483–487
63. Koussevitzky S, Nott A, Mockler TC, Hong F, Sachetto-Martins G, Surpin M, Lim J, Mittler R, Chory J (2007) Signals from chloroplasts converge to regulate nuclear gene expression. Science 316:715–719 [Epub 2007 Mar 29]
64. Krieger-Liszkay A (2005) Singlet oxygen production in photosynthesis. J Exp Bot 56:337–346
65. Kropat J, Oster U, Rudiger W, Beck CF (1997) Chlorophyll precursors are signals of chloroplast origin involved in light induction of nuclear heat-shock genes. Proc Natl Acad Sci U S A 94:14168–14172
66. Kuchka MR, Goldschmidt-Clermont M, van Dillewijn J, Rochaix JD (1989) Mutation at the *Chlamydomonas* nuclear NAC2 locus specifically affects stability of the chloroplast psbD transcript encoding polypeptide D2 of PS II. Cell 58:869–876
67. Laloi C, Stachowiak M, Pers-Kamczyc E, Warzych E, Murgia I, Apel K (2007) Cross-talk between singlet oxygen- and hydrogen peroxide-dependent signaling of stress responses in *Arabidopsis thaliana*. Proc Natl Acad Sci U S A 104:672–677
68. Larkin RM, Alonso JM, Ecker JR, Chory J (2003) GUN4, a regulator of chlorophyll synthesis and intracellular signaling. Science 299:902–906
69. Lee I, Hong W (2004) RAP-a putative RNA-binding domain. Trends Biochem Sci 29:567–570
70. Lee KP, Kim C, Landgraf F, Apel K (2007) EXECUTER1- and EXECUTER2-dependent transfer of stress-related signals from the plastid to the nucleus of *Arabidopsis thaliana*. Proc Natl Acad Sci U S A 104:10270–10275
71. Leister D (2005) Genomics-based dissection of the cross-talk of chloroplasts with the nucleus and mitochondria in *Arabidopsis*. Gene 354:110–116
72. Link G (2003) Redox regulation of chloroplast transcription. Antioxid Redox Signal 5:79–87
73. Loiselay C, Gumpel NJ, Girard-Bascou J, Watson AT, Purton S, Wollman FA, Choquet Y (2008) Molecular identification and function of cis- and trans-acting determinants for petA transcript stability in *Chlamydomonas reinhardtii* chloroplasts. Mol Cell Biol 28:5529–5542
74. Malnoe P, Mayfield SP, Rochaix JD (1988) Comparative analysis of the biogenesis of photosystem II in the wild-type and Y-1 mutant of *Chlamydomonas reinhardtii*. J Cell Biol 106:609–616
75. Mayfield SP, Taylor WC (1984) Carotenoid-deficient maize seedlings fail to accumulate light-harvesting chlorophyll a/b binding protein (LHCP) mRNA. Eur J Biochem 144:79–84
76. Meierhoff K, Felder S, Nakamura T, Bechtold N, Schuster G (2003) HCF152, an *Arabidopsis* RNA binding pentatricopeptide repeat protein involved in the processing of chloroplast psbB-psbT-psbH-petB-petD RNAs. Plant Cell 15:1480–1495

77. Merchant S, Bogorad L (1987) Metal ion regulated gene expression: use of a plastocyanin-less mutant of *Chlamydomonas reinhardtii* to study the Cu(II)-dependent expression of cytochrome c-552. EMBO J 6:2531–2535
78. Merendino L, Perron K, Rahire M, Howald I, Rochaix JD, Goldschmidt-Clermont M (2006) A novel multifunctional factor involved in trans-splicing of chloroplast introns in *Chlamydomonas*. Nucleic Acids Res 34:262–274 (Print 2006)
79. Meskauskiene R, Apel K (2002) Interaction of FLU, a negative regulator of tetrapyrrole biosynthesis, with the glutamyl-tRNA reductase requires the tetratricopeptide repeat domain of FLU. FEBS Lett 532:27–30
80. Meskauskiene R, Nater M, Goslings D, Kessler F, op den Camp R, Apel K (2001) FLU: a negative regulator of chlorophyll biosynthesis in *Arabidopsis thaliana*. Proc Natl Acad Sci U S A 98:12826–12831
81. Minai L, Wostrikoff K, Wollman FA, Choquet Y (2006) Chloroplast biogenesis of photosystem II cores involves a series of assembly-controlled steps that regulate translation. Plant Cell 18:159–175
82. Mittler R, Kim Y, Song L, Coutu J, Coutu A, Ciftci-Yilmaz S, Lee H, Stevenson B, Zhu JK (2006) Gain- and loss-of-function mutations in Zat10 enhance the tolerance of plants to abiotic stress. FEBS Lett 580:6537–6542
83. Mochizuki N, Brusslan JA, Larkin R, Nagatani A, Chory J (2001) *Arabidopsis* genomes uncoupled 5 (GUN5) mutant reveals the involvement of Mg-chelatase H subunit in plastid-to-nucleus signal transduction. Proc Natl Acad Sci U S A 98:2053–2058
84. Mochizuki N, Tanaka R, Tanaka A, Masuda T, Nagatani A (2008) The steady-state level of Mg-protoporphyrin IX is not a determinant of plastid-to-nucleus signaling in *Arabidopsis*. Proc Natl Acad Sci U S A 105:15184–15189 [Epub 2008 Sep 25]
85. Mochizuki N, Tanaka R, Grimm B, Masuda T, Moulin M, Smith AG, Tanaka A, Terry MJ (2010) The cell biology of tetrapyrroles: a life and death struggle. Trends Plant Sci 15:488–498
86. Morin RD, Aksay G, Dolgosheina E, Ebhardt HA, Magrini V, Mardis ER, Sahinalp SC, Unrau PJ (2008) Comparative analysis of the small RNA transcriptomes of *Pinus contorta* and *Oryza sativa*. Genome Res 18:571–584
87. Moulin M, McCormac AC, Terry MJ, Smith AG (2008) Tetrapyrrole profiling in *Arabidopsis* seedlings reveals that retrograde plastid nuclear signaling is not due to Mg-protoporphyrin IX accumulation. Proc Natl Acad Sci U S A 105:15178–15183 [Epub 2008 Sep 25]
88. Mullineaux PM, Karpinski S, Baker NR (2006) Spatial dependence for hydrogen peroxide-directed signaling in light-stressed plants. Plant Physiol 141:346–350
89. Nakamura T, Meierhoff K, Westhoff P, Schuster G (2003) RNA-binding properties of HCF152, an *Arabidopsis* PPR protein involved in the processing of chloroplast RNA. Eur J Biochem 270:4070–4081
90. Nickelsen J, van Dillewijn J, Rahire M, Rochaix J-D (1994) Determinants for stability of the chloroplast *psbD* RNA are located within its short leader region in *Chlamydomonas reinhardtii*. EMBO J 13:3182–3191
91. Nott A, Jung HS, Koussevitzky S, Chory J (2006) Plastid-to-nucleus retrograde signaling. Annu Rev Plant Biol 57:739–759
92. Oelmüller R, Levitan I, Bergfeld R, Rajasekhar VK, Mohr H (1986) Expression of nuclear genes is affecetd by treatments acting on the plastids. Planta 168:482–492
93. Ogrzewalla K, Piotrowski M, Reinbothe S, Link G (2002) The plastid transcription kinase from mustard (*Sinapis alba L.*). A nuclear-encoded CK2-type chloroplast enzyme with redox-sensitive function. Eur J Biochem 269:3329–3337
94. Okuda K, Shikanai T (2012) A pentatricopeptide repeat protein acts as a site-specificity factor at multiple RNA editing sites with unrelated cis-acting elements in plastids. Nucleic Acids Res 40:5052–5064
95. op den Camp RG et al (2003) Rapid induction of distinct stress responses after the release of singlet oxygen in *Arabidopsis*. Plant Cell 15:2320–2332

96. Ossenbuhl F, Nickelsen J (2000) cis- and trans-acting determinants for translation of psbD mRNA in *Chlamydomonas reinhardtii*. Mol Cell Biol 20:8134–8142
97. Ostersetzer O, Cooke AM, Watkins KP, Barkan A (2005) CRS1, a chloroplast group II intron splicing factor, promotes intron folding through specific interactions with two intron domains. Plant Cell 17:241–255
98. Ostheimer GJ, Williams-Carrier R, Belcher S, Osborne E, Gierke J, Barkan A (2003) Group II intron splicing factors derived by diversification of an ancient RNA-binding domain. EMBO J 22:3919–3929
99. Perron K, Goldschmidt-Clermont M, Rochaix JD (1999) A factor related to pseudouridine synthases is required for chloroplast group II intron trans-splicing in *Chlamydomonas reinhardtii*. EMBO J 18:6481–6490
100. Pesaresi P et al (2009) *Arabidopsis* STN7 kinase provides a link between short- and long-term photosynthetic acclimation. Plant Cell 21:2402–2423
101. Pfalz J, Liere K, Kandlbinder A, Dietz KJ, Oelmuller R (2006) pTAC2, -6, and −12 are components of the transcriptionally active plastid chromosome that are required for plastid gene expression. Plant Cell 18:176–197
102. Pfalz J, Bayraktar OA, Prikryl J, Barkan A (2009) Site-specific binding of a PPR protein defines and stabilizes 5' and 3' mRNA termini in chloroplasts. EMBO J 28:2042–2052
103. Pfannschmidt T, Nilsson A, Allen JF (1999) Photosynthetic control of chloroplast gene expression. Nature 397:625–628
104. Pfannschmidt T, Schutze K, Brost M, Oelmuller R (2001) A novel mechanism of nuclear photosynthesis gene regulation by redox signals from the chloroplast during photosystem stoichiometry adjustment. J Biol Chem 276:36125–36130
105. Pogson BJ, Woo NS, Forster B, Small ID (2008) Plastid signalling to the nucleus and beyond. Trends Plant Sci 13:602–609
106. Prikryl J, Rojas M, Schuster G, Barkan A (2011) Mechanism of RNA stabilization and translational activation by a pentatricopeptide repeat protein. Proc Natl Acad Sci U S A 108:415–420
107. Puthiyaveetil S (2011) A mechanism for regulation of chloroplast LHC II kinase by plastoquinol and thioredoxin. FEBS Lett 585:1717–1721 (Pesaresi, 2009 #63)
108. Puthiyaveetil S et al (2008) The ancestral symbiont sensor kinase CSK links photosynthesis with gene expression in chloroplasts. Proc Natl Acad Sci U S A 105:10061–10066
109. Puthiyaveetil S, Ibrahim IM, Jelicic B, Tomasic A, Fulgosi H, Allen JF (2010) Transcriptional control of photosynthesis genes: the evolutionarily conserved regulatory mechanism in plastid genome function. Genome Biol Evol 2:888–896
110. Rahire M, Laroche F, Cerutti L, Rochaix JD (2012) Identification of an OPR protein involved in the translation initiation of the PsaB subunit of photosystem I. Plant J 72:652–661
111. Ramundo S, Rahire M, Schaad O, Rochaix JD (2013) Repression of essential chloroplast genes reveals new signaling pathways and regulatory feedback loops in *Chlamydomonas*. Plant Cell 25(1):167–186
112. Rapp JC, Mullet JE (1991) Chloroplast transcription is required to express the nuclear genes rbcS and cab. Plant Mol Biol 17:813–823
113. Raynaud C, Loiselay C, Wostrikoff K, Kuras R, Girard-Bascou J, Wollman FA, Choquet Y (2007) Evidence for regulatory function of nucleus-encoded factors on mRNA stabilization and translation in the chloroplast. Proc Natl Acad Sci U S A 104:9093–9098
114. Rivier C, Goldschmidt-Clermont M, Rochaix JD (2001) Identification of an RNA-protein complex involved in chloroplast group II intron trans-splicing in *Chlamydomonas reinhardtii*. EMBO J 20:1765–1773
115. Rochaix JD (2013) Redox regulation of thylakoid protein kinases and photosynthetic gene expression. Antioxid Redox Signal 18(16):2184–2201
116. Saini G, Meskauskiene R, Pijacka W, Roszak P, Sjogren LL, Clarke AK, Straus M, Apel K (2011) 'happy on norflurazon' (hon) mutations implicate perturbance of plastid homeostasis with activating stress acclimatization and changing nuclear gene expression in norflurazon-treated seedlings. Plant J 65:690–702

117. Schmitz-Linneweber C, Small I (2008) Pentatricopeptide repeat proteins: a socket set for organelle gene expression. Trends Plant Sci 13:663–670
118. Schmitz-Linneweber C, Williams-Carrier RE, Williams-Voelker PM, Kroeger TS, Vichas A, Barkan A (2006) A pentatricopeptide repeat protein facilitates the trans-splicing of the maize chloroplast rps12 pre-mRNA. Plant Cell 18:2650–2663
119. Schroter Y, Steiner S, Matthai K, Pfannschmidt T (2010) Analysis of oligomeric protein complexes in the chloroplast sub-proteome of nucleic acid-binding proteins from mustard reveals potential redox regulators of plastid gene expression. Proteomics 10:2191–2204
120. Schwarz C, Elles I, Kortmann J, Piotrowski M, Nickelsen J (2007) Synthesis of the D2 protein of photosystem II in *Chlamydomonas* is controlled by a high molecular mass complex containing the RNA stabilization factor Nac2 and the translational activator RBP40. Plant Cell 19:3627–3639
121. Schwarz C, Bohne AV, Wang F, Cejudo FJ, Nickelsen J (2012) An intermolecular disulfide-based light switch for chloroplast psbD gene expression in *Chlamydomonas reinhardtii*. Plant J 72(3):378–389
122. Schweer J, Turkeri H, Kolpack A, Link G (2010) Role and regulation of plastid sigma factors and their functional interactors during chloroplast transcription-recent lessons from *Arabidopsis thaliana*. Eur J Cell Biol 89:940–946
123. Schweer J, Turkeri H, Link B, Link G (2010) AtSIG6, a plastid sigma factor from *Arabidopsis*, reveals functional impact of cpCK2 phosphorylation. Plant J 62:192–202
124. Shimizu M, Kato H, Ogawa T, Kurachi A, Nakagawa Y, Kobayashi H (2010) Sigma factor phosphorylation in the photosynthetic control of photosystem stoichiometry. Proc Natl Acad Sci U S A 107:10760–10764
125. Small ID, Peeters N (2000) The PPR motif-a TPR-related motif prevalent in plant organellar proteins. Trends Biochem Sci 25:46–47
126. Somanchi A, Barnes D, Mayfield SP (2005) A nuclear gene of *Chlamydomonas reinhardtii*, Tba1, encodes a putative oxidoreductase required for translation of the chloroplast psbA mRNA. Plant J 42:341–352
127. Sosso D, Canut M, Gendrot G, Dedieu A, Chambrier P, Barkan A, Consonni G, Rogowsky PM (2012) PPR8522 encodes a chloroplast-targeted pentatricopeptide repeat protein necessary for maize embryogenesis and vegetative development. J Exp Bot 63:5843–5857
128. Steiner S, Dietzel L, Schroter Y, Fey V, Wagner R, Pfannschmidt T (2009) The role of phosphorylation in redox regulation of photosynthesis genes psaA and psbA during photosynthetic acclimation of mustard. Mol Plant 2:416–429
129. Stern DB, Goldschmidt-Clermont M, Hanson MR (2010) Chloroplast RNA metabolism. Annu Rev Plant Biol 61:125–155
130. Strand A, Asami T, Alonso J, Ecker JR, Chory J (2003) Chloroplast to nucleus communication triggered by accumulation of Mg-protoporphyrinIX. Nature 421:79–83
131. Sullivan JA, Gray JC (1999) Plastid translation is required for the expression of nuclear photosynthesis genes in the dark and in roots of the pea lip1 mutant. Plant Cell 11:901–910
132. Sun X, Feng P, Xu X, Guo H, Ma J, Chi W, Lin R, Lu C, Zhang L (2011) A chloroplast envelope-bound PHD transcription factor mediates chloroplast signals to the nucleus. Nat Commun 2:477
133. Surzycki R, Cournac L, Peltier G, Rochaix JD (2007) Potential for hydrogen production with inducible chloroplast gene expression in *Chlamydomonas*. Proc Natl Acad Sci U S A 104:17548–17553 [Epub 2007 Oct 19]
134. Susek RE, Ausubel FM, Chory J (1993) Signal transduction mutants of *Arabidopsis* uncouple nuclear CAB and RBCS gene expression from chloroplast development. Cell 74:787–799
135. Tanaka K, Hanaoka M (2013) The early days of plastid retrograde signaling with resepct to replication and transcription. Front Plant Sci 3:301
136. Thomas J, Weinstein JD (1990) Measurement of heme efflux and heme content in isolated developing chloroplasts. Plant Physiol 94:1414–1423
137. Tian Q, Taupin J, Elledge S, Robertson M, Anderson P (1995) Fas-activated serine/threonine kinase (FAST) phosphorylates TIA-1 during Fas-mediated apoptosis. J Exp Med 182:865–874

138. Till B, Schmitz-Linneweber C, Williams-Carrier R, Barkan A (2001) CRS1 is a novel group II intron splicing factor that was derived from a domain of ancient origin. RNA 7:1227–1238
139. Tiller K, Link G (1993) Phosphorylation and dephosphorylation affect functional characteristics of chloroplast and etioplast transcription systems from mustard (*Sinapis alba L*). EMBO J 12:1745–1753
140. Trebitsh T, Levitan A, Sofer A, Danon A (2000) Translation of chloroplast psbA mRNA is modulated in the light by counteracting oxidizing and reducing activities. Mol Cell Biol 20:1116–1123
141. Vaistij FE, Boudreau E, Lemaire SD, Goldschmidt-Clermont M, Rochaix JD (2000) Characterization of Mbb1, a nucleus-encoded tetratricopeptide-like repeat protein required for expression of the chloroplast psbB/psbT/psbH gene cluster in *Chlamydomonas reinhardtii*. Proc Natl Acad Sci U S A 97:14813–14818
142. von Gromoff ED, Schroda M, Oster U, Beck CF (2006) Identification of a plastid response element that acts as an enhancer within the *Chlamydomonas* HSP70A promoter. Nucleic Acids Res 34:4767–4779
143. von Gromoff ED, Alawady A, Meinecke L, Grimm B, Beck CF (2008) Heme, a plastid-derived regulator of nuclear gene expression in *Chlamydomonas*. Plant Cell 20:552–567
144. Voss B, Meinecke L, Kurz T, Al-Babili S, Beck CF, Hess WR (2011) Hemin and magnesium-protoporphyrin IX induce global changes in gene expression in *Chlamydomonas reinhardtii*. Plant Physiol 155:892–905
145. Vranova E, Inze D, Van Breusegem F (2002) Signal transduction during oxidative stress. J Exp Bot 53:1227–1236
146. Wagner D et al (2004) The genetic basis of singlet oxygen-induced stress responses of *Arabidopsis thaliana*. Science 306:1183–1185
147. Wang X, Zamore PD, Hall TM (2001) Crystal structure of a Pumilio homology domain. Mol Cell 7:855–865
148. Wang X, McLachlan J, Zamore PD, Hall TM (2002) Modular recognition of RNA by a human pumilio-homology domain. Cell 110:501–512
149. Williams PM, Barkan A (2003) A chloroplast-localized PPR protein required for plastid ribosome accumulation. Plant J 36:675–686
150. Woodson JD, Chory J (2012) Organelle signaling: how stressed chloroplasts communicate with the nucleus. Curr Biol 22:R690–R692
151. Woodson JD, Perez-Ruiz JM, Chory J (2011) Heme synthesis by plastid ferrochelatase I regulates nuclear gene expression in plants. Curr Biol 21:897–903
152. Wostrikoff K, Stern D (2007) Rubisco large-subunit translation is autoregulated in response to its assembly state in tobacco chloroplasts. Proc Natl Acad Sci U S A 104:6466–6471
153. Wostrikoff K, Girard-Bascou J, Wollman FA, Choquet Y (2004) Biogenesis of PSI involves a cascade of translational autoregulation in the chloroplast of *Chlamydomonas*. EMBO J 23:2696–2705 [Epub 2004 Jun 10]
154. Xiao Y et al (2012) Retrograde signaling by the plastidial metabolite MEcPP regulates expression of nuclear stress-response genes. Cell 149:1525–1535
155. Yabuta Y, Maruta T, Yoshimura K, Ishikawa T, Shigeoka S (2004) Two distinct redox signaling pathways for cytosolic APX induction under photooxidative stress. Plant Cell Physiol 45:1586–1594
156. Yamaguchi K, Prieto S, Beligni MV, Haynes PA, McDonald WH, Yates JR 3rd, Mayfield SP (2002) Proteomic characterization of the small subunit of *Chlamydomonas reinhardtii* chloroplast ribosome: identification of a novel S1 domain-containing protein and unusually large orthologs of bacterial S2, S3, and S5. Plant Cell 14:2957–2974
157. Ye Q, Steudle E (2006) Oxidative gating of water channels (aquaporins) in corn roots. Plant Cell Environ 29:459–470
158. Yin L et al (2007) Rev-erbalpha, a heme sensor that coordinates metabolic and circadian pathways. Science 318:1786–1789

159. Yohn CB, Cohen A, Danon A, Mayfield SP (1998) A poly(A) binding protein functions in the chloroplast as a message-specific translation factor. Proc Natl Acad Sci U S A 95:2238–2243
160. Yohn CB, Cohen A, Rosch C, Kuchka MR, Mayfield SP (1998) Translation of the chloroplast psbA mRNA requires the nuclear-encoded poly(A)-binding protein, RB47. J Cell Biol 142:435–442
161. Zhang L, Hach A (1999) Molecular mechanism of heme signaling in yeast: the transcriptional activator Hap1 serves as the key mediator. Cell Mol Life Sci 56:415–426
162. Zhao Q, Wang J, Levichkin IV, Stasinopoulos S, Ryan MT, Hoogenraad NJ (2002) A mitochondrial specific stress response in mammalian cells. EMBO J 21:4411–4419

Part II
Plastid Differentiation

Chapter 4
An Overview of Chloroplast Biogenesis and Development

Barry J. Pogson and Veronica Albrecht-Borth

Abstract During evolution plants have incorporated chloroplasts into their cells. Within the chloroplast energy is produced during photosynthesis but also hormones and metabolites essential for plant metabolism. As a consequence, chloroplast biogenesis and development needs to be coordinated with seedling growth to ensure optimal rates of photosynthesis without causing oxidative damage upon seedling emergence. The importance of chloroplast development during germination for plant vitality, seed set and growth was shown in numerous studies. Thus, it is necessary to understand the regulation and mechanism of chloroplast development. This chapter will focus on how chloroplasts are formed with emphasis on recent findings about proteins and processes required for chloroplast development as well as into regulatory and molecular factors for chloroplast biogenesis and development gained from genetic studies.

Keywords Chloroplast biogenesis · Plant development · Regulatory processes

Abbreviations

ABA	Abscisic acid
ARC5	Accumulation and replication of chloroplasts 5
CAB	Chlorophyll a/b-binding proteins
CLB	Chloroplast biogenesis
cop/det/fus	Constitutive photomorphogenic/de-etiolated/fusca
GLK	Golden2-like proteins
NEP	Nuclear-encoded polymerase
PDI	Protein disulfide isomerases
PEP	Plastid-encoded polymerase
PIFs	Phytochrome interacting factors
PLB	Prolamellar body
POR	Protochlorophyllide oxidoreductase
PPRs	Pentatricopeptide repeat
ROS	Reactive oxygen species
SCO	Snowy cotyledon

B. J. Pogson (✉) · V. Albrecht-Borth
Australian National University, Canberra, Australia
e-mail: Barry.Pogson@anu.edu.au

S.M. Theg, F.-A. Wollman (eds.), *Plastid Biology,* Advances in Plant Biology 5,
DOI 10.1007/978-1-4939-1136-3_4,

STY	Serine, threonine, and tyrosine
THF1	Thylakoid formation 1
TIC	Translocon at inner envelope of chloroplast
TOC	Translocon at outer envelope of chloroplast
VIPP1	Vesicle inducing plastid protein

4.1 Introduction

Within the chloroplast energy is produced during photosynthesis as well as hormones and metabolites essential for plant metabolism. As a consequence, chloroplast biogenesis and development needs to be coordinated with seedling growth to ensure optimal rates of photosynthesis without causing oxidative damage upon seedling emergence. Indeed, that chloroplast development during germination is important for plant vitality, seed set and growth was shown in numerous studies. Thus, it is necessary to understand the regulation and mechanism of chloroplast development. The following chapter will focus on how chloroplasts are formed with emphasis on recent findings about proteins, processes and regulation required for chloroplast development as well as insight gained from genetic studies.

4.2 Chloroplast Biogenesis

Chloroplast biogenesis and development differs between dicotyledonous and monocotyledonous plants. In dicotyledonous seedlings chloroplast biogenesis can be described as the differentiation process from the plastid progenitor, a proplastid, to a mature chloroplast, whether direct or via the dark-grown intermediate form known as an etioplast (Fig. 4.1). In developed leaves, chloroplasts are further propagated by fission in a manner similar to that observed in bacteria. In contrast, chloroplast differentiation in monocotyledonous plants can be observed along the blade of the leaf, starting with proplastids in the first centimeter from the leaf ligula and progressing until they are fully developed after the middle of the leaf blade (Fig. 4.2). During this differentiation process no etioplast formation can be observed [31]. Furthermore, intensive analyses in maize leaves revealed a more complex process of chloroplast biogenesis, since this occurs differently in the mesophyll cells (serving for CO_2 fixation) and bundle sheath cells (where the Calvin cycle for CO_2 assimilation occurs), with the cells and chloroplasts close to the cell wall connected by plasmodesmata to exchange the metabolites. Thus, different chloroplast differentiation steps can be observed in monocotyledonous plants during their entire life cycle, whereas in the dicotyledonous plants it is restricted to a short timeframe. The focus on chloroplast biogenesis in this chapter will emphasize data from dicotyledonous plants.

Chloroplasts are found in all photosynthetic tissues, ranging from cotyledons and leaves to certain stems, fruits and floral organs. However, rather than being

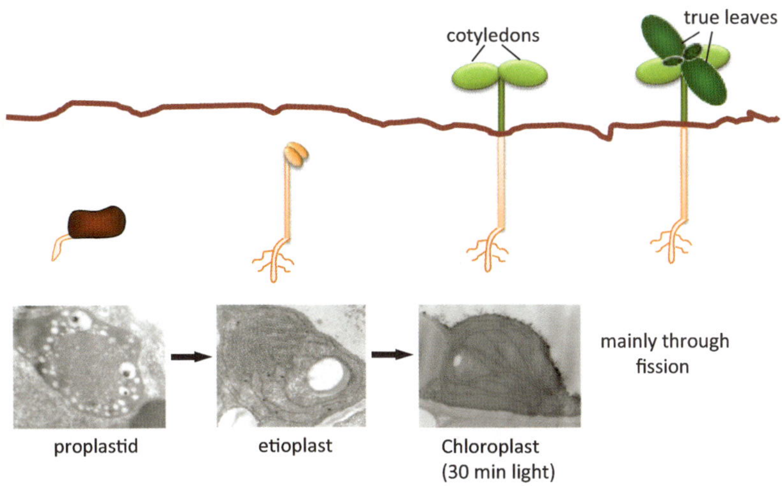

Fig. 4.1 Chloroplast biogenesis and development in dicotyledonous seedlings

Fig. 4.2 Chloroplast biogenesis and development in monocotyledonous seedlings. Example shown for a maize plant. *P* proplastid, *m* mesophyll cell, *bs* bundle sheath cell. (TEM pictures generously provided by Klaas van Wijk)

a uniform process, the biogenesis and development of the chloroplasts differs between organs as well as between species. For example, there is a need to consider if the seedlings exhibit epigaeic growth, having cotyledons which first serve as storage organs but become photosynthetically active, or alternatively exhibit hypogaeic growth, where the cotyledons only serve as a storage organ and are not photosynthetic. Herein, we only consider chloroplast biogenesis and development in epigaeic seedlings.

That chloroplast development proceeds differently in cotyledons and true leaves is demonstrated by genetic studies. That is, chloroplast mutants have been described that have the phenotype restricted to one leaf organ, such as chlorotic true leaves but green cotyledons, as in the *variegated* (*var*) and *immutans* (*im*) mutants [30]. Conversely, the *snowy cotyledon* (*sco*) mutant group displays chlorotic or bleached cotyledons, but green true leaves [1–3, 39]. Here, the observed impairment in chloroplast development in cotyledons seems to be due to a combination of factors: (1) a differential chloroplast development process in the plant organs (2) functional redundancy of different proteins in the true leaves and (3) different transport and/or assembly processes between cotyledons and true leaves. A third class of mutants have a white cotyledon phenotype, but require sucrose to develop green true leaves. Once green leaves have developed the plants can grow without sucrose and can be transferred to soil for propagation. To this group belongs, among others *plastid type I signal peptidase, plsp1,* which will be discussed later [41].

So how do plastids develop in germinating seedlings? In light grown seedlings the undifferentiated proplastid develops directly into a chloroplast. However, in dark-germinated seedlings the proplastid forms an etioplast. The etioplast is defined by the prolamellar body (PLB), a membranous structure that contains a limited number of lipids, metabolites and proteins required for photosynthesis. From this lattice-like structure prothylakoids emanate into the plastid stroma. The PLB contains the precursor of chlorophyll, protochlorophyllide, bound to its reducing enzyme Protochlorophyllide Oxidoreductase (POR), NADPH, lipids, a few proteins and typically two carotenoids, lutein and violaxanthin. In angiosperms, light is required for POR enzyme activity whereas gymnosperms additionally have a light-independent POR enzyme which catalyses the same reaction; hence they can synthesize chlorophyll in the dark [15]. Upon illumination chlorophyll is produced and integrated into the newly assembled photosystems for light absorption for photosynthesis. The photosystems are located in the thylakoids, which are the major internal membranous structure of the chloroplasts. In some cases, chloroplasts can also develop from other plastids such as chromoplasts, although our focus here will be only on chloroplast differentiation from proplastids.

There is coordination between plastids and the nucleus at the level of transcription, translation, import, protein turnover and metabolite flux—all of which is largely controlled by plastid and nuclear factors (Fig. 4.3 and Sect. 4.3). This facilitates stoichiometric assembly of nuclear-encoded and plastidic-encoded proteins together with chlorophylls and carotenoids: this is essential for photosynthesis, both with respect to limiting oxidative damage and ensuring optimal rates of protein synthesis. However, photosynthesis proceeds within the context of the environment, and the developmental stages of the cells and leaves. Accordingly, environmental and cytosolic processes also influence chloroplast biogenesis, an aspect outlined in Fig. 4.4 and discussed in Sect. 4.4.

4.3 Plastidic and Nuclear Factors Required for Chloroplast Biogenesis

The following is a summary of biogenic processes identified by mutations in chloroplastic and nuclear proteins as shown in Fig. 4.3.

4.3.1 Nuclear Transcription

Upon illumination one third of the nuclear gene transcription profile changes [8], including many that encode chloroplast-targeted proteins. This is not surprising given that the vast majority of the thousands of chloroplast proteins are nuclear-encoded. The perception of light requires the activation of the phytochrome photoreceptors, such as phyA and phyB. The photomorphogenic pathways that control gene transcription, chlorophyll biosynthesis, and protein degradation has been explored in the *cop/det/fus (constitutive photomorphogenic/de-etiolated/fusca)* mutants [49]. The *cop* mutants de-repress photomorphogenesis in the dark enabling the onset of chloroplast biogenesis in etiolated seedlings, and encode components of the so-called cytosolic COP-signalosome, which is involved in ubiquitin-dependent protein degradation [49]. COP1 is cryptochrome-regulated: a blue-light sensor that

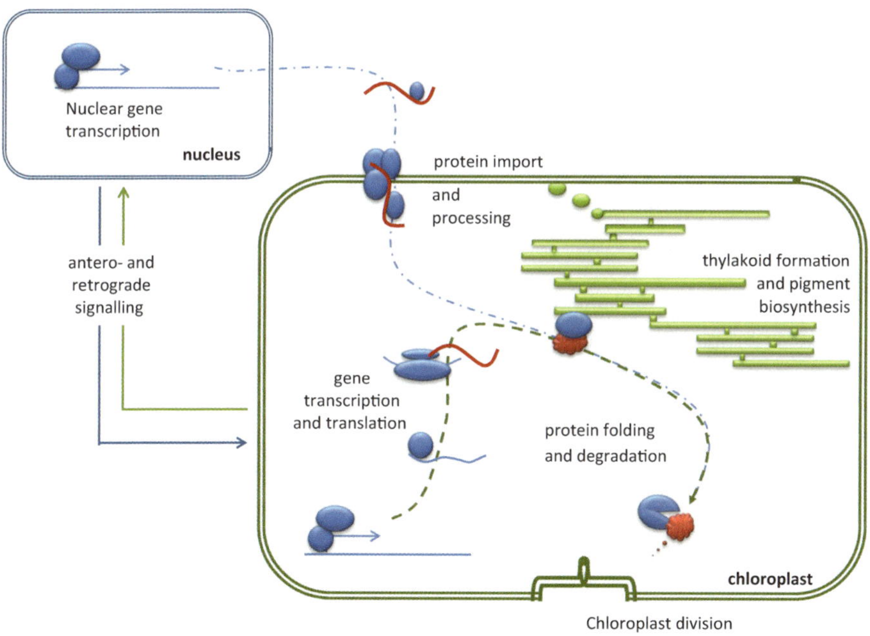

Fig. 4.3 Processes required for chloroplast biogenesis and development

deactivates the signalosome, thereby inhibiting further ubiquitination and degradation of proteins involved in light-signalling such as the transcription factor HY5 (Long Hypocotyl 5) [49]. *P*hytochrome *i*nteracting *f*actors (PIFs) are transcription factors that respond to activated phytochromes to initiate light-mediated transcription, such that mutations in PIF1 or PIF3 impair chloroplast biogenesis [33]. Interestingly, loss of four PIF proteins (PIF1, PIF3, PIF4, and PIF5) results in chloroplast biogenesis in dark-grown seedlings due to up-regulation of genes for chloroplast function, a situation comparable to the *cop* mutants [40].

HEMERA/pTAC12 was described to be localized both in the chloroplast transcriptional apparatus as well as in the nucleus [8]. Interestingly, in the nucleus it seems to be involved in the regulation of PIF and phyA protein degradation after phyB activation rather than directly promoting transcription. Loss of this protein results in an albino phenotype that can only be rescued when both the chloroplast and the nuclear targeting signals are accessible [8].

Another class of transcription factors required for chloroplast development which are directly involved in the transcription of photosynthesis genes are the *G*olden2-*lik*e proteins (GLK). Loss of both GLK transcription factors in Arabidopsis leads to a pale mutant phenotype which can be complemented by overexpression of either GLK [48].

4.3.2 Chloroplast Gene Transcription and Translation

Chloroplast gene transcription, RNA maturation, and protein translation and modification include both nuclear-encoded polymerase (NEP) and plastid-encoded polymerase (PEP) machinery. Nuclear encoded RNA polymerases, such as rpoT, have been shown to affect chloroplast biogenesis [4]. The plastidic RNA polymerase *sig*ma factors (SIG) are required as transcription factors for chloroplast genes, with the loss of *sig2* and *sig6* delaying chloroplast biogenesis [9, 22].

Interestingly, the aforementioned HEMERA/pTAC12 has been shown to be dual targeted to the chloroplast and nucleus [8]. HEMERA/pTAC12 interacts with another protein of the pTAC family, pTAC14, and is involved in the regulation of PEP-dependent plastid gene transcription [16].

The chloroplast genome only encodes about 100 genes, but most are essential to the chloroplast, with mutations that affect transcription and RNA processing, such as RNA editing, leading to impairment of photosynthetic function [12]. Indeed, there is a large class of nuclear-encoded proteins known as PPRs (*p*entatrico*p*eptide *r*epeat) that are required for RNA processing, splicing, editing, stability, maturation and translation in the chloroplast. In many instances PPR mutations are seedling-lethal. Interestingly, most PPR proteins have specific roles, be it RNA editing or transcript stability of a particular chloroplast-encoded gene [7, 9, 34].

The chloroplast ribosomal complex that translates mRNA into protein is more closely related to bacterial than eukaryotic ribosomes, reflecting the endosymbiotic origins of the plastid. Not surprisingly, loss of essential proteins of the plastid ribosomal complex is embryo-lethal, as the plastid provides crucial functions for cells in addition to photosynthesis. Phenotypes associated with a reduced functionality due to leaky point mutations in proteins of the ribosomal complex can be described as *virescent*, or somewhat green, as in the case of *sco1* or the rice mutant *Osvir2* [1, 43]. Such mutations result in either reduced protein translation, as has been shown for the *sco1* mutant in which the binding of chloroplast elongation factor G to the ribosomal complex is impaired, or a reduced amount of the ribosomal complex itself, as described for the loss of the DNA and RNA binding protein *Whirly1* in maize [37].

4.3.3 Protein Import and Processing

The majority of chloroplast-targeted proteins are imported into chloroplasts by the TOC and TIC translocon complexes (*T*ranslocon at *o*uter/*i*nner envelope of *c*hloroplast), respectively; others are targeted via the ER [23]. Proteins with a chloroplast targeting sequence are directed to the complex by HSP90 and TOC159. The chloroplast targeting sequence is removed by a stromal processing peptidase [11]. The absence of critical components of the translocon, such as TOC159, drastically impairs biogenesis, although plants are viable if minor components such as Toc 33 or Toc34 are absent [21, 23, 38]. The transport of some proteins to the chloroplasts seems to depend on the phosphorylation status of the transit peptide. Indeed, it has been found that the absence of some plant-specific STY kinases (STY stands for Serine, Threonine, and Tyrosine), specifically STY8, STY17, and STY46, results in delayed greening of the triple mutants after etiolation and transfer to light, as well as in the accumulation of transport vesicles in the chloroplast stroma, suggesting a delayed import of plastid proteins [28]. A detailed review on protein import into the chloroplast is reviewed elsewhere ([23] and Chap. 9 in this volume).

Chloroplast protein folding is mediated by chaperones, such as HSP70 and Cpn60, with the *hsp70–1* mutation leading to variegated cotyledons [29]. Another class of proteins, protein disulfide isomerases (PDI), have been shown to be required for protein folding by regulating the formation of disulphide bonds between cysteine residues [39]. Lesions in the SCO2/CYO1 protein disulphide isomerase result in pale green cotyledons due to their interaction with photosystem proteins [2, 39, 46].

Within the chloroplast, targeting systems integrate proteins into the thylakoid membranes and lumen space of the thylakoids. Thylakoid membrane proteins are integrated either spontaneously or via the cpSRP pathway which has been described for LHCB proteins [14]. Transport vesicles have been proposed to mediate lipid and protein targeting to developing thylakoids [20]. In the *sco2* mutant, accumulation of such transport vesicles has been observed. Indeed, the SCO2 protein disulphide isomerase interacts with specific photosystem proteins, such as LHCB1 [46]. Loss of function of either pathway, SRP and SCO2, results in impaired biogenesis and bleaching, though in varying degrees, depending on the tissue type.

Proteins described to be involved in the formation of transport vesicles and thus also in thylakoid formation are THF1 (*Th*ylakoid *F*ormation 1) and VIPP1 (*V*esicle *i*nducing *p*lastid *p*rotein) [27, 47]. Whereas the *thf1* mutant line accumulates transport vesicles in the white sections of the variegated true leaves, the *vipp1* mutant is not able to form these vesicles even at 4 °C, when these vesicles are usually observed in wild type. Both mutant lines are affected in normal thylakoid formation. Many mutations have been reported to affect thylakoid formation, although a direct involvement of the corresponding protein in thylakoid formation could not be attributed to their protein function, rather than to their essential presence in particular thylakoid protein complexes.

Transport of proteins into the thylakoid lumen is either ATP-dependent via the cpSec-pathway or ΔpH-dependent using the cpTAT mechanism. For both pathways, mutants have been described as being either seedling lethal or affecting thylakoid formation [11]. A thylakoidal processing peptidase, PLSP1, is required for removing the thylakoid-targeting sequence from thylakoidal proteins and has been shown to be involved in processing PsbO and PsbP proteins. Loss of the protein is seedling-lethal if no exogenous carbon source is provided [42].

The need for coordinated assembly and turnover of the photosystems is demonstrated by the observation that the rates of protein translation and degradation are important for chloroplast development. That is, mutations in the family of FtsH proteases, such as *variegated 2*, impairing Photosystem II D1 protein degradation, cause a white patched phenotype in true leaves, but not in cotyledons. Intriguingly, this bleaching can be complemented by decelerating protein biosynthesis by second site mutations in genes involved in protein translation [32, 51, 53]. The multi-subunit ClpPR proteinase complex also has a central role in degrading proteins and is required for chloroplast protein homeostasis. Loss of ClpP proteins are either embryo- or seedling-lethal [25].

4.3.4 Pigment Biosynthesis

Assembly of the photosystems requires a coordinated import of pigment-binding proteins such as chlorophyll a/b-binding proteins (CAB/LHCB) and the biosynthesis of the pigments. Thus, carotenoid and chlorophyll biosynthesis is tightly regulated during chloroplast development as free chlorophylls or their tetrapyrrole precursors are highly reactive oxidants [5]. Indeed, the ratio of different carotenoids in photosynthetic tissues is mediated in response to changes in photosystem ratios and is subjected to potential epigenetic regulation via a histone methyl transferase, SDG8 [6]. Therefore, it is not surprising, that screens for chloroplast development mutants have identified steps in these pathways, such as *clb4* and *clb6* (*chloroplast biogenesis),* which are lesions in enzymes in the chloroplast MEP pathway (2-C-methyl-D-erythritol 4-phosphate) [18, 19]. Furthermore, tetrapyrroles have been implicated in coordination of chloroplast biogenesis [52] and regulation of chloroplast nuclear signalling [35, 36].

4.3.5 Retrograde Signalling

The fact that most of the proteins needed for chloroplast function are encoded in the nucleus necessitates a tight regulation between the chloroplast functional status and nuclear gene transcription. Thus, signalling from the chloroplast is involved in regulating nuclear gene transcription via multiple pathways in a process referred to as retrograde signalling. Many factors have been suggested to be involved in the retrograde signalling control, ranging from ROS (reactive oxygen species) to metabolites and proteins that regulate or transduce the signals [35, 36]. Indeed, there must be multiple signalling pathways to coordinate the thousands of differentially regulated nuclear genes whose proteins are targeted to the chloroplast [35, 36]. However, identifying the actual signals has proven elusive till now. One recent finding is that a phosphonucleotide, PAP, whose levels are regulated by chloroplast-localised SAL1 can move between the chloroplast and nucleus, where it regulates the induction of a range of high light responsive genes presumably by inhibiting exoribonucleases [13].

A series of mutants that affect nuclear gene transcription in response to inhibitors of pigment biosynthesis and/or plastid transcription and translation are the *genomes uncoupled* (*gun*) mutants [10]. This series of mutants demonstrate a role for tetrapyrrole biosynthesis and a PPR protein, GUN1, in a retrograde signalling pathway, but the nature of this signalling cascade is still debated. Although the *gun1* mutation affects the signalling from chloroplast to nucleus, chloroplast biogenesis proceeds normally in the absence of inhibitors. This and other findings raise the question as to how much the different *gun* mutants directly affect signalling or result in a perturbation of plastidic processes and metabolic signatures that alters nuclear transcription [35]. Interestingly, overexpression of ABI4 in the *gun1* mutant background recovers the *gun* phenotype. Thus, the ABI4 4 transcription factor seems to integrate some of the GUN1-mediated signals from the chloroplasts by binding to light-regulatory elements [26]. But how are these particular signals relayed from the chloroplast to the nucleus? This is still unknown, although the recent finding of a chloroplast outer envelope-bound transcription factor, PTM (PHD type transcription factor with transmembrane domains), which after proteolysis transfers to the nucleus and binds to the promoter of ABI4, might offer some clues for elucidating this mechanism [44].

Of particular interest in the context of chloroplast development are signals related to biogenic control, that is, those that regulate nuclear transcription during plastid biogenesis [36]. The PSII associated proteins, EXECUTER1 and 2 (EX1, EX2), mediate singlet oxygen signalling pathways [24] and have been thought to be largely functional in operational control in response to excess light. Interestingly, with respect to biogenic control, is the observation that the double mutant *ex1ex2* exhibits white cotyledon regions which contain undifferentiated plastids that resemble proplastids. This chloroplast biogenesis defect can be overcome by growth of the seedlings on abscisic acid (ABA)-containing media [24].

4.3.6 Chloroplast Division

A typical higher plant cell contains between 80–120 chloroplasts, which necessitates organelle division, a process that has been shown to be largely independent of cell division. Chloroplast division is mediated by tubulin-like proteins known as FtsZ and dynamin-like proteins, such as ARC5 (*a*ccumulation and *r*eplication of *c*hloroplasts) that form concentric rings within and outside the chloroplast envelope, respectively [50]. Recent findings include a role for chaperones, cpn60, and PARC60 in coordinating and regulating division [17, 45]. However, although there are impacts on chloroplast size and function there are no substantive perturbations to chloroplast biogenesis.

4.4 Temporal and Cellular Factors Required for Chloroplast Biogenesis and Development

It is important to note that environmental, cellular and temporal factors also contribute to chloroplast biogenesis and development (Fig. 4.4), not just the actual processes and proteins within the plastid. Indeed, as noted above, the role of light mediated-signalling via the phytochromes and cryptochromes is essential. However, not only environmental factors influence chloroplast biogenesis but also other organelles and structures in the cell [3]. The SCO3 protein is a protein of unknown function required for chloroplast biogenesis with the *sco3* mutation impairing chloroplast and etioplast differentiation in seedlings. Unexpectedly, the SCO3 protein is targeted to the surface of peroxisomes and is associated with the microtubule cytoskeleton [3]. Furthermore, disrupting the cytoskeleton with inhibitors demonstrated a role for it, together with SCO3, in chloroplast biogenesis. One hypothesis is that transport and/or import of specific, but as yet unknown, proteins or metabolites might require interactions between the chloroplasts and cytoskeleton during plastid biogenesis in cotyledons.

Temporal factors also have an impact on chloroplast biogenesis in seedlings. Embryo maturation influences chloroplast development, that is, even though proplastids are present in the seed prior to germination and are largely undifferentiated, the way in which they are formed during embryogenesis affects chloroplast biogenesis. Mutants affected in chloroplast biogenesis in the embryonic leaves (cotyledons) of germinating seedlings, such as *sco* and *ex1ex2*, are not affected in chloroplast development during embryogenesis in the developing silique [2, 24]. Indeed, it could be shown that the chlorotic cotyledons of *sco2* and *ex1ex2* could be reverted to green if the mature green embryo was extracted from green seed and germinated precociously or if the seed was germinated from embryos that had matured in the dark, and not the light [2, 24]. The nature of this embryonic control on plastid development in germinating seedlings is as yet unknown, although ABA is proposed to be involved [24].

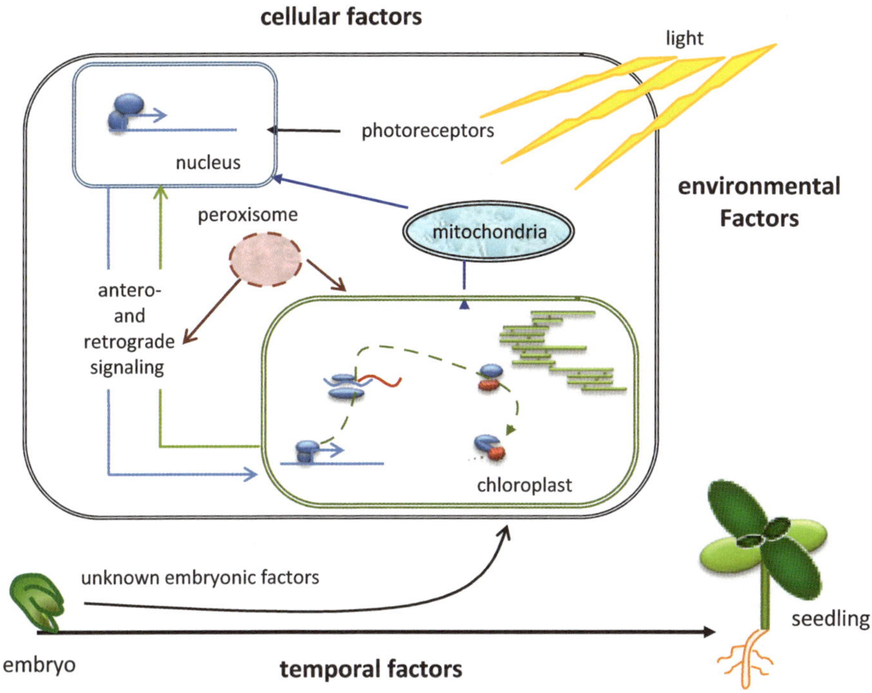

Fig. 4.4 Model of environmental, cellular and temporal factors that influence chloroplast biogenesis and development in seedlings. Environmental factors involved are among others light; cellular factors the communication between the organelles; and embryo development as one of the temporal factors, influencing chloroplast biogenesis in germinating seedlings

4.5 Future Directions

To date most studies of chloroplast development have identified chloroplast-localised proteins involved in import, chloroplast transcription/RNA maturation/protein translation, assembly and signalling. Temporal factors, with respect to tissue development and age, have to be taken into account. Indeed, insights into the effects of embryogenesis on photomorphogenesis reveal new directions for the analysis of the regulation of chloroplast biogenesis. Key questions that remain include: (1) what is the interaction between chloroplast development and photomorphogenesis; (2) what is the influence of other cellular compartments, such as the cytoskeleton and mitochondria on chloroplast biogenesis; (3) what are the actual retrograde signals that mediate chloroplast development; (4) what are the check points in chloroplast biogenesis and development and how do they interact with phytochrome-mediate signalling?

Acknowledgements This project has received support from the Australian Research Council Centre of Excellence in Plant Energy Biology (CE0561495).

References

1. Albrecht V, Ingenfeld A, Apel K (2006) Characterization of the snowy cotyledon 1 mutant of *Arabidopsis thaliana*: the impact of chloroplast elongation factor G on chloroplast development and plant vitality. Plant Mol Biol 60:507–518
2. Albrecht V, Ingenfeld A, Apel K (2008) Snowy cotyledon 2: the identification of a zinc finger domain protein essential for chloroplast development in cotyledons but not in true leaves. Plant Mol Biol 66:599–608
3. Albrecht V et al (2010) The cytoskeleton and the peroxisomal-targeted snowy cotyledon3 protein are required for chloroplast development in *Arabidopsis*. Plant Cell 22:3423–3438
4. Azevedo J, Courtois F, Hakimi MA, Demarsy E, Lagrange T, Alcaraz JP, Jaiswal P, Marechal-Drouard L, Lerbs-Mache S (2008) Intraplastidial trafficking of a phage-type RNA polymerase is mediated by a thylakoid RING-H2 protein. Proc Natl Acad Sci U S A 105:9123–9128
5. Cazzonelli CI, Pogson BJ (2010) Source to sink: regulation of carotenoid biosynthesis in plants. Trends Plant Sci 15:266–274
6. Cazzonelli CI, Cuttriss AJ, Cossetto SB, Pye W, Crisp P, Whelan J, Finnegan EJ, Turnbull C, Pogson BJ (2009) Regulation of carotenoid composition and shoot branching in *Arabidopsis* by a chromatin modifying histone methyltransferase, SDG8. Plant Cell 21:39–53
7. Chateigner-Boutin AL et al (2008) CLB19, a pentatricopeptide repeat protein required for editing of rpoA and clpP chloroplast transcripts. Plant J 56:590–602
8. Chen M, Galvao RM, Li M, Burger B, Bugea J, Bolado J, Chory J (2010) *Arabidopsis* HEMERA/pTAC12 initiates photomorphogenesis by phytochromes. Cell 141:1230–1240
9. Chi W, Mao J, Li Q, Ji D, Zou M, Lu C, Zhang L (2010) Interaction of the pentatricopeptide-repeat protein DELAYED GREENING 1 with sigma factor SIG6 in the regulation of chloroplast gene expression in *Arabidopsis cotyledons*. Plant J 64:14–25
10. Chory J (1991) Light signals in leaf and chloroplast development: photoreceptors and downstream responses in search of a transduction pathway. New Biol 3:538–548
11. Cline K, Dabney-Smith C (2008) Plastid protein import and sorting: different paths to the same compartments. Curr Opin Plant Biol 11:585–592
12. Delannoy E et al (2009) *Arabidopsis* tRNA adenosine deaminase arginine edits the wobble nucleotide of chloroplast tRNAArg(ACG) and is essential for efficient chloroplast translation. Plant Cell 21:2058–2071
13. Estavillo GM et al (2011) Evidence for a SAL1-PAP chloroplast retrograde pathway that functions in drought and high light signaling in *Arabidopsis*. Plant Cell 23:3992–4012
14. Falk S, Sinning I (2010) cpSRP43 is a novel chaperone specific for light-harvesting chlorophyll a, b-binding proteins. J Biol Chem 285:21655–21661
15. Forreiter C, Apel K (1993) Light-independent and light-dependent protochlorophyllide-reducing activities and two distinct NADPH-protochlorophyllide oxidoreductase polypeptides in mountain pine (*Pinus mugo*). Planta 190:536–545
16. Gao ZP, Yu QB, Zhao TT, Ma Q, Chen GX, Yang ZN (2011) A functional component of the transcriptionally active chromosome complex, *Arabidopsis* pTAC14, interacts with pTAC12/HEMERA and regulates plastid gene expression. Plant Physiol 157:1733–1745
17. Glynn JM, Yang Y, Vitha S, Schmitz AJ, Hemmes M, Miyagishima SY, Osteryoung KW (2009) PARC6, a novel chloroplast division factor, influences FtsZ assembly and is required for recruitment of PDV1 during chloroplast division in *Arabidopsis*. Plant J 59:700–711
18. Guevara-Garcia A, San Roman C, Arroyo A, Cortes ME, de la Luz Gutierrez-Nava M, Leon P (2005) Characterization of the *Arabidopsis* clb6 mutant illustrates the importance of posttranscriptional regulation of the methyl-D-erythritol 4-phosphate pathway. Plant Cell 17:628–643
19. Gutierrez-Nava Mde L, Gillmor CS, Jimenez LF, Guevara-Garcia A, Leon P (2004) CHLOROPLAST BIOGENESIS genes act cell and noncell autonomously in early chloroplast development. Plant Physiol 135:471–482
20. Hoober JK, Eggink LL, Chen M (2007) Chlorophylls, ligands and assembly of light-harvesting complexes in chloroplasts. Photosynth Res 94:387–400

21. Inoue H, Rounds C, Schnell DJ (2010) The molecular basis for distinct pathways for protein import into *Arabidopsis* chloroplasts. Plant Cell 22:1947–1960
22. Ishizaki Y et al (2005) A nuclear-encoded sigma factor, *Arabidopsis* SIG6, recognizes sigma-70 type chloroplast promoters and regulates early chloroplast development in cotyledons. Plant J 42:133–144
23. Kessler F, Schnell D (2009) Chloroplast biogenesis: diversity and regulation of the protein import apparatus. Curr Opin Cell Biol 21:494–500
24. Kim C, Lee KP, Baruah A, Nater M, Gobel C, Feussner I, Apel K (2009) (1)O2-mediated retrograde signaling during late embryogenesis predetermines plastid differentiation in seedlings by recruiting abscisic acid. Proc Natl Acad Sci U S A 106:9920–9924
25. Kim J, Rudella A, Ramirez Rodriguez V, Zybailov B, Olinares PD, van Wijk KJ (2009) Subunits of the plastid ClpPR protease complex have differential contributions to embryogenesis, plastid biogenesis, and plant development in *Arabidopsis*. Plant Cell 21:1669–1692
26. Koussevitzky S, Nott A, Mockler TC, Hong F, Sachetto-Martins G, Surpin M, Lim J, Mittler R, Chory J (2007) Signals from chloroplasts converge to regulate nuclear gene expression. Science 316:715–719
27. Kroll D, Meierhoff K, Bechtold N, Kinoshita M, Westphal S, Vothknecht UC, Soll J, Westhoff P (2001) VIPP1, a nuclear gene of *Arabidopsis thaliana* essential for thylakoid membrane formation. Proc Natl Acad Sci U S A 98:4238–4242
28. Lamberti G, Gugel IL, Meurer J, Soll J, Schwenkert S (2011) The cytosolic kinases STY8, STY17, and STY46 are involved in chloroplast differentiation in *Arabidopsis*. Plant Physiol 157:70–85
29. Latijnhouwers M, Xu XM, Moller SG (2010) Arabidopsis stromal 70-kDa heat shock proteins are essential for chloroplast development. Planta 232:567–578
30. Liu X, Yu F, Rodermel S (2010) *Arabidopsis* chloroplast FtsH, var2 and suppressors of var2 leaf variegation: a review. J Integr Plant Biol 52:750–761
31. Majeran W et al (2010) Structural and metabolic transitions of C4 leaf development and differentiation defined by microscopy and quantitative proteomics in maize. Plant Cell 22:3509–3542
32. Miura E, Kato Y, Matsushima R, Albrecht V, Laalami S, Sakamoto W (2007) The balance between protein synthesis and degradation in chloroplasts determines leaf variegation in *Arabidopsis* yellow variegated mutants. Plant Cell 19:1313–1328
33. Moon J, Zhu L, Shen H, Huq E (2008) PIF1 directly and indirectly regulates chlorophyll biosynthesis to optimize the greening process in *Arabidopsis*. Proc Natl Acad Sci U S A 105:9433–9438
34. Pfalz J, Bayraktar OA, Prikryl J, Barkan A (2009) Site-specific binding of a PPR protein defines and stabilizes 5′ and 3′ mRNA termini in chloroplasts. EMBO J 28:2042–2052
35. Pfannschmidt T (2010) Plastidial retrograde signalling-a true "plastid factor" or just metabolite signatures? Trends Plant Sci 15:427–435
36. Pogson BJ, Woo NS, Forster B, Small ID (2008) Plastid signalling to the nucleus and beyond. Trends Plant Sci 13:602–609
37. Prikryl J, Watkins KP, Friso G, van Wijk KJ, Barkan A (2008) A member of the Whirly family is a multifunctional RNA- and DNA-binding protein that is essential for chloroplast biogenesis. Nucleic Acids Res 36:5152–5165
38. Shi L-X, Theg SM (2010) A stromal heat shock protein 70 system functions in protein import into chloroplasts in the moss *Physcomitrella patens*. Plant Cell 22:205–220
39. Shimada H et al (2007) *Arabidopsis* cotyledon-specific chloroplast biogenesis factor CYO1 is a protein disulfide isomerase. Plant Cell 19:3157–3169
40. Shin J, Kim K, Kang H, Zulfugarov IS, Bae G, Lee CH, Lee D, Choi G (2009) Phytochromes promote seedling light responses by inhibiting four negatively-acting phytochrome-interacting factors. Proc Natl Acad Sci U S A 106:7660–7665
41. Shipman RL, Inoue K (2009) Suborganellar localization of plastidic type I signal peptidase 1 depends on chloroplast development. FEBS Lett 583:938–942

42. Shipman-Roston RL, Ruppel NJ, Damoc C, Phinney BS, Inoue K (2010) The significance of protein maturation by plastidic type I signal peptidase 1 for thylakoid development in *Arabidopsis* chloroplasts. Plant Physiol 152:1297–1308
43. Sugimoto H, Kusumi K, Noguchi K, Yano M, Yoshimura A, Iba K (2007) The rice nuclear gene, VIRESCENT 2, is essential for chloroplast development and encodes a novel type of guanylate kinase targeted to plastids and mitochondria. Plant J 52:512–527
44. Sun X, Feng P, Xu X, Guo H, Ma J, Chi W, Lin R, Lu C, Zhang L (2011) A chloroplast envelope-bound PHD transcription factor mediates chloroplast signals to the nucleus. Nat Commun 2:477
45. Suzuki K, Nakanishi H, Bower J, Yoder DW, Osteryoung KW, Miyagishima SY (2009) Plastid chaperonin proteins Cpn60 alpha and Cpn60 beta are required for plastid division in *Arabidopsis thaliana*. BMC Plant Biol 9:38
46. Tanz S, Kilian J, Johnsson C, Apel K, Small I, Harter K, Wanke D, Pogson B, Albrecht V (2012) The SCO2 protein disulphide isomerase is required for thylakoid biogenesis and interacts with LCHB1 chlorophyl a/b binding proteins which affects chlorophyll biosynthesis in *Arabidopsis* seedlings. Plant J 69:743–754
47. Wang Q, Sullivan RW, Kight A, Henry RL, Huang J, Jones AM, Korth KL (2004) Deletion of the chloroplast-localized Thylakoid formation1 gene product in *Arabidopsis* leads to deficient thylakoid formation and variegated leaves. Plant Physiol 136:3594–3604
48. Waters MT, Wang P, Korkaric M, Capper RG, Saunders NJ, Langdale JA (2009) GLK transcription factors coordinate expression of the photosynthetic apparatus in *Arabidopsis*. Plant Cell 21:1109–1128
49. Wei N, Serino G, Deng XW (2008) The COP9 signalosome: more than a protease. Trends Biochem Sci 33:592–600
50. Yang Y, Glynn JM, Olson BJ, Schmitz AJ, Osteryoung KW (2008) Plastid division: across time and space. Curr Opin Plant Biol 11:577–584
51. Yu F et al (2011) SUPPRESSOR OF VARIEGATION4, a new var2 suppressor locus, encodes a pioneer protein that is required for chloroplast biogenesis. Mol Plant 4:229–240
52. Zhang H, Li J, Yoo JH, Yoo SC, Cho SH, Koh HJ, Seo HS, Paek NC (2006) Rice Chlorina-1 and Chlorina-9 encode ChlD and ChlI subunits of Mg-chelatase, a key enzyme for chlorophyll synthesis and chloroplast development. Plant Mol Biol 62:325–337
53. Zhang D, Kato Y, Zhang L, Fujimoto M, Tsutsumi N, Sodmergen, Sakamoto W (2010) The FtsH protease heterocomplex in *Arabidopsis*: dispensability of type-B protease activity for proper chloroplast development. Plant Cell 22:3710–3725

Chapter 5
Dynamic Architecture of Plant Photosynthetic Membranes

Helmut Kirchhoff

Abstract The energy converting photosynthetic machinery in plants is harbored in the thylakoid membrane system inside the chloroplast. The structural flexibility of thylakoid membranes is a central feature that allows sessile plants to adapt to very different and highly fluctuating environmental conditions. This chapter surveys structural attributes of the thylakoid membrane system and its dynamics. It connects structural alterations to the functionality of the energy converting apparatus and its regulation. Two structural levels are addressed that cover different length scales. The first deals with changes in the overall membrane architecture (micrometer length scale) with a special emphasis on stacked grana regions. The second part focuses on the organization of many protein complexes in thylakoid membranes (100 nm length scale). It turns out that thylakoid membranes evolved a remarkable degree of plasticity that fine-tune different aspects of photosynthetic energy transformation.

Keywords Thylakoid membrane · Grana thylakoid · Macromolecular crowding · Protein arrays · Supramolecular level

Abbreviations

AFM	Atomic force microscopic
CLSM	Confocal laser scanning microscopy
cyt b6f	Cytochrome *b6f*
DLVO	Derjaguin-landau-verwey-overbeek
FRAP	Fluorescence recovery after photobleaching
LHCI	Light-harvesting complex I
LHCII	Light-harvesting complex II
OEC	Oxygen-evolving complex
PC	Plastocyanin
PMF	Protonmotive force
PSI	Photosystem I
PSII	Photosystem II
ROS	Reactive oxygen species
VDE	Violaxanthin deepoxidase

H. Kirchhoff (✉)
Institute of Biological Chemistry, Washington State University, Pullman, WA, USA
e-mail: kirchhh@wsu.edu

S.M. Theg, F.-A. Wollman (eds.), *Plastid Biology*, Advances in Plant Biology 5,
DOI 10.1007/978-1-4939-1136-3_5,

5.1 Introduction

The conversion of solar radiation into biologically useful energy by photosynthesis is confined to highly specialized membrane systems. In plants, the thylakoid membrane inside the chloroplast harbors the nanomachines that constitute the photosynthetic apparatus for sunlight-driven generation of $NADPH + H^+$ and ATP accompanied by water splitting and molecular oxygen production. Although the overall reaction equation is simple, the underlying molecular processes are extremely complex and fascinate thousands of researchers in many disciplines of natural sciences ranging from quantum mechanics to ecophysiology. Learning how nature optimizes and regulates photosynthetic energy transformation is a high priority research field with significant social impact because it might provide a key to solving global energy and food problems in a challenging climate. Understanding the primary processes in photosynthesis requires not only detailed knowledge of the molecular structure of the individual nanomachines, but it is also essential to know how the overall thylakoid system is structured and how its numerous protein complexes work together to establish functional networks for light-harvesting and electron transport. Recent work has given new insights into the structural plasticity of thylakoid membranes on the nanometer and micrometer length scale. This data reveals a high degree of structural flexibility in photosynthetic membranes, which is necessary to tackle the multiple challenges dictated by environmental changes that occur day-to-day on very different time scales.

A striking structural feature that is characteristic for the thylakoid system is that part of the membrane forms tightly stacked cylindrical domains called grana thylakoids [31, 90, 111]. Grana discs are interconnected by unstacked stroma lamellae forming a continuous membrane system that separates two aqueous reaction spaces: the chloroplast stroma and the intrathylakoid lumen space (see also Fig. 5.1). Embedded in the thylakoid membrane are six main transmembrane protein complexes that establish the energy transforming apparatus: photosystem II (PSII) with light-harvesting complex II (LHCII), photosystem I (PSI) with light-harvesting complex I (LHCI), the cytochrome *b6f* (cyt *b6f*) complex, and a F-type ATPase complex (ATPase). In addition to these major protein complexes, low abundance transmembrane proteins are present like the *psbS* protein or *stn* kinases. It is well established that differentiation into stacked and unstacked membrane regions leads to a lateral heterogeneity in protein complex distribution [2, 14, 35, 111]. The main part of PSII and LHCII is localized in stacked grana. In contrast, PSI with LHCI and the ATPase are excluded from stacked grana, most likely by steric hindrance because the stroma-facing protein moieties of PSI and ATPase are too large to enter the narrow gap between adjacent grana membranes (see Sect. 5.2.1.3). It is assumed that the cyt *b6f* complex is the only complex that is homogeneously distributed [111] but there is also evidence for its depletion in stacked grana [41, 122] as well as for its concentration in this subcompartment [2]. However, the lateral protein distribution within the thylakoid network is not static but highly dynamic. The dynamic redistribution of protein complexes within and between stacked and unstacked thylakoid regions could play a key role to adjust photosynthetic performance to environmental

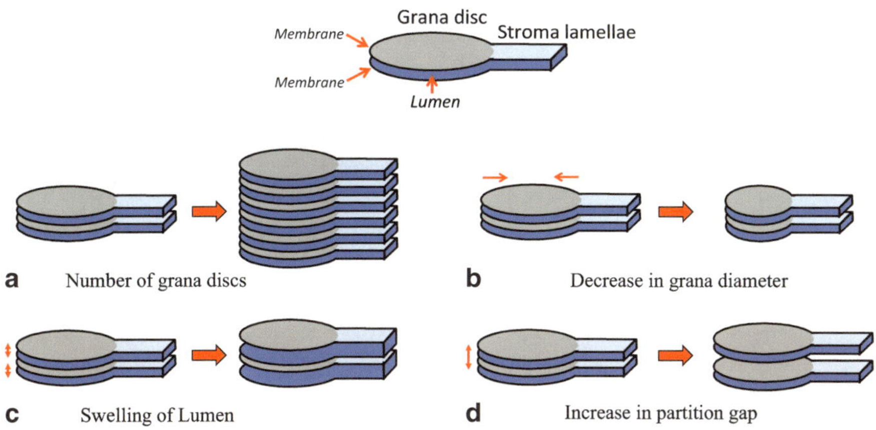

Fig. 5.1 Levels of protein organization in thylakoid membranes. Proteins can arrange themselves into different assembly levels. The example shows these levels for the LHCII-PSII complex. Different *psb*-subunits form the PSII (multi-subunit complex). This PSII complex associates to a dimer that binds two trimeric LHCII complexes (*red*) forming the LHCII-PSII supercomplex. Further trimeric LHCII attach to the supercomplex generating megacomplexes that can further form extended supramolecular networks. The different types of complexes are characterized by different binding strengths indicating a hierarchy in protein-protein interactions. For example, the two trimeric LHCII binds stronger to the dimeric PSII than the additional trimers in megacomplexes. As indicated by the arrows, the different assembly levels are in equilibrium and are inter-convertible. Intact thylakoid membranes contain all assembly levels [30] and the state of equilibrium is controlled by environmental factors. This inter-convertibility is critical for the PSII repair cycle (see text for further details)

changes. Thus, unraveling this interdependency is central to understanding the plasticity of the energy transforming machinery.

In the last few decades, considerable progress in solving the structure of photosynthetic protein complexes (for recent reviews see [8, 18, 21, 54, 121]) leads to an almost complete set of high-resolution models making the thylakoid membrane one of the best-characterized biomembranes. Furthermore, low-resolution data deduced from single-particle analysis of solubilized thylakoid membranes reveals that photosynthetic protein complexes form larger aggregates named supercomplexes (Fig. 5.2). It seems that protein complexes in thylakoid membranes are mainly organized into supercomplexes, i.e. dimeric cyt *b6f* complex, trimeric LHCII, PSI with four LHCI, and dimeric PSII with two trimeric LHCII [35]. In particular, the PSII supercomplex in grana thylakoids was extensively studied [25, 70, 91]. There is now an understanding that the dimeric PSII-LHCII supercomplex represents the structural building block in stacked grana (but see [115] for a different view). Furthermore, evidence exists [35] that in grana membranes isolated by mild-detergent treatments, supercomplexes can arrange into higher associations called megacomplexes (Fig. 5.2). It has to be clarified whether these higher associations represent a native state or that they are generated by the detergent treatment [88]. Under some conditions, the PSII-LHCII supercomplex can further associate to highly ordered semicrystalline supramolecular networks consisting of many proteins ([32, 35, 70],

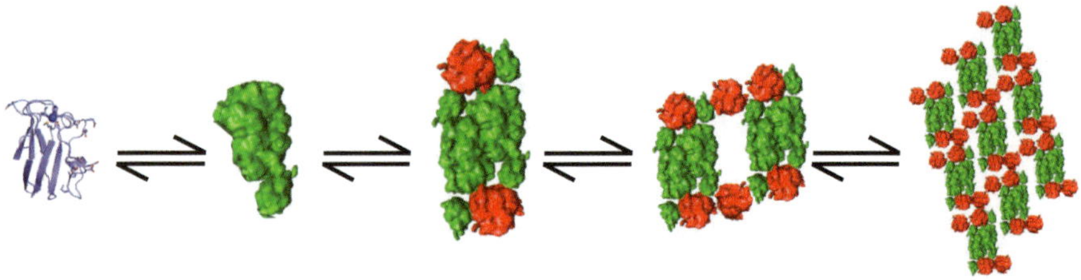

Fig. 5.2 Survey of geometrical changes in the overall grana architecture. **a** Light intensity is the major determinant that controls the number of grana thylakoid membranes constituting the grana cylinder. **b** The grana diameter seems to be a relative robust quantity. However, it was reported that it shrinks under certain unfavorable conditions. This leads to inter-conversion of stacked (*grey*) to unstacked thylakoid membrane regions (*light blue*). **c** Recent data suggests that the thylakoid lumen in grana (*dark blue*) swells in light that has impact on the molecular mobility of lumen-hosted proteins. **d** Although experimental evidence is weak, it was suggested that the distance between neighbored grana membranes on the stroma side (partition gap) can change under high-light stress

see Sect. 5.3.2). Since these semicrystalline arrays are found in intact thylakoid membranes that never had contact with detergents [32], it is safe to conclude that they represent an *in vivo* state of the photosynthetic machinery. Thus, a remarkable degree of order and hierarchy in protein organization is realized in particular in grana membranes (Fig. 5.2). Although this has been recognized for a long time, we are just now understanding the functional significance and dynamics of structural ordering in photosynthetic membranes. Interestingly, a similar complexity and hierarchy is realized for membrane proteins in the respiratory membranes in mitochondria that also form supercomplexes [37, 38, 40] and supramolecular rows [33, 38, 39, 112]. It seems that both bioenergetic membranes share common structural organization motifs indicating that learning the principles of protein organization in thylakoid membranes could be valuable for respiratory membranes too and *vice versa*. For example, electron transport in both membranes depends on diffusion of small electron carriers (plastoquinone, ubiquinone, plastocyanin, cytochrome c). Supercomplexes and supramolecular ordering as found photosynthetic as well as in respiratory membranes could have strong impact on the mobility of these carriers [53, 72, 59].

A current research challenge is to determine the supramolecular arrangement in native membranes and to understand how it dynamically responds to environmental changes. Furthermore, the overall thylakoid architecture is highly flexible and undergoes dynamic swelling and shrinkage processes that seem to be correlated with the adaptation of photosynthetic energy transformation. This review focuses on the flexibility of the whole thylakoid system (see Sect. 5.2) as well as on the supramolecular protein arrangement in thylakoid membranes (see Sect. 5.3) of higher plants. The reader is also referred to excellent review articles on related fields, i.e. on the supramolecular PSII organization [70] and on 3D models for the overall thylakoid membrane architecture derived from electron tomography [31, 90].

5.2 Overall Thylakoid Architecture

The structure of the thylakoid membrane system in chloroplasts was mainly studied by diverse electron microscopic techniques [86, 89, 111] and atomic force microscopy [55]. Most recently, electron tomography was established for visualizing the 3D thylakoid architecture [16, 32, 69, 87, 105]. An intensively discussed point is how the stroma lamellae insert into the grana stacks. Basically, two models have been proposed. One postulates that the stroma lamellae spiral up around the grana cylinder in the form of a right-handed helix and insert into the grana cylinder by slit like connections called "frets." The other model hypothesizes that bifurcation of stroma lamellae leads to grana stacks ("fork" model). This open aspect in the thylakoid structure is not addressed in this review and the interested reader is referred to [16, 31, 87, 90]. The following section focuses on the architecture of grana stacks that constitute about 80% of the thylakoid membrane [3].

5.2.1 Structural Organization of Grana Thylakoids

5.2.1.1 Grana Diameter

The 3D ultrastructural models derived from electron microscopy reveal the complexity but also the high degree of organization realized in thylakoid membranes. For example, it is striking that the diameters of the numerous grana cylinders in a chloroplast (typically 40–60) are in a narrow size range, i.e. between 300 and 600 nm [35, 111]. This size range seems to be a robust quantity because it does not change significantly under different growth conditions [59]. The grana diameter is particularly relevant for processes that are based on lateral diffusion between stacked and unstacked thylakoid membrane regions because it determines the diffusion distance and consequently the diffusion time. Diffusion dependent processes that are affected include the PSII repair cycle, *state transition* or electron shuttling between PSII and PSI by small electron carriers'. For a detailed discussion, see [59]. It is not clear why no broad statistical distribution in the grana diameters is realized, e.g. grana with a few tens of nm and μm sized grana. This points to physicochemical driving forces determining the size of stacked thylakoids. A possible driving force that is involved in grana organization is the phosphorylation level of PSII and/or LHCII. PSII and LHCII can be reversible phosphorylated by *stn7 and stn8* kinases [24, 42, 96, 116]. In *stn7/stn8* Arabidopsis double mutants with a very low PSII/LHCII phosphorylation level, the grana diameter is about 50% wider compared to wild-type plants [43]. Based on these findings, it was suggested that the phosphorylation level of grana hosted PSII determines the grana diameter [44]. This correlates with the hypothesis that stacking is a result of mutual electrostatic attraction between positive and negative surface charges localized on the stroma-facing N-terminal part of the LHCII complexes located on adjacent grana discs (for a model see [21]). In the context of this model, the introduction of negative charges by protein phosphorylation is expected to disturb the electrostatic balance leading to repulsion and destacking. The idea of stacking mediated by electrostatic interactions goes back to earlier models based on the *Derjaguin-Landau-Verwey-Overbeek* (DLVO) theory that combine electrostatic repulsion with van-der-Waals attraction [19]. In addition, at short distances (<2.5 nm), strong hydrostructural repulsion hinders further approximation of adjacent grana membranes [103]. In difference to this "surface-charge model," other models were postulated for explaining grana stacking including the molecular recognition of complementary protein surface regions [4, 5, 7] or entropic effects [26, 27]. It needs to be elucidated how all these factors work together for establishing tightly stacked grana. However, the observation that differences in the protein phosphorylation level of grana hosted protein complexes induce changes in grana diameter points to an important role of this post-translational protein modification for the structural organization of thylakoid membranes. A critical role of protein phosphorylation on the stromal LHCII site for thylakoid stacking is in accordance with studies showing that cleavage of a 2 kDa sized N-terminal domain of LHCII prevents stacking [81]. Since the activity of the *stn*-kinases are tightly controlled by the redox state of the photosynthetic electron transport chain (i.e. the plastoquinone pool), reversible protein phosphorylation allows dynamic adjustment of structural boundary conditions of the thylakoid membrane system to the needs determined by an ever-changing environment (see Sect. 5.2.2).

5.2.1.2 Grana Distribution in Chloroplasts

Another example for the high degree of organization found in the overall thylakoid architecture is that the many grana stacks in a chloroplast are remarkably homogenously distributed. We analyzed the grana distribution by confocal laser scanning microscopy (CLSM) in combination with mathematical analysis tools and found that the separation distance between adjacent grana stacks fall in a narrow range that varies between 450 and 750 nm with a mean at about 550 nm (unpublished results). As for the grana diameter, it is striking that a broad range of separation values is not realized. The non-random equal-distant separation by about ~550 nm of ~500 nm wide grana discs could be advantageous for absorption of visible light. This point has not been examined so far.

The constancy of the grana diameter and separation indicates that structural organizing forces are at work leading to a high level of order found in the overall thylakoid membrane architecture. Because cytoskeleton-like elements are very likely missing in chloroplasts, the capability to arrange the thylakoid network is most likely determined by physicochemical features of the membrane itself. The potential for self-organization can be demonstrated by *in vitro* destacking and restacking experiments of isolated stacked thylakoid membranes induced by changing salt concentrations [55, 65, 108, 109]. Destacking by low salt treatment not only leads to unfolding of grana stacks but also to randomization of the protein complexes

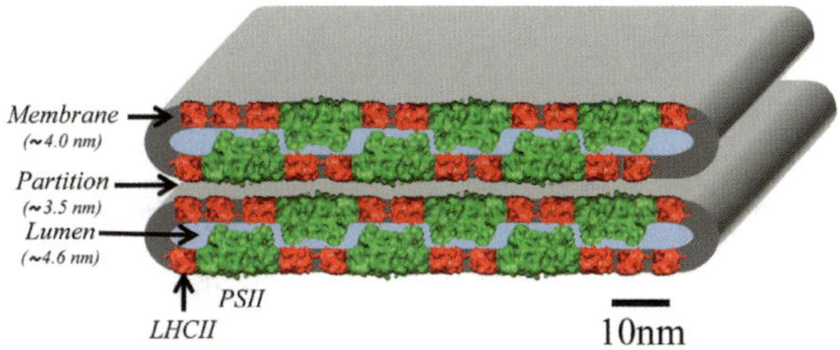

Fig. 5.3 In-scale model of a grana stack (*cross-section*). The model is based on recent cryo-EM data (see text for details). It represents the situation in dark- or dim light-adapted plants. The numbers change considerably in light-adapted plants. The model shows four grana membranes enclosing two lumen spaces and separated by one partition gap. The lateral grana dimension is not in-scale. Note that adjacent PSII complexes sharing the same lumen cannot adopt a face-to-face arrangement due to steric restrictions of the lumenal PSII protrusions

within the thylakoid membrane. Re-addition of salt to destacked thylakoids leads to almost full refolding to the complex grana network that is accompanied by the same inhomogeneous protein distribution between stacked and unstacked membrane regions as found for stacked control membranes. It follows that the complex folding and protein organization is self-organized.

5.2.1.3 Transversal Grana Structure

The transversal geometry of grana stacks recently came into focus of photosynthesis research since technical improvements (cryo-sample preparations and cryo-EM) allow determination of structural thylakoid attributes in near-native states. The pioneering work of Murakami and Packer [84] on thin-sections of room-temperature chemical fixed samples gives a first comprehensive quantitative picture of the transversal grana thylakoid structure. However, it was recognized that chemical fixation of thylakoids at ambient temperatures and using isolated thylakoids can be a source of artifacts [89]. Recently, cryo-EM tomography [32, 70] and cryo-EM [68] was applied to reexamine structural attributes of grana thylakoids. While the studies of Kouřil and co-workers were performed with isolated grana membranes, Daum et al. used ruptured chloroplasts and Kirchhoff et al. used intact leaf discs. The latter two studies gave a consistent picture of the transversal grana architecture summarized in Fig. 5.3. In difference to the values given in Fig. 5.3, Kouřil and co-workers found significant higher numbers for the luminal width (14–16 nm) and a variable width of the stroma partition gap. These differences may indicate that preparation of isolated grana lead to alterations in the native membrane organization and highlights the importance for analyzing membrane attributes with intact material.

Given that high-resolution information about photosynthetic protein complexes is available now, it is appealing to combine them with ultrastructurual data of stacked grana as shown in Fig. 5.3 for the LHCII-PSII supercomplex. These quantitative models give an intuitive access to the structural organization of the photosynthetic machinery and allow estimations of steric restrictions. From Fig. 5.3 it follows that the aqueous stroma space between two grana discs (partition gap) available for diffusion is less than 3 nm (taking into account that PSII extends ~1 nm into the stroma). It follows that the ATPase complex and PSI must be excluded from stacked grana by steric reasons because their stroma protrusions are too large (~10 nm for the ATPase [1] and ~3.5 nm for PSI [9, 55]). In turn, both PSII and LHCII have very flat stromal surfaces (Fig. 5.3) that enable tight grana stacking. The situation for the cyt *b6f* complex is less clear since stromal protrusions may cause steric hindrance and exclusion from stacked regions. These structural conclusions based on new EM data taken from material in a near native state correspond with earlier models [7, 35]. The overall picture derived in Sect. 5.2.1 represents the situation for dark-adapted samples or samples in dim light. However, as detailed in the next section the structural attributes of the thylakoid architecture are not static but can change significantly, in particular by illumination.

5.2.2 Flexibility of the Grana Architecture

5.2.2.1 Classification of Structural Changes in the Grana Architecture

Before summarizing the current knowledge on the structural flexibility of the thylakoid membrane system, it is worthwhile to classify possible geometrical changes for grana stacks. Figure 5.1 gives a schematic overview about these changes that are all realized in native thylakoid membranes. As detailed below, the four different types of structural changes have different functional consequences. Therefore, general statements like "destacking of grana" without specifying what exactly is meant by this may be not sufficient to understand the different functional implications associated with the different types of destacking.

The most obvious and best-documented change in the overall thylakoid architecture is alteration in the number of thylakoid membranes that constitute a grana stack (Fig. 5.1a). It has long been known that there is an inverse correlation between light intensity and the number of thylakoids per grana stack (for reviews see [11, 12, 111]). For example, shade plants form higher grana stacks (more membrane layers) than sun plants. An extreme example is the shade plant *Alocasia microrrhiza* growing in the deep shade in rainforests with more than 100 thylakoid membranes per grana stack [46]. Increasing the number of membrane layers could be a strategy to increase the probability for absorbing the limiting number of light quanta that reach the plant in the shadow [12]. A simple example illustrates this. A single (grana) membrane layer with a chlorophyll concentration of ~0.3 M [61] and a thickness of 4 nm (Fig. 5.3) would absorb only about 1 % of the incident light of 680 nm (extinction coefficient for chlorophylls at 680 nm is 53,000 mM^{-1} m^{-1}, [119]). (The wavelength of 680 nm was arbitrarily chosen). Although this is a rough estimate because the optical properties in intact membranes are much more complex, the number of 1 % is in agreement with more elaborate models [93]. Due to this low probability to harvest sunlight by one membrane layer, the stacking of many membrane layers to grana could be an important optimization for collecting limiting light quanta in shaded habitats. Besides changing the number of membranes per grana stack, other geometrical alterations in response to environmental cues are realized that are summarized in the following sections.

5.2.2.2 Lateral Changes in Grana Diameter (Fig. 5.1b)

As discussed in Sect. 5.2.1.1, the diameter of the grana cylinder seems to be fairly constant in non-stressed plants. However, under certain conditions, the diameter of grana stacks can change. An example is *state transition*. The broad outline of *state transition* [4, 6, 56, 75] is that it dynamically adjusts the functional antenna size of PSII and PSI (number of chlorophylls coupled to a photosystem reaction-center) to balance the energy distribution between them and in that way to synchronize the photochemical rates of the two photosystems. Under conditions where PSII receives more light energy than PSI, the imbalance is sensed by a higher reduction level of the intersystem PQ pool. The reduced PQ pool activates the *stn7* kinase that phosphorylates LHCII subunits (LHCII-P). LHCII-P uncouples from PSII and migrates to unstacked thylakoid regions where it serves as light harvester for PSI. This process is reversible, i.e. if the PQ pool gets oxidized again, LHCII-P is dephosphorylated and redistributes back from PSI to PSII. State 1 is defined when LHCII preferentially binds to PSII, state 2 when it binds to PSI. State transition is an important mechanism to optimize energy conversion under low light, i.e. under conditions where each light quantum should be used for photochemistry. Under high light, the regulation by state transition is turned off because the *stn7* kinase is inactivated by thiol-modulation [97]. Comparative EM micrographs of pea thylakoids in state 1 and 2 reveal that the grana diameter shrinks in state 2, leading to a 23 % decrease in the amount of stacked membranes and consequently to a corresponding increase in unstacked membranes [72, 110]. This lateral grana shrinkage under state 2 was recently confirmed [28].

What could be advantages to reduce the grana diameter? As discussed in detail in 5.3.1, grana thylakoids are heavily crowded by proteins that challenge lateral transport of membrane integral protein complexes. This interferes with the redistribution of LHCII between PSII in stacked grana to PSI in unstacked thylakoid parts. In this respect, decreasing the grana diameter has several advantages. First, LHCII localized in grana areas that are subjected to destacking reaches unstacked regions without long-range diffusion. Second, the diffusion distance of proteins that are still in stacked regions to reach unstacked membrane parts is shortened. Consequently, their diffusion time will be reduced. Third, the grana perimeter to area ratio increases after shrinkage of the grana diameter. Therefore, it is expected that the contact zones between stacked and unstacked membranes increases. In summary, lateral destacking by reducing the grana area could be relevant to improve lateral protein traffic between stacked and unstacked thylakoid membranes as required for *state transition*.

Another prime example where lateral protein traffic is essential is the PSII repair cycle. Plants have to deal with the problem that primary photochemical processes in the reaction center of PSII have an intrinsic probability for producing toxic reactive oxygen species (ROS) that mainly damage the D1 subunit of PSII [73, 79, 83, 92]. This cannot be completely avoided and becomes a severe problem under high-light or temperature stress. Plants addressed this challenge by the evolution of a sophisticated PSII repair cycle that is one of the fastest repair machineries in nature [73, 79, 83]. An open question concerning the PSII repair cycle is how damaged PSII in stacked grana becomes mobilized and can escape from crowded grana to find its repair machinery in distant (several 100 nm) stroma lamellae. As discussed above for *state transition*, reduction of the grana diameter could be an elegant way to solve this problem because it facilitates diffusion dependent steps in the PSII repair cycle. Recently, we analyzed structural grana attributes by mathematical analysis of CLSM images of dark-adapted and light-stressed *Arabidopsis* protoplasts (unpublished results). A key finding is that high-light stress causes a shrinkage of the grana diameter by about 21 % (from 380 nm to 300 nm). This contrasts with EM studies in which no change in grana diameter was observed by high light stress [43]. However, the error bars in the EM study are in the order of 35 %. Thus, it could be that the more subtle changes determined by CLSM were simply not detected in the EM study because the error bars are too large. It is striking that both *state transition* and the PSII repair cycle includes reversible phosphorylation of LHCII or PSII subunits (D1, D2, CP43, psbH) and lateral destacking of grana. This supports the concept that the protein phosphorylation level of grana hosted proteins determines the grana diameter (see Sect. 5.2.1.1).

5.2.2.3 Swelling of Thylakoid Lumen (Fig. 5.1c)

Until recently, the accepted view was that the thylakoid lumen shrinks in light compared to dark-adapted samples. The concept of a light-induced shrinkage mainly goes back to observations made by Murakami and Packer on samples prepared by classical room temperature fixation techniques [84]. In contrast to this mainstream concept, we could show in a collaborative study with Dr. Ziv Reich by applying two different cryo-EM techniques on intact leaf discs that the lumen expands in light-adapted Arabidopsis leaves from about 4.6 nm to about 9.2 nm, i.e. by about 100 % [68]. We interpreted the discrepancy of our study to the work of Murakami and Packer by differences in sample preparations and by working with intact leaf material instead of isolated thylakoids (see [88] for a detailed discussion). Therefore, recent data indicates that the lumen expands in light. A controlled swelling and shrinkage of the lumen introduces interesting types of regulation of electron transport reactions as well as photoprotective and protein degradation processes that are localized in the lumen.

What are the functional implications of a swelling or shrinkage of the thylakoid lumen? Before addressing this question it is essential to appreciate the high protein density in this narrow reaction space. In dark-adapted samples, a luminal width of only ~4.6 nm excludes the possibility that the protruding luminal parts of PSII complexes localized in adjacent grana thylakoid membranes adopt a face-to-face arrangement and therefore must be staggered (Fig. 5.3). This is because the height of the luminal protrusion of 4–5 nm [91] is too large to allow a face-to-face organization. From a staggered arrangement, it follows that the area occupied by luminal PSII protrusions in the middle of the lumen is doubled compared to a face-to-face arrangement. Taking measured PSII densities and mid-resolution models of PSII, we generated molecular landscapes of the lumen of dark-adapted Arabidopsis plants [68]. These models reveal that the lumen in grana is an extremely crowded space. About 70 % of the area is occupied by OECs. Estimation of the available diffusion space of the electron carrier plastocyanin (PC) shows that it is trapped in small diffusion microdomains. PC can escape from these microdomains only by rearrangements of the overall protein network in grana, which is a very slow process [62]. Consequently, long-range diffusion of PC in dark-adapted thylakoids is expected to be slow; this was supported by functional electron transport measurements [68]. In contrast, the significant swelling of the lumen in light-adapted thylakoids leads to a switch from highly localized diffusion (in microdomains) to long-range diffusion [68]. Thus, light-induced dynamic swelling and shrinkage of the thylakoid lumen can control photosynthetic electron transport by controlling the diffusion radius of plastocyanin.

The highly restricted diffusion in dark-adapted plants observed for PC is likely to hold also for other lumenal proteins. Steric restrictions for other proteins can even be more pronounced because they are often larger than PC. For example, the violaxanthin deepoxidase (VDE) is about 40 kDa [102]. It is assumed that the functional VDE-form is a dimer. Thus, the physical size (diameter) of functional dimeric VDE is ~twice that of PC $((80/10.5\ \text{kDa})^{1/3})$ assuming a spherical shape of both molecules. It follows that there is hardly any space in the lumen of stacked grana to accommodate VDE or that the enzyme is trapped in a few places in stacked grana only. VDE catalyzes the conversion of the xanthophyll violoxanthin to zeaxanthin [50]. The latter is an important activator of photoprotective high-energy quenching (qE). Size exclusion of VDE from stacked grana or a highly localized VDE within grana implies that the xanthophylls in grana have to migrate through the lipid membrane phase to and from the enzyme. Since the membrane is also a highly crowded environment, it is expected that this diffusion is slow, as was shown as well for plastoquinone [60, 74]. Taken together, it could be that activation of qE by zeaxanthin formation is kinetically limited by slow diffusion of xanthophylls through crowded grana to a tethered VDE in dark-adapted samples. This restriction could be reversed in the light by luminal swelling that could lead to acceleration of qE due to higher VDE mobility. Similar conclusions as for VDE can be drawn for luminal *Deg* proteases (molecular weight ~35 kDa [102], ~1.5 times larger than PC) that are involved in degradation of photodamaged PSII (see also FtsH proteases in Sect. 5.2.2.4). Further experiments have to prove these possibilities of dynamic restricted accesses to stacked grana and/or trapping in microdomains of lumen hosted VDE and proteases.

Another process that could be influenced by changes in the luminal width is the assembly/disassembly of the oxygen-evolving complex (OEC) of PSII. It has been postulated that changes in the luminal width can control the proper assembly of the four subunits that constitute the OEC for catalyzing water splitting [13]. Although the authors assumed shrinkage of the lumen in the light (based on the work of Murakami and Packer, 84], the concept of a dynamic switch in OEC activity controlled by swelling/shrinkage processes is an interesting possibility to regulate PSII activity and in consequence linear electron transport from water to PSI. A related mechanism is the disassembly of photodamaged PSII. It is known that the damaged LHCII-PSII holocomplex in grana dismantles, including the removal of the luminal OEC subunits. This removal may be required to mobilize PSII in stacked grana to make it accessible for its repair machinery in stroma lamellae (see Sect. 5.2.2.2 above). It is likely that this disassembly is facilitated by the expansion of the lumen. So far, this expansion was seen only for moderate light intensities (500 μmol quanta m^{-2} s^{-1}). It has yet to be determined whether this also takes place in high-light stressed plants.

At this point, we can only speculate about the factors that determine changes of the luminal width. A plausible scenario is that light-induced protonmotive force (*pmf*) drives an influx of chloride anions that in turn leads to osmotic swelling of the lumen. Thylakoid membranes contain at least two chloride channels [107]. At least one of these channels is voltage-gated [101]. Thus, the electrical potential gradient ($\Delta\varphi$) generated in the light could activate voltage-gated chloride channels leading to a *pmf*-driven influx of chloride anions into the lumen and to osmotic swelling. The important consequence of this chloride-channel regulated swelling is that the luminal width is not directly light-dependent but rather controlled by $\Delta\varphi$. Any factors that either manipulate the magnitude of the total *pmf* or that alter the partition between the chemical and electrical *pmf* parts would also control the luminal width. For example, metabolic control of the ATPase activity by thiol-modulation of the γ-subunit [69] would not only regulate the magnitude of the trans-thylakoid *pmf* but also structural attributes of the thylakoid system, i.e. the width of the lumen. Thus, there could be an indirect link between regulation of ATPase activity and control of PC-mediated electron transport and VDE-dependent photoprotection.

5.2.2.4 Vertical Destacking of Grana (Fig. 5.1d)

Vertical destacking (i.e. increase in the partition gap) was recently discussed in the context of the PSII repair cycle. It was postulated that widening of the stromal gap allows FtsH proteases to enter the grana region enabling degradation of damaged PSII [43, 123]. The height of the stromal protrusion of the hexameric FtsH proteases is ~6.5 nm [113]. This large protrusion may exclude the FtsH proteases from stacked regions in dark-adapted samples by steric hindrances (Fig. 5.3). Therefore, a significant widening of the stromal gap in light-stressed plants could allow FtsH proteases to access stacked grana regions for swift degradation of damaged PSII [123, 124]. It was also suggested that transversal destacking could be a factor to mobilize damaged PSII in stacked grana to reach its repair machinery in distant stroma lamellae [43]. The effect of vertical destacking on PSII mobility in grana could be two-fold. First, an increase in the stromal partition gap lowers attractive van-der-Waals forces between PSII in adjacent membrane discs that holds PSII in grana [20]. Second, transversal separation also diminishes steric hindrances since the stromal part of PSII sticks out ~1 nm from the membrane surface. The "effective" protrusion could be even larger due to structured water layers. Thus, steric hindrances on the stroma side could restrict PSII mobility in dark-adapted samples where the partition gap is only 3–4 nm (Fig. 5.3). These considerations make plausible that transversal destacking could facilitate the degradation of damaged PSII localized in stacked grana. According to the "surface charge model" of grana stacking (see Sect. 5.2.1.1) the driving force for separation of grana membranes is an increase in electrostatic repulsion by introduction of negative charges by protein phosphorylation of PSII (and/or LHCII) subunits. Although these ideas describing how widening of the stromal gap can facilitate accessibilities of FtsH proteases and the mobility of PSII are sound, the experimental evidence for a high light induced widening of the stromal gap in stacked grana is weak. There is evidence that grana partially destack under photoinhibitory conditions [43, 58]. But as mentioned above, grana destacking can mean different things and direct evidence for an increase in partition gap under light stress is missing. For example, no cryo-EM data exists on thylakoids from high-light stressed plants.

Another aspect of dynamic changes in grana partition gap concerns the excitonic energy transfer between protein-bound pigments. It was hypothesized that grana stacking allows for transversal energy transfer between LHCII-pigments localized in opposite grana membranes [7]. The significance of this vertical energy transfer from one grana disc across the stromal partition gap to the adjacent membrane disc for light harvesting by PSII is unclear. Experimental evidence supports [118] or contradicts [63] a significant contribution for vertical exciton transfer in grana. Theoretically, it is expected that lateral energy transfer between pigments in the same grana membrane is faster (and consequently more efficient) than the transversal transfer, because the pigment-pigment distances are shorter [63]. Within the membrane, the mean chlorophyll-chlorophyll distance is expected to be 2 nm or shorter as was estimated for adjacent trimeric LHCII in thylakoid membranes [21]. By contrast, the closest transversal chlorophyll-chlorophyll distance in dark-adapted plants is ~4 nm [estimated from [21, 32]. Assuming that inter-protein energy transfer can be described by Förster theory and assuming that pigment-pigment orientation factors and refractive indexes are the same for transversal and lateral exciton transfer pathways, the difference between 2 and 4 nm would reduce the energy transfer rate more than 60 fold. However, we do not have reliable information for the pigment orientation factors and for the refractive index (in particular for energy transfer across the aqueous partition gap) that can change this number considerably. Our recent cryo-EM studies show a tendency for a reduction in the partition gap in plants illuminated under non-photoinhibitory conditions [68]. Thus, there is a possibility for a light induced switch in transversal energy transfer, i.e. that is activated in illuminated samples by narrowing of the adjacent grana discs. This interesting possibility has to be proved by e.g. analyzing the partition gap under different light intensities.

Finally, the vertical stromal distance in grana thylakoids could control the sublocalization of the cyt *b6f* complex. The stromal protrusion of this complex is 1.5–2.0 nm (Dr. William Cramer, personal information). This is an interesting value because it is between the protrusion of PSII (~1 nm) and PSI (~3.5 nm). It could be that small changes in the partition gap have significant impact on the sublocalization of the cyt *b6f* complex in thylakoid membranes, i.e. a widening of the gap would give access to stacked grana whereas a narrowing would exclude the cyt *b6f* from this region. This in turn could determine whether the cyt *b6f* complex is involved in linear electron transport (localization in stacked grana) or in cyclic flow around PSI (localization in unstacked thylakoid regions).

5.3 The Supramolecular Level

Structural flexibility is not only apparent for the whole thylakoid membrane system but also realized on the supramolecular level, i.e. for the arrangement of many proteins in membranes. It is interesting that photosynthetic membranes in more ancient autotrophic organisms are less dynamic. For example, LHCs and reaction centers in purple bacteria form relative rigid highly ordered supramolecular assemblies [17]. It is likely that supramolecular flexibility in photosynthetic membranes was an important evolutionary selection criterion to adapt land plants to highly-dynamic terrestrial habitats [82]. For example, sunlight can be an elusive energy source for land plants because its intensity can change by orders of magnitude on time scales ranging from seconds to months. Photosynthetic energy conversion must compensate for these fluctuations. In low light, the absorption of light quanta and its conversion into chemical energy must be optimized to energetically fuel the cellular metabolism of the plant. In contrast, in high light, it could be necessary to switch the system to an energy dissipating state that minimizes severe damage by toxic side reactions of the primary photochemical reactions. The situation in nature is even more complex because plants are integrated in a highly dynamic network of different biotic and abiotic factors that determine different requirements on photosynthetic energy conversion. Thus, flexibility of photosynthetic functionality is central for plants. Recent finding suggest that a main part of the capability for adjustment of photosynthetic performance is realized by dynamic rearrangements of the protein network in grana thylakoids. Before describing the current knowledge on the supramolecular flexibility, it is essential to understand that photosynthetic membranes are densely packed by protein complexes.

5.3.1 Macromolecular Crowding

An important and long known tool to study supramolecular protein arrangements in photosynthetic membranes is EM with freeze-fractured or freeze-etched membranes [111]. In freeze-fracture EM, the lipid bilayer is split into the two monolayers.

Transmembrane protein complexes visibly stick out of the lipid monolayer as small knobs. In early freeze-fracture and freeze-etched EM studies, it was recognized that thylakoid membranes must be densely packed with photosynthetic protein complexes, especially in grana membranes (reviewed in [109]). Recently, the protein packing density was quantified for grana thylakoids [61, 62]. The study showed that ~70% of the membrane area in grana belongs to proteins and only ~30% to the lipid bilayer. Visualization of 70% protein packing density show that there is hardly any lipid space left. Indeed, both theoretical [61] as well as EPR studies with spin-labeled lipid probes [76, 77] indicate that a high fraction of thylakoid lipids (>50%) have direct protein contact (so-called boundary lipids). Thus, photosynthetic energy conversion takes places in an extremely crowded environment. The consequences of macromolecular crowing on the functionality of photosynthetic membranes are less considered so far, although it is known from theoretical considerations (such as percolation theory) that physicochemical membrane properties can be significantly affected by high protein packing densities [98–100]. Some aspects of high protein densities on photosynthetic energy conversion are discussed below. Again, it is interesting to look on related biomembrane systems. The protein/lipid ratio (as a measure of protein crowding) is 0.34 (w/w) in respiratory membranes [15, 29] that is almost identical to thylakoid membranes [61, 77, 85]. Furthermore, EM studies on photosynthetic membranes in cyanobacteria, red algae, purple bacteria, or green algae indicate a similar high protein packing density as in thylakoid membranes of higher plants. It appears that macromolecular crowding is a common structural feature of bioenergetic membranes.

Why are thylakoid membranes so densely packed with proteins? An obvious advantage is that it allows a tight packing of light-harvesting pigments bound to the proteins. On the molecular scale, even full sunlight is a dilute energy source [23]. Therefore, a high concentration of light-absorbing chromophores is a prerequisite for efficient photosynthetic energy conversion. This is realized by macromolecular crowding. However, the situation is more complex since the high chlorophyll concentration in thylakoid membranes (0.3 M) can lead to detrimental excimer formation that leads to an almost complete conversion of collected sunlight energy into heat if the pigments are randomly organized [22]. Energy quenching by high pigment concentrations is avoided by exact chlorophyll positioning in rigid protein scaffolds in LHCs and PSs. The evolutionary optimization of pigment-pigment distances, angles, and dielectric constants within the hydrophobic regions of light-harvesting proteins prevents unwanted energy losses and ensures ultrafast and thus very efficient energy transfer.

Recently, we could identify a further advantage of macromolecular crowding in grana thylakoid membranes [49]. The organization of the PS-II light-harvesting system in grana is modular. That means that multiple layers of LHCII complexes with different binding strengths surround the PSII core [35, 70]. In detail, the PSII core contains two core antenna complexes (CP43 and CP47) that are complemented by minor LHC (CP26, CP29) and one trimeric major LHCII forming the dimeric LHCII-PSII supercomplex. This supercomplex is named C2S2 (C=PSII *c*ore, S=*s*trongly bound trimeric LHCII). Further LHCII-trimers with the minor LHCII

CP24 can be attached to the C2S2 supercomplex forming a C2S2M2 supercomplex (M=*m*edium bound trimeric LHCII) that can further bind up to eight additional weakly attached major LHCII complexes. The consequence of this modular antenna organization is that it critically depends on efficient intermolecular energy transfer between LHCIIs. In our study [49], we diluted the natural high protein packing density by fusing isolated grana membranes with protein-free lipid-liposomes with the natural lipid composition. The data shows that dilution leads to uncoupling of peripheral LHCII trimers not tightly bound to the C2S2 supercomplex and consequently to a reduction in the functional antenna size of PSII. This result indicates that high protein packing densities in stacked grana are required to bring weakly interacting LHCII and PSII in close contact to enable efficient intermolecular exciton energy transfer. Thus, protein crowding is advantageous for light-harvesting by PSII surrounded by its multi-layer light-harvesting system. In contrast to advantages of protein crowding, there can be serious problems associated with high protein packing densities addressed in the next chapters.

5.3.2 Disordered Versus Ordered

Early EM studies in the 1960s on thylakoid membranes revealed that proteins in stacked grana can arrange themselves in highly ordered semicrystalline arrays [94, 95]. In the following decades, numerous publications described this remarkable self-ordering by using electron microscopy (reviewed in [35, 109]). Very recently, semicrystalline PSII arrays were visualized also by atomic force microscopy ([114] and Fig. 5.4). A deeper structural analysis of the protein arrays in grana reveals that they are composed of C2S2 supercomplex [32, 35]. Furthermore, it was recognized that there is variability in the crystal structure. The crystals can have different lattice constants and can also be composed of C2S2M or C2S2M2 supercomplexes [35].

Semicrystaline protein arrays are often seen under non-optimal growth conditions. This includes low temperatures, low light, high sugar concentrations, or special buffers [45, 64, 78, 80, 104, 106, 120]. Under these conditions, the arrays can cover a whole grana disc. The observation that different abiotic factors induce a similar switch from disordered to ordered is significant because it indicates that changes in the protein ordering level in grana could be a universal response to non-optimal growth condition that highlights its physiological relevance. Recently, EM tomography on ruptured chloroplast shows that PSII arrays are also formed under non-stressed conditions but with low abundance [32]. It correlates with the observation that in grana with disordered protein organization, a tendency for a parallel PSII alignment is already apparent [59, 66]. These findings are indicative for a dynamic equilibrium between disordered and semicrystalline states controlled by the environment. This equilibrium could represent a metastable state as suggested recently [70], i.e. small changes can trigger large-scale reorganizations from disordered to ordered and *vice versa*.

The factors that determine the equilibrium between disordered and ordered protein organizations in grana are unknown. One possibility is that properties

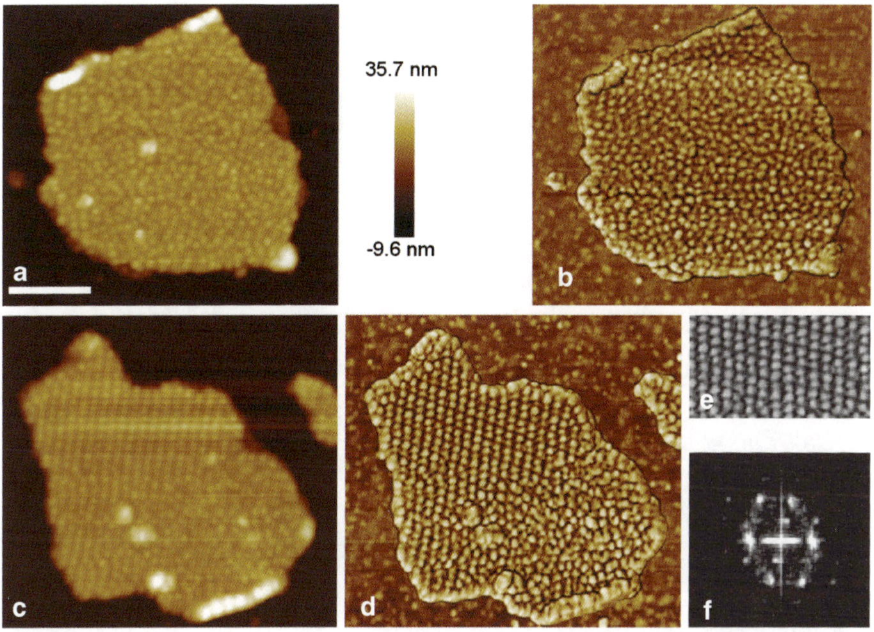

Fig. 5.4 Atomic force microscopic (AFM) micrograph of isolated grana membranes (unpublished results). The micrographs show the topography (**a**, **c**) and the phase image (**b**, **d**) of grana isolated from Arabidopsis. The numerous (*whitish*) particles represent the lumenal protrusion of PSII (mainly the water-splitting complex). Particles are better identifiable in the phase images. PSII rarely organizes into semicrystalline arrays (**c**, **d**, **e**). E is a zoom into the arrayed region. The lattice constants for the semicrystalline arrays can be derived from the Fourier transform (**f**). Scale bar 200 nm

of the lipid bilayer (hydrophobicity, hydrophobic matching, curvature forces) could govern the supramolecular protein organization. A lipid-controlled switch between disordered and ordered shines new light on the xanthophyll cycle (see Sect. 5.2.2.3) because the catalytic formation of zeaxanthin from violoxanthin changes physicochemical properties of the lipid bilayer phase (reviewed in [50]). This would enable control of the xanthophyll-cycle on the supramolecular protein arrangement in grana. Another possibility is that proteins could alter membrane properties. For example, it was suggested that the amount of the *psbS* protein controls the abundance of semicrystalline protein arrays in grana, i.e. the more *psbS* present in thylakoid membranes the more the proteins in grana adopt a disordered configuration [48, 57]. Although this 'inhibitory' effect of *psbS* on semicrystalline array formation is apparent, no mechanistic suggestions how *psbS* controls the supramolecular protein arrangement are around. Interestingly, both zeaxanthin and the *psbS* protein play key roles in photoprotective qE. It was also shown that qE is accompanied by large-scale protein reorganizations in grana [51]. Thus, changes in xanthophylls or *psbS* could trigger supramolecular switches required for photoprotection.

5.3.3 Functional Significance of Protein Ordering

Although macromolecular crowding seems to be required for efficient light-harvesting by the modular organized PSII complex (see Sect. 5.3.1), it can have severe impacts on all processes that require lateral diffusion in the thylakoid membrane system. The mobility of components of the photosynthetic machinery in thylakoid membranes is of vital importance. Not only the small electron carrier PQ has to diffuse through crowded grana to functionally connect PSII and the cyt *b6f* complex (for PC see Sect. 5.2.2.3), but *state transition* and the PSII repair cycle also require lateral traffic of proteins. Finally, thylakoids membranes remodel their protein composition in response to environmental changes. The site of new protein insertion is the stroma lamellae (ribosomes are size-excluded from stacked grana see Fig. 5.3). Therefore, changing the protein composition in grana requires migration over several 100 nm. Although it is intuitively evident that it is easier to laterally transport a molecule in a pure lipid bilayer compared to a bilayer fully packed with diffusion obstacles (proteins), the details of the correlation between obstacle density, size or shape, and diffusion can be very complex. Percolation theory describes this interdependency [98–100] and a main outcome of this theory is that at obstacle area occupations of ~70 %, small enclosed microdomains are formed that restrict long-range diffusion processes. Monte Carlo computer simulations on PQ diffusion [117] and PSII diffusion [62] reveal severe retardation of long-range diffusion for both molecules in accordance with percolation theory. This highlights the potential problem of macromolecular crowding.

An interesting discrepancy is that measurements of protein diffusion in grana by fluorescence recovery after photobleaching (FRAP) is heterogeneous [47, 67] whereas the Monte Carlo simulation show a homogenous very slow diffusion. In FRAP measurements, 15–25 % of granal proteins are moving relatively fast (e.g. they can cross the whole grana diameter in a few seconds). A main difference between simulation and experiment is that the simulation assumes a pure random protein distribution. Thus, it is likely that ordering of the proteins in grana can explain why a certain fraction is diffusing faster than others. It is important to recognize that even in disordered grana, proteins are not purely randomly arranged as demonstrated by mathematical analysis of freeze-fractured EM micrographs [62]. Very recently, we could show that lipid diffusion is faster in a mutant which constitutively forms semicrystalline arrays in grana (unpublished results). These observations point to a role of protein ordering for diffusion processes in grana thylakoids. Protein ordering could be a strategy to ensure high protein packing required for light harvesting by PSII and at the same time allows efficient protein/metabolite (PQ, xanthophylls) traffic required for ET function and regulation processes.

5.3.4 Stroma Lamellae

In contrast to grana thylakoids, stroma lamellae are less crowded by proteins as indicated by a higher lipid to protein ratio (0.64 for stroma lamellae, 0.16 for grana,

[85]). Furthermore, as discussed above, the protein composition is quite different, i.e. stroma lamellae are enriched in PSI-LHCI supercomplexes and ATPase and in addition contain cyt *b6f* complexes and a low amount PSII [2, 111]. In contrast to the pronounced protein heterogeneity between grana thylakoids and stroma lamellae, the four main acyl lipids (monogalactosyl-diacylglycerol, digalactosyl-diacylglycerol, sulfoquinovosyl-diacylglycerol, phosphatidyl-diacylglycerol) are evenly distributed between the two subcomparments [36]. So far, no ordered protein arrays were reported to exist in stroma lamellae indicating that physiochemical and/or structural properties of proteins in stroma lamellae (i.e. the PSI-LHCI supercomplex) do not facilitate organization in higher supramolecular assemblies. It seems that PSI-LHCI functions as an isolated entity in contrast to PSII-LHCII in grana thylakoids that form extended networks that exchange exciton energy (named connectivity, [52, 71]). Furthermore, compared to PSII in grana the antenna system of PSI is less modular, i.e. most of the 167 chlorophylls of the PSI-LHCI supercomplex are bound to the central *psaa/psab* subunits [10]. Thus, it could be that the missing (evolutionary) pressure to pack PSI and LHCI tightly (as realized for PSII/LHCII) to ensure efficient light harvesting (see Sect. 5.3.1) leads to lower protein packing densities that is advantageous for diffusion processes.

A striking difference exists for the organization of the ATPase in unstacked thylakoid regions compared to its mitochondrial counterpart. In respiratory membranes, the ATPase is organized as a dimer that in turn can form extended rows of dimers [33, 39, 112]. It has been discussed that this supramolecular ATPase arrangement is important for shaping the overall structure of the inner mitochondrial membrane (cristae formation) and for conversion of the protonmotive force to ATP [33]. No such structural or functional role is reported for the ATPase in stroma lamellae. The comparison between respiratory and photosynthetic membranes shows that very different structural principles are realized that govern the overall membrane architecture. In plants, grana are formed by electrostatic/van-der-Waals interactions between flat surfaces of LHCII (PSII). In respiratory membranes, ATPase dimer formation could exert a bending force to the lipid bilayer that causes cristae formation.

5.4 Outlook

After the first publication of the crystal structure of a photosynthetic protein complex in 1985 [34], the following two decades were dominated by unraveling the atomic structures of all photosynthetic complexes. We are now in the unique position to have an almost complete set of high-resolution structures. There are still essential tasks to do, i.e. no high-resolution structure of the PSII-LHCII supercomplex exists. Until appropriate PSII supercomplex crystals are available that allows generation of near atomic-resolution models, medium-resolution EM in combination with single-particle analysis has proved to be an excellent alternative. However, one challenge for the future is to study how many proteins interact in the intact membrane to form functional networks and to analyze the dynamics of these networks in response to environmental cues. The first glimpse of information we have on the supramolecular level reveals that thylakoid membranes, although they are densely packed with proteins, are very dynamic and can switch their protein organization from disordered to crystalline. What are the molecular mechanisms that determine these rearrangements? As suggested by Kouřil and co-workers [70], it could be small perturbations of structural or physicochemical parameters that trigger large-scale rearrangements. Since these reorganizations determine photoprotection, electron flow (by PQ and PC), adaption mechanisms (*state transition*), and repair processes (PSII repair cycle), it is a prime task to unravel the determinants for these protein rearrangements.

A further important research field for the future is the development of dynamic supramolecular models for photosynthetic membranes that ideally allow the tracking of individual molecules. Available methods related to this challenge have specific advantages and drawbacks. EM produces high-resolution molecular images but they are static, i.e., they give only a snapshot in time. It is expected that EM tomography will develop in the near future to allow drawing protein landscapes within whole thylakoid membranes. However, EM tomography will not give access to the dynamics of the protein assemblies. Conventional and high-speed AFM can visualize molecular mobility with a certain time resolution. However, AFM probes only surfaces and does not allow to track proteins in stacked grana of intact thylakoids. Methods that measure protein dynamics (diffusion) like FRAP do not have the resolution to visualize single molecules. A further drawback with FRAP is that it is established only for measuring bulk chlorophylls that does not allow to distinguish different protein types, i.e. between LHCII and PSII. Therefore one future task will be the design of specific labels for individual protein complexes that can be applied to high-resolution light microscopy, e.g. single particle tracking. However, although establishing this advanced light microscopic methods for photosynthetic membranes would be a technological breakthrough, they do not allow tracking of a whole protein assemble. A possible way to solve these limitations is to combine ultrastructural data (EM, AFM) and diffusion data in a dynamic computer simulation program. If the output parameters of these simulations are chosen carefully, i.e. that they can be tested by measurements, the combination between simulation and experiment could be extremely powerful.

The enormous progress made in the last few years with EM tomography are very encouraging and promise to study the thylakoid membrane architecture in intact chloroplasts. This tool would be very valuable to address the multiple questions raised in this review on the overall changes in grana architecture triggered by environmental challenges. The combination of ultrastructural data with functional measurement will allow unraveling dynamic structural attributes of the thylakoid system and how it affects photosynthetic performance. This will be the basis for designing strategies to improve plants for producing food and energy.

Acknowledgements The author is indebted to Drs. Egbert J. Boekema, William A. Cramer, Werner Kühlbrandt and Ziv Reich for critical reading of the manuscript and their very helpful comments and suggestions. The author is supported by Washington State University, the Washington State Agricultural Research Center, the National Science Foundation (MCB-1158571), and the United States-Israel Binational Agricultural Research and Development Fund (US-4334–10). Robert Yarbrough is acknowledged for proofreading the manuscript.

References

1. Abrahams J, Leslie AGW, Lutter R, Walker JE (1994) Structure at 2.8 Å resolution of F1-ATPase from bovine heart mitochondria. Nature 370:621–628
2. Albertsson PA (2001) A quantitative model of the domain structure of the photosynthetic membrane. Trends Plant Sci 6:349–354
3. Albertsson PA, Andreasson E (2004) The constant proportion of grana and stroma lamellae in plant chloroplast. Physiol Plant 121:334–342
4. Allen JF (1992a) Protein phosphorylation in regulation of photosynthesis. Biochim Biophys Acta 1098:275–335
5. Allen JF (1992b) How does protein phosphorylation regulate photosynthesis? Trends Biochem Sci 17:12–17
6. Allen JF (2003) State transitions-a question of balance. Science 299:1530–1532
7. Allen JF, Forsberg J (2001) Molecular recognition in thylakoid structure and function. Trends Plant Sci 6:317–326
8. Amunts A, Nelson N (2008) Functional organization of plant Photosystem I: evolution of a highly efficient photochemical machine. Plant Physiol Biochem 46:228–237
9. Amunts A, Nelson N (2009) Plant photosystem I design in the light of evolution. Structure 17:637–650
10. Amunts A, Drory O, Nelson N (2007) The structure of a plant photosystem I supercomplex at 3.4 Å resolution. Nature 447:58–63
11. Anderson JM (1986) Photoregulation of the composition, function and structure of thylakoid membranes. Annu Rev Plant Physiol 37:93–136
12. Anderson JM (1999) Insights into the consequences of grana stacking of thylakoid membranes in vascular plants: a personal perspective. Aust J Plant Physiol 26:625–639
13. Anderson JM, Chow WS, De Las Rivas J (2008) Dynamic flexibility in the structure and function of photosystem II in higher plant thylakoid membranes: the grana enigma. Photosynth Res 98:575–587
14. Andersson B, Anderson JM (1980) Lateral heterogeneity in the distribuition of chlorophyll-protein complexes of the thylakoid membranes of spinach. Biochim Biophys Acta 593:427–440
15. Ardail D, Lerme F, Louisot P (1990) Mitochondrial contact sites. Lipid composition and dynamics. J Biol Chem 265:18797–18802
16. Austin JR 2nd, Staehelin LA (2011) Three-dimensional architecture of grana and stroma thylakoids of higher plants as determined by electron tomography. Plant Physiol 155:1601–1611
17. Bahatyrova S, Frese RN, Siebert CA, Olsen JD, Van Der Werf KO, Van Grondelle R, Niederman RA, Bullough PA, Otto C, Hunter CN (2004) The native architecture of a photosynthetic membrane. Nature 430:1058–1062
18. Baniulis D, Yamashita E, Zhang H, Hasan SS, Cramer WA (2008) Structure-function of the Cytochrome b6f complex. Photochem Photobiol 84:1349–1358
19. Barber J (1980) Membrane surface charges and potentials in relation to photosynthesis. Biochim Biophys Acta 594:253–308
20. Barber J (1982) Influence of surface charges on thylakoid structure and function. Annu Rev Plant Physiol 33:261–295
21. Barros T, Royant A, Standfuss J, Dreuw A, Kühlbrandt W (2009) Crystal structure of plant light-harvesting complex shows the active, energy-transmitting state. EMBO J 28:298–306
22. Beddard GS, Porter G (1976) Concentration quenching in chlorophyll. Nature 260:366–367
23. Blankenship EB (2002) Molecular mechanisms of photosynthesis. Blackwell Science, Oxford
24. Bonardi V et al (2007) Photosystem II core phosphorylation and photosynthetic acclimation require two different protein kinases. Nature 437:1179–1182
25. Caffarri S, Kouril R, Kereiche S, Boekema EJ, Croce R (2009) Functional architecture of higher plant photosystem II supercomplexes. EMBO J 28:3052–3063

26. Chow WS (1999) Grana formation: entropy-assisted local order in chloroplasts? Aust J Plant Physiol 26:641–664
27. Chow WS, Kim EH, Horton P, Anderson JM (2005) Granal stacking of thylakoid membranes in higher plant chloroplasts: the physicochemical forces at work and the functional consequences that ensue. Photochem Photobiol Sci 4:1081–1090
28. Chuartzman SG, Nevo R, Shimoni E, Charuvi D, Kiss V, Ohad I, Brumfeld V, Reich Z (2008) Thylakoid membrane remodeling during state transitions in Arabidopsis. Plant Cell 20:1029–1039
29. Colbeau A, Nachbaur J, Vignais PM (1971) Enzymatic characterization and lipid composition of rat liver subcellular membranes. Biochim Biophys Acta 249:462–492
30. Danielsson R, Suorsa M, Paakkarinen V, Albertsson PA, Styring S, Aro EM, Mamedov F (2006) Dimeric and monomeric organization of photosystem II. J Biol Chem 281:14241–14249
31. Daum B, Kühlbrandt W (2011) Electron tomography of plant thylakoid membranes. J Exp Bot 62:2393–2402
32. Daum B, Nicastro D, Austin J II, McIntosh R, Kühlbrandt W (2010) Arrangement of photosystem II and ATP synthase in chloroplast membranes of spinach and pea. Plant Cell 22:1299–1312
33. Davies KM, Strauss M, Daum B, Kief JH, Osiewacz HD, Rycovska A, Zickermann V, Kühlbrandt W (2011) Maromolecular organization of ATP synthase and complex I n whole mitochrondria. Proc Natl Acad Sci U S A 108:14121–14126
34. Deisenhofer J, Epp O, Miki K et al (1985) Structure of the protein subunits in the photosynthetic reaction center of Rhodopseudomonas viridis at 3 Å resolution. Nature 318:618–624
35. Dekker JP, Boekema EJ (2005) Supramolecular organization of thylakoid membrane proteins in green plants. Biochim Biophys Acta 1706:12–39
36. Duchêne S, Siegenthaler PA (2000) Do glycerolipids display lateral heterogeneity in the thylakoid membrane? Lipids 35:739–744
37. Dudkina NV, Kouril R, Bultema JB, Boekema EJ (2010a) Imaging of organelles by electron microscopy reveals protein-protein interactions in mitochondria and chloroplasts. FEBS Lett 584:2510–2515
38. Dudkina NV, Kouril R, Peters K, Braun HP, Boekema EJ (2010b) Structure and function of mitochondrial supercomplexes. Biochim Biophys Acta 1797:664–670
39. Dudkina NV, Oostergetel GT, Lewejohann D, Braun H-P, Boekema EJ (2010c) Row-like organization of ATP synthase in intact mitochondria determined by cryo-electron tromography. Biochim Bioiphys Acta 1797:272–277
40. Dudkina NV, Kudryashev M, Stahlberg H, Boekema EJ (2011) Interaction of complexes I, III, and IV within the bovine respirasome by single particle cryoelectron tomography. Proc Natl Acad Sci U S A 108:15196–15200
41. Dunahay TG, Staehelin LA, Seibert M, Ogilvie PD, Berg SP (1984) Structural, biochemical and biophysical characterization of four oxygen-evolving photosystem II preparations from spinach. Biochim Biophys Acta 764:179–193
42. Fristedt R, Vener VV (2011) Highlight induced disassembly of photosystem II supercomplexes in Arabidopsis requires STN7-dependent phosphorylation of CP29. PloS ONE 6:e24565
43. Fristedt R, Willig A, Granath P, Crèvecoeur M, Rochaix J-D, Vener AV (2009) Phosphorylation of photosystem II controls functional macroscopic folding of photosynthetic membranes in Arabidopsis. Plant Cell 21:3950–3964
44. Fristedt R, Granath P, Vener AV (2010) A protein phosphorylation threshold for functional stacking of plant photosynthetic membranes. PLOS ONE 5:e10963
45. Garber MP, Steponkus PL (1976) Alterations in chloroplast thylakoids during cold acclimation. Plant Physiol 57:681–686
46. Goodchild DJ, Björkman O, Pyliotis NA (1972) Chloroplast ultrastructure, leaf anatomy, and soluble protein in rainforest species. Carnegie Inst Wash Yearb 71:102–107
47. Goral TK, Johnson MP, Kirchhoff H, Ruban AV, Mullineaux CW (2010) Visualizing the diffusion of chlorophyll-proteins in higher plant thylakoid membranes: effects of photoinhibition and protein phosphorylation. Plant J 62:948–959

48. Goral TK, Johnson MP, Duffy CDP, Brain APR, Ruban AV (2012) Light-harvesting antenna composition controls the macrostructure and dynamics of thylakoid membranes in Arabidopsis. Plant J 69:289–301
49. Haferkamp S, Haase W, Pascal AA, van Amerongen H, Kirchhoff H (2010) Efficient light harvesting by photosystem II requires an optimized protein packing density in grana thylakoids. J Biol Chem 285:17020–17028
50. Jahns P, Latowski D, Strzalka K (2009) Mechanism and regulation of the violaxanthin cycle: the role of antenna proteins and membrane lipids. Biochim Biophys Acta 1787:3–14
51. Johnson MP, Goral TK, Duffy CD, Brain AP, Mullineaux CW, Ruban AV (2011) Photoprotective energy dissipation involves the reorganization of photosystem II light-harvesting complexes in the grana membranes of spinach chloroplasts. Plant Cell 23:1468–1479
52. Joliot P, Joliot A (1964) Etudes cinetique de la reaction photochemique liberant l'oxygene au cours de la photosynthese. C R Acad Sci Paris 258:4622–4625
53. Joliot P, Verméglio A, Joliot A (1993) Supramolecular membrane protein assemblies in photosynthesis and respiration. Biochim Biophys Acta 1141:151–174
54. Junge W, Sielaff H, Engelbrecht S (2009) Torque generation and elastic power transmission in the rotary FoF1-ATPase. Nature 459:364–370
55. Kaftan D, Brumfeld V, Nevo R, Scherz A, Reich Z (2002) From chloroplasts to photosystems: *in situ* scanning force microscopy on intact thylakoid membranes. EMBO J 21:6146–6153
56. Kargul J, Barber J (2008) Photosynthetic acclimation: structural reorganization of light harvesting-antenna-role of redox-dependent phosphorylation of major and minor chlorophyll a/b binding proteins. FEBS Lett 275:1056–1068
57. Kereïche S, Kiss AZ, Kouril R, Boekema E, Horton P (2010) The PsbS protein controls the macro-organization of photosystem II complexes in the grana membranes of higher plant chloroplasts. FEBS Lett 584:754–764
58. Khatoon M, Inagawa K, Pospísil P, Yamashita A, Yoshioka M, Lundin B, Horie J, Morita N, Jajoo A, Yamamoto Y, Yamamoto Y (2009) Quality control of photosystem II. J Biol Chem 284:2543–2552
59. Kirchhoff H (2008) Molecular crowding and order in photosynthetic membranes. Trends Plant Sci 13:201–207
60. Kirchhoff H, Horstmann S, Weis E (2000) Control of the photosynthetic electron transport by PQ diffusion microdomains in thylakoids of higher plants. Biochim Biophys Acta 1459:148–168
61. Kirchhoff H, Mukherjee U, Galla HJ (2002) The molecular architecture of the thylakoid membrane: the lipidic diffusion space for plastoquinone. Biochemistry 41:4872–4882
62. Kirchhoff H, Tremmel I, Haase W, Kubitscheck U (2004a) Supramolecular photosystem II organization in grana thylakoid membranes: evidence for a structured arrangement. Biochemistry 43:9204–9213
63. Kirchhoff H, Borinski M, Lenhert S, Chi L, Büchel C (2004b) Transversal and lateral exciton energy transfer in grana thylakoids of spinach. Biochemistry 43:14508–14516
64. Kirchhoff H, Haase W, Wegner S, Danielsson R, Ackermann R, Albertsson PA (2007a) Low-light-induced formation of semicrystalline photosystem II arrays in higher plant chloroplasts. Biochemistry 46:11169–11176
65. Kirchhoff H, Haase W, Haferkamp S, Schott T, Borinski T, Kubitscheck U, Rögner M (2007b) Structural and functional self-organization of photosystem II in grana thylakoids. Biochim Biophys Acta 1167:1180–1188
66. Kirchhoff H, Lenhert S, Büchel C, Chi L, Nield J (2008a) Probing the organization of photosystem II in photosynthetic membranes by atomic force microscopy. Biochemistry 47:431–440
67. Kirchhoff H, Haferkamp S, Allen JF, Epstein, D Mullineaux CW (2008b) Significance of macromolecular crowding for protein diffusion in thylakoid membranes of chloroplasts. Plant Physiol 146:1571–1578
68. Kirchhoff H, Hall C, Wood M, Herbstová M, Tsabari O, Nevo R, Charuvi D, Shimoni E, Reich Z (2011) Dynamic control of protein diffusion within the granal thylakoid lumen. Proc Natl Acad Sci U S A 108:20248–20253

69. Kohzuma K, Dal Bosco C, Kanazawa A, Dhingdra A, Nitschke W, Meurer J, Kramer DM (2012) Thioredoxin-insensitive plastid ATP synthase that performs moonlight functions. Proc Nat Acad Sci U S A 109:2393–3298
70. Kouřil R, Oostergetel GT, Boekema EJ (2011) Fine structure of grana thylakoid membrane organization using cryo electron tomography. Biochim Biophys Acta 1807:368–374
71. Kramer DM, Johnson GN, Kiirats O, Edwards GE (2004) New fluorescence parameters for the determination of QA redox status and excitation energy fluxes. Photosynth Res 79:209–218
72. Kyle DJ, Staehelin LA, Arntzen CJ (1983) Lateral mobility of the light-harvesting complex in chloroplast membranes controls excitation energy distribution in higher plants. Arch Biochem Biophys 222:527–541
73. Kyle DJ, Ohad I, Arntzen CJ (1984) Membrane protein damage and repair: selective loss of quinone-protein function in chloroplast membranes. Proc Natl Acad Sci U S A 181:4070–4074
74. Lavergne J, Joliot P (1996) Dissipation in bioenergetic electron transfer chains. Photosynth Res 48:127–138
75. Lemeille S, Rochaix JD (2010) State transitions at the crossroad of thylakoid signaling pathways. Photosynth Res 106:33–46
76. Li G, Knowles PF, Murphy DJ, Nishida I, Marsh D (1989) Spin-label ESR studies of lipid-protein interactions in thylakoid membranes. Biochemistry 28:7446–7452
77. Li G, Knowles PF, Murphy DJ, Marsh D (1990) Lipid-protein interactions in stacked and destacked thylakoids membranes and the influence of phosphorylation and illumination. Spin label ESR studies. Biochim Biophys Acta 1024:278–284
78. Marr KM, McFeeters RL, Lyon MK (1996) Isolation and structural anaylsis of two-dimensional crystals of photosystem II from Hordeum viridis zb. J Struct Biol 117:86–98
79. Melis A (1999) Photosystem-II damage and repair cycle in chloroplasts: what modulates the rate of photodamage *in vivo*? Trends Plant Sci 4:130–135
80. Miller KR, Miller GJ, McIntyre KR (1976) The light-harvesting chlorophyll-protein complex of photosystem II. J Cell Biol 71:624–638
81. Mullet JE (1983) The amino acid sequence of the polypeptide segment which regulates membrane adhesion (grana stacking) in chloroplasts. J Biol Chem 258:9941–9948
82. Mullineaux CW (2005) Function and evolution of grana. Trends Plant Sci 10:521–525
83. Mulo P, Sirpio S, Suorsa M, Aro EM (2008) Auxiliary proteins involved in the assembly and sustenance of photosystem II. Photosynth Res 98:489–501
84. Murakami S, Packer S (1970) Protonation and chloroplast membrane structure. J Cell Biol 47:332–351
85. Murphy DJ (1986) The molecular organisation of the photosynthetic membranes of higher plants. Biochim Biophys Acta 864:33–94
86. Mustárdy L (1996) Development of thylakoid membrane stacking. In: Ort DR (ed) Oxygenic photosynthesis: the light reactions. Kluwer Academic, Dodrecht, pp 59–68
87. Mustárdy L, Buttle K, Steinbach G, Garab G (2008) The three-dimensional network of the thylakoid membranes in plants: quasihelical model of the granum-stroma assembly. Plant Cell 20:2552–2557
88. Nelson N, Yocum CF (2006) Structure and function of photosystem I and II. Annu Rev Plant Biol 57:521–565
89. Nevo R, Chuartzman SG, Tsabari O, Reich Z (2009) Architecture and plasticity of thylakoid membrane networks. In: Wada H, Murata N (eds.) Lipids in photosynthesis. Springer, New York, pp 295–328
90. Nevo R, Charuvi D, Tsabari O, Reich Z (2012) Composition, architecture and dynamics of the photosynthetic apparatus in higher plants. Plant J (in press)
91. Nield J, Barber J (2006) Refinement of the structural model for the photosystem II supercomplex of higher plants. Biochim Biophys Acta 1757:353–361
92. Ohad I, Kyle DJ, Arntzen CJ (1984) Membrane protein damage and repair: removal and replacement of inactivated 32-kilodalton polypeptide in chloroplast membranes. J Cell Biol 99:481–485

93. Paillotin G, Dobek A, Breton J, Leibl W, Trissl H-W (1993) Why does the light-gradient photovoltage from photosynthetic organelles show a wavelength-dependent polarity? Biophys J 65:379–385
94. Park RB, Biggins J (1964) Quantasome: size and composition. Science 144:1009–1011
95. Park RB, Pfeifhofer AO (1969) Ultrastructural observations on deep-etched thylakoids. J Cell Sci 5:299–311
96. Pesaresi P, Pribil M, Wunder T, Leister D (2011) Dynamics of reversible protein phosphorylation in thylakoids of flowering plants: the roles of STN7, STN8 and TAP38. Biochim Biophys Acta 1807:887–896
97. Rintamäki E, Martinsuo P, Pursiheimo S, Aro E-M (2000) Cooperative regulation of light-harvesting complex II phosphorylation via the plastoquinol and ferrodoxin-thioredoxin system in chloroplasts. Proc Natl Acad Sci U S A 97:11644–11649
98. Saxton MJ (1982) Lateral diffusion in an archipelago. Effects of impermeable patches on diffusion in a cell membrane. Biophys J 39:165–173
99. Saxton MJ (1987) Lateral diffusion in an archipelago. The effect of mobile obstacles. Biophys J 52:989–997
100. Saxton MJ (1989) Lateral diffusion in an archipelago. Distance dependence of the diffusion coefficient. Biophys J 56:615–622
101. Schönknecht G, Hedrich R, Junge W, Raschke K (1988) A voltage-dependent choride channel in the photosynthetic membrane of higher plants. Nature 336:589–592
102. Schubert M, Petersson UA, Haase BJ, Funk C, Schröder WP, Kieselbach T (2002) Proteome map of the chloroplast lumen of Arabidopsis thaliana. J Biol Chem 277:8354–8365
103. Sculley MJ, Duniec JT, Thorne SW, Chow WS, Boardman NK (1980) The stacking of chloroplast thylakoids-Quantitative analysis of the balance of forces between thylakoid membranes of chloroplasts, and the role of divalent cations. Arch Biochem Biophys 201: 339–346
104. Semenova GA (1995) Particle regularity on thylakoid fracture faces is influenced by storage conditions. Can J Bot 73:1676–1682
105. Shimoni E, Rav-Hon O, Ohad I, Brumfeld V, Reich Z (2005) Three-dimensional organization of higher-plant chloroplast thylakoid membranes revealed by electron tomography. Plant Cell 17:2580–2586
106. Simpson DJ (1978) Freeze-fracture studies on barley plastid membranes II. Wild-type chloroplasts. Carlsberg Res Commun 43:365–389
107. Spetea C, Schoefs B (2010) Solute transporter in plant thylakoid membranes. Commun Integr Biol 3:122–129
108. Staehelin LA (1976) Reversible particle movements associated with unstacking and restacking of chloroplast membranes *in vitro*. J Cell Biol 71:136–158
109. Staehelin LA (1986) Chloroplast structure and supramolecular organization of photosynthetic membranes. In: Staehelin LA, Arntzen CJ (eds) Photosynthesis III: photosynthetic membranes and light-harvesting systems. Springer, Berlin, pp 1–84
110. Staehelin LA, Arntzen CJ (1983) Regulation of chloroplast membrane function: protein phosphorylation changes the spatial organization of membrane components. J Cell Biol 97:1327–1337
111. Staehelin LA, van der Staay GWM (1996) Structure, composition, functional organization and dynamic properties of thylakoid membranes. In: Ort DA, Yocum CF (eds) Oxygenic photosynthesis: the light reactions. Kluwer, Netherlands, pp 11–30
112. Strauss M, Hofhaus G, Schröder RR, Kühlbrandt W (2009) Dimer ribbons of ATP synthase shape the inner mitochondrial membrane. EMBO J 27:1154–1160
113. Suno R, Niwa H, Tsuchiya D, Zhang X, Yoshida M, Morikawa K (2006) Structure of the whole cytosolic region of ATP-dependent protease FtsH. Mol Cell 22:575–585
114. Sznee K, Dekker JP, Dame RT, van Roon H, Wuite GJ, Frese RN (2012) Jumping mode atomic force microscopy on grana membranes from spinach. J Biol Chem 286:39164–39171
115. Takahashi T, Inoue-Kashino N, Ozawa S, Takahashi Y, Kashino Y, Satoh K (2009) Photosystem II complex *in vivo* is a monomer. J Biol Chem 284:15598–15606

116. Tikkanen M, Aro EM (2011) Thylakoid protein phosphorylation in dynamic regulation of photosystem II in higher plants. Biochim Biophys Acta 1817:232–238
117. Tremmel I, Kirchhoff H, Weis E, Farquhar GD (2003) Dependence of the plastoquinone diffusion coefficient on the shape, size, density of integral thylakoid proteins. Biochim Biophys Acta 1607:97–109
118. Trissl HW, Breton J, Deprez J, Leibl W (1987) Primary electrogenic reactions of photosystem II as probed by the light gradient method. Biochim Biophys Acta 893:305–319
119. Trissl H-W, Hecks B, Wulf K (1993) Invariable trapping times in photosystem I upon excitation of minor long-wavelength absorbing pigments. Photochem Photobiol 57:552–568
120. Tsvetkova NM, Apostolova EL, Brain APR, Williams WP, Quinn PJ (1995) Factors influencing PSII particle array formation in Arabidopsis thaliana chloroplasts and the relationship of such arrays to the thermostability of PSII. Biochim Biophys Acta 1228:201–210
121. Umena Y, Kawakami K, Shen J-R, Kamiya N (2011) Crystal structure of oxygen-evolving photosystem II at a resolution of 1.9 Å. Nature 473:55–61
122. van Roon H, van Breemen JF, de Weerd FL, Dekker JP, Boekema EJ (2000) Solubilization of green plant thylakoid membranes with n-dodecyl-alpha, D-maltoside. Implications for the structural organization of the potosystem II, photosystem I, ATPase sunthase and cytochrome b6f complexes. Photosynth Res 64:155–166
123. Yoshioka M, Yamamoto Y (2011) Quality control of photosystem II: where and how does the degradation of the D1 protein by FtsH proteases start under light stress?–Facts and hypotheses. J Photochem Photobiol B 104:229–235
124. Yoshioka M, Nakayama Y, Yoshida M, Ohashi K, Morita N, Kobayashi H, Yamamoto Y (2010) Quality control of photosystem II: FtsH hexamers are localized near photosystem II at grana for swift repair of damage. J Biol Chem 285:41972–41981

Chapter 6
Plastid Division

Jodi Maple-Grødem and Cécile Raynaud

Abstract Plastids evolved from free-living cyanobacteria that were engulfed by a host cell. One critical element for the stable establishment of endosymbiosis was the acquisition of the ability of the endosymbiont to multiply inside the host cell. In all plants, plastids divide by binary fission via the assembly and subsequent constriction of a multiproteic scaffold called the plastid division ring. This highly complex structure associates proteins that have been retained from the bacterial ancestor, such as FtsZ that forms a ring inside the plastid, with proteins brought by the host cell, that form a ring outside the plastid. Over the past 20 years, dramatic progress has been made in our understanding of the mechanisms underlying plastid division, thanks to the combination of various approaches including direct and reverse genetics, electron microscopy and biochemistry. In this review we describe in detail the different steps of plastid division, from the choice of the division site to the sequential assembly of all constituents of the plastid division ring. We also discuss the current knowledge of the regulation of this process, which is still in its infancy but raises fascinating questions for future research.

Keywords Plastid division ring · FtsZ · Min

Abbreviations

CGA1	Cytokinin-responsive GATA1
CJD1	Chloroplast J-like domain 1
CLMP1	Clumped chloroplast 1
CLS-8	Crinkled leaves 8
CpDNA	Chloroplast DNA
Cpn60	Chaperonin 60
CRL	Crumpled leaf
DGDG	Digalactosyldiacylglycerol
DRP5B	Dynamin-related protein 5B

C. Raynaud (✉)
Institut de Biologie des Plantes, Paris-Sud University, 91405 Orsay, France
e-mail: cecile.raynaud@u-psud.fr

J. Maple-Grødem
Centre for Organelle Research, University of Stavanger, 4036 Stavanger, Norway

Centre for Movement Disorders, Stavanger University Hospital, 4068 Stavanger, Norway

S.M. Theg, F.-A. Wollman (eds.), *Plastid Biology*, Advances in Plant Biology 5,
DOI 10.1007/978-1-4939-1136-3_6,

FP	Fluorescent protein
FRP	FtsZ regulatory proteins
FtsZ	Filamenting temperature-sensitive mutant Z
GED	GTPase effector domain
HP1	High pigment 1
IMS	Inter membrane space
MCD1	Multiple chloroplast division site 1
MD	Middle domain
MGDG	Monogalactosyldiacylglycerol
Min	Mini-cell
MORN	Membrane Occupation and Recognition Nexus
MSC	Mechano-sensitive channel
nDNA	Nuclear DNA
PARC6	Paralog of ARC6
PD	Plastid dividing
PDR1	Plastid dividing ring 1
PH	Pleckstrin homology
RNR	Ribonucleotide reductase
TEM	Transmission electron microscopy

6.1 Introduction

Although the hypothesis that plastid and mitochondria derived from a free living prokaryote was first formulated at the end of the nineteenth century, the question of the biogenesis of plastids in plant cells was still debated in the 1960s: some authors postulated that plastids could arise from small submicroscopic vesicles called proplastid precursors, even though a wealth of microscopic observations provided evidence for plastid division by binary fission [118]. It is now clear that plant cells cannot generate plastids *de novo*, and that binary fission is the main process allowing the multiplication of these organelles, although budding has been reported in ripening tomato fruits [30]. When observed under the microscope, the chloroplast division process can be described as follows: initially spherical chloroplasts become more ovoid, and constriction becomes visible at chloroplast midpoint. As constriction proceeds chloroplasts become dumbbell shaped and finally deeply constricted with a narrow neck between the two future organelles. At the final stage of division the two chloroplasts can twist around this neck and finally become completely separated. Observation of dividing chloroplasts was reported in the nineteenth century, but only in the last 20 years have we come to understand how plastid division actually takes place.

The study of plastid division has highly benefited from three types of approach: transmission electron microscopy (TEM), forward genetics and reverse genetics. TEM images of dividing *Cyanidium caldarium* cells revealed the existence of an electron dense structure at the division site of chloroplasts [90]. This observation was subsequently generalized to all plants, demonstrating that a complex structure

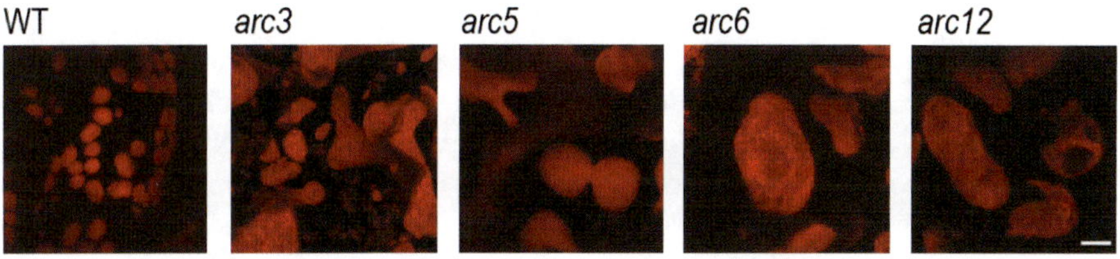

Fig. 6.1 Leaf mesophyll cells from fully expanded leaves of wild-type Arabidopsis (WT) and *arc* mutants *arc3*, *arc5*, *arc6* and *arc12*. Chlorophyll autofluorescence was detected, and extended focus images were generated. Scale bar = 10 μm

is responsible for plastid division. Genetic studies were then required to identify the constituents of this structure. In the early 1990s, K. Pyke and R. Leech isolated several mutants specifically affected for plastid division which were called *arc* for Accumulation and Replication of Chloroplasts [122, 123]. A total of 12 *arc* mutants have been described, most of which show a reduction of chloroplast number compensated by an increase in chloroplast size (Fig. 6.1). Because positional cloning is a long-term approach, especially when dealing with mutants whose phenotype must be studied under the microscope, more than 10 years passed before the first *ARC* gene (*ARC5*) was identified [40]. In the meantime, complete sequencing of the Arabidopsis genome allowed K. Osteryoung and E. Vierling to discover homologues of the *FtsZ* gene [113]; the bacterial FtsZ protein is distantly related to tubulin and can polymerize in a GTP-dependent manner. Its ability to form contractile rings allows cell division by binary fission. Since then, several studies have demonstrated that plastid division does require homologues of FtsZ and involves several other proteins of prokaryotic origin, but also proteins that were encoded by the genome of the host cell. The resulting machinery is extremely complex, and how this multi-proteic scaffold functions at the molecular level is only beginning to be elucidated.

In this chapter, we will focus mainly on chloroplast division in higher plants, because it is the best described model in many respects, but it is generally assumed that the division mechanisms are common to all plastid types. Over the past 15 years, our knowledge of the mechanisms allowing plastid division has made considerable progress. By contrast, very little is known at present regarding the regulation of plastid division. However, a few recent studies provide insights into this regulation and can now allow new interpretation of older reports regarding the factors that influence plastid division. In the last paragraph of this chapter we will therefore try to summarize these reports, illustrating how the recent discoveries about plastid division can allow a new understanding of early microscopic analyses.

6.2 Placement of the Plastid Division Site

Our knowledge of bacterial cell division has greatly contributed to the understanding of all aspects of chloroplast division, including division ring placement. To date, the best characterized model is *Escherichia coli* in which two distinct mechanisms allow the division machinery to assemble at midcell: the nucleoid occlusion system prevents FtsZ ring formation in the close vicinity of chromosomes (i.e. all sites except midcell and cell poles) and the Min system avoids FtsZ polymerization at cell poles [24]. In the absence of Min proteins, division frequently takes place at cell poles, giving rise to minicells. MinC is the inhibitor of FtsZ polymerization, while MinD and MinE regulate its location in the cell and thereby its activity. To affect FtsZ polymerization, MinC needs to be attached to the membrane, and this association requires MinD, an ATPase which forms dimers and engages its C-terminus into the membrane upon ATP binding. After ATP hydrolysis and release of ADP, MinD dissociates from the membrane. MinCD complexes oscillate from pole to pole, owing to the topological factor MinE. MinE forms a ring that appears to cap the Min polar zone and prevent it from extending past midcell: MinE stimulates the ATPase activity of MinD, thus promoting its dissociation from the membrane. The MinE ring then moves towards the pole, until most proteins are released from the membrane and assemble again at the opposite pole [11]. In other rod-shaped bacteria such as *Bacillus substilis*, the Min system is conserved, but MinCD proteins and their topology factors DivIVA and MinJ do not appear to oscillate from pole to pole. They rather seem to accumulate at the poles at early steps of division and to be dynamically localized to the mature division apparatus, thereby labeling the new cell poles after division is complete [11]. The Min system is also conserved in cyanobacteria, which are the closest relatives of chloroplasts, but their dynamic behavior during the division process has not been elucidated [85].

6.2.1 Many Proteins are Involved in the Choice of the Division Site

Homologues of MinD and MinE have been identified in Arabidopsis and appear to have antagonistic functions towards FtsZ polymerization. Mesophyll cells of AtMinE over-expressers and *atminD/arc11* mutants contain chloroplasts of heterogeneous size, showing aberrant Z-ring positioning and multiple Z-rings in one chloroplast [20, 34, 55, 77]. By contrast, AtMinD overexpressers and *atmine/arc12* show inhibition of plastid division [45]. These opposite effects of AtMinD and AtMinE on chloroplast division are due to their impact on FtsZ polymerization: FtsZ proteins form multiple rings in *atminD* mutants and AtMinE overexpressers whereas they form short filaments in *atminE* mutants and AtMinD overexpressers, indicating that AtMinD inhibits FtsZ polymerization whereas AtMinE promotes it [37, 45, 147]. Proper balance between AtMinD and AtMinE activities is therefore required for Z-ring formation at midplastid [35]. Molecular data further supports the notion that the Min system operates in chloroplasts in a similar way as in bacterial cells. As observed for their bacterial homologues, AtMinD and AtMinE can form homo or heterodimers [79]. Furthermore, the ATPase activity of AtMinD seems to be important for its proper localization in chloroplasts, and AtMinE, like its bacterial counterpart, stimulates this activity [4]. Finally, neither AtMinD nor AtMinE seem to interact with FtsZ proteins, suggesting that as observed in bacteria, a third component is required to regulate Z-ring positioning.

Although MinC is conserved in cyanobacteria [85] and the chloroplast division machinery is sensitive to overexpression of the *E. coli* MinC protein [143], plant genomes lack *MinC* homologues. ARC3 has been suggested to replace MinC functionally. The phenotype of *arc3* mutants and over-expressers is similar to *arc11* and AtMinD over-expressers respectively [80, 134], providing evidence for an inhibitory function of ARC3 on Z-ring assembly away from the mid-plastid point. ARC3 interacts with AtMinE and AtMinD via its central domain and with AtFtsZ1 via its N-terminal FtsZ-like domain, although this domain lacks the tubulin signature and amino acids crucial for GTP hydrolysis [80]. Taken together with the division site misplacement in the *arc3* mutant, the presence of ARC3 in an FtsZ/Min complex suggests that ARC3 plays a central role in division site placement.

Two more Arabidopsis proteins participate in division site placement, bringing further complexity to the system compared with bacterial cell division: MCD1 [100] and PARC6/CDP1 [46, 155]. PARC6 is a paralogue of the ARC6 protein which destabilizes FtsZ ring (see below). PARC6 is conserved from algae to land plants whereas MCD1 is specific to land plants. Both proteins appear to be negative regulators of Z-ring formation, but how they achieve this function is not completely clear. In *mcd1* mutants, FtsZ proteins form long filaments, and overexpression of AtMinD has little effect on FtsZ polymerization. In this background the AtMinD protein accumulates at wild-type level, but shows diffuse localization in chloroplasts; taken together, these results suggest that MCD1 may enhance the inhibitory effect of AtMinD on FtsZ polymerization by modulating its intra-plastidial localization [100]. PARC6/CDP1 negatively regulates FtsZ polymerization, possibly by interacting with ARC3 [46, 155]. Interestingly, this interaction appears to be mediated by the C-terminal MORN domain of ARC3 [46]. Mesophyll cells of *parc6* mutants contain chloroplasts of heterogeneous sizes suggesting that PARC6 participates in the choice of the division site, but also constricted chloroplasts blocked at a later stage of division, pointing at additional roles of PARC6 ([46], see below).

Finally, mechano-sensitive channel (MSC) proteins appear to be involved in division site positioning [49]. A systematic functional analysis of MSC of the S family (Msc-S Like) led to the observation that MSL2 and MSL3 play redundant roles in chloroplast division: *msl2* and *msl3* mutants are identical to the wild-type whereas *msl2msl3* double mutants are variegated and show abnormal chloroplast size and shape [49]. Further genetic analyses revealed that MSL2 and 3 proteins function as inhibitors of Z-ring formation in a Min/ARC3 dependent manner, although their molecular mechanism of action is not clear [149]. Strikingly, in *B. subtilis*, the MinE functional equivalent DivIVA can recognize curved membrane regions and has therefore been suggested to have the intrinsic capacity to bind and accumulate at negatively curved membranes [11]. MSL proteins, by regulating ion fluxes to or from the chloroplast may affect membrane curvature and in this way influence the polarity of the organelle.

6.2.2 How Do Proteins Involved in the Choice of the Division Site Work Together?

Although many players of the Z-ring placement have been identified in the past 10 years, our understanding of the underlying mechanism remains very incomplete and the molecular functions of all the identified players are not fully elucidated yet. Based on interaction data (summarized in Fig. 6.2), and on the effect of these proteins on FtsZ polymerization, AtMinD, AtMinE and ARC3 may be the functional equivalent of the bacterial MinCDE complex. However, it is not clear whether AtMinD engages the chloroplast envelope upon ATP binding and dimerization, like its bacterial counterpart. By contrast, MCD1 is a membrane spanning protein which seems to be required for proper AtMinD localization and function [100]. It is therefore tempting to speculate that MCD1 may be required to attach the MinDE/ARC3 complex to the chloroplast inner envelope. PARC6 may also play a similar role since it is a transmembrane protein and can interact with ARC3 [46]. All these proteins may hence form a complex attached to the chloroplast inner membrane as pictured on Fig. 6.2a, but it is not clear whether a single complex harboring all proteins or several sub-complexes differing by their composition exist in chloroplasts.

Even more elusive is how and where in the chloroplast all these proteins function. This gap in our knowledge can be attributed to the lack of live imaging of proteins regulating Z-ring assembly, and to conflicting results regarding their localization in dividing chloroplasts. Depending on the technique used for localization experiments, two types of distributions have been reported for division site placement machinery. According to fluorescent protein (FP) fusions and BiFC assays, AtMinD and AtMinE co-localize into large dots at chloroplast poles [77, 79]. The same localization was reported for MSL2 and MSL3 [49]. PARC6 [46] was also found in pole localized dots, but also at the division site, where it seems to function at later steps of the division process. The distribution of ARC3 seems more complex since it was found to co-localize with FtsZ1 in ring-like structures and to form discrete dots containing MinD [80], or multiple dots in which it interacts with PARC6/CDP1 [155]. By contrast, immuno-fluorescence experiments suggest that MCD1 [100] and AtMinD [37] are localized in ring-like structures at the chloroplast midpoint as well as punctuate structures dispersed throughout the envelope, whereas ARC3 is localized at the division ring during early and middle steps of chloroplast division [134]. Several hypotheses can explain these discrepancies. In some of these studies, localization of FP-fusion proteins was performed in transient expression assays using tobacco leaves in which chloroplast division may have stopped, and the localization of these proteins might therefore be altered. Furthermore, Fujiwara et al reported that the AtMinD-YFP fusion cannot complement the *arc11/minD* mutant [36], suggesting that the reported localization of the fusion protein may not reflect the proper distribution of AtMinD. Reciprocally, some of the observed immunofluorescence signals are very faint, and the inferred distribution of the target protein may not be accurate.

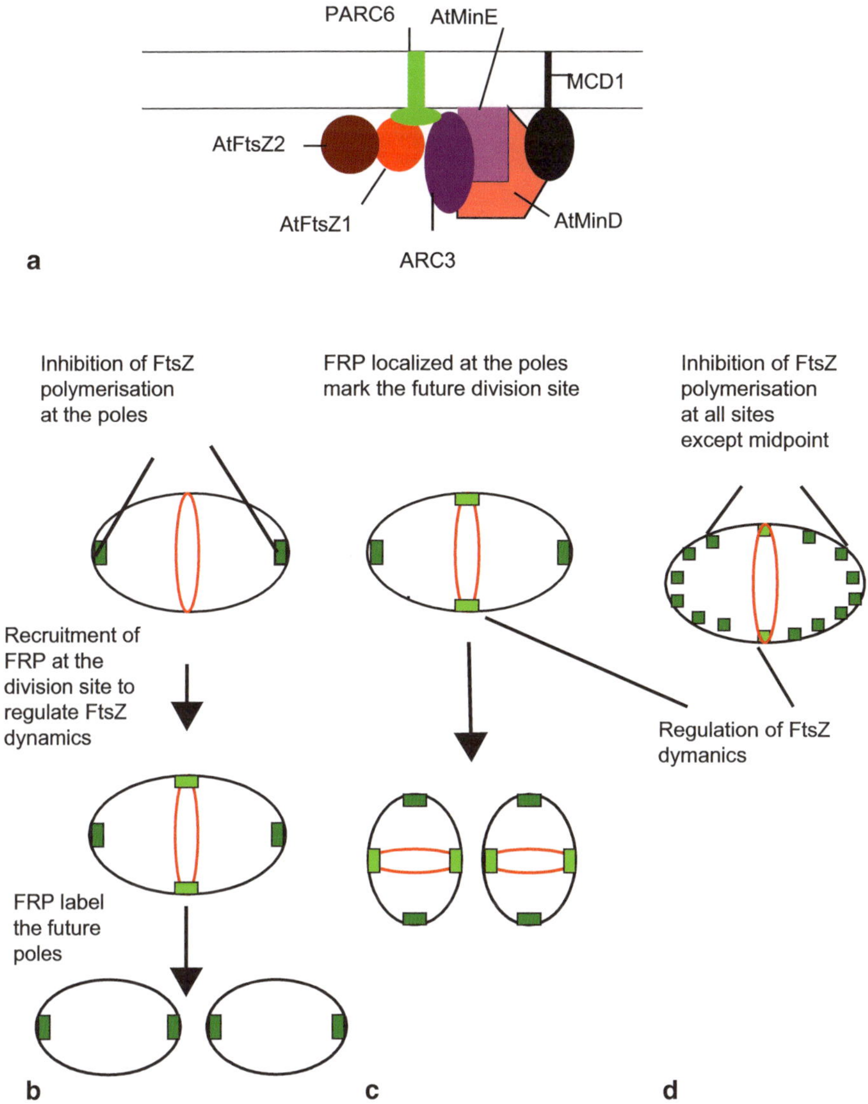

Fig. 6.2 Regulation of Z-ring positioning during chloroplast division. **a** The Z-ring regulatory complex. This picture summarizes the reported interactions between proteins and their membrane association. It is not clear however whether such a complex exists in chloroplasts, or several subcomplexes differing by their composition co-exist. **b–d** Three alternative models for Z-ring positioning based on localization data. *Dark green* and *light green* boxes represent complexes of FtsZ regulatory proteins. *Dark green* boxes are for a combination of proteins that completely inhibit FtsZ polymerization, whereas *light green* boxes represent a complex capable to modulate FtsZ dynamics at the division site. **b** FtsZ regulatory proteins (FRP) operate predominantly at chloroplast poles to inhibit ectopic Z-ring formation. After Z-ring assembly (*red* ring), a different set of FRP may be recruited at the division site and regulate FtsZ dynamics at the division site. FRP could also be required to label the newly formed poles, thereby maintaining chloroplast polarity. **c** Alternatively, the polar localization of FRP may mark the site of the future division site. **d** FRP inhibit Z-ring formation at all sites except midpoint. At the division site, another regulatory complex differing from the previous one by its composition regulates the dynamics of the Z-ring

Based on these localization experiments, several models can be proposed for the choice of chloroplast division site (Fig. 6.2b–d). All these models assume that the Z-ring regulatory complex is present at least transiently at the division site, implying that a minimum of two complexes differing by their proteic composition must exist in chloroplasts, one of them being an inhibitor of Z-ring formation (Fig. 6.2b–d; dark green boxes), and the other a regulator of FtsZ dynamics (Fig. 6.2b–d; light green boxes). One possibility would be that inhibitors of FtsZ polymerization act predominantly at the poles, as suggested by FP-fusion experiments (Fig. 6.2b, c). Recognition of the poles by this inhibitory complex may involve membrane curvature and require MSL proteins. In addition, accumulation of proteins involved in the choice of the division site at the poles may reflect their role in the regulation of chloroplast polarity [44]. Indeed, in *B. subtilis*, Min proteins are associated to the cell poles, and are recruited at the division site to mark the newly formed poles after division (Fig. 6.2b). Another possibility would be that Min-containing dots mark the site of the future division (Fig. 6.2c). Indeed, the cyanobacteria *Synechocystis sp* divides in alternating perpendicular planes [85]. Finally, as suggested by immuno-fluorescence analysis, inhibitors of FtsZ polymerisation may be distributed both in discrete dots dispersed throughout the inner membrane and at the division site (Fig. 6.2d). If this were the case, the discrete dots would represent inhibitory complexes preventing ectopic Z-ring formation, whereas the inhibitory action of AtMinD, ARC3, MCD1 and PARC6 would somehow be alleviated by AtMinE or another yet unidentified factor at the division site (Fig. 6.2d). Furthermore, the presence of ARC3 and PARC6 at the division site may be required to modulate the dynamics of AtFtsZ polymerization.

Models illustrated by Fig. 6.2c, d have the advantage that they may better account for the choice of the division site. Indeed, unlike bacteria in which the nucleoid occlusion system prevents the formation of the division apparatus everywhere except at midcell and at the poles, chloroplasts contain numerous small nucleoids that are distributed throughout the stroma [67]. Therefore, preventing Z-ring assembly at the poles may not be sufficient to ensure symmetric division of plastids. To clarify the model for Z-ring positioning, time-course analysis of the localization of all the above mentioned proteins during chloroplast division is absolutely indispensable.

6.3 Assembly and Dynamics of the FtsZ Ring

FtsZ is a stable protein and in bacteria the intracellular concentration changes little during cell division, therefore, rapid responses required for regulation of the Z-ring occur through control over FtsZ assembly. Recent evidence indicates that regulation of Z-ring assembly/maintenance in plastids may be achieved through the opposing effects of ARC6 and PARC6 and also through the action of the plastid chaperone system. Further complexity is added to the potential regulatory mechanisms in plants because at variance with prokaryotes, plant genomes encode several FtsZ isoforms.

6.3.1 Z-ring Assembly Requires Two Isoforms of FtsZ

In prokaryotes, FtsZ is usually encoded by a single gene even though exceptions have been reported [82]. By contrast, three homologues of FtsZ distributed in two families are present in Arabidopsis (AtFtsZ1, AtFtsZ2-1 and AtFtsZ2-2). Divergence between the two FtsZ families occurred early during evolution, probably before the divergence between Chlorophycean and Charophycean [138], and the number of isoforms in each family seems to vary from one plant to another. For example, only one FtsZ1 isoform can be found in *Arabidopsis thaliana* versus four in *Nicotiana tabacum*. AtFtsZ1 and AtFtsZ2 are assumed to play distinct role in chloroplast division because mutants cannot be complemented by overexpression of a member of the other family [132] and AtFtsZ1 and AtFtsZ2 differ by their ability to interact with other plastid division proteins: ARC3 interacts specifically with AtFtsZ1 [80] whereas ARC6 binds only to AtFtsZ2 ([147] see below). In Arabidopsis, *ftsz2-2* null mutants show only mild plastid division defects, and this phenotype can be complemented by expression of AtFtsZ2-1. Reciprocally, the more severe phenotype of *ftsz2-1* mutants can be rescued by AtFtsZ2-2 expression, suggesting that the two proteins are functionally redundant but differ by their expression level [132]. However, the two FtsZ2 isoforms may differ in their role in thylakoid development [59], see below).

AtFtsZ1 and AtFtsZ2-1 proteins co-localize and form rings at chloroplast midpoint (Fig. 6.3) [79, 86]. FtsZ proteins are structural homologues of tubulins, and can form both longitudinal and lateral interactions. Longitudinal interactions are stimulated by GTP binding and allow the formation of proto-filaments that consist of a head-to-tail linear polymer of FtsZ. However, unlike tubulin, FtsZ proto-filaments do not associate into microtubules but instead form bundles or sheets. Whether FtsZ1 and FtsZ2 proteins form distinct proto-filaments or hetero-multimeric ones is not clear. In chloroplasts, AtFtsZ1 and AtFtsZ2-1 can form curved filaments in the absence of the other protein, even if these structures are often disordered and more numerous than in the wild-type [146]. Purification of FtsZ complexes from Arabidopsis suggests that Z-rings contain both AtFtsZ1 and AtFtsZ2 at a 1:2 ratio [87], and the stoichiometry between the two proteins is probably important to their function because overexpression of either isoform severely inhibits chloroplast division [114]. Interestingly, AtFtsZ1 and AtFtsZ2-1 can form hetero-polymers *in vitro* and FtsZ1 promotes co-assembly, suggesting it can favor lateral interactions between proto-filaments [109]. Finally, phosphorylation of AtFtsZ2 proteins has recently been reported to somewhat affect their ability to interact with AtFtsZ1 and ARC6 and to form rings *in vivo*, providing evidence for an additional layer of complexity in the mechanisms governing Z-ring assembly [41].

6.3.2 Opposing Roles of ARC6 and PARC6 on Z-Ring Formation

One major difference between AtFtsZ1 and AtFtsZ2 is that only AtFtsZ2 possesses a C-terminal extension harboring a CORE domain allowing its interaction with ARC6 (Fig. 6.3) [79]. *arc6* mutants are the most severely affected of all *arc* mutants,

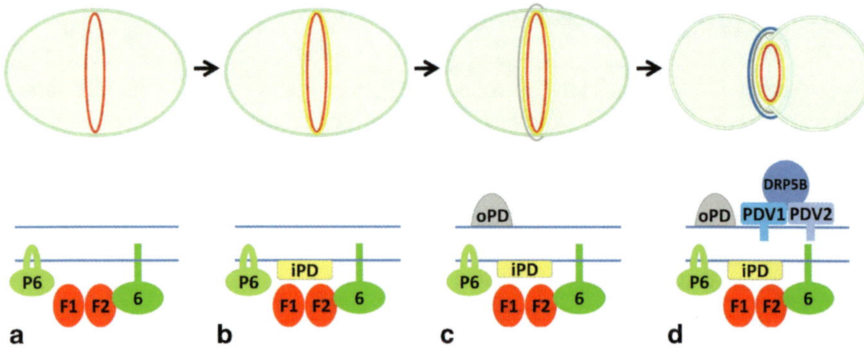

Fig. 6.3 The stromal and cytoplasmic division machineries of chloroplasts are assembled in a step-wise manner. **a** In Arabidopsis plastid division is initiated by the assembly of AtFtsZ1-1 (F1), AtFtsZ2-1 and AtFtsZ2-2 (F2) at the centre of chloroplasts to form the Z-ring. Z-ring assembly/ maintenance may be achieved through the opposing effects of ARC6 (6) and PARC6 (P6). **b** After completion of the Z-ring the inner PD ring (iPD) assembles on the inside of the Z-ring. This is followed by the middle PD ring in the inter membrane space, if present (not shown). **c** Subsequently the outer PD ring (oPD) forms on the cytosolic face of the chloroplast. In *C. merolae* the oPD is composed of PRD1 and polyglucan filaments but the composition in higher plants is unknown. **d** Finally dynamin is recruited to the division site. In Arabidopsis this is achieved by the combined action of PDV1 and PDV2. DRP5B is faintly detected on un-constricted but only forms ring-like structures after constriction has begun. The timings of the assembly of the rings are inferred from microscopy studies in both Arabidopsis and *C. merolae*

containing only 2 chloroplasts per mesophyll cell on average [129]. ARC6 is an integral membrane protein homologous to the Cyanobacterial protein Ftn2 [64, 147], which appears to affect FtsZ polymerization positively: FtsZ2 forms short filaments in *arc6* mutants whereas it forms long polymers in ARC6 over-expressing lines [147]. In this respect, ARC6 seems to function similarly to AtMinE. Nevertheless, chloroplast division is almost completely abolished by ARC6 overexpression or loss of function, and asymmetric division was never observed: ARC6 is therefore a component of the chloroplast division machinery and not involved in the choice of the division site. ARC6 has been reported to form a discontinuous ring at the chloroplast division site [79], and is likely to function as a Z-ring anchor, stabilizing FtsZ polymers and associating them with the inner envelope of the chloroplast, This is similar to the role played by Zip and FtsA in *E. coli* which interact with FtsZ through the CORE domain and play a role in maintaining/stabilizing the Z-ring [117].

Unlike ARC6, PARC6 lacks a functional J-domain and subsequent associated co-chaperone activity [46]. This observation has led to the suggestion that PARC6 arose in vascular plants as a new functional class of plastid division proteins and evidence suggests that PARC6 carries out at least two, potentially overlapping functions, both playing a role in Z-ring placement (see above) and influencing Z-ring assembly and/or maintenance. PARC6 appears to act antagonistically to ARC6 in its role in Z-ring stability; in *arc6* chloroplasts FtsZ filaments are short, indicating that ARC6 acts to stabilize the Z-ring, but in contrast to this in *parc6* chloroplasts FtsZ filaments are relatively long and appear as multiple rings or spirals, indicating that

PARC6 inhibits Z-ring assembly [46, 147]. These data point to ARC6 and PARC6 having opposing effects on the formation/stability of the Z-ring, despite the high levels of homology between them.

ARC6 is proposed to mediate its effects on the Z-ring by anchoring the ring to the membrane through a direct interaction of the N-terminal conserved domain of ARC6 with AtFtsZ2-1 and AtFtsZ2-2 [79, 132]. Given the high levels of similarity between the N-terminal conserved domains of ARC6 and PARC6, the possibility that PARC6 directly interacted with AtFtsZ1 or AtFtsZ2-1 was investigated by yeast two-hybrid assays, but no interactions were detected [46]. Unlike ARC6, PARC6 is predicted to contain two trans-membrane domains, which would also orientate the C-terminal domain of PARC6 in the stroma and it will be important to test if the C-terminal domain can interact with any of the plastid FtsZ proteins to determine if PARC6 can directly influence the Z-ring.

The N-terminal stromal domain of PARC6 has been found to interact with ARC3. This interaction is dependent on the C-terminal MORN domain of ARC3, which is the same domain of ARC3 which inhibits the interaction of ARC3 with AtFtsZ1 in yeast two-hybrid studies [46, 80]. If PARC6 exerts the destabilizing effects on the Z-ring through ARC3, the availability of the MORN domains could play a vital regulatory role. The interaction of PARC6 with ARC3 could sequester the MORN domain, thus enabling ARC3 to interact with the Z-ring and render the Z-ring unstable. In the absence of the PARC6-ARC3 interaction, the MORN domain could inhibit the interaction of ARC3 with AtFtsZ1, thus rendering the Z-ring more stable, as seen in the *parc6* mutant [46]. In an additional level of complexity the availability of ARC3 to interact with PARC6 could also be influenced by the other interacting partners of ARC3, including MinD, MinE, and MCD1 [80, 100], and much work will be required to unravel the role of PARC6 on the placement and assembly of the Z-ring.

6.3.3 Chaperone-Regulation of Z-Ring Assembly/Maintenance

Recent evidence has shed light on the possible role of molecular chaperones in the assembly and/or maintenance of the Z-ring. Chaperones assist in the folding or unfolding of proteins and the assembly or disassembly of protein complexes. The first chaperones shown to play a direct role in the process of plastid division were members of the chaperonin 60 (cpn60, hsp60) family in Arabidopsis [142]. The Cpn60 family of chaperonins is highly conserved and found in bacteria, mitochondria and plastids [51]. The best characterized members include the GroEL complex in *E. coli*, which together with the cochaperonin GroES forms a large (~1 MDa) chaperone complex termed GroE. In *E. coli* GroE function is required for the formation of the cell division machinery and cells with impaired GroE activity exhibit filamentous cell morphology, characteristic of cell division inhibition [33, 104]. The filamentous phenotypes are also observed in GroE-depleted *Streptococcus mutans* and *Caulobacter crescentus*, suggesting that GroE plays a universal role in cell division in bacteria [72, 141].

In Arabidopsis at least two members of the Cpn60 family, ptCpn60α1 and ptCpn60β1, are required for chloroplast division [142]. A null mutation in *ptCpn60α1* or the double *ptcpn60β1-1 ptcpn60β2* mutant results in albino and dwarf seedlings harboring small, colorless plastids, indicating Cpn60 function is vital for plant development [5, 142]. However a moderate reduction in the levels of activity of ptCpn60βB (caused by a premature stop codon in the ptCpn60β1) or of ptCpn60α1 (caused by a missense mutation, Ala-342-Val), results in the modest impairment of plastid division and mesophyll cells that harbor fewer but larger chloroplasts than those in wild type cells [142]. In *E. coli*, treatment of cells with penicillin has been shown to trigger the localization of GroE to possible division sites in an FtsZ dependent manner [104], although insults with other stresses did not replicate these results [16]. ptCpn60β is localized as speckles throughout the chloroplasts in Arabidopsis [142] and there is no evidence to date of localization to the division machinery. However In both the ptCpn60α1 and ptCpn60β1 mutants FtsZ forms long, disorganized filaments, similar to those observed in Arabidopsis plants in which ARC6 is over-expressed, suggesting that reduced ptCpn60 levels result in excessively stable FtsZ filaments.

How changes in the levels of Cpn60 function might alter FtsZ filament stability in plastids is not yet clear. In bacteria proteome-wide analysis of GroE substrates have revealed several cell division proteins as putative targets of GroE [15, 52, 62]. To date GroE has been shown to assist in the folding of the FtsE [33]. FtsE interacts with FtsZ and works together with FtsX to support septal ring assembly and most likely ring constriction [6, 21]. FtsE is absent in plant and algal genomes and in Arabidopsis FtsZ is properly imported and processed in the ptCpn60β1-1 mutant, suggesting that the plastid Cpn60 system targets a different plastid division substrate(s) [142]. However, it is attractive to speculate that akin to bacterial system, the impaired chloroplast division phenotype and excessively stable FtsZ filament phenotype observed in the *cpn60* mutants are as a result of changes in the stability of a regulator of Z-ring formation and identification of possible targets of Cpn60 chaperonin activity will be important.

Given the striking similarity in the structure of the long, disorganized FtsZ filaments observed the *cpn60* mutants and in Arabidopsis plants overexpressing ARC6, it is interesting to observe that ARC6 might also play a role in the chloroplast chaperone systems. ARC6 harbors a J domain characteristic of co-chaperones that interact with the Hsp70 family of chaperone proteins [17, 147, 148]. J domains can associate with unfolded polypeptide chains and deliver them to Hsp70 chaperones for processing and also regulate the activity of the Hsp70 chaperones [148]. Most J domains harbor a central His-Pro-Asp motif, but in ARC6 and its orthologues only the central Pro is conserved uniformly. The His-Pro-Asp motif is commonly crucial for interaction with the Hsp70 chaperone partners, although exceptions exist [50]. Z-ring formation in *E. coli* is known to be influenced by Hsp70 [145] and it is possible that in addition to the proposed role in anchoring the Z-ring, ARC6 may have additional functions.

Whilst it remains to be shown whether ARC6 can interact with a chaperone through its atypical J-domain, recently it was shown that ARC6 can interact with a second chloroplast protein harboring a J domain, Chloroplast J-like Domain 1 (CJD1) [3]. Like ARC6, CJD1 is an inner envelope protein and harbors an N-terminal J-like domain that resides in the stroma. The J-domain of CJD1 interacts with the ARC6 J-like domain and the adjacent conserved region. Interestingly, during mitochondrial protein import in yeast the dynamic interaction of two J-domain containing proteins (the J co-chaperone Tim14 and the J-like protein Tim16) has been shown to regulate the co-chaperone activity of Tim14 on mtHsp70, which plays an important regulatory role [98] and it is possible that the ARC6-CJD1 interaction has a similar regulatory role.

CJD1 plants were identified as containing altered levels of the leaf galactolipids monogalactosyldiacylglycerol (MGDG) and digalactosyldiacylglycerol (DGDG) [3]. These two galactolipids commonly constitute up to 80 % of the total chloroplast lipids and play an essential role in chloroplast membrane biogenesis and are indispensable for correct protein import into chloroplasts [31]. Mutant analysis failed to reveal evidence for a direct link between the function of CJD1 and ARC6, however in the *tgd-1-1, fad6-1* and *tgd-4-3, fad6-1* double mutants, a strong reduction in the levels of MGDG and DGDG is coupled with a reduction in the number of chloroplasts per cell [29]. The authors suggest that an inadequate lipid supply for membrane proliferation during chloroplast division is responsible. This hypothesis is supported by the recent finding that impairment of very-long-chain fatty acids biosynthesis via genetic of pharmacologic approaches affects plastid division [103]. Whilst the picture is complicated by the lack of strong linkage between lipid mutants and chloroplast division mutants phenotypes, it will be interesting to determine the functional significance of interaction between CJD1 and ARC6.

6.4 Assembly of the Plastid Division Machinery

In bacteria the Z-ring is known to act as a scaffold protein, playing an essential role in the sequential and coordinated recruitment of up to nine additional cell division proteins to the division site (FtsX, FtsE, FtsK, FtsQ, FtsL, FtsB, FtsI, FtsW and FtsN) to form the divisome or septal ring (reviewed in [24]). Studies in Arabidopsis and algae have also shown that assembly of the chloroplast division machinery appears to occur in a linear order. After the formation of the Z-ring the timed addition of division machinery components results in assembly of up to four additional ring-like structures at the division site (Fig. 6.3).

6.4.1 The Plastid Dividing Rings

Many algae have just one chloroplast per cell and have been used extensively for the study of the coordinated assembly of the chloroplast division machinery because, unlike in Arabidopsis in which each cell can contain as many as 100 chloroplasts that divide asynchronously, algal cell and organelle division can be highly synchronized by light/dark cycles. The completion of Z-ring formation is followed by the appearance of ring like structures known as plastid dividing (PD) rings. PD rings were first identified as fuzzy plaques in electron microscopy images in young wheat leaves [71]. This fuzzy plaque was subsequently shown to be an electron-dense ring-like structure in the alga *C. caldarium* [90]. Studies in both *Avena sativa* and *Cyanidioschyzon merolae* further resolved this ring to be composed of two or three electron-dense rings: an inner ring on the stromal face of the inner envelope, an outer ring on the cytosolic face of the outer envelope, and occasionally a middle ring in the intermembrane space. PD rings have subsequently been detected in many plant and algae species and are thought to be a universal feature of plastid division machineries [47, 68, 90–92, 94].

The first PD ring to assemble after the completion of the Z-ring is the inner PD ring, which forms between the inner envelope membrane and the Z-ring [93, 94]. This is followed by the middle PD ring in the inter membrane space (IMS), if present [91]. The components of the inner and middle PD ring are as yet unidentified. Finally the outer PD ring, which is composed of a bundle of 5–7 nm filaments, forms on the outer chloroplast envelope (Fig. 6.3) [95].

The identity of the components of the PD rings remained elusive for many years. It was found that isolated PD machineries from *C. merolae* could not be fully decomposed with proteases, indicating a non-proteic component, and electron dense deposits appeared on the PD rings after staining with periodic acid–horseradish peroxidase, indicating that it is saccharic in nature [153]. Proteomic analysis of the isolated PD machinery identified more than 20 putative components, one of which was a glycogenin-like protein and was designated Plastid Dividing Ring 1 (PDR1) [153]. PDR1 forms a ring that is recruited to the division site after the formation of the Z-ring and down-regulation of *C. merolae PRD1* by antisense suppression prevents the formation of the PDR1 ring, but does not affect the assembly of the Z-ring. PDR1 harbors a glycosyltransferase domain in the C-terminal region, putatively associated with the transfer of a sugar moiety from an activated donor sugar onto an acceptor molecule. Immunoblot analysis of synchronized cell cultures revealed that during chloroplast division the band corresponding to PDR1 shifted downwards before decreasing in intensity. The authors propose that PDR1 synthesizes the sugars that constitute the outer PD ring and is glycosylated during chloroplast division before decomposing [153].

The precise role of PRD1 and the polyglucan filaments is unclear, but down-regulation of *PRD1* leads to a decrease in the frequency of chloroplast division in *C. merolae* compared with control cells [153], indicating that the correct formation of the PD ring is essential for chloroplast division. PDR1 is widely conserved from red alga to land plants, and in Arabidopsis there is a family of 6 PDR1 proteins (PGSIP1-6). PGSIP1 and PGSIP3 are located in the Golgi [105], but the remaining proteins are yet to be investigated. It will be interesting to determine if any members of the family are involved in chloroplast division and if the outer PD ring is also a bundle of PDR1-mediated polyglucan filaments in higher plants.

6.4.2 PDV1, PDV2 and Dynamin Recruitment to the Division Site

Although the outer PD ring is not well characterized in higher plants, microscopy studies have revealed that the formation of the Z-ring is followed by the recruitment of two novel proteins, Plastid Division 1 and Plastid Division 2 (PDV1 and PDV2) to the division site on the outer envelope [97]. PDV1 and PDV2 are integral outer envelope proteins that show no significant sequence similarity to known proteins. PDV1 and PDV2 have similar domain structure and are composed of a large N-terminal domain residing in the cytosol, a single transmembrane domain and a small C-terminal domain residing in the in the IMS [44, 97]. The cytosolic domains harbor two short conserved regions and a coiled-coil domain, whilst the IMS domains both terminate in a conserved C-terminal glycine. The PDV1 family of proteins is found in only higher plants, whilst the PDV2 family is found in higher plants and in moss [97]. The primary defining feature between the two PDV families is the presence of a 28 amino acid extension in the C-terminal domain of the PDV2 proteins, that is absent in all PDV1 proteins [44].

Both PDV1 and PDV2 localize to patches on the outer membrane of the chloroplasts and are recruited to ring-like structures at the division site after Z-ring formation. The analysis of the localization of PDV1 and PDV2 in the *pdv2* and *pdv1* mutants, respectively, reveals that both proteins are recruited to the division site independently of each other [46]. PDV2 is recruited to the division site by a direct interaction of the IMS domain of PDV2 with the IMS domain of ARC6 [44]. Disruption of this interaction, by deletion of the conserved C-terminal glycine of PDV2 ($PDV2_{G307D}$) destroys PDV function *in vivo*, as demonstrated by the inability of $PDV2_{G307D}$ to complement the *pdv2* mutant [44]. The IMS of PDV1 does not interact with ARC6 and it is possible that the unique 28 amino acid extension present in the PDV2 family of proteins mediates this interaction [44]. The recruitment of PDV1 to the division site is dependent on PARC6, although no direct interaction has been detected between the IMS domains of PDV1 and PARC6 [46]. Consequently the mechanism by which PDV1 is recruited to the division site is not clear.

A mutation of the C-terminal conserved glycine residue of PDV1 ($PDV1_{G272D}$) does not disrupt the insertion of PDV1 into the outer chloroplast membrane but, $PDV1_{G272D}$ is localized dispersed throughout the membrane and is unable to be recruited to the division site [97]. It will be important to determine the protein interacting partner that mediates the ability of PDV1 to identify and/or localize to the division site through the C-terminal Gly. Interestingly, analysis of a truncated form of ARC6 lacking the IMS domain ($ARC6_{\Delta IMS}$) found that $ARC6_{\Delta IMS}$ is able to insert into the inner chloroplast envelope and to correctly localize to the division site [44]. Furthermore, whilst the ARC6 ring is detected in both the *pdv1* and *pdv2* mutant background, neither PDV1 nor PDV2 are recruited to the division site in *arc6* mutant [44]. These data indicate that ARC6 acts upstream of PDV1 and PDV2 and is necessary for the membrane recruitment of both proteins. The link between the PDV proteins and ARC6 (and potentially PARC6) is the first clue as to how the cytosolic and stromal division machineries can be coordinated across the inner and outer chloroplast envelope. In addition to conveying topological information regarding the location of the division site, it may also act to convey signals concerning the status of the stromal to the cytosolic division machineries.

Disruption or deletion of either *PDV1* or *PDV2* results in mesophyll cells containing chloroplasts that are fewer in number and larger than those in the wild type, and which frequently show prevalent constrictions [97]. This phenotype suggests that plants lacking functional PDV proteins are able to initiate constriction of the division sites but are unable to complete division. A strikingly similar phenotype is observed in Arabidopsis plants expressing a truncated form of DRP5B (Dynamin-related protein 5B; ARC5) [40], the last chloroplast division protein known to be recruited to the division site, and recent work has shown these three proteins work together as part of the cytosolic chloroplast division machinery.

DRP5B was first identified through cloning of the *arc5* mutant and around the same time a dynamin-like protein in *C. merolae* (CmDnm2) was also identified as a late stage chloroplast division protein [40, 96]. The dynamin and dynamin-like protein superfamily of self-assembling GTPases are multidomain proteins and are well documented to participate in fission and fusion events of intracellular membrane structures (such as endocytosis, intracellular vesicle trafficking, cytokinesis, mitochondrial fusion and fission and peroxisome division) reviewed in [22]. How dynamin proteins function to bring about such a diverse range of events has been the focus of much study.

There are 16 dynamin-like proteins in Arabidopsis and only two in *C. merolae*. CmDnm2 and AtDRP5B, along with AtDRP5A which is involved in cytokinesis, represent a novel plant-specific group of dynamin-related proteins. This group harbor the three domains central to dynamin proteins: N-terminal GTPase domain, a middle domain (MD) and a C-terminal GTPase effector domain (GED), thought to interact directly with the GTPase domain and to mediate self-assembly [22]. Both AtDRP5B and CmDnm2 also have a pleckstrin homology (PH) domain, which have been shown to mediate lipid binding of other dynamin-like proteins. DRP5B is expressed as two alternatively spliced forms in Arabidopsis, with the shorter protein (714 amino acids) harboring a shorter PH domain than the longer variant (777 amino acids), although the functional significance of this is not yet apparent [40].

Both CmDnm2 and DRP5B relocate from the cytosol, where they localize as patches, to the division site where they form a ring like structure on the surface of the chloroplast [40, 96]. In Arabidopsis this dynamin ring is faintly detected in un-constricted chloroplasts but in *C. merolae* immunoblot analysis has shown that the dynamin ring only associates with chloroplasts during the division phase [96]. DRP5B is recruited to the division site by the combined action of PDV1 and PDV2. DRP5B is observed to localize to the division site in both the single *pdv1* and *pdv2* mutants, but not in the double *pdv1,pdv2* mutant, indicating that PDV1 and PDV2 function independently to recruit DRP5B to the division site [97]. The precise mechanism of recruitment is unclear and no interaction has been detected between DRP5B and PDV1 or PDV2 [97, 107]. However, both PDV proteins harbor cytoplasmic coiled-coil domains and there are several examples of proteins that recruit dynamin proteins to the mitochondrial division site via a direct interaction with their coiled-coil domain, for example Mdv1p recruits Dnm1p during mitochondrial division [144].

In *C. merolae* the PDV families of proteins have not been identified, indicating that a novel mechanism may exist to recruit CmDnm2 to the chloroplasts division site in algae. Immunofluorescence microscopy demonstrated that PDR1 appears earlier than CmDnm2 in the cytoplasm and forms a ring from the beginning to the end of chloroplast division. CmDnm2 forms a ring between the outer most rings of the PD1/polyglucan filaments, leading to the hypothesis that PD ring plays a role in organizing the dynamin ring [153].

Recently it was found that DRP5B is also required for peroxisome division in Arabidopsis [154]. Akin to the phenotype observed in the chloroplasts of the *arc5* mutant, these plants exhibit enlarged peroxisomes and peroxisomes that have undergone membrane constriction but failed to complete fission and therefore are unable to separate from each other [154]. Two other dynamin-like proteins, two dynamin-related proteins (DRP3A and DRP3B) are also implicated in peroxisome division [32] and it will be interesting to see both how these they are coordinated with DRP5B and also how peroxisome division is coordinated with chloroplast division.

To date the dynamin-like proteins are the final component known to be recruited to the chloroplast division machinery, however it is doubtful that we have identified all of the components of the stromal and cytosolic division machineries. In the future the collection of *arc* mutants may provide novel insights into chloroplast division, as the mutations in at least four of the original *arc* mutants are yet to be identified. Furthermore, disruptions of several Arabidopsis genes have been demonstrated to lead to defects in chloroplast division, for example GC1/AtSulA, CRL (crumpled leaf) and ARTEMIS, although a direct link to the division machinery has yet to be demonstrated [8, 39, 78, 126, 140]. Likewise, the negative regulator of chloroplast division ARC1 has recently been identified as FtsHi, but does not affect accumulation of chloroplast division proteins or Z-ring formation [58]. Further characterization of these proteins will be important to identify the full complement of chloroplast division machinery proteins.

6.5 Constriction and Separation of Chloroplasts

After formation of the division machinery on the inner and outer envelope, microscopic observations reveal that constriction of the plastid is initiated and the septum progressively tightens to eventually separate the two new daughter plastids. Whilst events mediating the placement and the temporal assembly of the machinery are beginning to be elucidated, less clear is how the chloroplast division machinery is initiated to begin constriction, how the force for constriction is generated and which molecules/mechanisms mediate the final scission event.

The analysis of plants in which the mechanism of chloroplast division is disrupted, either through the absence of a functional form of a division protein or through overexpression of a division protein, has revealed distinct classes of chloroplast division phenotypes. Most striking are the presence of either (i) enlarged chloroplasts which are globular/round in shape or (ii) enlarged chloroplasts which are frequently constricted and have a dumbbell-shaped appearance. The first category are commonly caused by diminution of functional stromal ring division components, such as ARC6 or FtsZ [86, 125, 139], whilst the second class are only caused by diminution of the cytosolic division components [40, 97] (Fig. 6.1). These phenotypic differences suggest that the early and late stages of plastid constriction are governed by different forces, generated by either the stromal or the cytosolic division machineries.

6.5.1 Force Generation by the Stromal Plastid Division Machinery

It is thought that FtsZ is necessary and sufficient to generate the force for constriction of the septum during bacterial cell division [9, 76]. In support of this in some bacteria the sole cell division protein is FtsZ, suggesting that FtsZ alone is sufficient to bring about cell division, furthermore FtsZ protofilaments are observed to undergo a conformational change from straight to curved that would be capable of generating a force [75, 81]. This hypothesis was strengthened when membrane tethered FtsZ successfully assembled into Z-rings on the inside of liposomes [110, 111]. The Z-rings formed were dynamic and capable of constricting the liposomes in a GTP dependent manner, proving that FtsZ alone is capable of self-assembly and force generation. How FtsZ might generate the force for constriction in bacteria is still controversial but years of genetic, biochemical and structural studies have revealed many properties of the protein.

FtsZ has the ability to bind GTP and FtsZ proteins from at least seven bacteria, as well as both AtFtsZ1 and AtFtsZ2-1 from Arabidopsis, have been shown to be functional GTPase enzymes, although both AtFtsZ1 and AtFtsZ2-1 hydrolyze GTP at a slower rate than their bacterial counterparts [25, 74, 109, 137]. Monomeric FtsZ hydrolyses GTP very slowly but can self-assemble into filaments. When assembled into a protofilaments the bottom interface of one FtsZ makes contact with the GTP pocket of the subunit below [131].

FtsZ filaments have three preferred conformations; straight, intermediate curved and highly curved [28]. Whilst the structure of the Z-ring in cell or chloroplast division is not yet clear, it is thought to be formed from small, overlapping subunits of curved FtsZ filaments that are tethered to the membrane via interaction with FtsA, ZipA or ARC6 [79, 117]. GTP hydrolysis is not necessary for FtsZ self-assembly and FtsZ polymers from bacteria have been shown to contain a substantial amount of GTP, suggesting that hydrolysis takes place with some lag after polymer formation [28, 108]. Despite many years of research it is not known what happens following GTP hydrolysis in a protofilament. It is known that GDP substantially destabilizes the protofilaments and one suggestion is that after hydrolysis the protofilament fragments, however it has been possible to detect FtsZ protofilaments that contain a 1:1 ratio of GTP:GDP and many believe that the polymers fragment only at the ends [18, 130].

Several mathematical and physical models have been proposed to explain how the physical properties of FtsZ can generate the force necessary to constrict a membrane. It is proposed that the membrane tethered Z-ring generates force by the protofilaments exerting a bending force on the membrane as they are induced into a curved conformation. In support of this elegant experiments were performed in which Z-rings were reconstituted in liposomes by either targeting the C-terminal end of FtsZ (FtsZ-c) to the membrane (to mimic the *in vivo* orientation of FtsZ tethered to the membrane) or by targeting the N-terminal end of FtsZ (FtsZ-n) to the membrane (to reverse the orientation). As described above, the Z-rings reconstituted from FtsZ-c were capable in inducing concave constrictions of the liposomes. In stark contrast, the Z-rings formed from FtsZ-n caused convex bulges to form on the liposomes, indicating that the bending force was generated in the opposing direction [111, 112]. It will be extremely interesting to determine if each of the three Arabidopsis FtsZ proteins is capable of generating such bending forces alone and whether they act cooperatively.

If FtsZ is not (solely) responsible for generating the force necessary to initiate constriction of the chloroplast division site, what other proteins could play a role? To date the best candidates are the components of the PD rings. During constriction the inner PD ring has been observed to remain a constant thickness and loses components as constriction proceeds, disassembling late in constriction, indicating that any role played in force generation is not related to final constriction/scission [93]. Identification of the components of these rings and analysis of the phenotype of mutants lacking functional components will be vital in determining possible roles.

6.5.2 Force Generation by the Cytosolic Plastid Division Machinery

The principal candidates to generate force for constriction on the outer envelope are the chloroplast division dynamin-like proteins. Dynamin-like proteins are well established to be essential for constriction during the division of organelles in many species. For example mitochondrial division is known to employ dynamin-like proteins in mammals (Drp1/Dlp1), yeast (Dnm1), higher plants (ADL1 and ADL2) and *C. merolae* (CmDnm1) (reviewed in [69]), and the involvement of dynamin-related proteins in peroxisome division has been demonstrated in Arabidopsis (DRP5B, DRP3A, DRP3B), humans (DLP1,Fis1) and yeast (Vps1, Dnm1) (reviewed in [61]). Best characterized is the role of Dnm1p/Drp1 in mitochondrial division and current models leave no doubt that dynamin is the key player mediating membrane scission.

The study of how the dynamin chloroplast division proteins function during division is in its relative infancy but many parallels can be drawn between the mechanisms of chloroplast division and mitochondrial division. Like the chloroplast dynamin division proteins, Dnm1p/Drp1 interacts with other proteins to form the mitochondrial division machinery (reviewed in [116]). Furthermore, Dnm1p/Drp1 are believed to be involved in late stages of division: *C. elegans* with mutant *drp-1* have constrictions in the outer mitochondrial membranes but fail to complete division, a phenotype reminiscent of the chloroplasts in the *arc5* mutant, which are frequently constricted and dumbbell shaped [70]. Also, the majority of Dnm1p/Drp1 is present in the cytosol and is recruited to the mitochondrial outer membrane during the late stage of division [136]. Similarly, CmDnm2 and DRP5B also relocate from cytosolic patches to associate with the chloroplast division site [40, 101].

Drp1 exists as small oligomers (dimers/tetramers) that can assemble into larger multimeric structures at the mitochondrial outer membrane [38, 135]. DRP5B can interact with itself *in vivo* [154], suggesting DRP5B can form higher order structures on chloroplasts. Drp1 can polymerize *in vivo* into spirals with a diameter of approximately 100 nm, which is considerably smaller than the diameter of mitochondria (400 nm) [54] and it is thought that as higher order structures of dynamin are only observed at later stages of mitochondrial division, they may only form once the membrane has undergone some constriction. In *C. merolae* small dynamin patches are discontinuously localized at the division site at the onset of constriction and only form a continuous ring at later stages, suggesting that the same mechanism may be in place in chloroplast division [96].

Once assembled, dynamin can undergo GTPase activity, which is stimulated by interactions between the GTPase domain, the MD and the GED [14, 54]. The GTPase activity is stimulated once a 'critical mass' is reached and the resulting activity resembles a chain reaction. When dynamin hydrolyses GTP there is a conformation change in the protein which results in considerable bending of the filament, sufficient to bring about constriction of the membrane [88]. Although the identified dynamin-like proteins from Arabidopsis and *C. merolae* are yet to be proven to be GTPases, given the high levels of conservation in the domain architecture it is likely that they are functional enzymes. Additionally the membrane-free PD machineries isolated from *C. merolae* formed super-twisted rings, circular rings and spirals which is attributed to the motive force generated for contraction by CmDnm2 [152].

What is the role of the outer PD ring? PDR1 is only glycosylated during division and after constriction has been initiated the outer PD ring is observed to widen and thicken, and remains attached until division has been completed [94, 153]. It is possible that the PDR1 and sugar ring is required to organize the dynamin ring throughout the division process, but it could also play a role in force generation.

The mechanism behind the final scission of chloroplasts remains unknown. It is possible that dynamin is necessary and sufficient to bring about this stage, but other factors may also be required. After constriction is completed, plastids are separated and can move inside the cell via the actin cytoskeleton [66, 133]. Very recently, the CLUMPED CHLOROPLAST 1 protein (CLMP1) has been identified as an important factor for chloroplast separation [151]. In *clmp1* mutants, chloroplasts remain interconnected and form clumps instead of being distributed throughout the cells, resulting in the formation of some aplastidic cells. This phenotype is transient and almost completely disappears as leaves age. Interestingly, both early and late steps of chloroplast division appear to occur normally since FtsZ and DRP5B localization is normal in this mutant. Ultrastructure analysis revealed that some chloroplasts within the clumps are held together by thin membranous-connections, similar in structure to the isthmus, suggesting that only the final scission of chloroplasts is impaired in this mutant [151]. CLMP1 does not localize at the division site, and further analysis will be required to fully understand possible roles in scission and how its activity affects chloroplast separation.

6.6 Regulation of Plastid Division

It is now clear that all plastids originate from the division of a pre-existing organelle, and fusion between plastids appears to occur very rarely if at all [7]. Because plastid number is maintained constant depending on the species and the cell type, plastid division must be tightly regulated to achieve the correct number of organelles in a given cell. Exogenous cues such as light, temperature or nutrition have also been reported to affect chloroplast division (reviewed in [118]), but the underlying mechanisms remain largely unknown.

6.6.1 Regulation During Cell Proliferation

Regulation of chloroplast division during the cell cycle has been extensively studied in unicellular algae. In such model organisms cells contain one or a few chloroplasts, and the requirement for a strict regulation of chloroplast division during the cell cycle is obvious: if chloroplast division did not occur prior to cell division, aplastidic daughter cells would be generated. This can be observed in the case of *Hatena sp*, a protist that hosts an endosymbiotic green algal partner but inherits it unevenly because the algae does not divide inside the host cell [106]. The ability of the host cell to promote and control the division of the endosymbiont has hence clearly been a major step during the evolution of chloroplasts. In the red algae *C. merolae*, accumulation of chloroplast division proteins CmFtsZ1 and CmFtsZ2 precedes accumulation of tubulin during mitosis [102]. In the brown algae *Seminavis robusta*, chloroplast division has been shown to occur during the S/G2 phase of the cell cycle, and this regulation may involve transcriptional regulation of *FtsZ* genes expression [43]. In *Chlamydomonas reinhardtii*, expression of *FtsZ* and *Min* genes is also cell cycle regulated [2, 53].

In higher plants, the relationship between cell and plastid division is far more complex because (i) cells contain numerous plastids, and a strict coordination between cell cycle and plastid division is therefore no longer required to avoid the formation of aplastidic cells, and (ii) plastid division can occur in two contrasting cellular contexts: in proliferating cells or in differentiating cells. Indeed, mature

Arabidopsis mesophyll cells contain on average 120 chloroplasts in the L-*er* background, whereas meristematic cells contain about 10 proplastids [83]. Hence, proplastid division is required to keep pace with cell division, and in developing leaves, differentiated chloroplasts can divide both in dividing and in expanding cells.

The existence of mechanisms regulating plastid division during the cell cycle in higher plants is debated. Early observations of dividing chloroplasts in various plant species do not provide evidence for synchrony of chloroplast division in a given cell. Furthermore, chloroplast division can occur independently of the cell cycle in wheat leaves where the maximum rate of chloroplast division is observed in a region of the leaf where cell division has ceased [10]. However, plastid division might be regulated depending on nuclear DNA (nDNA) content (or synthesis) because there is a correlation between chloroplast number and cell ploidy: chloroplast number is increased in a polyploid plant compared to a diploid plant of the same species, and chloroplast number in a given cell type has also been proposed to vary according to endoreduplication (a particular type of cell cycle during which the cell undergoes subsequent phases of DNA replication without mitosis) [12]. In agreement with this hypothesis, in tomato fruit, in which endoreduplication is particularly high, chromoplast number can be up to 1000 per cell [121].

To date, data is lacking to firmly establish that chloroplast division is cell cycle regulated in higher plants, or to provide a molecular mechanism underlying this regulation, but some studies have provided insights into this process. In algae, cell cycle-dependent regulation of FtsZ seems to be the rule, but is not so clear in higher plants: it was observed in BY-2 cells [27], but not in micro-arrays on synchronized Arabidopsis cells [89]. Furthermore, expression of genes encoding proteins involved at early steps of plastid division such as FtsZ or ARC6 does not seem to vary much in developing leaves [107]. In Arabidopsis, silencing of the *AtCDT1a* gene, encoding a protein involved in nuclear DNA replication licensing, resulted in an inhibition of cell cycle progression and chloroplast division. AtCDT1a was also found to interact with ARC6, suggesting that this factor may not only be a key factor for nuclear DNA replication but also play a direct role in plastid division, and providing evidence for a common regulatory pathway to both processes at the onset of S-phase [127]. In tomato, Caspi et al (2008) reported that the *high pigment 1* (*hp1*) mutation, affecting the DDB1 protein, results in increased chloroplast number per cell [13]. Interestingly, DDB1 is a sub-unit of an ubiquitin-E3 ligase complex and CDT1 is one of its targets; this result may hence support the view of a CDT1-mediated coordination of cell cycle and chloroplast division at the onset of S-phase.

6.6.2 Plastid Division and Plastid DNA Replication

To date, it is not clear whether plastid DNA replication and plastid division are strictly coupled processes. Plastids contain multiple copies of their genome, and replication of plastid DNA is therefore not a pre-requisite to plastid division. In some instances, chloroplast division can occur independently of cpDNA replication since in the latest stages of leaf development DNA copy number per chloroplast decreases [48]. However, in dividing chloroplasts, nucleoid replication usually occurs before chloroplast division [121]. The phenotype of the *crinkled leaves 8* mutant (*cls-8*) may provide further evidence for a link between cpDNA replication and chloroplast division. These plants are deficient for the large sub-unit of ribonucleotide reductase (RNR), but display severe inhibition of chloroplast division. Despite a clear reduction in dTTP and dATP pools, little effect was observed on nDNA replication whereas chloroplast DNA (CpDNA) copy number was drastically reduced [42]. Finally, the YlmG protein was recently identified in cyanobacteria and the red alga *C. merolae* for its putative role in nucleoid partitioning [57]. Its overexpression inhibits chloroplast division whereas loss of function only affects nucleoid localization, suggesting that partitioning of chloroplast DNA may somehow be related to the chloroplast division process.

As illustrated by the above paragraphs, several lines of evidence suggest that plastid division may be influenced by nuclear and plastidial DNA replication, although further work is needed to further support this hypothesis and to decipher the molecular mechanisms involved.

6.6.3 Plastid Division Is Regulated by Cell Differentiation, and Cell Expansion

Plastid number strikingly differs depending on the cell type: for example in Arabidopsis, meristematic cells contain about 10 proplastids [83] whereas stomata guard cells contain only 4 [129], petal cells 15–20 [124], and mesophyll cells more than 100 [83]. How plastid division is regulated according to cell type remains largely unknown, but cytokinins are probably responsible for the stimulation of chloroplast division in young developing leaves. Early reports supporting this hypothesis came from the isolation of the *PC22* mutant in the moss *Physcomitrella patens*. In this mutant, chloroplast division is severely impaired, but can be partially restored by exogenously applied cytokinins [1, 60]. A positive effect of cytokinins on chloroplast division was also reported in bean leaves [99] and *Brassica rapa* leaf disks [150], but not in spinach leaf disk [119]. The molecular basis for this regulation is beginning to be elucidated and is mediated at least partly by PDV proteins. Indeed, unlike FtsZ or ARC6, accumulation of PDV1 and PDV2 is higher in young leaves, where chloroplast division is most active, than in older leaves, and their overexpression is sufficient to stimulate chloroplast division, like exogenous cytokinin application. Furthermore, the cytokinin response factor CRF2 was shown to stimulate *PDV* genes expression and chloroplast division [107]. More recently, two cytokinin responsive transcription factors, GATA Nitrate-inducible Carbon metabolism involved (GNC) and Cytokinin-Responsive GATA1 (CGA1) have also been involved in the cytokinin-dependent stimulation of chloroplast division downstream of B-type ARRs [19]. Whether these factors regulate chloroplast division via PDV proteins or via an independent mechanism remains unclear, but PDV proteins nevertheless appear to be key regulators of chloroplast division in developing leaves. The mechanisms regulating plastid division in other organs such as roots, flowers, tuber or seeds have not been investigated, but it is tempting to speculate that other phytohormones such as gibberellins in the case of flowers [23] may play a role in regulating plastid division in non-green tissues.

Additional regulatory factors probably affect chloroplast division in leaves. Indeed in mesophyll cells, chloroplast number clearly depends on cell area: in Arabidopsis, a very nice correlation is observed between cell size and chloroplast content in the mesophyll cells [123]. This correlation is retained between different species: for example cocoa mesophyll cells contain on average 3 chloroplasts but are very small [121]. Isolation of *arc* mutants led to the conclusion that total chloroplast area, rather than chloroplast number is under tight control in mesophyll cells. Diminution of chloroplast number due to impaired chloroplast division is compensated by an increase in chloroplast size, so that the total chloroplast plan area remains almost constant [123]. Reciprocally, in the few *arc* mutants characterized by an increased chloroplast number, chloroplast size is reduced [120]. Finally, comparison of mesophyll cell size, chloroplast number and chloroplast size in different species demonstrated that total chloroplast area in cells is strictly related to cell size over a tenfold range of cell sizes [121]. Because cell size is generally related to cell ploidy [65], some authors postulate that chloroplast division is regulated directly by cell size and independently of DNA replication. In agreement with this hypothesis, chloroplast number still correlates with cell size in transgenic plants over-expressing an inhibitor of cell cycle regulation, resulting in increased cell size and reduced cell ploidy [56].

As shown by the above summary, several studies have tackled the issue of the regulation of plastid division during development, but little is known yet regarding the exogenous control of this cellular process. Due to their sessile nature, external conditions affect many aspects of plant growth and development, and in this respect are also likely to affect chloroplast division, but evidence for such regulatory pathways is scarce.

6.6.4 Chloroplast Division Is Regulated by Exogenous Cues

The most obvious external signal that could affect chloroplast division is probably light. Evidence for light regulation of chloroplast division has been provided long ago by Hashimoto and Possingham [48], in spinach leaf disks they found chloroplast generation time to be more than 2 times longer in the dark. Interestingly, the late phases of chloroplast division are most affected: chloroplasts remain constricted (at the dumbbell stage) for 22 h instead of 3.5 h [48]. Recently *DRP5B* expression has been shown to be regulated by FHY3, a transcription factor involved in phytochrome A signaling [115]. This result fits nicely with early microscopic observations since chloroplasts in *arc5* mutants are arrested at the dumbbell stage of division [83]. FHY3 is specifically involved in far-red signaling and appears to stimulate *DRP5B* expression in far-red light. In agreement with this observation, mutants deficient for FHY3 or its homolog are affected in chloroplast division, and accumulation of DRP5B at the division site is no longer detected in *fhy3*. Chloroplast division is even more severely impaired in *fhy3 far1* double mutants, demonstrating that FHY3 and FAR1 act partially redundantly to promote chloroplast division in response to far-red light. Nevertheless, other light dependent signaling pathways regulating chloroplast division may exist. Indeed, the chloroplast division phenotype of the *PC22* mutant in *P. patens* can be reverted by blue light treatment [128], suggesting that other photoreceptors such as cryptochromes may play a role in regulating chloroplast division.

Capture of light energy by chloroplasts and active photosynthesis can modify the redox status of the plastoquinone pool in the chloroplast, and the redox state of PQ is likely to regulate nuclear gene expression [63]. In addition, chloroplasts generate high amounts of reactive oxygen species, especially when submitted to excess light, which can result in severe photo-oxidative stress. Because the chloroplast redox balance is most important for the survival and function of plant cells, redox regulation of chloroplast division may well occur. In agreement with such a hypothesis, *ntrc* mutants deficient for the chloroplast NADPH-thioredoxin reductase appear to accumulate fewer chloroplasts than the wild-type during their first month of growth [73], but further work is needed to confirm this observation and determine how this regulation may function.

Finally, many other factors have been reported to affect chloroplast division including temperature [119] or mineral nutrition [118], but it is not clear whether these effects are direct or indirect, and the underlying molecular mechanisms have yet to be investigated.

Now that the mechanism for chloroplast division is better understood, an increasing number of studies have investigated its regulation, and are providing molecular basis accounting for early cytological observations. Regulation of plastid division by external cues is only beginning to be unraveled but the role of internal cues is more thoroughly described. Two contrasting situations can be distinguished: in unicellular algae containing one or a few chloroplasts, cell cycle regulation of chloroplast division appears to be the rule, and seems to involve transcriptional regulation of genes involved at early steps of the process such as *FtsZ* or *Min*. By contrast in higher plants, cell cycle regulation of plastid division is still debated, an although several lines of evidence suggest that plastid division may be modulated by nuclear DNA replication, the only consensus is that total chloroplast area is modulated according to cell area, but the molecular mechanisms for this regulation have yet to be described. One striking feature of plastid division in higher plants is that expression of FtsZ proteins or ARC6 does not seem to vary a lot and proteins involved at later stages of division such as ARC5 or PDV seem to be targets for regulatory pathways. This may point to additional functions for FtsZ proteins unrelated to plastid division. Indeed, in Arabidopsis, FtsZ proteins are associated with thylakoid membranes [26, 59], and *fsZ2-2* mutants show defects in chloroplast shape and thylakoid development [59]. Likewise, in *P. patens*, the five FtsZ isoforms appear to play distinctive roles in cell patterning, chloroplast shaping, plant development and gravity sensing [84] suggesting that in multi-cellular plants, the functions of FtsZ proteins may have diversified.

Acknowledgements Jodi Maple was supported by a grant from the Norwegian Research Council (Grant no. 204761/F20), and Cécile Raynaud by a grant from the French National Research Agency (ANR 2010 JCJC1207 01). We thank Daniela Gargano for critical reading of the manuscript and Ingrid Holtsmark for help with microscopy.

References

1. Abel W, Knebel W, Koop H, Marienfeld J, Quader H, Reski R, Schnepf E, Spörlein B (1989) A cytokinin-sensitive mutant of the mos, Physcomitrella patens, defective in chloroplast division. Protoplasma 152:1–13
2. Adams S, Maple J, Moller SG (2008) Functional conservation of the MIN plastid division homologues of Chlamydomonas reinhardtii. Planta 227:1199–1211
3. Ajjawi I, Coku A, Froehlich JE, Yang Y, Osteryoung KW, Benning C, Last RL (2011) A j-like protein influences fatty acid composition of chloroplast lipids in Arabidopsis. PLoS ONE 6:e25368
4. Aldridge C, Moller SG (2005) The plastid division protein AtMinD1 is a Ca2+-ATPase stimulated by AtMinE1. J Biol Chem 280:31673–31678
5. Apuya NR, Yadegari R, Fischer RL, Harada JJ, Zimmerman JL, Goldberg RB (2001) The Arabidopsis embryo mutant schlepperless has a defect in the chaperonin-60alpha gene. Plant Physiol 126:717–730
6. Arends SJ, Kustusch RJ, Weiss DS (2009) ATP-binding site lesions in FtsE impair cell division. J Bacteriol 191:3772–3784
7. Arimura S, Yamamoto J, Aida GP, Nakazono M, Tsutsumi N (2004) Frequent fusion and fission of plant mitochondria with unequal nucleoid distribution. Proc Natl Acad Sci U S A 101:7805–7808
8. Asano T, Yoshioka Y, Kurei S, Sakamoto W, Machida Y (2004) A mutation of the CRUMPLED LEAF gene that encodes a protein localized in the outer envelope membrane of plastids affects the pattern of cell division, cell differentiation, and plastid division in Arabidopsis. Plant J 38:448–459
9. Bi EF, Lutkenhaus J (1991) FtsZ ring structure associated with division in Escherichia coli. Nature 354:161–164
10. Boffey SA, Ellis JR, Sellden G, Leech RM (1979) Chloroplast division and DNA synthesis in light-grown wheat leaves. Plant Physiol 64:502–505
11. Bramkamp M, van Baarle S (2009) Division site selection in rod-shaped bacteria. Curr Opin Microbiol 12:683–688
12. Butterfass T (1973) Control of plastid division by means of nuclear DNA amount. Protoplasma 76:167–195
13. Caspi N, Levin I, Chamovitz DA, Reuveni M (2008) A mutation in the tomato DDB1 gene affects cell and chloroplast compartment size and CDT1 transcript. Plant Signal Behav 3:641–649
14. Chang CR, Blackstone C (2010) Dynamic regulation of mitochondrial fission through modification of the dynamin-related protein Drp1. Ann N Y Acad Sci 1201:34–39
15. Chapman E et al (2006) Global aggregation of newly translated proteins in an *Escherichia coli* strain deficient of the chaperonin GroEL. Proc Natl Acad Sci U S A 103:15800–15805
16. Charbon G, Wang J, Brustad E, Schultz PG, Horwich AL, Jacobs-Wagner C, Chapman E (2011) Localization of GroEL determined by *in vivo* incorporation of a fluorescent amino acid. Bioorg Med Chem Lett 21:6067–6070
17. Cheetham ME, Caplan AJ (1998) Structure, function and evolution of DnaJ: conservation and adaptation of chaperone function. Cell Stress Chaperones 3:28–36
18. Chen Y, Erickson HP (2009) FtsZ filament dynamics at steady state: subunit exchange with and without nucleotide hydrolysis. Biochemistry 48:6664–6673

19. Chiang YH, Zubo YO, Tapken W, Kim HJ, Lavanway AM, Howard L, Pilon M, Kieber JJ, Schaller GE (2012) Functional characterization of the GATA transcription factors GNC and CGA1 reveals their key role in chloroplast development, growth, and division in Arabidopsis. Plant Physiol 160:332–348
20. Colletti KS, Tattersall EA, Pyke KA, Froelich JE, Stokes KD, Osteryoung KW (2000) A homologue of the bacterial cell division site-determining factor MinD mediates placement of the chloroplast division apparatus. Curr Biol 10:507–516
21. Corbin BD, Wang Y, Beuria TK, Margolin W (2007) Interaction between cell division proteins FtsE and FtsZ. J Bacteriol 189:3026–3035
22. Danino D, Hinshaw JE (2001) Dynamin family of mechanoenzymes. Curr Opin Cell Biol 13:454–460
23. Davis SJ (2009) Integrating hormones into the floral-transition pathway of *Arabidopsis thaliana*. Plant Cell Environ 32:1201–1210
24. de Boer PA (2010) Advances in understanding *E. coli* cell fission. Curr Opin Microbiol 13:730–737
25. de Boer P, Crossley R, Rothfield L (1992) The essential bacterial cell-division protein FtsZ is a GTPase. Nature 359:254–256
26. El-Kafafi el S, Mukherjee S, El-Shami M, Putaux JL, Block MA, Pignot-Paintrand I, Lerbs-Mache S, Falconet D (2005) The plastid division proteins, FtsZ1 and FtsZ2, differ in their biochemical properties and sub-plastidial localization. Biochem J 387:669–676
27. El-Shami M, El-Kafafi S, Falconet D, Lerbs-Mache S (2002) Cell cycle-dependent modulation of FtsZ expression in synchronized tobacco BY2 cells. Mol Genet Genomics 267:254–261
28. Erickson HP, Anderson DE, Osawa M (2010) FtsZ in bacterial cytokinesis: cytoskeleton and force generator all in one. Microbiol Mol Biol Rev 74:504–528
29. Fan J, Xu C (2011) Genetic analysis of Arabidopsis mutants impaired in plastid lipid import reveals a role of membrane lipids in chloroplast division. Plant Signal Behav 6:458–460
30. Forth D, Pyke KA (2006) The suffulta mutation in tomato reveals a novel method of plastid replication during fruit ripening. J Exp Bot 57:1971–1979
31. Froehlich JE, Benning C, Dormann P (2001) The digalactosyldiacylglycerol (DGDG) synthase DGD1 is inserted into the outer envelope membrane of chloroplasts in a manner independent of the general import pathway and does not depend on direct interaction with monogalactosyldiacylglycerol synthase for DGDG biosynthesis. J Biol Chem 276:31806–31812
32. Fujimoto M et al (2009) Arabidopsis dynamin-related proteins DRP3A and DRP3B are functionally redundant in mitochondrial fission, but have distinct roles in peroxisomal fission. Plant J 58:388–400
33. Fujiwara K, Taguchi H (2007) Filamentous morphology in GroE-depleted *Escherichia coli* induced by impaired folding of FtsE. J Bacteriol 189:5860–5866
34. Fujiwara MT, Nakamura A, Itoh R, Shimada Y, Yoshida S, Moller SG (2004) Chloroplast division site placement requires dimerization of the ARC11/AtMinD1 protein in Arabidopsis. J Cell Sci 117:2399–2410
35. Fujiwara MT, Hashimoto H, Kazama Y, Abe T, Yoshida S, Sato N, Itoh RD (2008) The assembly of the FtsZ ring at the mid-chloroplast division site depends on a balance between the activities of AtMinE1 and ARC11/AtMinD1. Plant Cell Physiol 49:345–361
36. Fujiwara MT, Li D, Kazama Y, Abe T, Uno T, Yamagata H, Kanamaru K, Itoh RD (2009) Further evaluation of the localization and functionality of hemagglutinin epitope- and fluorescent protein-tagged AtMinD1 in *Arabidopsis thaliana*. Biosci Biotechnol Biochem 73:1693–1697
37. Fujiwara MT, Sekine K, Yamamoto YY, Abe T, Sato N, Itoh RD (2009) Live imaging of chloroplast FtsZ1 filaments, rings, spirals, and motile dot structures in the AtMinE1 mutant and overexpressor of *Arabidopsis thaliana*. Plant Cell Physiol 50:1116–1126
38. Fukushima NH, Brisch E, Keegan BR, Bleazard W, Shaw JM (2001) The GTPase effector domain sequence of the Dnm1p GTPase regulates self-assembly and controls a rate-limiting step in mitochondrial fission. Mol Biol Cell 12:2756–2766
39. Fulgosi H, Gerdes L, Westphal S, Glockmann C, Soll J (2002) Cell and chloroplast division requires ARTEMIS. Proc Natl Acad Sci USA 99:11501–11506

40. Gao H, Kadirjan-Kalbach D, Froehlich JE, Osteryoung KW (2003) ARC5, a cytosolic dynamin-like protein from plants, is part of the chloroplast division machinery. Proc Natl Acad Sci U S A 100:4328–4333
41. Gargano D, Maple-Grodem J, Moller SG (2012) *In vivo* phosphorylation of FtsZ2 in *Arabidopsis thaliana*. Biochem J 446:517–521
42. Garton S, Knight H, Knight MR, Thorlby GJ (2007) Nucleotide depletion and chloroplast division. Plant Signal Behav 2:197–198
43. Gillard J et al (2008) Physiological and transcriptomic evidence for a close coupling between chloroplast ontogeny and cell cycle progression in the pennate diatom Seminavis robusta. Plant Physiol 148:1394–1411
44. Glynn JM, Froehlich JE, Osteryoung KW (2008) Arabidopsis ARC6 coordinates the division machineries of the inner and outer chloroplast membranes through interaction with PDV2 in the intermembrane space. Plant Cell 20:2460–2470
45. Glynn JM, Miyagishima SY, Yoder DW, Osteryoung KW, Vitha S (2007) Chloroplast division. Traffic 8:451–461
46. Glynn JM, Yang Y, Vitha S, Schmitz AJ, Hemmes M, Miyagishima SY, Osteryoung KW (2009) PARC6, a novel chloroplast division factor, influences FtsZ assembly and is required for recruitment of PDV1 during chloroplast division in Arabidopsis. Plant J 59:700–711
47. Hashimoto H (1986) Double ring structure around the constricting neck of dividing plastids of *Avena sativa*. Protoplasma 135:166–172
48. Hashimoto H, Possingham JV (1989) Effect of light on the chloroplast division cycle and DNA synthesis in cultured leaf discs of spinach. Plant Physiol 89:1178–1183
49. Haswell ES, Meyerowitz EM (2006) MscS-like proteins control plastid size and shape in *Arabidopsis thaliana*. Curr Biol 16:1–11
50. Hennessy F, Cheetham ME, Dirr HW, Blatch GL (2000) Analysis of the levels of conservation of the J domain among the various types of DnaJ-like proteins. Cell Stress Chaperones 5:347–358
51. Hill JE, Hemmingsen SM (2001) *Arabidopsis thaliana* type I and II chaperonins. Cell Stress Chaperones 6:190–200
52. Houry WA, Frishman D, Eckerskorn C, Lottspeich F, Hartl FU (1999) Identification of *in vivo* substrates of the chaperonin GroEL. Nature 402:147–154
53. Hu Y, Chen Z-W, Liu W-Z, Liu X-L, He Y-K (2008) Chloroplast division is regulated by the circadian expression of FTSZ and MIN genes in *Chlamydomonas reinhardtii*. Eur J Phycol 43:207–215
54. Ingerman E, Perkins EM, Marino M, Mears JA, McCaffery JM, Hinshaw JE, Nunnari J (2005) Dnm1 forms spirals that are structurally tailored to fit mitochondria. J Cell Biol 170:1021–1027
55. Itoh R, Fujiwara M, Nagata N, Yoshida S (2001) A chloroplast protein homologous to the eubacterial topological specificity factor minE plays a role in chloroplast division. Plant Physiol 127:1644–1655
56. Jasinski S, Saraiva Leite C, Domenichini S, Stevens R, Raynaud C, Perennes C, Bergounioux C, Glab N (2003) NtKIS2, a novel tobacco cyclin-dependent kinase inhibitor is differentially expressed during the cell cycle and plant development. Plant Physiol Biochem 41:667–676
57. Kabeya Y, Nakanishi H, Suzuki K, Ichikawa T, Kondou Y, Matsui M, Miyagishima SY (2010) The YlmG protein has a conserved function related to the distribution of nucleoids in chloroplasts and cyanobacteria. BMC Plant Biol 10:57
58. Kadirjan-Kalbach DK, Yoder DW, Ruckle ME, Larkin RM, Osteryoung KW (2012) FtsHi1/ARC1 is an essential gene in Arabidopsis that links chloroplast biogenesis and division. Plant J 72(5):856–867
59. Karamoko M, El-Kafafi el S, Mandaron P, Lerbs-Mache S, Falconet D (2011) Multiple FtsZ2 isoforms involved in chloroplast division and biogenesis are developmentally associated with thylakoid membranes in Arabidopsis. FEBS Lett 585:1203–1208
60. Kasten B, Buck F, Nuske J, Reski R (1997) Cytokinin affects nuclear- and plastome-encoded energy-converting plastid enzymes. Planta 201:261–272

61. Kaur N, Hu J (2009) Dynamics of peroxisome abundance: a tale of division and proliferation. Curr Opin Plant Biol 12:781–788
62. Kerner MJ et al (2005) Proteome-wide analysis of chaperonin-dependent protein folding in *Escherichia coli*. Cell 122:209–220
63. Kleine T, Voigt C, Leister D (2009) Plastid signalling to the nucleus: messengers still lost in the mists? Trends Genet 25:185–192
64. Koksharova OA, Wolk CP (2002) A novel gene that bears a DnaJ motif influences cyanobacterial cell division. J Bacteriol 184:5524–5528
65. Kondorosi E, Roudier F, Gendreau E (2000) Plant cell-size control: growing by ploidy? Curr Opin Plant Biol 3:488–492
66. Krzeszowiec W, Rajwa B, Dobrucki J, Gabrys H (2007) Actin cytoskeleton in *Arabidopsis thaliana* under blue and red light. Biol Cell 99:251–260
67. Kuroiwa T (1991) The replication, differentiation, and inheritance of plastids with emphasis on the concept of organelle nuclei. Int Rev Cytol 128:1–62
68. Kuroiwa H, Mori T, Takahara M, Miyagishima SY, Kuroiwa T (2002) Chloroplast division machinery as revealed by immunofluorescence and electron microscopy. Planta 215:185–190
69. Kuroiwa T, Nishida K, Yoshida Y, Fujiwara T, Mori T, Kuroiwa H, Misumi O (2006) Structure, function and evolution of the mitochondrial division apparatus. Biochim Biophys Acta 1763:510–521
70. Labrousse AM, Zappaterra MD, Rube DA, van der Bliek AM (1999) *C. elegans* dynamin-related protein DRP-1 controls severing of the mitochondrial outer membrane. Mol Cell 4:815–826
71. Leech RM, Thomson WW, Platt-Aloika KA (1981) Observations on the mechanism of chloroplast division in higher plants. New Phytologist 87:1–9
72. Lemos JA, Luzardo Y, Burne RA (2007) Physiologic effects of forced down-regulation of dnaK and groEL expression in *Streptococcus mutans*. J Bacteriol 189:1582–1588
73. Lepisto A, Kangasjarvi S, Luomala EM, Brader G, Sipari N, Keranen M, Keinanen M, Rintamaki E (2009) Chloroplast NADPH-thioredoxin reductase interacts with photoperiodic development in Arabidopsis. Plant Physiol 149:1261–1276
74. Lu CL, Stricker J, Erickson HP (1998) FtsZ from *Escherichia coli*, Azotobacter vinelandii, and Thermotoga maritima-Quantitation, GTP hydrolysis, and assembly. Cell Motil Cytoskeleton 40:71–86
75. Lu C, Reedy M, Erickson HP (2000) Straight and curved conformations of FtsZ are regulated by GTP hydrolysis. J Bacteriol 182:164–170
76. Lutkenhaus J (1993) FtsZ ring in bacterial cytokinesis. Mol Microbiol 9:403–409
77. Maple J, Chua NH, Moller SG (2002) The topological specificity factor AtMinE1 is essential for correct plastid division site placement in Arabidopsis. Plant J 31:269–277
78. Maple J, Fujiwara MT, Kitahata N, Lawson T, Baker NR, Yoshida S, Moller SG (2004) GIANT CHLOROPLAST 1 is essential for correct plastid division in Arabidopsis. Curr Biol 14:776–781
79. Maple J, Aldridge C, Moller SG (2005) Plastid division is mediated by combinatorial assembly of plastid division proteins. Plant J 43:811–823
80. Maple J, Vojta L, Soll J, Moller SG (2007) ARC3 is a stromal Z-ring accessory protein essential for plastid division. EMBO Rep 8:293–299
81. Margolin W (2001) Spatial regulation of cytokinesis in bacteria. Curr Opin Microbiol 4:647–652
82. Margolin W, Long SR (1994) Rhizobium meliloti contains a novel second homolog of the cell division gene ftsZ. J Bacteriol 176:2033–2043
83. Marrison JL, Rutherford SM, Robertson EJ, Lister C, Dean C, Leech RM (1999) The distinctive roles of five different ARC genes in the chloroplast division process in Arabidopsis. Plant J 18:651–662
84. Martin A, Lang D, Hanke ST, Mueller SJ, Sarnighausen E, Vervliet-Scheebaum M, Reski R (2009) Targeted gene knockouts reveal overlapping functions of the five *Physcomitrella patens* FtsZ isoforms in chloroplast division, chloroplast shaping, cell patterning, plant development, and gravity sensing. Mol Plant 2:1359–1372

85. Mazouni K, Domain F, Cassier-Chauvat C, Chauvat F (2004) Molecular analysis of the key cytokinetic components of cyanobacteria: FtsZ, ZipN and MinCDE. Mol Microbiol 52:1145–1158
86. McAndrew RS, Froehlich JE, Vitha S, Stokes KD, Osteryoung KW (2001) Colocalization of plastid division proteins in the chloroplast stromal compartment establishes a new functional relationship between FtsZ1 and FtsZ2 in higher plants. Plant Physiol 127:1656–1666
87. McAndrew RS, Olson BJ, Kadirjan-Kalbach DK, Chi-Ham CL, Vitha S, Froehlich JE, Osteryoung KW (2008) In vivo quantitative relationship between plastid division proteins FtsZ1 and FtsZ2 and identification of ARC6 and ARC3 in a native FtsZ complex. Biochem J 412:367–378
88. Mears JA, Lackner LL, Fang S, Ingerman E, Nunnari J, Hinshaw JE (2011) Conformational changes in Dnm1 support a contractile mechanism for mitochondrial fission. Nat Struct Mol Biol 18:20–26
89. Menges M, Murray JA (2002) Synchronous Arabidopsis suspension cultures for analysis of cell-cycle gene activity. Plant J 30:203–212
90. Mita T, Kanbe T, Tanaka K, Kuroiwa T (1986) A ring structure around the dividing plane of the Cyanidium caldarium chloroplast. Protoplasma 130:211–213
91. Miyagishima S-y, Itoh R, Toda K, Takahashi H, Kuroiwa H, Kuroiwa T (1998) Identification of a triple ring structure involved in plastid division in the primitive red alga Cyanidioschyzon merolae. J Electron Microsc 47:269–272
92. Miyagishima S, Itoh R, Aita S, Kuroiwa H, Kuroiwa T (1999) Isolation of dividing chloroplasts with intact plastid-dividing rings from a synchronous culture of the unicellular red alga cyanidioschyzon merolae. Planta 209:371–375
93. Miyagishima SY, Takahara M, Mori T, Kuroiwa H, Higashiyama T, Kuroiwa T (2001) Plastid division is driven by a complex mechanism that involves differential transition of the bacterial and eukaryotic division rings. Plant Cell 13:2257–2268
94. Miyagishima S, Kuroiwa H, Kuroiwa T (2001) The timing and manner of disassembly of the apparatuses for chloroplast and mitochondrial division in the red alga Cyanidioschyzon merolae. Planta 212:517–528
95. Miyagishima S, Takahara M, Kuroiwa T (2001) Novel filaments 5 nm in diameter constitute the cytosolic ring of the plastid division apparatus. Plant Cell 13:707–721
96. Miyagishima SY, Nishida K, Mori T, Matsuzaki M, Higashiyama T, Kuroiwa H, Kuroiwa T (2003) A plant-specific dynamin-related protein forms a ring at the chloroplast division site. Plant Cell 15:655–665
97. Miyagishima SY, Froehlich JE, Osteryoung KW (2006) PDV1 and PDV2 mediate recruitment of the dynamin-related protein ARC5 to the plastid division site. Plant Cell 18:2517–2530
98. Mokranjac D, Bourenkov G, Hell K, Neupert W, Groll M (2006) Structure and function of Tim14 and Tim16, the J and J-like components of the mitochondrial protein import motor. EMBO J 25:4675–4685
99. Momotani E, Aoki K, Tsuji H (1991) Effect of benzyladenine on diurnal changes in DNA content per chloroplast and chloroplast replication in intact bean leaves. J Exp Bot 42: 1287–1293
100. Nakanishi H, Suzuki K, Kabeya Y, Miyagishima SY (2009) Plant-specific protein MCD1 determines the site of chloroplast division in concert with bacteria-derived MinD. Curr Biol 19:151–156
101. Nishida K, Takahara M, Miyagishima SY, Kuroiwa H, Matsuzaki M, Kuroiwa T (2003) Dynamic recruitment of dynamin for final mitochondrial severance in a primitive red alga. Proc Natl Acad Sci U S A 100:2146–2151
102. Nishida K, Yagisawa F, Kuroiwa H, Nagata T, Kuroiwa T (2005) Cell cycle-regulated, microtubule-independent organelle division in *Cyanidioschyzon merolae*. Mol Biol Cell 16:2493–2502
103. Nobusawa T, Umeda M (2012) Very-long-chain fatty acids have an essential role in plastid division by controlling Z-ring formation in *Arabidopsis thaliana*. Genes Cells 17:709–719

104. Ogino H, Wachi M, Ishii A, Iwai N, Nishida T, Yamada S, Nagai K, Sugai M (2004) FtsZ-dependent localization of GroEL protein at possible division sites. Genes Cells 9:765–771
105. Oikawa A, Joshi HJ, Rennie EA, Ebert B, Manisseri C, Heazlewood JL, Scheller HV (2010) An integrative approach to the identification of Arabidopsis and rice genes involved in xylan and secondary wall development. PLoS ONE 5:e15481
106. Okamoto N, Inouye I (2005) A secondary symbiosis in progress? Science 310:287
107. Okazaki K, Kabeya Y, Suzuki K, Mori T, Ichikawa T, Matsui M, Nakanishi H, Miyagishima SY (2009) The PLASTID DIVISION1 and 2 components of the chloroplast division machinery determine the rate of chloroplast division in land plant cell differentiation. Plant Cell 21:1769–1780
108. Oliva MA, Huecas S, Palacios JM, Martin-Benito J, Valpuesta JM, Andreu JM (2003) Assembly of archaeal cell division protein FtsZ and a GTPase-inactive mutant into double-stranded filaments. J Biol Chem 278:33562–33570
109. Olson BJ, Wang Q, Osteryoung KW (2010) GTP-dependent heteropolymer formation and bundling of chloroplast FtsZ1 and FtsZ2. J Biol Chem 285:20634–20643
110. Osawa M, Anderson DE, Erickson HP (2008) Reconstitution of contractile FtsZ rings in liposomes. Science 320:792–794
111. Osawa M, Anderson DE, Erickson HP (2009) Curved FtsZ protofilaments generate bending forces on liposome membranes. EMBO J 28:3476–3484
112. Osawa M, Erickson HP (2011) Inside-out Z rings-constriction with and without GTP hydrolysis. Mol Microbiol 81:571–579
113. Osteryoung KW, Vierling E (1995) Conserved cell and organelle division. Nature 376:473–474
114. Osteryoung KW, Stokes KD, Rutherford SM, Percival AL, Lee WY (1998) Chloroplast division in higher plants requires members of two functionally divergent gene families with homology to bacterial ftsZ. Plant Cell 10:1991–2004
115. Ouyang X et al (2011) Genome-wide binding site analysis of FAR-RED ELONGATED HYPOCOTYL3 reveals its novel function in Arabidopsis development. Plant Cell 23:2514–2535
116. Palmer CS, Osellame LD, Stojanovski D, Ryan MT (2011) The regulation of mitochondrial morphology: intricate mechanisms and dynamic machinery. Cell Signal 23:1534–1545
117. Pichoff S, Lutkenhaus J (2002) Unique and overlapping roles for ZipA and FtsA in septal ring assembly in *Escherichia coli*. EMBO J 21:685–693
118. Possingham JV (1980) Plastid replication and development in the life cycle of higher plants. Annu Rev Plant Physiol 31:113–129
119. Possingham J, Smith J (1972) Factors affecting chloroplast replication in spinach J Exp Bot 23:1050–1059
120. Pyke K (1997) The genetic control of plastid division in higher plants. Am J Bot 84:1017
121. Pyke KA (1999) Plastid division and development. Plant Cell 11:549–556
122. Pyke KA, Leech RM (1991) Rapid image analysis screening procedure for identifying chloroplast number mutants in mesophyll cells of *Arabidopsis thaliana* (L.) Heynh. Plant Physiol 96:1193–1195
123. Pyke KA, Leech RM (1992) Chloroplast division and expansion is radically altered by nuclear mutations in *Arabidopsis thaliana*. Plant Physiol 99:1005–1008
124. Pyke KA, Page AM (1998) Plastid ontogeny during petal development in Arabidopsis. Plant Physiol 116:797–803
125. Pyke KA, Rutherford SM, Robertson EJ, Leech RM (1994) arc6, A fertile Arabidopsis mutant with only two mesophyll cell chloroplasts. Plant Physiol 106:1169–1177
126. Raynaud C, Cassier-Chauvat C, Perennes C, Bergounioux C (2004) An Arabidopsis homolog of the bacterial cell division inhibitor SulA is involved in plastid division. Plant Cell 16:1801–1811
127. Raynaud C, Perennes C, Reuzeau C, Catrice O, Brown S, Bergounioux C (2005) Cell and plastid division are coordinated through the prereplication factor AtCDT1. Proc Natl Acad Sci U S A 102:8216–8221

128. Reski R, Wehe M, Hadeler B, Marienfeld J, Abel WO (1991) Cytokinin and light quality interact at the molecular level in the chloroplast-mutant PC22 of the moss *Physcomitrella patens*. J Plant Physiol 138: 236–243
129. Robertson EJ, Pyke KA, Leech RM (1995) arc6, an extreme chloroplast division mutant of Arabidopsis also alters proplastid proliferation and morphology in shoot and root apices. J Cell Sci 108(Pt 9):2937–2944
130. Scheffers DJ, den Blaauwen T, Driessen AJ (2000) Non-hydrolysable GTP-gamma-S stabilizes the FtsZ polymer in a GDP-bound state. Mol Microbiol 35:1211–1219
131. Scheffers DJ, de Wit JG, den Blaauwen T, Driessen AJ (2002) GTP hydrolysis of cell division protein FtsZ: evidence that the active site is formed by the association of monomers. Biochemistry 41:521–529
132. Schmitz AJ, Glynn JM, Olson BJ, Stokes KD, Osteryoung KW (2009) Arabidopsis FtsZ2-1 and FtsZ2-2 are functionally redundant, but FtsZ-based plastid division is not essential for chloroplast partitioning or plant growth and development. Mol Plant 2:1211–1222
133. Sheahan MB, Rose RJ, McCurdy DW (2004) Organelle inheritance in plant cell division: the actin cytoskeleton is required for unbiased inheritance of chloroplasts, mitochondria and endoplasmic reticulum in dividing protoplasts. Plant J 37:379–390
134. Shimada H, Koizumi M, Kuroki K, Mochizuki M, Fujimoto H, Ohta H, Masuda T, Takamiya K (2004) ARC3, a chloroplast division factor, is a chimera of prokaryotic FtsZ and part of eukaryotic phosphatidylinositol-4-phosphate 5-kinase. Plant Cell Physiol 45:960–967
135. Shin HW, Takatsu H, Mukai H, Munekata E, Murakami K, Nakayama K (1999) Intermolecular and interdomain interactions of a dynamin-related GTP-binding protein, Dnm1p/Vps1p-like protein. J Biol Chem 274:2780–2785
136. Smirnova E, Griparic L, Shurland DL, van der Bliek AM (2001) Dynamin-related protein Drp1 is required for mitochondrial division in mammalian cells. Mol Biol Cell 12:2245–2256
137. Smith AG, Johnson CB, Vitha S, Holzenburg A (2010) Plant FtsZ1 and FtsZ2 expressed in a eukaryotic host: GTPase activity and self-assembly. FEBS Lett 584:166–172
138. Stokes KD, Osteryoung KW (2003) Early divergence of the FtsZ1 and FtsZ2 plastid division gene families in photosynthetic eukaryotes. Gene 320:97–108
139. Stokes KD, McAndrew RS, Figueroa R, Vitha S, Osteryoung KW (2000) Chloroplast division and morphology are differentially affected by overexpression of FtsZ1 and FtsZ2 genes in Arabidopsis. Plant Physiol 124:1668–1677
140. Sugita C, Kato Y, Yoshioka Y, Tsurumi N, Iida Y, Machida Y, Sugita M (2012) CRUMPLED LEAF (CRL) homologs of *Physcomitrella patens* are involved in the complete separation of dividing plastids. Plant Cell Physiol 53:1124–1133
141. Susin MF, Baldini RL, Gueiros-Filho F, Gomes SL (2006) GroES/GroEL and DnaK/DnaJ have distinct roles in stress responses and during cell cycle progression in *Caulobacter crescentus*. J Bacteriol 188:8044–8053
142. Suzuki K, Nakanishi H, Bower J, Yoder DW, Osteryoung KW, Miyagishima SY (2009) Plastid chaperonin proteins Cpn60 alpha and Cpn60 beta are required for plastid division in *Arabidopsis thaliana*. BMC Plant Biol 9:38
143. Tavva VS, Collins GB, Dinkins RD (2006) Targeted overexpression of the *Escherichia coli* MinC protein in higher plants results in abnormal chloroplasts. Plant Cell Rep 25:341–348
144. Tieu Q, Nunnari J (2000) Mdv1p is a WD repeat protein that interacts with the dynamin-related GTPase, Dnm1p, to trigger mitochondrial division. J Cell Biol 151:353–366
145. Uehara T, Matsuzawa H, Nishimura A (2001) HscA is involved in the dynamics of FtsZ-ring formation in *Escherichia coli* K12. Genes Cells 6:803–814
146. Vitha S, McAndrew RS, Osteryoung KW (2001) FtsZ ring formation at the chloroplast division site in plants. J Cell Biol 153:111–120
147. Vitha S, Froehlich JE, Koksharova O, Pyke KA, van Erp H, Osteryoung KW (2003) ARC6 is a J-domain plastid division protein and an evolutionary descendant of the cyanobacterial cell division protein Ftn2. Plant Cell 15:1918–1933
148. Walter S, Buchner J (2002) Molecular chaperones-cellular machines for protein folding. Angew Chem Int Ed Engl 41:1098–1113

149. Wilson ME, Jensen GS, Haswell ES (2011) Two mechanosensitive channel homologs influence division ring placement in Arabidopsis chloroplasts. Plant Cell 23:2939–2949
150. Yagisawa F, Mori T, Higashiyama T, Kuroiwa H, Kuroiwa T (2003) Regulation of *Brassica rapa* chloroplast proliferation *in vivo* and in cultured leaf disks. Protoplasma 222:139–148
151. Yang Y, Sage TL, Liu Y, Ahmad TR, Marshall WF, Shiu SH, Froehlich JE, Imre KM, Osteryoung KW (2011) CLUMPED CHLOROPLASTS 1 is required for plastid separation in Arabidopsis. Proc Natl Acad Sci U S A 108:18530–18535
152. Yoshida Y, Kuroiwa H, Misumi O, Nishida K, Yagisawa F, Fujiwara T, Nanamiya H, Kawamura F, Kuroiwa T (2006) Isolated chloroplast division machinery can actively constrict after stretching. Science 313:1435–1438
153. Yoshida Y et al (2010) Chloroplasts divide by contraction of a bundle of nanofilaments consisting of polyglucan. Science 329:949–953
154. Zhang X, Hu J (2010) The Arabidopsis chloroplast division protein DYNAMIN-RELATED PROTEIN5B also mediates peroxisome division. Plant Cell 22:431–442
155. Zhang M, Hu Y, Jia J, Li D, Zhang R, Gao H, He Y (2009) CDP1, a novel component of chloroplast division site positioning system in Arabidopsis. Cell Res 19:877–886

Chapter 7
Stromules

Amutha Sampath Kumar, Savithramma P. Dinesh-Kumar and Jeffrey L. Caplan

Abstract Stomules are dynamic tubular structures that extend from plastid's outer surface. Although stromule-like structures were reported in the 1930s, only recently the dynamic nature of stromules is beginning to emerge with the advent of live cell imaging using fluorescent proteins. In this chapter we describe historical observations on stromule-like structures and modern cell biological approaches that are being used to visualize stromules. We summarize current knowledge about stromule formation, regulation, and their interaction with other cellular organelles. Finally, we discuss potential role for stromules during biotic and abiotic stress responses in plants.

Keywords Plastids · Chloroplast · Stromule · Inter-organelle communication · Fluorescent proteins · Virus · Actin · Microtubule · Abiotic and biotic stress responses · Signaling

Abbreviations

FP	Fluorescent protein
TEM	Transmission electron microscopy
DIC	Differential interference contrast
GFP	Green fluorescent protein
CT-GFP	Chloroplast targeted GFP
OEP14	Outer envelope protein
FtsZ	Filamenting temperature sensitive Z
NRIP1	N-receptor interacting protein 1

S. P. Dinesh-Kumar (✉)
Department of Plant Biology and The Genome Center, College of Biological Sciences, University of California, Davis, CA 95616, USA
e-mail: spdineshkumar@ucdavis.edu

A. S. Kumar · J. L. Caplan
Department of Plant and Soil Sciences, Delaware Biotechnology Institute, University of Delaware, Newark, DE 19711, USA
e-mail: amutha@udel.edu; jcaplan@udel.edu

S.M. Theg, F.-A. Wollman (eds.), *Plastid Biology*, Advances in Plant Biology 5,
DOI 10.1007/978-1-4939-1136-3_7,

cpHSC70-1:YFP	The *C*-terminal part of the nuclear encoded *p*lastid targeted *H*eat *S*hock protein *C*ognate 70-1 fused with the *Y*ellow *F*luorescent *P*rotein
BiFC	Bimolecular fluorescence complementation
MP	Movement protein
AM	Arbuscular mychorrhizal
ER	Endoplasmic reticulum
FRAP	Fluorescence recovery after photobleaching

7.1 Introduction

Stromules are stroma-filled tubular projections that extend from plastids outer surface. Although stromule-like structures were described over a hundred years ago, recent use of fluorescent protein (FP) markers and live cell imaging has contributed towards better understanding of their morphology and dynamic movement. Stromules have received increased attention recently owing to their rapid extension and retraction rates within the cell. In this chapter, we provide a comprehensive summary of the characteristics and functions of stromules in plants.

7.2 Initial Observations of Stromule-Like Structures in Plants

The description of stromule-like structures dates back to [24], when the Austrian botanist Haberlandt, while working with the alga *Selaginella* found chains of chloroplasts linked together by thin connections that measured about 30 μm. These structures were believed to be the interconnections due to incomplete separation of the plastids. Later, Senn [73] in his monograph illustrated the presence of a mobile jacket called the peristromium surrounding the chlorophyll- containing chloroplasts of the mosses *Funaria* and *Mnium*, the alga *Bryopsis* and in the liverwort *Marchantia*. Several connecting filaments measuring 0.4–1.0 μm in diameter were found to arise from the peristromium which connected chloroplasts. Following these observations there have been a number of reports where the presence of thin connecting filaments between chloroplasts has been catalogued. In the 1930s many groups reported the existence of these connecting structures in the amyloplasts of the castor bean roots, french bean, squash and onion epidermal cells [12, 21, 28]. Rapid extension and the retraction of the "pseudopodium" in the chloroplasts of *Colchicum bornmulleri* and *Allium angularis* and also the amoeboid movement of the chloroplasts in the alga *Selaginella* has been documented [28]. Similar structures were observed in the chloroplasts of sugar beet with dynamic changes in the shape over a time period of 15 min [12]. Wildman et al. [83] showed that the chlorophyll lacking outer jacket constantly changed shape wherein long protuberances extended into the cytoplasm which further segmented into particles resembling a mitochondrion. Most of these initial observations coincided with the development of compound lens microscopes in the early 1900s. The structures described in these earlier studies correspond to what are now known as "stromules" [47].

7.2.1 Stromules in Different Cells and Tissues

Stromules have been observed in various cell types such as the guard cells, epidermal cells, spongy parenchyma, trichomes, roots, flower petals and mesophyll cells [41]. They are tubular extensions from the plastids that range from 0.5–10 μm in length and exist in both monocots and dicots [41]. Compared to chlorophyll containing plastids, the non-chlorophyll containing plastids, amyloplasts and chromoplasts, show a wide variation in the shape, length and the abundance of stromules [6, 51, 63, 77]. The existence of stromules in the non-chlorophyll cells indicates putative functions other than photosynthesis. In roots, the size and number of the stromules differ in various cell types present at the different zones [41]. In meristematic zones, stromules are more densely packed, shorter and arranged around the nucleus more frequently, whereas in the elongation zone, the stromules are elongated. In the mature region of the roots, stromules are more widely spaced but shorter in length with occasional arrangement around the nucleus. Stromules are short in non-dividing callus tissue; however, in actively dividing cells, they are fairly long, reaching 40 μm and are highly dynamic, with most of them positioned around the nucleus and extending long distances ramifying the cytoplasm [41].

7.2.2 Initial Observations of Stromules in Virus Infected Tissue

Virus infection results in the production of stromules and was first described by Esau [12]. Stromules were abundant in the yellowish chlorotic regions of beet (*Beta vulgaris*) leaves when infected with the beet mosaic virus, but were absent in the green areas. Later, Shalla [74] reported that when tomato leaves were infected with the tobacco mosaic virus, "formation of projections which rapidly extended and contracted" from plastids were observed. However the studies by Shalla [74] failed to provide concrete evidence to support stromule induction during virus infection. After 8–10 days of infection, numerous plastids appeared distorted with long filamentous structures which were occasionally associated with the mitochondria as visualized using phase contrast light microscopy and transmission electron microscopy (TEM). The formation of vacuoles in the mesophyll chloroplasts during TMV infection was observed as a general phenomenon; however, the rapid extension and engulfment of cytosol by stromules was considered as the mode of formation of vacuoles. Due to this engulfment, plastids also contained the virus particles.

7.2.3 Stromules are Dynamic

Stromules with varying diameter from 0.35–1.5 μm have been documented for a number of plant species [1, 42, 51, 75, 80]. Stromules vary considerably because they retract and extend constantly. Lengths measuring 15 μm have been observed in tobacco and petunia [42, 75], 9 μm in *Arabidopsis* [80] and 20 μm in wheat [51]. The longest stromule measuring 200 μm in length was observed in the mesocarp of the tomato fruit [19, 82].

7.2.4 Extension and Retraction Rates

Extension and retraction rates of the stromules have been best observed in the leucoplasts of sub-epidermal peels of the midribs in *Iris unguicularis* [13, 23]. The retraction rates ranged from 0.23 μm/s to as low as 0.05 μm/s. Usually the retraction rates were faster than the outgrowth and the outgrowth of stromules was frequented with stops and starts accompanied with temporary retractions. Looping was commonly observed with the retraction of the stromules and these loops produced new stromules besides those arising from the other regions of the chloroplasts [23]. Stromules form almost 35 % of the surface area of chloroplasts, suggesting a high rate of membrane formation and disposal. Frequently, the un-anchored regions of the stromules show intense cytoplasmic streaming activity with continuous change in directions. Besides retraction and extension, stromules show transient positioning and anchoring which in turn is coordinated with the line of moving organelles. Depending on the movement of the organelles, the free end of the stromules position themselves transiently within the cell. Live virtual cytoplasmic streaming, retention and extension were captured using the differential interference contrast (DIC) microscopy in the tomato trichome [23]. Together with retraction and extension, stromules also orchestrate a wide variety of branching and anchoring within the cells. The branch points also moved along the cytoplasmic stream or along the length of the stromules. Besides shedding of the stromule tip, bridging of stromules was also observed as stromules connected two plastids. This bridge was reported to be thicker than the typical stromule thickness, however, the bridge moved out of the field of view in a tandem fashion after maintaining connection with the stromules.

Primarily, stromule movement is dependent upon the actin cytoskeleton [19, 30, 44]. The power for the movement is derived from the ATPase activity of myosin XI [59, 69]. Recently it was shown that the plastid filamentation is reversible using chemicals such as antimycin. The root cortical cells of *A. thaliana* showed reversible plastid filamentation when treated with the respiratory inhibitor compound antimycin A. Instead of imparting changes in the mitochondria, the inhibitor caused extreme filamentation of the plastids in the mature cortical cells of the roots. Filamentation was rapid and was observed as early as 5 min and was completed within an hour. However, the filaments were different from the stromules. The plastid body actually changed to a filamentous structure rather than as a separate filament emerging from the plastid's main body. The filamentation reverted back to the original condition when the chemical compound was washed away [35].

7.3 Visualization of Stromules

7.3.1 Differential Interference Contrast (DIC) Imaging of Stromules

Conventional light microscopy in the early nineteenth century paved the way for the visualization of stromules. Filamentous structures measuring ≥9 μm in length and less than 0.5 μm in diameter were observed with compound light microscopes. Though light microscopy provided ample evidence supporting the existence of stromules, electron microscopic images banished several of the light microscopic observations of stromules as artifacts [3]. Though evidence for the existence of stromules accumulated, until the early 1990s, many questioned the mere existence of stromules as a physiological entity of plastids. However, the advent of the confocal microscopy together with fluorescent protein markers revolutionized the visualization and observation of the dynamics of stromules in living plant cells [42, 58].

Prior to the advent of green fluorescent protein (GFP), phase contrast light microscopy was the most wide spread way of imaging plastid protuberances and the dynamic nature of stromules. Major breakthroughs in the observation of the mobility of the stromal contents between chloroplasts were made by phase contrast imaging in the trichomes of tobacco and tomato [83] and in spinach [32, 76]. Live capture of the breakage of small organelles from the chloroplast protuberances, and the association of mitochondria-like structures fusing with the plastid filaments raised widespread speculation regarding the authenticity of the findings due to limitation on the resolving power. However, to date there is no definite support to authenticate the findings of the mitochondria like structures with the stromules as recorded by Wildman et al. [83]. Time-lapsed microscopy was used by Hongladarom et al. [32] to show real time dynamism of the stromal contents, the shape changes and the extension of the chloroplast protuberances in the mesophyll cells of spinach. These protuberances were long (15 μm), branched and extended all over the cytosol. Following which Suzuki et al. [79] observed plastids with "elongated and string like" structures using phase contrast microscopy in suspension cultured *Nicotiana tabacum* BY-2 cells.

7.3.2 Fluorescent Protein Markers and Confocal Microscopy

Stromules and their existence have been widely accepted only after the advent of the use of GFP as a marker to track stromules [19]. Stromule identification and observation has been difficult due to the very low frequency of stromules in mature leaves. However, labeling of the stromules found in all cell types using the GFP or other types of FPs has made it possible to easily observe stromules under *in vivo* conditions. Kohler et al. [42] were the first to show the presence of the stromules more abundantly in the epidermal cells and the roots of the plants using the transgenic tobacco plants expressing GFP in the plastids. The fusion of the GFP with the RecA–transit peptide (CT-GFP, Chloroplast targeted GFP) under the control of *35S CaMV* promoter helped visualize the stromules as the CT-GFP specifically associated with the soluble stroma fraction [42]. These findings were replicated in other labs and Kohler and Hanson [41] coined the term "stromules" to describe stroma filled tubules projecting from plastid surface. Since then other labs have provided proof of existence of stromules and they are now considered an integral part of the plastids [1, 18, 42, 57, 75, 80]. By developing transgenic plants expressing GFP, it was possible to observe stromules in different tissue types as the GFP was localized in the plastids. The important advantage of transgenic plants expressing GFP is that the presence of GFP does not alter the physiology of the plastids. Thus GFP made it possible to determine developmental variations in size, shape, structure and position of the stromules in cell types otherwise not amenable to conventional light microscopy.

GFP has been used to study the connections that occur between plastids and other organelles, such as the nucleus. Using photobleaching techniques in root cells, it has been shown that proteins could transfer between plastids through stromule connections [42]. However, such transfer was not observed in suspension cells [42] nor in the leucoplasts of roots [70]. Though shedding of the stromule tips has been previously shown through phase contrast or DIC microscopy [22, 23, 32, 83], none of the studies using GFP could provide evidence for the release of these free-streaming organelles except where GFP-containing vesicles have been observed in the epidermal cells of tobacco [1] and in the mesocarp cells of the tomato fruit [66, 82]. However, in these studies proof that these vesicles originated from the GFP expressing chloroplasts is lacking.

7.3.3 Transmission Electron Microscopy (TEM) and Cryo-fixation of Stromules

Before the advent of confocal microscopy, electron microscopy was the preferred method to observe stromules. Though conventional fixation and electron microscopy provides high resolution micrographs of stromules, artifacts and the inability to protect the sensitive structures caused a decline in their use. To better preserve stromules, high pressure freezing and freeze substitution was used as an alternative method by [3] in mesophyll cells of rice leaves. Rapid freezing is considered a robust method to maintain the original structure without the formation of structure-disrupting crystals of ice [37, 39]. The cells are fixed close to their native state. Stromules were well preserved using the high pressure freezing technique; however, breakages in the cell wall and the plasma membrane were common and the cell membranes were poorly stained. Nonetheless, these techniques allowed stromules to be more accurately measured as thin filaments measuring 0.475 μm in mean diameter and maximum length of 4.1 μm.

7.4 Origin, Formation and Regulation of Stromule Formation

7.4.1 Chloroplast Envelope

Stromules consist of both the outer and the inner envelope of the chloroplasts. Though attempts were made to separate the inner and the outer envelope of the stromules by isotonic shock [8], Scott and Theg [72] showed that both the membranes were separate and suggested that occasionally the membranes might to be fused during protein translocation with the help of translocases. By marking the inner and the outer membrane envelopes of the stromules with different colored FPs, stromules were confirmed to be extensions of the double membrane of the plastids present in higher plants thus corroborating the previous studies [19, 26]. The presence of the double layer in the stromules was confirmed by the fusion of the GFP with the OEP14 (Outer Envelope Protein) [52] and the GFP fused with the phosphate translocator which targets the protein to the inner envelope [40]. Intact existence of both the membranes through electron micrographs were described by Holzinger et al. [29, 30]. This was later confirmed by the distinctive labeling of the inner and the outer membranes of the stromules with the fluorescent proteins [19, 26]. However, the functional and biological significance behind the formation of these dynamic structures are yet to be unraveled. Hanson and Sattarzadeh [27] speculated that the stromules emerge due to the pressure from within the plastid. Recently, a unique class of the Filamenting temperature sensitive Z (FtsZ) proteins were proposed to play a role in the initiation and formation of stromules. Two families, FtsZ-1 and FtsZ-2 participate in plastid division and are capable of forming the compacted protofilaments under *in vitro* conditions [62]. These FtsZ proteins are considered to be the volunteers in the cytoskeletal structures involved in stromule formation [68]. However, Itoh [35] have reported that FtsZ1 proteins are not involved in the antimycin-induced plastidic filamentation. Apart from these proteins, *arc* (accumulation and replication of chloroplast) genes involved in the chloroplast division are also considered to play a role in the formation of the stromules. The stromule lengths in the *arc3* and *arc6* mutants were longer than those observed in wild type plants [31].

7.4.2 Actin and Microtubule Requirements

Plant cells are known for their ability to undergo rapid cytoplasmic streaming and plastids have been shown to be involved in this highly dynamic process [84]. Microfilaments play a pivotal role in the orientation and the positioning of the chloroplasts in plants [11, 38, 56, 85]. Several lines of evidence suggest a close association between the plastids (both green and non-green) and the microfilaments [9, 36, 38]. Stromules have been found in close association with the microfilaments, as confirmed through DIC imaging [22, 23, 45] and through CLSM in the epidermal cells of the hypocotyls in *Arabidopsis* [45].

Plastid movement was affected in treatments involving the actin microfilament inhibitors. Using latrunculin B that depolymerizes microfilaments and 2,3-butanedione 2-monoxime, which inhibits myosin ATPase activity, it was shown that stromules move along the actin microfilaments using the myosin motors [19]. Using a microfilament inhibitor cytochalasin, but not by the microtubule inhibitors aminoprophosmethyl or chloroisopropyl-N-phenylcarbamate, it was shown that the movement could be stopped [19, 44]. In contrast, microtubule inhibitors, oryzalin and amiprophosmethyl in onion epidermal cells or tobacco hypocotyl epidermal cells had no effect on stromule movement and in fact decreased stromule length by 75 % [19, 44]. Recently Natesan et al. [59] noticed a decrease in the length and the number of stromules after treating the cells with the myosin ATPase inhibitor 2,3-butanedion 2-monoxime. The treatment not only abolished cytoplasmic streaming but also the movement of stromules supporting the fact that myosin motors play a significant role in the mobility of stromules. In agreement with these findings, silencing of myosin XI in *Nicotiana benthamiana* using tobacco rattle virus based silencing vector [53] altered the positioning of the plastids and affected stromule formation [69]. In an independent study, the inverted repeat-mediated silencing of myosin XI in transgenic *N. benthamiana* plants reduced stromule number in the tobacco leaf epidermal cells [59].

GFP was used as the primary marker to study the active interaction of plastids and microfilaments without the need for the microinjection and immunolocalization [45]. The GFP protein was fused with hTalin to study the interactions between the stromules and microfilaments. Both confocal microscopy as well as DIC was used to study the plastids, stromules and their interaction with the microfilaments, especially in the dark grown plants. Predominantly the plastids were engulfed in a network of microfilaments. Simultaneously the authors also found that the expression of the GFP-hTalin in the plastids did not disrupt the normal activity of the stromules or the plastids. Time lapse studies revealed that lower regions of the stromules are bound by the microfilaments that run in parallel, while the top portion remains swaying in the cytoplasmic matrix and is constantly bombarded by the vesicles whose movement are themselves guided by microfilaments. While the microfilaments help in the anchoring, they also aid in the movement of the stromules. Though bound on either side by the intersecting thick microfilaments, while the plastids move, the intersection point of the microfilaments also moved. The *in vivo* studies of Kwok and Hanson [45] also revealed that the mitochondria and the plastids colocalized and both were found in close association with the same microfilaments. Collectively, these findings illustrate that stromules utilize the available cell motility system within the plant cells.

7.5 Induction and Regulation of Stromules

7.5.1 Abiotic Stress Responses

Stromules were initially considered to be structures associated with stress responses in plants. Drought and salt stress in wheat and barley were shown to result in plastids with protrusions [16, 33, 86]. In angiosperms, stromules were detected in desiccated mosses after hydration [64]. The extensions in the moss *Polytrichum formosum* [65] were even more prominent in the leaf lamella after hydration of the desiccated tissues. Most of these observations were recorded using electron microscopy, which raises serious concerns about the original structure of the stromules as most of them are visible only after serial sectioning through the tissues. Recently, Gray et al. [20] used confocal microscopy to show that stromules are significantly induced upon abiotic stress treatments in *N. tabacum* seedlings, and the monocot, wheat. The increase was proportional to the concentration of PEG or mannitol used to induce abiotic stress. Furthermore, addition of the salts NaCl and KCl increased stromule number. Light also plays a major role in the number of stromules in hypocotyl epidermal cells. Light grown seedlings had fewer stromules compared to the dark grown seedlings [44]. Stromule formation was also shown to be sensitive to blue light rather than red or far-red light [20].

Holzinger et al. [29] have conducted detailed studies on the effect of temperature on the induction of stromules in *A. thaliana*. Using a temperature controlled chamber fitted to the light microscope, they showed that stromule formation in the plastids of the *A. thaliana* mesophyll cells was temperature dependent. At lower temperatures (5–15 °C), there was no stromule induction whereas as the temperature increased, the authors observed stromule induction. Previously "chloroplast proliferations" were observed in arctic-alpine plant *Ranunculus glacialis* [54, 55]. At increasing temperatures more protrusions from the plastids were shown to occur. At 35 °C, long thylakoid free stroma-filled tubules extended from the plastid resulting in an irregular shape of the chloroplasts. Between 35–45 °C, numerous plastids were shown to possess the protrusions also referred to as the "mobile jackets" of the chloroplasts. Though numerous protrusions were observed, not all protrusions extended as stromules [76]. Buchner et al. [4] initiated an in depth understanding of the temperature-mediated effects on the formation of the stroma-filled protrusions in the leaf mesophyll cells of nine different alpine plants. Effects of stepwise increase in temperatures on the protrusions were analyzed. In seven out of the nine plants tested, temperatures over 20 °C increased chloroplast protrusions with light intensities having no effect on stromule induction.

7.5.2 Biotic Stress Defense Response

Stromules were also observed in plants during biotic stress. Stromule formation during defense responses in plants was reported by Caplan et al. [7]. The authors observed stromule induction in the presence of the Tobacco Mosaic Virus effector protein p50. Through fluorescent proteins, they also showed that the in the presence of the p50 viral effector, the N-receptor interacting protein 1 (NRIP1) which localizes in the chloroplasts was distributed in the cytoplasm and the nucleus, suggesting a possible role of stromules in this process.

During the infection of the *A. thaliana* plants by the abutilon mosaic virus, "strings of pearls" were found attached to the plastids [50]. The cpHSC70-1:YFP (the *C*-terminal part of the nuclear encoded *p*lastid targeted *H*eat *S*hock protein *C*ognate 70-1 fused with the *Y*ellow *F*luorescent *P*rotein) was found in these strings of pearls as identified using the bimolecular fluorescence complementation (BiFC). A movement protein (MP) from the Abutilon Mosaic Virus and the nuclear/plastid-targeted heat shock cognate protein were shown to co-localize in the chloroplasts and in the periphery of the thin filaments extending from the chloroplasts. In some cells these structures extended from the plastid to the cell periphery, especially in the virus infected cells, while some plastids had short-length filaments (in uninfected cells). These structures were referred to as "stromules". Though the study did not target the elucidation of the functions of the stromules, the presence of the stromules with the labeled target protein during virus infection suggests the functional significance of the stromules in the distribution of the labeled protein and the involvement of the stromules in the distribution of the cpHSC70-1 oligomers throughout the cell during a virus infection [49].

Infection with root arbuscular mycorrhizae led to an increase in stromule proliferation in the root cortical cells [14, 15, 78]. Stromule-like structures were observed to protrude from the plastids in the root cells of *Alnus glutinosa* after infection by the actinomycete *Frankia* [17]. Numerous "elongate" plastids were observed in root cortical cells colonized by the arbuscular mychorrhizal (AM) fungi. Upon colonization, there occurred a massive increase in the number of plastids as shown by GFP expression in the plastids of transgenic *N. tabacum* cortical cells with AM fungi. Surprisingly, the authors found that the plastids were interconnected by numerous thin structures, which were referred to as "stromules". The nuclei of the cortical cells showed numerous plastids around them resembling an "octopus-like" structure as observed by Kohler and Hanson [41] and Kwok and Hanson [48]. The authors speculated that the cortical cell nuclei depended on the plastid derived nucleotides, and suggested that the plastid extensions might play a major role in transport [14, 15]. These observations were similar to those observed by Dexheimer et al. [10] in which long curved plastids were seen after AM colonization of the roots of *Prunus avium* and *Pirus malus*. The close association of the tubular network in the cells with the arbuscules also led to speculation that there might be terpenoid secretion due to the plant defense response.

7.5.3 ABA and H_2O_2-Mediated Induction of Stromules

By using ABA biosynthesis inhibitors, it was shown that ABA is directly linked to stromule formation in the presence of mannitol [67]. Since ABA was primarily involved in stromule formation, the authors further supported their findings by including methyl jasmonic acid, salicylic acid, ethylene and H_2O_2, suggesting that stromule formation responds to herbivore and pathogen attack. Among these, salicylic acid treatment had a negative effect on stromule induction. The treatment of the light-grown wheat seedlings with 0.1 mM H_2O_2 resulted in stromule induction in root hair cells, and the induction was sensitive to abamine, an ABA biosynthesis inhibitor. The presence of sugars such as glucose and sucrose in the apoplast increased the formation of the stromules in epidermal cells and the palisade parenchyma chloroplasts [71]. The authors concluded that epidermal cells are an excellent choice for the study of the physiological conditions influencing the stromule induction.

7.6 Putative Functions of Stromules

7.6.1 Increased Chloroplast Surface Area

Given that stromules can account for up to 50% of the plastid surface area it is logical to contemplate the importance of the large surface area in exchanging the materials with the cytosol. Extensions from the plastids increase the surface area of the envelope membrane and may facilitate transport of substances in and out of the plastids. Besides, there exists an inverse relationship between the total number of the plastids present and the number of stromules. Cells with fewer numbers of plastids (epidermal and the root) possess more stromules than do mesophyll cells [41, 82]. With approximately 15% of the volume of the plastid material, it is apparent that stromules could maximize the area for vital signal transduction and movement of the substrates [19]. More research into the morphometric analysis of membrane surface area would provide more definitive answers regarding the presence of increased stromule numbers in certain cell types. Though there is an increase in the surface area, minimal changes occur with respect to the volume. Thus one of the prominent functions of stromules may be to extend as much as possible into the cytosol to facilitate metabolite exchange as plastid metabolic functions are connected with other cellular organelles such as mitochondria, peroxisome and the endoplasmic reticulum (ER). It has been documented that, GFP moves more slowly in the stroma due to its viscous nature than in the cytosol [27]. Taken together, (i) the encircling of the cytosol along with the ER, ribosomes, and mitochondria by the plastids in the root tips cells [60], (ii) the more organized arrangement of the stromules and the plastids around the nuclei in the cell suspensions of tobacco [42, 48], and (iii) the large network of the stromules within the cells after AM infection all suggest that the stromules not only aid in metabolite exchange but may also contain transporters [14, 25]. The extensive framework of the stromules from the plastids of the AM infected cells also suggests that several metabolites might move from the plastids to the arbuscules during this symbiotic interaction. Mycorrhiza induces the accumulation of carotenoids, and stromules might be involved in the transport and exchange of the plastid-derived terpenoids, amino acids and nucleosides necessary for the AM fungi development. In all the above cases, stromules were not observed as a general phenomenon but only under infection regimes and unique culture conditions.

7.6.2 Intra- and Inter-organelle Trafficking

Despite size exclusion limits, it is possible that macromolecules and higher order structures such as chloroplast polysomal complexes, RNA, ribosomes, and protein complexes might make it through the stromules. Bourett et al. [3] showed the movement of 550 kDa Rubisco (Ribulose bis-phosphate carboxylase) through the protuberances of the chloroplasts in the rice mesophyll cells. During transfer, thylakoid components are excluded from the stromules which suggests that only stromal components can move through the stromules. Though plastid DNA is restricted to nucleoids, recombination facilitates transfer of DNA between the plastids during protoplast fusion. However, these initial observations still await confirmation by other labs.

The first insights into the inter-organelle trafficking by the stromules was put forth by Kohler et al. [42] via photo bleaching experiments (FRAP, Fluorescence Recovery after Photobleaching). The interconnected plastids were targeted for the experiments wherein one of the connected plastid was photo-bleached and after a 7 s recovery time period, the other plastid fluoresced confirming that stromules allow the movement of a 27 kDa globular protein between plastids through the stromules. Simultaneously these findings were also corroborated by Tirlapur et al. [80] and Kohler et al. [43]. Kohler et al. [43] used two-photon excitation fluorescence correlation spectroscopy; the concentration and movement of the small numbers of fluorescent molecules were tracked and were reported to occur as batches with 10–100 fold increase in brightness than the background. These findings were supported by using the active transport inhibitors and uncouplers of the oxidative phosphorylation. However, the mere presence of the GFP protein did result in the movement across the stromules as GFP proteins are not plastid specific. The typical illustration by Kohler et al. [43] suggests that there occurs no movement of GFP even after photo bleaching of the selected plastids in case of the tobacco suspension cells where the plastids encircle the nucleus with interconnections. Recent findings of Schattat et al. [70] refuted the idea that exchange of ions, molecules or any other metabolite takes place via the stromules connecting two plastids. Using differential coloring of the individual plastids through expression of photoswitchable fluorescent proteins, it was shown that the plastids maintained their identity even when the stromules from the plastids formed networks. This showed that the transfer did not take place even if the stromules were present. Besides, there are several reports that indicate that presence of stromules does not guarantee transport of macromolecules across stromules. However, other studies suggest that transfer of contents between plastids do occur.

Recently, Schattat and Klosgen [71] have highlighted the importance of stromules in cellular transport in the upper epidermal cells of *A. thaliana* leaves. Stromule numbers were shown to significantly increase in the apoplast of the leaves in the presence of glucose or sucrose, but not of fructose, sorbitol or mannitol. Plastids present in the epidermal cells are photosynthetically less active and they import large amounts of the glucose-6-phosphate for their energy and basic metabolism. In return, the triose phosphates produced are transported back to the cytosol. Thus there exists a constant transport of the metabolites and this correlates with the increased abundance of stromules in the presence of the apoplastic sugars. Though normally the chloroplasts of the green tissues do not produce stromules, in the presence of sugars the source activity is suppressed and the sink activity is initiated. Consequently, stromules are more abundant in ripening tomato fruits than non-green tomatoes [82]. Thus stromules are also seen to play an important role in carbohydrate metabolism.

Newell et al. [61] used GFP-labeled plastid DNA and ribosomes to study their movement through the stromules in tobacco and *Arabidopsis*. Though GFP fluorescence could be detected in the plastids, the stromules did not show the presence of the GFP, suggesting that the transfer of genetic information through the stromules does not take place.

7.6.3 Signaling

The close association of the stromules with the nuclear [46, 48] and the plasma membranes [80] suggests that signaling and exchange of biophysical signals takes place at the interface. However, it is yet to be experimentally demonstrated what types of signals could move across the point of contact between stromules and the target organelles. Caplan et al. [7] suggested that stromules may be actively involved in chloroplast-cytoplasm or chloroplast-nucleus signaling when the plant mounts a defense response. The tobacco N immune receptor that detects and activates defense against TMV does so by interacting with the chloroplastic protein, NRIP1. NRIP1 normally resides in the chloroplast and is recruited to the cytoplasm and the nucleus by TMV. The N immune receptor recognizes the TMV-NRIP1 pre-immune complex to activate successful defense response. Caplan et al. [7] speculated that NRIP1 could be released from chloroplast through stromules into the cytoplasm and nucleus. Recently, some proteins present in the stroma were shown to be involved in the sensing of the pathogens and the presence of sugars, but whether this occurs via stromules need to be addressed [34, 49, 81].

7.6.4 Starch Granule Formation

Apart from signaling and macromolecular transport, stromules are suggested to play a role in the formation of the B-type starch granule in wheat endosperm [51]. Using TEM, Buttrose [5, 6] and Parker [63] observed stromule like protrusions during the B-type starch granule formation. Langevald et al. [51] used CLSM to show that these protrusions arise from the amyloplasts. CLSM showed the presence of the connections between the amyloplasts and B-type granules in the protrusions, and time-lapse movies showed the movement of these protrusions. Betchel and Wilson [2] later studied the development and maturation of wheat starch post-flowering using TEM. The sequential observations of wheat starch show that A-type amyloplasts form protrusions within which the B-type granules develop. The degree of occurrence of these protrusions and their duration could not be determined. However, these studies concluded that the protrusions occurred at a specific time during the endosperm development. The TEM sections of wheat endosperm tissue provided strong evidence that the division of the amyloplasts occurred in the area where plastid protrusions were found. The division of the protrusions into individual amyloplasts is considered to be a unique form of plastid division in wheat. Due to the presence of the large starch granule within the plastid, these protrusions are the only viable areas for amyloplast division. The occurrence of the protrusions was also synchronized with time during which the B and C-type starch granule formed in wheat starch. The protrusions were prevalent during the entire 10–12 days of development of the wheat starch granule after flowering, and they were observed before the onset of the C-type starch granule formation i.e., 17 days after flowering.

7.7 Future Insights

Stromules are dynamic structures that are predicted to have multifarious functions in different cell types and under physiological conditions. An extensive literature indicates that they share an intimate relationship with several vital cellular organelles involved in the transport, signaling and diffusion of metabolites. However experimental evidence is lacking regarding functional significance of stromules in specific biological processes. In addition, several important questions remain to be answered: What triggers the formation of the stromules? What components or signals could move across the membranes and through the stromules? Are stromules formed in response to increased pressure within the plastids as projections or do the projections form as a result of extensive pulling forces from outside? Are there specific types of stromules for specific functions? It is challenging to capture the complex interplay and movement of the proteins and other defense related compounds via stromules as these structures are highly sensitive to the conventional fixation techniques and laser exposure. With the wealth of information on the existence of stromules as a physical entity of the chloroplasts, it would be interesting to probe into their origin and functions using genetic and molecular approaches. To date genes that control the formation of stromules have not been identified. The spontaneous existence of these thin tubular structures in different plant cell types and their highly sensitive nature complicates simple methods used for stromule studies. Available advanced techniques such as fixation by high pressure freezing, TEM tomography and combining light and EM via correlative microscopy, will make it possible to unravel the various intricate mechanisms involved in stromule formation and their connection with other intra-cellular organelles.

Acknowledgments Stromule research in JC and SPD-K laboratory is supported by NIH grant GM097587.

References

1. Arimura S-I, Hirai A, Tsutsumi N (2001) Numerous and highly developed tubular projections from plastids observed in tobacco epidermal cells. Plant Sci 160:449–454
2. Bechtel DB, Wilson JD (2003) Amyloplast formation and starch grain development in hard red winter wheat. Cereal Chem 80:175–183
3. Bourett TM, Czymmek KJ, Howard RJ (1999) Ultrastructure of chloroplast protuberances in rice leaves preserved by high-pressure freezing. Planta 208:472–479
4. Buchner O, Holzinger A, Lütz C (2007) Effects of temperature and light on the formation of chloroplast protrusions in leaf mesophyll cells of high alpine plants. Plant Cell Environ 30:1347–1356
5. Buttrose MS (1960) Submicroscopic development and structure of starch granules in cereal endosperms. J Ultrastruct Res 4:231–257
6. Buttrose MS (1963) Ultrastructure of the developing wheat endosperm. Aus J Biol Sci 16:305–317
7. Caplan JL, Mamillapalli P, Burch-Smith TM, Czymmek K, Dinesh-Kumar SP (2008) Chloroplastic protein NRIP1 mediates innate immune receptor recognition of a viral effector. Cell 132:449–462
8. Cline K, Andrews J, Mersey B, Newcomb EH, Keegstra K (1981) Separation and characterization of inner and outer envelope membranes of pea chloroplasts. Proc Natl Acad Sci U S A 78:3595–3599
9. Collings DA, Zsuppan G, Allen NS, Blancaflor EB (2001) Demonstration of prominent actin filaments in the root columella. Planta 212:392–403
10. Dexheimer J, Gérard J, Bougarda K, Jeanmaire C (1990) Les plastes des cellules-hô tes des mycorrhizes à vésicules et arbuscules. Can J Bot 68:50–55
11. Dong X-J, Ry J-H, Takagi S, Nagai R (1996) Dynamic changes in the organization of microfilaments associated with the photocontrolled motility of chloroplasts in epidermal cells of *Vallisneria*. Protoplasma 195:18–24
12. Esau K (1944) Anatomical and cytological studies on beet mosaic. J Agric Res 69:95–117
13. Faull AF (1935) Elaioplasts in iris: a morphological study. J Arnold Arboretum 16:225–267

14. Fester T, Strack D, Hause B (2001) Reorganization of tobacco root plastids during arbuscule development. Planta 213:864–868
15. Fester T, Lohse S, Halfmann K (2007) "Chromoplast" development in arbuscular mycorrhizal roots. Phytochemistry 68:92–100
16. Freeman T, Duysen M (1975) The effect of imposed water stress on the development and ultrastructure of wheat chloroplasts. Protoplasma 83:131–145
17. Gardner IC, Abbas H, Scott A (1989) The occurrence of amoeboid plastids in the actinorhizal root nodules of *Alnus glutinosa* (L.) Gaertn. Plant Cell Environ 12:205–211
18. Gray JC, Hibberd JM, Linley PJ, Uijtewaal B (1999) GFP movement between chloroplasts. Nat Biotechnol 17:1146
19. Gray JC, Sullivan JA, Hibberd JM, Hansen MR (2001) Stromules: mobile protrusions and interconnections between plastids. Plant Biol 3:223–233
20. Gray JC, Hansen MR, Shaw DJ, Graham K, Dale R, Smallman P, Natesan SKA, Newell CA (2012) Plastid stromules are induced by stress treatments acting through abscisic acid. Plant J 69:387–398
21. Guilliermond A (1934) Les Consituants Morphologiques du Cytoplasme: Le Chondriome. Hermann, Paris
22. Gunning BES (2004) Plant cell biology on CD: information for students and a resource for teachers. BES Gunning Pty Ltd, Canberra
23. Gunning BE (2005) Plastid stromules: video microscopy of their outgrowth, retraction, tensioning, anchoring, branching, bridging, and tip-shedding. Protoplasma 225:33–42
24. Haberlandt G (1888) Die Chlorophyllkörper der Selaginellen. Flora 71:291–308
25. Hans J, Hause B, Strack D, Walter MH (2004) Cloning, characterization, and immunolocalization of a mycorrhiza-inducible 1-deoxy-d-xylulose 5-phosphate reductoisomerase in arbuscule-containing cells of maize. Plant Physiol 134:614–624
26. Hanson MR, Sattarzadeh A (2008) Dynamic morphology of plastids and stromules in angiosperm plants. Plant Cell Environ 31:646–657
27. Hanson MR, Sattarzadeh A (2011) Stromules: recent insights into a long neglected feature of plastid morphology and function. Plant Physiol 155:1486–1492
28. Heitz E (1937) Untersuchungen über den Bau der Plastiden I. Die gerichteten Chlorophyllscheiben der Chloroplasten. Planta 26:134–163
29. Holzinger A, Buchner O, Lütz C, Hanson MR (2007a) Temperature-sensitive formation of chloroplast protrusions and stromules in mesophyll cells of *Arabidopsis thaliana*. Protoplasma 230:23–30
30. Holzinger A, Wasteneys GO, Lütz C (2007b) Investigating cytoskeletal function in chloroplast protrusion formation in the arctic-alpine plant *Oxyria digyna*. Plant Biol 9:400–410
31. Holzinger A, Kwok EY, Hanson M (2008) Effects of *arc3*, *arc5* and *arc6* mutations on plastid morphology and stromule formation in green and nongreen tissues of *Arabidopsis thaliana*. Photochem Photobiol 84:1324–1335
32. Hongladarom T, Honda S, Wildman SG (1964) Organelles in living plant cells. Videotape available from University of California Extension Center for Media and Independent Learning, Berkeley, California
33. Huang CX, van Steveninck RFM (1990) Salinity induced structural changes in meristematic cells of barley roots. New Phytol 115:17–22
34. Huang J, Taylor JP, Chen JG, Uhrig JF, Schnell DJ, Nakagawa T, Korth KL, Jones AM (2006) The plastid protein THYLAKOID FORMATION1 and the plasma membrane G-protein GPA1 interact in a novel sugar-signaling mechanism in *Arabidopsis*. Plant Cell 18:1226–1238
35. Itoh RD, Yamasaki H, Septiana A, Yoshida S, Fujiwara MT (2010) Chemical induction of rapid and reversible plastid filamentation in *Arabidopsis thaliana* roots. Physiol Plant 139:144–158
36. Kadota A, Wada M (1992) Photoinduction of formation of circular structures by microfilaments on chloroplasts during intracellular orientation in protonemal cells of the fern *Adiantum capillus-veneris*. Protoplasma 167:97–107

37. Kaeser W, Koyro H-W, Moor H (1989) Cryofixation of plant tissues without pretreatment. J Microsc 154:279–288
38. Kandasamy M, Meagher R (1999) Actin-organelle interaction: association with chloroplast in *Arabidopsis* leaf mesophyll cells. Cell Motil Cytoskeleton 44:110–118
39. Kiss JZ, Staehelin LA (1995) High pressure freezing. In: Severs NJ, Shotton DM (eds) Rapid freezing, freeze fracture and deep etching. Wiley-Liss, Somerset, pp 89–104
40. Knight JS, Gray JC (1995) The N-terminal hydrophobic region of the mature phosphate translocator is sufficient for targeting to the chloroplast inner envelope membrane. Plant Cell 7:1421–1432
41. Köhler RH, Hanson MR (2000) Plastid tubules of higher plants are tissue-specific and developmentally regulated. J Cell Sci 113:81–89
42. Köhler RH, Cao J, Zipfel WR, Webb WW, Hanson MR (1997) Exchange of protein molecules through connections between higher plant plastids. Science 276:2039–2042
43. Köhler RH, Schwille P, Webb WW, Hanson MR (2000) Active protein transport through plastid tubules: velocity quantified by fluorescence correlation spectroscopy. J Cell Sci 113:3921–3930
44. Kwok EY, Hanson MR (2003) Microfilaments and microtubules control the morphology and movement of non-green plastids and stromules in *Nicotiana tabacum*. Plant J 35:16–26
45. Kwok EY, Hanson MR (2004a) In vivo analysis of interactions between GFP-labeled microfilaments and plastid stromules. BMC Plant Biol 4:2
46. Kwok EY, Hanson MR (2004b) GFP-labeled Rubisco and aspartate aminotransferase are present in plastid stromules and traffic between plastids. J Exp Bot 55:595–604
47. Kwok EY, Hanson MR (2004c) Stromules and the dynamic nature of plastid morphology. J Microsc 214:124–137
48. Kwok EY, Hanson MR (2004d) Plastids and stromules interact with the nucleus and cell membranes in vascular plants. Plant Cell Rep 23:188–195
49. Krenz B, Windeisen V, Wege C, Jeske H, Kleinow T (2010) A plastid-targeted heat shock cognate 70 kDa protein interacts with the Abutilon mosaic virus movement protein. Virology 401:6–17
50. Krenz B, Neugart F, Kleinow T, Jeske H (2011) Self-interaction of Abutilon mosaic virus replication initiator protein (Rep) in plant cell nuclei. Virus Res 161:194–197
51. Langeveld SMJ, van Wijk R, Stuurman N, Kijne JW, dePater S (2000) B-type granule containing protrusions and interconnections between amyloplasts in developing wheat endosperm revealed by transmission electron microscopy and GFP expression. J Exp Bot 51:1357–1361
52. Li H-M, Moore T, Keegstra K (1991) Targeting of proteins to the outer envelope membrane uses a different pathway than transport into chloroplasts. Plant Cell 3:709–717
53. Liu Y, Schiff M, Marathe R, Dinesh-Kumar SP (2002) Tobacco Rar1, EDS1 and NPR1/NIM1 like genes are required for N-mediated resistance to tobacco mosaic virus. Plant J 30:415–429
54. Lutz C (1987) Cytology of high alpine plants II. Microbody activity in leaves of *Ranunculus glacialis* L. Cytologia 52:679–686
55. Lutz C, Engel L (2007) Changes in chloroplast ultrastructure in some high-alpine plants: adaptation to metabolic demands and climate? Protoplasma 231:183–192
56. Menzel D, Schliwa M (1986) Motility in the siphonous green alga *Bryopsis*. II. Chloroplast movement requires oraganized arrays of both microtubules and actin filaments. Eur J Cell Biol 40:286–293
57. Millen RS, Olmstead RG, Adams KL, Palmer JD, Lao NT, Heggie L, Kavanagh TA, Hibberd JM, Gray JC, Morden CW, Calie PJ, Jermiin LS, Wolfe KH (2001) Many parallel losses of infA from chloroplast DNA during angiosperm evolution with multiple independent transfers to the nucleus. Plant Cell 13:645–658
58. Natesan SK, Sullivan JA, Gray JC (2005) Stromules: a characteristic cell-specific feature of plastid morphology. J Exp Bot 56:787–797
59. Natesan SK, Sullivan JA, Gray JC (2009) Myosin XI is required for actin-associated movement of plastid stromules. Mol Plant 2:1262–1272

60. Newcomb EH (1967) Fine structure of protein-storing plastids in bean root tips. J Cell Biol 33:143–163
61. Newell CA, Natesan SK, Sullivan JA, Jouhet J, Kavanagh TA, Gray JC (2012) Exclusion of plastid nucleoids and ribosomes from stromules in tobacco and *Arabidopsis*. Plant J 69:399–410
62. Olson BJ, Wang Q, Osteryoung KW (2010) GTP-dependent heteropolymer formation and bundling of chloroplast FtsZ1 and FtsZ2. J Biol Chem 285:20634–20643
63. Parker ML (1985) The relationship between A-type and B-type starch granules in the developing endosperm of wheat. J Cereal Sci 3:271–278
64. Pressel S, Duckett JG (2010) Cytological insights into the desiccation biology of a model system: moss protonema. New Phytol 185:944–963
65. Proctor MCF, Ligrone R, Duckett JG (2007) Desiccation tolerance in the moss *Polytrichum formosum*: physiological and fine-structural changes during desiccation and recovery. Ann Bot 99:75–93
66. Pyke KA, Howells CA (2002) Plastid and stromule morphogenesis in tomato. Ann Bot 90:559–566
67. Raab S, Toth Z, De Groot C, Stamminger T, Hoth S (2006) ABA-responsive RNA- binding proteins are involved in chloroplast and stromule function in *Arabidopsis* seedlings. Planta 224:900–914
68. Reski R (2009) Challenges to our current view on chloroplasts. Biol Chem 390:731–738
69. Sattarzadeh A, Krahmer J, Germain AD, Hanson M (2009) A myosin XI tail domain homologous to the yeast myosin vacuole-binding domain interacts with plastids and stromules in *Nicotiana benthamiana*. Mol Plant 2:1351–1358
70. Schattat MH, Griffiths S, Mathur N, Barton K, Wozny MR, Dunn N, Greenwood JS, Mathur J (2012) Differential coloring reveals that plastids do not form networks for exchanging macromolecules. Plant Cell 24:1465–1477
71. Schattat MH, Klösgen RB (2011) Induction of stromule formation by extracellular sucrose and glucose in epidermal leaf issue of *Arabidopsis thaliana*. BMC Plant Biol 11:115
72. Scott SV, Theg SM (1996) A new intermediate on the chloroplast protein import pathway reveals distinct translocation machineries in the two envelope membranes: energetics and mechanistic implications. J Cell Biol 132:63–75
73. Senn G (1908) Die Gestalts-und Lageveränderung der Pflanzen-Chromatophoren. Engelmann, Leipzig
74. Shalla TA (1964) Assembly and aggregation of tobacco mosaic virus in tomato leaves. J Cell Biol 21:253–264
75. Shiina T, Hayashi K, Ishii N, Morikawa K, Toyoshima Y (2000) Chloroplast tubules visualized in transplastomic plants expressing green fluorescent protein. Plant Cell Physiol 41:367–371
76. Spencer D, Wildman SG (1962) Observations on the structure of grana-containing chloroplasts and a proposed model of chloroplast structure. Aust J Biol Sci 15:599–610
77. Steffen K (1964) Chromoplastenstudien. I. Der amöboide Chromoplastentyp. Planta 60: 506–522
78. Strack D, Fester T (2006) Isoprenoid metabolism and plastid reorganization in arbuscular mycorrhizal roots. New Phytol 172:22–34
79. Suzuki T, Kawano S, Sakai A, Fujie M, Kuroiwa H, Nakamura H, Kuroiwa T (1992) Preferential mitochondrial and plastid DNA synthesis before multiple cell divisions in *Nicotiana tabacum*. J Cell Sci 103:831–837
80. Tirlapur UK, Dahse I, Reiss B, Meurer J, Oelmüller R (1999) Characterization of the activity of a plastid-targeted green fluorescent protein in *Arabidopsis*. Eur J Cell Biol 78:233–240
81. Wangdi T, Uppalapati SR, Nagaraj S, Ryu CM, Bender CL, Mysore K (2010) A virus- induced gene silencing screen identifies a role for thylakoid formation1 in *Pseudomonas syringae* pv *tomato* symptom development in tomato and *Arabidopsis*. Plant Physiol 152:281–292
82. Waters MT, Fray RG, Pyke KA (2004) Stromule formation is dependent upon plastids size, plastid differentiation status and the density of plastids within the cell. Plant J 39:655–667

83. Wildman SG, Hongladarom T, Honda SI (1962) Chloroplasts and mitochondria in living plant cells: cinephotomicrographic studies. Science 138:434–436
84. Williamson RE (1993) Organelle movements. Annu Rev Plant Physiol 44:181–202
85. Witztum A, Parthasarathy MV (1985) Role of actin in chloroplast clustering and banding in leaves of Egeria, Elodea and Hydrilla. Eur J Cell Biol 39:21–26
86. Yan XX (1995) NaCl-induced amoeboid plastids and mitochondria in meristematic cells of barley roots. Biol Plant 37:363–369

Chapter 8
The Apicoplast: A Parasite's Symbiont

Lilach Sheiner and Boris Striepen

Abstract The endosymbiotic capture of a red alga brought photosynthesis to a previously heterotrophic protist, and marked the birth of a now very diverse new branch of the eukaryotic tree of life. Among the many plastid-bearing descendants of this event are the Apicomplexa, a phylum of obligate animal parasites. These include the causative agents of important diseases like malaria and toxoplasmosis. The apicomplexan plastid, or apicoplast, has experienced dramatic changes in function, organization and protein content as Apicomplexa adapted from photosynthesis to parasitism. In this chapter we outline the broad strokes of the organelle's remarkable evolutionary history and follow how these changes shaped its biology and metabolism.

Keywords Apicoplast · Plastid · Toxoplasma · Malaria · Chromalveolates · Endosymbiosis · Heme · Isoprenoid precursors · FASI · Import

Abbreviations

ACP	Acyl carrier protein
ALA	Aminolevulinate
ALAD	Aminolevulinate dehydratase
ALAS	Aminolevulinate synthase
CMK	4-diphosphocytidyl-2-*C*-methyl-D-erythritol kinase
CMS	4-diphosphocytidyl-2-*C*-methyl-D-erythritol synthase
Cox2	Cytochrome oxidase subunit 2
CPO	Coproporphyrinogen III oxidase
DMAPP	Dimethylallyl pyrophosphate
DOXP	1-deoxy-D-xylulose-5-phosphate
DOXPRI	DOXP-reductoisomerase
ER	Endoplasmic reticulum
ERAD	Endoplasmatic reticulum associated degradation

B. Striepen (✉) · L. Sheiner
Center for Tropical and Emerging Global Diseases & Department of Cellular Biology, University of Georgia, 500 D.W. Brooks Drive, Athens, GA 30602, USA
e-mail: striepen@uga.edu

L. Sheiner
e-mail: lilash@uga.edu

S.M. Theg, F.-A. Wollman (eds.), *Plastid Biology*, Advances in Plant Biology 5,
DOI 10.1007/978-1-4939-1136-3_8, © Springer Science+Business Media New York 2014

FC	Ferrochelatase
GPTs	Glucose 6-phosphate/phosphate transporter
IPP	Isopentenyl pyrophosphate
LipA	Lipoic acid synthase
MECS	2-*C*-methyl-D-erythritol 2,4-cyclodiphosphate synthase
PB	Porphobilinogen
PI3P	Phosphatidylinositol 3-monophosphate
PPO	Protoporphyrinogen IX oxidase
PPT	Phosphoenolpyruvate phosphate/phosphate transporters
PPTs	Plastid phosphate translocators
Tic	Translocon of the inner chloroplast membrane
Toc	Translocon of the outer chloroplast membrane
TP	Transit peptide
TPT	Triose phosphate/phosphate transporters
UROD	Uroporphyrinogen decarboxylase
UROS	Uroporphyrinogen-III synthase
XPTs	Xylulose 5-phosphate/phosphate transporter
3*PGA*	3-phosphoglyceraldehyde

8.1 Introduction

8.1.1 What are Apicomplexans?

Apicomplexa are a protozoan phylum encompassing some 5000 described species, whose members are parasites. The name is based on the presence of an apical complex of cytoskeletal and secretory organelles found in the infectious stages of all phylum members. This complex typically consists of a conoid, a polar ring, subpellicular microtubules and the secretory organelles known as micronemes, rhoptries and dense granules [33], not all elements are present in all species. Most Apicomplexa are obligate intracellular parasites, namely they can only replicate and survive inside a host cell, and the apical organelles play a role in the indispensable process of host cell invasion. Over their long evolutionary history Apicomplexa have adapted to infecting a wide range of animals and to thrive in a variety of cell types that provide different metabolic opportunities and challenges.

Members of the phylum are the etiological agents of diseases of medical and veterinary importance. The phylum includes several human pathogens such as *Plasmodium*, which cause malaria; *Toxoplasma gondii*, the causative agent of toxoplasmosis, which threatens immunocompromised individuals and may cause blindness or encephalitis, and *Cryptosporidium*, a waterborne pathogen responsible for severe early childhood diarrhea. Other members cause diseases across a range of livestock and include *Eimeria*, *Neospora*, *Sarcocystis*, and *Theileria*. The best studied apicomplexans are *Plasmodium* and *Toxoplasma* and this chapter is based largely on data established in those systems.

8.1.2 Obligate Intracellular Parasites with a Plastid

Until the early 1990s most Apicomplexa researchers might have never considered a photosynthetic ancestry to their favorite phylum. However molecular evidence accumulated, specifically between 1991 and 1996, describing an extrachromosomal DNA element whose sequence suggested a "plant connection" for apicomplexans [6, 25, 38, 50–54, 97, 115, 165–167]. In fact, reports of an unusual organelle of complex structure where emerging as early as the 60s as part of the era of electron microscopy studies and ultrastructuaral characterization of phylum members. The main characteristic observed for these organelles were their numerous delimiting membranes and their localization within the cell, which resulted in various names highlighting these morphological characteristics (reviewed in [129]). It was only in 1996 that these mysterious multi-membrane organelles and the plastid-like extrachromosomal DNA were linked by experimental evidence, and it became unambiguously clear that these parasites possess a non-photosynthetic chloroplast-like organelle [81, 97], **the apicoplast**. Since then numerous studies have aimed to identify the origin of this plastid, to map its metabolic capabilities and to understand its biogenesis. In this chapter we briefly describe the evolutionary history of the apicoplast, and then we outline the current insights into apicoplast metabolism and cell biology, highlighting some of the research strategies used to approach these problems. We conclude with remarks on the challenges and opportunities we are now facing to deepen our understanding of this fascinating organelle.

8.2 Apicoplast Origin and Evolution

8.2.1 Metamorphosis of an Alga

It has been proposed that the apicomplexans plastid resulted from two consecutive endosymbiotic events (Fig. 8.1a), in a similar manner to the evolution of the plastids of chromista algae [18]. The first event was the endosymbiosis between a cyanobacterium and a eukaryote that gave rise to the primary plastids found in glaucophytes, red algae and green algae (as reviewed in [60, 99]). The second event was a love affair between an auxotrophic single celled protist and a single celled alga. Both partners progressively came to depend on each other. Specifically, the host came to rely on metabolites provided by the symbiont's pathways (in particular its photosynthetic carbon fixation). On the other hand genetic material was transferred from the symbiont to the host nucleus turning inheritance and protein synthesis over to the host. Based on this hypothetical scenario, the apicomplexan ancestor was likely a photosynthetic organism with a secondary plastid, an alga that abandoned its photosynthetic machinery and turned to parasitism. Numerous studies have gradually built substantial support for this hypothesis: the evidence is based on morphology, the phylogenetic analyses of apicoplast genomes and their genes, and the many aspects of organelle function and biogenesis shared between the apicoplast and algal

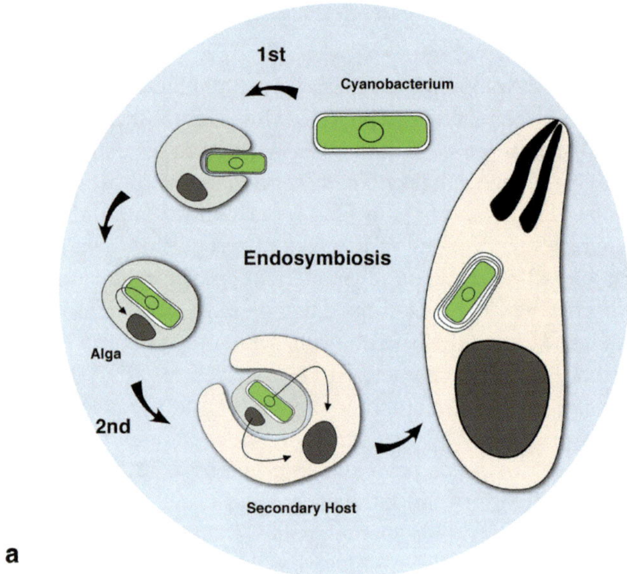

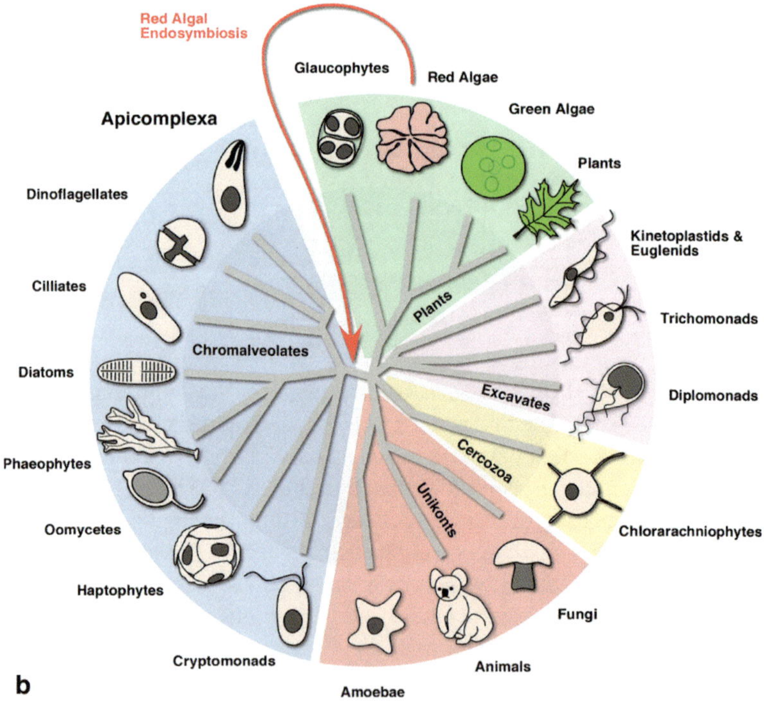

Fig. 8.1 Secondary endosymbiosis and the origin of the phylum Apicomplexa. **a** Schematic representation of the two consecutive events of endosymbiosis that gave rise to primary and secondary plastids. A cyanobacterium was engulfed by a heterotrophic eukaryote giving rise to the plastids of red and green algae and glaucophytes. In a second event, an alga was taken up by another eukaryote

plastids. We review two prominent debates: the affiliation of the apicoplast with the red or the green lineage of plastids, and the evolution of organisms with secondary plastids of red origin.

8.2.2 The Origin of the Apicoplast: The Red Versus Green Debate

An important consideration in understanding the evolutionary root of the apicoplast is to establish whether it is part of the green or red lineage of chloroplasts. These lineages are typically distinguished by the differences among their photosynthetic pigments. However, apicoplasts have completely lost photosynthesis and neither the associated pigments, nor any of the genes encoding the proteins required for the light reaction or pigment synthesis has remained in the genome. As a consequence, apicoplasts lack some of the features most frequently used to resolve plastid phylogeny. Initially the *Toxoplasma* apicoplast was described as a plastid of green algal origin [81]. This was based on phylogenetic analysis using the sequence of *TufA,* a gene present on the organellar genome, encoding the protein synthesis elongation factor Tu. *TufA*, was used previously for construction of molecular phylogenies, and served well to solidify the placement of the apicoplast genome within the plastids (as opposed to the mitochondria lineages) [81]. Phylogenetic analysis of the sequences of a nuclear-encoded gene, cytochrome oxidase subunit 2 (cox2), also supported green ancestry. Specifically, in chlorophytes, legumes and apicomplexans, this gene split into two (Cox2 a and b), whereas in other organisms it remained a single gene [48, 49]. However, the conclusion of green algal ancestry was quickly challenged. The school supporting a red algal origin pointed to evidence suggesting that the cox2 split happened repeatedly in evolution and is not suitable to track the apicoplast origin [162]. A recent phylogentic analysis performed with a larger set of *TufA* homologs questioned its suitability to address the ancestry of the apicoplast [110].

Currently, evidence favors for a red lineage for the apicoplast. The arrangement and phylogeny of the apicoplast genome ribosomal RNA genes resembles that found in red algae [11, 100, 136, 175]. Phylogenetic analysis of several nuclear-encoded plastid-targeted proteins lent further support to this conclusion [7, 37, 64, 113]. Lastly, much of what was learned about apicoplast cell biology and metabolism, detailed later in this chapter, shows strong resemblance with organisms that carry secondary plastids of red algal origin.

(note that this occurred multiple independent times). The resulting complex plastids are surrounded by three or four membranes. Narrow arrows indicate gene transfer from symbiont to host genome. **b** Schematic tree of eukaryotes drawn after phylogenetic analyses summarized by Keeling and colleagues [79]. The chromalveolate theory is shown by the uptake of a red alga in their ancestor branch. Diversification and adaptation to different ecological niches led to subsequent loss of photosynthesis (as in Apicomplexa) or of entire plastids (as in ciliates or oomycetes). Reproduced with slight modification from [1]

8.2.3 *Was the Red Algal Acquisition a Unique Event in the History of Evolution?*

Many algal taxa appear to have acquired their plastids through secondary endosymbiosis, but how many times in evolution has such secondary acquisition taken place? In the case of the green plastids, phylogenetic analysis indicated that the two main lineages, euglenids and chlorarachniophytes, originated independently, namely from two separate events of secondary endosymbiosis [122]. On the other hand, the origin of red plastids found in a larger range of eukaryotic organisms is less well resolved. The chromalveolate hypothesis suggests that the acquisition of a red alga and its enslavement into a secondary plastid occurred only once in evolution [18]. This implies a common ancestor of cryptophytes, alveolates (including dinoflagellates, cilliates and apicomplexans), stramenopiles (heterokonts), and haptophytes [18, 19] (Fig. 8.1b). Its parsimony makes this hypothesis conceptually appealing: a complex new protein import system, widely considered the biggest hurdle on the way to stable endosymbiosis, had to evolve only once. Numerous molecular and phylogenetic studies have tested this hypothesis over the last decade and in general support remains strong (reviewed in [78]).

Plastid gene phylogenies produced robust support for a common origin of stramenopiles, cryptophytes and haptophytes [80, 122, 172]. Including the apicoplast and the plastids of dinoflagellates in these analyzes was difficult, due to their divergent plastid genomes. However, this could be overcome by analyzing nuclear-encoded plastid proteins. In this context, an important breakthrough was the finding that apicomplexans and other chromalveolates have substituted their cyanobacterial-type plastid glyceraldehyde-3-phosphate dehydrogenase (GAPDH) with a duplication of a eukaryotic GAPDH [37], supporting a single origin for chromalveolates.

It should be noted, however, that this hypothesis is yet to become a consensus, and recent studies continue to challenge the chromalveolate monophyly. Specifically, studies claim to detect evidence for potential polyphyly among the hosts in the chromalveolates sub-groups [39, 62, 65]. This might be attributed to the recruitment of rhodophyte translocation machinery, that it may have occurred more than once, which led some to a potentially more complex model of the origin of chromalveolates [39].

8.2.4 *Chromera velia, a (No Longer) Missing Link*

A breakthrough in the understanding of apicoplast evolution came from the recent discovery of a closely related photosynthetic sister of Apicomplexa, *Chromera velia* [101]. The sequence of *Chromera*'s plastid genome provided a compelling link between the plastids of apicomplexans, dinoflagellates, and stramenopiles and further solidified the red origin of the apicoplast [71]. *Chromera* is a photosynthetic symbiont of corals [101]. This tight association with a marine animal provides interesting food for speculation on the potential route that Apicomplexa might have taken from free-living autotrophy, via symbiosis towards parasitism [110]. Apicomplexan animal parasitism is likely ancient starting with simple invertebrates predating the main radiation and diversification of animals. In other words: when the first Devonian fish crawled to land in all likelihood they were already infected with an apicomplexan. This scenario is supported by the notion that the acquisition of red secondary plastids and the branching of stramenopiles and alveolates occurred between 767 and 1072 million years ago [32], long before the appearance of land vertebrates.

8.3 Apicoplast Biogenesis

8.3.1 Import of Nuclear-encoded Proteins to various Apicoplast Destinations

Due to the massive gene transfer from the endosymbiont genome to the host, the vast majority of the organelle's proteins are now encoded in the nucleus. These proteins are hence translated in the cytosol from where they travel to the apicoplast lumen, one of its four membranes, or to the various intermembrane spaces. Some of the signals and machineries involved in this import process have been characterized and we review them below.

8.3.2 Targeting Signals

Similar to nuclear-encoded components of secondary chloroplasts in algae [1, 12], most stromal apicoplast proteins studied to date possess an N-terminal bipartite targeting signal. The first part of this sequence shares characteristics with the signal peptide present in secretory proteins [160], while the second part, the transit peptide (TP), shows features similar to the transit peptides of primary chloroplast proteins. Studies in *Plasmodium falciparum* (*P. falciparum*) and *Toxoplasma gondii* (*T. gondii*) confirmed that the N-termini of several luminal apicoplast proteins are necessary and sufficient to target a GFP reporter to the lumen of the organelle [46, 160, 161]. It was also experimentally demonstrated that the TP part of a *T. gondii* apicoplast ribosomal protein can direct import into pea chloroplasts [28]. In addition it was found that deletion of the TP results in the targeting of a GFP reporter to the parasitophorous vacuole, the default secretory route in *T. gondii*, while deletion of the signal peptide results in cytosolic or mitochondrial localization of the reporter [28, 63, 161]. The amino acid composition of these bipartite signals was studied in detail showing that an overall positive charge at the N-terminus is essential for targeting to the apicoplast stroma, however that their exact position can vary [46, 63, 146].

The two parts of the bipartite leader are thought to guide a two-step sequential localization, where the signal peptide dictates cotranslational insertion of the nascent peptide into the endoplasmic reticulum (ER). The SP seems to be rapidly removed [160] and this cleavage results in exposing the TP. The TP directs targeting from the ER to the apicoplast stroma where it is cleaved to release the mature protein. The presence of TP-containing pre-protein and the TP-processed mature form are detectable by western blot, and were observed both in *T. gondii* and *P. falciparum* [158, 160, 161, 174]. Furthermore the kinetics of this processing in *P. falciparum* was measured by pulse-chase labeling experiments and was demonstrated to take 45 minutes to an hour from the moment of translation [151]. A homolog of the plant chloroplast transit peptide peptidase was identified in the nuclear genome of *T. gondii* and *P. falciparum* [66, 151], however its role in cleaving the TP is yet to be experimentally demonstrated.

While most apicoplast proteins adhere to the rule just described, some seem to possess a non-canonical targeting signal. Several proteins do not have a signal peptide at the immediate N- terminus but rather show a recessed hydrophobic patch [3, 109, 128]. Others have no obvious signal peptide but contain multi-membrane spanning domains [30, 44, 75–77]. How the different N-terminal structures affect the mode of targeting or the final destination among the apicoplast sub-compartment remains to be established.

8.3.3 Crossing the Borders

8.3.3.1 The Two Innermost Membranes

Different hypotheses were initially considered as to how proteins may cross the four apicoplast membranes [147]. One of the models suggests that distinct translocons are found in the different membranes, and that their ancestry corresponds to the evolutionary origin of each respective membrane. According to this model the two inner membranes were predicted to have import machineries derived from those found in the red algal chloroplast. In ultrastructural support for this idea, the two inner membrane of the apicoplast show tight physical apposition, as described for primary chloroplast, and present large intramembranous structures that were hypothesized to represent translocons [143]. The similarity of the transit peptide part of the bipartite signal to plant chloroplast transit peptides is consistent with this idea.

In primary chloroplasts protein import is mediated by two protein complexes, the translocon of the outer chloroplast membrane (Toc) and the translocon of the inner chloroplast membrane (Tic). These were studied in detail in green chloroplast of land plants, and were shown to be composed of numerous proteins, some of which are inherited from the cyanobacterial ancestor (see [132] and Chap. 9 in this volume for a thorough review). Bioinformatics searches for homologs of these components in the genomes of red algae and of *Plasmodium* identified only a small subset of these proteins [1, 99, 159]. Those that have been found show significant sequence divergence, which may implicate modest sequence conservation as the reason for the relatively small number of identified factors. A homolog of Tic20 is the most extensively characterized among those. In the chloroplast, Tic20 is an integral membrane protein shown to function as a part of the protein pore that facilitates protein transport through the inner membrane [21, 82]. Van Dooren and coworkers, confirmed that in *T. gondii*, Tic20 is an integral membrane protein of the apicoplast [152]. Using split-GFP reporters they could also demonstrate its localization in the inner most membrane [152]. A conditional knock-out of *TgTic20* showed that this gene is essential for *T. gondii* survival. While the *Toxoplasma* plastid could not be isolated in significant quantity at the time, several pulse-chase assays were developed to measure protein import *in vivo* by following posttranslational modifications of reporter proteins. These analyses performed with the *TgTic20* mutant showed a profound loss of protein import followed by organelle loss and cell death. These experiment supported that TgTic20 acts in import [152]. A second putative component of the Tic complex, the soluble protein Tic22, was identified in the apicoplast of *P. falciparum* and of *T. gondii* [73]. Loss of this protein in a conditional mutant produces a phenotype comparable to that observed for *TgTic20* [56]. Interestingly, a homolog of ClpC (Hsp93), a stromal chaperone component of the Tic complex, is encoded on the apicoplast genome both in *P. falciparum* and *T. gondii* [81, 167]. At this time genetic manipulation of the organellar genome is not established and the role of ClpC in import awaits confirmation.

Some homologs of the Toc components were also identified in the red lineage. One of the critical components of this complex in plants is the receptor protein Toc34 that recognizes the transit peptide and directs proteins in transit to the pore in the outer chloroplast membrane, Toc75. A Toc34 homolog was identified in the genomes of green, red and secondary plastid-containing algae and is thought to have been acquired after the cyanobacterial uptake and before the split to red and green [99]. A putative Toc34 homolog was reported for the genome of *P. falciparum* [159], but was not confirmed. A Toc75 homolog was not immediately identifiable in the genomes of organisms containing secondary plastid of red origin. Recently however, Bullmann and coworkers have identified a homolog of the cyanobacterial Omp85 in the genome of the diatom *Phaeodactylum tricornutum (P. tricornutum*) [14]. This gene encodes a protein with a bipartite signal and affinity to the Toc75 group in phylogenetic studies [14]. Electrophysiological experiments using recombinant PtOmp85 confirmed that it can act as a pore and that it shares biochemical characteristics with plant Toc75 [14].

Homologs of the *PtOpm85* are found in the genome of *Plasmodium* and *Toxoplasma* [1, 14] and our preliminary studies indicate that loss of this protein in Toxoplasma results in a pronounced apicoplast protein import defect (Sheiner, Agrawal, Brooks and Striepen unpublished). Taken together these observations make a compelling case for Tic and Toc derived translocons as mechanisms to cross the two innermost membranes of the apicoplast.

8.3.3.2 The Second Outermost Membrane

The second outermost membrane is thought to be derived from the plasma membrane of the red algal endosymbiont. Initially this membrane represented the most stringent barrier between symbiont and host and the establishment of exchange of metabolites and proteins across this barrier represent pivotal early events in the evolution of the organelle. Key to the discovery of the machinery that breaches this barrier was the sequencing of the nucleomorph genome of the cryptomonad *G. theta*. The nucleomorph is a highly reduced remnant of the algal symbiont's nucleus. This cellular "fossil" is found between the second and third membrane of some secondary plastids of the red lineage, in a compartment derived from the symbiont's cytoplasm now referred to as the periplastid compartment [31] (note that a nucleomorph is also present in some secondary plastids of the green lineage, e.g. in *Chlorarachnion*, where it is found between the first and second of three membranes). Analysis of the *G. theta* nucleomorph revealed that it encoded core elements of the endoplasmatic reticulum associated degradation (ERAD) system [133]. In the ER, ERAD is responsible for retro-translocation of miss-folded proteins across the ER membrane into the cytoplasm, to be degraded by the proteasome. In light of the apparent absence of an ER in the *G. theta* plastid, this finding inspired Sommer and colleagues to formulate a hypothesis according to which the ERAD translocon had been retooled to import proteins into complex plastids [133] (see also Chap. 11 of this volume). This hypothesis gained broader support by the identification of duplicate sets of genes encoding ERAD components in additional organisms with complex red plastids such as diatoms and Apicomplexa [3, 73, 133, 134]. These organisms lack a nucleomorph, but the nucleus encodes both ER and plastid targeted ERAD components [73, 133, 134]. Electron microscopy in *T. gondii* and split-GFP experiments performed with the diatom *P. tricornutum* showed that these proteins localize to the periphery of the complex plastid and more specifically to the second outermost membrane and the periplastid compartments [2, 68].

The elements of the plastid ERAD machinery that have been characterized so far are the membrane protein Der-1, the AAA-ATPase, Cdc48, and its cofactor, Ufd-1. ER-Der-1 is essential for retro-translocation of miss-folded luminal ER proteins in yeast and human cells. It is hypothesized by some that Der-1 forms the translocation channel [170], while others suggest that its role is that of a receptor aiding interaction with the ubiquitin ligase HRD1 (see [131] for a recent review of the ERAD translocon). In this system, proteins are marked for degradation by conjugation of ubiquitin resulting in poly-ubiquitin chains. This appears to occur during translocation on the cytoplasmic side of the membrane. Cdc48 then extracts these substrate proteins from the pore with the help of its cofactors, the Ufd-1–Npl4 complex [170]. After confirming the localization of these three core components to the periphery of the apicoplast of *T. gondii*, Agrawal and colleagues generated a conditional mutant of the apicoplast Der1. Ablation of Der1 results in ablation of apicoplast protein import, as measured using a variety of biochemical assays demonstrating a direct role of the ERAD system in import [3], validating the initial hypothesis by Sommer and colleagues. Importantly, the apicoplast ERAD machinery is derived from the ERAD system of the symbiont and does not represent a duplication of the host system [3, 39]. This is consistent with a model under which the symbiont grants access to host derived proteins by modifying its membranes. Overall, this may suggest that the symbiont played an active role in establishing a stable relationship with the host rather than being subjugated to do the bidding of the host, as implied by the endosymbiosis as enslavement narrative.

ERAD associated ubiquitination is required not only for the subsequent degradation of the protein but also appears critical to the translocation step across the ER itself [131]. For example, deletion of the Ufd-1 amino terminus, known to bind polyubiquitin, results in disruption of protein translocation across the ER membrane in yeast [112, 163]. Moreover, overexpression of the yeast ubiquitin E3 ligase (Hard1p) bypasses the need of other ER components such as Der1, and result in ER-luminal protein degradation that depends only on its own active ubiquitin ligase activity and on Cdc48 [17]. This model appears also to apply to secondary plastids of the red lineage including the apicoplast: a series of putative E1, E2 and E3 ubiquitination enzymes has been described to either target to the apicoplast or to contain suitable leaders for apicoplast targeting, and functional data linking their enzymatic activity to protein import is emerging ([68, 114, 134]; Fellows et al. unpublished).

Recently we have identified two new *T. gondii* periplastid proteins that are conserved among the red algal lineage and are also encoded in the nucleomorphs of cryptomonads. A conditional mutant of one of the two, PPP1, results in a drastic apicoplast import defect leading to apicoplast demise and ultimately to cell death [128]. Exactly how PPP1 integrates into the ERAD model of crossing the periplastid compartment awaits further clarification.

8.3.3.3 The Outermost Membrane

Three translocons have been identified that collectively provide transport across the three inner membranes of the apicoplast, but how do apicoplast proteins find and cross the outermost apicoplast membrane to begin this journey? Since these proteins enter the ER courtesy of their N-terminal signal peptide, it seemed intuitive that like other secretory proteins they may travel from the ER to the Golgi and then on to the apicoplast. However, the transport of apicoplast targeted GFP reporters was shown to be resistant to the action of the fungal toxin Brefeldin A, a potent disruptor of the Golgi apparatus [29, 148]. This points to the possibility of direct ER to apicoplast transport. Note that in several algal systems the outer membrane of the complex plastid is continuous with the nuclear envelope and the ER [55]. Such constant direct connection is not evident in Apicomplexa, yet recent electron microscopy and tomographic studies suggested that the two organelles come into close contact, which may reflect functional interaction [143, 144]. Alternative to the direct contact model, vesicles could ferry proteins from the ER to the apicoplast, side-stepping the Golgi. Several groups have described such vesicles using light and electron microscopy [30, 76, 152, 153]. These vesicles become more apparent in apicoplast import mutants [153]. Two new studies report the presence of the lipid phosphatidylinositol 3-monophosphate (PI3P) in the apicoplast of *P. falciparum* and *T. gondii* [140, 141]. Interference with PI3P in *T. gondii* through drug treatment or the overexpression of a heterologous PI3P binding protein leads to profound and complex plastid biogenesis defects [141]. Interestingly, in these mutant parasites apicoplast proteins again accumulate in a vesicular compartment outside of the apicoplast. The role of PI3P in the regulation of endosomal trafficking is well established [89], and the observation of PI3P in the apicoplast suggests a potential connection between the apicoplast and the endosome. The precise mechanistic role of PI3P in this process remains to be defined.

In summary, the mechanism used to traffic nuclear-encoded proteins to the stroma appears conserved among organisms with secondary red algal plastids. Initial synthesis appears to be ER-associated, followed by a likely vesicle-mediated transfer to the outermost compartment of the plastid. Modified ERAD, Tic and Toc complexes then transport proteins across the remaining membranes. Numerous mechanistic questions remain open, including how the targeting machinery distinguishes stromal proteins from those that remain in the outer compartments of the apicoplast and how the trafficking of membrane proteins differs from that of soluble proteins.

8.3.4 Organelle Replication and Division—Drug Targets Beyond Metabolism

Like other plastids, the apicoplast possess its own genome. Genome replication and partitioning are a prerequisite to organelle inheritance. As noted for protein import, the origin of the proteins that govern organelle replication is dictated by the origin of the compartment that they are found in. The apicoplast genome is found in the stroma and is replicated and maintained by a prokaryote-type machinery derived from cyanobacteria, while components involved in organelle fission and segregation have evolved from factors originally localized to the cytoplasm of the host.

8.3.4.1 Apicoplast Genome Replication and Gene Translation

The highly reduced apicoplast genome (35 kb) is dedicated almost entirely to encoding RNAs and proteins for its own transcription and translation, in addition to two known and seven hypothetical genes that may serve other functions [167]. In some cases, like for the apicoplast RNA-polymerase, further nuclear-encoded components are imported to join forces with apicoplast encoded proteins [8]. Phylogenetic analysis suggests that regardless of the current location of the gene (nucleus or plastid), this machinery is largely of bacterial origin, and this is consistent with the sensitivity of Apicomplexa to antibiotics that typically interfere with bacterial transcription and translation. Rifampin, an inhibitor of eubacterial RNA polymerases is one example [10, 26, 96]. A number of additional antibiotics that show some efficacy against apicomplexans typically target the prokaryotic ribosome. These include clindamycin [16, 42], doxycycline [26], thiostrepton, and azithromycin [23, 130]. Analysis of the apicoplast ribosomal RNA sequence predicts the apicoplast as their target and mutants selected for antibiotic resistance show mutation of the apicoplast ribosomal RNA genes [98].

The apicoplast genome does not, however, appear to encode components involved in its replication, and those are presumably imported. Indeed, several nuclear-encoded homologs of such components were identified and localized to the apicoplast. Those include Prex (plastid-DNA replication enzyme complex) [127], gyrase subunits A and B, homologs of the bacterial strand relaxation topoisomerase subunits [5, 27, 116], and a group of hypothetical proteins with domains involved in DNA repair such as gamma integrase and helicases [128]. Interestingly, the apicoplast genome is associated with a bacterial histone-like HU protein [117, 120]. The bacterial nature of the apicoplast genome replication machinery is also support by the sensitivity of the apicoplast to antibiotics such as ciprofloxacin [42, 164].

8.3.4.2 Division of the Apicoplast Organelle

The apicoplast shows a unique mode of division when compared to other plastids. Early microscopy-based studies reported a surprisingly tight association of apicoplast and nuclear division. The cell and division cycle shows significant diversity in Apicomplexa (see details in [138]) and nuclear division occurs in different stereotypic forms, importantly, the apicoplast follows the nucleus precisely through these complex processes. Studies in *Toxoplasma* and *Sarcocystis* confirmed that apicoplast division coincides with nuclear division and further demonstrated its coordination by physical association between the apicoplast and the centrosome of the spindle [137, 150]. In *Toxoplasma* the apicoplast associates with the recently divided centrosomes early in M-phase and is subsequently suspended between the two poles of the spindle. As mitosis and budding proceeds both the nucleus and the apicoplast appear U-shaped, and this U elongates until fission occurs concurrently with budding of the new daughter cells [137, 153]. Hitching a ride on the host centrosomes ensures faithful segregation of the organelle and may potentially also be involved in equal segregation of its genome. Note that in those apicomplexans that replicate nuclear DNA multiple times before cytokinetic partitioning into new invasive stages, the apicoplast develops into an impressive tubular network [41, 135, 150]. Segregation and partitioning again appears to be controlled by centrosome tethering as initially shown for *Toxoplasma*.

How is apicoplast fission accomplished? Essentially all plastids studied so far, including secondary plastids, divide using a mechanism that at its core uses bacterial elements, in particular the tubulin homolog FtsZ which drives constriction, and minD and minE which regulate FtsZ (Reviewed in [57]; cf. Chap. 6 in this volume). From the cytoplasmic site, constriction is aided by eukaryotic components, including the dynamin-related protein ARC5. Importantly, the process appears to be orchestrated from the inside out and a protein complex relays the position of the FtsZ ring to the outside recruiting ARC5 [58]. The apicoplast represents a unique exception, Apicomplexa lack FtsZ and any of its associated factors [149]. Nevertheless, the presence of a division ring was implied by a plastid constrictions observed in ultrastructural studies [40, 93]. The molecular components involved in this process were recently described in *Toxoplasma*. The first player emerged with the description of a constrictive cytokinetic ring identified by the repeat protein MORN1 [61]. The position of the MORN1 rings during budding coincides with constrictions in the apicoplast and nucleus. This lent support to a pull and cut model in which the centrosomes and budding ring cooperate in apicoplast division [149, 152]. Consistent with this model, MORN1 deficient *T. gondii* parasites display apicoplast segregation defect [67, 92]. While MORN1 is required for fission it is not sufficient. At least one more component is required to complete fission, the dynamin-related protein DrpA [153]. Genetic and cell biological studies using a dominant negative mutant demonstrated that DrpA is essential for apicoplast fission. In mutant parasites large elongated apicoplasts connect recently divided daughter cells [153]. The assembly of dynamin-related proteins into active rings is typically induced by an initial constriction of the target area [85]. We recently suggested a model combining these elements, according to which the centrosome and MORN1 rings generate opposing forces that result in constrictions. Those then become the site of DrpA activity [4, 153]. Interestingly, TgDrpA is phylogenetically distinct from the ARC5 dynamin involved in chloroplast division [85]. This suggests that dynamins have been recruited independently more than once to act in the fission of endosymbiont organelles.

8.4 Metabolic Functions of the Apicoplast

8.4.1 The Apicoplast's Function is Anabolic Metabolism

Among their secondary red plastid containing kin, Apicomplexa are unique in that they have maintained the plastid despite loss of photosynthesis. Which evolutionary forces drove apicomplexans to keep this seemingly pointless organelle? The key to understand this process is to understand the current cellular needs fulfilled by the apicoplast. Below we review the current understanding of apicoplast metabolism. Note that not all of the pathways described are found in all species (Fig. 8.2). Apicomplexa occupy different niches within their hosts and, as a result, their metabolic needs and capabilities vary dramatically (reviewed in [45]). *T. gondii*, who can infect a broad range of cells and tissues, has maintained the most elaborated metabolic network in its plastid. This includes pathways for the synthesis of fatty acid, heme, isoprenoid precursors and iron-sulfur cluster. *Cryptosporidium*, marks the other end of the spectrum, this parasite has lost the apicoplast and essentially all of its metabolic capabilities [45].

8.4.2 Type II Fatty Acid Synthesis (FASII)

Fatty acids are indispensable for cell growth and many pathogens have developed specialized strategies to salvage host fatty acids and fatty acid derived lipids. They

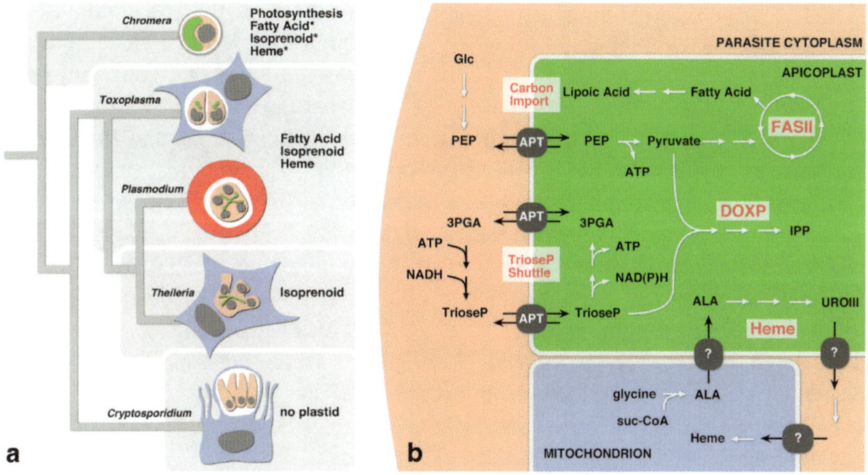

Fig. 8.2 The metabolic functions of the apicoplast. **a** Schematic tree representing the evolutionary relationships among different members of the Apicomplexa. Apicoplast metabolic functions are indicated and based on comparative genomic analyses. *Note that the functions indicated for Chromera at this time are speculative. **b** A simplified schematic overview of the metabolic pathways of the apicoplast of *Toxoplasma*. Three major pathways (FASII, DOXP, and heme) have been identified, and two of them depend on the activity of the apicoplast phosphate translocator (APT). Note that many steps were omitted for simplicity here, but are described in more detail in the text. *Glc* glucose, *PEP* phosphoenolpyruvate, 3*PGA* 3-phosphoglyceraldehyde, *IPP* isopentenyl pyrophosphate, *ALA* aminolevulinic acid, *UROIII* uroporphyrinogen-III

are the main building blocks of cellular membranes and numerous proteins undergo post-translational modifications by acylation and lipidation. Apicomplexa have a high turnover of membrane lipids as they rely on an elaborate secretory system that includes numerous apical organelles that are essential for host cell invasion. Moreover, fatty acid acylation of some components of the invasion and motility machinery, e.g. GAP45, is essential for its assembly and correct targeting [47]. In most organisms *de novo* synthesis of fatty acids occurs mainly via one of two pathways. FASI is found typically in animals and fungi and is executed by a large multi-domain polypeptide. The FASII pathway, on the other hand, depends on individually expressed enzymes and is more commonly found in prokaryotes and in the chloroplasts of plants.

Prior to the identification of the apicoplast, Apicomplexa, and more specifically *Plasmodium*, was presumed incapable of *de novo* fatty acid synthesis and thought to entirely rely on scavenging from the host [69, 121]. However, among the first identified nuclear-encoded apicoplast proteins were three fatty acid biosynthetic proteins: acyl carrier protein (ACP), β-ketoacyl-ACP synthase III (FabH), and β-hydroxyacyl-ACP dehydratase (FabZ) [160]. Homologs of these proteins are known to act in FASII in the chloroplasts of plants and algae. The FASII pathway consists of six enzymes, FabD, FabH, FabZ, FabI, FabB/F and FabG and the core component ACP. FabD first synthesizes a precursor from malonyl co-A on an ACP starter. In successive rounds the decarboxylative condensation of malonyl-CoA with the acyl chain by FabH yields a two carbon extensions. Upon each addition the new unit is then reduced, dehydrated and reduced again by FabG, FabZ and FabI. Ultimately, a fatty acid of 14 or 16 carbon chain length is released by FabB/F.

All FASII enzymes are encoded in the genomes of *Plasmodium* and *Toxoplasma* and there is now a wealth of kinetic, structural and pharmacological support based on the analysis of recombinant proteins (reviewed in [59, 94, 139]. However, not all Apicomplexa have maintained the FASII pathway, aside from *Cryptosporidium* who has lost the apicoplast and may instead rely on a cytosolic FASI [177, 178], *Theilaria* and *Babesia,* still possessing an apicoplast, have also lost FASII. Interestingly, this is to the exception of ACP, which is still encoded and was recently shown to be apicoplast localized in *Babesia* [15]. This may represent a molecular fossil as discussed by the authors, however, in bacteria ACP has multiple cellular roles, and it is therefore conceivable that its maintenance may be driven by its importance beyond FASII.

The FASII pathway is not only differentially maintained in different species, there are also interesting differences in its importance for parasite survival. In *T. gondii* for example, genetic interference with apicoplast FASII blocks the growth of the parasite in tissue culture and in animals [95], indicating that FASII is essential. In *Plasmodium*, similar genetic studies have shown that FASII is essential to the development of liver stages but not for the bloodstream and mosquito phase of its complex parasite lifecycle [156, 173]. This suggests that in parasites that infect red blood cells, FASII is dispensable, while it is required in those that live in nucleated cells, namely the importance of FASII may depend on the host cell or tissue environment.

What is then the role of FASII in the parasite metabolism? In plants, the chloroplast and its FASII are the sole sources of fatty acids. Surprisingly however, incorporation of radiolabelled acetate into fatty acids remained unchanged in the *T. gondii* FASII mutant. This initial observation seemed to suggest that in Apicomplexan FASII's role is more modest then in plants and solely focused on the synthesis of certain specialized lipids like the enzyme cofactor lipoic acid [142, 168]. Indeed, pharmacological [24] and genetic [95] disruption of *T. gondii* FASII results in the loss of lipoylation of plastid pyruvate dehydrogenase. To further test these possibilities we recently used a combination of metabolomic and genetic analysis. By labeling with ^{13}C-U-glucose followed by GC-MS analysis we showed that *T. gondii* engages in vigorous *de novo* fatty acid synthesis and that this activity is entirely dependent on FASII [118] (note that we found acetate not to be a suitable substrate for apicoplast FASII). Most (60–80 %) of the parasite's myristic and palmitic acids (C:14 and 16) originate from FASII activity. There is also good evidence that apicoplast-synthesized fatty acids are exported from the organelle and further modified by an elongation system in the ER. Overall it appears that, as with the chloroplast, the apicoplast is a significant source of fatty acids – however, the host niche governs whether this production is essential or can be replaced by lipid salvage from the host cell.

8.4.3 Heme Biosynthesis

Heme is a large and complex molecule in which a cyclic tetrapyrrole backbone coordinates a central iron. Heme functions as the prosthetic group of proteins that act in redox reactions. These include enzymes like catalase and most importantly, steps of the mitochondrial electron transfer chain, a key element of oxidative phosphorylation and aerobic metabolism. They also function in the transport and storage of gases. Given its many important roles of securing a steady supply either by synthesis or salvage, heme is essential for parasite growth. Curiously, in some apicomplexan that parasitize red blood cells and feast on their hemoglobin, heme can become too much of a good thing. *Plasmodium* parasites detoxify the overabundance of heme by polymerization into hemozoin, the malaria pigment, and interference with heme detoxification is the mode of action of several important antimalarials. Nonetheless, *Plasmodium* [104–107, 124, 155] and *Toxoplasma* [169] encode a complete heme biosynthesis pathway. Some researchers believe this pathway to be critical for the growth of the erythrocyte stage of *P. falciparum* [119, 139], but as discussed below, other studies point to isoprenoid precursor synthesis as the sole truly essential apicoplast pathway in that stage [109, 171]. Furthermore, several apicomplexans have apparently lost heme biosynthesis; these include *Theileria*, *Babesia*, and *Cryptosporidium* [126].

A comprehensive and elegant model for the evolution of the heme synthesis pathway was recently described in a review by van Dooren and colleagues [154]. Initially host and endsymbiont maintained independent pathways: one in the mitochondrion and the cytoplasm, one in the plastid. Establishment of transport for key substrates and products likely allowed the subsequent merger of both and the heavy reliance on plastid enzymes (a situation still evident in the photosynthetic apicomplexan *Chromera*). Upon loss of photosynthesis the balance shifted again and *Plasmodium* and *Toxoplasma* now rely on a pathway that begins and ends in the mitochondrion and represents a complex evolutionary patchwork (Fig. 8.2). The first step occurs in the mitochondrion where δ-aminolevulinate synthase (ALAS, also named HemA) utilizes glycine and succinyl-CoA to generate 5-aminolevulinate (ALA). The next three steps occur in the apicoplast and are executed by enzymes that cluster phylogenetically with plastid enzymes [169]. ALA is converted to porphobilinogen (PB) by δ-aminolevulinate dehydratase (ALAD or HemB). Typically the next steps are catalyzed by two enzymes, PB deaminase (PBGD or HemC) and uroporphyrinogen-III synthase (UROS or HemD), yielding hydroximethyl bilane and then uroporphynogenIII. However, in *Plasmodium,* HemC was shown to catalyze both steps [103]. Next the hydrophilic uroporphynogen III is converted to the hydrophobic coproporphinogen III by uroporphyrinogen decarboxylase (UROD or HemE). While in *Plasmodium* this is a plastid localized step [124], in *Toxoplasma* HemE appears to localize to the cytoplasm [126]. The next two enzymes, coproporphyrinogen III oxidase (CPO or HemF) and protoporphyrinogen IX oxidase (PPO or HemG), lack obvious targeting signals. Studies in *P. falciparum* confirmed that HemF is cytosolic, where it generates protoporphyrinogen III [107], whereas HemG is a mitochondrial enzyme that anaerobically catalyzes protoporphyrin IX [106].

The last step, the synthesis of heme from protoporphyrin IX by ferrochelatase (FC or HemH), also occurs in the mitochondrion [105].

There are a number of important remaining questions. How does ALA get from the mitochondrion to the apicoplast? Suitable transporters will have to breach numerous membranes in different organelles. Similarly how do products shuttle back out from the apicoplast and how are later intermediates imported into the mitochondrion? Finally, how important is heme synthesis in different parasite and host settings?

8.4.4 Iron–sulfur Clusters and Isoprenoids

Iron–sulfur clusters/ISCs [Fe–S] consist of iron and sulfur ions that are coordinated by cysteine or histidine residues in a variety of configurations. ISCs are important metal cofactors for redox reactions and are involved in electron transfer in metabolism, cellular regulation and homeostasis, and in stress response.

[Fe–S]-proteins appear to require assembly *in situ* and eukaryotes possess several synthesis pathways for different compartments. The major players in plastids are homologs of the bacterial Suf system. In bacteria this system is largely used to assemble [Fe–S] clusters under environmental conditions such as oxidative stress or iron starvation [111]. Plant chloroplasts rely on this pathway likely because of its lower sensitivity to oxygen.

Sulfide is generated from L-cysteine by the desulfurase SufS, in complex with sulfide transferase SufE [84]. The source for iron ions and their donor in the organelle remain elusive. The sulfide is then passed to the scaffold protein(s) for assembly of a transient [Fe–S]. Several scenarios have been proposed to occur downstream of this step. The latest model proposed for *E. coli* suggests that SufA, the SUF member of the A-type carriers class [157], acts as a shuttle and transfers the [Fe–S] from a scaffold complex (that consists of SufB, C, and D) to the apo-proteins [20, 84].

Importantly, a SufB homolog is encoded by the apicoplast genome [70]. Homologs of Suf members are also found in the genome of several apicomplexans and their products are predicted to target to the apicoplast [36, 125], however, their localization and function have little experimental support. Kumar and coworkers have established that the nuclear-encoded SufC is an active ATPase that resides in the apicoplast, and demonstrated its interaction with the plastome encoded PfSufB [83]. We have shown that the nuclear homolog of NFU in *T. gondii* encodes an apicoplast resident, however we found that it is dispensable for tachyzoites grown in culture [128]. This data suggests that scaffold proteins may play a redundant role, or that the apo-protein subjected to transfer from NFU is not essential under these growth conditions.

The apicoplast is home for several proteins whose homologs typically require ISCs for function. These include Ferredoxin (Fd), LipA (lipoic acid synthase) and MiaB (tRNA methylthiotransferase) and two enzymes involved in isoprenoid synthesis in the apicoplast, LytB and HDS. The genes encoding these proteins are conserved in all plastid containing Apicomplexa studied to date. It is tempting to hypothesize that the two pathways were conserved together, one to generate an essential metabolite, and the other for supporting that function. However, the genomes of *Theileria* and *Babesia* seem to encode only two components of the ISC pathway, a SufS-like desulfurase gene and an NFU-like [126]. This suggests that, if active, their pathways act very differently from that of *Plasmodium* and *Toxoplasma*.

Isoprenoids are a diverse group of biological compounds that are derived from isopentenyl pyrophosphate (IPP) or from its isomeric form dimethylallyl pyrophosphate (DMAPP). There are two major routes of IPP synthesis, the mevalonate pathway is common in animal and fungi, and the 1-deoxy-D-xylulose-5-phosphate (DOXP, or nonmevalonate pathway), is found mainly in eubacteria. Plants are home to both a mevalonate pathway in the cytoplasm and a DOXP pathway in their plastids. Apicomplexa lack the mevalonate pathway but have maintained the DOXP pathway in the apicoplast [22, 72].

DOXP-synthase catalyzes the condensation of pyruvate and glyceraldehyde-3-phosphate to initiate isoprenoid synthesis. This yields DOXP which is subsequently rearranged and reduced to 2-C-methyl-D-erythritol-4-phosphate (MEP) by the enzyme DOXP-reductoisomerase (DOXPRI). The activity of three additional enzymes, 4-diphosphocytidyl-2-*C*-methyl-D-erythritol synthase (CMS), 4-diphosphocytidyl-2-*C*-methyl-D-erythritol kinase (CMK), and 2-*C*-methyl-D-erythritol 2,4-cyclodiphosphate synthase (MECS), results in the formation of a cyclic diphosphate. The latter is transformed to yield either IPP or DMAPP in two additional steps mediated by (*E*)-4-hydroxy-3-methyl-but-2-enyl pyrophosphate (HMB-PP) synthase (HDS) and HMD-PP reductase (HDR) (reviewed in [35]). In the apicoplast, the last steps of producing IPP and DMAPP are performed by a single enzyme, LytB [123].

One reason that apicoplast isoprenoid synthesis has garnered particular attention is the fact that the antibiotic, fosmidomycin, an inhibitor of DOXPRI, shows robust activity against *Plasmodium* and *Babesia in vitro* and in animal models [72]. Recent analysis has suggested that the drug may have a second target in the same pathway potentially delivering a double blow [176]. In combination with other drug, fosmidomycin has also been shown to cure uncomplicated malaria in people [86]. Yeh and DeRisi showed that the toxicity of fosmidomycin for *P. falciparum* blood stage parasites can be overcome by chemical complementation with IPP, supporting the specificity of the drug's activity [171]. In this remarkable study, they further demonstrated that supplementation of growth media with IPP allows to derive, and more importantly, continuously culture plastid-less parasites [171]. This is very strong support for the notion that IPP is the most critical apicoplast-derived metabolite (at least under these growth conditions). Genetic studies in *T. gondii* come to the same conclusion. Mutants in the DOXP pathway have severe growth defects that mirror complete ablation of apicoplast metabolism and that are more immediate than those that ablate apicoplast fatty acid synthesis [13, 109]. This is consistent with the model that dependency on plastid synthesized IPP may drive maintenance of the apicoplast [108].

Keeping the IPP dependence model in mind, it was very surprising that fosmidomycin showed little or no effect on the growth of numerous apicomplexans, including *Eimeria*, *Theileria* or *Toxoplasma* [22, 90, 91, 109]. Two studies recently resolved this mystery by demonstrating that the DOXP pathway in *T. gondii* is susceptible to fosmidomycin but that the drug cannot reach its target in these parasites [9, 109]. Parasites engineered to express a bacterial transporter protein capable of fosmidomycin import are rendered fully susceptible, pointing to important differences in the transport of metabolites in different parasites [109].

While it is now clear that IPP production is essential, it is less clear why. There is ample evidence for the salvage of various isoprenoids from the host cell, yet it appears that IPP has to be supplied *in situ*. May this be, again, the consequence of lack of a suitable transport system?

There are several downstream metabolites and metabolite-dependent functions that may require IPP. Some are localized to the apicoplast itself (i.e. ispoprenylation of tRNAs). Other likely candidates are found outside of the apicoplast e.g. providing dolichol to the ER glycosylation machinery, or ubiquinone to maintain the mitochondrial electron transport chain. Apicomplexa also appear to produce carotenoids [145] and their derivatives [102]. Carotenoids where proposed to be used, as in plants, as antioxidants or as the starting compounds to generate plant hormones [145]. Enzymes involved in the synthesis of plant hormones have been proposed to be encoded in the genome of *T. gondii*, and the presence of one such hormone, abscisic acid, was demonstrated in this parasite [102]. While the precise role of these molecules is still elusive, it has been speculated that they play a signaling role in host cell egress [102, 145].

8.4.5 Transport of Metabolites

Photosynthesis and carbon fixation are the prime function of chloroplasts. Chloroplasts heavily export anabolic metabolites in addition to a number of sugar phosphate compounds that funnel into and fuel the central carbon metabolism of the plant cell. The numerous anabolic activities of the apicoplast require robust sources of carbon, reduction power and energy – but it has lost photosynthesis as a provider and now has to be fueled from the parasite cytoplasm. Chloroplast are known to use an arsenal of plastid phosphate translocators (PPTs) to balance the carbon and energy needs between organelle and cytoplasm: triose phosphate/phosphate transporters (TPT), phosphoenolpyruvate phosphate/phosphate transporters (PPT; [43]), glucose 6-phosphate/phosphate transporter (GPTs; [74]) and xylulose 5-phosphate/phosphate transporter (XPTs; [34]). Comparatively little is known about transport in Apicomplexa [87]. *Theileria* and *Toxoplasma* encode a single PPT homolog while two genes are found in *Plasmodium* [44, 75, 102]. This difference appears to reflect localization rather than function. The two *Plasmodium* proteins are differentially targeted to outer and inner membranes while the *Toxoplasma* protein is found in both compartments. Heterologous expression and reconstitution of transport *in vitro* showed the apicoplast transporters to be unique among phosphate translocators in exchanging phosphate against phosphoenolpyruvate, glyceraldehyde-3-phosphate and triose phosphate [13, 88] (chloroplast use at least two independent transporters to achieve this). A single transporter thus provides the substrates for at least two apicoplast pathways (FASII and DOXP) and establishes a reverse triose shuttle to supply the apicoplast with energy and reduction power (Fig. 8.2).

In agreement with this central role, genetic ablation of the *APT* gene in *Toxoplasma* has a dramatic impact on apicoplast metabolism and results in swift parasite demise [13]. Additional transporters likely operate to feed other metabolic needs of the apicoplast (i.e. heme intermediates, nucleotides for RNA and DNA synthesis, iron for assembly of ISC), and the continuous efforts to complete the apicoplast proteome will hopefully unravel those components.

8.5 Conclusion

Apicoplast functions similar to other plastids in the dark, it acts as an anabolic hub and does not engage in photosynthesis. The importance of apicoplast-made metabolites varies and depends on the host environments faced by the various parasite species, and their life stages. Understanding these niches will be key to exploiting the apicoplast as a drug target.

The function of the organelle has changed dramatically over time and was accompanied by significant gain and loss of genes. Some of these changes rendered pathways redundant or dispensable, or most interestingly, resulted in chimerical pathways whose components are of various origins. Many of the players in apicoplast metabolism and biogenesis have been identified based on sequence similarity with their chloroplast counterparts, however this strategy may be exhausted. Recent efforts have developed tools to discover new proteins independent of similarity. These include organelle purification schemes followed by proteomics, forward genetic screens, and functional genomic analyses. These methods will lead to novel insights into the biology of this fascinating symbiont organelle.

References

1. Agrawal S, Striepen B (2010) More membranes, more proteins: complex protein import mechanisms into secondary plastids. Protist 161:672–687
2. Agrawal S, van Dooren GG, Beatty WL, Striepen B (2009) Genetic evidence that an endosymbiont-derived endoplasmic reticulum-associated protein degradation (ERAD) system functions in import of apicoplast proteins. J Biol Chem 284:33683–33691
3. Agrawal S, van Dooren GG, Beatty WL, Striepen B (2009) Genetic evidence that an endosymbiont-derived ERAD system functions in import of apicoplast proteins. J Biol Chem 284:33683–33691
4. Agrawal S, Nair S, Sheiner L, Striepen B, (2010) The apicoplast: an ancient algal endosymbiont of apicomplexa. In: De Souza W (ed) Structures and organelles in pathogenic protists. Springer, Berlin
5. Ahmed A, Sharma YD (2008) Ribozyme cleavage of *Plasmodium falciparum* gyrase A gene transcript affects the parasite growth. Parasitol Res 103:751–763

6. Allison LA, Simon LD, Maliga P (1996) Deletion of rpoB reveals a second distinct transcription system in plastids of higher plants. EMBO J 15:2802–2809
7. Arenas AF, Escobar AJ, Gomez-Marin JE (2008) Evolutionary origin of the protozoan parasites histone-like proteins (HU). Silico Biol 8:15–20
8. Bahl A et al (2003) PlasmoDB: the plasmodium genome resource. A database integrating experimental and computational data. Nucleic Acids Res 31:212–215
9. Baumeister S et al (2011) Fosmidomycin uptake into plasmodium and Babesia-infected erythrocytes is facilitated by parasite-induced new permeability pathways. PLoS ONE 6:e19334
10. Beckers CJ, Roos DS, Donald RG, Luft BJ, Schwab JC, Cao Y, Joiner KA (1995) Inhibition of cytoplasmic and organellar protein synthesis in *Toxoplasma gondii*. Implications for the target of macrolide antibiotics. J Clin Invest 95:367–376
11. Blanchard JL, Hicks JS (1999) The non-photosynthetic plastid in malarial parasites and other apicomplexans is derived from outside the green plastid lineage. J Eukaryot Microbiol 46:367–375
12. Bolte K, Bullmann L, Hempel F, Bozarth A, Zauner S, Maier UG (2009) Protein targeting into secondary plastids. J Eukaryot Microbiol 56:9–15
13. Brooks CF, Johnsen H, van Dooren GG, Muthalagi M, Lin SS, Bohne W, Fischer K, Striepen B (2009) The Toxoplasma apicoplast phosphate translocator links cytosolic and apicoplast metabolism and is essential for parasite survival. Cell Host Microbe 7:62–73
14. Bullmann L, Haarmann R, Mirus O, Bredemeier R, Hempel F, Maier UG, Schleiff E (2010) Filling the gap, evolutionarily conserved Omp85 in plastids of chromalveolates. J Biol Chem 285:6848–6856
15. Caballero MC, Pedroni MJ, Palmer GH, Suarez CE, Davitt C, Lau AO (2011) Characterization of acyl carrier protein and LytB in Babesia bovis apicoplast. Mol Biochem Parasitol 181:125–133
16. Camps M, Arrizabalaga G, Boothroyd J (2002) An rRNA mutation identifies the apicoplast as the target for clindamycin in *Toxoplasma gondii*. Mol Microbiol 43:1309–1318
17. Carvalho P, Stanley AM, Rapoport TA (2010) Retrotranslocation of a misfolded luminal ER protein by the ubiquitin-ligase Hrd1p. Cell 143:579–591
18. Cavalier-Smith T (1999) Principles of protein and lipid targeting in secondary symbiogenesis: euglenoid, dinoflagellate, and sporozoan plastid origins and the eukaryote family tree. J Eukaryot Microbiol 46:347–366
19. Cavalier-Smith T (2004) Only six kingdoms of life. Proc Biol Sci 271:1251–1262
20. Chahal HK, Dai Y, Saini A, Ayala-Castro C, Outten FW (2009) The SufBCD Fe-S scaffold complex interacts with SufA for Fe-S cluster transfer. Biochemistry 48:10644–10653
21. Chen X, Smith MD, Fitzpatrick L, Schnell DJ (2002) In vivo analysis of the role of atTic20 in protein import into chloroplasts. Plant Cell 14:641–654
22. Clastre M, Goubard A, Prel A, Mincheva Z, Viaud-Massuart MC, Bout D, Rideau M, Velge-Roussel F, Laurent F (2007) The methylerythritol phosphate pathway for isoprenoid biosynthesis in coccidia: presence and sensitivity to fosmidomycin. Exp Parasitol 116:375–384
23. Clough B, Rangachari K, Strath M, Preiser PR, Wilson RJ (1999) Antibiotic inhibitors of organellar protein synthesis in *Plasmodium falciparum*. Protist 150:189–195
24. Crawford MJ, Thomsen-Zieger N, Ray M, Schachtner J, Roos DS, Seeber F (2006) *Toxoplasma gondii* scavenges host-derived lipoic acid despite its de novo synthesis in the apicoplast. EMBO J 25:3214–3222
25. Creasey A, Mendis K, Carlton J, Williamson D, Wilson I, Carter R (1994) Maternal inheritance of extrachromosomal DNA in malaria parasites. Mol Biochem Parasitol 65:95–98
26. Dahl EL, Shock JL, Shenai BR, Gut J, DeRisi JL, Rosenthal PJ (2006) Tetracyclines specifically target the apicoplast of the malaria parasite *Plasmodium falciparum*. Antimicrob Agents Chemother 50:3124–3131
27. Dar MA, Sharma A, Mondal N, Dhar SK (2007) Molecular cloning of apicoplast-targeted Plasmodium falciparum DNA gyrase genes: unique intrinsic ATPase activity and ATP-independent dimerization of PfGyrB subunit. Eukaryot Cell 6:398–412

28. DeRocher A, Hagen CB, Froehlich JE, Feagin JE, Parsons M (2000) Analysis of targeting sequences demonstrates that trafficking to the *Toxoplasma gondii* plastid branches off the secretory system. J Cell Sci 113(Pt 22):3969–3977
29. DeRocher A, Gilbert B, Feagin JE, Parsons M (2005) Dissection of brefeldin A-sensitive and -insensitive steps in apicoplast protein targeting. J Cell Sci 118:565–574
30. DeRocher AE, Coppens I, Karnataki A, Gilbert LA, Rome ME, Feagin JE, Bradley PJ, Parsons M (2008) A thioredoxin family protein of the apicoplast periphery identifies abundant candidate transport vesicles in *Toxoplasma gondii*. Eukaryot Cell 7:1518–1529
31. Douglas S et al (2001) The highly reduced genome of an enslaved algal nucleus. Nature 410:1091–1096
32. Douzery EJ, Snell EA, Bapteste E, Delsuc F, Philippe H (2004) The timing of eukaryotic evolution: does a relaxed molecular clock reconcile proteins and fossils? Proc Natl Acad Sci U S A 101:15386–15391
33. Dubremetz JF, Garcia-Reguet N, Conseil V, Fourmaux MN (1998) Apical organelles and host-cell invasion by Apicomplexa. Int J Parasitol 28:10071013
34. Eicks M, Maurino V, Knappe S, Flugge UI, Fischer K (2002) The plastidic pentose phosphate translocator represents a link between the cytosolic and the plastidic pentose phosphate pathways in plants. Plant Physiol 128:512522
35. Eisenreich W, Bacher A, Arigoni D, Rohdich F (2004) Biosynthesis of isoprenoids via the non-mevalonate pathway. Cell Mol Life Sci 61:1401–1426
36. Ellis KE, Clough B, Saldanha JW, Wilson RJ (2001) Nifs and Sufs in malaria. Mol Microbiol 41:973–981
37. Fast NM, Kissinger JC, Roos DS, Keeling PJ (2001) Nuclear-encoded, plastid-targeted genes suggest a single common origin for apicomplexan and dinoflagellate plastids. Mol Biol Evol 18:418–426
38. Feagin JE, Werner E, Gardner MJ, Williamson DH, Wilson RJ (1992) Homologies between the contiguous and fragmented rRNAs of the two *Plasmodium falciparum* extra chromosomal DNAs are limited to core sequences. Nucleic Acids Res 20:879–887
39. Felsner G, Sommer MS, Gruenheit N, Hempel F, Moog D, Zauner S, Martin W, Maier UG (2011) ERAD components in organisms with complex red plastids suggest recruitment of a preexisting protein transport pathway for the periplastid membrane. Genome Biol Evol 3:140–150
40. Ferguson DJ, Henriquez FL, Kirisits MJ, Muench SP, Prigge ST, Rice DW, Roberts CW, McLeod RL (2005) Maternal inheritance and stage-specific variation of the apicoplast in *Toxoplasma gondii* during development in the intermediate and definitive host. Eukaryot Cell 4:814–826
41. Ferguson DJ et al (2007) Enzymes of type II fatty acid synthesis and apicoplast differentiation and division in *Eimeria tenella*. Int J Parasitol 37:3351
42. Fichera ME, Roos DS (1997) A plastid organelle as a drug target in apicomplexan parasites. Nature 390:407–409
43. Fischer K, Kammerer B, Gutensohn M, Arbinger B, Weber A, Hausler RE, Flugge UI (1997) A new class of plastidic phosphate translocators: a putative link between primary and secondary metabolism by the phosphoenolpyruvate/phosphate antiporter. Plant Cell 9:453–462
44. Fleige T, Fischer K, Ferguson DJ, Gross U, Bohne W (2007) Carbohydrate metabolism in the *Toxoplasma gondii* apicoplast: localization of three glycolytic isoenzymes, the single pyruvate dehydrogenase complex, and a plastid phosphate translocator. Eukaryot Cell 6:984–996
45. Fleige T, Limenitakis J, Soldati-Favre D (2010) Apicoplast: keep it or leave it. Microbes Infect 12:253–262
46. Foth BJ, Ralph SA, Tonkin CJ, Struck NS, Fraunholz M, Roos DS, Cowman AF, McFadden GI (2003) Dissecting apicoplast targeting in the malaria parasite *Plasmodium falciparum*. Science 299:705–708
47. Frenal K, Polonais V, Marq JB, Stratmann R, Limenitakis J, Soldati-Favre D (2010) Functional dissection of the apicomplexan glideosome molecular architecture. Cell Host Microbe 8:343–357

48. Funes S, Davidson E, Reyes-Prieto A, Magallon S, Herion P, King MP, Gonzalez-Halphen D (2002) A green algal apicoplast ancestor. Science 298:2155
49. Funes S, Reyes-Prieto A, Perez-Martinez X, Gonzalez-Halphen D (2004) On the evolutionary origins of apicoplasts: revisiting the rhodophyte vs. chlorophyte controversy. Microbes Infect 6:305–311
50. Gardner MJ, Feagin JE, Moore DJ, Spencer DF, Gray MW, Williamson DH, Wilson RJ (1991) Organisation and expression of small subunit ribosomal RNA genes encoded by a 35-kilobase circular DNA in *Plasmodium falciparum*. Mol Biochem Parasitol 48:77–88
51. Gardner MJ, Williamson DH, Wilson RJ (1991) A circular DNA in malaria parasites encodes an RNA polymerase like that of prokaryotes and chloroplasts. Mol Biochem Parasitol 44:115–123
52. Gardner MJ, Feagin JE, Moore DJ, Rangachari K, Williamson DH, Wilson RJ (1993) Sequence and organization of large subunit rRNA genes from the extra chromosomal 35 kb circular DNA of the malaria parasite *Plasmodium falciparum*. Nucleic Acids Res 21:1067–1071
53. Gardner MJ, Goldman N, Barnett P, Moore PW, Rangachari K, Strath M, Whyte A, Williamson DH, Wilson RJ (1994) Phylogenetic analysis of the rpoB gene from the plastid-like DNA of *Plasmodium falciparum*. Mol Biochem Parasitol 66:221–231
54. Gardner MJ, Preiser P, Rangachari K, Moore D, Feagin JE, Williamson DH, Wilson RJ (1994) Nine duplicated tRNA genes on the plastid-like DNA of the malaria parasite *Plasmodium falciparum*. Gene 144:307–308
55. Gibbs SP (1981) The chloroplasts of some algal groups may have evolved from endosymbiotic eukaryotic algae. Ann N Y Acad Sci 361:193–208
56. Glaser S, van Dooren GG, Agrawal S, Brooks CF, McFadden GI, Striepen B, Higgins MK (2012) Tic22 is an essential chaperone required for protein import into the apicoplast. J Biol Chem 287(47):39505–39512. doi: 10.1074/jbc.M112.405100
57. Glynn JM, Miyagishima SY, Yoder DW, Osteryoung KW, Vitha S (2007) Chloroplast division. Traffic 8:451–461
58. Glynn JM, Froehlich JE, Osteryoung KW (2008) Arabidopsis ARC6 coordinates the division machineries of the inner and outer chloroplast membranes through interaction with PDV2 in the intermembrane space. Plant Cell 20:2460–2470
59. Goodman CD, McFadden GI (2007) Fatty acid biosynthesis as a drug target in apicomplexan parasites. Curr Drug Targets 8:15–30
60. Gray MW (1993) Origin and evolution of organelle genomes. Curr Opin Genet Dev 3:884–890
61. Gubbels MJ, Vaishnava S, Boot N, Dubremetz JF, Striepen B (2006) A MORN-repeat protein is a dynamic component of the *Toxoplasma gondii* cell division apparatus. J Cell Sci 119:2236–2245
62. Hackett JD, Yoon HS, Li S, Reyes-Prieto A, Rummele SE, Bhattacharya D (2007) Phylogenomic analysis supports the monophyly of cryptophytes and haptophytes and the association of rhizaria with chromalveolates. Mol Biol Evol 24:1702–1713
63. Harb OS, Chatterjee B, Fraunholz MJ, Crawford MJ, Nishi M, Roos DS (2004) Multiple functionally redundant signals mediate targeting to the apicoplast in the apicomplexan parasite *Toxoplasma gondii*. Eukaryot Cell 3:663–674
64. Harper JT, Keeling PJ (2003) Nucleus-encoded, plastid-targeted glyceraldehyde-3-phosphate dehydrogenase (GAPDH) indicates a single origin for chromalveolate plastids. Mol Biol Evol 20:1730–1735
65. Harper JT, Waanders E, Keeling PJ (2005) On the monophyly of chromalveolates using a six-protein phylogeny of eukaryotes. Int J Syst Evol Microbiol 55:487–496
66. He CY, Striepen B, Pletcher CH, Murray JM, Roos DS (2001) Targeting and processing of nuclear-encoded apicoplast proteins in plastid segregation mutants of *Toxoplasma gondii*. J Biol Chem 276:28436–28442
67. Heaslip AT, Dzierszinski F, Stein B, Hu K (2010) TgMORN1 is a key organizer for the basal complex of *Toxoplasma gondii*. PLoS Pathog 6:e1000754

68. Hempel F, Bullmann L, Lau J, Zauner S, Maier UG (2009) ERAD-derived preprotein transport across the second outermost plastid membrane of diatoms. Mol Biol Evol 26:1781–1790
69. Holz GG Jr (1977) Lipids and the malarial parasite. Bull World Health Organ 55:237–248
70. Howe CJ, Purton S (2007) The little genome of apicomplexan plastids: its raison d'etre and a possible explanation for the 'delayed death' phenomenon. Protist 158:121–133
71. Janouskovec J, Horak A, Obornik M, Lukes J, Keeling PJ (2010) A common red algal origin of the apicomplexan, dinoflagellate, and heterokont plastids. Proc Natl Acad Sci U S A 107:10949–10954
72. Jomaa H et al (1999) Inhibitors of the nonmevalonate pathway of isoprenoid biosynthesis as antimalarial drugs. Science 285:1573–1576
73. Kalanon M, Tonkin CJ, McFadden GI (2009) Characterization of two putative protein translocation components in the apicoplast of *Plasmodium falciparum*. Eukaryot Cell 8:1146–1154
74. Kammerer B, Fischer K, Hilpert B, Schubert S, Gutensohn M, Weber A, Flugge UI (1998) Molecular characterization of a carbon transporter in plastids from heterotrophic tissues: the glucose 6-phosphate/phosphate antiporter. Plant Cell 10:105–117
75. Karnataki A, Derocher A, Coppens I, Nash C, Feagin JE, Parsons M (2007) Cell cycle-regulated vesicular trafficking of Toxoplasma APT1, a protein localized to multiple apicoplast membranes. Mol Microbiol 63:1653–1668
76. Karnataki A, Derocher AE, Coppens I, Feagin JE, Parsons M (2007) A membrane protease is targeted to the relict plastid of *Toxoplasma* via an internal signal sequence. Traffic 8:1543–1553
77. Karnataki A, DeRocher AE, Feagin JE, Parsons M (2009) Sequential processing of the *Toxoplasma apicoplast* membrane protein FtsH1 in topologically distinct domains during intracellular trafficking. Mol Biochem Parasitol 166:126–133
78. Keeling PJ (2010) The endosymbiotic origin, diversification and fate of plastids. Philos Trans R Soc Lond B Biol Sci 365:729–748
79. Keeling PJ, Burger G, Durnford DG, Lang BF, Lee RW, Pearlman RE, Roger AJ, Gray MW (2005) The tree of eukaryotes. Trends Ecol Evol 20:670–676
80. Khan H, Parks N, Kozera C, Curtis BA, Parsons BJ, Bowman S, Archibald JM (2007) Plastid genome sequence of the cryptophyte alga *Rhodomonas salina* CCMP1319: lateral transfer of putative DNA replication machinery and a test of chromist plastid phylogeny. Mol Biol Evol 24:1832–1842
81. Kohler S, Delwiche CF, Denny PW, Tilney LG, Webster P, Wilson RJ, Palmer JD, Roos DS (1997) A plastid of probable green algal origin in Apicomplexan parasites. Science 275:1485–1489
82. Kouranov A, Chen X, Fuks B, Schnell DJ (1998) Tic20 and Tic22 are new components of the protein import apparatus at the chloroplast inner envelope membrane. J Cell Biol 143:991–1002
83. Kumar B, Chaubey S, Shah P, Tanveer A, Charan M, Siddiqi MI, Habib S (2011) Interaction between sulphur mobilisation proteins SufB and SufC: evidence for an iron-sulphur cluster biogenesis pathway in the apicoplast of *Plasmodium falciparum*. Int J Parasitol 41:991–999
84. Layer G, Gaddam SA, Ayala-Castro CN, Ollagnier-de Choudens S, Lascoux D, Fontecave M, Outten FW (2007) SufE transfers sulfur from SufS to SufB for iron-sulfur cluster assembly. J Biol Chem 282:13342–13350
85. Legesse-Miller A, Massol RH, Kirchhausen T (2003) Constriction and Dnm1p recruitment are distinct processes in mitochondrial fission. Mol Biol Cell 14:1953–1963
86. Lell B et al (2003) Fosmidomycin, a novel chemotherapeutic agent for malaria. Antimicrob Agents Chemother 47:735–738
87. Lim L, McFadden GI (2010) The evolution, metabolism and functions of the apicoplast. Philos Trans R Soc Lond B Biol Sci 365:749–763
88. Lim L, Linka M, Mullin KA, Weber AP, McFadden GI (2010) The carbon and energy sources of the non-photosynthetic plastid in the malaria parasite. FEBS Lett 584:549–554

89. Lindmo K, Stenmark H (2006) Regulation of membrane traffic by phosphoinositide 3-kinases. J Cell Sci 119:605–614
90. Ling Y, Sahota G, Odeh S, Chan JM, Araujo FG, Moreno SN, Oldfield E (2005) Bisphosphonate inhibitors of *Toxoplasma gondi* growth: in vitro, QSAR, and in vivo investigations. J Med Chem 48:3130–3140
91. Lizundia R, Werling D, Langsley G, Ralph SA (2009) Theileria apicoplast as a target for chemotherapy. Antimicrob Agents Chemother 53:1213–1217
92. Lorestani A, Sheiner L, Yang K, Robertson SD, Sahoo N, Brooks CF, Ferguson DJ, Striepen B, Gubbels MJ (2010) A *Toxoplasma* MORN1 null mutant undergoes repeated divisions but is defective in basal assembly, apicoplast division and cytokinesis. PLoS ONE 5:e12302
93. Matsuzaki M, Kikuchi T, Kita K, Kojima S, Kuroiwa T (2001) Large amounts of apicoplast nucleoid DNA and its segregation in *Toxoplasma gondii*. Protoplasma 218:180–191
94. Mazumdar J, Striepen B (2007) Make it or take it: fatty acid metabolism of apicomplexan parasites. Eukaryot Cell 6:1727–1735
95. Mazumdar J EHW, Masek K CAH, Striepen B (2006) Apicoplast fatty acid synthesis is essential for organelle biogenesis and parasite survival in *Toxoplasma gondii*. Proc Natl Acad Sci U S A 103:13192–13197
96. McConkey GA, Rogers MJ, McCutchan TF (1997) Inhibition of *Plasmodium falciparum* protein synthesis. Targeting the plastid-like organelle with thiostrepton. J Biol Chem 272:2046–2049
97. McFadden GI, Reith ME, Munholland J, Lang-Unnasch N (1996) Plastid in human parasites. Nature 381:482
98. McFadden DC, Camps M, Boothroyd JC (2001) Resistance as a tool in the study of old and new drug targets in *Toxoplasma*. Drug Resist Updat 4:7984
99. McFadden GI, van Dooren GG (2004) Evolution: red algal genome affirms a common origin of all plastids. Curr Biol 14:514–516
100. McFadden GI, Waller RF (1997) Plastids in parasites of humans. Bioessays 19:1033–1040
101. Moore RB et al (2008) A photosynthetic alveolate closely related to apicomplexan parasites. Nature 451:959–963
102. Nagamune K, Hicks LM, Fux B, Brossier F, Chini EN, Sibley LD (2008) Abscisic acid controls calcium-dependent egress and development in *Toxoplasma gondii*. Nature 451:207–211
103. Nagaraj VA, Arumugam R, Gopalakrishnan B, Jyothsna YS, Rangarajan PN, Padmanaban G (2008) Unique properties of *Plasmodium falciparum* porphobilinogen deaminase. J Biol Chem 283:437–444
104. Nagaraj VA, Arumugam R, Chandra NR, Prasad D, Rangarajan PN, Padmanaban G (2009) Localisation of Plasmodium falciparum uroporphyrinogen III decarboxylase of the heme-biosynthetic pathway in the apicoplast and characterisation of its catalytic properties. Int J Parasitol 39:559–568
105. Nagaraj VA, Prasad D, Rangarajan PN, Padmanaban G (2009) Mitochondrial localization of functional ferrochelatase from *Plasmodium falciparum*. Mol Biochem Parasitol 168:109–112
106. Nagaraj VA, Arumugam R, Prasad D, Rangarajan PN, Padmanaban G (2010) Protoporphyrinogen IX oxidase from *Plasmodium falciparum* is anaerobic and is localized to the mitochondrion. Mol Biochem Parasitol 174:44–52
107. Nagaraj VA, Prasad D, Arumugam R, Rangarajan PN, Padmanaban G (2010) Characterization of coproporphyrinogen III oxidase in *Plasmodium falciparum* cytosol. Parasitol Int 59:121–127
108. Nair SC, Striepen B (2011) What do human parasites do with a chloroplast anyway? PLoS Biol 9(8):e1001137
109. Nair SC et al (2011) Apicoplast isoprenoid precursor synthesis and the molecular basis of fosmidomycin resistance in *Toxoplasma gondii*. J Exp Med 208:1547–1559

110. Obornik M, Janouskovec J, Chrudimsky T, Lukes J (2009) Evolution of the apicoplast and its hosts: from heterotrophy to autotrophy and back again. Int J Parasitol 39(1):1–12
111. Outten FW, Djaman O, Storz G (2004) A suf operon requirement for Fe-S cluster assembly during iron starvation in *Escherichia coli*. Mol Microbiol 52:861–872
112. Park S, Isaacson R, Kim HT, Silver PA, Wagner G (2005) Ufd1 exhibits the AAA-ATPase fold with two distinct ubiquitin interaction sites. Structure 13:995–1005
113. Patron NJ, Rogers MB, Keeling PJ (2004) Gene replacement of fructose-1,6-bisphosphate aldolase supports the hypothesis of a single photosynthetic ancestor of chromalveolates. Eukaryot Cell 3:1169–1175
114. Ponts N, Saraf A, Chung DW, Harris A, Prudhomme J, Washburn MP, Florens L, Le Roch KG (2011) Unraveling the ubiquitome of the human malaria parasite. J Biol Chem 286(46):40320–40330
115. Preiser P, Williamson DH, Wilson RJ (1995) tRNA genes transcribed from the plastid-like DNA of *Plasmodium falciparum*. Nucleic Acids Res 23:43294336
116. Raghu Ram EV, Kumar A, Biswas S, Chaubey S, Siddiqi MI, Habib S (2007) Nuclear gyrB encodes a functional subunit of the *Plasmodium falciparum* gyrase that is involved in apicoplast DNA replication. Mol Biochem Parasitol 154:30–39
117. Ram EV, Naik R, Ganguli M, Habib S (2008) DNA organization by the apicoplast-targeted bacterial histone-like protein of *Plasmodium falciparum*. Nucleic Acids Res 36:5061–5073
118. Ramakrishnan S et al (2012) Apicoplast and endoplasmic reticulum cooperate in fatty acid biosynthesis in apicomplexan parasite *Toxoplasma gondii*. J Biol Chem 287:4957–4871
119. Ramya TN, Mishra S, Karmodiya K, Surolia N, Surolia A (2007) Inhibitors of nonhousekeeping functions of the apicoplast defy delayed death in *Plasmodium falciparum*. Antimicrob Agents Chemother 51:307–316
120. Reiff SB, Vaishnava S, Striepen B (2012) The HU protein is important for apicoplast genome maintenance and inheritance in *Toxoplasma gondii*. Eukaryot Cell 11:905–915
121. Rock RC (1971) Incorporation of 14 C-labelled fatty acids into lipids of rhesus erythrocytes and *Plasmodium knowlesi* in vitro. Comp Biochem Physiol B 40:893–906
122. Rogers MB, Gilson PR, Su V, McFadden GI, Keeling PJ (2007) The complete chloroplast genome of the chlorarachniophyte Bigelowiella natans: evidence for independent origins of chlorarachniophyte and euglenid secondary endosymbionts. Mol Biol Evol 24:54–62
123. Rohrich RC et al (2005) Reconstitution of an apicoplast-localised electron transfer pathway involved in the isoprenoid biosynthesis of *Plasmodium falciparum*. FEBS Lett 579:6433–6438
124. Sato S, Clough B, Coates L, Wilson RJ (2004) Enzymes for heme biosynthesis are found in both the mitochondrion and plastid of the malaria parasite *Plasmodium falciparum*. Protist 155:11725
125. Seeber F (2002) Biogenesis of iron-sulphur clusters in amitochondriate and apicomplexan protists. Int J Parasitol 32:1207–1217
126. Seeber F, Soldati-Favre D (2010) Metabolic pathways in the apicoplast of apicomplexa. Int Rev Cell Mol Biol 281:161–228
127. Seow F, Sato S, Janssen CS, Riehle MO, Mukhopadhyay A, Phillips RS, Wilson RJ, Barrett MP (2005) The plastidic DNA replication enzyme complex of *Plasmodium falciparum*. Mol Biochem Parasitol 141:145–153
128. Sheiner L, Demerly JL, Poulsen N, Beatty WL, Lucas O, Behnke MS, White MW, Striepen B (2011) A systematic screen to discover and analyze apicoplast proteins identifies a conserved and essential protein import factor. PLoS Pathog 7:e1002392
129. Siddall ME (1992) Hohlzylinders. Parasitol Today 8:90–91
130. Sidhu AB, Sun Q, Nkrumah LJ, Dunne MW, Sacchettini JC, Fidock DA (2007) In vitro efficacy, resistance selection, and structural modeling studies implicate the malarial parasite apicoplast as the target of azithromycin. J Biol Chem 282:2494–2504
131. Smith MH, Ploegh HL, Weissman JS (2011) Road to ruin: targeting proteins for degradation in the endoplasmic reticulum. Science 334:1086–1090
132. Soll J, Schleiff E (2004) Protein import into chloroplasts. Nat Rev Mol Cell Biol 5:198–208

133. Sommer MS, Gould SB, Lehmann P, Gruber A, Przyborski JM, Maier UG (2007) Der1-mediated preprotein import into the periplastid compartment of chromalveolates? Mol Biol Evol 24:918–928
134. Spork S, Hiss JA, Mandel K, Sommer M, Kooij TW, Chu T, Schneider G, Maier UG, Przyborski JM (2009) An unusual ERAD-like complex is targeted to the apicoplast of *Plasmodium falciparum*. Eukaryot Cell 8:1134–1145
135. Stanway RR, Mueller N, Zobiak B, Graewe S, Froehlke U, Zessin PJ, Aepfelbacher M, Heussler VT (2011) Organelle segregation into *Plasmodium* liver stage merozoites. Cell Microbiol 13:1768–1782
136. Stoebe B, Kowallik KV (1999) Gene-cluster analysis in chloroplast genomics. Trends Genet 15:344–347
137. Striepen B, Crawford MJ, Shaw MK, Tilney LG, Seeber F, Roos DS (2000) The plastid of *Toxoplasma gondii* is divided by association with the centrosomes. J Cell Biol 151:1423–1434
138. Striepen B, Jordan CN, Reiff S, van Dooren GG (2007) Building the perfect parasite: cell division in apicomplexa. PLoS Pathog 3:e78
139. Surolia N, Padmanaban G (1992) de novo biosynthesis of heme offers a new chemotherapeutic target in the human malarial parasite. Biochem Biophys Res Commun 187:744–750
140. Tawk L, Chicanne G, Dubremetz JF, Richard V, Payrastre B, Vial HJ, Roy C, Wengelnik K (2010) Phosphatidylinositol 3-phosphate, an essential lipid in *Plasmodium*, localizes to the food vacuole membrane and the apicoplast. Eukaryot Cell 9:1519–1530
141. Tawk L et al (2011) Phosphatidylinositol 3-monophosphate is involved in *Toxoplasma apicoplast* biogenesis. PLoS Pathog 7(2):e1001286
142. Thomsen-Zieger N, Schachtner J, Seeber F (2003) Apicomplexan parsites contain a single lipoic acid synthase located in the plastid. FEBS Lett 547:80–86
143. Tomova C, Geerts WJ, Muller-Reichert T, Entzeroth R, Humbel BM (2006) New comprehension of the apicoplast of Sarcocystis by transmission electron tomography. Biol Cell 98:535–545
144. Tomova C, Humbel BM, Geerts WJ, Entzeroth R, Holthuis JC, Verkleij AJ (2009) Membrane contact sites between apicoplast and ER in *Toxoplasma gondii* revealed by electron tomography. Traffic 10:1471–1480
145. Tonhosolo R et al (2009) Carotenoid biosynthesis in intraerythrocytic stages of *Plasmodium falciparum*. J Biol Chem 284:9974–9985
146. Tonkin CJ, Roos DS, McFadden GI (2006) N-terminal positively charged amino acids, but not their exact position, are important for apicoplast transit peptide fidelity in *Toxoplasma gondii*. Mol Biochem Parasitol 150:192–200
147. Tonkin CJ, Kalanon M, McFadden GI (2008) Protein targeting to the malaria parasite plastid. Traffic 9:166–175
148. Tonkin CJ, Struck NS, Mullin KA, Stimmler LM, McFadden GI (2006) Evidence for Golgi-independent transport from the early secretory pathway to the plastid in malaria parasites. Mol Microbiol 61:614–630
149. Vaishnava S, Striepen B (2006) The cell biology of secondary endosymbiosis-how parasites build, divide and segregate the apicoplast. Mol Microbiol 61:1380–1387
150. Vaishnava S, Morrison DP, Gaji RY, Murray JM, Entzeroth R, Howe DK, Striepen B (2005) Plastid segregation and cell division in the apicomplexan parasite Sarcocystis neurona. J Cell Sci 118:3397–3407
151. van Dooren GG, Su V, D'Ombrain MC, McFadden GI (2002) Processing of an apicoplast leader sequence in *Plasmodium falciparum* and the identification of a putative leader cleavage enzyme. J Biol Chem 277:23612–23619
152. van Dooren GG, Tomova C, Agrawal S, Humbel BM, Striepen B (2008) *Toxoplasma gondii* Tic20 is essential for apicoplast protein import. Proc Natl Acad Sci U S A 105:13574–13579
153. van Dooren GG, Reiff SB, Tomova C, Meissner M, Humbel BM, Striepen B (2009) A novel dynamin-related protein has been recruited for apicoplast fission in *Toxoplasma gondii*. Curr Biol 19:267–276

154. van Dooren GG, Kennedy AT, McFadden GI (2012) The use and abuse of heme in Apicomplexan parasites. Antioxid Redox Signal 17:634–656
155. Varadharajan S, Dhanasekaran S, Bonday ZQ, Rangarajan PN, Padmanaban G (2002) Involvement of delta-aminolaevulinate synthase encoded by the parasite gene in de novo haem synthesis by *Plasmodium falciparum*. Biochem J 367:321–327
156. Vaughan AM, O'Neill MT, Tarun AS, Camargo N, Phuong TM, Aly AS, Cowman AF, Kappe SH (2009) Type II fatty acid synthesis is essential only for malaria parasite late liver stage development. Cell Microbiol 11:506–520
157. Vinella D, Brochier-Armanet C, Loiseau L, Talla E, Barras F (2009) Iron-sulfur (Fe/S) protein biogenesis: phylogenomic and genetic studies of A-type carriers. PLoS Genet 5:e1000497
158. Vollmer M, Thomsen N, Wiek S, Seeber F (2001) Apicomplexan parasites possess distinct nuclear-encoded, but apicoplast-localized, plant-type ferredoxin-NADP+reductase and ferredoxin. J Biol Chem 276:5483–5490
159. Waller RF, McFadden GI (2005) The apicoplast: a review of the derived plastid of apicomplexan parasites. Curr Issues Mol Biol 7:57–79
160. Waller RF et al (1998) Nuclear-encoded proteins target to the plastid in *Toxoplasma gondii* and *Plasmodium falciparum*. Proc Natl Acad Sci U S A 95:12352–12357
161. Waller RF, Reed MB, Cowman AF, McFadden GI (2000) Protein trafficking to the plastid of *Plasmodium falciparum* is via the secretory pathway. EMBO J 19:1794–1802
162. Waller RF, Keeling PJ, van Dooren GG, McFadden GI (2003) Comment on "A green algal apicoplast ancestor". Science 301:49 (author reply 49)
163. Walters KJ (2005) Ufd1 exhibits dual ubiquitin binding modes. Structure 13:943–947
164. Weissig V, Vetro-Widenhouse TS, Rowe TC (1997) Topoisomerase II inhibitors induce cleavage of nuclear and 35-kb plastid DNAs in the malarial parasite *Plasmodium falciparum*. DNA Cell Biol 16:1483–1492
165. Williamson DH, Gardner MJ, Preiser P, Moore DJ, Rangachari K, Wilson RJ (1994) The evolutionary origin of the 35 kb circular DNA of *Plasmodium falciparum*: new evidence supports a possible rhodophyte ancestry. Mol Gen Genet 243:249–252
166. Wilson RJ, Williamson DH, Preiser P (1994) Malaria and other Apicomplexans: the "plant" connection. Infect Agents Dis 3:29–37
167. Wilson RJ et al (1996) Complete gene map of the plastid-like DNA of the malaria parasite *Plasmodium falciparum*. J Mol Biol 261:155–172
168. Wrenger C, Muller S (2004) The human malaria parasite *Plasmodium falciparum* has distinct organelle-specific lipoylation pathways. Mol Microbiol 53:103–113
169. Wu B Heme biosynthetic pathway in Apicomplexan parasites. University of Pennsylvania. http://repositoryupennedu/dissertations/AAI3246256/2006
170. Ye Y, Shibata Y, Yun C, Ron D, Rapoport TA (2004) A membrane protein complex mediates retro-translocation from the ER lumen into the cytosol. Nature 429:841–847
171. Yeh E, DeRisi JL (2011) Chemical rescue of malaria parasites lacking an apicoplast defines organelle function in blood-stage *Plasmodium falciparum*. PLoS Biol 9(8):e1001138
172. Yoon HS, Hackett JD, Pinto G, Bhattacharya D (2002) The single, ancient origin of chromist plastids. Proc Natl Acad Sci U S A 99:15507–155012
173. Yu M et al (2008) The fatty acid biosynthesis enzyme FabI plays a key role in the development of liver-stage malarial parasites. Cell Host Microbe 4:567–578
174. Yung S, Unnasch TR, Lang-Unnasch N (2001) Analysis of apicoplast targeting and transit peptide processing in *Toxoplasma gondii* by deletional and insertional mutagenesis. Mol Biochem Parasitol 118:11–21
175. Zhang Z, Green BR, Cavalier-Smith T (2000) Phylogeny of ultra-rapidly evolving dinoflagellate chloroplast genes: a possible common origin for sporozoan and dinoflagellate plastids. J Mol Evol 51:2640
176. Zhang B, Watts KM, Hodge D, Kemp LM, Hunstad DA, Hicks LM, Odom AR (2011) A second target of the antimalarial and antibacterial agent fosmidomycin revealed by cellular metabolic profiling. BioChemistry 50:35707

177. Zhu G, Li Y, Cai X, Millership JJ, Marchewka MJ, Keithly JS (2004) Expression and functional characterization of a giant Type I fatty acid synthase (CpFAS1) gene from *Cryptosporidium parvum*. Mol Biochem Parasitol 134:127135
178. Zhu G, Shi X, Cai X (2010) The reductase domain in a Type I fatty acid synthase from the apicomplexan *Cryptosporidium parvum*: restricted substrate preference towards very long chain fatty acyl thioesters. BMC Biochem 11:46

Part III
Biogenesis of Chloroplast Proteins

Chapter 9
Mechanisms of Chloroplast Protein Import in Plants

Paul Jarvis and Felix Kessler

Abstract Most chloroplast proteins are encoded as preproteins by the nuclear genome. Their import into chloroplasts occurs post-translationally. An N-terminal pre-sequence, the transit peptide, contains the organellar targeting information. It is specifically recognized by receptor components at the chloroplast surface. These receptors are components of the TOC (*t*ranslocon at the *o*uter envelope membrane of *c*hloroplasts) complex. Together with the TIC (*t*ranslocon at the *i*nner envelope membrane of *c*hloroplasts) machinery, this mediates the import of proteins into chloroplasts. In addition to the receptors, these complexes incorporate channel, motor and regulatory functions. Many putative or actual components have been identified. Multiple isoforms of the TOC receptors (and possibly of some other components) constitute the molecular basis of separate import pathways with distinct client preferences. This perhaps reduces competition effects between highly abundant and less abundant preproteins. Client preferences of different import pathways might also facilitate the differentiation of various plastid types. In addition to the canonical TOC/TIC-mediated import routes, alternative, mechanistically distinct pathways of protein transport to chloroplasts have been identified; one of these passes through the endoplasmic reticulum and Golgi apparatus. Other work has revealed several protein targeting pathways leading to the envelope membranes.

Keywords Chloroplast envelope · Chloroplast protein import · Plastid biogenesis · Protein targeting · Protein transport · TOC/TIC machinery · Translocon

Abbreviations

CAH1	Carbonic anhydrase 1
ceQORH	Chloroplast envelope quinone oxidoreductase homolog
cpHsc70	Chloroplast stromal Hsp70
MGD1	Monogalactosyldiacylglycerol synthase 1

P. Jarvis (✉)
Department of Plant Sciences, University of Oxford, South Parks Road, Oxford OX1 3RB, UK
e-mail: paul.jarvis@plants.ox.ac.uk

F. Kessler
Laboratoire de Physiologie Végétale, Université de Neuchâtel, Rue Emile-Argand 11, 2000 Neuchâtel, Switzerland
e-mail: felix.kessler@unine.ch

S.M. Theg, F.-A. Wollman (eds.), *Plastid Biology*, Advances in Plant Biology 5,
DOI 10.1007/978-1-4939-1136-3_9,

OEP7/14	Outer envelope protein, 7/14 kD
PPI1	Plastid protein import 1
SPP	Stromal processing peptidase
SSU	Small subunit of Rubisco
TIC	Translocon at the inner envelope membrane of chloroplasts
TOC	Translocon at the outer envelope membrane of chloroplasts
TROL	Thylakoid rhodanese-like protein
UPS	Ubiquitin-proteasome system

9.1 Introductory Topics

9.1.1 Background

Plastids are the prototypical plant organelles. The chloroplast is the most prominent representative of the family. Chloroplasts derived from cyanobacteria by an endosymbiotic process during which a primordial photosynthetic bacterium was engulfed by a eukaryotic host cell. Over time, most of the cyanobacterial genetic material was transferred to host nucleus, both transforming the host genome and progressively reducing the chloroplast genome. The cyanobacterial model *Synechocystis* sp. PCC6803 [99] has 3168 predicted protein coding genes, whereas the chloroplast genome of *Arabidopsis thaliana*, the eudicot model system, retains 87 protein coding genes [178]. However, the number of chloroplast proteins has been estimated at around 1500 in recent proteomics studies [59, 113, 198]. The large difference is explained by the nuclear-encoded origin of the vast majority of chloroplast proteins. Therefore, a mechanism for importing proteins synthesized in the cytosol into chloroplasts is required [94, 105, 136, 187]. In this chapter, we will address how these nuclear-encoded chloroplast constituents are targeted to the organelle and translocated across the outer and inner envelope membranes.

9.1.2 Transit Peptides

Some of the first studies on chloroplast protein import in the late 70s and early 80s of the past century focused on one of the most abundant and best known chloroplast proteins, the small subunit of Rubisco (SSU), establishing it as a model chloroplast import substrate [52, 67]. In cell-free translation experiments of isolated *Chlamydomonas* mRNA, Dobberstein and colleagues in 1977 observed that SSU was synthesized as a protein larger than that present in chloroplasts. They hypothesized that this was an extrachloroplastic form and that the additional amino acid sequence may be required for its transfer into the chloroplast. Since then many chloroplast proteins have been identified and studied, and as it turned out the majority of these are synthesized with N-terminal extensions as "preproteins", abbreviated as pSSU in the case of SSU. Furthermore, it has been demonstrated that the N-terminal extensions (known as transit peptides) are required for the targeting and translocation of preproteins into the chloroplast. Interestingly, transit peptides are not conserved in primary structure, and their length is also quite variable from approximately 30 to 70 amino acids [33, 34]. Certain physicochemical characteristics, however, such as a preponderance of hydroxylated and basic amino acids together with the under-representation of acidic amino acids are typical of transit peptides. It has also been found that transit peptides have a tendency to take on a natively unfolded structure [223]. Such an unfolded structure may facilitate the engagement of components of the chloroplast protein import machinery.

9.1.3 Energetics and Stages of Import

Before any of the components of the chloroplast protein import machinery were identified, the in vitro import system using isolated pea chloroplasts was extensively studied in terms of energetics and differentiable import stages. Early on it was recognized that preprotein import requires energy: light could be used as it leads to the production of ATP via the electron transport chain and the ATP synthase. Exogenous ATP is required when chloroplasts are depleted of energy in the dark and by the dissolution of the proton gradient using nigericin. Depending on the ATP concentration, at least three different stages of import were distinguished. In the absence of added ATP ("energy-independent stage"), the preprotein interacted with components at the chloroplast surface in a reversible, unstable way that could be trapped using covalent chemical crosslinking [115, 143, 163]. At low concentrations of ATP (100 micromolar), and GTP, the preprotein inserted across the outer envelope membrane in a stable way and was isolated as the so-called "early translocation intermediate" [106, 159, 228]. Later it was demonstrated that at this stage the preprotein is exposed to the cytoplasm but has also already engaged components at the inner envelope membrane. Complete translocation of the preprotein across both envelope membranes required "high" concentrations of ATP (1-5 millimolar) [160, 202]. At this stage the preprotein arrives in the chloroplast stroma where the transit peptide is cleaved by the stromal processing peptidase (SPP) (see Sect. 9.3.4). The preprotein can be arrested at this stage by chilling the chloroplasts on ice resulting in the "late translocation intermediate" [85, 133].

9.1.4 Identification of Translocon Components

The identification of translocon components long proved challenging: Eventually, three separate approaches turned out to be successful. A study using chemical crosslinking of pSSU in an in vitro import assay in the absence of added ATP at the "energy-independent stage" resulted in the cross-linking of an 86 kD protein [163]. At a later stage of import, in the presence of ATP, an additional 75 kD protein

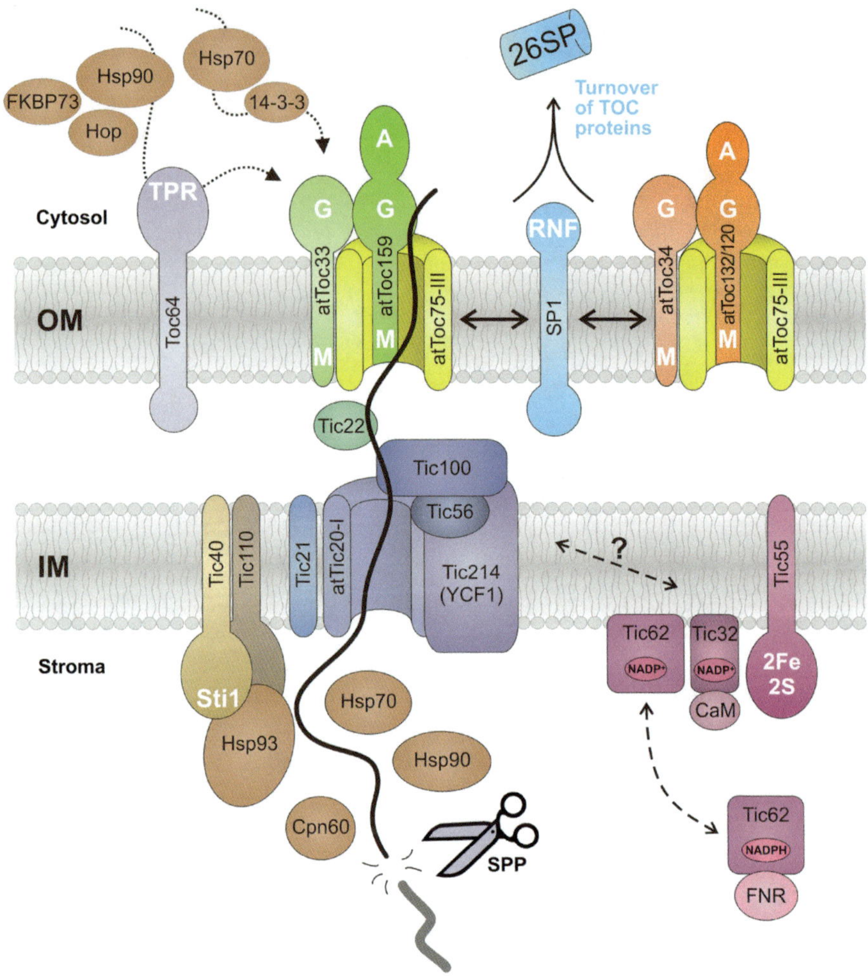

Fig. 9.1 The TOC and TIC complexes of the chloroplast protein import machinery. The TOC and TIC translocons in the outer and inner envelope membranes (*OM* and *IM*, respectively) are shown, as is a translocating preprotein (*black line*). Individual translocon components are identified by their molecular weights (*black text*), while some key functional domains are indicated (*white text*). Toc159, Toc34 and Toc75 together form the core TOC complex in the outer membrane (Toc159 is represented here by the *Arabidopsis* isoforms atToc159, atToc132 and atToc120; and Toc34 is represented by atToc33 and atToc34). While Toc159 and Toc34 are responsible for preprotein recognition, Toc75 (atToc75-III in *Arabidopsis*) forms the outer envelope channel. Different receptor isoforms enable the formation of different TOC complexes, and thus the operation of different import pathways with distinct client preferences. The RING finger (*RNF*) ubiquitin E3 ligase SP1 mediates the ubiquitination of TOC components, leading to their turnover by the 26S proteasome (*26SP*); this enables the dynamic reorganization of the protein import machinery. Various cytosolic chaperones and their cofactors (Hsp70 and 14-3-3; Hsp90, Hop and FKBP73) are proposed to interact with unfolded preproteins (forming so-called "guidance complexes"), to maintain their import competence and direct them to the Toc34 or Toc64 (OEP64) receptors. Tic22 is thought to provide a link between the TOC and TIC complexes, facilitating preprotein passage through the intermembrane space. Tic20 (atTic20-I) may participate in inner membrane channel formation, as

was crosslinked. These results indicated that preprotein import proceeds through a sequence of interactions from the cytosol to the chloroplast stroma. However, this study did not molecularly identify either the 86 or the 75 kD components [163]. In a separate study, IgG directed against the 86 kD protein at the outer membrane was shown to inhibit preprotein import into the chloroplast [73]. This finding suggested that the 86 kD protein functioned as a preprotein import receptor. In a third study, recombinant purified pSSU fused to IgG-binding of ProteinA (pSSU-ProtA) was incubated on a large scale with isolated chloroplasts either in the presence of "low" concentrations of ATP to produce the early intermediate or with "high" concentrations of ATP to produce the late intermediate [106, 182]. Subsequently, the reactions were stopped on ice. The envelope membranes were isolated, solubilized and subjected to IgG-affinity chromatography. In the case of the early intermediate, this resulted in the co-isolation of three chloroplast envelope proteins (86, 75 and 34 kD) together with un-processed pSSU-ProtA. For the late intermediate, the corresponding experiment resulted in the co-isolation of five envelope membrane proteins (the same 86, 75, 34 kD bands and additional bands at 110 and 36 kD) together with the mature SSU-ProtA.

In hindsight, these three seminal studies together yielded the first evidence for components of the chloroplast protein import machinery. The components at the outer membrane were termed Toc (*t*ranslocon at the *o*uter membrane of the *c*hloroplast) [183]. The core of the TOC translocon consists of an apparently stable complex of Toc159 (of which the 86 kD protein is a fragment, see below), Toc75 and Toc34 that correspond to the proteins that were identified in the initial studies (Fig. 9.1) [107, 181]. One of the additional components that co-isolated with the late intermediate was later identified as Tic110 (*t*ranslocon at the *i*nner membrane of the *c*hloroplast) and was the first known component at the inner membrane [103, 140]. The 36 kD component remained unidentified.

At the early intermediate stage of translocation the preprotein is inserted across the outer membrane and already makes contact with components of the inner membrane. The components at the inner membrane are Tic20 and Tic22, which were both identified by covalent crosslinking to the trapped intermediate [115, 116]. Tic20 is an integral protein of the inner membrane, while Tic22 is one of a few known intermembrane space proteins and is only peripherally associated with the outer face of the inner envelope membrane (see Sect. 9.3).

---

part of a large TIC complex which also incorporates Tic214, Tic100, Tic56 and Tic21, the latter being only loosely associated with the complex. Thereafter, Tic110 functions together with Tic40 in the recruitment and regulation of stromal chaperones, such as Hsp93, Hsp90, Hsp70 and Cpn60, which may form motor complexes for protein import propulsion, facilitate protein folding, or aid intraorganellar routing. SPP removes the transit peptide (*grey line*) upon its arrival in the stroma. A redox-regulator, comprising Tic62, Tic55 and Tic32, might be involved in fine-tuning the import process, working in conjunction with FNR and calmodulin (*CaM*)

9.2 Translocation Across the Outer Membrane

9.2.1 Components of the TOC Complex

The three main components of TOC core complex have highly intriguing primary structures: Toc159 and Toc34 turned out to be homologous GTP-binding proteins of the septin class [106, 134]. While Toc34 consists of the conserved GTP-binding domain followed by a C-terminal transmembrane segment, Toc159 has an extensive N-terminal acidic (A-) domain and a large C-terminal membrane-anchoring (M-) domain [31, 40]. Toc75 is a beta-barrel type membrane protein of the Omp85 family, members of which are involved in the insertion of outer membrane proteins in bacteria [57, 63, 209]. Both Toc159 and Toc34 are located at the chloroplast surface and are accessible to exogenous protease. Crosslinking at the earliest stages of protein import as well as antibody inhibition suggested that Toc159 constitutes the primary import receptor at the chloroplast surface [73, 163]. But there is now ample evidence that Toc159 and Toc34 co-operate in the recognition of preproteins at the chloroplast surface although the exact sequence of events occurring at the chloroplast surface is not known [22, 104, 138]. Toc75 is resistant to exogenously added protease indicating that it is deeply buried in the outer membrane. Based on its topology and primary structure Toc75 was proposed to function as the preprotein-conducting channel in the outer membrane [182, 209]. Indeed, Toc75 has channel properties in electrophysiological setups [70]. In most graphic models, Toc159 and -34 are represented as GTP-regulated, preprotein-specific receptors providing access to Toc75 channel. This view is supported biochemically by the finding that non- or slowly-hydrolyzable GTP analogs block preprotein import into isolated chloroplasts [106, 228]. However, more recent reverse genetic studies in the *Arabidopsis thaliana* model system point to a very complex mechanism of GTP-regulation that we do not completely understand (see below) [1, 9, 129, 226]. A reconstitution study also suggested that the three components are sufficient to accomplish translocation in vitro [180]. Because of their central role in import and stable association, Toc159, -34 and -75 together were termed the TOC core complex (Fig. 9.1) [182].

9.2.2 TOC Components in **Arabidopsis thaliana**

The research described so far was carried out using the pea (*Pisum sativum*) chloroplast in vitro system. This system allows large-scale chloroplast isolation, which in turn is useful for all sorts of biochemical experimentation. But up to now pea has not been amenable to genetic techniques. Here, *Arabidopsis thaliana* has proven incredibly powerful. A first mutant was identified in a forward screen of T-DNA insertion lines to find pale-green chloroplast biogenesis mutants—this was the *plastid protein import 1* (*ppi1*) mutant [95]. The PPI1 protein (atToc33) turned out to be highly homologous to pea Toc34, thereby providing the first insight into its role in chloroplast biogenesis. Moreover, the *Arabidopsis* genome contained a close homolog of atToc33, atToc34, that was able to functionally complement the *ppi1* mutant. Indeed, the atToc33 and -34 proteins share considerable redundancy, with the double knock-out being embryo lethal [50]. This finding indicates that the two proteins together fulfil an essential role in plastid development, presumably by their participation in protein import. Since then, many more details of a complex TOC system in *Arabidopsis* have emerged: For instance, a total of four Toc159 homologs exist (atToc159, -132, -120 and -90). This GTPase subfamily shares the A-, G- and M-domain features of pea Toc159, with the exception of atToc90 that lacks the A-domain [19, 68].

Inevitably, questions regarding the possibility of distinct functions of these homologs emerged. Evidence for such a scenario stems from the analysis of the *ppi2* mutant in which the *atTOC159* gene is disrupted by a T-DNA insertion [19]. The *ppi2* mutant has a tell-tale seedling-lethal, albino phenotype pointing to a defect in chloroplast biogenesis. Moreover, a strong reduction in the levels of many, but not all photosynthesis-associated proteins was observed. This suggested that atToc159 is a major receptor required for the massive accumulation of photosynthesis-associated proteins. While this may hold true, the *ppi2* mutation also affects the accumulation of proteins that are not associated with photosynthesis [29]. Therefore, the functional boundaries of atToc159 are not as clear cut as originally believed. The single T-DNA insertion mutants of the other Toc159 homologs have milder (Toc132) or no phenotypes (Toc120/-90) [69, 91, 124]. However, the *toc132 toc120* double mutant resulted in a very severe phenotype, indicating redundancy. Because these two genes are expressed predominantly in non-photosynthetic tissues such as roots, this suggested that the two homologs together may have a central role in the import of non-photosynthesis-associated, "house-keeping" proteins [89, 91, 124]. Preferential assignment of the Toc34 isoforms to different import pathways has also been proposed (Fig. 9.1) [50, 91, 123]. While along rather general lines, the respective functional assignments of Toc159 and Toc132/-120 to specific groups of preproteins are probably more or less correct; nonetheless, a much more differentiated view must be developed.

Just recently, it has been demonstrated that chloroplast preproteins fall into three age-dependent classes, with the optimal import efficiency of each correlating with different chloroplast ages [204]. Intriguingly, the molecular determinants of age-dependent import lie within the transit peptide. In light of these findings, the question arises as to what extent age-dependent import pathways correspond to the separate import pathways that have been identified in the context of TOC receptor-dependent client-specificity.

9.2.3 TOC Complexes

In the absence of genome sequence information, the situation in pea appears simple as a sole TOC core complex consisting of three components is known; however, inspection of the available sequenced plant genomes indicates that *Arabidopsis* is not an unusual case, and that the aforementioned TOC receptor diversity is common.

Regardless, exactly how many of each one of these proteins is present in the individual complexes is still a matter of debate. A megadalton complex containing the components has been identified, which suggests that some of them must be present in more than just one copy [39, 107, 180].

In *Arabidopsis*, only Toc75 is encoded by a unique gene (it has homologs but these are either inactive or are not directly implicated in the translocation of transit peptide containing preproteins) [14, 55, 80]. Given the existence of small families of Toc GTPases, a variety of TOC core complexes may exist. Experimental evidence suggests that complexes consisting of atToc159, atToc33 and atToc75 are predominantly present in green, chloroplast-containing tissues and transport mostly photosynthesis-associated genes. Complexes consisting of atToc132/-120, atToc34 and atToc75 are predominantly present in non-green tissues and transport mostly the "house-keeping" proteins [91]. The evidence, however, is not as clear cut as it may appear: atToc34 can complement the absence of atToc33 in *Arabidopsis* suggesting that at least in this case the function of one component can be taken over by another and also replace it in the TOC core complex [50, 95].

Earlier results had shown that non- or slowly-hydrolyzable analogs of GTP inhibit chloroplast protein import in vitro [106, 228]. Therefore, the GTP-binding motifs in the Toc receptor family appeared to be interesting targets for further analysis. Initial experiments analyzed the effects of the GTPase mutants. A triple mutant in the GTP-binding site of atToc159 was non-functional and did not assemble into the TOC complex in planta [20]. But other single mutants that functionally disrupted GTP binding and/or hydrolysis apparently fully complemented the albino *ppi2* mutant, although import activity in some of the mutants was reduced [1, 226]. Similar results were obtained in the case of atToc33 [9, 129]. A particularly interesting result was obtained with the atToc159 A864R mutant that binds GTP strongly but is unable to hydrolyze GTP: this mutant not only rescued the albino phenotype of *ppi2* but increased the preprotein import efficiency in vitro [226]. These results suggest that, unexpectedly, GTP-binding to the TOC receptors is not essential but important for regulation of import activity levels.

9.2.4 Cytosolic Factors and Regulators of Chloroplast Protein Import

Before preproteins destined for the chloroplast reach the TOC complex they have to interact with cytosolic targeting factors (Fig. 9.1) [61, 132]. These include the chaperones Hsp70 and -90 that serve to prevent the newly synthesized preproteins from aggregating, maintaining their import-competent state. It has been shown that Hsp70 interacts with many preproteins immediately after their synthesis. Some preproteins may be phosphorylated at their transit peptide which promotes the binding of a 14-3-3 dimer [58, 148, 153]. Together, with Hsp70, the 14-3-3 dimer forms a so-called guidance complex that accompanies the preprotein to the receptors of the TOC complex and hands over the preprotein to Toc34. A set of cytosolic kinases (STY8-, 14 and -46) have been identified that are implicated in the phosphorylation of the transit peptide [126, 147]. Yet other preproteins have been shown to form complexes with Hsp90 [58, 166]. Preprotein/Hsp90 complexes interact with Toc64, a tetratricopeptide repeat (TPR)-containing protein, loosely associating with the TOC complex. In addition to Hsp90, Hsp70/Hsp90-organizing protein (Hop) and the immunophilin FKBP73 may be players in the HSP90 pathway [58]. Surprisingly, however, Toc64 is essential neither in *Physcomitrella* nor *Arabidopsis* suggesting that this component of the pathway can be bypassed [8, 76].

Phosphorylation does not only play a role at the level of the transit peptide. It has been shown that Toc34 is a target of phosphorylation, and that in its phosphorylated state it is unable to bind preproteins or GTP and needs to be activated by a phosphatase [7, 96, 200]. Toc159 is highly phosphorylated at its A-domain, and this is most likely due to cytosolic casein kinase II [2]. The role of the phosphorylation is likely of regulatory nature, but exactly how it works still needs to be clarified. The A-domain also exists as a separate, soluble protein. This finding may explain why Toc159 was originally identified as an 86 kD protein lacking the A-domain [31, 40]. Again, the cleavage of the A-domain may have a regulatory purpose, such as activating the protein. More research in this area is required to clarify the role of phosphorylation and other mechanisms in chloroplast protein import.

Interestingly, cytosolic preproteins are rarely observed in living plants. This suggests that preprotein synthesis is tightly coupled to translocation so that very few preproteins remain in the cytosol. In this context, an interesting regulatory mechanism was discovered by which "un-imported" preproteins are degraded by the ubiquitin-proteasome system (UPS) [130]. Thus, both tight coupling of preprotein synthesis and import and the UPS may contribute to efficient accumulation inside the chloroplast and removal of preproteins from the cytosol.

Recently, a second exciting role for the UPS was discovered. A screen for second-site suppressors of the *Arabidopsis ppi1* mutant identified SP1 (*s*uppressor of *p*pi1 locus *1*) [139]. SP1 is a chloroplast outer membrane E3 ubiquitin ligase that directly interacts with components of the TOC core complex. Furthermore, SP1 was shown to ubiquitinate the TOC components in vivo as well as in vitro. The *sp1* mutant showed defects in plastid differentiation; i.e., the etioplast-to-chloroplast and chloroplast-to-gerontoplast (old chloroplasts in aging leaves) transitions. These results suggest that the UPS controls changes in the composition of TOC complexes to accommodate different sets of preproteins according to the needs of the developing plastid type (Fig. 9.1) [139]. Two homologs of SP1 were also identified, and it will be of great interest to see what their respective roles in the chloroplast may be.

9.3 Translocation Across the Inner Membrane

9.3.1 Arrival and Conductance

The Tic22 protein resides in the intermembrane space, and is peripherally associated with the inner membrane [115, 116]. It is perhaps the first TIC component to be encountered by translocating preproteins, and it may facilitate their passage from TOC to TIC, possibly functioning in association with other intermembrane space components [21, 167]. It might also play a role in the formation of so-called TOC-TIC supercomplexes [3, 116, 158], enabling simultaneous transport across the two membranes. Tic22 homologs exist in cyanobacteria, and so this component was likely acquired with the endosymbiont. Its function is essential in cyanobacteria and apicomplexan parasites (where it is localized in the apicoplast), and structural analyses suggest that it may act as a chaperone [64, 211]. However, Tic22 is not essential in plants, as *Arabidopsis* mutants lacking both canonical Tic22 isoforms exhibit only moderate defects in greening and preprotein import [102, 176].

Tic110 is one of the most abundant TIC components [220]. It is encoded by a single-copy gene in *Arabidopsis*, and is essential [84, 118]. Based on electrophysiological analyses, it was proposed to form a cation-selective, β-barrel channel with a pore diameter of 15-31 Å [66]. However, another study showed that it is composed mainly of α-helices, and that it is anchored in the inner membrane by two N-terminal transmembrane spans [83]. In the latter topology, a large hydrophilic domain is oriented towards the stroma and is thought to recruit stromal chaperones for import propulsion [83, 92, 103]. A later study aimed to resolve these discrepancies, and concluded that the hydrophilic part contains four amphipathic helices that contribute to the channel [15].

Another component that has been proposed to form the TIC channel is Tic20 [116]. This protein possesses four α-helical transmembrane domains, similar to the mitochondrial inner membrane channel components Tim17, Tim22 and Tim23 [98, 101, 116], and interacts with preproteins at a slightly later stage than Tic22 [115, 143]. In *Arabidopsis*, deficiency of the main Tic20 isoform (atTic20-I) causes defects in chloroplast biogenesis and protein import [41], while complete loss causes severe albinism and seedling lethality [71, 101, 108, 203]. In fact, there are four Tic20 genes in *Arabidopsis* that fall into two distinct, evolutionarily-conserved groups: the Group 1 proteins (atTic20-I and atTic20-IV) are demonstrably important for chloroplast biogenesis, whereas the Group 2 proteins are dispensable [101, 206]. The atTic20-I protein seems to be important for the import of photosynthesis-associated preproteins in shoots, while atTic20-IV may deliver mainly non-photosynthetic, "house-keeping" preproteins in roots [71, 108]. Embryos lacking both Group 1 proteins are not viable [71, 101]. It has been suggested that the localization of Tic20 proteins is not restricted to the chloroplast inner envelope membrane [144].

Blue native PAGE analysis indicated that Tic20 exists in a large, 1 MD complex together with Tic21 (see below) and translocating preprotein [108]; Tic21 is only loosely associated with the complex, while Tic20 appears to be a core component. Very recently, the purified complex was found to contain two additional nucleus-encoded proteins (Tic56 and -100) and, surprisingly, the elusive YCF1 protein (Tic214) encoded by the chloroplast genome (Fig. 9.1) [109]. The 1 MD complex reconstituted in a planar lipid bilayer had channel activity, and was therefore proposed to form a general TIC translocon. Electrophysiological analysis also revealed that Tic20 alone is able to form a channel, with cation selectivity and a pore size of 8–14 Å [121]. Notably, Tic110 was absent from the 1 MD complex, and instead was present in a smaller 200–300 kD complex. Thus, Tic20 may form a large channel complex (including Tic21 and the other components), whereas Tic110 may act later in the import mechanism as part of a distinct motor complex, or in other stromal events such as protein folding (see Sect. 9.3.2) [108, 109]. However, Tic20 protein was reported to be considerably less abundant than other translocon components, and so its candidacy as the main TIC channel has been questioned [121, 220].

While Tic110, Tic22 and Tic20 were all identified through biochemical analyses of isolated pea chloroplasts, Tic21 (or CIA5, for *c*hloroplast *i*mport *a*pparatus *5*) was found genetically in *Arabidopsis*, by screening for plants defective in the chloroplast import of a selectable marker [203]. *Arabidopsis* Tic21 knockout mutants are albino, and display similar defects in the import of photosynthetic preproteins to *tic20-I* mutants [108]. Interestingly, *tic21 tic20-I* double mutants do not exhibit phenotypic additivity, supporting the notion that the two proteins function together [203]. It was suggested that Tic20 might act early in plant development, with Tic21 taking over later on [203], but this seems inconsistent with the fact that the two proteins have been found together in the same complex [108].

Tic21 was also reported to act in iron transport, and thus given the alternative name of PIC1 (for *p*ermease *i*n *c*hloroplasts *1*) [54]. *Arabidopsis* PIC1/Tic21 mutants accumulated ferritin (a protein which binds iron to prevent iron loss or oxidative stress caused by free iron ions) in chloroplasts, and displayed up-regulated expression of ferritin and other factors related to iron stress and metabolism, while plants overexpressing PIC1 accumulated free iron ions in the stroma. Moreover, a yeast iron uptake mutation could be complemented using PIC1 [54]. It is conceivable that a block in iron uptake could affect protein import indirectly, accounting for some of the results linking the protein to import. However, genes related to iron homeostasis are also up-regulated in other pale mutants with defects in chloroplast biogenesis [108], and so further work is needed to determine the causal relationship between the iron homeostasis and protein import defects in *pic1/tic21* mutants. An alternative possibility is that PIC1/Tic21 has a dual role, acting in both processes [65].

9.3.2 Import Propulsion

The Tic110 C-terminus projects, at least partly, into the stroma and can bind transit peptides upon their emergence from the TIC channel [15, 83, 92]. It also recruits molecular chaperones, and these are believed to consume the ATP that is needed to drive preprotein import, and to assist the folding of newly-imported proteins [3, 61, 103, 158]. In mitochondria, a matrix Hsp70 (mtHsp70) delivers the energy for preprotein import [156], but until recently it has generally been thought that an Hsp100 protein, ClpC/Hsp93, is the principal component of the TIC motor [3, 158]. However, an important role for chloroplast stromal Hsp70 (cpHsc70) in the import mechanism has now been established [186, 197], while a stromal Hsp90 was also recently found to play a role (Fig. 9.1) [90].

Hsp93 (or ClpC) is an Hsp100-type AAA+ATPase. In addition to its function in preprotein import, it also forms part of the Clp protease complex in chloroplasts [61, 184]. It is believed to assemble into hexamers, and to act by threading clients (either importing preproteins or proteins to be degraded) through the resulting axial pore, towards either the stroma or the Clp proteolytic core [93, 179]. Reflecting its different roles, Hsp93 partitions between the envelope and stroma, and recent work identified its N-terminus as an important determinant of envelope association [48]. There are two Hsp93 isoforms in *Arabidopsis*, called atHsp93-V (ClpC1) and atHsp93-III (ClpC2). The former is expressed at much higher levels than the latter, while *hsp93-V* knockouts are pale and exhibit reduced preprotein import efficiency; *hsp93-III* knockout mutants are indistinguishable from wild type [49, 118, 119, 191]. Because double mutants are embryo lethal, and because the mature domains share ~91 % identity [119], the two isoforms are believed to have largely redundant functions.

Tic40 (previously named Com44/Cim44) can be crosslinked to Tic110 via a disulfide bridge under oxidizing conditions [194], and its loss causes a pale phenotype and inefficient chloroplast protein import in *Arabidopsis* [46, 118]. It is anchored in the inner membrane by a single, N-terminal transmembrane span, and it projects a large C-terminal domain into the stroma, much like Tic110 [46, 194]. This stromal region contains a putative TPR domain (whether it is truly a TPR was recently questioned [16]), and a tightly-folded Sti1 domain of the type found in eukaryotic Hip/Hop co-chaperones [24, 46, 47, 100]. Tic40 associates with Tic110 and Hsp93, and these three proteins appear to function at similar times in the import mechanism [46]. It is proposed that Tic40 binds to Tic110 (via its putative TPR domain) when the transit peptide binding site of the latter is occupied [47, 83]. Upon binding of Tic40, the transit peptide is released from Tic110 and passed to Hsp93. The chaperone then draws the preprotein into the stroma at the expense of ATP hydrolysis, which is stimulated by the Tic40 Sti1 domain. Curiously, the Sti1 domain of Tic40 can be functionally replaced with that of mammalian Hip (*H*sp70-*i*nteracting *p*rotein), for which an ATPase-stimulating function was not previously proposed [24].

Early attempts to identify stromal Hsp70 in import complexes failed, seemingly due to the lack of a suitable antibody [3, 158, 197]. However, this issue was recently overcome, while new genetic evidence also supports a role for this chaperone in preprotein import. Two stromal Hsp70 isoforms exist in *Arabidopsis* (cpHsc70-1 and cpHsc70-2), and plants lacking either one exhibit defective preprotein import [196, 197]; double mutants lacking both isoforms are embryo lethal, implying that the proteins share redundant functions [197]. Interestingly, the *cphsc70-1 hsp93-V* and *cphsc70-1 tic40* double mutants are phenotypically more severe than the corresponding single mutants [197], suggesting that cpHsc70's import function acts in parallel to the Tic40/Hsp93 system. The *cphsc70-1 tic40* genotype is lethal, whereas *hsp93-V tic40* causes only a pale phenotype, suggesting that cpHsc70, but not Hsp93, becomes essential and limiting in the *tic40* background [118, 197]. In the moss *Physcomitrella patens*, cpHsc70-deficient mutants similarly display inefficient chloroplast protein import, as do other mutants with a deficiency in the stromal co-chaperone CGE (*c*hloroplast *G*rp*E* homolog) [186]. Related GrpE proteins promote nucleotide exchange at Hsp70 in prokaryotic systems, and play a well-established role in mitochondrial protein import in conjunction with mtHsp70 [156]. Finally, immunoprecipitation studies in both moss and pea showed that cpHsc70 associates with preproteins and translocon components such as Tic110, Hsp93 and Tic40 [186, 197].

Most recently, a chloroplast Hsp90 protein (Hsp90C) was implicated in import [90]. This chaperone was identified in import intermediates, and was co-purified with translocon components including Tic110, Tic40, Toc75 and Tic22, as well as Hsp93 and cpHsp70. Moreover, an inhibitor of Hsp90 ATPase activity, radicicol, reversibly inhibited the import of several preproteins during inner envelope translocation. Insertion mutations affecting the single Hsp90C gene in *Arabidopsis* are embryo lethal, indicating an essential role for this chaperone, presumably as part of a stromal chaperone complex that facilitates membrane translocation during protein import [90].

Conceivably, the different stromal chaperones implicated in import may act sequentially in the process, or exhibit selectivity towards different preprotein clients. The unusual complexity of the import-associated chaperone network in chloroplasts suggests that the chaperones do not simply function as components of a translocation motor, but perhaps participate in a series of events necessary for efficient import.

9.3.3 Redox Regulation

Chloroplast redox signals inform many important regulatory mechanisms [13], and so it is not surprising that chloroplast protein import is also a proposed target of redox control. The TIC translocon in particular is a proposed target for regulation by chloroplast redox status (Fig. 9.1) [17, 120]. In maize chloroplasts, precursors of different isoforms of ferredoxin and ferredoxin-NADP$^+$reductase (FNR) are imported differentially under light and dark conditions [72]: photosynthetic isoforms are similarly imported by light- and dark-exposed chloroplasts, but non-photosynthetic isoforms are missorted to the intermembrane space in the light. Conceivably, the non-photosynthetic isoforms might interfere with photosynthesis, and so perhaps systems evolved to prevent their import under light conditions.

Two TIC proteins, Tic32 and Tic62, are proposed to modulate preprotein import in response to changing stromal NADP$^+$/NADPH ratios [17, 77, 78, 125]. In fact, their association with the TIC apparatus is dependent on such ratios, with binding and dissociation occurring under oxidizing and reducing conditions, respectively [44, 120, 195]. Both proteins have an N-terminal NADP(H)-binding site and are members of the short-chain dehydrogenase/reductase family, and they associate with Tic110 and other translocon components at the stromal side of the inner membrane [77, 125, 195]. Tic32 also has a binding site for calmodulin, and the inhibition of preprotein import by ophiobolin A and ionomycin, which both disrupt calcium signalling, has been linked to Tic32 [43, 44]. Calmodulin and NADPH binding to Tic32 are mutually exclusive, suggesting that calcium signals are relayed via calmodulin only under oxidizing conditions when Tic32 is associated with the TIC machinery [44].

Like TROL (*t*hylakoid *r*hodanese-*l*ike protein), which tethers FNR to thylakoids for the reduction of $NADP^+$ in photosynthetic electron transport, Tic62 has a C-terminal FNR-binding site [4, 97, 125]. Under reducing conditions, Tic62 breaks its peripheral association with the envelope and moves into the stroma (unlike Tic32, which behaves as an integral membrane protein), thereby increasing its affinity for FNR and enabling its association with the thylakoids [26, 195]. However, the Tic62-bound FNR appears not to be involved in photosynthetic electron transport, even when attached to the thylakoids [26]. Whether Tic62 has different functions dependent on its location (in the envelope, stroma or thylakoids), or the capacity to relay thylakoid signals to the TIC machinery, remains to be seen.

A third component implicated in the redox-regulation of preprotein import is Tic55, which was identified in a complex with Tic110, translocating preproteins, and other translocon components [36]. Tic55 is anchored in the inner membrane by two C-terminal transmembrane spans, has a Rieske-type iron-sulfur centre and a mononuclear iron-binding site, and is proposed to act via an electron transfer process or as a sensor of oxidative stress [36, 120]. Tic55 can also bind to thioredoxins and contains conserved cysteine residues that have the potential to form disulfide bridges [18]. However, an *Arabidopsis* Tic55 knockout mutant displays neither visible abnormalities nor defects in chloroplast protein import [30]. Doubts over the participation of Tic55 in preprotein import were also raised when two groups failed to detect the protein in import complexes [116, 170].

9.3.4 Transit Peptide Cleavage

Upon emergence from the TIC machinery, the transit peptide of a translocating preprotein is quickly removed by the stromal processing peptidase (SPP) (Fig. 9.1) [174, 201]. SPP is a zinc-binding metalloendopeptidase of the M16 family, which also includes the mitochondrial processing peptidase MPP, the presequence protease PreP (see below), and *Escherichia coli* pitrilysin [171, 217]. A stretch of 10–15 residues near the C-terminus of the transit peptide, where basic residues tend to be concentrated, is recognized by SPP, and cleavage occurs at a loosely-conserved site dependent upon physicochemical properties of the sequence [56, 172, 175, 230]. Interestingly, SPP is encoded by a single gene in *Arabidopsis*, and so the protein must accommodate a wide range of transit peptides with highly variable sequences [34, 171]. Following release of the newly-processed protein, SPP terminates its interaction with the transit peptide by a second cleavage event [172]. The peptide fragments are then degraded by the presequence protease, PreP [151, 172, 173]. The SPP protein is evolutionarily well conserved, as related sequences are found in various algae, apicomplexan parasites, and even cyanobacteria, suggesting that an ancestral activity was probably inherited with the endosymbiont [174].

Suppression of SPP expression in *Arabidopsis* or tobacco plants causes various abnormal phenotypes, ranging from albinism to seedling lethality, and is associated with ultrastructural defects and reduced numbers of chloroplasts [225, 232].

Similarly, a point mutation affecting a conserved glutamate residue of SPP causes chlorosis and small, abnormal chloroplasts in rice [229]. Interestingly, SPP-deficiency leads to reduced chloroplast protein import efficiency [225, 232]; this may reflect the fact that most TIC components, and Toc75, have a transit peptide and so depend on SPP for proper maturation, or indicate that transit peptide cleavage is an integrated step in the import mechanism. *Arabidopsis* SPP knockout mutations are embryo lethal, further emphasizing the importance of this protein for organelle and plant development [212].

9.4 Targeting to the Envelope Membranes

9.4.1 Sorting to the Outer Membrane

Outer envelope membrane proteins typically do not have transit peptides, but instead are targeted by intrinsic, non-cleavable signals. There are several different pathways for outer membrane protein insertion [74], and perhaps the best understood of these is that used by signal-anchored proteins such as OEP7/14 (*o*uter *e*nvelope *p*rotein, *7*/*14* kD) and Toc64/OEP64 [132]. In these proteins, targeting information is linked to the amino-terminal transmembrane domain, which is superficially similar to signal peptides that direct proteins to the ER [74, 127]. Flanking the transmembrane domain there is a positively-charged region that, together with the hydrophobicity of the transmembrane region itself, plays a critical role in ensuring that such proteins are targeted to chloroplasts rather than the ER [127, 131].

Despite early suggestions that signal-anchored proteins insert spontaneously, it is now clear that their targeting involves proteinaceous cofactors and the consumption of nucleoside triphosphates [75, 214]. In fact, competition, cross-linking and reconstitution results indicate that Toc75 is involved, and that in this role it may function without assistance from the TOC receptors, Toc34 and Toc159 [215]. Involvement of Toc75 parallels the situation in mitochondria, where the equivalent import channel, Tom40, is similarly employed [169]. More recently, AKR2A (*ank*yrin *r*epeat-containing protein *2A*) was identified as a cytosolic sorting factor in this pathway [11, 23]. In conjunction with its cofactor, Hsp17.8, a member of the small heat shock protein family, AKR2A is proposed to act as a chaperone, preventing the aggregation of its clients and guiding them to the envelope [110]. Interestingly, AKR2A also mediates protein insertion into the peroxisomal membrane [185], suggesting that it acts in the targeting of a broad class of membrane proteins [231].

Similar intrinsic information directs the targeting of Toc34, but in this case the relevant transmembrane domain lies at the C-terminus (i.e., it is a tail-anchored protein). As with signal-anchored proteins, insertion requires both envelope proteins and an energy source [213]. Indeed, competition results suggest that Toc34, OEP7/14 and Toc64/OEP64 may all follow the same pathway [74, 75, 214]. However, Toc34 insertion was also reported to depend on previously-inserted Toc34, as well as on membrane lipids, and to follow a different pathway from that used by another tail-anchored protein [51, 165]. Further complexity arises from the fact that some outer membrane proteins are dual-targeted to mitochondria and chloroplasts [199].

Toc159 employs a different targeting mechanism, perhaps due to its large, atypical M-domain. Its insertion is thought to involve a homotypic G-domain interaction with resident Toc34, controlled by guanine nucleotide status, as well as Toc75 [20, 193, 224]. Nonetheless, the M-domain itself seems to possess targeting information [128, 142]. The M-domain has no typical transmembrane spans, and so its insertion most likely depends on the TOC complex. That said, a short hydrophobic segment near the C-terminus may interface with the lipid bilayer [82].

Unusually, Toc75 has a cleavable, bipartite targeting signal at its N-terminus: the N-terminal part is a standard transit peptide, while the second part directs intraorganellar sorting [208, 209]. The latter contains a poly-glycine stretch that enables disengagement from the translocon and membrane integration [86]. The transit peptide is cleaved by SPP, whereas the second domain is removed by an envelope-localized type I signal peptidase (which additionally resides in thylakoids for the maturation of thylakoidal proteins) [79, 88, 188, 189]. How Toc75 becomes integrated into the outer membrane is unclear. In bacteria and mitochondria, the biogenesis of similar β-barrels is assisted by proteins of the Omp85 superfamily [205], and a related protein in chloroplasts, OEP80, was proposed to play a similar role [55, 87]. Supporting this idea, OEP80 is an essential protein in *Arabidopsis* (like Toc75), while its depletion affects Toc75 accumulation in vivo [80, 161]. Phylogenetic data are also consistent with the notion that OEP80 has retained an ancestral function [207].

9.4.2 Sorting to the Intermembrane Space and Inner Membrane

Unlike most outer membrane proteins, those destined for the intermembrane space or inner membrane typically have cleavable, N-terminal targeting information. Sorting to the intermembrane space has been studied for two proteins that follow different pathways: Tic22 and MGD1 (*m*onogalactosyl*d*iacylglycerol synthase *1*) [117, 221]. Both proteins have a targeting sequence, but only that of MGD1 is cleaved by SPP. Along with the energetic requirements for its import, this indicates that MGD1 partially enters the stroma. In contrast, Tic22 is processed by an unknown protease in the intermembrane space, suggesting that it does not enter the TIC channel. There is also uncertainty over the participation of the TOC apparatus in Tic22 sorting.

Inner membrane proteins typically follow one of two routes: the stop-transfer and post-import pathways. The requirement for a transit peptide in both cases implies involvement of the TOC/TIC apparatus [114, 141, 194]. In the stop transfer pathway, a hydrophobic transmembrane domain arrests preprotein transport in the channel, enabling lateral exit into the membrane [32, 60, 114, 210]. This pathway may be particularly important for hydrophobic or polytopic proteins that are prone to aggregation. Recent work on the APG1 (*a*lbino or *p*ale *g*reen mutant *1*) protein, a stop-transfer client, revealed that membrane targeting information lies in the transmembrane domain, which is sufficient to direct stop-transfer insertion even in the context of heterologous passenger proteins [218].

In the post-import pathway, proteins undergo complete translocation into the stroma, where they form soluble intermediates, prior to membrane insertion [137, 141, 210]. Tic40 and Tic110 are clients of this pathway, and both are anchored in the inner membrane by N-terminal helices and have large stromal domains. Interference with their membrane integration leads to the accumulation of SPP-processed forms in the stroma [24, 84]. Tic40 possesses a bipartite targeting sequence, but the role of the second domain is unclear as a serine/proline-rich region of the mature sequence and the adjacent transmembrane domain control insertion [137, 210]. The latter two may cooperate to form a membrane insertion loop, while in Tic110 the two transmembrane domains may create an equivalent structure. Efficacy of Tic40's targeting information is influenced by context within the protein sequence, implying that post-import signals are complex, which might be necessary to avoid stop-transfer insertion and an incorrect topology [218]. Stromal events in the post-import pathway may involve Hsp93 [222], while integration depends on proteinaceous membrane components [137].

In mitochondria, sorting to the inner membrane employs stop-transfer and conservative sorting pathways [157]. The latter is similar to the post-import pathway of chloroplasts and, as its name suggests, it is at least partly of prokaryotic origin. Bearing this in mind, it is intriguing that a second Sec translocase (in addition to the well-known thylakoidal system) was recently identified in chloroplast envelopes [192]. There is also evidence that resident Tic40 (and possibly Tic110) acts in the integration of other proteins into the membrane [45, 84].

9.5 Alternative Protein Import Pathways

9.5.1 Dual-Targeting

Although most chloroplast proteins are targeted specifically to plastids, a significant number (> 100) are transported to more than one location [37, 145, 190]. Transport to chloroplasts and mitochondria is the most common form of dual-targeting, but there are also proteins that exist in the nucleus, ER or peroxisomes as well as in chloroplasts [122, 135, 177]. Such multi-destination transport implies that protein targeting is rather flexible, and is supportive of a model for the relocation of organellar genes to the nucleus that depends on the "minor mistargeting" of many proteins to multiple locations [146]. Dual-targeted proteins tend to have highly-conserved functions that are easily shared, including nucleic acid and protein synthesis or processing, and cellular stress response [37, 145]. A particularly striking example occurs amongst the aminoacyl-tRNA synthetases, where 17 of the 24 organellar proteins in *Arabidopsis* are targeted to both chloroplasts and mitochondria [53]; some are even targeted to all three of the compartments that possess translational machinery.

Dual-targeting to chloroplasts and mitochondria typically involves one of two mechanisms [162]. In the first of these, alternative splicing and/or differential transcriptional or translational initiation leads to the production of protein variants with different N-terminal leader sequences and distinct targeting properties. In the second mechanism, a single protein is produced that possesses an ambiguous leader, competent for sorting to both chloroplasts and mitochondria. Alternatively, dual-targeting information may be, either wholly or partly, an intrinsic feature of the mature protein [12, 216].

Ambiguous transit peptides for dual-targeting to endosymbiotic organelles have been scrutinized, and in general they have properties intermediate between those that target either organelle specifically [27, 28, 162, 164]. In the N-terminal region, serine content is more similar to that of chloroplast transit peptides, while arginine content is more similar to that of mitochondrial presequences. They show enrichment of phenylalanine and leucine residues and, while certain segments are more important for transport to one or the other organelle in some cases, they do not share a common functional-domain architecture [27, 28]. Dual-targeting is also influenced by the mature domain of the preprotein, and by developmental factors [38, 145]. Software for the predication of ambiguous targeting peptides suggests that as many as ~400–500 proteins may be dual-targeted to chloroplasts and mitochondria [150]. Competition data indicate that dual-targeted proteins utilize the same import machineries as organelle-specific proteins [28].

9.5.2 Non-canonical Protein Transport

In recent years it has become apparent that transit peptide-dependent import is not the only sorting pathway to the chloroplast interior [94, 112, 168]. One study estimated that more than 10% of plastid proteins lack a typical transit peptide [6]. The ceQORH (*c*hloroplast *e*nvelope *q*uinone *o*xido*r*eductase *h*omolog) protein was identified through proteomics, and found to associate with the inner envelope membrane even though it lacks a transit peptide. An internal sequence of ~40 residues controls its localization, and while its import does require proteinaceous machinery and ATP, the TOC/TIC apparatus is not involved [149]. Another inner membrane protein, Tic32/IEP32 (*i*nner *e*nvelope *p*rotein, *32* kD), similarly lacks a transit peptide, and it too localizes independently of the TOC translocon [152]. Competition results imply that ceQORH and Tic32 utilize different import pathways [149].

Proteomic analysis also led to the identification of a large number of chloroplast proteins with predicted signal peptides for ER translocation [112]. Chloroplast protein traffic through the endomembrane system is well documented in organisms that have complex plastids with more than two bounding membranes, such as algae and apicomplexan parasites [155], but was not thought to occur in plants. That said, physical and functional links between the ER and the outer envelope membrane have long been known [5, 25, 227], while glycoproteins and proteins with apparent signal peptides were detected in plastids [10, 42, 62]. The breakthrough came from analyses of *Arabidopsis* CAH1 (*c*arbonic *anh*ydrase *1*) [219]. This stromal protein has a signal peptide and is imported and processed by ER microsomes, but not by chloroplasts. Glycosylated forms of CAH1 and other proteins are present in chloroplasts, while brefeldin A (a chemical that disrupts traffic through the Golgi) interferes with their localization, indicating passage through the Golgi en route to chloroplasts [154, 219]. Other data suggest that this sorting pathway, and the glycosylation it enables, are functionally important [35, 81].

Exactly how proteins are directed through the endomembrane system to chloroplasts remains unclear. Some data suggest that the signal peptide provides the necessary sorting information [42], while others argue that surface characteristics of the mature protein are important [111]. The proteins may be released into the intermembrane space upon vesicle fusion with the outer membrane, before entering an unknown translocon, the TIC apparatus, or vesicles that pinch off from the inner membrane [168].

9.6 Concluding Remarks

Our understanding of chloroplast protein import, and of the molecular events that underlie the process, has improved significantly in recent years. Nonetheless, important unanswered questions remain, while several inconsistencies in the literature need to be resolved. Even though our knowledge concerning TOC receptor GTPase function has expanded, the precise mode-of-action of these receptors is hotly debated, and a consensus model is lacking. It is generally accepted that client-specific protein import pathways operate in chloroplasts, but the molecular basis for TOC receptor (and possibly also TIC channel) selectivity requires further work. Chloroplast protein import must be tightly regulated, and indeed recent research has unveiled direct control of the TOC machinery by the ubiquitin-proteasome system, while other work suggests redox-regulation at the TIC apparatus. Nonetheless, the mechanistic details behind such regulatory systems are largely unknown. We have learnt much about the functions of putative inner envelope channel components, but there is now a need to reconcile the different hypotheses that have been proposed. Identification of a large TIC complex containing Tic20, but excluding Tic110 and Tic40, indicates that a more comprehensive framework is required that integrates the roles of these proteins. Furthermore, the confounding complexity of stromal chaperone complexes involved in chloroplast protein import needs to be unravelled. We expect that future research focusing on these fascinating questions will bring us closer to a full understanding of this essential process.

Acknowledgements Research in the authors' laboratories is supported by the Biotechnology and Biological Sciences Research Council (BBSRC) and the Gatsby Charitable Foundation (to P.J.), and the University of Neuchâtel and the Swiss National Science Foundation (to F.K.). We thank Jocelyn Bédard for helpful comments on the manuscript.

References

1. Agne B et al (2009) A Toc159 import receptor mutant, defective in hydrolysis of GTP, supports preprotein import into chloroplasts. J Biol Chem 284:8670–8679
2. Agne B et al (2010) The acidic A-domain of Arabidopsis Toc159 occurs as a hyperphosphorylated protein. Plant Physiol 153:1016–1030
3. Akita M, Nielsen E, Keegstra K (1997) Identification of protein transport complexes in the chloroplastic envelope membranes via chemical cross-linking. J Cell Biol 136:983–994
4. Alte F, Stengel A, Benz JP, Petersen E, Soll J, Groll M, Bolter B (2010) Ferredoxin: NADPH oxidoreductase is recruited to thylakoids by binding to a polyproline type II helix in a pH-dependent manner. Proc Natl Acad Sci U S A 107:19260–19265
5. Andersson MX, Goksör M, Sandelius AS (2007) Optical manipulation reveals strong attracting forces at membrane contact sites between endoplasmic reticulum and chloroplasts. J Biol Chem 282:1170–1174
6. Armbruster U et al (2009) Chloroplast proteins without cleavable transit peptides: rare exceptions or a major constituent of the chloroplast proteome? Mol Plant 2:1325–35
7. Aronsson H, Combe J, Patel R, Jarvis P (2006) In vivo assessment of the significance of phosphorylation of the *Arabidopsis* chloroplast protein import receptor, atToc33. FEBS Lett 580:649–655
8. Aronsson H, Boij P, Patel R, Wardle A, Töpel M, Jarvis P (2007) Toc64/OEP64 is not essential for the efficient import of proteins into chloroplasts in *Arabidopsis thaliana*. Plant J 52:53–68
9. Aronsson H, Combe J, Patel R, Agne B, Martin M, Kessler F, Jarvis P (2010) Nucleotide binding and dimerization at the chloroplast pre-protein import receptor, atToc33, are not essential in vivo but do increase import efficiency. Plant J 63:297–311
10. Asatsuma S, Sawada C, Itoh K, Okito M, Kitajima A, Mitsui T (2005) Involvement of alpha-amylase I-1 in starch degradation in rice chloroplasts. Plant Cell Physiol 46:858–869
11. Bae W, Lee YJ, Kim DH, Lee J, Kim S, Sohn EJ, Hwang I (2008) AKR2A-mediated import of chloroplast outer membrane proteins is essential for chloroplast biogenesis. Nat Cell Biol 10:220–227
12. Bahaji A et al (2011) Dual targeting to mitochondria and plastids of AtBT1 and ZmBT1, two members of the mitochondrial carrier family. Plant Cell Physiol 52:597–609
13. Baier M, Dietz KJ (2005) Chloroplasts as source and target of cellular redox regulation: a discussion on chloroplast redox signals in the context of plant physiology. J Exp Bot 56: 1449–1462
14. Baldwin A, Wardle A, Patel R, Dudley P, Park SK, Twell D, Inoue K, Jarvis P (2005) A molecular-genetic study of the Arabidopsis Toc75 gene family. Plant Physiol 138:715–733
15. Balsera M, Goetze TA, Kovács-Bogdán E, Schürmann P, Wagner R, Buchanan BB, Soll J, Bölter B (2009) Characterization of Tic110, a channel-forming protein at the inner envelope membrane of chloroplasts, unveils a response to $Ca^{2+}$and a stromal regulatory disulfide bridge. J Biol Chem 284:2603–2616
16. Balsera M, Soll J, Buchanan BB (2009) Protein import in chloroplasts: an emerging regulatory role for redox. Adv Bot Res 52:277–332
17. Balsera M, Soll J, Buchanan BB (2010) Redox extends its regulatory reach to chloroplast protein import. Trends Plant Sci 15:515–521
18. Bartsch S, Monnet J, Selbach K, Quigley F, Gray J, von Wettstein D, Reinbothe S, Reinbothe C (2008) Three thioredoxin targets in the inner envelope membrane of chloroplasts function in protein import and chlorophyll metabolism. Proc Natl Acad Sci U S A 105:4933–4938
19. Bauer J, Chen K, Hiltbrunner A, Wehrli E, Eugster M, Schnell D, Kessler F (2000) The major protein import receptor of plastids is essential for chloroplast biogenesis. Nature 403: 203–207
20. Bauer J, Hiltbrunner A, Weibel P, Vidi PA, Alvarez-Huerta M, Smith MD, Schnell DJ, Kessler F (2002) Essential role of the G-domain in targeting of the protein import receptor atToc159 to the chloroplast outer membrane. J Cell Biol 159:845–854

21. Becker T, Hritz J, Vogel M, Caliebe A, Bukau B, Soll J, Schleiff E (2004) Toc12, a novel subunit of the intermembrane space preprotein translocon of chloroplasts. Mol Biol Cell 15:5130–5144
22. Becker T, Jelic M, Vojta A, Radunz A, Soll J, Schleiff E (2004) Preprotein recognition by the Toc complex. EMBO J 23:520–530
23. Bédard J, Jarvis P (2008) Green light for chloroplast outer-membrane proteins. Nat Cell Biol 10:120–122
24. Bédard J, Kubis S, Bimanadham S, Jarvis P (2007) Functional similarity between the chloroplast translocon component, Tic40, and the human co-chaperone, Hsp70-interacting protein (Hip). J Biol Chem 282:21404–21414
25. Benning C (2009) Mechanisms of lipid transport involved in organelle biogenesis in plant cells. Annu Rev Cell Dev Biol 25:71–91
26. Benz JP et al (2009) *Arabidopsis* Tic62 and ferredoxin-NADP(H) oxidoreductase form light-regulated complexes that are integrated into the chloroplast redox poise. Plant Cell 21: 3965–3983
27. Berglund AK, Pujol C, Duchene AM, Glaser E (2009) Defining the determinants for dual targeting of amino acyl-tRNA synthetases to mitochondria and chloroplasts. J Mol Biol 393:803–814
28. Berglund AK, Spanning E, Biverstahl H, Maddalo G, Tellgren-Roth C, Maler L, Glaser E (2009) Dual targeting to mitochondria and chloroplasts: characterization of Thr-tRNA synthetase targeting peptide. Mol Plant 2:1298–1309
29. Bischof S et al (2011) Plastid proteome assembly without Toc159: photosynthetic protein import and accumulation of N-acetylated plastid precursor proteins. Plant Cell 23:3911–3928
30. Boij P, Patel R, Garcia C, Jarvis P, Aronsson H (2009) In vivo studies on the roles of Tic55-related proteins in chloroplast protein import in *Arabidopsis thaliana*. Mol Plant 2:1397–1409
31. Bölter B, May T, Soll J (1998) A protein import receptor in pea chloroplasts, Toc86, is only a proteolytic fragment of a larger polypeptide. FEBS Lett 441:59–62
32. Brink S, Fischer K, Klösgen RB, Flügge UI (1995) Sorting of nuclear-encoded chloroplast membrane proteins to the envelope and the thylakoid membrane. J Biol Chem 270: 20808–20815
33. Bruce BD (2000) Chloroplast transit peptides: structure, function and evolution. Trends Cell Biol 10:440–447
34. Bruce BD (2001) The paradox of plastid transit peptides: conservation of function despite divergence in primary structure. Biochim Biophys Acta 1541:2–21
35. Burén S et al (2011) Importance of post-translational modifications for functionality of a chloroplast-localized carbonic anhydrase (CAH1) in *Arabidopsis thaliana*. PLoS ONE 6:e21021
36. Caliebe A, Grimm R, Kaiser G, Lubeck J, Soll J, Heins L (1997) The chloroplastic protein import machinery contains a Rieske-type iron- sulfur cluster and a mononuclear iron-binding protein. EMBO J 16:7342–7350
37. Carrie C, Small I (2013) A reevaluation of dual-targeting of proteins to mitochondria and chloroplasts. 1833: 253-259
38. Carrie C, Giraud E, Whelan J (2009) Protein transport in organelles: dual targeting of proteins to mitochondria and chloroplasts. FEBS J 276:1187–1195
39. Chen KY, Li HM (2007) Precursor binding to an 880-kDa Toc complex as an early step during active import of protein into chloroplasts. Plant J 49:149–158
40. Chen K, Chen X, Schnell DJ (2000) Initial binding of preproteins involving the Toc159 receptor can be bypassed during protein import into chloroplasts. Plant Physiol 122:813–822
41. Chen X, Smith MD, Fitzpatrick L, Schnell DJ (2002) In vivo analysis of the role of atTic20 in protein import into chloroplasts. Plant Cell 14:641–654
42. Chen MH, Huang LF, Li HM, Chen YR, Yu SM (2004) Signal peptide-dependent targeting of a rice alpha-amylase and cargo proteins to plastids and extracellular compartments of plant cells. Plant Physiol 135:1367–1377
43. Chigri F, Soll J, Vothknecht UC (2005) Calcium regulation of chloroplast protein import. Plant J 42:821–831

44. Chigri F, Hörmann F, Stamp A, Stammers DK, Bölter B, Soll J, Vothknecht UC (2006) Calcium regulation of chloroplast protein translocation is mediated by calmodulin binding to Tic32. Proc Natl Acad Sci U S A 103:16051–16056
45. Chiu CC, Li HM (2008) Tic40 is important for reinsertion of proteins from the chloroplast stroma into the inner membrane. Plant J 56:793–801
46. Chou ML, Fitzpatrick LM, Tu SL, Budziszewski G, Potter-Lewis S, Akita M, Levin JZ, Keegstra K, Li HM (2003) Tic40, a membrane-anchored co-chaperone homolog in the chloroplast protein translocon. EMBO J 22:2970–2980
47. Chou ML, Chu CC, Chen LJ, Akita M, Li HM (2006) Stimulation of transit-peptide release and ATP hydrolysis by a cochaperone during protein import into chloroplasts. J Cell Biol 175:893–900
48. Chu CC, Li HM (2012) The amino-terminal domain of chloroplast Hsp93 is important for its membrane association and functions in vivo. Plant Physiol 158:1656–1665
49. Constan D, Froehlich JE, Rangarajan S, Keegstra K (2004) A stromal Hsp100 protein is required for normal chloroplast development and function in Arabidopsis. Plant Physiol 136:3605–3615
50. Constan D, Patel R, Keegstra K, Jarvis P (2004) An outer envelope membrane component of the plastid protein import apparatus plays an essential role in *Arabidopsis*. Plant J 38:93–106
51. Dhanoa PK, Richardson LG, Smith MD, Gidda SK, Henderson MP, Andrews DW, Mullen RT (2010) Distinct pathways mediate the sorting of tail-anchored proteins to the plastid outer envelope. PLoS ONE 5:e10098
52. Dobberstein B, Blobel G, Chua NH (1977) In vitro synthesis and processing of a putative precursor for the small subunit of ribulose-1,5-bisphosphate carboxylase of *Chlamydomonas reinhardtii*. Proc Natl Acad Sci U S A 74:1082–1085
53. Duchêne AM, Giritch A, Hoffmann B, Cognat V, Lancelin D, Peeters NM, Zaepfel M, Marechal-Drouard L, Small ID (2005) Dual targeting is the rule for organellar aminoacyl-tRNA synthetases in *Arabidopsis thaliana*. Proc Natl Acad Sci U S A 102:16484–16489
54. Duy D, Wanner G, Meda AR, von Wiren N, Soll J, Philippar K (2007) PIC1, an ancient permease in *Arabidopsis* chloroplasts, mediates iron transport. Plant Cell 19:986–1006
55. Eckart K, Eichacker L, Sohrt K, Schleiff E, Heins L, Soll J (2002) A Toc75-like protein import channel is abundant in chloroplasts. EMBO Rep 3:557–562
56. Emanuelsson O, Nielsen H, von Heijne G (1999) ChloroP, a neural network-based method for predicting chloroplast transit peptides and their cleavage sites. Protein Sci 8:978–984
57. Ertel F, Mirus O, Bredemeier R, Moslavac S, Becker T, Schleiff E (2005) The evolutionarily related beta-barrel polypeptide transporters from *Pisum sativum* and *Nostoc* PCC7120 contain two distinct functional domains. J Biol Chem 280:28281–28289
58. Fellerer C, Schweiger R, Schongruber K, Soll J, Schwenkert S (2011) Cytosolic HSP90 cochaperones HOP and FKBP interact with freshly synthesized chloroplast preproteins of *Arabidopsis*. Mol Plant 4:1133–1145
59. Ferro M et al (2010) AT_CHLORO, a comprehensive chloroplast proteome database with subplastidial localization and curated information on envelope proteins. Mol Cell Proteomics 9:1063–1084
60. Firlej-Kwoka E, Strittmatter P, Soll J, Bolter B (2008) Import of preproteins into the chloroplast inner envelope membrane. Plant Mol Biol 68:505–519
61. Flores-Pérez U, Jarvis P (2013) Molecular chaperone involvement in chloroplast protein import. Biochim Biophys Acta 1833:332–340
62. Gaikwad A, Tewari KK, Kumar D, Chen W, Mukherjee SK (1999) Isolation and characterisation of the cDNA encoding a glycosylated accessory protein of pea chloroplast DNA polymerase. Nucleic Acids Res 27:3120–3129
63. Gentle I, Gabriel K, Beech P, Waller R, Lithgow T (2004) The Omp85 family of proteins is essential for outer membrane biogenesis in mitochondria and bacteria. J Cell Biol 164:19–24
64. Glaser S, van Dooren GG, Agrawal S, Brooks CF, McFadden GI, Striepen B, Higgins MK (2012) Tic22 is an essential chaperone required for protein import into the apicoplast. J Biol Chem 287:39505–39512

65. Gross J, Bhattacharya D (2009) Revaluating the evolution of the Toc and Tic protein translocons. Trends Plant Sci 14:13–20
66. Heins L, Mehrle A, Hemmler R, Wagner R, Küchler M, Hörmann F, Sveshnikov D, Soll J (2002) The preprotein conducting channel at the inner envelope membrane of plastids. EMBO J 21:2616–2625
67. Highfield PE, Ellis RJ (1978) Synthesis and transport of the small subunit of chloroplast ribulose bisphosphate carboxylase. Nature 271:420–424
68. Hiltbrunner A, Bauer J, Alvarez-Huerta M, Kessler F (2001) Protein translocon at the *Arabidopsis* outer chloroplast membrane. Biochem Cell Biol 79:629–635
69. Hiltbrunner A, Grunig K, Alvarez-Huerta M, Infanger S, Bauer J, Kessler F (2004) AtToc90, a new GTP-binding component of the *Arabidopsis* chloroplast protein import machinery. Plant Mol Biol 54:427–440
70. Hinnah SC, Wagner R, Sveshnikova N, Harrer R, Soll J (2002) The chloroplast protein import channel Toc75: pore properties and interaction with transit peptides. Biophys J 83:899–911
71. Hirabayashi Y, Kikuchi S, Oishi M, Nakai M (2011) *In vivo* studies on the roles of two closely related *Arabidopsis* Tic20 proteins, AtTic20-I and AtTic20-IV. Plant Cell Physiol 52:469–478
72. Hirohashi T, Hase T, Nakai M (2001) Maize non-photosynthetic ferredoxin precursor is missorted to the intermembrane space of chloroplasts in the presence of light. Plant Physiol 125:2154–2163
73. Hirsch S, Muckel E, Heemeyer F, von Heijne G, Soll J (1994) A receptor component of the chloroplast protein translocation machinery. Science 266:1989–1992
74. Hofmann NR, Theg SM (2005) Chloroplast outer membrane protein targeting and insertion. Trends Plant Sci 10:450–457
75. Hofmann NR, Theg SM (2005) Protein- and energy-mediated targeting of chloroplast outer envelope membrane proteins. Plant J 44:917–927
76. Hofmann NR, Theg SM (2005) Toc64 is not required for import of proteins into chloroplasts in the moss *Physcomitrella patens*. Plant J 43:675–687
77. Hörmann F, Küchler M, Sveshnikov D, Oppermann U, Li Y, Soll J (2004) Tic32, an essential component in chloroplast biogenesis. J Biol Chem 279:34756–34762
78. Hörmann F, Kuchler M, Sveshnikov D, Oppermann U, Li Y, Soll J (2009) Tic32, an essential component in chloroplast biogenesis (correction). J Biol Chem 284:29240
79. Hsu SC, Endow JK, Ruppel NJ, Roston RL, Baldwin AJ, Inoue K (2011) Functional diversification of thylakoidal processing peptidases in *Arabidopsis thaliana*. PLoS ONE 6:e27258
80. Huang W, Ling Q, Bédard J, Lilley K, Jarvis P (2011) In vivo analyses of the roles of essential Omp85-related proteins in the chloroplast outer envelope membrane. Plant Physiol 157:147–159
81. Hummel E, Osterrieder A, Robinson DG, Hawes C (2010) Inhibition of Golgi function causes plastid starch accumulation. J Exp Bot 61:2603–2614
82. Inaba T, Schnell DJ (2008) Protein trafficking to plastids: one theme, many variations. Biochem J 413:15–28
83. Inaba T, Li M, Alvarez-Huerta M, Kessler F, Schnell DJ (2003) atTic110 functions as a scaffold for coordinating the stromal events of protein import into chloroplasts. J Biol Chem 278:38617–38627
84. Inaba T, Alvarez-Huerta M, Li M, Bauer J, Ewers C, Kessler F, Schnell DJ (2005) *Arabidopsis* Tic110 is essential for the assembly and function of the protein import machinery of plastids. Plant Cell 17:1482–1496
85. Inoue H, Akita M (2008) Three sets of translocation intermediates are formed during the early stage of protein import into chloroplasts. J Biol Chem 283:7491–7502
86. Inoue K, Keegstra K (2003) A polyglycine stretch is necessary for proper targeting of the protein translocation channel precursor to the outer envelope membrane of chloroplasts. Plant J 34:661–669
87. Inoue K, Potter D (2004) The chloroplastic protein translocation channel Toc75 and its paralog OEP80 represent two distinct protein families and are targeted to the chloroplastic outer envelope by different mechanisms. Plant J 39:354–365

88. Inoue K, Baldwin AJ, Shipman RL, Matsui K, Theg SM, Ohme-Takagi M (2005) Complete maturation of the plastid protein translocation channel requires a type I signal peptidase. J Cell Biol 171:425–430
89. Inoue H, Rounds C, Schnell DJ (2010) The molecular basis for distinct pathways for protein import into *Arabidopsis* chloroplasts. Plant Cell 22:1947–1960
90. Inoue H, Li M, Schnell DJ (2013) An essential role for chloroplast heat shock protein 90 (Hsp90C) in protein import into chloroplasts. Proc Natl Acad Sci U S A 110:3173–3178
91. Ivanova Y, Smith MD, Chen K, Schnell DJ (2004) Members of the Toc159 import receptor family represent distinct pathways for protein targeting to plastids. Mol Biol Cell 15: 3379–3392
92. Jackson DT, Froehlich JE, Keegstra K (1998) The hydrophilic domain of Tic110, an inner envelope membrane component of the chloroplastic protein translocation apparatus, faces the stromal compartment. J Biol Chem 273:16583–16588
93. Jackson-Constan D, Akita M, Keegstra K (2001) Molecular chaperones involved in chloroplast protein import. Biochim Biophys Acta 1541:102–113
94. Jarvis P (2008) Targeting of nucleus-encoded proteins to chloroplasts in plants (Tansley Review). New Phytol 179:257–285
95. Jarvis P, Chen LJ, Li H, Peto CA, Fankhauser C, Chory J (1998) An *Arabidopsis* mutant defective in the plastid general protein import apparatus. Science 282:100–103
96. Jelic M, Soll J, Schleiff E (2003) Two Toc34 homologues with different properties. Biochemistry 42:5906–5916
97. Juric S et al (2009) Tethering of ferredoxin:$NADP^+$ oxidoreductase to thylakoid membranes is mediated by novel chloroplast protein TROL. Plant J 60:783–794
98. Kalanon M, McFadden GI (2008) The chloroplast protein translocation complexes of *Chlamydomonas reinhardtii*: a bioinformatic comparison of Toc and Tic components in plants, green algae and red algae. Genetics 179:95–112
99. Kaneko T et al (1996) Sequence analysis of the genome of the unicellular cyanobacterium *Synechocystis* sp. strain PCC6803. II. Sequence determination of the entire genome and assignment of potential protein-coding regions. DNA Res 3:109–136
100. Kao YF, Lou YC, Yeh YH, Hsiao CD, Chen C (2012) Solution structure of the C-terminal NP-repeat domain of Tic40, a co-chaperone during protein import into chloroplasts. J Biochem 152:443–451
101. Kasmati AR, Töpel M, Patel R, Murtaza G, Jarvis P (2011) Molecular and genetic analyses of Tic20 homologues in *Arabidopsis thaliana* chloroplasts. Plant J 66:877–889
102. Kasmati AR, Töpel M, Khan NZ, Patel R, Ling Q, Karim S, Aronsson H, Jarvis P (2013) Evolutionary, molecular and genetic analyses of Tic22 homologues in *Arabidopsis thaliana* chloroplasts. PLoS ONE 8:e63863
103. Kessler F, Blobel G (1996) Interaction of the protein import and folding machineries of the chloroplast. Proc Natl Acad Sci U S A 93:7684–7689
104. Kessler F, Schnell DJ (2004) Chloroplast protein import: solve the GTPase riddle for entry. Trends Cell Biol 14:334–338
105. Kessler F, Schnell DJ (2006) The function and diversity of plastid protein import pathways: a multilane GTPase highway into plastids. Traffic 7:248–257
106. Kessler F, Blobel G, Patel HA, Schnell DJ (1994) Identification of two GTP-binding proteins in the chloroplast protein import machinery. Science 266:1035–1039
107. Kikuchi S, Hirohashi T, Nakai M (2006) Characterization of the preprotein translocon at the outer envelope membrane of chloroplasts by blue native PAGE. Plant Cell Physiol 47:363–71
108. Kikuchi S, Oishi M, Hirabayashi Y, Lee DW, Hwang I, Nakai M (2009) A 1-megadalton translocation complex containing Tic20 and Tic21 mediates chloroplast protein import at the inner envelope membrane. Plant Cell 21:1781–1797
109. Kikuchi S, Bédard J, Hirano M, Hirabayashi Y, Oishi M, Imai M, Takase M, Ide T, Nakai M (2013) Uncovering the protein translocon at the chloroplast inner envelope membrane. Science 339:571–574

110. Kim DH, Xu ZY, Na YJ, Yoo YJ, Lee J, Sohn EJ, Hwang I (2011) Small heat shock protein Hsp17.8 functions as an AKR2A cofactor in the targeting of chloroplast outer membrane proteins in *Arabidopsis*. Plant Physiol 157:132–146
111. Kitajima A et al (2009) The rice alpha-amylase glycoprotein is targeted from the Golgi apparatus through the secretory pathway to the plastids. Plant Cell 21:2844–2858
112. Kleffmann T, Russenberger D, von Zychlinski A, Christopher W, Sjolander K, Gruissem W, Baginsky S (2004) The *Arabidopsis thaliana* chloroplast proteome reveals pathway abundance and novel protein functions. Curr Biol 14:354–362
113. Kleffmann T, Hirsch-Hoffmann M, Gruissem W, Baginsky S (2006) plprot: a comprehensive proteome database for different plastid types. Plant Cell Physiol 47:432–436
114. Knight JS, Gray JC (1995) The N-terminal hydrophobic region of the mature phosphate translocator is sufficient for targeting to the chloroplast inner envelope membrane. Plant Cell 7:1421–1432
115. Kouranov A, Schnell DJ (1997) Analysis of the interactions of preproteins with the import machinery over the course of protein import into chloroplasts. J Cell Biol 139:1677–1685
116. Kouranov A, Chen X, Fuks B, Schnell DJ (1998) Tic20 and Tic22 are new components of the protein import apparatus at the chloroplast inner envelope membrane. J Cell Biol 143:991–1002
117. Kouranov A, Wang H, Schnell DJ (1999) Tic22 is targeted to the intermembrane space of chloroplasts by a novel pathway. J Biol Chem 274:25181–25186
118. Kovacheva S, Bédard J, Patel R, Dudley P, Twell D, Ríos G, Koncz C, Jarvis P (2005) *In vivo* studies on the roles of Tic110, Tic40 and Hsp93 during chloroplast protein import. Plant J 41:412–428
119. Kovacheva S, Bédard J, Wardle A, Patel R, Jarvis P (2007) Further *in vivo* studies on the role of the molecular chaperone, Hsp93, in plastid protein import. Plant J 50:364–379
120. Kovacs-Bogdan E, Soll J, Bölter B (2010) Protein import into chloroplasts: the Tic complex and its regulation. Biochim Biophys Acta 1803:740–747
121. Kovacs-Bogdan E, Benz JP, Soll J, Bolter B (2011) Tic20 forms a channel independent of Tic110 in chloroplasts. BMC Plant Biol 11:133
122. Krause K, Krupinska K (2009) Nuclear regulators with a second home in organelles. Trends Plant Sci 14:194–199
123. Kubis S, Baldwin A, Patel R, Razzaq A, Dupree P, Lilley K, Kurth J, Leister D, Jarvis P (2003) The *Arabidopsis ppi1* mutant is specifically defective in the expression, chloroplast import, and accumulation of photosynthetic proteins. Plant Cell 15:1859–1871
124. Kubis S et al (2004) Functional specialization amongst the *Arabidopsis* Toc159 family of chloroplast protein import receptors. Plant Cell 16:2059–2077
125. Küchler M, Decker S, Hörmann F, Soll J, Heins L (2002) Protein import into chloroplasts involves redox-regulated proteins. EMBO J 21:6136–6145
126. Lamberti G, Gugel IL, Meurer J, Soll J, Schwenkert S (2011) The cytosolic kinases STY8, STY17, and STY46 are involved in chloroplast differentiation in *Arabidopsis*. Plant Physiol 157:70–85
127. Lee YJ, Kim DH, Kim YW, Hwang I (2001) Identification of a signal that distinguishes between the chloroplast outer envelope membrane and the endomembrane system in vivo. Plant Cell 13:2175–2190
128. Lee KH, Kim SJ, Lee YJ, Jin JB, Hwang I (2003) The M domain of atToc159 plays an essential role in the import of proteins into chloroplasts and chloroplast biogenesis. J Biol Chem 278:36794–36805
129. Lee J, Wang F, Schnell DJ (2009) Toc receptor dimerization participates in the initiation of membrane translocation during protein import into chloroplasts. J Biol Chem 284: 31130–31141
130. Lee S, Lee DW, Lee Y, Mayer U, Stierhof YD, Jurgens G, Hwang I (2009) Heat shock protein cognate 70-4 and an E3 ubiquitin ligase, CHIP, mediate plastid-destined precursor degradation through the ubiquitin-26S proteasome system in *Arabidopsis*. Plant Cell 21:3984–4001

131. Lee J, Lee H, Kim J, Lee S, Kim DH, Kim S, Hwang I (2011) Both the hydrophobicity and a positively charged region flanking the C-terminal region of the transmembrane domain of signal-anchored proteins play critical roles in determining their targeting specificity to the endoplasmic reticulum or endosymbiotic organelles in *Arabidopsis* cells. Plant Cell 23:1588–1607
132. Lee DW, Jung C, Hwang I (2013) Cytosolic events involved in chloroplast protein targeting. Biochim Biophys Acta 1833:245–252
133. Leheny EA, Theg SM (1994) Apparent inhibition of chloroplast protein import by cold temperatures is due to energetic considerations not membrane fluidity. Plant Cell 6:427–437
134. Leipe DD, Wolf YI, Koonin EV, Aravind L (2002) Classification and evolution of P-loop GTPases and related ATPases. J Mol Biol 317:41–72
135. Levitan A, Trebitsh T, Kiss V, Pereg Y, Dangoor I, Danon A (2005) Dual targeting of the protein disulfide isomerase RB60 to the chloroplast and the endoplasmic reticulum. Proc Natl Acad Sci U S A 102:6225–6230
136. Li HM, Chiu CC (2010) Protein transport into chloroplasts. Annu Rev Plant Biol 61:157–180
137. Li M, Schnell DJ (2006) Reconstitution of protein targeting to the inner envelope membrane of chloroplasts. J Cell Biol 175:249–259
138. Li HM, Kesavulu MM, Su PH, Yeh YH, Hsiao CD (2007) Toc GTPases. J Biomed Sci 14:505–508
139. Ling Q, Huang W, Baldwin A, Jarvis P (2012) Chloroplast biogenesis is regulated by direct action of the ubiquitin-proteasome system. Science 338:655–659
140. Lübeck J, Soll J, Akita M, Nielsen E, Keegstra K (1996) Topology of IEP110, a component of the chloroplastic protein import machinery present in the inner envelope membrane. EMBO J 15:4230–4238
141. Lübeck J, Heins L, Soll J (1997) A nuclear-coded chloroplastic inner envelope membrane protein uses a soluble sorting intermediate upon import into the organelle. J Cell Biol 137:1279–1286
142. Lung SC, Chuong SD (2012) A transit peptide-like sorting signal at the C terminus directs the *Bienertia sinuspersici* preprotein receptor Toc159 to the chloroplast outer membrane. Plant Cell 24:1560–1578
143. Ma Y, Kouranov A, LaSala SE, Schnell DJ (1996) Two components of the chloroplast protein import apparatus, IAP86 and IAP75, interact with the transit sequence during the recognition and translocation of precursor proteins at the outer envelope. J Cell Biol 134:315–327
144. Machettira AB, Gross LE, Sommer MS, Weis BL, Englich G, Tripp J, Schleiff E (2011) The localization of Tic20 proteins in *Arabidopsis thaliana* is not restricted to the inner envelope membrane of chloroplasts. Plant Mol Biol 77:381–390
145. Mackenzie SA (2005) Plant organellar protein targeting: a traffic plan still under construction. Trends Cell Biol 15:548–554
146. Martin W (2010) Evolutionary origins of metabolic compartmentalization in eukaryotes. Phil Trans R Soc Lond B 365:847–855
147. Martin T, Sharma R, Sippel C, Waegemann K, Soll J, Vothknecht UC (2006) A protein kinase family in *Arabidopsis* phosphorylates chloroplast precursor proteins. J Biol Chem 281:40216–40223
148. May T, Soll J (2000) 14-3-3 proteins form a guidance complex with chloroplast precursor proteins in plants. Plant Cell 12:53–64
149. Miras S, Salvi D, Piette L, Seigneurin-Berny D, Grunwald D, Reinbothe C, Joyard J, Reinbothe S, Rolland N (2007) Toc159- and Toc75-independent import of a transit sequence-less precursor into the inner envelope of chloroplasts. J Biol Chem 282:29482–29492
150. Mitschke J, Fuss J, Blum T, Hoglund A, Reski R, Kohlbacher O, Rensing SA (2009) Prediction of dual protein targeting to plant organelles. New Phytol 183:224–235
151. Moberg P, Stahl A, Bhushan S, Wright SJ, Eriksson A, Bruce BD, Glaser E (2003) Characterization of a novel zinc metalloprotease involved in degrading targeting peptides in mitochondria and chloroplasts. Plant J 36:616–628

152. Nada A, Soll J (2004) Inner envelope protein 32 is imported into chloroplasts by a novel pathway. J Cell Sci 117:3975–3982
153. Nakrieko KA, Mould RM, Smith AG (2004) Fidelity of targeting to chloroplasts is not affected by removal of the phosphorylation site from the transit peptide. Eur J Biochem 271:509–516
154. Nanjo Y et al (2006) Rice plastidial *N*-glycosylated nucleotide pyrophosphatase/phosphodiesterase is transported from the ER-golgi to the chloroplast through the secretory pathway. Plant Cell 18:2582–2592
155. Nassoury N, Morse D (2005) Protein targeting to the chloroplasts of photosynthetic eukaryotes: getting there is half the fun. Biochim Biophys Acta 1743:5–19
156. Neupert W, Brunner M (2002) The protein import motor of mitochondria. Nat Rev Mol Cell Biol 3:555–565
157. Neupert W, Herrmann JM (2007) Translocation of proteins into mitochondria. Annu Rev Biochem 76:723–749
158. Nielsen E, Akita M, Davila-Aponte J, Keegstra K (1997) Stable association of chloroplastic precursors with protein translocation complexes that contain proteins from both envelope membranes and a stromal Hsp100 molecular chaperone. EMBO J 16:935–946
159. Olsen LJ, Keegstra K (1992) The binding of precursor proteins to chloroplasts requires nucleoside triphosphates in the intermembrane space. J Biol Chem 267:433–439
160. Pain D, Blobel G (1987) Protein import into chloroplasts requires a chloroplast ATPase. Proc Natl Acad Sci U S A 84:3288–3292
161. Patel R, Hsu S, Bédard J, Inoue K, Jarvis P (2008) The Omp85-related chloroplast outer envelope protein OEP80 is essential for viability in *Arabidopsis*. Plant Physiol 148:235–245
162. Peeters N, Small I (2001) Dual targeting to mitochondria and chloroplasts. Biochim Biophys Acta 1541:54–63
163. Perry SE, Keegstra K (1994) Envelope membrane proteins that interact with chloroplastic precursor proteins. Plant Cell 6:93–105
164. Pujol C, Marechal-Drouard L, Duchene AM (2007) How can organellar protein N-terminal sequences be dual targeting signals? In silico analysis and mutagenesis approach. J Mol Biol 369:356–367
165. Qbadou S, Tien R, Soll J, Schleiff E (2003) Membrane insertion of the chloroplast outer envelope protein, Toc34: constrains for insertion and topology. J Cell Sci 116:837–846
166. Qbadou S, Becker T, Mirus O, Tews I, Soll J, Schleiff E (2006) The molecular chaperone Hsp90 delivers precursor proteins to the chloroplast import receptor Toc64. EMBO J 25:1836–1847
167. Qbadou S, Becker T, Bionda T, Reger K, Ruprecht M, Soll J, Schleiff E (2007) Toc64- a preprotein-receptor at the outer membrane with bipartide function. J Mol Biol 367:1330–1346
168. Radhamony RN, Theg SM (2006) Evidence for an ER to Golgi to chloroplast protein transport pathway. Trends Cell Biol 16:385–387
169. Rapaport D (2005) How does the TOM complex mediate insertion of precursor proteins into the mitochondrial outer membrane? J Cell Biol 171:419–423
170. Reumann S, Keegstra K (1999) The endosymbiotic origin of the protein import machinery of chloroplastic envelope membranes. Trends Plant Sci 4:302–307
171. Richter S, Lamppa GK (1998) A chloroplast processing enzyme functions as the general stromal processing peptidase. Proc Natl Acad Sci U S A 95:7463–7468
172. Richter S, Lamppa GK (2002) Determinants for removal and degradation of transit peptides of chloroplast precursor proteins. J Biol Chem 277:43888–43894
173. Richter S, Lamppa GK (2003) Structural properties of the chloroplast stromal processing peptidase required for its function in transit peptide removal. J Biol Chem 278: 39497–39502
174. Richter S, Zhong R, Lamppa G (2005) Function of the stromal processing peptidase in the chloroplast import pathway. Physiol Plant 123:362–368
175. Rudhe C, Clifton R, Chew O, Zemam K, Richter S, Lamppa G, Whelan J, Glaser E (2004) Processing of the dual targeted precursor protein of glutathione reductase in mitochondria and chloroplasts. J Mol Biol 343:639–647

176. Rudolf M et al (2012) In vivo function of Tic22, a protein import component of the intermembrane space of chloroplasts. Mol Plant 6:817–829
177. Sapir-Mir M, Mett A, Belausov E, Tal-Meshulam S, Frydman A, Gidoni D, Eyal Y (2008) Peroxisomal localization of *Arabidopsis* isopentenyl diphosphate isomerases suggests that part of the plant isoprenoid mevalonic acid pathway is compartmentalized to peroxisomes. Plant Physiol 148:1219–1228
178. Sato S, Nakamura Y, Kaneko T, Asamizu E, Tabata S (1999) Complete structure of the chloroplast genome of *Arabidopsis thaliana*. DNA Res 6:283–290
179. Schirmer EC, Glover JR, Singer MA, Lindquist S (1996) HSP100/Clp proteins: a common mechanism explains diverse functions. Trends Biochem Sci 21:289–296
180. Schleiff E, Jelic M, Soll J (2003) A GTP-driven motor moves proteins across the outer envelope of chloroplasts. Proc Natl Acad Sci U S A 100:4604–4609
181. Schleiff E, Soll J, Küchler M, Kuhlbrandt W, Harrer R (2003) Characterization of the translocon of the outer envelope of chloroplasts. J Cell Biol 160:541–551
182. Schnell DJ, Kessler F, Blobel G (1994) Isolation of components of the chloroplast protein import machinery. Science 266:1007–1012
183. Schnell DJ, Blobel G, Keegstra K, Kessler F, Ko K, Soll J (1997) A consensus nomenclature for the protein-import components of the chloroplast envelope. Trends Cell Biol 7:303–304
184. Shanklin J, DeWitt ND, Flanagan JM (1995) The stroma of higher plant plastids contain ClpP and ClpC, functional homologs of *Escherichia coli* ClpP and ClpA: an archetypal two-component ATP-dependent protease. Plant Cell 7:1713–1722
185. Shen G, Kuppu S, Venkataramani S, Wang J, Yan J, Qiu X, Zhang H (2010) ANKYRIN REPEAT-CONTAINING PROTEIN 2A is an essential molecular chaperone for peroxisomal membrane-bound ASCORBATE PEROXIDASE3 in *Arabidopsis*. Plant Cell 22:811–831
186. Shi LX, Theg SM (2010) A stromal heat shock protein 70 system functions in protein import into chloroplasts in the moss *Physcomitrella patens*. Plant Cell 22:205–220
187. Shi LX, Theg SM (2013) The chloroplast protein import system: from algae to trees. Biochim Biophys Acta 1833: 314–331
188. Shipman RL, Inoue K (2009) Suborganellar localization of plastidic type I signal peptidase 1 depends on chloroplast development. FEBS Lett 583:938–942
189. Shipman-Roston RL, Ruppel NJ, Damoc C, Phinney BS, Inoue K (2010) The significance of protein maturation by plastidic type I signal peptidase 1 for thylakoid development in *Arabidopsis* chloroplasts. Plant Physiol 152:1297–1308
190. Silva-Filho MC (2003) One ticket for multiple destinations: dual targeting of proteins to distinct subcellular locations. Curr Opin Plant Biol 6:589–595
191. Sjögren LL, MacDonald TM, Sutinen S, Clarke AK (2004) Inactivation of the *clpC1* gene encoding a chloroplast Hsp100 molecular chaperone causes growth retardation, leaf chlorosis, lower photosynthetic activity, and a specific reduction in photosystem content. Plant Physiol 136:4114–4126
192. Skalitzky CA, Martin JR, Harwood JH, Beirne JJ, Adamczyk BJ, Heck GR, Cline K, Fernandez DE (2011) Plastids contain a second sec translocase system with essential functions. Plant Physiol 155:354–369
193. Smith MD, Hiltbrunner A, Kessler F, Schnell DJ (2002) The targeting of the atToc159 preprotein receptor to the chloroplast outer membrane is mediated by its GTPase domain and is regulated by GTP. J Cell Biol 159:833–843
194. Stahl T, Glockmann C, Soll J, Heins L (1999) Tic40, a new "old" subunit of the chloroplast protein import translocon. J Biol Chem 274:37467–37472
195. Stengel A, Benz P, Balsera M, Soll J, Bölter B (2008) Tic62- redox-regulated translocon composition and dynamics. J Biol Chem 283:6656–6667
196. Su PH, Li HM (2008) *Arabidopsis* stromal 70-kD heat shock proteins are essential for plant development and important for thermotolerance of germinating seeds. Plant Physiol 146:1231–1241
197. Su PH, Li HM (2010) Stromal Hsp70 is important for protein translocation into pea and *Arabidopsis* chloroplasts. Plant Cell 22:1516–1531

198. Sun Q, Zybailov B, Majeran W, Friso G, Olinares PD, van Wijk KJ (2009) PPDB, the plant proteomics database at Cornell. Nucleic Acids Res 37:D969–D974
199. Sun F, Carrie C, Law S, Murcha MW, Zhang R, Law YS, Suen PK, Whelan J, Lim BL (2012) AtPAP2 is a tail-anchored protein in the outer membrane of chloroplasts and mitochondria. Plant Signal Behav 7:927–932
200. Sveshnikova N, Soll J, Schleiff E (2000) Toc34 is a preprotein receptor regulated by GTP and phosphorylation. Proc Natl Acad Sci U S A 97:4973–4978
201. Teixeira PF, Glaser E (2013) Processing peptidases in mitochondria and chloroplasts. Biochim Biophys Acta 1833:360–370
202. Theg SM, Bauerle C, Olsen LJ, Selman BR, Keegstra K (1989) Internal ATP is the only energy requirement for the translocation of precursor proteins across chloroplastic membranes. J Biol Chem 264:6730–6736
203. Teng YS, Su YS, Chen LJ, Lee YJ, Hwang I, Li HM (2006) Tic21 is an essential translocon component for protein translocation across the chloroplast inner envelope membrane. Plant Cell 18:2247–2257
204. Teng YS, Chan PT, Li HM (2012) Differential age-dependent import regulation by signal peptides. PLoS Biol 10:e1001416
205. Tommassen J (2010) Assembly of outer-membrane proteins in bacteria and mitochondria. Microbiology 156:2587–2596
206. Töpel M, Jarvis P (2011) The Tic20 gene family: phylogenetic analysis and evolutionary considerations. Plant Signal Behav 6:1046–1048
207. Töpel M, Ling Q, Jarvis P (2012) Neofunctionalization within the Omp85 protein superfamily during chloroplast evolution. Plant Signal Behav 7:161–164
208. Tranel PJ, Keegstra K (1996) A novel, bipartite transit peptide targets OEP75 to the outer membrane of the chloroplastic envelope. Plant Cell 8:2093–2104
209. Tranel PJ, Froehlich J, Goyal A, Keegstra K (1995) A component of the chloroplastic protein import apparatus is targeted to the outer envelope membrane via a novel pathway. EMBO J 14:2436–2446
210. Tripp J, Inoue K, Keegstra K, Froehlich JE (2007) A novel serine/proline-rich domain in combination with a transmembrane domain is required for the insertion of AtTic40 into the inner envelope membrane of chloroplasts. Plant J 52:824–838
211. Tripp J et al (2012) Structure and conservation of the periplasmic targeting factor Tic22 protein from plants and cyanobacteria. J Biol Chem 287:24164–24173
212. Trösch R, Jarvis P (2011) The stromal processing peptidase of chloroplasts is essential in *Arabidopsis*, with knockout mutations causing embryo arrest after the 16-cell stage. PLoS ONE 6:e23039
213. Tsai LY, Tu SL, Li HM (1999) Insertion of atToc34 into the chloroplastic outer membrane is assisted by at least two proteinaceous components in the import system. J Biol Chem 274:18735–18740
214. Tu SL, Li HM (2000) Insertion of OEP14 into the outer envelope membrane is mediated by proteinaceous components of chloroplasts. Plant Cell 12:1951–1960
215. Tu SL, Chen LJ, Smith MD, Su YS, Schnell DJ, Li HM (2004) Import pathways of chloroplast interior proteins and the outer-membrane protein OEP14 converge at Toc75. Plant Cell 16:2078–2088
216. Ueda M, Nishikawa T, Fujimoto M, Takanashi H, Arimura S, Tsutsumi N, Kadowaki K (2008) Substitution of the gene for chloroplast RPS16 was assisted by generation of a dual targeting signal. Mol Biol Evol 25:1566–1575
217. VanderVere PS, Bennett TM, Oblong JE, Lamppa GK (1995) A chloroplast processing enzyme involved in precursor maturation shares a zinc-binding motif with a recently recognized family of metalloendopeptidases. Proc Natl Acad Sci U S A 92:7177–7181
218. Viana AA, Li M, Schnell DJ (2010) Determinants for stop-transfer and post-import pathways for protein targeting to the chloroplast inner envelope membrane. J Biol Chem 285:12948–12960
219. Villarejo A et al (2005) Evidence for a protein transported through the secretory pathway en route to the higher plant chloroplast. Nat Cell Biol 7:1124–1131

220. Vojta A, Alavi M, Becker T, Hörmann F, Küchler M, Soll J, Thomson R, Schleiff E (2004) The protein translocon of the plastid envelopes. J Biol Chem 279:21401–21405
221. Vojta L, Soll J, Bölter B (2007) Protein transport in chloroplasts—targeting to the intermembrane space. FEBS J 274:5043–5054
222. Vojta L, Soll J, Bölter B (2007) Requirements for a conservative protein translocation pathway in chloroplasts. FEBS Lett 581:2621–2624
223. von Heijne G, Nishikawa K (1991) Chloroplast transit peptides. The perfect random coil? FEBS Lett 278:1–3
224. Wallas TR, Smith MD, Sanchez-Nieto S, Schnell DJ (2003) The roles of Toc34 and Toc75 in targeting the Toc159 preprotein receptor to chloroplasts. J Biol Chem 278:44289–44297
225. Wan J, Bringloe D, Lamppa GK (1998) Disruption of chloroplast biogenesis and plant development upon down-regulation of a chloroplast processing enzyme involved in the import pathway. Plant J 15:459–468
226. Wang F, Agne B, Kessler F, Schnell DJ (2008) The role of GTP binding and hydrolysis at the atToc159 preprotein receptor during protein import into chloroplasts. J Cell Biol 183:87–99
227. Whatley JM, McLean B, Juniper BE (1991) Continuity of chloroplast and endoplasmic reticulum membranes in *Phaseolus vulgaris*. New Phytol 117:209–217
228. Young ME, Keegstra K, Froehlich JE (1999) GTP promotes the formation of early-import intermediates but is not required during the translocation step of protein import into chloroplasts. Plant Physiol 121:237–244
229. Yue R, Wang X, Chen J, Ma X, Zhang H, Mao C, Wu P (2010) A rice stromal processing peptidase regulates chloroplast and root development. Plant Cell Physiol 51:475–485
230. Zhang XP, Glaser E (2002) Interaction of plant mitochondrial and chloroplast signal peptides with the Hsp70 molecular chaperone. Trends Plant Sci 7:14–21
231. Zhang H, Li X, Zhang Y, Kuppu S, Shen G (2010) Is AKR2A an essential molecular chaperone for a class of membrane-bound proteins in plants? Plant Signal Behav 5:1520–1522
232. Zhong R, Wan J, Jin R, Lamppa G (2003) A pea antisense gene for the chloroplast stromal processing peptidase yields seedling lethals in *Arabidopsis*: survivors show defective GFP import *in vivo*. Plant J 34:802–812

Chapter 10
Protein Routing Processes in the Thylakoid

Carole Dabney-Smith and Amanda Storm

Abstract Thylakoids are complex sub-organellar membrane systems whose role in photosynthesis makes them critical to life. To function properly, thylakoids require the coordinated incorporation of both nuclear- and plastid-encoded proteins allowing rapid response to changing environmental conditions. Protein trafficking to thylakoids is complex; the processes occurring in thylakoids result in an exceptionally protein-dense membrane in which some proteins experience rapid turn-over. Protein transport in thylakoids is accomplished via an intriguing mix of conserved ancestral translocases with novel adaptations to a sub-organellar location. This chapter describes the four known transport pathways into the thylakoid membrane and the thylakoid lumen, namely the chloroplast general secretory system (cpSec), signal recognition particle (cpSRP), twin arginine transport (cpTat), and spontaneous insertion pathways as well as a potential secondary Sec system proposed to be in the inner chloroplast envelope. An overview is provided of known aspects of translocase components, energy requirements, and mechanisms with a focus on recent discoveries. Some of the most exciting new studies have been in determining the structure and binding features of the translocase components and substrate proteins. This chapter highlights the connection between structural and biochemical data and how these complementary avenues of study allow for a more detailed understanding and confirmation of mechanistic models and a means to imagine new areas of pursuit. The cpTat system is of particular interest because it transports folded protein domains using only the proton motive force for energy. Recent structural data has contributed to rapid progress in studying the individual cpTat components by translating their structural features into mechanistic functions.

Keywords Chloroplast sec · Chloroplast tat · Chloroplast SRP · Thylakoid protein translocation · Protein routing in chloroplasts

Abbreviations

BPTI	Bovine pancreatic trypsin inhibitor
CD	Chromodomain
cpSec	Chloroplast secretory system

C. Dabney-Smith (✉) · A. Storm
Department of Chemistry and Biochemistry, Miami University, Oxford, OH 45056, USA
e-mail: adabney@MiamiOH.edu

S.M. Theg, F.-A. Wollman (eds.), *Plastid Biology*, Advances in Plant Biology 5,
DOI 10.1007/978-1-4939-1136-3_10,

cpSRP	Chloroplast signal recognition particle
cpTat	Chloroplast twin arginine transport
LHCPs	Light-harvesting chlorophyll-binding proteins
PMF	Protonmotive force
SRP	Signal recognition particle
TAT	Twin arginine transport
TMD	Transmembrane domains

10.1 Overview of Conservative Protein Sorting to the Thylakoid Membrane and Lumen

Chloroplasts contain three membranes and three aqueous compartments, giving rise to six possible destinations for proteins: the chloroplast envelope outer membrane, the intermembrane space, chloroplast envelope inner membrane, the stroma, the thylakoid, and the thylakoid lumen. Chloroplasts also contain up to 3000 unique proteins, of which ~90 % are nuclear-encoded and most of those are stromal localized [79, 104]. The thylakoid contains ~100 proteins of which 50 % are encoded by plastid genes and the thylakoid lumen contains ~150 proteins, all of which are nuclear-encoded [58]. This chapter will focus on routing to interior locations of the chloroplast, namely the thylakoid and thylakoid lumen. The reader is referred to other chapters in this volume for import into the chloroplast with localization to the outer or inner membranes and stroma. Many proteins destined for the thylakoid and all of the proteins destined for the thylakoid lumen are encoded by nuclear genes and are synthesized in the cytoplasm as higher molecular weight precursors containing bipartite N-terminal amino acid extensions, transit peptides, which target the precursor to the chloroplast. Once import into the chloroplast has occurred, the stromal targeting domain of the bipartite transit peptide is cleaved to reveal a second targeting sequence, which subsequently targets the protein to the thylakoid or the thylakoid lumen. The mechanisms for transport of a protein across the thylakoid or to insert it into the thylakoid are evolutionarily ancient and can be found still in the plasma membranes of many prokaryotes. Two chloroplast (cp) pathways are found to transport proteins into the thylakoid lumen, the cpSec and the cpTat; while the cpSec, cpSRP,and the spontaneous insertion pathways are capable of integrating proteins into the thylakoid. This chapter will look at what is known about these pathways and the roles they play in thylakoid biogenesis.

10.2 The Sec and SRP Transport Pathways in Chloroplasts

10.2.1 cpSec Transport

Sec pathways are evolutionarily conserved protein translocation machines found in the eukaryotic endoplasmic reticulum [77], the archaeal plasma membrane [84],

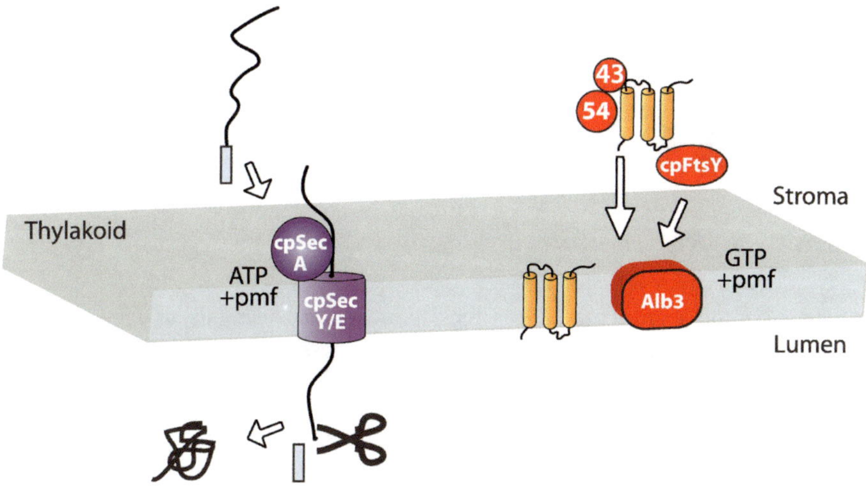

Fig. 10.1 cpSec and cpSRP protein translocation machineries. *Left,* a fixed channel composed of cpSecY/E (*purple barrel*) allows passage across the thylakoid of an unfolded precursor containing a signal peptide (*gray box*). cpSecA (*purple sphere*) is an ATPase that serves as a translocation motor to drive the translocation event. *Right,* the only known protein substrates of the cpSRP system are members of the light harvesting chlorophyll binding proteins, LHCP, which contain three transmembrane spanning domains (*orange*). cpSRP54 and cpSRP43 bind LHCP and guide it to the membrane localized cpSRP receptor, cpFtsY. Interaction with cpFtsY allows engaging the insertase, Alb3, which serves to insert LHCP into the membrane. See text for more information

bacterial plasma membrane [81] and the thylakoid of plant and algal chloroplasts. For more information the reader is referred to several excellent reviews on bacterial and eukaryotic endoplasmic reticulum Sec transport [77, 81, 111].

A common feature of all Sec transport systems is the translocation of precursor proteins in an unfolded conformation through a permanent protein-conducting channel of defined size. Most of the work to describe the Sec pathway has been done in bacteria and eukaryotes and most models of the cpSec system are derived from those. In thylakoid the cpSec system is composed of three components: cpSecA [110] and a membrane complex of cpSecY/E [40, 89]. The cpSecA-cpSecY/E system appears to be a minimal Sec system and operates much like the bacterial system in which the SecY/E forms a channel through which unfolded protein substrates pass, while the cpSecA ATPase serves as a translocation motor (Fig. 10.1, *left panel*). While cpSecA is homologous to *E. coli* SecA, its ATPase activity is specific for thylakoid signal peptides as the activity is not stimulated by binding *E. coli* signal peptides [97]. In addition, the ATPase activity of cpSecA is also affected by the lipid composition of the membrane, having maximal activity in the presence of thylakoid specific lipids [97]. cpSecA is required for development of photosynthetic complexes in *Arabidopsis* because its absence led to severe defects in thylakoid structure and function [61]. Sec systems are used in other organisms for integration of protein via a cotranslation translocation mechanism whereby the ribosome nascent polypeptide complex bind the Sec translocon throught interactions with a signal recognition particle (SRP, see below). Integration of membrane proteins encoded by the plastid genome may use a similar mechanism, although involvement of the chloroplast SRP system has not been shown. Integration of the photosystem II reaction center D1 protein seems to be dependent upon cpSecY/E, but not cpSecA [103]. However, understanding the role that cpSec has in the cotranslational translocation of plastid encoded genes has been difficult to study and very little is known.

10.2.2 cpSRP Transport

Though all domains of life have some form of signal recognition particle (SRP) transport pathway, the chloroplast SRP (cpSRP) is fundamentally distinct from the prokaryotic and cytosolic eukaryotic systems. Prokaryotic systems are composed minimally of a conserved 54 kDa subunit (SRP54) and a RNA component (SRP-RNA). These form the SRP complex, and with additional subunits in eukaryotic systems recognize and bind to the signal sequence of substrates as the nascent proteins are just exiting the ribosome. The SRP-ribosome-nascent protein transit complex is targeted to the specified membrane where it further interacts with the SRP receptor (SR) and engages the Sec translocase. An integral protein in bacterial systems, YidC, that often acts in conjunction with the Sec system to aid in transmembrane protein insertion also appears to be able to engage the transit complex and independently act as an insertase [36, 56, 102]. Transfer of the ribosome-nascent protein to the translocase requires the coordinated hydrolysis of GTP by SRP54 and the SR, which also releases the SRP components for future transport cycles. For a more detailed review of SRP that focuses on the more canonical pathways the reader is referred to Akopian et al. [1] and a thorough review of the cpSRP system can be found in Richter et al. [82].

The chloroplast SRP has adapted to perform a specific role in thylakoid transport resulting in a number of major mechanistic changes. First, even though the cpSRP54 subunit is well conserved from the bacterial systems, it is unable to bind to the SRP-RNA component [10, 93]. Whereas SRP-RNA is otherwise strictly conserved and required for accelerating interaction between SRP54 with the SR, it has been replaced in the cpSRP system with a completely novel protein cpSRP43. The unique features of cpSRP43 that allow it to replace SRP-RNA as the central component of the SRP complex have been the subject of much recent work as discussed below. Although the SRP-RNA is not found in the genome of higher land plants, it has been detected in the plastome of some red and basal green algae [74, 88]. Of additional interest is that cpSRP-RNA and cpSRP43 were recently determined to occur simultaneously in many green algae and nonspermatophyte land plants, strongly suggesting a co-evolution of the cpSRP43 and SRP-RNA as opposed to cpSRP43 replacing the SRP-RNA [99].

Another major difference is that the cpSRP system functions primarily post-translationally as opposed to co-translationally in coordination with the ribosome as in the other systems. There is evidence that cpSRP54 is involved in co-translational integration of some plastid-encoded proteins but very little is known about this area of function [73, 82]. Current knowledge of the system restricts the cpSRP system to a single class of substrates, the light-harvesting chlorophyll-binding proteins

(LHCPs), which are responsible for binding photosynthetic pigments. LHCPs are three-span integral thylakoid proteins that are translated in the cytosol with a stromal signal peptide, which is removed following import through the Toc/Tic translocase so that recognition by SRP components occurs via integral recognition sequences, including a requisite 18 residue motif, L18, located adjacent to TM3 [31, 101].

After recognition and binding of the substrate by cpSRP43 and cpSRP54, the cpSRP transit complex engages the membrane anchored cpSRP receptor, cpFtsY, at the thylakoid membrane, likely through interactions of the GTPase domains of cpSRP54 and cpFtsY in their nucleotide bound state (Fig. 10.1, *right panel*) [35]. Binding between cpSRP54 and cpFtsY serves to reciprocally activate their respective GTPase ability as well as tether the complex to the thylakoid membrane [82]. The cpSRP-cpSR-substrate complex is then targeted to the chloroplast insertase, Alb3, a homolog of the bacterial YidC and mitochondrial Oxa1p proteins. Docking with Alb3 is believed to be mediated primarily by cpSRP43 due to its known interaction with Alb3 C-tail domain, though the details of docking interactions and how substrate is transferred to the insertase *in vivo* remains uncertain [34, 38, 59, 65].

Understanding the novel chloroplast post-transport SRP mechanism largely depends on deciphering the role of the cpSRP exclusive component cpSRP43. This protein is composed entirely of known protein-protein interaction domains, containing an N-terminal chromodomain (CD) followed by four ankyrin repeats (Ank) and two C-terminal CDs. CpSRP43 appears to directly contact all SRP pathway participants except FtsY and functions in both recognition and regulation of SRP interactions [45, 53]. Given its centrality it is not surprising that a number of recent publications have focused on obtaining both structural and biochemical data on cpSRP43 interactions. These studies have provided new insights into how cpSRP43 specifically engages the LHCP substrate, cpSRP54, and Alb3. For example, the available structures have indicated the substrate binding domain of cpSRP [96], which was then further detailed through crosslinking studies between substrate and cpSRP43 [15]. In addition, structural studies have provided a wealth of information about the interaction between cpSRP54 and cpSRP43 (PDB 3UI2) [47] and have confirmed biochemical studies demonstrating interactions between the CD2 domain of cpSRP43 and the chloroplast-specific C-tail extension of cpSRP54 [41, 46]. This interaction involves a unique twinned cage architecture that permits specific reading of the twin arginines in the RRKR C-tail motif [47]. Chromodomains are almost exclusively found in nuclear proteins involved in chromatin remodeling where CD domains selectively recognize methylated Lys residues by means of aromatic cages. The cpSRP43 CD domain appears to have adapted to sensing the positively charged arginines by replacement of some of the aromatic residues in the cages with charged groups able to form salt bridges. The CD2 domain has two consecutive modified cages that enable it to selectively bind with the two initial arginines in the RRKR motif [47]. The CD2 domain, in concert with CD3, has also been implicated in binding the C-tail of Alb3. Two conserved motifs in Alb3 C-tail, AKRS, and SKRS, where found to be required for cpSRP43 binding and the similarity with the cpSRP54 motif suggests the cage architectures in CD2 are likewise able to read Alb3 motifs to promote selective targeting of the SRP complex to the Alb3 insertase [37].

Before a true picture of Alb3-cpSRP43 interaction can be established, however, more biochemical and structural date will be needed to resolve some differences between recent studies. Both the ankayrin repeat domains and CD2 have been proposed as the primary binding site on cpSRP43 and, although two conserved motifs have been identified in Alb3 C-tail domain, another study using full-length Alb3 found only one to be required for cpSRP43 binding and identified an additional binding site on the lumenal side of Alb3 TM5 [34, 37, 59]. The additional binding site was detected due to the use of full-length Alb3 integrated into proteoliposomes as opposed to earlier studies that screened binding of different cpSRP43 constructs with isolated Alb3 C-tail domain. The existence of an intramembrane binding site presents intriguing questions about the mechanism of cpSRP43-Alb3 interaction. A possible explanation that has been proposed is a model in which Alb3 dimerizes to form a pore, reminiscent of the SecYEG channel structure [34]. Binding of cpSRP43 could trigger pore formation and allow it to partially insert, making contact with the luminal portion of Abl3 TM5 and potentially leading to partial insertion of LHCP cargo. A similar model has been proposed for YidC and Oxa1p, which were shown to form dimers stabilized by interaction of the ribosome with their respective C-tail domains [55].

10.2.3 A Second Sec System

There are a number of integral proteins of the inner membrane of chloroplasts for which there is no known translocase [100]. These 'orphan' proteins, such as Tic21, Tic40, and Tic110, have been shown to have soluble stromal intermediates, indicating that they are completely transported through the chloroplast envelope but then somehow become integrated back into the inner membrane [20, 60, 100, 106]. Chloroplasts are an intriguing mixture of both conservative and novel protein transport systems. All translocases of the thylakoid into the lumen or membrane are homologous to bacterial export systems [17, 22]. Movement of proteins out of the stroma, into or across the inner membrane would also be analogous to export out of bacterial cytoplasm. If there is a conservative sorting pathway in the inner membrane the Sec system is a likely candidate as it is the primary export translocase in the bacterial plasma membrane [72, 111].

The first evidence of dual Sec systems was recently discovered in plants. Skalitzky et al. [95] identified genes in *Arabidopsis thaliana* encoding a second set of SecY and SecA homologs that they classified as Scy2 and SecA2, respectively. Using phenotypic analysis of loss-of-function mutants, they were able to show that these two proteins perform essential functions distinct from the previously characterized thylakoid-localized Scy1 and SecA1. Loss of either Scy2 or SecA2 results in early embryo lethality at the globular stage of development as compared to the slow-growing, sucrose-dependent, albino phenotype evidenced in plants lacking one of the thylakoid Sec components, Scy1, SecA1, or SecE. Scy1 and Scy2 were unable to supplement for loss of the other homolog even when their respective promoter sequences were swapped, indicating that loss of a distinct protein not differences in expression produced the observed phenotypes.

Both *in vivo* immunogold localization and *in vitro* chloroplast import and fractionation assays pointed to Scy2 being preferentially envelope-localized, though a small fraction was also detected in thylakoid membrane fractions. *In vitro* import and fractionation also showed that SecA2 was competent for chloroplast import and was localized to the stroma. The third essential component of a Sec translocase system, SecE, remains unidentified for the Scy2 system. Arabidopsis only encodes a single plastid-localized SecE homolog and phenotypic analysis indicated that this protein functions solely with the thylakoid Sec system. However, Scy2 and SecA2 are most closely related to bacterial Sec systems and some bacterial accessory Sec systems contain proteins with distant homology to SecE [83]. If there is a third component associated with Scy2 it is possible that there is insufficient sequence homology to known SecE proteins to be detected by BLAST searches [95].

Beyond the existence of 'orphan' precursors lacking a known targeted translocase, the cellular utility of an inner membrane translocase is echoed by the discovery of Sec systems in both the thylakoid and plasma membrane of cyanobacteria and in the envelope and thylakoid membrane of cyanelles, a primitive plastid that occupies a bridge position between cyanobacteria and chloroplast [71, 112], although it appears that both Sec systems in these organisms come from a single gene whose product is dual localized. Although it remains to be determined whether Scy2 and SecA2 compose a functional translocase, a body of evidence suggests this system as a strong candidate for the missing post-import transport pathway in chloroplast inner membranes. The additional Sec system found in Arabidopsis, Scy2 and SecA2, is plastid-localized with Scy2 preferentially associating with the inner membrane. Both proteins obviously perform essential functions given the severe loss-of-function phenotypes. The fact that embryo abortion occurs at the globular stage, prior to biogenesis of the thylakoid membrane, further supports an envelope-localized functionality. It is of additional interest that knocking out Tic110, one of the 'orphan' precursor proteins that would be a potential Scy2 substrate, produces the same phenotype as Scy2 or SecA2 mutants [51, 57]. Finally, participation of Scy2 and SecA2 in protein transport is supported by their conservation of key sequence elements identified as essential for functionality in the E. coli Sec translocase [95].

It will be especially exciting to follow the future discoveries concerning this putative new translocase as it may be a key to unraveling the chicken-or-egg dilemma posed by the integration of multi-spanning membrane components of thylakoid translocases. For example, previous work in chloroplasts eliminated all known thylakoid translocation pathways as being necessary or involved in the integration of the Tat component cpTatC [64]. The discovery of a new translocase, particularly one that may be involved in thylakoid membrane biogenesis, opens a new avenue of investigation.

10.3 The Twin Arginine Transport (Tat) Pathway in Chloroplasts

10.3.1 Introduction

The presence of a Sec-independent transport system was first identified in plant thylakoid in the early 1990's [26] due to its independence from NTP hydrolysis and dependence upon the pH gradient across the thylakoid for energy [26, 69]. This system was originally called the ΔpH-dependent transport system because of its dependence upon the H^+ gradient. However, it has since been shown to rely upon the protonmotive force (PMF) of which the ΔpH is a component [14]. More recently the system has been referred to as the chloroplast Tat (cpTat) system to reflect the requirement for a twin arginine motif in the signal peptide. The first Tat component required for this system was identified in 1997 in a maize mutant with *h*igh *c*hlorophyll *f*luorescence (*hcf106*) [94]. This was quickly followed by the identification of an operon in *E. coli* containing homologs to Hcf106 [12, 90]. The presence of Hcf106 homologous genes and twin arginine signal peptides in sequenced organisms indicates that the Tat system is widely represented among prokaryotes and prokaryote-derived organelles, but absent from fungi and animals, an exception being in the mitochondrial genome of certain sponges [33, 42, 76]. All plants and algae genomes sequenced so far indicate the presence of Tat systems localized to their chloroplasts. Interestingly, one Tat component, TatC, is often encoded in plant and algal mitochondrial genomes, although a functional Tat system in mitochondria has not been characterized or demonstrated. In eubacteria, Tat substrates represent a small subset of the total secretome (less than 10% in *E. coli*) and are frequently metal cofactor-containing proteins that use cytosolic machinery for cofactor insertion. Of the ~50 predicted substrates of the thylakoid Tat system, only a few possess cofactors. One likely reason for maintaining this system in chloroplasts is that imported Tat substrates are folded rapidly in the stroma making them difficult to unfold by the thylakoid Sec system.

Currently both chloroplasts and prokaryotes, especially *E. coli*, provide insight into mechanistic capabilities of the system, exposing the similarities and nuanced differences as adapted to their individual environments. Work in *E. coli* has been especially useful for investigating structural and physical aspects of the Tat system components, as well as *in vivo* function and genetic dissection of the system. Plant thylakoids, however, provide a robust way to characterize the energetic requirements and individual component contributions. Recently however, a blurring between the systems has occurred such that either system can provide valuable insight into function of the Tat system. The following discussion will summarize results directed toward a mechanistic understanding of the thylakoid system, with reference to similarities and differences between thylakoids and bacterial Tat systems.

10.3.2 Capabilities and Requirements of the Chloroplast Tat system

10.3.2.1 Transport of Folded and Unfolded Proteins

The thylakoid Tat system is capable of transporting both tightly folded and unfolded proteins. Transport of folded proteins was directly demonstrated by following the efficient transport of an internally cross-linked bovine pancreatic trypsin inhibitor (BPTI) fused to the C-terminus of precursor to OE17 (pOE17) [21], and efficient transport of DHFR, locked in a folded conformation by methotrexate binding, fused to the C-terminus of precursor to OE23 (pOE23) [50]. Further confirmation was shown by localization of functional GFP to thylakoid lumen only when transport by the Tat system [63]. cpTat substrates are predicted range in mass from ~2 kDa to over 60 kDa, correlating to a potential spherical size of 1 nm to over 5 nm in diameter, however the folded state of most of these substrates is unknown [78]. The *E. coli* Tat, however, can transport folded substrates up to about 7 nm particle [8, 32, 85], suggesting the chloroplast system may have similar capabilities.

Unlike the *E. coli* Tat system, which rejects unfolded proteins [32], cpTat appears capable of transporting unfolded or unstructured proteins. On the one hand, a C-terminally truncated pOE23-DHFR or a pOE23-DHFR that was translated with amino acid analogues to destabilize its conformation was efficiently transported [50]. Alternatively use of a (Gly_4Ser) polylinker, which lacks structure, was used to show that cpTat could transport unstructured domains [23].

10.3.2.2 Biochemical Requirements: Soluble Factors and Energetics

While it is likely that Tat pathway precursors interact with stromal chaperones after import into the chloroplast, these interactions appear unnecessary for protein translocation per se. This is supported by many observations of efficient transport of thylakoid lumen proteins using isolated, washed thylakoids in either the presence or absence of stromal extract [2, 4, 27, 70, 98]. In addition, efficient cpTat transport occurs without added NTPs or even in the presence of non-hydrolyzable NTP analogs [26]. Further indirect evidence supporting the lack of NTPs for cpTat transport is that the identified Tat components lack nucleotide-binding motifs [26]. In this respect, Tat systems differ from virtually all other protein translocation systems.

Initial experiments demonstrated the dependence upon light for cpTat transport. Detailed work on the energetics of the Tat system demonstrated that there is a threshold ΔpH for transport that is likely different for each precursor [2]. Additionally, transport of the precursor causes counter proton flow at a rate of roughly 80,000 H^+ per protein translocated [2]. Interestingly, the authors also demonstrated that over 90 % of the H^+ counter flow was tied to protein transport process [2]. These findings imply that the cpTat system contains a proton well, a membrane-imbedded species that carries protons as part of the reaction mechanism (see [25] for discussion).

Further studies have also demonstrated that the electric potential, $\Delta\psi$, could also support translocation of precursor [14] and that ion-sensitive cpTat transport occurred even after a ΔpH was no longer detectable [13]. Therefore thylakoid Tat transport requires energy in the form of the thylakoidal PMF, which is usually generated by light and photosynthetic electron transfer, and can also be generated by the reverse action of the chloroplast ATP synthase [26] but *in vitro* is mostly parsed as the ΔpH.

What is not completely understood is how the PMF is used for cpTat transport. Various experiments have demonstrated the requirement of the PMF for the assembly of Tha4 to the precursor bound receptor complex (see Section 10.4). Indeed, physiologically, precursors bind the receptor only weakly in de-energized thylakoids, but become more deeply inserted and bound to the receptor upon energization of the thylakoid [44]. Bacterial Tat transport has been shown to require the $\Delta\psi$ at two distinct steps of the translocation process: a large $\Delta\psi$ early in the process presumably for assembly of components, and a smaller $\Delta\psi$ during the translocation event [5]. The contribution to the translocation event by the PMF in thylakoids has not directly been demonstrated and leaves open the question of whether the PMF is required for the translocation event or rather required to set up the translocation through assembly of components in a priming step.

10.3.2.3 Biochemical Requirements: Components of the Tat machinery

The cpTat system or its α-proteobacterial homolog Tat system is composed of three transmembrane proteins that work together to form the active translocase capable of converting electrochemical energy into the mechanical energy [98] needed to translocate folded proteins, likely of varying diameters, as well as unstructured or unfolded proteins (Fig. 10.2). In thylakoid, these proteins are cpTatC, Hcf106, and Tha4, and in *E. coli* the corresponding orthologues are TatC, TatB, and TatA. In pea chloroplasts, Tat components have a ratio of ~21:4:1 (Tha4:Hcf106:cpTatC) [16], which is different from that found in *E. coli* 75:2.5:1 (TatA:TatB:TatC) [9, 52, 91]. Most studies of the chloroplast Tat system are in pea chloroplasts or isolated thylakoids.

cpTatC

cpTatC is encoded by a nuclear gene as a precursor containing a transit peptide. The pea isoform of mature cpTatC is a 303 residue, 33.3 kD protein. The protein is a multispanning membrane protein containin six transmembrane domains (TMD) with both the N and C termini facing the stroma [68]. cpTatC is found almost exclusively as a heterodimer with Hcf106. cpTatC serves as the primary receptor for signal sequence binding.

Recently the crystal structure of the cpTatC homologue from *Aquifex aeolicus* was solved by two groups [80, 87]. Both groups present similar structures, which

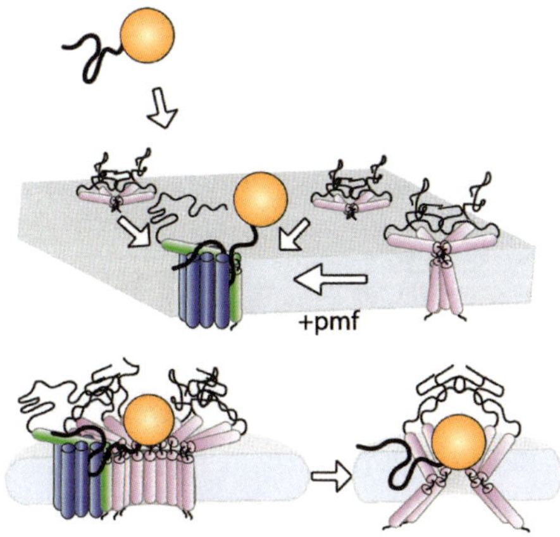

Fig. 10.2 cpTat pathway translocation machinery. Three membrane components, Tha4 (*pink*), Hcf106 (*green*), and cpTatC (*blue*) work together to promote transport of precursors (*orange*) containing signal peptides with obligate twin arginines. Precursors bind a cpTatC/Hcf106 receptor and in the presence of a pmf triggers assembly of Tha4 with the precursor-bound receptor complex (*upper panel*). Once assembly occurs transport of precursor occurs. How the precursor navigates across the membrane is an area of active research, but likely employs a transient channel or pore composed largely of Tha4

are suggestive of a glove with a concave pocket on one side. Based on those structures, cpTatC has four stroma facing regions (N terminus, stromal loop 1 between TMD 2 and 3, stromal loop 2 between TMD 4 and 5, and very short C terminus) when placed in the membrane. Mutational analysis of cpTatC demonstrates that substrate preferentially binds to amino acids in the stromal proximal portions of TMD 1 and 2 [62]. Further analysis indicated that mutations in TMD 5 cpTatC affected complex assembly [62]. Understanding the assembly of the receptor complex and interactions with substrate precursors is an active area of investigation and should yield a greater understanding of not only how the complexes are organized in the membrane but potentially how Tha4 assembles to promote translocation.

Tha4 and Hcf106

Hcf106 (TatB) and Tha4 (TatA) are similar proteins that likely arose from a gene duplication event [109]. Both are nuclear-encoded and are synthesized as higher molecular weight precursors containing transit peptides. Mature Tha4 from pea is an 82-residue, 8.9 kD protein, whereas mature Hcf106 is an 176-residue, 19 kD protein. They each contain N-terminal transmembrane domains with an obligate glutamate in the transmembrane region followed by a hinge region containing a consensus sequence of FGPK and a stromal-facing amphipathic helix and a loosely structured C-tail. The amino acid compositions of the transmembrane regions are very similar and can substitute for each other [29]. The sequence identity is approximately 43 % overall with the transmembrane and amphiphathic regions greater than 50% identity. However, Hcf106 contains extended amphipathic helix and C-tail regions than those in Tha4. Cysteine substitution of amino acid residues in Tha4 has at most minor effects on function of Tha4 with one exception. Substitution of the transmembrane obligate glutamate E10 resulted in nonfunctional protein [29, 39]. Further E10 was shown to be essential for Tha4 assembly into the transclocase.

Recently, the structure was solved for TatAd from *Bacillus subtilis*, the homologous protein to Tha4 using solution NMR [49] and solid state NMR [107]. Both structures demonstrate a short transmembrane region of ~14 amino acids and extensive inter-helical contacts, which may have implications on how Tha4 is found in the membrane as discussed below (Section 10.4). Further, the TatA structure from *E. coli* shows similar structural features [86], although none of the structures solved contain the carboxy-tail region. While Hcf106 and Tha4 are structurally quite similar, they have distinct function [29]. Hcf106 is found in pea thylakoids in two separate complexes when the membranes are subject to detergent solubilization and blue native polyacrylamide gel electrophoresis (BN-PAGE). One complex migrating as a 700 kD complex also contains cpTatC and the other is a smaller complex containing only Hcf106.

However, some gram-positive organisms, such as *Bacillus subtilis*, or cyanobacteria, such as *Synechocystis* PCC6803, have minimal Tat pathways lacking a TatB component [33, 54]. The TatA in these systems is bifunctional and capable of fulfilling both roles. Indeed in studies of *E. coli* Tat transport, bifunctionality of TatA could be restored by specific point mutations in the N terminus of the protein [7, 11], providing further evidence of a gene duplication event leading to distinct function of the thylakoid Tha4 and Hcf106 proteins.

10.4 Operation of the Tat System

Tat transport can be biochemically divided into discrete steps such as precursor binding, Tha4 assembly, precursor translocation, and Tha4 disassembly [24, 66]. Through studies of component and precursor interactions during these stages, researchers have tried to define the roles of the different components.

10.4.1 Precursor Binding

Recent studies suggest that the precursor engages the thylakoid membrane initially through direct interaction with lipids and can then encounter the protein receptor complex, cpTatC/Hcf106. For example, a chimera of two oxygen evolving subunits, OE16 and OE23, containing the signal peptide from OE16 and the mature domain of OE23 demonstrated an initial interaction with the thylakoid membrane prior to transport [48]. Further, in *E. coli* the precursor has been shown to be able engage the membrane in the absence of Tat components and that interaction is predominantly through the signal sequence [6]. In both of these examples, the interactions were predominantly, if not exclusively, through the targeting sequence, which is known to contain a hydrophobic sequence fully capable of engaging lipids [105]. In addition, the signal peptide/lipid interaction may be due to supersaturating amounts of the precursor present [6, 48]. However, more recent studies failed to detect transport of lipid-bound precursors [16]. The true relevance of precursors interacting with the lipids prior to engaging the receptor complex remains to be demonstrated *in vivo*, although physiologically for the plant, this could allow synthesis of the precursor in the cytoplasm and localization at the thylakoid membrane at anytime, while the actual transport process would occur during the daytime or whenever the PMF was sufficient.

Regardless of whether the precursor engages lipids prior to the cpTat transport components, in order for translocation to occur the precursor must engage the receptor complex comprising cpTatC and Hcf106 [24]. The precursor through its signal peptide primarily engages the cpTatC N terminus and first stromal loop [43, 62]. The receptor complex in thylakoid, which when isolated from detergent solubilized membranes and separated by BN-PAGE migrates at ~700 kD, appears to contain only cpTatC and Hcf106 in a 1:1 molar ratio, suggesting that each complex contains ~8 dimers of cpTatC and Hcf106 [24]. Interestingly, recent binding studies indicate that each cpTatC in the complex is capable of binding precursor in a non-cooperative manner [16], thus allowing for staging of up to eight precursors at each receptor complex.

10.4.2 Tha4 Assembly and Interaction with Precursor

Studies have shown that translocation does not occur without the participation of Tha4 [29, 39, 66]. Furthermore, inactivation of endogenous Tha4 can be overcome by integration of recombinant Tha4 [29, 39], thus allowing a mutagenic approach to investigate the role of Tha4 in cpTat transport. Data from mutagenic studies suggest a major role for Tha4 in the translocation event. First, Tha4 is found in molar excess over the other cpTat components. An earlier report placed Tha4 at a roughly 8-fold excess over cpTatC [67], but more recently it was shown that a roughly 25-fold excess is more likely [16]. Second, Tha4 demonstrates reversible assembly with the precursor-bound receptor complex [66], and that assembly is only triggered in the presence of the protonmotive force (PMF). Third, Tha4 is found in the thylakoid in homotetramers through interactions between the TMD of the monomer Tha4 [28]. These homotetramers then assemble in the presence of the precursor bound to the receptor complex and the PMF to form large complexes through the C tail region [28, 30]. Fourth, Tha4 undergoes localized conformational changes during the transport process [4], but does not undergo major topology change such as flipping into the membrane, as was suggest for *E. coli* TatA [18, 19]. Lastly, Tha4 was shown to directly interact with the mature domain of the precursor, only after it was bound to the receptor [75]. This interaction with precursor is significant because the interactions were detected between the mature domain of the precursor and Tha4, but if the signal peptide was cleaved the interactions did not occur [75], suggesting that the interaction occurs when precursor is bound to the receptor complex, possibly before transport, clearly before cleavage of the signal peptide.

10.4.3 Models for Tat Translocation and Future Prospects

So how do these data fit into a model for protein transport by the cpTat pathway? The transport process seems to occur in a cyclical and transient fashion, all the while maintaining the impermeability of the membrane to ions. Precursors in the stroma find their way to the thylakoid and either surf along the thylakoid surface until they reach a receptor complex (cpTatC/Hcf106) or bind directly to the receptor complex (Fig. 10.2). After precursor binding and in the presence of a PMF, Tha4 assembles with the complex. At this point transport of the precursor occurs. After translocation or concomitant with it, Tha4 disassembles from the complex, thus resetting the system for subsequent rounds of translocation. These aspects of the model are based upon biochemical analysis using a robust *in vitro* isolated thylakoid assay and provide the general scheme for translocation of a precursor. Major questions such as how translocation occurs (e.g., through a pore composed of Tha4 homooligomer or a heteroligomer composed of mostly Tha4 with Hcf106 and cpTatC, or not through a pore at) still remain unanswered although they are extremely active areas of research.

10.5 Spontaneous Insertion of Thylakoid Proteins

There are a few thylakoid proteins for which insertion into the thylakoid appears to occur unassisted or in a spontaneous manner, not requiring the presence of other proteins or energy. Some of these proteins are single transmembrane spanning proteins, such as CF_0II of the ATP synthase and the photosystem II proteins, PsbX and PsbW [108], while others, such as PsaG and PsaK contain two transmembrane spans with a short loop in between (reviewed in [3, 92]). Two of the thylakoid Tat components, Tha4 and Hcf106, also insert spontaneously [39]. In each of these instances, the thylakoids in the reactions had been treated to be free of energy sources or to inactivate known thylakoids translocases or insertases.

References

1. Akopian D, Shen K, Zhang X, Shan SO (2013) Signal recognition particle: an essential protein-targeting machine. Ann Rev Biochem 82:693–721
2. Alder NN, Theg SM (2003) Energetics of protein transport across biological membranes. a study of the thylakoid DeltapH-dependent/cpTat pathway. Cell 112:231–242
3. Aldridge C, Cain P, Robinson C (2009) Protein transport in organelles: protein transport into and across the thylakoid membrane. FEBS J 276:1177–1186
4. Aldridge C, Storm A, Cline K, Dabney-Smith C (2012) The chloroplast twin arginine transport (tat) component, tha4, undergoes conformational changes leading to tat protein transport. J Biol Chem 287:34752–34763
5. Bageshwar UK, Musser SM (2007) Two electrical potential-dependent steps are required for transport by the *Escherichia coli* Tat machinery. J Cell Biol 179:87–99

6. Bageshwar UK, Whitaker N, Liang FC, Musser SM (2009) Interconvertibility of lipid- and translocon-bound forms of the bacterial Tat precursor pre-SufI. Mol Microbiol 74:209–226
7. Barrett CM, Freudl R, Robinson C (2007) Twin arginine translocation (Tat)-dependent export in the apparent absence of TatABC or TatA complexes using modified *Escherichia coli* TatA subunits that substitute for TatB. J Biol Chem 282:36206–36213
8. Berks BC, Sargent F, Palmer T (2000) The Tat protein export pathway. Mol Microbiol 35:260–274
9. Berks BC, Palmer T, Sargent F (2003) The Tat protein translocation pathway and its role in microbial physiology. Adv Microb Physiol 47:187–254
10. Berthelmann F, Mehner D, Richter S, Lindenstrauss U, Lunsdorf H, Hause G, Brüser T (2008) Recombinant expression of tatABC and tatAC results in the formation of interacting cytoplasmic TatA tubes in Escherichia coli. J Biol Chem 283:25281–25289
11. Blaudeck N, Kreutzenbeck P, Muller M, Sprenger GA, Freudl R (2005) Isolation and characterization of bifunctional *Escherichia coli* TatA mutant proteins that allow efficient tat-dependent protein translocation in the absence of TatB. J Biol Chem 280:3426–3432
12. Bogsch EG, Sargent F, Stanley NR, Berks BC, Robinson C, Palmer T (1998) An essential component of a novel bacterial protein export system with homologues in plastids and mitochondria. J Biol Chem 273:18003–18006
13. Braun NA, Theg SM (2008) The chloroplast Tat pathway transports substrates in the dark. J Biol Chem 283:8822–8828
14. Braun NA, Davis AW, Theg SM (2007) The chloroplast Tat pathway utilizes the transmembrane electric potential as an energy source. Biophys J 93:1993–1998
15. Cain P, Holdermann I, Sinning I, Johnson AE, Robinson C (2011) Binding of chloroplast signal recognition particle to a thylakoid membrane protein substrate in aqueous solution and delineation of the cpSRP43-substrate interaction domain. Biochem J 437:149–155
16. Celedon J, Cline K (2012) Stoichiometry for binding and transport by the twin arginine translocation system. J Cell Biol 197:523–534
17. Celedon JM, Cline K (2013) Intra-plastid protein trafficking: how plant cells adapted prokaryotic mechanisms to the eukaryotic condition. BBA-Mol Cell Res 1833:341–351
18. Chan CS, Zlomislic MR, Tieleman DP, Turner RJ (2007) The TatA subunit of *Escherichia coli* twin-arginine translocase has an N-in topology. BioChemistry 46:7396–7404
19. Chan CS, Haney EF, Vogel HJ, Turner RJ (2011) Towards understanding the Tat translocation mechanism through structural and biophysical studies of the amphipathic region of TatA from *Escherichia coli*. Biochim Biophys Acta 1808:2289–2296
20. Chiu CC, Li HM (2008) Tic40 is important for reinsertion of proteins from the chloroplast stroma into the inner membrane. Plant J 56:793–801
21. Clark SA, Theg SM (1997) A folded protein can be transported across the chloroplast envelope and thylakoid membranes. Mol Biol Cell 8:923–934
22. Cline K, Dabney-Smith C (2008) Plastid protein import and sorting: different paths to the same compartments. Curr Opin Plant Biol 11:585–592
23. Cline K, McCaffery M (2007) Evidence for a dynamic and transient pathway through the TAT protein transport machinery. EMBO J 26:3039–3049
24. Cline K, Mori H (2001) Thylakoid DeltapH-dependent precursor proteins bind to a cpTatC-Hcf106 complex before Tha4-dependent transport. J Cell Biol 154:719–729
25. Cline K, Theg SM (2007) The Sec and Tat protein translocation pathways in chloroplasts. In: Dalbey RE, Koehler CM, Tamanoi F (eds) The enzymes: molecular machines involved in protein transport across cellular membranes. Academic, New York, pp 463–492
26. Cline K, Ettinger WF, Theg SM (1992) Protein-specific energy requirements for protein transport across or into thylakoid membranes—two lumenal proteins are transported in the absence of ATP. J Biol Chem 267:2688–2696
27. Cline K, Henry R, Li C, Yuan J (1993) Multiple pathways for protein transport into or across the thylakoid membrane. EMBO J 12:4105–4114
28. Dabney-Smith C, Cline K (2009) Clustering of C-terminal stromal domains of Tha4 homo-oligomers during translocation by the Tat protein transport system. Mol Biol Cell 20:2060–2069

29. Dabney-Smith C, Mori H, Cline K (2003) Requirement of a Tha4-conserved transmembrane glutamate in thylakoid Tat translocase assembly revealed by biochemical complementation. J Biol Chem 278:43027–43033
30. Dabney-Smith C, Mori H, Cline K (2006) Oligomers of Tha4 organize at the thylakoid Tat translocase during protein transport. J Biol Chem 281:5476–5483
31. DeLille J, Peterson EC, Johnson T, Moore M, Kight A, Henry R (2000) A novel precursor recognition element facilitates posttranslational binding to the signal recognition particle in chloroplasts. Proc Natl Acad Sci U S A 97:1926–1931
32. DeLisa MP, Tullman D, Georgiou G (2003) Folding quality control in the export of proteins by the bacterial twin-arginine translocation pathway. Proc Natl Acad Sci U S A 100:6115–6120
33. Dilks K, Rose RW, Hartmann E, Pohlschroder M (2003) Prokaryotic utilization of the twin-arginine translocation pathway: a genomic survey. J Bacteriol 185:1478–1483
34. Dunschede B, Bals T, Funke S, Schunemann D (2011) Interaction studies between the chloroplast signal recognition particle subunit cpSRP43 and the full-length translocase Alb3 reveal a membrane-embedded binding region in Alb3 protein. J Biol Chem 286:35187–35195
35. Egea PF, Shan SO, Napetschnig J, Savage DF, Walter P, Stroud RM (2004) Substrate twinning activates the signal recognition particle and its receptor. Nature 427:215–221
36. Facey SJ, Neugebauer SA, Krauss S, Kuhn A (2007) The mechanosensitive channel protein MscL is targeted by the SRP to the novel YidC membrane insertion pathway of *Escherichia coli*. J Mol Biol 365:995–1004
37. Falk S, Sinning I (2010) The C terminus of Alb3 interacts with the chromodomains 2 and 3 of cpSRP43. J Biol Chem 285:le25–le26; author reply le26–le28
38. Falk S, Ravaud S, Koch J, Sinning I (2010) The C terminus of the Alb3 membrane insertase recruits cpSRP43 to the thylakoid membrane. J Biol Chem 285:5954–5962
39. Fincher V, Dabney-Smith C, Cline K (2003) Functional assembly of thylakoid deltapH-dependent/Tat protein transport pathway components in vitro. Eur J Biochem 270:4930–4941
40. Froderberg L, Rohl T, van Wijk KJ, de Gier JW (2001) Complementation of bacterial SecE by a chloroplastic homologue. FEBS Lett 498:52–56
41. Funke S, Knechten T, Ollesch J, Schunemann D (2005) A unique sequence motif in the 54-kDa subunit of the chloroplast signal recognition particle mediates binding to the 43-kDa subunit. J Biol Chem 280:8912–8917
42. Gazave E, Lapebie P, Renard E, Vacelet J, Rocher C, Ereskovsky AV, Lavrov DV, Borchiellini C (2010) Molecular phylogeny restores the supra-generic subdivision of homoscleromorph sponges (Porifera, Homoscleromorpha). PLoS ONE 5:e14290
43. Gérard F, Cline K (2006) Efficient twin arginine translocation (Tat) pathway transport of a precursor protein covalently anchored to its initial cpTatC binding site. J Biol Chem 281:6130–6135
44. Gérard F, Cline K (2007) The thylakoid proton gradient promotes an advanced stage of signal peptide binding deep within the Tat pathway receptor complex. J Biol Chem 282:5263–5272
45. Goforth RL, Peterson EC, Yuan J, Moore MJ, Kight AD, Lohse MB, Sakon J, Henry RL (2004) Regulation of the GTPase cycle in post-translational signal recognition particle-based protein targeting involves cpSRP43. J Biol Chem 279:43077–43084
46. Hermkes R, Funke S, Richter C, Kuhlmann J, Schunemann D (2006) The alpha-helix of the second chromodomain of the 43 kDa subunit of the chloroplast signal recognition particle facilitates binding to the 54 kDa subunit. FEBS Lett 580:3107–3111
47. Holdermann I, Meyer NH, Round A, Wild K, Sattler M, Sinning I (2012) Chromodomains read the arginine code of post-translational targeting. Nat Struct Mol Biol 19:260–263
48. Hou B, Frielingsdorf S, Klosgen RB (2006) Unassisted membrane insertion as the initial step in DeltapH/Tat-dependent protein transport. J Mol Biol 355:957–967
49. Hu Y, Zhao E, Li H, Xia B, Jin C (2010) Solution NMR structure of the TatA component of the twin-arginine protein transport system from gram-positive bacterium *Bacillus subtilis*. J Am Chem Soc 132:15942–15944

50. Hynds PJ, Robinson D, Robinson C (1998) The sec-independent twin-arginine translocation system can transport both tightly folded and malfolded proteins across the thylakoid membrane. J Biol Chem 273:34868–34874
51. Inaba T, Alvarez-Huerta M, Li M, Bauer J, Ewers C, Kessler F, Schnell DJ (2005) Arabidopsis tic110 is essential for the assembly and function of the protein import machinery of plastids. Plant Cell 17:1482–1496
52. Jack RL, Sargent F, Berks BC, Sawers G, Palmer T (2001) Constitutive expression of *Escherichia coli* tat genes indicates an important role for the twin-arginine translocase during aerobic and anaerobic growth. J Bacteriol 183:1801–1804
53. Jonas-Straube E, Hutin C, Hoffman NE, Schunemann D (2001) Functional analysis of the protein-interacting domains of chloroplast SRP43. J Biol Chem 276:24654–24660
54. Jongbloed JD, Grieger U, Antelmann H, Hecker M, Nijland R, Bron S, van Dijl JM (2004) Two minimal Tat translocases in *Bacillus*. Mol Microbiol 54:1319–1325
55. Kohler R, Boehringer D, Greber B, Bingel-Erlenmeyer R, Collinson I, Schaffitzel C, Ban N (2009) YidC and Oxa1 form dimeric insertion pores on the translating ribosome. Mol Cell 34:344–353
56. Kolkman MA, van der Ploeg R, Bertels M, van Dijk M, van der Laan J, van Dijl JM, Ferrari E (2008) The twin-arginine signal peptide of *Bacillus subtilis* YwbN can direct either Tat- or Sec-dependent secretion of different cargo proteins: secretion of active subtilisin via the *B. subtilis* Tat pathway. Appl Environ Microbiol 74:7507–7513
57. Kovacheva S, Bedard J, Patel R, Dudley P, Twell D, Rios G, Koncz C, Jarvis P (2005) In vivo studies on the roles of Tic110, Tic40 and Hsp93 during chloroplast protein import. Plant J 41:412–428
58. Leister D (2003) Chloroplast research in the genomic age. Trends Genet 19:47–56
59. Lewis NE, Marty NJ, Kathir KM, Rajalingam D, Kight AD, Daily A, Kumar TK, Henry RL, Goforth RL (2010) A dynamic cpSRP43-Albino3 interaction mediates translocase regulation of chloroplast signal recognition particle (cpSRP)-targeting components. J Biol Chem 285:34220–34230
60. Li M, Schnell DJ (2006) Reconstitution of protein targeting to the inner envelope membrane of chloroplasts. J Cell Biol 175:249–259
61. Liu D et al (2010) cpSecA, a thylakoid protein translocase subunit, is essential for photosynthetic development in *Arabidopsis*. J Exp Bot 61:1655–1669
62. Ma X, Cline K (2013) Mapping the signal peptide binding and oligomer contact sites of the core subunit of the pea twin arginine protein translocase. Plant Cell 25:999–1015
63. Marques JP, Dudeck I, Klosgen RB (2003) Targeting of EGFP chimeras within chloroplasts. Mol Genet Genom 269:381–387
64. Martin JR, Harwood JH, McCaffery M, Fernandez DE, Cline K (2009) Localization and integration of thylakoid protein translocase subunit cpTatC. Plant J 58:831–842
65. Moore M, Goforth RL, Mori H, Henry R (2003) Functional interaction of chloroplast SRP/FtsY with the ALB3 translocase in thylakoids: substrate not required. J Cell Biol 162:1245–1254
66. Mori H, Cline K (2002) A twin arginine signal peptide and the pH gradient trigger reversible assembly of the thylakoid [Delta]pH/Tat translocase. J Cell Biol 157:205–210
67. Mori H, Summer EJ, Ma X, Cline K (1999) Component specificity for the thylakoidal Sec and Delta pH-dependent protein transport pathways. J Cell Biol 146:45–56
68. Mori H, Summer EJ, Cline K (2001) Chloroplast TatC plays a direct role in thylakoid (Delta) pH-dependent protein transport. FEBS Lett 501:65–68
69. Mould RM, Robinson C (1991) A proton gradient is required for the transport of two lumenal oxygen-evolving proteins across the thylakoid membrane. J Biol Chem 266:12189–12193
70. Musser SM, Theg SM (2000) Characterization of the early steps of OE17 precursor transport by the thylakoid DeltapH/Tat machinery. Eur J Biochem 267:2588–2598
71. Nakai M, Sugita D, Omata T, Endo T (1993) Sec-Y protein is localized in both the cytoplasmic and thylakoid membranes in the cyanobacterium *Synechococcus* PCC7942. Biochem Biophys Res Commun 193:228–234

72. Natale P, Bruser T, Driessen AJ (2008) Sec- and Tat-mediated protein secretion across the bacterial cytoplasmic membrane-distinct translocases and mechanisms. Biochim Biophys Acta 1778:1735–1756
73. Nilsson R, van WKJ (2002) Transient interaction of cpSRP54 with elongating nascent chains of the chloroplast-encoded D1 protein; 'cpSRP54 caught in the act'. FEBS Lett 524:127–133
74. Packer JC, Howe CJ (1998) Algal plastid genomes encode homologues of the SRP-associated RNA. Mol Microbiol 27:508–510
75. Pal D, Fite K, Dabney-Smith C (2013) Direct interaction between a precursor mature domain and transport component Tha4 during twin arginine transport of chloroplasts. Plant Physiol 161:990–1001
76. Palmer T, Berks BC (2012) The twin-arginine translocation (Tat) protein export pathway. Nat Rev Microbiol 10:483–496
77. Park E, Rapoport TA (2012) Mechanisms of Sec61/SecY-mediated protein translocation across membranes. Ann Rev Biophys 41:21–40
78. Peltier JB et al (2002) Central functions of the lumenal and peripheral thylakoid proteome of *Arabidopsis* determined by experimentation and genome-wide prediction. Plant Cell 14:211–236
79. Peltier JB, Friso G, Kalume DE, Roepstorff P, Nilsson F, Adamska I, van Wijk KJ (2000) Proteomics of the chloroplast: systematic identification and targeting analysis of lumenal and peripheral thylakoid proteins. Plant Cell 12:319–341
80. Ramasamy S, Abrol R, Suloway CJ, Clemons WM Jr (2013) The glove-like structure of the conserved membrane protein TatC provides insight into signal sequence recognition in twin-arginine translocation. Structure 21:777–888
81. Rapoport TA (2007) Protein translocation across the eukaryotic endoplasmic reticulum and bacterial plasma membranes. Nature 450:663–669
82. Richter CV, Bals T, Schunemann D (2010) Component interactions, regulation and mechanisms of chloroplast signal recognition particle-dependent protein transport. Eur J Cell Biol 89:965–973
83. Rigel NW, Braunstein M (2008) A new twist on an old pathway—accessory secretion systems. Mol Microbiol 69:291–302
84. Robinson C, Bolhuis A (2004) Tat-dependent protein targeting in prokaryotes and chloroplasts. Biochim Biophys Acta 1694:135–147
85. Rodrigue A, Chanal A, Beck K, Muller M, Wu LF (1999) Co-translocation of a periplasmic enzyme complex by a hitchhiker mechanism through the bacterial tat pathway. J Biol Chem 274:13223–13228
86. Rodriguez F, Rouse SL, Tait CE, Harmer J, De RA, Timmel CR, Sansom MS, Berks BC, Schnell JR (2013) Structural model for the protein-translocating element of the twin-arginine transport system. Proc Natl Acad Sci U S A 110:E1092–E1101
87. Rollauer SE et al (2012) Structure of the TatC core of the twin-arginine protein transport system. Nature 492:210–214
88. Rosenblad MA, Zwieb C, Samuelsson T (2004) Identification and comparative analysis of components from the signal recognition particle in protozoa and fungi. BMC Genomics 5:5
89. Roy LM, Barkan A (1998) A SecY homologue is required for the elaboration of the chloroplast thylakoid membrane and for normal chloroplast gene expression. J Cell Biol 141:385–395
90. Sargent F, Bogsch EG, Stanley NR, Wexler M, Robinson C, Berks BC, Palmer T (1998) Overlapping functions of components of a bacterial Sec-independent protein export pathway. EMBO J 17:3640–3650
91. Sargent F, Gohlke U, De Leeuw E, Stanley NR, Palmer T, Saibil HR, Berks BC (2001) Purified components of the *Escherichia coli* Tat protein transport system form a double-layered ring structure. Eur J Biochem 268:3361–3367
92. Schleiff E, Klosgen RB (2001) Without a little help from 'my' friends: direct insertion of proteins into chloroplast membranes? Biochim Biophys Acta 1541:22–33

93. Schuenemann D, Amin P, Hoffman NE (1999) Functional divergence of the plastid and cytosolic forms of the 54-kDa subunit of signal recognition particle. Biochem Biophys Res Commun 254:253–258
94. Settles AM, Yonetani A, Baron A, Bush DR, Cline K, Martienssen R (1997) Sec-independent protein translocation by the maize Hcf106 protein. Science 278:1467–1470
95. Skalitzky CA, Martin JR, Harwood JH, Beirne JJ, Adamczyk BJ, Heck GR, Cline K, Fernandez DE (2011) Plastids contain a second sec translocase system with essential functions. Plant Physiol 155:354–369
96. Stengel KF, Holdermann I, Cain P, Robinson C, Wild K, Sinning I (2008) Structural basis for specific substrate recognition by the chloroplast signal recognition particle protein cpSRP43. Science 321:253–256
97. Sun C, Rusch SL, Kim J, Kendall DA (2007) Chloroplast SecA and *Escherichia coli* SecA have distinct lipid and signal peptide preferences. J Bacteriol 189:1171–1175
98. Teter SA, Theg SM (1998) Energy-transducing thylakoid membranes remain highly impermeable to ions during protein translocation. Proc Natl Acad Sci U S A 95:1590–1594
99. Trager C et al (2012) Evolution from the prokaryotic to the higher plant chloroplast signal recognition particle: the signal recognition particle RNA is conserved in plastids of a wide range of photosynthetic organisms. Plant Cell 24:4819–4836
100. Tripp J, Inoue K, Keegstra K, Froehlich JE (2007) A novel serine/proline-rich domain in combination with a transmembrane domain is required for the insertion of AtTic40 into the inner envelope membrane of chloroplasts. Plant J 52:824–838
101. Tu CJ, Peterson EC, Henry R, Hoffman NE (2000) The L18 domain of light-harvesting chlorophyll proteins binds to chloroplast signal recognition particle 43. J Biol Chem 275:13187–13190
102. van derLM, Bechtluft P, Kol S, Nouwen N, Driessen AJ (2004) F1F0 ATP synthase subunit c is a substrate of the novel YidC pathway for membrane protein biogenesis. J Cell Biol 165:213–222
103. van Wijk KJ, Knott TG, Robinson C (1995) Evidence for SecA- and delta pH-independent insertion of D1 into thylakoids. FEBS Lett 368:263–266
104. van Wijk KJ, Peltier JB, Giacomelli L (2007) Isolation of chloroplast proteins from *Arabidopsis thaliana* for proteome analysis. Methods Mol Biol 355:43–48
105. van't Hof R, van Klompenburg W, Pilon M, Kozubek A, de Korte-KoolG, Demel R, Weisbeek PJ, de Kruijff B (1993) The transit sequence mediates the specific interaction of the precursor of ferredoxin with chloroplast envelope membrane lipids. J Biol Chem 268:4037–4042
106. Vojta L, Soll J, Bolter B (2007) Requirements for a conservative protein translocation pathway in chloroplasts. FEBS Lett 581:2621–2624
107. Walther TH, Grage SL, Roth N, Ulrich AS (2010) Membrane alignment of the pore-forming component TatA(d) of the twin-arginine translocase from *Bacillus subtilis* resolved by solid-state NMR spectroscopy. J Am Chem Soc 132:15945–15956
108. Woolhead CA, Thompson SJ, Moore M, Tissier C, Mant A, Rodger A, Henry R, Robinson C (2001) Distinct Albino3-dependent and -independent pathways for thylakoid membrane protein insertion. J Biol Chem 276:40841–40846
109. Yen MR, Tseng YH, Nguyen EH, Wu LF, Saier MH Jr (2002) Sequence and phylogenetic analyses of the twin-arginine targeting (Tat) protein export system. Arch Microbiol 177:441–450
110. Yuan J, Henry R, McCaffery M, Cline K (1994) SecA homolog in protein transport within chloroplasts: evidence for endosymbiont-derived sorting. Science 266:796–798
111. Yuan J, Zweers JC, van DJM, Dalbey RE (2010) Protein transport across and into cell membranes in bacteria and archaea. CMLS 67:179–199
112. Yusa F, Steiner JM, Loffelhardt W (2008) Evolutionary conservation of dual Sec translocases in the cyanelles of *Cyanophora paradoxa*. BMC Evol Biol 8:304

Chapter 11
Protein Transport into Plastids of Secondarily Evolved Organisms

Franziska Hempel, Kathrin Bolte, Andreas Klingl, Stefan Zauner and Uwe-G. Maier

Abstract Secondary or complex plastids arose by the engulfment of photosynthetically active eukaryotes by eukaryotic host cells. Co-evolution of the host cell and the endosymbiont led to the establishment of complex plastids, which are surrounded by additional membranes in comparison to chloroplasts from land plants. Plastid proteins, encoded by the genome of the host cell have to be imported from the host cytoplasm into the complex plastid thereby crossing up to four plastid surrounding membranes. This resulted in an increased complexity of targeting signals as well as transport- and sorting machineries. Here we summarize current knowledge about protein transport into different types of complex plastids, indicating that pre-existing mechanisms were often reused and altered to fulfill new requirements.

Keywords Complex plastids · Protein translocation · Secondary endosymbiosis · Targeting sequences

Abbreviations

BTS	Bipartite targeting signal
CASH	Cryptophytes, alveolates, stramenopiles, and haptophytes
EGT	Endosymbiotic gene transfer
ER	Endoplasmatic reticulum
ERAD	ER-associated degradation
PPC	Periplastidal compartment
SELMA	Symbiont-specific ERAD-like machinery
SP	Signal peptide
TAT	Twin arginine targeting
TIC	Translocon of the inner chloroplast membrane

U.-G. Maier (✉) · F. Hempel · K. Bolte · S. Zauner
Laboratory for Cell Biology, Philipps University of Marburg, Karl-von-Frisch Strasse 8, 35032 Marburg, Germany
e-mail: maier@biologie.uni-marburg.de

A. Klingl · F. Hempel · U.-G. Maier
LOEWE-Zentrum für Synthetische Mikrobiologie (Synmikro), Hans-Meerwein-Strasse, 35032 Marburg, Germany

S.M. Theg, F.-A. Wollman (eds.), *Plastid Biology,* Advances in Plant Biology 5,
DOI 10.1007/978-1-4939-1136-3_11,

TOC	Translocon of the outer chloroplast membrane
TP	Transit peptide
TTDs	Thylakoid targeting domains

11.1 Evolution of Complex Plastids

During cellular evolution, major transitions were catalyzed by endosymbiotic events leading to new cellular entities. Here, either a prokaryote or a eukaryote was integrated into a eukaryotic host cell and reduced to an organelle, thereby creating a cellular chimera with new capacities. Beside the generally known organelles of prokaryotic origin—the chloroplasts, the mitochondrion as well as mitochondrion-related organelles (such as hydrogenosomes and mitosomes) [51]—eukaryotic endosymbionts are known from many algae and some intracellular parasites. These organisms arose in a process called secondary endosymbiosis, when either green or red algae were integrated into unicellular eukaryotes [22]. Here, the co-evolution of the two eukaryotic cells led to the conversion of the intracellular symbiont into an organelle. This co-evolution includes endosymbiotic gene transfer (EGT) as well as gene loss and the reduction of redundant endosymbiotic cellular structures [36, 37, 61]. The established new organelle, a secondary or complex plastid, provides for the cellular merger—at least initially—a new light driven machinery which is, in contrast to primary plastids, surrounded by additional membranes.

In the case of organisms with a red algal endosymbiont, the secondary plastids are surrounded by four membranes in cryptophytes, heterokontophytes and haptophytes (Figs. 11.1a, b and 11.2). This is also true for the apicoplast, the plastid of apicomplexa, which is photosynthetically inactive in the parasitic lines (Fig. 11.1c). However, in peridinin-containing dinoflagellates, major contributors to marine biomass, the plastids are surrounded by only three membranes (Fig. 11.1e).

In the case of organisms with a complex plastid of green algal origin two lines can be distinguished due to their plastid architecture. Whereas plastids of chlorarachniophytes are surrounded by four membranes, the plastids of euglenophytes possess only three surrounding membranes (Fig. 11.1d, e).

Phylogenetic relationships of organisms with complex plastids are currently highly debated [12, 39, 67]. In any case, organisms harboring a secondary or complex plastid evolved at least three times independently, but probably additional serial and potentially tertiary endosymbiotic events took place [3, 15, 22, 46, 48]. One of the key steps in the transition of an intracellular symbiont into an organelle is the evolution and development of protein transport systems, which allow correct import of nucleus-encoded proteins into the organelle by crossing all surrounding membranes. Here we summarize the current knowledge of the protein transport systems embedded within the membranes of complex plastids and show parallel evolution of transporters, which are re-cycled from pre-existing cellular transport mechanisms.

11.2 Protein Transport into Complex Plastids Surrounded by Four Membranes

Although of different phylogenetic origin, targeting mechanisms for nucleus-encoded plastid proteins share basic principles in organisms with complex plastids. Nucleus-encoded pre-proteins carry a bipartite targeting signal (BTS) at the N-terminus consisting of a signal peptide followed by a transit peptide-like sequence [5, 20, 24, 57]. The signal peptide mediates co-translational transport via the Sec61 complex into the ER, which is in case of cryptophytes, heterokontophytes and haptophytes equivalent to the translocation across the outermost plastid membrane as in these organisms the outermost plastid membrane is continuous with the ER membrane of the host cell [9, 17] (Fig. 11.1a, b). In apicomplexa and chlorarachniophytes the outermost membrane is located close to the endomembrane system although a fusion of the membranes does not exist [9] (Fig. 11.1c, d). Consequently, pre-protein transport from the endomembrane system to the outermost plastid membrane is mediated by vesicles, which seem to traverse the cytoplasm directly from the ER (i.e. without passing the Golgi apparatus) to the complex plastid in case of apicomplexa [11, 62]. Deletion of the transit peptide-like sequences results in secretion of the protein demonstrating the necessity of this part of the targeting signal for the targeting step [16, 63]. Further details regarding receptor proteins and other transport factors inside the ER-lumen are not known yet. A different model postulating that the apicoplast, the complex plastid of apicomplexa, itself lies within the ER lumen is described in [64]. Unfortunately, data providing details on pre-protein transport from the ER to the outermost plastid membrane of chlorarachniophytes are not available yet; however a vesicle-based transport is postulated [28–30].

After pre-protein transport across the first outermost plastid membrane the signal peptide is presumably cleaved off, thereby exposing the transit peptide-like sequence, which is essential for targeting across the second, third and fourth plastid membrane. The second outermost plastid membrane encloses the periplastidal compartment (PPC), the former symbiotic cytosol [38]. The complex plastids of cryptophytes and chlorarachniophytes still harbor a nucleomorph inside the PPC, representing the remnant nucleus of the red respectively green algal endosymbiont, which encodes a number of proteins that are expressed by a symbiont-specific machinery [13, 18, 34, 35, 60]. In all other groups the nucleus of the secondary endosymbiont was completely lost. The second outermost plastid membrane probably traces back to the former plasma membrane of the red or green algal endosymbiont respectively, which did not possess a protein translocation machinery by natural means. Hence, mechanisms for transporting pre-proteins across that membrane had to be evolved during secondary endosymbiosis. How nature solved that issue was cryptic for a long time. A first experimentally accessible hypothesis came from *in silico* analyses of the nucleomorph genome of the cryptophyte *Guillardia theta* revealing that core components of the red algal ERAD (ER-associated degradation) translocation machinery are still encoded on this minimized genome [54]. ERAD

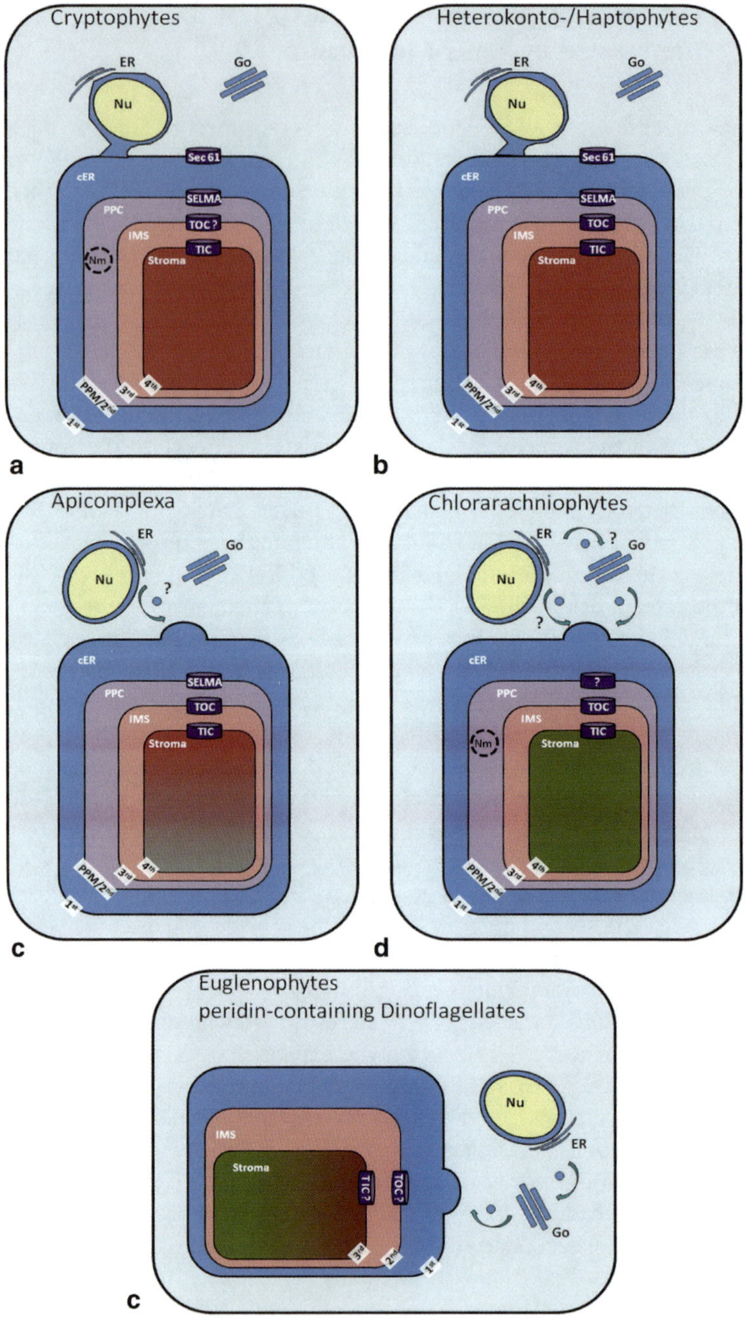

Fig. 11.1 Illustration of organisms harboring complex plastids with different membrane architectures and pre-protein transport systems. In a process called secondary endosymbiosis either a red alga (cryptophytes (**a**), heterokonto- and haptophytes (**b**), apicomplexa (**c**), peridinin-containing

in general represents a quality control system transporting misfolded proteins from the ER back into the cytosol, where such proteins get degraded [4]. Interestingly, the ER within the periplastidal compartment was completely eliminated raising the question why parts of this machinery were still retained. Further analyses revealed that also other organisms harboring complex plastids, namely heterokontophytes, haptophytes and apicomplexa, retained such symbiont-specific ERAD components in addition to the ERAD machinery of the host [15, 32, 54–56]. The ERAD translocation system fulfils all criteria postulated for a possible protein transporter within the second outermost membrane, leading to the hypothesis that the ERAD machinery of the red algal endosymbiont might have been recycled during evolution and relocated from the symbiotic ER to the second outermost membrane to fulfill a new function in pre-protein translocation [54]. First experimental proof for that hypothesis came from analyses in the diatom *Phaeodactylum tricornutum* demonstrating that core components of this symbiont-specific ERAD-like machinery (SELMA) are indeed localized within the postulated membrane and form an oligomeric complex which might possibly be part of a translocation channel [25]. Further *in vivo* analyses revealed that these proteins (sDer-1-1 and sDer-1-2) specifically interact with the transit peptide-like sequences of pre-proteins destined for the PPC [25]. Interestingly, transit peptide-like sequences of stroma-specific proteins (which have to be further transported across membranes three and four) showed no interaction with these SELMA-factors, which might be a discrimination mechanism for both protein populations. Furthermore, conditional knock out mutants for a sDer1 protein in the apicomplexan parasite *Toxoplasma gondii* demonstrated that pre-protein transport into the apicoplast is strictly dependent on this SELMA factor [1]. Additionally, it was shown very recently that ubiquitination of nucleus-encoded proteins is essential for protein import into the apicoplast [2] as it is postulated for *P. tricornutum* as well [26, 56]. Phylogenetic analyses showed that SELMA indeed originates from the ERAD system of the red algal endosymbiont [15]. Hence, the endogenous ERAD translocation system of the red alga symbiont was recycled and modified during secondary endosymbiosis giving an elegant example of "making new out of old" [6].

dinoflagellates (**e**)) or a green alga (chlorarachniophytes (**d**), euglenophytes (**e**)) was engulfed by a eukaryotic host cell. This led to organisms with complex plastids and numerous pre-protein import mechanisms due to a variety of membrane architectures. **a**–**b** For those plastids which are surrounded by four membranes and the outermost membrane stays in continuum with the ER-membrane of the host cell, pre-protein translocation across the first membrane is mediated by the Sec61 complex. In apicomplexa (**c**) and chlorarachniophytes (**d**) transport to the four membrane-bound plastid is mediated by the endomembrane system which can additionally involve the Golgi apparatus as in both groups no connection of the first plastid membrane to the ER of the host cell could be observed. **e** Peridinin-containing dinoflagellates and euglenophytes possess plastids which are surrounded by three membranes. Protein transport into these plastids also starts at the endomembrane system. Transport mechanisms across the remaining membranes are still unclear. *cER* chloroplast ER, *ER* endoplasmatic reticulum, *Go* Golgi apparatus, *IMS* intermembrane space, *Nu* Nucleus, *PPC* periplastidal compartment, *SELMA* symbiont-specific ERAD-like machinery, *TIC* translocon of the inner chloroplast membrane, *TOC* translocon of the outer chloroplast membrane

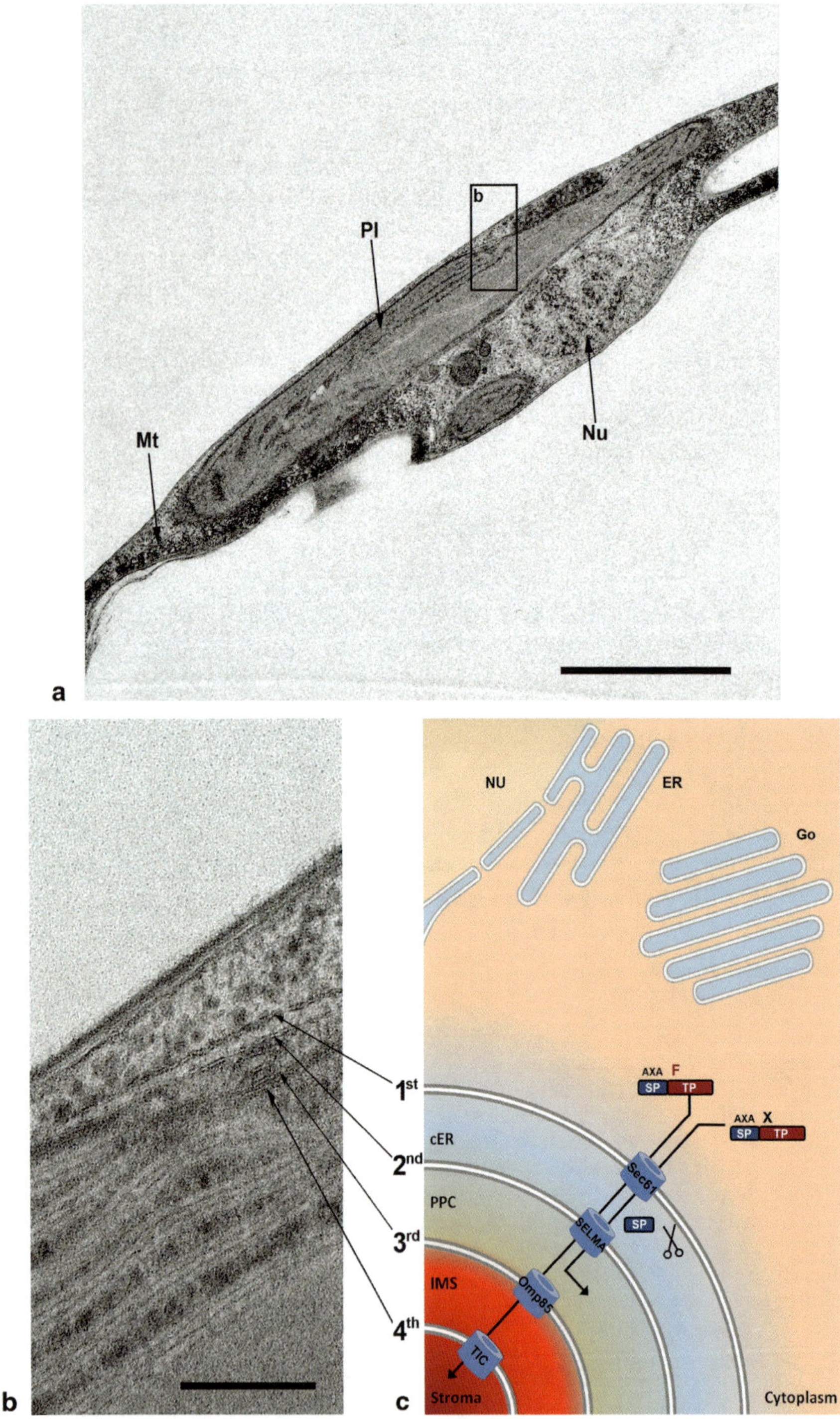

Fig. 11.2 Illustration of the membrane system surrounding the complex plastid of one member of the heterokontophytes, *Phaeodactylum tricornutum*. **a** Electron micrograph of a high-pressure frozen and ultrathin sectioned cell of *P. tricornutum*. Freeze-substitution, embedding and ultrathin

In chlorarachniophytes, which evolved by the engulfment of a green alga, little is known about pre-protein transport across the second outermost membrane, even though recent studies revealed novel targeting signals for discriminating PPC and stromal protein populations [28, 29]. As no protein transport machinery similar to SELMA could be identified so far [30], it is of huge interest how pre-protein transport across that membrane is managed.

Proteins destined for the plastids stroma and the thylakoids have to cross two further membranes, the third outermost and the innermost membrane of four membrane-bound plastids. These two membranes are generally accepted to be homologous to the plastid envelope of primary plastids. Again the transit peptide-like sequence, the second part of the BTS, is essential for this targeting step, mediating transport across membrane three and four (Fig. 11.2c). In contrast to periplastidal proteins, which are processed within the PPC, transit peptide-like sequences of stromal specific proteins carry an aromatic amino acid or a leucin at the +1 position at least in cryptophytes and heterokontophytes, which mostly occur in an AXA-F context [19, 23, 33]. Up to now, the+1 position of the transit peptide-like sequence is the only factor known for discriminating PPC-resistant proteins from stromal proteins and mutations at this position (from a non-aromatic amino acid into an aromatic one or vice versa) result in targeting the protein to the stroma or retaining the protein within the PPC, respectively [20].

Import of proteins into primary plastids is well studied in land plants, leading to a detailed knowledge on the mechanistical and structural characteristics of pre-protein translocation [50]. Traversing these two membranes requires two different translocons, TOC (Translocon of the outer chloroplast membrane) and TIC (translocon of the inner chloroplast membrane), which are embedded into the respective chloroplast membranes of primary plastids [31, 53].

Scanning the genomes of organisms with complex plastids for protein translocons which might be embedded into the two innermost membranes resulted in the identification of homologs of several components of the TIC translocon [13, 18, 30,

sectioning was performed according to [27]. When examining the indicated area in (**a**) at higher magnification, the four plastid membranes can be easily distinguished (**b**). In dependence on this electron micrograph, the protein transport in the complex plastid is depicted schematically (**c**). The outermost plastid membrane is connected with the ER membrane of the host cell. Nucleus-encoded plastid proteins possess a bipartite targeting signal (*BTS*) composed of a signal peptide (*SP*) followed by a transit peptide-like sequence (*TP*). Proteins destined for the plastid stroma usually contain a phenylalanine (*F*) at the +1 position of the transit peptide-like sequence, whereas targeting into the periplastidal compartment (*PPC*), requires a non-aromatic amino acid (*X*) at this position. The signal peptide mediates co-translational transport into the ER lumen, i.e. crossing the first outermost plastid membrane is mediated via the Sec61 complex. Further transport across the remaining membranes is directed by the transit peptide-like sequence. Transport across the second outermost membrane is facilitated by an ERAD-derived translocation system (SELMA). Finally, an Omp85/TIC-system is responsible for protein translocation across the third and fourth membrane. *cER* chloroplast ER, *ER* endoplasmatic reticulum, *Go* Golgi apparatus, *IMS* intermembrane space, *Mt* mitochondrium, *Nu* nucleus, *Omp85* outer membrane protein, *Pl* plastid, *PPC* periplastidal compartment, *SELMA* symbiont-specific ERAD-like machinery, *SP* signal peptide, *TIC* translocon of the inner chloroplast membrane, *TP* transit peptide-like sequence. Scale bars indicate 2 μm (**a**) and 200 nm (**b**)

32, 47, 65]. The significance of these results was shown by analyzing one TIC component of the apicomplexa demonstrating the protein localization within the innermost membrane and its necessity for pre-protein translocation [65]. In chlorarachniophytes, a nucleomorph-encoded Toc75 homolog—the core component of the TOC-complex—was identified in addition to a recently discovered nucleus encoded version [18, 30]. Data bases from organisms of the red lineages, however, initially refused to highlight a protein translocon component for the third outermost plastid membrane. Not until 2007, Wunder and co-workers mentioned a putative Omp85 homolog in diatoms [68]. This protein was subsequently characterized in detail in the diatom *Phaeodactylum tricornutum* [8]. *In vivo* localization studies demonstrated that the protein, forming a beta barrel like protein with two POTRA domains at the N-terminus, is indeed localized within the third outermost plastid membrane. Electrophysiological characteristics of ptOmp85 showed that core parameters like pore-size and cation selectivity are nearly equal to plant Toc75 proteins. Finally, it could be shown that the intracellular targeting mechanism of ptOmp85 into the third outermost membrane is as complicated as for Toc75 proteins in land plants i.e., the protein is transported first into the plastids stroma across all four plastid surrounding membranes and in a second step targeted to the respective membrane [8]. All together these data show overwhelming similarities to the Toc75 translocon of higher plants [8]. Scanning the genomes of further secondary evolved organisms revealed ptOmp85 homolog proteins in haptophytes and apicomplexa as well [8]. Altogether these studies suggest that protein transport across the two innermost membranes might be basically conserved in plastids surrounded by either two or four membranes.

After crossing the two innermost plastid surrounding membranes, the transit peptide-like sequence is cleaved off and stromal proteins are folded to their mature conformation. Although only initial work was done, especially in the identification of plastid specific peptidases [10], one would expect that mechanisms of folding and processing of stromal proteins are mechanistically comparable in complex plastids and primary plastids. Proteins with a final destination in the thylakoid lumen have to cross one additional membrane and possess therefore an additional targeting signal [7, 21]. These proteins harbor an N-terminal tripartite targeting sequence with either a Sec targeting domain or a twin arginine targeting (TAT) motif located C-terminal to the BTS. Thus, in complex plastids protein transport across the thylakoid membrane is managed for unfolded and folded proteins, the latter in a proposed picky pack mechanism [7, 21]. In any case, these results show that pre-existing mechanisms, here for thylakoid import, are still used in secondary plastids.

In 2005 it was shown for the first time that in plastids of higher plants proteins exist, which get glycosylated in the ER and are not imported via the classical TOC/TIC pathway but use an alternative vesicle mediated route [66]. Recently, it was demonstrated that also in complex plastids glycoproteins exist. For the diatom *P. tricornutum*, it was shown that nucleus-encoded proteins get N-glycosylated after transport across the outermost plastid membrane and are subsequently transported across all three further membranes into the plastid stroma [45]. This observation revealed new questions on the operating mode of transport systems like the TOC/TIC machinery, which is so far not believed to transport such bulky proteins, and uncovered new pressure on the evolution of translocation machineries. Also in other organisms with four membrane bound plastids N-glycosylated plastid proteins are predicted and in future it will be very interesting to elucidate mechanistic details on glycoprotein transport into complex plastids [44, 45].

11.3 Protein Transport into Complex Plastids Surrounded by Three Membranes

In comparison to the above-mentioned groups, the complex plastids of peridinin-containing dinoflagellates and photosynthetic active euglenophytes are surrounded by only three instead of four membranes [9]. Although both groups of organisms share a comparable membrane topology they are phylogenetically not related to each other. While peridinin-containing dinoflagellates harbor a plastid of red algal origin, the euglenophytes contain one of green algal origin. As in apicomplexa and chlorarachniophytes the outermost plastid membrane is not fused to the ER of the host cell, consequently pre-protein transport from the secretory system to the plastid has to be carried out via vesicles. In both groups there is evidence suggesting that those vesicles bud from the Golgi apparatus [41, 42, 52, 58, 59] and are thought to fuse with the outermost plastid membrane. The analogy between peridinin-containing dinoflagellates and euglenophytes in terms of three membrane topology and Golgi-mediated transport also led to comparable characteristics of the targeting signals of nuclear encoded plastid proteins. The pre-proteins are grouped in three classes in respect to their targeting information. All of them contain a bipartite targeting signal connatural to that of the above described groups of organisms with four membrane bound plastids. Additional to that, class I pre-proteins possess a so called stop-transfer domain in direct proximity to the transit peptide like part of the BTS [14, 43]. This stop-transfer halts co-translational import into the ER lumen, therefore causing functional domains of mature proteins to have a cytosolic localization for the duration of transport to the plastid [40, 59]. In contrast, class II proteins contain no such hydrophobic stop-transfer sequences and are expected to be fully translated into the ER lumen and are therefore encapsulated by vesicles and the Golgi apparatus during transport to the outermost plastid membrane [14, 40]. To differentiate them from stromal proteins, proteins with thylakoid targeting domains (TTDs) have been termed class III proteins in dinoflagellates [43] or class IB in euglenophytes [14].

Although the transport across the second and third innermost membrane could also be accomplished in the same manner as in primary plastids there is still no evidence for this. The main reason for this is the lack of large scale genome information which makes it difficult to identify subunits or candidates for a possible translocon of one of these membranes. In any case, further investigations are needed to close the gaps in knowledge on protein transport across the three membranes of the plastids of peridinin-containing dinoflagellates and euglenophytes. By doing so, it also could be possible to determine the origin of the two innermost plastid membranes, which might have originated either from a eukaryotic, a prokaryotic membrane or by fusions of membranes.

11.4 Conclusions

In the last decade substantial progress was made in understanding protein import into complex plastids. One common theme certainly is that, although of different phylogenetic origin, organisms tend to recycle already established transport systems for new duties instead of inventing completely new mechanisms. Nevertheless, this still might be only the tip of the iceberg and further investigations will surely reveal mechanistic details and might highlight additional mechanisms that evolved during secondary endosymbiosis as seen by recent contributions [45, 49].

References

1. Agrawal S, van Dooren GG, Beatty WL, Striepen B (2009) Genetic evidence that an endosymbiont-derived endoplasmic reticulum-associated protein degradation (ERAD) system functions in import of apicoplast proteins. J Biol Chem 284:33683–33691
2. Agrawal S et al (2013) An apicoplast localized ubiquitylation system is required for the import of nuclear-encoded plastid proteins. PLoS Pathog 9:e1003426
3. Archibald JM (2009) The puzzle of plastid evolution. Curr Biol 19:R81–R88
4. Bagola K, Mehnert M, Jarosch E, Sommer T (2011) Protein dislocation from the ER. Biochim Biophys Acta 1808:925–936
5. Bolte K, Bullmann L, Hempel F, Bozarth A, Zauner S, Maier UG (2009) Protein targeting into secondary plastids. J Eukaryot Microbiol 56:9–15
6. Bolte K, Gruenheit N, Felsner G, Sommer MS, Maier UG, Hempel F (2011) Making new out of old: recycling and modification of an ancient protein translocation system during eukaryotic evolution. Mechanistic comparison and phylogenetic analysis of ERAD, SELMA and the peroxisomal importomer. Bioessays 33:368–376
7. Broughton MJ, Howe CJ, Hiller RG (2006) Distinctive organization of genes for light-harvesting proteins in the cryptophyte alga Rhodomonas. Gene 369:72–79
8. Bullmann L, Haarmann R, Mirus O, Bredemeier R, Hempel F, Maier UG, Schleiff E (2010) Filling the gap, evolutionarily conserved Omp85 in plastids of chromalveolates. J Biol Chem 285:6848–6856
9. Cavalier-Smith T (2003) Genomic reduction and evolution of novel genetic membranes and protein-targeting machinery in eukaryote-eukaryote chimaeras (meta-algae). Philos Trans R Soc Lond B Biol Sci 358:109–133 (discussion 133–144)
10. Chaal BK, Ishida K, Green BR (2003) A thylakoidal processing peptidase from the heterokont alga *Heterosigma akashiwo*. Plant Mol Biol 52:463–472
11. DeRocher A, Gilbert B, Feagin JE, Parsons M (2005) Dissection of brefeldin A-sensitive and -insensitive steps in apicoplast protein targeting. J Cell Sci 118:565–574
12. Deschamps P, Moreira D (2012) Reevaluating the green contribution to diatom genomes. Genome Biol Evol 4:683–688
13. Douglas S et al (2001) The highly reduced genome of an enslaved algal nucleus. Nature 410:1091–1096
14. Durnford DG, Gray MW (2006) Analysis of *Euglena gracilis* plastid-targeted proteins reveals different classes of transit sequences. Eukaryot Cell 5:2079–2091
15. Felsner G, Sommer MS, Gruenheit N, Hempel F, Moog D, Zauner S, Martin W, Maier UG (2010) ERAD components in organisms with complex red plastids suggest recruitment of a preexisting protein transport pathway for the periplastid membrane. Genome Biol Evol 3:140–150

16. Foth BJ, Ralph SA, Tonkin CJ, Struck NS, Fraunholz M, Roos DS, Cowman AF, McFadden GI (2003) Dissecting apicoplast targeting in the malaria parasite *Plasmodium falciparum*. Science 299:705–708
17. Gibbs SP (1979) The route of entry of cytoplasmically synthesized proteins into chloroplasts of algae possessing chloroplast ER. J Cell Sci 35:253–266
18. Gilson PR, Su V, Slamovits CH, Reith ME, Keeling PJ, McFadden GI (2006) Complete nucleotide sequence of the chlorarachniophyte nucleomorph: nature's smallest nucleus. Proc Natl Acad Sci U S A 103:9566–9571
19. Gould SB, Sommer MS, Kroth PG, Gile GH, Keeling PJ, Maier UG (2006) Nucleus-to-nucleus gene transfer and protein retargeting into a remnant cytoplasm of cryptophytes and diatoms. Mol Biol Evol 23:2413–2422
20. Gould SB, Sommer MS, Hadfi K, Zauner S, Kroth PG, Maier UG (2006) Protein targeting into the complex plastid of cryptophytes. J Mol Evol 62:674–681
21. Gould SB, Fan E, Hempel F, Maier UG, Klosgen RB (2007) Translocation of a phycoerythrin alpha subunit across five biological membranes. J Biol Chem 282:30295–30302
22. Gould SB, Waller RF, McFadden GI (2008) Plastid evolution. Annu Rev Plant Biol 59: 491–517
23. Gruber A, Vugrinec S, Hempel F, Gould SB, Maier UG, Kroth PG (2007) Protein targeting into complex diatom plastids: functional characterisation of a specific targeting motif. Plant Mol Biol 64:519–530
24. Hempel F, Bozarth A, Sommer MS, Zauner S, Przyborski JM, Maier UG (2007) Transport of nuclear-encoded proteins into secondarily evolved plastids. Biol Chem 388:899–906
25. Hempel F, Bullmann L, Lau J, Zauner S, Maier UG (2009) ERAD-derived preprotein transport across the second outermost plastid membrane of diatoms. Mol Biol Evol 26:1781–1790
26. Hempel F, Felsner G, Maier UG (2010) New mechanistic insights into pre-protein transport across the second outermost plastid membrane of diatoms. Mol Microbiol 76:793–801
27. Hempel F, Bozarth AS, Lindenkamp N, Klingl A, Zauner S, Linne U, Steinbuchel A, Maier UG (2011) Microalgae as bioreactors for bioplastic production. Microb Cell Fact 10:81
28. Hirakawa Y, Nagamune K, Ishida K (2009) Protein targeting into secondary plastids of chlorarachniophytes. Proc Natl Acad Sci U S A 106:12820–12825
29. Hirakawa Y, Gile GH, Ota S, Keeling PJ, Ishida K (2010) Characterization of periplastidal compartment-targeting signals in chlorarachniophytes. Mol Biol Evol 27:1538–1545
30. Hirakawa Y, Burki F, Keeling PJ (2012) Genome-based reconstruction of the protein import machinery in the secondary plastid of a chlorarachniophyte alga. Eukaryot Cell 11:324–333
31. Inaba T, Schnell DJ (2008) Protein trafficking to plastids: one theme, many variations. Biochem J 413:15–28
32. Kalanon M, Tonkin CJ, McFadden GI (2009) Characterization of two putative protein translocation components in the apicoplast of *Plasmodium falciparum*. Eukaryot Cell 8:1146–1154
33. Kilian O, Kroth PG (2005) Identification and characterization of a new conserved motif within the presequence of proteins targeted into complex diatom plastids. Plant J 41:175–183
34. Lane CE, van den Heuvel K, Kozera C, Curtis BA, Parsons BJ, Bowman S, Archibald JM (2007) Nucleomorph genome of *Hemiselmis andersenii* reveals complete intron loss and compaction as a driver of protein structure and function. Proc Natl Acad Sci U S A 104:19908–19913
35. Maier UG, Zauner S, Woehle C, Bolte K, Hempel F, Allen JF, Martin WF (2013) Massively convergent evolution for ribosomal protein gene content in plastid and mitochondrial genomes. Genome Biol Evol 5:2318–2329
36. Martin W (2003) Gene transfer from organelles to the nucleus: frequent and in big chunks. Proc Natl Acad Sci U S A 100:8612–8624
37. Martin W, Stoebe B, Goremykin V, Hapsmann S, Hasegawa M, Kowallik KV (1998) Gene transfer to the nucleus and the evolution of chloroplasts. Nature 393:162–165
38. Moog D, Stork S, Zauner S, Maier UG (2011) In silico and in vivo investigations of proteins of a minimized eukaryotic cytoplasm. Genome Biol Evol 3:375–382

39. Moustafa A, Beszteri B, Maier UG, Bowler C, Valentin K, Bhattacharya D (2009) Genomic footprints of a cryptic plastid endosymbiosis in diatoms. Science 324:1724–1726
40. Nassoury N, Cappadocia M, Morse D (2003) Plastid ultrastructure defines the protein import pathway in dinoflagellates. J Cell Sci 116:2867–2874
41. Nassoury N, Wang Y, Morse D (2005) Brefeldin a inhibits circadian remodeling of chloroplast structure in the dinoflagellate *Gonyaulax*. Traffic 6:548–561
42. Osafune T, Schiff JA, Hase E (1991) Stage-dependent localization of LHCP II apoprotein in the Golgi of synchronized cells of *Euglena gracilis* by immunogold electron microscopy. Exp Cell Res 193:320–330
43. Patron NJ, Waller RF, Archibald JM, Keeling PJ (2005) Complex protein targeting to dinoflagellate plastids. J Mol Biol 348:1015–1024
44. Peschke M, Hempel F (2013) Glycoprotein import: a common feature of complex plastids? Plant Signal Behav 8:e26050
45. Peschke M, Moog D, Klingl A, Maier UG, Hempel F (2013) Evidence for glycoprotein transport into complex plastids. Proc Natl Acad Sci U S A 110:10860–10865
46. Petersen J, Ludewig AK, Michael V, Bunk B, Jarek M, Baurain D, Brinkmann H (2014) Chromera velia, endosymbioses and the rhodoplex hypothesis-plastid evolution in cryptophytes, alveolates, stramenopiles, and haptophytes (CASH lineages). Genome Biol Evol 6:666–684
47. Rogers MB, Gilson PR, Su V, McFadden GI, Keeling PJ (2007) The complete chloroplast genome of the chlorarachniophyte *Bigelowiella natans*: evidence for independent origins of chlorarachniophyte and euglenid secondary endosymbionts. Mol Biol Evol 24:54–62
48. Sanchez-Puerto MV, Delwiche CF (2008) A hypothesis for plastid evolution in chromalveolates. J Phycol 44:1097–1107
49. Sheiner L, Demerly JL, Poulsen N, Beatty WL, Lucas O, Behnke MS, White MW, Striepen B (2011) A systematic screen to discover and analyze apicoplast proteins identifies a conserved and essential protein import factor. PLoS Pathog 7:e1002392
50. Shi LX, Theg SM (2013) The chloroplast protein import system: from algae to trees. Biochim Biophys Acta 1833:314–331
51. Shiflett AM, Johnson PJ (2010) Mitochondrion-related organelles in eukaryotic protists. Annu Rev Microbiol 64:409–429
52. Slavikova S, Vacula R, Fang Z, Ehara T, Osafune T, Schwartzbach SD (2005) Homologous and heterologous reconstitution of Golgi to chloroplast transport and protein import into the complex chloroplasts of Euglena. J Cell Sci 118:1651–1661
53. Soll J, Schleiff E (2004) Protein import into chloroplasts. Nat Rev Mol Cell Biol 5:198–208
54. Sommer MS, Gould SB, Lehmann P, Gruber A, Przyborski JM, Maier UG (2007) Der1-mediated preprotein import into the periplastid compartment of chromalveolates? Mol Biol Evol 24:918–928
55. Spork S, Hiss JA, Mandel K, Sommer M, Kooij TW, Chu T, Schneider G, Maier UG, Przyborski JM (2009) An unusual ERAD-like complex is targeted to the apicoplast of *Plasmodium falciparum*. Eukaryot Cell 8:1134–1145
56. Stork S, Moog D, Przyborski JM, Wilhelmi I, Zauner S, Maier UG (2012) Distribution of the SELMA translocon in secondary plastids of red algal origin and predicted uncoupling of ubiquitin-dependent translocation from degradation. Eukaryot Cell 11:1472–1481
57. Stork S, Lau J, Moog D, Maier UG (2013) Three old and one new: protein import into red algal-derived plastids surrounded by four membranes. Protoplasma 250:1013–1023
58. Sulli C, Schwartzbach SD (1995) The polyprotein precursor to the Euglena light-harvesting chlorophyll a/b-binding protein is transported to the Golgi apparatus prior to chloroplast import and polyprotein processing. J Biol Chem 270:13084–13090
59. Sulli C, Fang Z, Muchhal U, Schwartzbach SD (1999) Topology of Euglena chloroplast protein precursors within endoplasmic reticulum to Golgi to chloroplast transport vesicles. J Biol Chem 274:457–463

60. Tanifuji G, Onodera NT, Wheeler TJ, Dlutek M, Donaher N, Archibald JM (2011) Complete nucleomorph genome sequence of the nonphotosynthetic alga *Cryptomonas paramecium* reveals a core nucleomorph gene set. Genome Biol Evol 3:44–54
61. Timmis JN, Ayliffe MA, Huang CY, Martin W (2004) Endosymbiotic gene transfer: organelle genomes forge eukaryotic chromosomes. Nat Rev Genet 5:123–135
62. Tonkin CJ, Struck NS, Mullin KA, Stimmler LM, McFadden GI (2006) Evidence for Golgi-independent transport from the early secretory pathway to the plastid in malaria parasites. Mol Microbiol 61:614–630
63. Tonkin CJ, Roos DS, McFadden GI (2006) N-terminal positively charged amino acids, but not their exact position, are important for apicoplast transit peptide fidelity in *Toxoplasma gondii*. Mol Biochem Parasitol 150:192–200
64. Tonkin CJ, Kalanon M, McFadden GI (2008) Protein targeting to the malaria parasite plastid. Traffic 9:166–175
65. van Dooren GG, Tomova C, Agrawal S, Humbel BM, Striepen B (2008) *Toxoplasma gondii* Tic20 is essential for apicoplast protein import. Proc Natl Acad Sci U S A 105:13574–13579
66. Villarejo A et al (2005) Evidence for a protein transported through the secretory pathway en route to the higher plant chloroplast. Nat Cell Biol 7:1224–1231
67. Woehle C, Dagan T, Martin WF, Gould SB (2011) Red and problematic green phylogenetic signals among thousands of nuclear genes from the photosynthetic and apicomplexa-related *Chromera velia*. Genome Biol Evol 3:1220–1230
68. Wunder T, Martin R, Loffelhardt W, Schleiff E, Steiner JM (2007) The invariant phenylalanine of precursor proteins discloses the importance of Omp85 for protein translocation into cyanelles. BMC Evol Biol 7:236

Chapter 12
Processing and Degradation of Chloroplast Extension Peptides

Kentaro Inoue and Elzbieta Glaser

Abstract Most chloroplast proteins are synthesized as larger precursors with cleavable extension peptides. These extensions include import signals called transit peptides, export signals for thylakoid transfer, and the C-terminal extension of the chloroplast-encoded D1 subunit of the photosystem II. Transit peptides are necessary for transport of nuclear-encoded proteins from the cytoplasm across the double-membrane envelope, and are cleaved off by Stromal Processing Peptidase (SPP) in the stroma. Further degradation of transit peptides involves SPP and Presequence Protease (PreP). Thylakoid-transfer sequences are required for correct intraorganellar protein sorting and cleaved by Thylakoidal Processing Peptidase (TPP) in the thylakoid lumen. The C-terminal extension of the D1 protein is not required for precursor targeting and integration into the protein complex; however its removal by Carboxyl-terminal peptidase called CtpA in the thylakoid lumen is needed for proper formation of the photosystem II Mn_4CaO_5 cluster. Biochemical studies in the 1980s–1990s defined basic properties of SPP, TPP and CtpA, while PreP was discovered in the early 2000s. Recent molecular genetic studies demonstrated physiological importance as well as some unprecedented functions of these enzymes. This chapter gives a comprehensive survey on processing and degradation of chloroplast extension peptides. The emphasis is on biochemical, molecular and evolutionary aspects of proteases. The significances of the presence and processing of these extension peptides are also discussed.

Keywords C-terminal extension · Export signal · Import signal · Plastidic type I signal peptidase (Plsp) · Presequence protease (PreP) · Stromal processing peptidase (SPP) · The D1 carboxyl-terminal peptidase (CtpA) · Thylakoidal processing peptidase (TPP) · Thylakoid-transfer sequence · Transit peptide

K. Inoue (✉)
Department of Plant Sciences, University of California, One Shields Avenue, Davis, CA 95616, USA
e-mail: kinoue@ucdavis.edu

E. Glaser
Department of Biochemistry and Biophysics, Stockholm University, SE-106 91 Stockholm, Sweden

S.M. Theg, F.-A. Wollman (eds.), *Plastid Biology*, Advances in Plant Biology 5,
DOI 10.1007/978-1-4939-1136-3_12,

Abbreviations

Aβ	Amyloid beta peptide
ACP	Acyl Carrier Protein
AD	Alzheimer's disease
cpSEC	Chloroplastic SEC
cpSRP	Chloroplastic Signal Recognition Particle
cpTAT	Chloroplastic Twine-Arginine-Transport
CtpA	The D1 Carboxyl-terminal peptidase
GFP	Green Fluorescent Protein
HSP21	Heat Shock Protein 21
MPP	Mitochondrial Processing Peptidase
MS	Mass Spectrometry
OEC	Oxygen Evolving Complex
Plsp	Plastidic type I signal peptidase
PreP	Presequence Protease
RBCS	Small subunit of Rubisco
ROS	Reactive Oxygen Species
Rubisco	Ribulose-1,5-bisphosphate carboxylase/oxygenase
S2P	Site-2-Protease
SPase I	Type I signal peptidase
SPP	Stromal Processing Peptidase
SStpPs	Transit peptide of the RBCS precursor
TIC	Transocon at the Inner-envelope-membrane of Chloroplasts
TOC	Translocon at the Outer-envelope-membrane of Chloroplasts
TPP	Thylakoidal Processing Peptidase

12.1 Introduction

In nearly all cells, numerous proteins are synthesized as larger precursors with cleavable extension peptides. Most of these extensions are localized N-terminally and function as targeting signals, which are necessary for secretion, intracellular targeting, or intraorganellar sorting in the case of eukaryotic proteins [68]. Some other extension peptides are not required for targeting but their removal is essential for the proteins to become active (e.g., [4]). Hence, the presence and removal of extension peptides are important processes for protein functions in general. Three types of extension peptides are known for chloroplast proteins: import signals, export signals and the C-terminal extension of the chloroplast-encoded D1 protein. Processing and degradation of these signals inside the organelle are the subject of this chapter.

The chloroplast import signals are called transit peptides and are present in the N termini of most nuclear-encoded precursor proteins (Fig. 12.1a). Transit peptides vary in length and no consensus motif can be identified among them. However, they share several common characteristics in that they are overall rich in hydroxylated residues, contain relatively large number of small residues and hydrophobic residues near the N and C termini, respectively, and are largely deficient in acidic amino acids [11, 84]. Transit peptides are necessary and sufficient for TOC/TIC-dependent protein transport across the double-membrane of the chloroplast envelope (see Chap. 9), and are cleaved off the precursor proteins in the stroma by Stromal Processing Peptidase (SPP) (Fig. 12.1b). The released peptides are further degraded by SPP and the organellar peptidasome called Presequence Protease (PreP) (Fig. 12.1b).

Protein transport from the chloroplast stroma to the thylakoid lumen is considered as export [68]. This is because the stroma and lumen are homologous to the cytosol and extracytosolic compartment, respectively, as indicated by the direction of proton transport during the electron transfer and by the presence of homologous protein transport systems in thylakoid and bacterial plasma membranes. Thylakoid-transfer sequences are necessary for proper intraorganeller sorting of precursor proteins via the chloroplastic SEC (cpSEC), chloroplastic Twine-Arginine-Transport (cpTAT) or non-assisted spontaneous pathways. These sequences are processed in the lumen by Thylakoidal Processing Peptidase (TPP) (Fig. 12.1c). All known proteins in the thylakoid lumen such as oxygen evolving complex (OEC) subunits, plastocyanin, DEG protease and the D1 Carboxyl-terminal peptidase (CtpA) are nuclear-encoded [41, 54, 70]. These and some but not all nuclear-encoded integral membrane proteins in thylakoids, such as FtsH proteases and the CFoII subunit of ATP synthase, carry N-terminal cleavable transport signals, which consist of a stroma-targeting transit peptide followed by a thylakoid-transfer sequence [61]. Thylakoid-transfer sequences show typical characteristics of export signals, comprising three domains: the N-terminal positively charged domain (n), the central hydrophobic core (h), and more polar C domain (c) ending with the conserved cleavage site (Fig. 12.1a). Some but not all thylakoid-transfer sequences contain an additional acidic domain preceding the N domain [28]. A thylakoid-transfer sequence is also found in the chloroplast-encoded cytochrome f, which uses the cpSEC pathway for thylakoid transport and is anchored to the membrane by its C-terminal transmembrane domain [62]. Unlike nuclear-encoded thylakoid proteins, the cytochrome f precursor does not carry a transit peptide for an obvious reason: it is synthesized in the chloroplast stroma and thus does not need to be imported via the TOC/TIC machinery.

The D1 protein forms the reaction center of the photosystem II, binding to the Mn_4Ca cluster to catalyze water oxidation. In higher plants, this protein is encoded in the chloroplast genome with a C-terminal extension of 8–9 residues [67] (Fig. 12.1a), and appears to be inserted into the thylakoid membrane co-translationally by the cpSRP pathway [50]. The C-terminal extension does not affect membrane insertion and further incorporation of the D1 precursor into the photosystem II, but its removal in the lumen by CtpA (Fig. 12.1c) is required for proper formation of the Mn_4Ca cluster, thus water oxidation [51].

This chapter surveys what we know about processing and degradation of the chloroplast extension peptides. We focus on properties of the four enzymes involved in these processes (SPP, PreP, TPP and CtpA) (Fig. 12.1) and discuss significances of the presence and cleavage of the extension sequences.

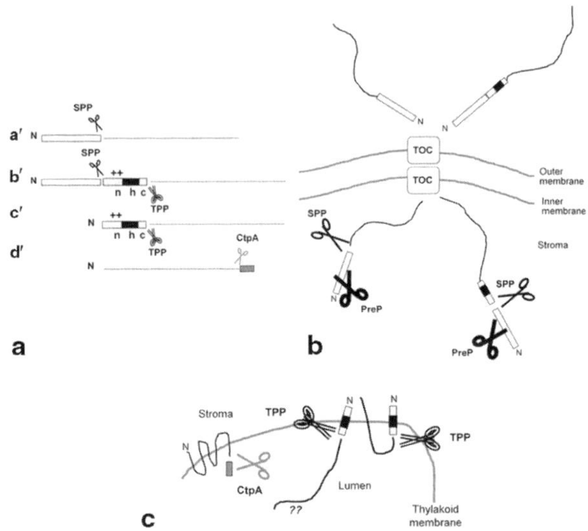

Fig. 12.1 Chloroplast extension peptides and their processing enzymes. **a** Major chloroplast precursor proteins with extension peptides. (a′) A typical nuclear-encoded chloroplast precursor protein. Its N-terminal import signal is called transit peptide (*white box*), which is cleaved by Stromal Processing Peptidase (SPP). (b′) A nuclear-encoded precursor protein sorted to the thylakoid lumen. Its N-terminal transport signal consists of a transit peptide (*white box*) followed by a thylakoid-transfer sequence, which is cleaved by Thylakoidal Processing Peptidase (*TPP*). The thylakoid-transfer sequence comprises the N-terminal positive domain (*n*), the central hydrophobic core (*h*, indicated as a *black box*), and more polar C-terminal domain (*c*). (c′) The cytochrome f precursor. Its N terminus functions as a thylakoid-transfer sequence. (d′) The D1 protein precursor. Its C terminus is removed by a carboxyl-terminal peptidase called CtpA. **b** Processing and degradation of the chloroplast extension peptides in the stroma. Nuclear-encoded precursor proteins that carry transit peptides are translocated across the double-membrane envelope by TOC and TIC complexes. In the stroma, the import signals are cleaved off the precursors by SPP. Further degradation of the cleaved signals involves SPP and PreP. **c** Processing of the chloroplast extension peptides in the thylakoid lumen.The export signals are cleaved off in the thylakoid lumen by TPP. Whether the TPP processing occurs during or after the completion of the protein translocation remains unknown. The C-terminal extension of the D1 subunit of the photosystem II is cleaved off in the lumen by CtpA. The fates of the released extension peptides remain unknown

12.2 Processing and Degradation of Import Signals by SPP

Newly imported chloroplast precursor proteins containing transit peptides, independently whether they are destined to stroma, thylakoid membrane or thylakoid lumen, are proteolytically processed in the stroma by SPP (Fig. 12.1b). SPP was originally identified in the pea (*Pisum sativum*) chloroplast which cleaved the precursor for the major light-harvesting chlorophyll binding protein [52]. The involvement of SPP in maturation of the small subunit of Rubisco (RBCS) and the acyl carrier protein (ACP) was also demonstrated by immunodepletion assays [52]. Further studies revealed that the enzyme processes various precursor proteins destined for plastids of both photosynthetic and non-photosynthetic tissues, suggesting that it is a key component of the import machinery [57, 83]. SPP from pea is a 124 kDa soluble protein and its activity is highly sensitive to metal chelators such as 1,10-phenanthroline and EDTA. Indeed, the enzyme contains a zinc-binding motif (His-X-X-Glu-His) in the catalytic site, classifying it to the pitrilysin metalloprotease family (MEROPS metallopeptidase family M16), which includes *Escherichia coli* protease III, insulin-degrading enzyme, MPP (Mitochondrial Processing Peptidase) and PreP (Presequence Protease; see below) [57–60, 83].

SPP has been shown to use a single endoproteolytic step to cleave transit peptides off a wide variety of precursor proteins such as those of RBCS, Rubisco activase, the $CF_1\gamma$-subunit of ATP synthase, OE33 (the 33-kD subunit of OEC), plastocyanin, ACP1, 3-deoxy-d-arabino-heptulosonate 7-phosphate synthase, ferredoxin, HSP21 (heat shock protein 21) [57]. After cleavage, the mature protein is released, but the transit peptide remains bound to SPP and is further trimmed to a subfragment form without its original C terminus. The trimming of the transit peptide by SPP triggers the release of the peptide from SPP and the subfragment has been proposed to be further degraded in the stroma by an ATP-dependent metallopeptidase [58, 59]. SPP itself is also synthesized as a large precursor with a transit peptide and cleaves off its transit peptide in *trans* [57].

Examination of binding and processing characteristics of several chloroplast precursor proteins was shown to be dependent on specific interactions of SPP with the region consisting of the C-terminal 10–15 residues of the transit peptides, independent of the amino acid sequence or length of the transit peptide [59]. In silico analysis of processing determinants identified a loosely conserved sequence consensus for the SPP cleavage, (Val/Ile)-X-(Ala/Cys)↓Ala, in which the Val at −3 and the Ala at −1 to the cleavage site (↓) seemed to be most conserved [23]. However, the prediction analysis was based on the N-terminal amino acid sequence of the endogenous mature chloroplast proteins. Hence it did not consider possible N-terminal modifications, including the N-terminal processing after transit peptide cleavage, important for stability of proteins and chloroplast viability. A neural network-based analysis called ChloroP that used a set of 62 chloroplast precursor proteins revealed a clear difference between the predicted and the SWISS-PROT annotated cleavage site, essentially shifted by one residue, and proposed a motif Val-Arg↓Ala-Ala-Ala-Val-X-X, which gave 60% correct predictions [18]. Large-scale mass spectrometry (MS) analysis of the Arabidopsis (*Arabidopsis thaliana*) chloroplast proteome identified some new features of the cleavage sites [93]. In this study, N-terminal acetylation was identified and taken as evidence for an authentic N terminus of the mature protein. The amino acid sequences from 47 N-acetylated proteins revealed conservation of a motif [Val/Ile]-X-[Ala/Cys/Ser]↓[Ala/Val/Ser]-[Ala/Ser/Val/Leu]-[Ser/Thr/Ala/Val] that likely represents a combination of a consensus motif for the SPP and the N-acetylase. This motif was essentially similar to a motif deciphered from N-terminal peptides of 62 unmodified proteins [93]. Creating a sequence logo of the predicted transit peptides from a dataset of 898 annotated Arabidopsis chloroplast proteins, a consensus with an enriched Arg at the position −1 to the cleavage site was observed, although the experimental data do not support such a strong preference for Arg [93].

There is also a group of proteins that are targeted to both mitochondria and chloroplasts using an ambiguous targeting sequence recognized by both organelles. Investigation of determinants for processing of such a dually targeted protein, pea glutathione reductase, revealed that the 60 amino acid-long N-terminal targeting peptide was recognized by separate information patterns by the MPP in mitochondria and by SPP in chloroplasts [65]. Numerous single mutations of amino acid residues in proximity of the cleavage site did not affect processing by SPP, whereas processing by MPP was severely inhibited. This showed that processing by SPP has a low level of sensitivity to single mutations on the targeting peptide and led to the conclusion that recognition of precursors by SPP is likely to involve recognition of the physiochemical properties of the sequence in the vicinity of the cleavage site rather than a requirement for specific amino acid residues [65].

Studies of the role of SPP *in vivo* demonstrated that SPP is essential for chloroplast biogenesis and plant survival [92]. When SPP cDNA antisense constructs were introduced into Arabidopsis, a large percentage of the transgenic plants were seedling lethal. Surviving plants exhibited slower growth, aberrant leaf morphology, abnormal pigmentation and not fully developed chloroplasts with accumulated starch granules. Protein import into plastids was also affected in antisense plants; a transit peptide-GFP (Green Fluorescent Protein) fusion construct was not imported into chloroplasts, but remained in the cytosol. This could indicate that SPP cleavage is required for progression of the protein import or proper assembly of TOC/TIC import machinery because all known TIC subunits are synthesized with transit peptides [92]. A rice mutant having a Glu deletion in the highly conserved C terminus of SPP exhibited chlorosis associated with small, underdeveloped chloroplasts as well as defective root development [89]. A recent study on T-DNA mutagenized Arabidopsis further demonstrated that SPP plays an essential role in embryo development: the *spp*-null Arabidopsis mutant embryos exhibited delayed development, with cell divisions not completing properly beyond the 16-cell stage [81].

12.3 Degradation of Import Signals by PreP

In 1999, Richter and Lamppa showed that the chloroplast extract rapidly degraded both the ferredoxin and HSP21 transit peptides and their respective subfragments created after trimming of the transit peptides by SPP [58]. The degradation was suggested to be catalyzed by a soluble ATP-dependent metallopeptidase. The proteolytic activity did not show any sequence specificity, but had some preference for larger oligopeptides and it was effectively inhibited by both the N-and C-terminal regions of the transit peptide, as well as other unrelated oligopeptides [59]. However, molecular identity of the peptidase remains unknown.

In 2002, a novel organellar peptidasome called the Presequence Protease (PreP) was isolated from potato (*Solanum tuberosum*) as a metalloprotease degrading presequences in the mitochondrial matrix [76]. Its molecular identification was achieved by electrospray ionization-MS/MS and database search. Its homolog in

Arabidopsis, *At*PreP (At3g19170), was found to contain an inverted zinc-binding motif (His-X-X-Glu-His) and thus belongs to the pitrilysin protease family (subfamily M16C) [76]. Interestingly, recombinant *At*PreP was shown to degrade not only mitochondrial presequences, but also other unstructured peptides of 10–65 amino acid residues including the transit peptide of the RBCS precursor (SStpPs) *in vitro*. PreP had no substrate specificity, but showed a preference for positively charged amino acids at P1′ position (+1 to the cleavage site) and small, nonpolar residues or Ser at P1 position (−1 to the cleavage site) [49, 77]. The presence of PreP in the spinach (*Spinacia oleracea*) chloroplast stroma was demonstrated by immunoblotting using the antibody against *At*PreP and the assay for the SStpPs degradation activity, which was completely abolished by the *At*PreP antibody [49]. Interestingly, however, in contrast to the result of Richter et al. [59], degradation of SStpPs was not dependent on ATP. There are two isoforms of PreP in Arabidopsis, *At*PreP1 and *At*PreP2, both of which contain 85 amino acid-long N-terminal cleavable targeting peptides. *In vitro* and *in vivo* import studies of GFP-fused *At*PrePs demonstrated that both isoenzymes are dually targeted to mitochondria and chloroplasts using an ambiguous targeting signal [9, 10]. Furthermore, the N-terminal targeting peptide of the *At*PreP1 precursor was shown to be organized in domains with an N-terminal domain required for the mitochondrial import and the C-terminal domain sufficient for the chloroplast import.

Both *AtPreP1* and *AtPreP2* genes are expressed in all tissues, but *AtPreP1* is expressed to a much higher level than *AtPreP2*. The single atprep1-knockout and the double knockout mutations resulted in a chlorotic phenotype, especially during early plant development. The mutant plants exhibited a slower growth rate. The accumulated biomass was 40% lower all through the development in the mutant plants in comparison to wild type. Both mitochondria and chloroplasts exhibited altered morphology. Chloroplasts contained less grana stacking and less starch granules, and chlorophyll a and b content was diminished. Mitochondria were variable in size, partially uncoupled and the respiratory rates were lower [14]. These results demonstrated the importance of PreP for efficient organellar functions and normal plant growth and development.

Notably, the 3D crystal structure of *At*PreP has been solved at 2.1 Å resolution in a closed conformation that enabled for the first time understanding of the proteolytic mechanism of this type of a protease from the pitrilysin family [39]. *At*PreP consists of bowl-shaped halves, connected by a hinge region that create a large internal chamber of about 10,000 $Å^3$, where the active site resides. The active site is formed by the inverted zinc-binding motif (77-His X-X-Glu-His) with a distal Glu-177 located in the N-terminal half and also by Arg-848 and Tyr-854 located in the C-terminal half of the enzyme, that are essential for the catalysis. The proteolysis occurs only in the closed conformation and a mechanism involving hinge-bending motions causing opening and closing of the enzyme in response to substrate binding has been proposed [39]. The crystal structure also revealed two non-catalytic Mg^{2+}binding sites. Mutation studies demonstrated that one of these sites located inside the proteolytic chamber close to the active site was essential for the enzyme activity [5].

Homologues of PreP are present in all species except Archaea. Deletion of the PreP homolog called Mop112/Cym1 in *Saccharomyces cerevisiae* results in a severe phenotype that can be complemented by *At*PreP, suggesting functional conservation of the yeast and plant homologues [1]. Interestingly, the human PreP homologue (hPreP) has been shown to degrade amyloid beta peptide (Aβ) associated with Alzheimer's disease (AD). Immunological studies showed that hPreP is the protease responsible for degradation of Aβ in mitochondria [20]. Recent studies showed reduced hPreP activity in the mitochondrial matrix of AD brains and in AD transgenic mouse models compared with controls that correlated with increased reactive oxygen species (ROS) production [2]. The decreased PreP proteolytic activity in concert with enhanced ROS production contributes to Aβ accumulation in mitochondria, leading to the mitochondrial toxicity and neuronal death that is exacerbated in AD. Clearance of mitochondrial Aβ by PreP may thus be not only essential for sustenance of mitochondrial and chloroplastic functions but also of importance in the pathology of AD [24].

12.4 Processing of Thylakoid-Transfer Sequences by TPP

Thylakoidal Processing Peptidase (TPP) activity was first demonstrated in the mid-1980s using a radiolabeled precursor of the lumenal protein plastocyanin from white campion (*Silene pratensis*) as a substrate and the thylakoid preparation from pea seedlings [25]. Extensive biochemical studies since then, up until early 1990s, revealed properties of TPP, including its membrane topology and catalytic properties [26, 27, 38, 42, 43, 71, 75, 85]. TPP associates integrally to the stroma lamellae, without tightly associating to major thylakoid protein complexes, and faces its catalytic site in the lumen (Fig. 12.1c) [43]. Detailed biochemical characterization revealed that TPP belongs to a group of membrane-bound serine proteases called the type I signal peptidase (SPase I) family [27]. SPases I are enzymes found in both prokaryotes and eukaryotes, using a Ser/Lys (or Ser/His) catalytic dyad to cleave export signals [53]. Prokaryotic SPases I, which are often called leader peptidases, are located in the plasma membrane and process the N-terminal export signals from a number of proteins in the periplasmic space. In eukaryotes, in addition to TPP, two distinct SPases I exist in the endoplasmic reticulum and mitochondria inner membrane, respectively. SPase I homologs from different sources show similar catalytic activities. They recognize several short sequence motifs in the substrate proteins, especially small hydrophobic residues, mainly Ala, at the −3 and −1 positions to the processing site. Interestingly, TPP was shown to have more stringent requirements for this "Ala-X-Ala rule" than other types of SPases I *in vitro* [71]. Nonetheless, a prokaryotic SPase I could process a thylakoid-transfer sequence and TPP could cleave bacterial export signals *in vitro* [27], and both bacterial and thylakoidal enzymes were inhibited by the penem compounds [7]. These findings not only established the prokaryotic origin of TPP, but also provided one of the first indications that the thylakoid protein transport is homologous to the bacterial protein export.

By the time the biochemical properties were established in the early 1990s, however, the study on TPP appeared to peak and start simmering down. Nonetheless, TPP remained unforgotten. In 1998, Chaal et al. performed a homology search using the sequence of a cyanobacterial leader peptidase and obtained the first plant TPP cDNA from Arabidopsis [15]. This clone, *At2g30440*, encodes a protein of 340 amino acid residues which consists of a potential transit peptide in the N terminus, a single putative transmembrane domain and catalytic residues conserved among SPases I. The C-terminal soluble portion of the At2g30440 protein was produced as a recombinant protein in *E. coli,* and the antibody against it reacted with a 30-kD protein in the thylakoid membrane. The catalytic activity of the recombinant TPP was demonstrated against the 23-kD subunit of OEC (OE23) from wheat [15]. However, this activity was very low compared to that of the catalytic domain of the *E. coli* leader peptidase [29], which appeared to hamper further characterization of the enzyme. In the meantime, completion of the Arabidopsis genome sequencing allowed identification of two additional TPP homologs (At1g06870 and At3g24590) in this model plant [29].

TPP was brought to light unexpectedly in 2005 by a study that was aimed to identify an enzyme that processes Toc75 [36]. Toc75 exists in all known plant species and functions as the protein import channel in the chloroplast outer envelope membrane (see Chap 9). Toc75 orthologs are the only known outer membrane proteins in chloroplasts or mitochondria which carry an N-terminal cleavable extension peptide [80]. This extension is necessary for proper targeting of Toc75 and consists of two parts. Its N-terminal portion acts as a canonical transit peptide [35, 79], and the C-terminal portion, which contains the unique polyglycine stretch, prevents further protein transport across the inner envelope membrane [6, 34, 79]. Removal of the N-terminal portion of the Toc75 transit peptide was shown to be catalyzed by SPP [79], whereas the enzyme responsible for the second processing had remained unknown. In 2003, it was demonstrated that two Ala at −3 and −1 to the second processing site are conserved among Toc75 orthologs from different plant species, and mutating these two residues in pea Toc75 did not disrupt proper targeting but resulted in preventing the proper maturation during an *in vitro* import assay [34]. These results led to the hypothesis that SPase I is involved in the second processing, which was further supported by the *in vitro* study showing that the *E. coli* leader peptidase cleaved Toc75 precursor at the second processing site [36]. Hence Inoue et al. identified At3g24590 as a candidate for the Toc75 processing enzyme and named it as Plsp1 (plastidic type I signal peptidase 1). Although its catalytic activity has not been demonstrated *in vitro*, results of a genetic study indicated that Plsp1 is involved in Toc75 maturation *in vivo*: disruption of the *PLSP1* gene by T-DNA insertion was found to cause accumulation of incompletely matured Toc75; localization of Plsp1 in the envelope and thylakoids was demonstrated first by *in vitro* import assay [36] and later by electron microscopy-immunolocalization study [73]. Interestingly, *plsp1*-null plants are seedling-lethal and accumulate not only Toc75 intermediates but also the intermediate of lumenal proteins OE33, OE23 and plastocyanin [36, 74]. These results indicate that Plsp1 may be the main TPP which cleaves both cpSEC and cpTAT substrates at thylakoids, and is also involved in Toc75 processing at the envelope membrane.

Hsu et al. attempted to define the significance of the presence of the three TPP homologs in Arabidopsis [30]. In their work, At1g06870 and At2g30440 were named as Plsp2A and Plsp2B, respectively, because they are more similar to each other than to Plsp1. Interestingly, most cyanobacteria contain two distinct SPase I homologs, named lepB1 and lepB2, respectively. In *Synechocystis* PCC6803, lepB1 is necessary for photoautotrophic growth, whereas lepB2 is essential for cell viability [91]. This and other results suggest that lepB2 is located in the plasma membrane to cleave proteins to be transported to periplasmic and extracellular space, whereas lepB1 may process thylakoidal proteins. Hsu et al.'s phylogenetic analysis revealed that plant TPPs are more similar to lepB1 than to lepB2, indicating that TPP may have originated before the endosymbiotic event. It was also found that the duplication leading to the two groups of TPP, one including Plsp1 and another including Plsp2A and Plsp2B, occurred before the divergence of gymnosperms and angiosperms (Fig. 12.2). Furthermore, transcript analysis revealed distinct expression patterns of the *TPP* genes: two *PLSP2* genes are co-expressed in both photosynthetic tissues and roots, whereas the *PLSP1* transcript accumulates mainly in photosynthetic tissues. The expression level of the two *PLSP2* genes in aerial parts of the *plsp1*-null plants was comparable to that in wild type plants. The phenotypic defects of *plsp1*-null plants could be rescued by expression of a cDNA encoding Plsp1 but not by that for other two TPP isoforms. These results indicate that Plsp1 and Plsp2 evolved to take on different functions [30]. Although the importance of Plsp1 for proper assembly of photosynthetic membranes has been demonstrated, if and how Plsp2 plays a role in chloroplast development remains to be examined.

12.5 Significance of the Cleavage of Targeting Signals

A transport signal ensures the engagement of a protein to a specific transport pathway. Hence such an extension is essential for proper localization and, thus, the function of the protein. Now a question arises: why does the signal need to be removed? Is the signal cleavage required for correct folding and/or assembly of the mature protein? Surprisingly, few studies have addressed these questions. Nonetheless, the SPP precursor, which contains a transit peptide, was shown to be proteolytically inactive and its affinity to the substrate was significantly lower than that of the mature form *in vitro* [60]. A similar result has been obtained with the PreP precursor containing its targeting peptide, which was also proteolytically inactive (Teixeira and Glaser, unpublished results). Furthermore, SPP was shown to be strongly inhibited by C-terminal portions of transit peptides, suggesting the need for complete degradation of transit peptides [59]. The embryo-lethal phenotype of the *spp*-null plant may be due to the accumulation of unfolded, thus nonfunctional, proteins in the chloroplast stroma [81], although this possibility needs to be tested. Additionally, *At*PreP double knockouts in Arabidopsis displayed chloroplastic and mitochondrial dysfunctions such as chlorosis due to lower levels of chlorophylls, less grana and starch in chloroplasts, lower respiratory rates, partially uncoupled mitochondria

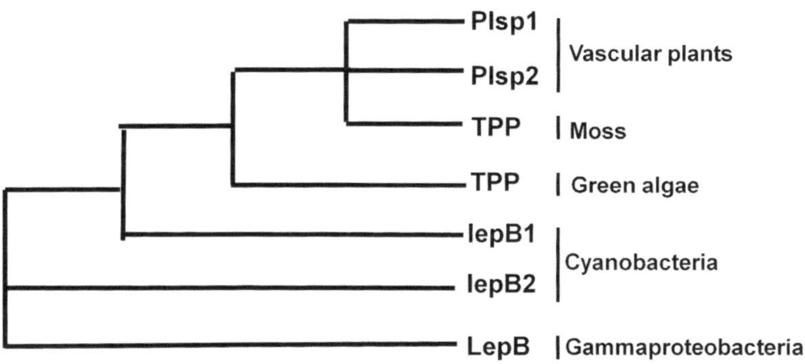

Fig. 12.2 A schematic showing the evolution of TPP. Cyanobacteria have two leader peptidases (lepB1 and lepB2). Cyanobacterial lepB2 appears to function as a type I signal peptidase that cleaves off the export signals at the plasma membrane, similar to its homologs in gammaproteobacteria. lepB1 in cyanobacteria appears to share the common ancestor with TPP from photosynthetic eukaryotes. Green algae and moss contain a single TPP isoform, whereas vascular plants have two, namely Plsp1 and Plsp2. Duplication leading to Plsp1 and Plsp2 occurred before gymnosperm-angiosperm diversification [30]

and aberrant organelle morphology early in plant development [14]. Mitochondrial presequences possess membrane interacting capacity and were shown to uncouple mitochondrial membranes [31]. Similarly, the released chloroplast transit peptides might disrupt the H^+gradient at the thylakoid membrane. All these effects indicate an important role of removal of import signals [45].

In the case of a lumenal protein, its thylakoid-transfer sequence may act as a membrane anchor via its H-domain (Fig. 12.1a). Hence, one may predict that removal of the export signal by TPP is required for either proper protein folding or the release of the mature protein from the membrane. Interestingly, inhibition of the TPP processing by site-directed mutagenesis and chloroplast transformation did not disrupt assembly of the membrane-anchored cytochrome f into the cytochrome b_6f complex *in vivo*, although the resultant unprocessed precursor and misprocessed forms appeared to decrease the redox potential of the hemes [8]. In addition, Popelkova et al. produced in bacteria the precursor of spinach OE33 which contained both the transit peptide and thylakoid-transfer sequence, and showed that it was assembled into the functional PSII with wild type-level water oxidation activity [55]. Hence, the cleavage of the targeting signals is not essential for proper folding and assembly of OE33 *in vitro,* although its significance *in vivo* has not been addressed.

More recent studies provided another perspective on the significance of the processing of thylakoid-transfer sequences. Unprocessed cpTAT substrates (OE23 and the 17-kD subunit of OEC, OE17), which were generated by introducing mutations around the TPP cleavage site, were found to associate with the thylakoid membrane instead of localizing in the lumen *in vivo* [17] and *in vitro* [21]. The *in vitro* study also showed that the unprocessed cpTat substrates may be present free in lipid bilayers, suggesting that the cleavage of the thylakoid-transfer sequence may occur after translocation [21]. Similarly, it was shown that unprocessed OE33 accumulates in the peripheral area of the "ballooned thylakoids" of the *plsp1*-null plastids by an electron microscopy-immunolocalization assay [74]. In the *plsp1*-null plastids, maturation of proteins not only in thylakoids but also in the envelope is affected. Shipman-Roston et al. used a genetic complementation assay with the embryo-lethal *toc75*-null mutant to show that disrupting proper maturation of the envelope substrate Toc75 did not cause the severe phenotypic defect seen in the *plsp1*-null mutant [74]. Hence, it was hypothesized that complete maturation of lumenal proteins is necessary to maintain proper thylakoid assembly [19]. It remains to be elucidated whether or not processing of any specific proteins is important for proper flattening of thylakoids.

12.6 Maturation of the D1 Subunit of the Photosystem II by CtpA

The D1 protein exists in all known organisms that perform oxygenic photosynthesis. Because of its central role in the oxygenic photosynthetic electron transport, the D1 protein has been studied extensively in its structural, biophysical and biochemical properties. Consequently, we have a good understanding, although not complete yet, of its conformation, localization within the photosystem II complex, and targeting and assembly pathways [51]. With a few exceptions seen in Euglenophyta, Chlorarachniophyta and Dinophyta, the D1 protein is synthesized as a larger precursor with a C-terminal extension [67]. This extension does not interfere with the membrane targeting and the initial integration of the D1 protein into the photosystem II complex. However, removal of the C-terminal extension is essential for proper assembly of the Mn_4CaO_5 cluster. There are some differences between cyanobacteria and higher plants in the extension peptides and their processing. In cyanobacteria, the extension is usually 12–16 amino acids long and is processed in two steps. By contrast, higher plant D1 precursors contain shorter extensions of 8–9 amino acid residues and are processed in one step. Nonetheless, most of the D1 precursors, regardless of their origin, contain a conserved Leu–Asp–Leu–Ala↓Ala/Ser at −4 to+1 to the final cleavage site, supporting the idea that the bacterial and eukaryotic D1 protein carboxyl terminal peptidases (CtpAs) are homologous (for a detailed review on the D1 carboxyl-terminal processing, see [67]).

Biochemical characterization in late 1980s to 1990s [13, 22, 32, 78] facilitated molecular cloning of eukaryotic CtpAs [33, 82], while genetic studies led to identification of the *ctpA* gene from the cyanobacterium *Synechocystis* PCC6803 [3, 72]. Extensive studies in the early 2000s, including the X-ray crystal structure analysis at 1.8 Å resolution [47], demonstrated that CtpA contains a Ser-Lys catalytic dyad although it is not inhibited by known Ser protease inhibitors, including the penem compounds [87]. In higher plants, CtpA is encoded in the nuclear genome and synthesized in the cytoplasm as a larger precursor with an N-terminal extension [33, 82]. This extra sequence consists of a transit peptide and a thylakoid-transfer sequence. Hence the higher plant CtpA is processed by SPP in the stroma and TPP in the thylakoid lumen before it gets matured. The cyanobacterial CtpA also carries an N-terminal extension, but it comprises only an export signal [72]. Interestingly, the prokaryotic CtpA is located in the periplasm, instead of the thylakoid lumen [90], although its export signal was shown to act as a thylakoid-transfer sequence in an *in vitro* assay using spinach chloroplasts [40].

Recent molecular genetic studies revealed the involvement of proteins other than CtpA in the maturation of the D1 protein. These include PratA [44, 69] and Psb27 [48, 63, 64] from *Synechocystis* PCC6803, and a Psb27 homolog called LPA19 from Arabidopsis [16, 86]. However, it remains unknown if they are directly involved in the D1 processing. Indeed, available data suggest that it is likely that these proteins play roles in proper assembly of the D1 protein in the photosystem II.

In contrast to the case of transport signals as discussed in the previous section, the significance of the D1 carboxyl-terminal processing is evident. However, the function of this extension peptide is rather enigmatic. Nonetheless, results of molecular genetic studies using *Synechocystis* sp. PCC 6803 demonstrated that the C-terminal extension increases general fitness [37] and effectiveness of the photosystem II repair under increased irradiance [46]. It would be interesting to compare assembly and repair of photosynthetic machineries under various light conditions in organisms that produce the D1 with the C-terminal extension and those that don't.

12.7 Conclusion

The chloroplast depends for its biogenesis on posttranslational import of nuclear-encoded proteins from the cytoplasm. Many of these proteins need N-terminal cleavable signals for the transport across the double-membrane envelope. Proteins sorted to the thylakoid lumen require another type of N-terminal extensions, which function as export signals. Removal of the C-terminal extension of the D1 subunit precursor is necessary for assembly of the catalytic site for water oxidation in the photosystem II. Biochemical properties of the enzymes that cleave these extension peptides were largely established in the 1980s–1990s. The discovery of PreP in early 2000s added a novel aspect of the metabolism of extension peptides. With the help of recent molecular studies, our understanding of the processing of the chloroplast extension peptides appears to be quite comprehensive now. Nonetheless, there are several intriguing questions that remain to be answered.

The first of such questions concerns the fates of the peptides after cleavage. In the stroma, transit peptides released from the imported precursors are further processed by SPP and PreP. Involvement of additional enzymes including yet unidentified ATP-dependent stromal metallopeptidase [58, 59] and a membrane-bound metallopeptidase [14] for sequential degradation of these peptides is very likely, and the interaction of these peptidases would be an interesting problem. In the thylakoid, thylakoid-transfer sequences may be liberated into the membrane after cleavage by TPP because of their hydrophobic core (Fig. 12.1c). In bacterial plasma membranes, site-2 proteases (S2Ps), which contain zinc and act as intermembrane-cleaving enzymes, directly process export signals after these sequences are released from the precursor proteins by the leader peptidase [66]. Two potential S2P homologs are encoded in the Arabidopsis genome, *AT1G05140* and *AT2G32480*. It is tempting to speculate that either one of them, or both, is involved in degradation of thylakoid-transfer sequences after they are cleaved by TPP. Interestingly, however, a previous genetic study showed that At2g32480, which was found to be essential for plant viability, exists only in the chloroplast envelope [12]. Whether At1g05140, or another unrelated protease, plays a role in processing of thylakoid-transfer sequences remains to be examined. Finally, the C-terminal extension of the D1 protein should be released in the thylakoid lumen, where DegP proteases are known to play role in the degradation of the D1 protein (see Chap. 13). Whether the released D1 C-terminal extension is further processed by these or other enzymes, or whether it possesses some biological functions on its own remains to be examined.

Another intriguing unanswered question is evolution of the processing peptidases. SPP appears to have evolved from proteins in non-photosynthetic bacteria although the details of the evolutionary process remain elusive [56]. By contrast, the cyanobacterial origin of TPP is evident [15, 29, 30]. Nonetheless, the significance of its duplication in land plants leading to Plsp1 and Plsp2 remains unknown (Fig. 12.2). Interestingly, similar to the case of TPP, CtpA also forms a small gene family in Arabidopsis, namely *AT3G57680*, *AT4G17740* and *AT5G46390*. Disruption of the *AT3G57680* gene by a T-DNA insertion did not cause accumulation of the D1 precursor [88]. Whether this is due to functional redundancy among the three gene products or whether they have evolved to take on distinct functions, similar to the case of TPP, remains to be defined.

In summary, studies on the chloroplast enzymes that cleave and degrade extension peptides have established the mechanism and importance of peptide processing. Addressing the emerging questions should not only advance our understanding of the biological significance of these extension peptides, but also contribute greatly to our knowledge of the general biological processes of protein trafficking, protein homeostasis, and molecular evolution.

Acknowledgements The work on TPP in the Inoue laboratory has been supported by the Division of Chemical Sciences, Geosciences, and Biosciences, Office of Basic Energy Sciences of the U.S. Department of Energy through Grant DE-FG02-08ER15963. The work on PreP in the Glaser laboratory was supported by research grants from the Swedish Research Council (VR-NT 2009-5641), Carl Tryggers Foundation and Alzheimerfonden.

References

1. Alikhani N, Berglund AK, Engmann T, Spanning E, Vogtle FN, Pavlov P, Meisinger C, Langer T, Glaser E (2011) Targeting capacity and conservation of PreP homologues localization in mitochondria of different species. J Mol Biol 410:400–410
2. Alikhani N, Guo L, Yan S, Du H, Pinho CM, Chen JX, Glaser E, Yan SS (2011) Decreased proteolytic activity of the mitochondrial amyloid-beta degrading enzyme, PreP peptidasome, in Alzheimer's disease brain mitochondria. J Alzheimers Dis 27:75–87

3. Anbudurai PR, Mor TS, Ohad I, Shestakov SV, Pakrasi HB (1994) The ctpA gene encodes the C-terminal processing protease for the D1 protein of the photosystem II reaction center complex. Proc Natl Acad Sci U S A 91:8082–8086
4. Artenstein AW, Opal SM (2011) Proprotein convertases in health and disease. New Eng J Med 365:2507–2518
5. Backman HG, Pessoa J, Eneqvist T, Glaser E (2009) Binding of divalent cations is essential for the activity of the organellar peptidasome in Arabidopsis thaliana, AtPreP. FEBS Lett 583:2727–2733
6. Baldwin AJ, Inoue K (2006) The most C-terminal tri-glycine segment within the polyglycine stretch of the pea Toc75 transit peptide plays a critical role for targeting the protein to the chloroplast outer envelope membrane. FEBS J 273:1547–1555
7. Barbrook AC, Packer JC, Howe CJ (1996) Inhibition by penem of processing peptidases from cyanobacteria and chloroplast thylakoids. FEBS Lett 398:198–200
8. Baymann F, Zito F, Kuras R, Minai L, Nitschke W, Wollman FA (1999) Functional characterization of Chlamydomonas mutants defective in cytochrome f maturation. J Biol Chem 274:22957–22967
9. Bhushan S, Lefebvre B, Stahl A, Wright SJ, Bruce BD, Boutry M, Glaser E (2003) Dual targeting and function of a protease in mitochondria and chloroplasts. EMBO Rep 4:1073–1078
10. Bhushan S et al (2005) Catalysis, subcellular localization, expression and evolution of the targeting peptides degrading protease, AtPreP2. Plant Cell Physiol 46:985–996
11. Bhushan S, Kuhn C, Berglund AK, Roth C, Glaser E (2006) The role of the N-terminal domain of chloroplast targeting peptides in organellar protein import and miss-sorting. FEBS Lett 580:3966–3972
12. Bolter B, Nada A, Fulgosi H, Soll J (2006) A chloroplastic inner envelope membrane protease is essential for plant development. FEBS Lett 580:789–794
13. Bowyer JR, Packer JC, McCormack BA, Whitelegge JP, Robinson C, Taylor MA (1992) Carboxyl-terminal processing of the D1 protein and photoactivation of water-splitting in photosystem II. Partial purification and characterization of the processing enzyme from Scenedesmus obliquus and Pisum sativum. J Biol Chem 267:5424–5433
14. Cederholm SN, Backman HG, Pesaresi P, Leister D, Glaser E (2009) Deletion of an organellar peptidasome PreP affects early development in Arabidopsis thaliana. Plant Mol Biol 71:497–508
15. Chaal BK, Mould RM, Barbrook AC, Gray JC, Howe CJ (1998) Characterization of a cDNA encoding the thylakoidal processing peptidase from Arabidopsis thaliana. Implications for the origin and catalytic mechanism of the enzyme. J Biol Chem 273:689–692
16. Chen H, Zhang D, Guo J, Wu H, Jin M, Lu Q, Lu C, Zhang L (2006) A Psb27 homologue in Arabidopsis thaliana is required for efficient repair of photodamaged photosystem II. Plant Mol Biol 61:567–575
17. Di Cola A, Robinson C (2005) Large-scale translocation reversal within the thylakoid Tat system in vivo. J Cell Biol 171:281–289
18. Emanuelsson O, Nielsen H, von Heijne G (1999) ChloroP, a neural network-based method for predicting chloroplast transit peptides and their cleavage sites. Protein Sci 8:978–984
19. Endow JK, Ruppel NJ, Inoue K (2010) Keep the balloon deflated: the significance of protein maturation for thylakoid flattening. Plant Signal Behav 5:721–723
20. Falkevall A et al (2006) Degradation of the amyloid beta-protein by the novel mitochondrial peptidasome, PreP. J Biol Chem 281:29096–29104
21. Frielingsdorf S, Klosgen RB (2007) Prerequisites for terminal processing of thylakoidal Tat substrates. J Biol Chem 282:24455–24462
22. Fujita S, Inagaki N, Yamamoto Y, Taguchi F, Matsumoto A, Satoh K (1995) Identification of the carboxyl-terminal processing protease for the D1 precursor protein of the photosystem-ii reaction-center of spinach. Plant Cell Physiol 36:1169–1177
23. Gavel Y, von Heijne G (1990) A conserved cleavage-site motif in chloroplast transit peptides. FEBS Lett 261:455–458

24. Glaser E, Alikhani N (2010) The organellar peptidasome, PreP: a journey from Arabidopsis to Alzheimer's disease. Biochim Biophys Acta 1797:1076–1080
25. Hageman J, Robinson C, Smeekens S, Weisbeek P (1986) A thylakoid processing protease is required for complete maturation of the lumen protein plastocyanin. Nature 324:567–569
26. Hageman J, Baecke C, Ebskamp M, Pilon R, Smeekens S, Weisbeek P (1990) Protein import into and sorting inside the chloroplast are independent processes. Plant Cell 2:479–494
27. Halpin C, Elderfield PD, James HE, Zimmermann R, Dunbar B, Robinson C (1989) The reaction specificities of the thylakoidal processing peptidase and Escherichia coli leader peptidase are identical. EMBO J 8:3917–3921
28. Henry R, Carrigan M, McCaffrey M, Ma X, Cline K (1997) Targeting determinants and proposed evolutionary basis for the Sec and the Delta pH protein transport systems in chloroplast thylakoid membranes. J Cell Biol 136:823–832
29. Howe CJ, Floyd KA (2001) Chloroplast and mitochondrial type I signal peptidases. In: Dalbey RE, Sigman DS (eds) Co- and posttranslational proteolysis of proteins, Academic, San Diego, pp 101–125
30. Hsu SC, Endow JK, Ruppel NJ, Roston RL, Baldwin AJ, Inoue K (2011) Functional diversification of thylakoidal processing peptidases in Arabidopsis thaliana. PLoS ONE 6:e27258
31. Hugosson M, Andreu D, Boman HG, Glaser E (1994) Antibacterial peptides and mitochondrial presequences affect mitochondrial coupling, respiration and protein import. Eur J Biochem 223:1027–1033
32. Inagaki N, Fujita S, Satoh K (1989) Solubilization and partial-purification of a thylakoidal enzyme of spinach involved in the processing of D1-protein. FEBS Lett 246:218–222
33. Inagaki N, Yamamoto Y, Mori H, Satoh K (1996) Carboxyl-terminal processing protease for the D1 precursor protein: cloning and sequencing of the spinach cDNA. Plant Mol Biol 30:39–50
34. Inoue K, Keegstra K (2003) A polyglycine stretch is necessary for proper targeting of the protein translocation channel precursor to the outer envelope membrane of chloroplasts. Plant J 34:661–669
35. Inoue K, Demel R, de Kruijff B, Keegstra K (2001) The N-terminal portion of the preToc75 transit peptide interacts with membrane lipids and inhibits binding and import of precursor proteins into isolated chloroplasts. Eur J Biochem 268:4036–4043
36. Inoue K, Baldwin AJ, Shipman RL, Matsui K, Theg SM, Ohme-Takagi M (2005) Complete maturation of the plastid protein translocation channel requires a type I signal peptidase. J Cell Biol 171:425–430
37. Ivleva NB, Shestakov SV, Pakrasi HB (2000) The carboxyl-terminal extension of the precursor D1 protein of photosystem II is required for optimal photosynthetic performance of the cyanobacterium Synechocystis sp PCC 6803. Plant Physiol 124:1403–1411
38. James HE, Bartling D, Musgrove JE, Kirwin PM, Herrmann RG, Robinson C (1989) Transport of proteins into chloroplasts—import and maturation of precursors to the 33-kDa, 23-kDa, and 16-kDa proteins of the photosynthetic oxygen-evolving complex. J Biol Chem 264:19573–19576
39. Johnson KA, Bhushan S, Stahl A, Hallberg BM, Frohn A, Glaser E, Eneqvist T (2006) The closed structure of presequence protease PreP forms a unique 10,000 Angstroms3 chamber for proteolysis. EMBO J 25:1977–1986
40. Karnauchov I, Herrmann RG, Pakrasi HB, Klosgen RB (1997) Transport of CtpA protein from the cyanobacterium Synechocystis 6803 across the thylakoid membrane in chloroplasts. Eur J Biochem 249:497–504
41. Kieselbach T, Schroder WP (2003) The proteome of the chloroplast lumen of higher plants. Photosynth Res 78:249–264
42. Kirwin PM, Elderfield PD, Robinson C (1987) Transport of proteins into chloroplasts - partial-purification of a thylakoidal processing peptidase involved in plastocyanin biogenesis. J Biol Chem 262:16386–16390
43. Kirwin PM, Elderfield PD, Williams RS, Robinson C (1988) Transport of proteins into chloroplasts - organization, orientation, and lateral distribution of the plastocyanin processing peptidase in the thylakoid network. J Biol Chem 263:18128–18132

44. Klinkert B, Ossenbuhl F, Sikorski M, Berry S, Eichacker L, Nickelsen J (2004) PratA, a periplasmic tetratricopeptide repeat protein involved in biogenesis of photosystem II in Synechocystis sp PCC 6803. J Biol Chem 279:44639–44644
45. Kmiec B, Glaser E (2012) A novel mitochondrial and chloroplast peptidasome, PreP. Physiol Plant 145:180–186
46. Kuvikova S, Tichy M, Komenda J (2005) A role of the C-terminal extension of the photosystem II D1 protein in sensitivity of the cyanobacterium Synechocystis PCC 6803 to photoinhibition. Photoch Photobiol Sci 4:1044–1048
47. Liao DI, Qian J, Chisholm DA, Jordan DB, Diner BA (2000) Crystal structures of the photosystem II D1 C-terminal processing protease. Nat Struct Biol 7:749–753
48. Liu HJ, Roose JL, Cameron JC, Pakrasi HB (2011) A genetically tagged psb27 protein allows purification of two consecutive photosystem ii (psii) assembly intermediates in Synechocystis 6803, a cyanobacterium. J Biol Chem 286:24865–24871
49. Moberg P, Stahl A, Bhushan S, Wright SJ, Eriksson A, Bruce BD, Glaser E (2003) Characterization of a novel zinc metalloprotease involved in degrading targeting peptides in mitochondria and chloroplasts. Plant J 36:616–628
50. Nilsson R, van Wijk KJ (2002) Transient interaction of cpSRP54 with elongating nascent chains of the chloroplast-encoded D1 protein; 'cpSRP54 caught in the act'. FEBS Lett 524:127–133
51. Nixon PJ, Michoux F, Yu JF, Boehm M, Komenda J (2010) Recent advances in understanding the assembly and repair of photosystem II. Ann Bot 106:1–16
52. Oblong JE, Lamppa GK (1992) Identification of two structurally related proteins involved in proteolytic processing of precursors targeted to the chloroplast. EMBO J 11:4401–4409
53. Paetzel M, Karla A, Strynadka NC, Dalbey RE (2002) Signal peptidases. Chem Rev 102:4549–4580
54. Peltier JB et al (2002) Central functions of the lumenal and peripheral thylakoid proteome of Arabidopsis determined by experimentation and genome-wide prediction. Plant Cell 14:211–236
55. Popelkova H, Im MM, D'Auria J, Betts SD, Lydakis-Simantiris N, Yocum CF (2002) N-terminus of the photosystem II manganese stabilizing protein: effects of sequence elongation and truncation. Biochemistry 41:2702–2711
56. Reumann S, Inoue K, Keegstra K (2005) Evolution of the general protein import pathway of plastids. Mol Membr Biol 22:73–86
57. Richter S, Lamppa GK (1998) A chloroplast processing enzyme functions as the general stromal processing peptidase. Proc Natl Acad Sci U S A 95:7463–7468
58. Richter S, Lamppa GK (1999) Stromal processing peptidase binds transit peptides and initiates their ATP-dependent turnover in chloroplasts. J Cell Biol 147:33–43
59. Richter S, Lamppa GK (2002) Determinants for removal and degradation of transit peptides of chloroplast precursor proteins. J Biol Chem 277:43888–43894
60. Richter S, Lamppa GK (2003) Structural properties of the chloroplast stromal processing peptidase required for its function in transit peptide removal. J Biol Chem 278:39497–39502
61. Robinson C, Mant A (2005) Biogenesis of the thylakoid membrane. In Möller SG (ed) Plastids, Blackwell: Oxford, pp 180–213
62. Rohl T, van Wijk KJ (2001) In vitro reconstitution of insertion and processing of cytochrome f in a homologous chloroplast translation system. J Biol Chem 276:35465–35472
63. Roose JL, Pakrasi HB (2004) Evidence that D1 processing is required for manganese binding and extrinsic protein assembly into photosystem II. J Biol Chem 279:45417–45422
64. Roose JL, Pakrasi HB (2008) The Psb27 protein facilitates manganese cluster assembly in photosystem II. J Biol Chem 283:4044–4050
65. Rudhe C, Clifton R, Whelan J, Glaser E (2002) N-terminal domain of the dual-targeted pea glutathione reductase signal peptide controls organellar targeting efficiency. J Mol Biol 324:577–585
66. Saito A, Hizukuri Y, Matsuo E, Chiba S, Mori H, Nishimura O, Ito K, Akiyama Y (2011) Post-liberation cleavage of signal peptides is catalyzed by the site-2 protease (S2P) in bacteria. Proc Natl Acad Sci U S A 108:13740–13745

67. Satoh K, Yamamoto Y (2007) The carboxyl-terminal processing of precursor D1 protein of the photosystem II reaction center. Photosynth Res 94:203–215
68. Schatz G, Dobberstein B (1996) Common principles of protein translocation across membranes. Science 271:1519–1526
69. Schottkowski M, Gkalympoudis S, Tzekova N, Stelljes C, Schunemann D, Ankele E, Nickelsen J (2009) Interaction of the Periplasmic PratA Factor and the PsbA (D1) Protein during biogenesis of photosystem ii in Synechocystis sp PCC 6803. J Biol Chem 284:1813–1819
70. Schubert M, Petersson UA, Haas BJ, Funk C, Schroder WP, Kieselbach T (2002) Proteome map of the chloroplast lumen of Arabidopsis thaliana. J Biol Chem 277:8354–8365
71. Shackleton JB, Robinson C (1991) Transport of proteins into chloroplasts. The thylakoidal processing peptidase is a signal-type peptidase with stringent substrate requirements at the -3 and -1 positions. J Biol Chem 266:12152–12156
72. Shestakov SV, Anbudurai PR, Stanbekova GE, Gadzhiev A, Lind LK, Pakrasi HB (1994) Molecular cloning and characterization of the ctpA gene encoding a carboxyl-terminal processing protease. Analysis of a spontaneous photosystem II-deficient mutant strain of the cyanobacterium Synechocystis sp. PCC 6803. J Biol Chem 269:19354–19359
73. Shipman RL, Inoue K (2009) Suborganellar localization of plastidic type I signal peptidase 1 depends on chloroplast development. FEBS Lett 583:938–942
74. Shipman-Roston RL, Ruppel NJ, Damoc C, Phinney BS, Inoue K (2010) The significance of protein maturation by plastidic type I signal peptidase 1 for thylakoid development in Arabidopsis chloroplasts. Plant Physiol 152:1297–1308
75. Smeekens S, Bauerle C, Hageman J, Keegstra K, Weisbeek P (1986) The role of the transit peptide in the routing of precursors toward different chloroplast compartments. Cell 46:365–375
76. Stahl A, Moberg P, Ytterberg J, Panfilov O, von Lowenhielm HB, Nilsson F, Glaser E (2002) Isolation and identification of a novel mitochondrial metalloprotease (PreP) that degrades targeting presequences in plants. J Biol Chem 277:41931–41939
77. Stahl A et al (2005) Two novel targeting peptide degrading proteases, PrePs, in mitochondria and chloroplasts, so similar and still different. J Mol Biol 349:847–860
78. Taylor MA, Packer JCL, Bowyer JR (1988) Processing of the D1 polypeptide of the photosystem-II reaction center and photoactivation of a low fluorescence mutant (Lf-1) of Scenedesmus-Obliquus. FEBS Lett 237:229–233
80. Tranel PJ, Keegstra K (1996) A novel, bipartite transit peptide targets OEP75 to the outer membrane of the chloroplastic envelope. Plant Cell 8:2093–2104
80. Tranel PJ, Froehlich J, Goyal A, Keegstra K (1995) A component of the chloroplastic protein import apparatus is targeted to the outer envelope membrane via a novel pathway. EMBO J 14:2436–2446
81. Trosch R, Jarvis P (2011) The stromal processing peptidase of chloroplasts is essential in arabidopsis, with knockout mutations causing embryo arrest after the 16-cell stage. PLoS ONE 6:e23039
82. Trost JT, Chisholm DA, Jordan DB, Diner BA (1997) The D1 C-terminal processing protease of photosystem II from Scenedesmus obliquus—protein purification and gene characterization in wild type and processing mutants. J Biol Chem 272:20348–20356
83. Vandervere PS, Bennett TM, Oblong JE, Lamppa GK (1995) A chloroplast processing enzyme involved in precursor maturation shares a zinc-binding motif with a recently recognized family of metalloendopeptidases. Proc Natl Acad Sci U S A 92:7177–7181
84. von Heijne G, Steppuhn J, Herrmann RG (1989) Domain-structure of mitochondrial and chloroplast targeting peptides. Eur J Biochem 180:535–545
85. Wallace TP, Robinson C, Howe CJ (1990) The reaction specificities of the pea and a cyanobacterial thylakoid processing peptidase are similar but not identical. FEBS Lett 272:141–144
86. Wei L, Guo J, Ouyang M, Sun X, Ma J, Chi W, Lu C, Zhang L (2010) LPA19, a Psb27 homolog in Arabidopsis thaliana, facilitates D1 protein precursor processing during PSII biogenesis. J Biol Chem 285:21391–21398

87. Yamamoto Y, Inagaki N, Satoh K (2001) Overexpression and characterization of carboxyl-terminal processing protease for precursor D1 protein: regulation of enzyme-substrate interaction by molecular environments. J Biol Chem 276:7518–7525
88. Yin SM, Sun XW, Zhang LX (2008) An Arabidopsis ctpA homologue is involved in the repair of photosystem II under high light. Chin Sci Bull 53:1021–1026
89. Yue RQ, Wang XF, Chen JY, Ma XX, Zhang HH, Mao CZ, Wu P (2010) A rice stromal processing peptidase regulates chloroplast and root development. Plant Cell Physiol 51:475–485
90. Zak E, Norling B, Maitra R, Huang F, Andersson B, Pakrasi HB (2001) The initial steps of biogenesis of cyanobacterial photosystems occur in plasma membranes. Proc Natl Acad Sci U S A 98:13443–13448
91. Zhbanko M, Zinchenko V, Gutensohn M, Schierhorn A, Klosgen RB (2005) Inactivation of a predicted leader peptidase prevents photoautotrophic growth of Synechocystis sp. strain PCC 6803. J Bacteriol 187:3071–3078
92. Zhong R, Wan JX, Jin RG, Lamppa G (2003) A pea antisense gene for the chloroplast stromal processing peptidase yields seedling lethals in Arabidopsis: survivors show defective GFP import in vivo. Plant J 34:802–812
93. Zybailov B, Rutschow H, Friso G, Rudella A, Emanuelsson O, Sun Q, van Wijk KJ (2008) Sorting signals, N-terminal modifications and abundance of the chloroplast proteome. PLoS ONE 3:e1994

Chapter 13
Molecular Chaperone Functions in Plastids

Raphael Trösch, Michael Schroda and Felix Willmund

Abstract Molecular chaperones play essential roles in a wide variety of cellular processes, from *de-novo* protein folding to protein disaggregation under stress conditions, unfolding and re-folding of misfolded proteins, protein degradation, protein transport and proteome remodeling during development. Almost all cell compartments contain chaperone activity to some extent, hence it is not surprising that a large number of chaperones also play essential roles in the plastid compartment. Here, the focus of chaperone activity is on protein targeting (protein import and assembly of complexes in target membranes) as well as protection from specific chloroplast-derived stresses. Moreover, chaperones play important roles in *de-novo* folding of plastid-encoded proteins, in the folding of soluble proteins after import and processing of the transit peptides, and in protein degradation. The four major groups of molecular chaperones, the chaperonin/Cpn60, Hsp70, Hsp90 and Hsp100 families of chaperones, are all present in plastids but many cofactors and co-chaperones have not yet been identified. Although chaperone function is generally conserved, it seems that plastid-localized chaperones have evolved some specific functions and mechanisms. Current research on plastid-localized chaperones focuses therefore on the specificities of chaperone function in the context of their plastid environment and requirements.

Keywords Molecular chaperone · Chaperonin · Heat shock protein · Co-chaperone · Stress · Protein targeting · Folding · Disaggregation · Protein import · Degradation

M. Schroda (✉) · F. Willmund
Department of Molecular Biotechnology & Systems Biology, TU Kaiserslautern, Paul-Ehrlich-Straße 23, 67663 Kaiserslautern, Germany
e-mail: schroda@biologie.uni-kl.de

R. Trösch
Institute of Biology, Humboldt University of Berlin, Chauseestraße 117, 10115 Berlin, Germany
e-mail: raphael.troesch@aol.de

F. Willmund
e-mail: willmund@rhrk.uni-kl.de

S.M. Theg, F.-A. Wollman (eds.), *Plastid Biology*, Advances in Plant Biology 5,
DOI 10.1007/978-1-4939-1136-3_13,

Abbreviations

ABRE	Abscisic acid responsive element
APG	Albino and pale green
CDJ	Chloroplast DnaJ-like protein
CGE	Chloroplast GrpE homolog
CPN	Chaperonin
ER	Endoplasmic reticulum
HEP	Hsp70 escort protein
HSP	Heat shock protein
PS	Photosystem
TAC	Transcriptionally active chromosome
VIPP1	Vesicle-inducing proteins in plastids 1
ZFHD1	Zinc-finger homeodomain 1

13.1 Introduction

Folding of newly synthesized polypeptides as well as protein maintenance and degradation of misfolded proteins are essential processes in cells. Misfolded protein species are often found to accumulate in protein aggregates with drastic consequences for cell viability [59, 192]. Over the last decades it became clear that both eukaryotic and prokaryotic organisms contain a whole set of conserved proteins dedicated to maintain a healthy cellular proteome. These factors are termed molecular chaperones and are defined as proteins that transiently bind other proteins to allow for their stabilization and acquisition of a functional conformation without being present in their final conformation [56, 58]. Cells harbor structurally and functionally distinct classes of chaperones that vary in size and complexity. The diversity ranges from chaperones that only bind to misfolded polypeptides and prevent their aggregation to those that recognize specific substrate proteins and facilitate their folding to the native state. Further, chaperones are involved in the translocation of partially unfolded proteins across membranes, complex assembly and disassembly, and many other regulatory processes within the cell [44, 57]. Many chaperones show increased expression under cellular stress situations and were thus annotated as heat shock proteins (HSPs). A large group of chaperones hydrolyze ATP, which allows for the marked conformational changes required for substrate processing. For the folding of some substrates, multiple rounds of ATP hydrolysis are necessary to produce a stable "native" conformation of the substrate protein (reviewed in [17, 67]).

Chaperone functions are best studied in bacteria and the cytosol of eukaryotic organisms. However, most members of the major chaperone classes are also found in subcompartments and organelles of eukaryotic cells such as the endoplasmic reticulum (ER), mitochondria and plastids. Multiple studies indicate an equally important function of these subcellular chaperone systems for protein biosynthesis and protein homeostasis (reviewed in [14, 160]). Despite their homology, interesting distinctions are observed among organellar chaperones and their cytosolic counterparts, indicating their specific adaptation to the need of the respective proteome.

This chapter summarizes current knowledge on molecular chaperones of the heat-shock protein family that are targeted to plastids. While many diverse factors are known to promote protein maturation and turnover in the chloroplast, we will focus here on the four major classes of ATP hydrolyzing molecular chaperones, i.e., Cpn60/Cpn10, Hsp70, Hsp90, and Hsp100.

13.2 Chaperonins/Hsp60s

13.2.1 Chaperonin Isoforms and Complex Compositions

Chaperonins/Hsp60s are remarkable chaperone machines of ~800–900 kDa sized complexes that enclose substrate proteins in their central cavities for folding. Group I chaperonins are found in eubacteria and organelles of bacterial origin such as mitochondria and chloroplasts. The most famous member of this chaperonin family is GroEL found in bacteria; members of the mitochondria are the Hsp60s, and members of chloroplast chaperonins are termed Cpn60. Chaperonins are structurally distinct from other chaperones, as the chaperonin folding machinery is composed of two stacked rings, each ring containing seven subunits of about 60 kDa. The functional unit is complemented by a heptameric cofactor complex that forms the lid of the folding cage. This cofactor complex consists of members of the Hsp10 family with GroES in bacteria and Hsp10 or Cpn10 in mitochondria and plastids, respectively (reviewed in [46, 59]). Group II chaperonins are found in archaea (Thermosome) and the eukaryotic cytosol (TRiC/CCT) and are structurally distinct from group I chaperonins as they consist of octomeric rings with built-in lids substituting the role of Hsp10 cofactors [178]. The individual subunits of Group I and II chaperonins share a similar structure with three distinct domains: an equatorial ATP-binding domain; an apical domain that is involved in substrate binding; and a central hinge domain that enables communication between the equatorial domain and the apical domain (reviewed in [59, 67]).

Most information about the mechanisms of the chaperonins was gained by studies on the group I GroEL/GroES system from *E. coli*. The folding of non-native proteins is accomplished by nucleotide-dependent cycling of the chaperonin between a binding-active state and a folding-active state. In the cycle, the two rings act in an asymmetric and anti-cooperative behavior, also termed a "two-stroke engine". Substrate binding takes place at the inner wall of the cavity and is mediated by exposed hydrophobic residues of the apical domains (in the *cis* ring). Folding is induced by binding of GroES to ATP-bound GroEL inducing conformational changes of GroEL that eventually lead to the formation of a cage with a highly hydrophilic net-negatively-charged cavity wall. The transition from a hydrophilic to a polar environment is thought to be an important factor promoting folding. The process of ATP-hydrolysis by GroEL provides about 10 seconds for substrate folding within the encapsulated cavity. Allosteric ATP binding in the opposite *trans* ring results in GroES dissociation and subsequent substrate and ADP release (reviewed in [67]).

In the chloroplast, chaperonins possess several intriguing features concerning complex composition and functional properties that are not shared by other group I chaperonin family members. From green algae to higher plants, both Cpn60 chaperonins and Cpn10 components comprise multiple isoforms instead of only one member in other phyla [64, 160]. Cpn60s are found in two major forms, alpha and beta, that share only about 50 % amino acid sequence identity. In the unicellular green alga *Chlamydomonas reinhardtii* three chloroplast-targeted Cpn60 members have been identified and are termed CPN60A, CPN60B1, and CPN60B2 [160]. Genomes of higher plants such as *Arabidopsis thaliana* encode six members, including three abundantly expressed Cpn60 subunits (Cpn60α1, Cpn60β1, and Cpn60β2) and three low abundant isoforms (Cpn60α2, Cpn60β3, and Cpn60β4) (reviewed in [207]).

Isolation of chaperonin oligomers from chloroplasts showed that complexes contained equal amounts of both isoforms [13, 112, 120]. In *E. coli* lysates of heterologously expressed Cpn60 isoforms and in *in vitro* reconstitution studies, Cpn60β was capable of auto-assembly into functional homo-oligomers, while Cpn60α needs Cpn60β to assemble into hetero-oligomers with a suggested stoichiometry of α7β7 [29, 33]. However, the structural organization of formed complexes has not been unequivocally determined yet [33].

The cofactors of the Hsp10 family from both, bacteria and mitochondria consist of heptameric rings of 10-kDa subunits [69, 197]. In contrast, plastidic cofactors are expressed as conventional Cpn10 isoforms and as proteins with two fused Cpn10 domains. The latter are termed Cpn20, as sequence analysis indicated a head-to-tail fusion of two Cpn10 domains, joined by a putative TDDVKD-linker sequence resulting in proteins with a molecular weight of ~21 kDa [8, 12]. Both Cpn10 and Cpn20 isoforms have been described in various plant species ranging from green algae to higher plants [64, 80, 160]. In contrast to Cpn10s, no genes encoding Cpn20s are found in cyanobacteria, suggesting that the likely gene fusion event must have occurred in the common ancestor of unicellular algae and higher plants after the endosymbiotic event. The reason for multiple isoforms in the chloroplast remains unclear. Possible differences in suborganellar localization have been suggested in the past [153].

Of the four cofactor genes identified in *C. reinhardtii*, two genes appear to encode for plastidic Cpn20 homologs (CPN20 and CPN23) and one for a Cpn10 isoform (termed CPN11). The fourth gene seems to code for the mitochondrial isoform [160]. The genome *of A. thaliana* most likely encodes only one plastidic Cpn20 cofactor and four Cpn10 isoforms two each located to chloroplasts and mitochondria [64, 80]. Interestingly, it has been reported recently that recombinant cofactors of *C. reinhardtii* do not form functional higher homo-oligomers by themselves but rather hetero-oligomeric forms of $CPN11_1/CPN20_3$ and $CPN11_1/CPN23_3$. Both assemblies constituted functional complexes with seven Cpn10 domains, thus matching the seven-fold symmetry of the Cpn60 ring [190]. For the *A. thaliana* cofactors, hetero-oligomeric Cpn10/Cpn20 complexes were also observed. However, also homo-tetrameric Cpn20 complexes were reported to serve as functional cofactors of Cpn60 rings [80, 168, 207]. A study with heterologously expressed AtCpn20 indicates that such AtCpn20$_4$ complexes (with 4×2 Cpn10 domains) achieve seven-fold symmetry by partial proteolytic cleavage of one Cpn20 protein, resulting in seven Cpn10 domains of this complex [190]. It has been suggested that a varying number of Cpn10/20 components in the hetero-oligomeric cofactor complex might correlate with different binding affinities to the Cpn60 complex and thus provide different folding kinetics for the respective substrates [207].

It will be interesting to examine how the Cpn60-Cpn10/20 hetero-oligomers are distributed in the chloroplast. Further, it would be interesting to analyze how subunits are arranged within hetero-oligomeric Cpn60 complexes. The unique composition of the plastidic chaperonins might be an important contributor for substrate specificity of a chaperonin complex and might reveal novel mechanisms for this folding machinery.

13.2.2 Functions of Chloroplast Localized Chaperonins

Chloroplast Cpn60 was initially identified as oligomeric protein complex that interacts with the large subunit of the ribulose-1,5-bisphosphate carboxylase (Rubisco) (reviewed in [14]). Besides this prominently studied chaperonin substrate, a few other chloroplast proteins were identified to associate with the complex [13, 75, 100, 109, 191]. Notably, several of these substrates require both the plastidic chaperonin and Hsp70 to assume the native state. Such substrates include the coupling factor CF1 [19], ferredoxin-NADP+reductase [191], and the Rieske protein [101].

Multiple studies indicate that different compositions of alpha and beta subunit complexes together with their respective cofactors serve specific tasks and substrates: The low abundance *A. thaliana* subunit Cpn60β_4 specifically participates in a hetero-oligomeric Cpn60 complex for the folding of the chloroplast protein NdhH, a subunit of the NADH dehydrogenase-like complex [135]. The other abundant Cpn60 subunits are thought to fulfill general housekeeping functions by facilitating the folding of other obligate substrates as indicated by the lethal phenotype upon disruption of the genes for Cpn60α1 and Cpn60β1-2 [5, 182]. An important housekeeping function of chloroplast chaperonins is also indicated by their up-regulation during heat shock, which was shown both at the mRNA and protein levels although with varying kinetics between organisms [41, 111, 157, 184, 186].

Further, Cpn60β was shown to protect Rubisco activase from thermal denaturation during heat stress [147], and both *A. thaliana* Cpn60α and Cpn60β isoforms are required for the proper assembly of the plastid division apparatus [182]. A non-protein-folding related function of CPN60α was observed in *C. reinhardtii*, where CPN60α was found to specifically interact with group II intron RNA, suggesting a specialized role as a general organellar mRNA splicing factor [7].

While many of these studies revealed different aspects of the chaperonin function, the broad substrate spectrum of the different Cpn60 isoforms remains elusive. It will be interesting to analyze how different complex compositions serve in general *de novo* protein folding of the chloroplast proteome. The fascinating differences among group I chaperonins highlight specific adaptations to its subcellular environment and underscore the importance of studying multiple model systems when attempting to formulate universal mechanisms for how a given chaperone conducts protein folding.

13.3 Hsp70

13.3.1 General Aspects of Hsp70 Chaperones and Cochaperones

Chaperones of the Hsp70 family are among the most highly conserved proteins known (reviewed in [105]). Their N-terminal ATPase domain regulates substrate binding and release at the C-terminal substrate-binding domain. Hsp70 substrate proteins expose hydrophobic amino acids flanked by basic residues [143], a characteristic feature of non-native, but also of native Hsp70 substrates. Binding of Hsp70 to hydrophobic regions prevents the formation of aggregates. In addition, the ATP-driven unfolding activity of Hsp70s may introduce conformational changes to bound substrates that eventually allow non-native proteins to assume the native state [169]. Thus, Hsp70s play a major role in the folding of nascent chains and in the renaturation of non-native proteins that have accumulated during stress. However, they are also involved in many highly specialized functions like the regulation of the general stress response in bacteria [187], the uncoating of clathrin-coated vesicles [193], or the translocation of proteins across membranes [74].

Hsp70s can be divided into DnaK-type and Hsp110-type Hsp70s. DnaK-type Hsp70s again can be subdivided into those which are regulated by GrpE-type nucleotide exchange factors (the main bacterial and organellar Hsp70s), those regulated by BAG-1 (Hsp70s in the eukaryotic cytosol), and those that do not need any nucleotide exchange factor (specialized bacterial and organellar Hsp70s) [15]. All GrpE homologs form dimers that interact with their Hsp70 partners in the ADP-state and catalyze the release of ADP to allow for rebinding of ATP [95].

In addition to nucleotide exchange factors, Hsp70s function in concert with J-domain cochaperones [31]. J-domain proteins stimulate the ATPase activity of their Hsp70 partner [95] and lock it onto specific substrates [53]. Thus, J-domain proteins mediate substrate specificity and thereby the function of their Hsp70 partner. The J domain has a length of about 70 amino acids and contains a conserved tripeptide of histidine, proline, and aspartate (HPD motif), which is essential for stimulating the ATPase activity of their Hsp70 partners [200]. The number of J-domain proteins an organism possesses exceeds the number of Hsp70 chaperones, indicating that one Hsp70 may be recruited by multiple J-domain proteins to different targets within a cellular compartment. For example, *Chlamydomonas reinhardtii* encodes only 7 DnaK-type Hsp70s (HSP70A-D, HSP70F, BIP1, and BIP2), but at least 63 J-domain proteins [165].

Hsp70 systems do not act in isolation from the other cellular chaperone systems. They cooperate with ClpBs in protein disaggregation and refolding [60], with GroEL/Hsp60 in protein folding [87], and with the Hsp90 system in protein folding and signal transduction [136, 204].

13.3.2 Plastidic Hsp70s

Hsp70s have been proposed to be present in four different plastid compartments: the outer envelope [79, 81, 216], the intermembrane space [10, 20, 103, 159], the stroma [37, 103, 201], and the thylakoid lumen [153].

Hsp70s detected in outer envelope fractions apparently are eukaryotic-type, cytosolic Hsp70s, of which subfractions are engaged in delivering preproteins to the outer envelope translocon (see below). The presence of Hsp70s in the thylakoid lumen has been confirmed by some proteomics studies [132, 133], but not by others [166]. The finding that a small subfraction of stromal Hsp70 is firmly associated with (thylakoid) membranes [88, 164, 201] suggests that Hsp70s detected in thylakoid lumen preparations are in fact stromal contaminants. Also the detection of an Hsp70 in the intermembrane space, termed hsp70-IAP (import intermediate associated protein) [20, 159] or imsHsp70 [10], is likely caused by contaminations of envelope fractions by soluble stromal Hsp70s [139].

In *Arabidopsis thaliana* two Hsp70s, termed cpHsc70-1 and cpHsc70-2 [181], were shown to be imported into the chloroplast stroma [139, 179]. Both proteins are 90.7% identical to each other and appear to accumulate to comparable levels, albeit transcript levels appear to be higher for *cpHsc70-1* in cotyledons, root tips and seeds [88, 179]. In the moss *Physcomitrella patens* three stromal Hsp70s exist that are ~80% identical to each other and were termed PpHsp70-1 to 3 [170]. *Chlamydomonas reinhardtii* harbours a major stromal Hsp70 dubbed HSP70B [37]. The recent completion of the Chlamydomonas genome revealed two additional Hsp70s termed HSP70D and HSP70F that contain putative chloroplast transit peptides and are most similar to cyanobacterial and plastidic Hsp70s. Interestingly, while mature HSP70B with 68 kDa is a typical Hsp70, predicted molecular masses of HSP70D (~57 kDa) and HSP70F (~42 kDa) suggest that they have truncations at their C-termini. Whether the *HSP70D* and *HSP70F* genes give rise to functional proteins is unclear. In general, Hsp70s in the chloroplast stroma are of prokaryotic-type and, when compared with the three DnaK isoforms normally present in cyanobacteria, exhibit the highest degree of homology with DnaK2 [119, 139, 144, 170, 179].

Taken together, chloroplasts appear to harbour Hsp70s only in the stroma, but stromal Hsp70s also associate with thylakoid and/or inner envelope membranes. Moreover, cytosolic Hsp70s appear to associate with the outer chloroplast envelope, presumably while delivering precursor proteins for import into the organelle.

13.3.3 Cofactors and Cochaperones of Plastidic Hsp70 Systems

As is true for bacteria, also stromal Hsp70s cooperate with cochaperones and cofactors. Cofactors specific for mitochondrial and plastidic Hsp70s are the so-called Hsp70 escort proteins (HEP) [172]. HEPs are L-shaped molecules, where one leg of the "L" is formed by two two-stranded, antiparallel β-sheets which, at their connecting loops, each contain two cysteines that together coordinate a Zn^{2+}ion [110]. Apparently, this zinc finger is required to stabilize the structure of the protein. HEP homologs are not found in cyanobacteria and bacteria and therefore must have evolved after endosymbiosis. *C. reinhardtii* contains single HEP homologs in mitochondria (HEP1) and chloroplasts (HEP2) [214], while *A. thaliana* contains a single homolog in mitochondria (ZR3) and two homologs in plastids (ZR1/ET1 and ZR2) [32, 78].

Other important cochaperones for stromal Hsp70s are the GrpE-type nucleotide exchange factors termed CGEs (chloroplast GrpE homologs). Like their stromal Hsp70 partners, also stromal CGEs have a cyanobacterial origin. *A. thaliana*, *P. patens* and *C. reinhardtii* encode two putative stromal CGEs, but a location to the chloroplast stroma has been demonstrated only for CGE1 from *C. reinhardtii* and for CGE1 and CGE2 from *P. patens* [164, 170]. While algal CGE1 and moss CGE1 and CGE2 are true homologs of bacterial GrpE with molecular masses of ~24 kDa, this is not the case for CGE2 from *C. reinhardtii*. The latter is predicted to be a protein of ~152 kDa with the GrpE-domain at the C-terminus and a long N-terminal sequence of yet unknown function.

Among the 89 J-domain proteins identified in *A. thaliana*, about 12 were predicted to be targeted to the chloroplast [107]. Similarly, up to 13 out of the at least 63 J-domain proteins in *C. reinhardtii* might be targeted to the chloroplast [121]. A localization to the chloroplast stroma was verified for true homologs of bacterial DnaJ, which are termed PCJ1 in pea [154], and CDJ1 in *C. reinhardtii* [213]. While *C. reinhardtii* encodes only a single chloroplast DnaJ homolog, *A. thaliana* encodes four, which are termed AtDjA24, AtDjA26, AtDjA52 and AtDjA54 [107, 213]. As judged from their high sequence similarity with cyanobacterial DnaJ proteins, the chloroplast homologs appear to be derived from the endosymbiont. Like the true DnaJ homologs, also the ARC6 DnaJ-like protein has a close relative in cyanobacteria, the Ftn2 protein [198]. ARC6 is located to the chloroplast inner envelope membrane and exposes a J domain into the stroma. The ARC6 J domain is particular, as only the proline of the conserved HPD motif within the J domain is conserved.

In contrast to the true DnaJ homologs and ARC6, many other J-domain proteins that are predominantly targeted to the stroma appear to have evolved after endosymbiosis, as they have no (cyano)bacterial orthologs [38]. These include: CDJ2/AtDjB42 that interact with the vesicle-inducing proteins in plastids 1 (VIPP1) [97]; CDJ3/CDJ4/AtDjC18 and CDJ5/AtDjC17 containing iron-sulfur clusters [34]; AtJ11, consisting of a J domain and N- and C-terminal extensions of only 24 and 21 amino acids, respectively [124]; AtJ8 (formerly and erroneously TOC12) that contains a conserved 60-amino acids domain C-terminally to the J domain [23]; and AtJ20, which like AtJ8 and AtJ11 is a small J-domain protein [21].

13.3.4 Specific Functions of Chloroplast Hsp70 Systems

13.3.4.1 Functions of Cochaperones and Cofactors of Plastidic Hsp70s

Hsp70 Escort Proteins In *C. reinhardtii*, the Hsp70 escort protein 2 (HEP2) is a constitutively (but weakly) expressed stromal protein with a calculated molecular weight of ~ 14 kDa and apparent molecular weight of ~ 21 kDa [214]. HEP2 appears to form dimers and interacts preferably in the ADP-bound state as a minor partner with stromal HSP70B. Active HSP70B, *i.e.*, HSP70B in a protease-resistant conformation that is capable of interacting with CGE1, could be quantitatively produced in *E. coli* only when co-expressed with HEP2. HEP2 binds to active and inactive HSP70B, but cannot activate inactive forms of HSP70B such as HSP70B expressed in *E. coli* without HEP2.

Hep1 was proposed to interact with mitochondrial Hsp70s to maintain them in a functional conformation by preventing their aggregation [110, 149, 172]. Apparently, the situation is different for chloroplast HSP70B-HEP2 from *C. reinhardtii.* Instead of aggregating, HSP70B expressed in the absence of HEP2 assumed a protease-sensitive configuration unable to interact with CGE1 [214]. Active HSP70B that was co-expressed with HEP2 remained active for 48 h in the absence of HEP2, suggesting that HEP2 was not involved in maintaining HSP70B in an active state. Moreover, some HSP70B expressed as a fusion protein containing an N-terminal extension became activated if HEP2 was present during cleavage of the fusion protein, suggesting that *in vivo* HEP2 might be required for *de novo* folding of HSP70B after transit peptide cleavage. This hypothesis is supported by a recent phylogenetic analysis of plastidic HEP homologs from several algal and higher plant species. That study revealed that plastidic HEP homologs are absent in red algae and glaucophytes, which still encode Hsp70s on their plastid genomes [78].

It turned out that the *etched* (*ET1*) gene from *Zea mays* encodes one of two chloroplast-targeted HEP homologs [78]. Hence, the virescent phenotype (delayed plastid development in seedlings) associated with the maize *etched 1* mutant [32] is likely caused by reduced levels of functional stromal Hsp70s.

CGE Nucleotide Exchange Factors The biochemically so far best characterized CGE is CGE1 from *C. reinhardtii*, which interacts firmly with chloroplast HSP70B in its ADP-bound state and thus appears to be its nucleotide exchange factor [158, 164]. HSP70B and CGE1 constitute about 0.19 % and 0.01 % of total cell protein, respectively, thus corresponding to a molar ratio of ~ 6.7: 1 [98, 164]. CGE1 is only 32 % identical to its *E. coli* homolog, GrpE, yet shares a number of important structural features with GrpE [212]. This includes the ability to form dimers, and a three-dimensional architecture consisting (from N- to C-terminus) of a paired α-helix, a four-helix-bundle, and a β-sheet domain [54]. Moreover, CGE1 complements the temperature sensitive growth phenotype of an *E. coli* strain lacking GrpE, and interacts with *E. coli* DnaK [164, 212]. Therefore, despite the low sequence conservation and substantial evolutionary distance between CGE1 and bacterial GrpE, the proteins are quite similar at both the structural and functional levels.

However, stromal CGE1 and bacterial GrpE also differ in some important aspects. First, CGE1 exists as two isoforms, a and b, which differ by an additional valine-glutamine dipeptide at positions 4 and 5 of the mature CGE1b protein absent in CGE1a [164]. This difference is due to the temperature-dependent alternative splicing of the *CGE1* transcript, with *CGE1b* transcript and protein levels increasing upon heat shock [212]. Curiously, the two isoforms have different affinities for HSP70B: the affinity of CGE1b is about 25 % higher than that of CGE1a, indicating that the CGE1 extreme N-terminus plays an important role in determining the affinity of the cochaperone for HSP70B. However, the functional significance of this finding is not yet understood. Moreover, alternative splicing of *CGE* transcripts from moss or higher plants has not been reported yet and therefore might be a species-specific phenomenon.

An additional important difference between CGE1 and GrpE relates to their N-termini and dimer formation. The N-terminus of CGE1 contains a coiled-coil motif as opposed to the unstructured N-terminus of GrpE. Deletion analyses revealed that the N-terminal coiled-coil of CGE1 is essential for dimer formation, while dimerization of *E. coli* GrpE is mediated by the four-helix bundle at the posterior part of the molecule [55, 212]. Hence, although general structural and functional properties of GrpE and CGE1 appear to be conserved, the proteins have clearly evolved somewhat differently.

Single *cge1* and *cge2* mutants in moss displayed slow growth phenotypes and were delayed in the development of leafy shoots [170]. While double mutants were not viable, a *cge1* knockout/*cge2* knockdown mutant accumulating 10–20 % CGE2 displayed a 20–40 % reduction in growth rate and smaller and fewer chloroplasts compared with the wild-type. Moreover, the efficiency of protein import into chloroplasts was reduced in *cge* mutants, thus suggesting that stromal Hsp70s play a role in the import process and for this require efficient nucleotide exchange by CGE protein [170] (see below).

13.3.4.2 Functions of Plastidic Hsp70s and J-domain Proteins

General Protein Folding A role for stromal Hsp70s in the maturation of chloroplast proteins has been well documented for the subunits of the coupling factor CF1 [19], ferredoxin-NADP+reductase [191], the Rieske protein [101], and phytoene desaturase [13]. In all these cases, Hsp70 and chaperonin systems were shown to act coordinately. The identification of a complex formed by the stromal HSP70B, CGE1, CDJ1, and HSP90C proteins in *C. reinhardtii* also suggests a cooperation of Hsp70 and Hsp90 systems in chloroplasts [213]. Possibly, similar to the "foldosome" consisting of components of the Hsp70 and Hsp90 systems in the eukaryotic cytosol [136], the chloroplast equivalent might also be involved in the maturation of specific client proteins involved in signal transduction [161].

For a thorough understanding of the roles of stromal Hsp70 in general protein folding in the chloroplast a comprehensive map of its substrates under diverse stress conditions is essential. Key to this will be novel methods based on immunoprecipitation combined with stable isotope labeling and mass spectrometry [62, 111, 158].

Redox Regulation and Redox Sensing Oxidative stress has been shown to cause multiple protein modifications [11] that lead to the increased expression of molecular chaperones and proteases [196]. However, in yeast and other organisms oxidative stress also results in a dramatic drop in cellular ATP levels, which precludes ATP-dependent folding by molecular chaperones [125, 215]. Accordingly, mammalian cytosolic Hsc70 in the nucleotide-free state performed significantly better in preventing protein aggregation when it was glutathionylated compared to its unmodified conformation [65]. In the presence of ATP the performance of glutathionylated and unmodified Hsc70 were similar. Recent studies revealed that stromal HSP70B from *C. reinhardtii* is a target for thioredoxin and glutathionylation [94, 106]. This finding suggests that the activity of stromal Hsp70s might be regulated by the redox-state of the chloroplast.

An interesting type of chloroplast-targeted Hsp70 cochaperones are those containing a J domain, a bacterial 4Fe-4S cluster, and a domain of unknown function [34]. Interestingly, genes encoding proteins with a J domain and the 4Fe-4S cluster, but lacking the domain of unknown function have also been found in mesophilic Crenarchaeota (or Thaumarchaeota). Most likely, the genes encoding the chloroplast proteins were transferred to the archaebacteria by horizontal gene transfer. The biochemically best-characterized homologs of these peculiar plastidic J-domain proteins are the CDJ3/4 proteins from *C. reinhardtii* [34]. The *CDJ3/4* genes are weakly expressed. While transcript levels increase in the light, they decrease after heat shock, suggesting that CDJ3/4 do not play roles in the folding of stress-denatured proteins. Redox-active Fe-S clusters are assembled on CDJ3/4 expressed in *E. coli*, and both proteins interact *in vitro* with purified HSP70B when it is in the ATP-bound state. CDJ3 is localized to the stroma, whereas CDJ4 appears to be associated with thylakoid membranes.

CDJ3 was shown to bind RNA [34], a function that might point to a chaperone-mediated remodelling of RNA-binding protein complexes that, for example, are involved in translation initiation/elongation or mRNA stability. Such complexes are found in the stroma and associated with thylakoids [102], where CDJ3/4 and HSP70B are also located. Well-studied examples for chaperone-mediated remodelling of replication initiation occur in *E. coli*, where DnaK and DnaJ monomerize RepA dimers and dissociate DnaB-helicase-Lambda P complexes to trigger replication of plasmid P1 and lambda phage, respectively [4, 209]. As post-transcriptional regulation of the expression of many chloroplast genes is strongly regulated by light [102], it is possible that CDJ3/4 represent nuclear-encoded factors that act as redox switches by recruiting stromal HSP70B for the reorganization of regulatory protein complexes.

VIPP1 (Dis)assembly The CDJ2 protein was shown to mediate the interaction between stromal HSP70B/CGE1 and the vesicle-inducing protein in plastids (VIPP1) in *C. reinhardtii* [97]. There, HSP70B, CDJ2, CGE1 and VIPP1 are present in a molar ratio of 17.5: 1: 2.6: 10.6 [98]. VIPP1 dimers form rings of up to 2 MDa that contain a central hole [6, 45, 98]. VIPP1 rings may further assemble into large rod-shaped tubules, whose assembly state is ATP-dependently dynamically altered by the HSP70B-CDJ2-CGE1 chaperones [98]. Presumably the chaperones introduce conformational changes into VIPP1 that facilitate assembly and disassembly.

C. reinhardtii strains in which VIPP1 was downregulated to ~25% of wild-type levels exhibited several defects in the organization of their thylakoid membranes when compared with wild-type [122]. This was evident by (i) swelling of thylakoid membranes in mutant cells grown on ammonium as nitrogen source and exposed to high light. Obviously, the thylakoid membranes in the mutants were unable to withstand the increased osmotic pressure generated by the accumulation of ammonium ions in the acidic lumen. (ii) Photosystem (PS)II in the mutants was more sensitive to heat shock. (iii) The maximum temperature of the Q-band was downshifted in the mutants, thus reflecting a lowering of the midpoint potential of the redox couple Q_A/Q_A^- in PSII [84]. (iv) Cooperativity between PSII complexes was reduced. (v) Levels of PSI, PSII, Cytb_6/*f*, and ATP synthase were reduced by up to 20%, while levels of LHCII were increased by ~30%. Strikingly, these phenotypes came along with aberrant structures resembling prolamellar bodies at positions within the chloroplast where multiple thylakoid membranes emerge. Electron microscopy data suggest that so-called thylakoid centers are located at such positions, which in cyanobacteria are located close to the plasma membrane and contain long tubules as a central component [86, 195]. In *C. reinhardtii*, VIPP1 was visualized by immunofluorescence and localized to distinct spots within the chloroplast [122]. Interestingly, using YFP fused to the C-termini of stromal cpHsc70-1 and 2, both were also localized to distinct spots within chloroplasts [88]. Thus, it is tempting to speculate that stromal Hsp70s accumulate at regions were VIPP1 oligomers are localized.

It was suggested recently that PSII biogenesis in cyanobacteria occurs at these thylakoid centers [117]. As the dimensions of the tubules within thylakoid centers fit exactly those recorded for VIPP1 tubules [86, 98], it may be possible that both are identical and a reduction in VIPP1 levels thus results in impaired function of thylakoid centers in the biogenesis of PSII, and perhaps also of other thylakoid membrane complexes [122]. In this case, HSP70B-CDJ2-CGE1, by mediating the dynamic interconversion between VIPP1 monomers and oligomers, would indirectly play a role in the biogenesis of thylakoid membrane complexes. This idea is supported by the finding that downregulation of the *C. reinhardtii* Alb3.2 protein, which is involved in photosystem assembly, resulted in the upregulation of VIPP1, HSP70B, and CDJ2 [49].

Protection of PSII from High-light Damage One of the earliest investigations of the stromal HSP70B in *C. reinhardtii* revealed that its gene is highly induced by light [199], resulting in an approximately two-fold increase in HSP70B protein levels [37]. This finding suggested a possible role for this chaperone in processes that help the cell to cope with photodamage. Accordingly, cells overexpressing HSP70B exhibited less severe damage to photosystem II and recovered photosystem II activity faster after photoinhibition, compared to wild-type cells. The opposite effect (more severe damage and slower recovery) was observed for cells underexpressing HSP70B [162]. It was hypothesized that HSP70B might facilitate a coordinated exchange of damaged D1 protein by *de novo*-synthesized D1 protein [163]. In support of this hypothesis, HSP70B in the green alga *Dunaliella salina* was found to be part of a ~320 kDa complex containing photodamaged D1, D2, and CP47 proteins [219].

Interestingly, knockout mutants of the chloroplast-targeted AtJ8, AtJ11, and AtJ20 proteins each displayed a reduced stability of PSII-LHCII supercomplexes and of PSII dimers in high light [21]. As these mutants showed impaired CO_2 assimilation rates, which correlated with ~20% lower levels of Rubisco activase, this effect might be caused by a reduced electron sink. In this case the positive effect of *C. reinhardtii* stromal HSP70B on PSII photoprotection [162] is an indirect one, as stromal HSP70 would cooperate with the small J-domain proteins in improving the efficiency of the dark reactions. The *atJ8*, *atJ11*, and *atJ20* mutants also accumulated higher levels of enzymes involved in the detoxification of H_2O_2, which suggests that they have elicited an oxidative stress response [21]. The finding that AtJ8 transcript and protein levels are high in dark and decline drastically with increasing light intensities led to the suggestion that the protein might be active only in the dark [22]. Obviously, more work is required to bring these findings into a functional context.

Protein Import into Plastids In the past, Hsp70 proteins have been associated with all stages of chloroplast protein import, although many ideas and models are heavily disputed [42]. Targeting of pre-proteins to the chloroplast may involve either an accumulation of the corresponding mRNA at the chloroplast surface, or the guidance of the fully translated protein in an unfolded, import-competent state to the chloroplast surface [104, 194]. In the latter case, cytosolic Hsp70s have been suggested to interact with transit peptides and 14-3-3 proteins to form a so-called guidance complex [104]. Cytosolic Hsp70 proteins can interact with chloroplast transit peptides, although this interaction does not seem to be necessary for protein import, but may rather have an influence on the degradation of accumulating precursors [70, 71, 93, 140, 141]. Similarly, mutations of putative 14-3-3 binding phosphopeptide motifs within transit peptides did not affect protein import [90, 116]. Therefore, the existence of a guidance complex and hence the involvement of cytosolic Hsp70 proteins in pre-protein guidance remains highly speculative.

Another unsolved problem concerns the observed NTP requirement in the intermembrane space which is necessary for pre-protein binding to the chloroplast surface [123]. As outlined above, the presence of an intermembrane space Hsp70 involved in protein import in pea has been proposed [10, 103]. However, of the three putative chloroplast homologues of *Arabidopsis* two are located in the stroma and one is not imported *in vitro* [139, 179]. Hence, the involvement of an intermembrane space Hsp70 in protein import is highly speculative as well.

More evidence is accumulating though for an active involvement of stromal Hsp70 in chloroplast protein import. In *Arabidopsis*, both stromal Hsp70 mutants *cphsp70-1* and *cphsp70-2* showed reduced protein import at an age of 14 and 24 days but not 30 days, suggesting an involvement of both isoforms in protein import at least during early plant development [180]. Interestingly, the *cphsp70-2* mutants have a wild-type phenotype despite the observed reduction of pre-RBCS import which rather suggests that the import defect is of a small scale [179, 180]. Still, cpHsp70 could be cross-linked to importing precursors as well as TIC and even TOC components, and the association with precursors and Tic110 could be detected even without added cross-linker [180]. Most strikingly, the *cphsp70-1 tic40* double mutant is embryonic lethal and the *cphsp70-2 tic40* and *cphsp70-1 hsp93-V* double mutants have severe additive phenotypes, proposing a model in which stromal Hsp70 acts in parallel to the Tic40-Hsp93 motor system [180]. An independent study in the moss *Physcomitrella patens* discovered that one of three chloroplast-localized Hsp70s is essential, and that the corresponding conditional, temperature-sensitive mutant was impaired in chloroplast protein import upon heat-shock of isolated chloroplasts [170]. Similar to the observations in *Arabidopsis*, moss stromal Hsp70 could be cross-linked to importing precursor, Hsp93 and Tic40 which suggests that involvement of Hsp70 in chloroplast protein import is conserved from moss to higher plants [170].

Only little is known about the contribution of chloroplast Hsp70 nucleotide exchange factors, CGEs, to protein import. The knockout of both chloroplast-localized CGEs in *Physcomitrella* is lethal, but the *cge1* null *cge2* knockdown has a reduced import capacity in most cases [170]. The depletion of CGE proteins leads to an upregulation of the essential Hsp70-2, thus possibly restoring protein import at least in a few cases. Nevertheless, this shows for the first time that these chloroplast nucleotide exchange factors may be involved in protein import, too, evidence which is still missing for *Arabidopsis*. About a possible influence of J-domain proteins on chloroplast protein import nothing is known to date; this question remains open for future research.

Plastid Development *cphsc70-1* knock-out mutants in *A. thaliana* display variegated cotyledons, vegetative leaves with irregular margins and small lesions, and show impaired growth rates [88, 179]. These phenotypes were much more severe in plants exposed to drought stress. In contrast, knock-out mutants of stromal *cphsc70-2* did not display any visible phenotypes even when exposed to drought stress. *cphsc70-1 cphsc70-2* double knock-out mutants were lethal due to reduced pollen transmission efficiency [179]. Plants in which levels of both stromal Hsc70s were strongly reduced by artificial microRNA or co-suppression were almost completely white [88]. In white tissues and in pale tissues of *cphsc70-1* mutants, no or only small chloroplasts were found that had an altered morphology. The latter contained few, unorganized thylakoid membranes and levels of photosystem core subunits were strongly reduced [88]. These strong developmental phenotypes are reminiscent of those observed with the maize *etched 1* mutant lacking an escort protein of stromal Hsp70s [32], and also with the moss *cge* mutants [170], i.e., mutants expected to affect the functionality of stromal Hsp70s. It will be interesting to see whether these phenotypes are due to reduced protein import rates [170, 180], due to a reduced capability to maintain/generate thylakoid membranes e.g. via VIPP1 [97, 98, 122, 162], the reduced folding of yet unknown substrates, or a combination of these scenarios.

13.4 Hsp90s

13.4.1 General Aspects of Hsp90 Chaperones and Cochaperones

Molecular chaperones of the Hsp90 family are highly conserved proteins and found in most organisms ranging from bacteria to mammals. In eukaryotic cells, Hsp90s are located in the cytosol, the endoplasmic reticulum (ER), and mitochondria [72]. Plants contain an additional member located in plastids. For instance, the *A. thaliana* genome encodes seven Hsp90s: four members (AtHsp90-1 to 4) are located in the cytosol, while AtHsp90-5, AtHsp90-6 and AtHsp90-7 are located in plastids, mitochondria and the ER, respectively [85]. In contrast, the *C. reinhardtii* genome encodes only three Hsp90s: cytosolic HSP90A, HSP90B of the ER, and plastidic HSP90C [160]. An interesting aspect concerning the evolution of subcellular Hsp90 isoforms was observed by a phylogenetic analysis of 77 different *HSP90* genes from various organisms. This study revealed a close relation between bacterial Hsp90 and mitochondrial chaperones. In contrast, plastidic Hsp90 seems to be more closely related to the ER isoform than to bacterial homologs, indicating that chloroplast Hsp90 is not derived from the cyanobacterial ancestor HtpG but rather by a gene duplication event of ER *HSP90* and subsequent acquisition of a chloroplast transit sequence [39]. In general, Hsp90s are essential components of the cellular heat shock response mechanism but their putative roles in cellular protein homeostasis go far beyond simple stress tolerance, as indicated by their abundant expression and their broad substrate spectrum [183].

All Hsp90s analyzed so far contain three conserved domains: the N-terminal ATPase domain, a structurally flexible middle domain and a C-terminal domain, which mediates homo-dimerization of two Hsp90 proteins (reviewed in [130]). The active form of Hsp90 functions as a dimeric "clamp" and both subunits undergo large dynamic conformational shifts driven by ATP binding and hydrolysis. In a simplified model of this substrate cycle, nucleotide-free Hsp90 is in an open conformation, only dimerized via the C-terminal domain. ATP binding leads to a closed conformation with additional dimer interactions at the N-terminus. Subsequent ATP-hydrolysis results in a conformational shift of the complex that is structurally not fully understood, and ADP-release restores the dimer in the open conformation (reviewed in [183]). Substrate binding occurs in the closed conformation and is mediated by multiple binding sites over the entire length of the Hsp90 structure.

In the best-understood cytosolic Hsp90 system of yeast, ATP-hydrolysis, cycling through the open and closed conformations, and substrate binding are tightly connected to a vast number of cofactors (twelve in *S. cerevisiae* and an even higher number in mammalian cells) (reviewed in [72]). Little is known about cofactors in other cellular compartments. It has been shown recently that the Hsp90 of the ER (Grp94) requires a cofactor termed Canopy 3 (CNPY3), which targets specific substrates to the chaperone for folding [99]. However, no cofactors have been identified yet for the plastidic and mitochondrial Hsp90 chaperones.

Substrates of cytosolic Hsp90 (termed clients) embrace a broad spectrum of cellular factors that are inherently unstable or require assistance to fold properly. Hsp90 clients are components of signal pathways such as kinases, transcription factors and steroid hormone receptors indicating its central function as a hub in protein homeostasis (for details see [183, 204]). In the process of client maturation, Hsp90 often cooperates in a successive folding cascade with Hsp40 and Hsp70, where Hsp90 folds client proteins after initial folding by Hsp70 [205]. In this folding complex, the interaction with Hsp70 is promoted by a well-studied cochaperone termed Hsp70-Hsp90 organizing protein, Hop [72, 137, 155]. Active cytosolic Hsp90 in this multi-chaperone complex exhibits a significantly higher ATP turnover than inactive Hsp90 [73].

An interesting aspect about eukaryotic Hsp90 function is its role in influencing phenotypic manifestation. Studies with multiple model organisms ranging from plants to vertebrates report that Hsp90 acts as capacitor of genetic variation. Impairment of Hsp90 function resulted in occurrence of a variety of phenotypes that were heritable and dependent on the genetic background. Some of the phenotypes had the capacity to be stable and independent of Hsp90 when enriched by selection [138, 145, 148, 218].

The central function of Hsp90 in cellular protein homeostasis makes it an important target for therapeutic targeting. Thus, a number of small molecule inhibitors such as Geldanamycin or 17-AAG have been identified over the last decade with potential to act as potent anti-tumor agents. Most of these components mimic the unique kinked conformation of ATP bound at the N-terminal ATPase domain [131].

13.4.2 Functions of Plastidic Hsp90s

Like cytosolic Hsp90, chloroplast Hsp90 is a highly abundant and constitutively expressed protein that is localized in stroma, low-density membranes (consisting of inner envelopes and transitory membranes between inner envelope and thylakoids), and thylakoid fractions [211]. Chloroplast HSP90C from *C. reinhardtii* shares common features with Hsp90s from other compartments: it forms dimers and displays a low ATP hydrolysis rate (Km = 48 μM; Kcat = 0.71/min), which is comparable with those of other Hsp90s [40, 127, 211]. Analysis of Hsp90s from *A. thaliana* indicated that the plastidic chaperone is critical for cellular stress response and tolerance [177], consistent with its moderate induction by heat stress and light [18, 156, 211].

It has been observed that chloroplast HSP90C of *C. reinhardtii* forms a multi-chaperone complex together with stromal HSP70B. This complex contains also cofactors of the Hsp40 family (CDJ1) and the nucleotide exchange factor CGE1, indicating an orthologous function in protein maturation like the cytosolic multi-chaperon complex [161, 213]. It has been further reported that HSP90C interacts with the vesicle-inducing protein in plastids 1 (VIPP1), which might indicate that HSP90C, in addition to HSP70B-CDJ2-CGE1, serves in the assembly/disassembly of VIPP1 oligomeric complexes [62] (see above). However, it remains elusive how the interaction between HSP90C and HSP70B is mediated. Both chaperones appear not to directly interact *in vitro* [211], suggesting the need for a coordinating cochaperone, as known from the cytosol [72]. In the cytosol, many cofactors contain TPR-domains that recognize dimeric Hsp90 via its C-terminal MEEVD acceptor motif. This motif is absent in plastidic Hsp90s, however, they all contain a distinct C-terminal DPW motif, which might serve an orthologous function [18, 156]. It would be interesting to determine, if this region is essential for plastidic Hsp90 function.

Although cyanobacterial HtpG is not directly related with plastidic Hsp90, some functional aspects might indicate related functions in the chloroplast. It was recently reported that HtpG interacts with a highly unstable linker polypeptide of phycobilisomes. Phycobilisomes are the major light harvesting complexes in cyanobacteria and the authors conclude that HtpG protects this polypeptide from thermal aggregation [150].

An interesting functional aspect regarding chloroplast Hsp90 was gained by analysis of an *A. thaliana* mutant carrying a point mutation in the dimerization domain of plastidic Hsp90 [18]. This mutant showed a yellow-green phenotype due to retarded development of chloroplasts, particularly in young leaves. In addition, the mutant exhibited reduced light-inducible expression of the *NR2, CAB* and *RBCS* genes and retarded deetiolation in red light [18, 96]. The authors concluded that chloroplast Hsp90 might exhibit a role in the transduction of light signals responsible for the regulation of a distinct set of photosynthesis-related genes. In *C. reinhardtii*, intermediates of chlorophyll biogenesis serve as signaling molecules between the chloroplast and the nucleus that mediate light induced gene expression [9]. Interestingly, in cyanobacteria, HtpG controls the activity of HemE, which is located at the first branching point of the tetrapyrrole biosynthetic pathway. Thus, these findings might point to an orthologous mechanism by which light induction of nuclear genes is influenced by plastidic Hsp90 [161, 202].

13.5 Hsp100s

13.5.1 The Hsp100 family of AAA+ATPases in Chloroplasts

The Hsp100 family of AAA+(ATPases associated with various cellular activities) proteins was originally discovered in bacteria. Although they are indeed associated with a wide variety of different functions within the cell and its compartments, they were found to employ a similar mode of action: they use ATP to induce changes in the folding and assembly of other proteins [151]. The Hsp100 family proteins are divided into two classes, with class 1 members having two conserved but fairly different nucleotide binding domains that arose from gene fusion, and class 2 members having only one nucleotide binding domain with homology to the second nucleotide binding domain from class 1 members [151]. The middle region between the two nucleotide binding domains varies in size between class 1 members, which therefore are further sub-divided in A-, B-, C- and D-type Hsp100 proteins. Similarly, class 2 members are further sub-divided in M-,N-, X-and Y-type Hsp100 proteins based on the longer N-terminus of M- than N-type proteins, the higher homology of M- and N-type proteins to class 1 members than X- and Y-type proteins, and an insertion into Y-type proteins between Walker A and B domains compared to X-type proteins [151].

Hsp100 family proteins can form ring-shaped hexamers upon binding of ATP, as determined initially for *E. coli* ClpA and yeast Hsp104 [151]. Yeast Hsp104, a ClpB protein, was found inefficient in preventing the aggregation of a chemically denatured model protein, a property which is otherwise characteristic for molecular chaperones [47]. Instead, it is essential in restoring already aggregated proteins along with other chaperones that assist the refolding of untangled proteins [47]. The hexameric structure of Hsp104 led to the suggestion that different aggregated components could be bound by several substrate recognition sites at the same time, and that a reversible conformational change upon ATP hydrolysis could pry the entangled components apart [48]. In this so called molecular crowbar model the primary function of Hsp104 would be to reduce the size of the aggregates by breaking them apart while other chaperones prevent re-aggregation [48]. The molecular ratchet model on the other hand explains disaggregation by the threading of individual components through the central pore of the Hsp104 hexamer [48]. In this model a free N- or C-terminus or a disordered loop would pass through the pore in one direction only while backsliding is prevented by residues, probably a tyrosine of a conserved GYV/IG domain, that are able to "shut" the pore upon backward movement of the emerging protein [48]. This latter model could equally well explain the ability of Hsp100 proteins to deliver individual proteins to proteases, which have a requirement for unfolded substrate, or the unidirectional pulling force exerted by Hsp100 proteins involved in the import of unfolded pre-proteins into cell compartments [30, 66].

The *Arabidopsis thaliana* genome revealed nine Hsp100 genes, namely *CLPB1*, *CLPB2*, *CLPB3*, *CLPB4*, *CLPC1*, *CLPC2*, *CLPD*, *CLPX1* and *CLPX2* [1, 92]. ClpB1 is a cytosolic protein also known as AtHsp101 which is crucial for thermotolerance [50]. The *CLPB2* gene is not expressed and encodes only a part of normal B-type proteins, and ClpB4, ClpX1 and ClpX2 are mitochondrial proteins [1, 92]. The remaining four proteins, ClpB3, ClpC1, ClpC2 and ClpD, are all chloroplast localized [92, 134, 221]. The genome of *Chlamydomonas reinhardtii* contains homologues of ClpB3, ClpC and ClpD which are putatively located in the chloroplast [165]. In rice, 3 ClpB, 4 ClpC and 2 ClpD proteins are encoded in the genome, of which two ClpC proteins are predicted and ClpB-c has been confirmed to localize in the chloroplast [175].

13.5.2 ClpB and its Role in Protein Disaggregation

While it was established for *E.coli* ClpA and ClpX proteins that they can interact with ClpP proteases and thus participate in protein degradation, the tripeptide motif IGF/L necessary for this interaction is absent in ClpB proteins [66, 76]. Therefore, ClpB proteins do not associate with ClpP and have likely no direct influence on proteolysis [151, 210]. Instead, they seem to play a role in disaggregating stress induced protein aggregates, as it was originally found for the yeast ClpB component Hsp104 [129]. Hsp104 and ClpB assemble into hexamers upon binding of ATP, with each protomer protruding a relatively large middle domain either towards the central pore or towards the surface [91, 208]. This large middle domain is unique for ClpB proteins and appears to be important for ClpB function as well as interaction of ClpB with other chaperones [60, 108, 152]. On the other hand, the shorter

N-terminal domain is not intrinsically important for ClpB function but might facilitate the interaction of ClpB with some substrates [26, 35, 68, 108]. It was shown that high concentrations of Hsp104 alone can disassemble Sup35 and Ure2 aggregates *in vitro*, but in most *in vivo* conditions it interacts with the Hsp70/Hsp40 system [47, 171]. Thereby, Hsp70 likely binds to unstructured regions of aggregated proteins before and after ClpB activity to induce substrate binding and prevent re-aggregation [36, 206, 222]. ClpB hexamers act by pulling unfolded protein ends or even unfolded loops through the central pore [61, 185, 206].

Many plant species have a cytosolic, a mitochondrial and a chloroplast ClpB with the organellar ClpB proteins being more closely related to each other than to the cytosolic ones [92, 174]. However, the high structural similarity of all Arabidopsis and rice ClpB proteins and the high homology to yeast Hsp104 suggests a common function for all ClpB proteins [174, 175]. Indeed, all three isoforms of ClpB in *Arabidopsis thaliana, Chlamydomonas reinhardtii* and rice are upregulated by heat stress, suggesting that ClpB may have a quite conserved role in the disaggregation of heat-induced aggregates [92, 111, 175]. A cytosolic ClpB from soybean can functionally compensate a yeast *hsp104* mutant and restore thermotolerance, again suggesting that ClpB proteins share a conserved function [89].

In Arabidopsis, ClpB3 was discovered as APG6 in a screen for albino and pale green (*apg*) mutants [113]. Chloroplast localization of ClpB3, which is encoded with a predicted transit peptide, has been confirmed by GFP tagging [92, 113]. The pale, seedling lethal phenotype of *apg6* mutants and the reduced chloroplast size and thylakoid network indicates a role for ClpB3 in normal chloroplast development and growth [92, 113]. In addition, homozygous mutant seeds showed a defect of *apg6* already in embryogenesis and seed development [113]. Interestingly, while heat shock leads to a general expression of a ClpB3 promoter-GUS fusion in all tissues, non-heat-shock treated plants showed highest expression of the construct in meristematic and post-mitotic tissues containing proplastids, which points towards an additional role of ClpB3 in proplastid differentiation and maturation [113]. In tomato, however, antisense lines of a chloroplast localized ClpB3 homologue had a wild-type phenotype even though both ClpB3 mRNA and protein were reduced below detection levels even after heat shock [217]. While these tomato antisense lines showed the expected heat sensitivity, in line with the notion of a generally conserved ClpB function, the Arabidopsis *apg6* mutants did not show an additional defect after heat shock [92, 217]. It was speculated that Arabidopsis *apg6* mutants might lack a thermosensitivity phenotype only because it is masked by the seedling lethal phenotype [92].

Recently, it was shown that chloroplast ClpB3 is not only heat shock induced. In maize, Friso et al. [43] showed that bundle sheath cells contain higher ClpB3 and Hsp90 levels than mesophyll cells [43]. This shows that certain cell types may use ClpB3 to remodel the chloroplast proteome for developmental purposes rather than as a result of stress [43]. In an Arabidopsis *clpr2* mutant background, which shows impaired proteolysis, ClpB3 is upregulated, presumably to disaggregate higher amounts of misfolded proteins that accumulate due to slower degradation [223]. Consequently, the *clpr2 clpb3* double mutant shows additive defects [223].

Finally, it was shown that even high irradiance can induce ClpB3 expression, although the reason for this might be an increased surface temperature caused by the high irradiation [2]. All in all, it seems that the function of all ClpB proteins, including chloroplastic ClpB3, is quite conserved, namely the disaggregation of protein aggregates. This function is almost always necessary as a consequence of heat stress but probably also other stresses and even developmental proteome rearrangements.

13.5.3 ClpC: Partitioning Between Protein Degradation and Chloroplast Protein Import

Chloroplast ClpC was originally identified as a functional homologue of *E. coli* ClpA that is able to facilitate the degradation of 3H-methylcasein by *E. coli* ClpP in an ATP-dependent fashion [167]. Both ClpC and the ClpP protease have been localized to the chloroplast stroma by subfractionation and immunolocalization, suggesting that they act together as a ClpC-ClpP protease [167]. The interaction of ClpC with ClpP was found to be dependent on ATP binding but not hydrolysis, while ATP hydrolysis was necessary for the degradation of beta-casein and OE33 [51, 52]. This degradation activity could be inhibited by phenylmethylsulfonyl fluoride, showing that the ClpC-ClpP protease is an ATP-dependent serine protease [52]. Interestingly, both pea ClpC mRNA and protein levels were increased by exposure of etiolated seedlings to light and decreased at temperatures higher or lower than ambient 25°C, indicating that, unlike other Hsp100 chaperones, ClpC has a role in the constitutive development of green leaves [126]. Similarly, a Synechocystis ClpC homologue was expressed higher under rapid growth conditions but lower under heat shock conditions, again implying a role of ClpC for general photosynthetic growth [28]. This finding was supported by Arabidopsis *clpc1/hsp93-V* mutants, ClpC1/Hsp93-V being the main isoform of two Arabidopsis ClpCs, which showed retarded growth, a chlorotic phenotype, impaired photosynthetic performance and a lower PSI and PSII content [30, 82, 176].

In Arabidopsis, the two stromal ClpC isoforms have been called Hsp93-V and Hsp93-III, designating their apparent molecular weight and the chromosome number on which the corresponding gene is located [30]. The two isoforms have likely overlapping functions because *hsp93-III* knockout mutants show no defect compared to wild-type while *hsp93-V hsp93-III* double mutants are embryo-lethal [30, 82, 83]. This suggests that ClpC has an essential function in chloroplast biogenesis and therefore plant survival. In fact, some reports indicate that the ClpC-ClpP protease might indirectly regulate thylakoid biogenesis: An *ffc/cpsrp54* mutant impaired in thylakoid protein targeting showed strong upregulation of ClpC1/Hsp93-V, and the *ffc clpc1* double mutants are seedling lethal, suggesting that ClpC1/Hsp93-V might be necessary for the turnover of non- or mis-targeted proteins [146]. Also, the variegation phenotype of *var2/ftsh2*, being defective in thylakoid protein turnover in white sectors, was shown to be suppressed by a *clpc2/hsp93-III* splice defect mutant [128]. As *ftsh2 ftsh8* double mutants are albino, it might be that FtsH8 levels and consequently thylakoid protein levels are regulated by ClpC2 proteolysis in *var2/ftsh2* mutants [128, 220].

ClpC is considered to play a crucial role in chloroplast protein import in addition to proteolysis. It was discovered to interact with importing protein precursors on their way into the chloroplast, as well as with translocon complexes of the TOC/TIC (translocon at the outer/inner chloroplast envelope membrane) machinery under limiting ATP conditions [3, 118]. High levels of ATP, however, destabilized this interaction pointing towards a short-lived association of ClpC with precursors and a fast ATP-dependent protein import process [118]. Recently, it was found that ClpC2/Hsp93-III, like ClpD, can bind to a transit peptide when it is fused N-terminally to GST, but neither when fused C-terminally to GST, nor when interleaved between GST and Ferredoxin-NADP+-reductase [16]. Additionally, a pool of ClpC1/Hsp93-V seems to be constantly associated with inner envelope membranes, and this association is dependent on the presence of the ClpC1/Hsp93-V N-terminus [27]. In fact, an N-terminal deletion construct of ClpC1/Hsp93-V could not complement the pale-green and protein import-defective phenotype of the *clpC1/hsp93-V* knockout mutant while ATPase activity and the degradation of the assumed ClpP target glutamine synthetase are not affected [27]. All these data strongly propose that ClpC has an important role as a motor protein involved in protein import propulsion.

In an import time-course experiment, Arabidopsis ClpC1/Hsp93-V was crosslinked to the TIC components Tic40 and Tic110 in the late stages of protein import, which implies that these components act together in the late, stromal stages of protein import [24]. Intriguingly, Tic40 is a cochaperone, which contains a TPR domain as well as domains with homology to human cochaperones Hop (Hsp70-Hsp90 organizing protein) and Hip (Hsp70 interacting protein), while Tic110 is the proposed protein channel at the inner envelope membrane [24, 63]. In a study using yeast two-hybrid and immunoprecipitations with constructs containing individual domains of Tic40 and Tic110 it was shown that the stromal Tic110 domain interacts with the Tic40 TPR but not the Hip/Hop domain and that this interaction is increased upon binding of RBCS precursors but not mature RBCS to the stromal domain of Tic110 [25]. The binding of Tic40 to Tic110 releases the precursors from Tic110, whereupon the Tic40 Hip/Hop domain stimulates ATP hydrolysis by ClpC1/Hsp93-V, a property which can be abolished by mutations in the Hip/Hop domain [25]. Indeed, three independent studies noted a decreased protein import rate in Arabidopsis *clpc1/hsp93-V* mutants, suggesting active participation of ClpC/Hsp93 in the import process, although a fourth study failed to show the same trend [27, 30, 82, 176]. On the other hand, while unfolded PC-DHFR precursor is imported at a higher rate than folded PC-DHFR (stabilized with methotrexate) in wild-type, the *hsp93-V hsp93-III-1* double mutant (where *hsp93-III-1* is a knockdown rather than a knockout) imported the stabilized PC-DHFR and the unfolded precursor equally well, which shows that ClpC/Hsp93 is not required for precursor unfolding [83]. Finally, the model of ClpC hexamers being able to thread an already unfolded precursor through its central pore and thus creating an importing force by ATP hydrolysis is in line with the proposed model for other Hsp100 proteins [185, 206].

13.5.4 Is ClpD a Stress and Senescence Related Chaperone?

ClpD was discovered in Arabidopsis in a cDNA pool from 1-hour-dehydrated plants and named ERD1 for *e*arly *r*esponse to *d*ehydration 1 [77]. The nucleotide sequence shows 38 % and 34 % identity with *E.coli* ClpA and ClpB, respectively, and has an N-terminal transit peptide [77]. Localization to the plastid was confirmed by fusing the N-terminal part of ClpD/ERD1 to GFP [115]. The ClpD/ERD1 protein is properly processed after import into the chloroplast and accumulates in the soluble fraction [203]. Expression of ClpD/ERD1 was strongly induced upon dehydration, but not upon any other stress treatments such as high or low temperature, growth hormone, heavy metal or starvation treatments [77]. Nakashima et al. [115] showed by northern blots and GUS reporter gene fusion that expression of ClpD/ERD1 was also induced by natural senescence and dark-induced etiolation [115]. Studies on the ClpD/ERD1 protein level in dark-induced etiolation and natural senescence have, however, produced quite conflicting results. While Nakabayashi et al. [114] report an increase of ClpD/ERD1, detected at 97 kDa, in both dark-induced etiolation and natural senescence, Weaver et al. [203] reported a decrease of the same protein, detected at about 100 kDa, again in both conditions [114, 203]. Finally, Zheng et al. [221] showed that ClpD/ERD1 protein levels are not increased upon short-term dehydration but rather upon long-term high-light and cold acclimation [221].

Using ClpD/ERD1 promoter luciferase fusions with various deletions, Simpson et al. [173] found two discrete ClpD/ERD1 promoter elements with two motifs each: a dehydration-responsive element with an *rps1* site-1-like motif and a MYC-like motif as well as an etiolation-responsive element with an abscisic acid responsive element (ABRE) motif and an ACGT motif [173]. Because the etiolation-induced luciferase expression could be inhibited by incubation in 1 % sucrose, it was suggested that the ClpD/ERD1 etiolation-responsive element is inhibited by endogenous levels of assimilates and activated upon starvation during etiolation or senescence, although earlier studies have shown that sucrose-starvation for 10 hours alone is not enough to induce ClpD/ERD1 expression [77, 173]. Using yeast one-hybrid and gel retardation assays, Tran et al. [188] found interactions of the NAC transcription factors ANAC019, ANAC055 and ANAC072 with the ClpD/ERD1 MYC-like promoter sequence *in vivo* and *in vitro*, and found that ClpD/ERD1 promoter GUS fusions are induced by all three factors [188]. ANAC072 expression is induced upon dehydration whereas ANAC019 and ANAC055 are rather high-salt-induced, but the fast and ABA-independent induction of all three factors suggest that they might act upstream of ClpD/ERD1 expression [188]. An additional factor, zinc-finger homeodomain 1 (ZFHD1), was found to bind to the *rps1* site-1-like motif using yeast one-hybrid assays and to each of the three NAC factors using yeast two-hybrid assays [189]. In fact, only co-expression of each NAC factor with ZHFD1 could induce ClpD/ERD1 expression, suggesting that the two factors act in concert during dehydration and salinity stress [188, 189].

In terms of functionality, ClpD/ERD1 has been shown to form hexamers upon addition of ATP like it is the case for other members of the Hsp100 chaperone family [142]. Although intrinsic ATPase activity under physiological conditions was found to be unlikely, ClpD/ERD1 was still able to disaggregate up to 20% of heat-denatured luciferase [142]. ClpD/ERD1 appears to be C-terminally processed, and this processed form is able to interact with a model transit peptide N-terminally fused to GST both *in vitro* and in isolated chloroplasts [142]. In contrast, no interaction of ClpD/ERD1 was found with the same transit peptide fused C-terminally to GST or interleaved between GST and FNR, suggesting that a free amino terminus of the transit peptide is required for interaction [16]. In addition to a potential role in protein import, the presence of the tripeptide domain IGF/L, which is required for association with the ClpP protease, could point towards a role of ClpD in protein degradation [76, 174]. Although the inconsistent results concerning the ClpD/ERD1 protein levels in dehydration-stressed and senescent plants need further investigation, it could be speculated that ClpD/ERD1 might play a role in protein import or degradation during certain stress conditions, senescence or high light and cold acclimation.

References

1. Adam Z et al (2001) Chloroplast and mitochondrial proteases in *Arabidopsis*. A proposed nomenclature. Plant Physiol 125:1912–1918
2. Adamiec M, Lucinski R, Jackowski G (2011) The irradiance dependent transcriptional regulation of AtCLPB3 expression. Plant Sci 181:449–456
3. Akita M, Nielsen E, Keegstra K (1997) Identification of protein transport complexes in the chloroplastic envelope membranes via chemical cross-linking. J Cell Biol 136:983–94
4. Alfano C, McMacken R (1989) Ordered assembly of nucleoprotein structures at the *bacteriophage lambda* replication origin during the initiation of DNA replication. J Biol Chem 264:10699–10708
5. Apuya NR, Yadegari R, Fischer RL, Harada JJ, Zimmerman JL, Goldberg RB (2001) The *Arabidopsis* embryo mutant schlepperless has a defect in the chaperonin-60alpha gene. Plant Physiol 126:717–30
6. Aseeva E, Ossenbühl F, Eichacker LA, Wanner G, Soll J, Vothknecht UC (2004) Complex formation of Vipp1 depends on its alpha-helical PspA-like domain. J Biol Chem 279:35535–35541
7. Balczun C, Bunse A, Schwarz C, Piotrowski M, Kuck U (2006) Chloroplast heat shock protein Cpn60 from *Chlamydomonas reinhardtii* exhibits a novel function as a group II intron-specific RNA-binding protein. FEBS Lett 580:4527–4532
8. Baneyx F, Bertsch U, Kalbach CE, van der Vies SM, Soll J, Gatenby AA (1995) Spinach chloroplast cpn21 co-chaperonin possesses two functional domains fused together in a toroidal structure and exhibits nucleotide-dependent binding to plastid chaperonin 60. J Biol Chem 270:10695–10702
9. Beck CF (2005) Signaling pathways from the chloroplast to the nucleus. Planta 222:743–756
10. Becker T, Hritz J, Vogel M, Caliebe A, Bukau B, Soll J, Schleiff E (2004) Toc12, a novel subunit of the intermembrane space preprotein translocon of chloroplasts. Mol Biol Cell 15:5130–5144
11. Berlett BS, Stadtman ER (1997) Protein oxidation in aging, disease, and oxidative stress. J Biol Chem 272:20313–20316
12. Bertsch U, Soll J, Seetharam R, Viitanen PV (1992) Identification, characterization, and DNA sequence of a functional "double" groES-like chaperonin from chloroplasts of higher plants. Proc Natl Acad Sci U S A 89:8696–8700

13. Bonk M, Tadros M, Vandekerckhove J, Al-Babili S, Beyer P (1996) Purification and characterization of chaperonin 60 and heat-shock protein 70 from chromoplasts of *Narcissus pseudonarcissus*. Plant Physiol 111:931–939
14. Boston RS, Viitanen PV, Vierling E (1996) Molecular chaperones and protein folding in plants. Plant Mol Biol 32:191–222
15. Brehmer D, Rudiger S, Gassler CS, Klostermeier D, Packschies L, Reinstein J, Mayer MP, Bukau B (2001) Tuning of chaperone activity of Hsp70 proteins by modulation of nucleotide exchange. Nat Struct Biol 8:427–432
16. Bruch EM, Rosano GL, Ceccarelli EA (2012) Chloroplastic Hsp100 chaperones ClpC2 and ClpD interact in vitro with a transit peptide only when it is located at the N-terminus of a protein. BMC Plant Biol 12:57
17. Bukau B, Weissman J, Horwich A (2006) Molecular chaperones and protein quality control. Cell 125:443–451
18. Cao D, Froehlich JE, Zhang H, Cheng CL (2003) The chlorate-resistant and photomorphogenesis-defective mutant *cr88* encodes a chloroplast-targeted HSP90. Plant J 33:107–118
19. Chen GG, Jagendorf AT (1994) Chloroplast molecular chaperone-assisted refolding and reconstitution of an active multisubunit coupling factor CF1 core. Proc Natl Acad Sci U S A 91:11497–11501
20. Chen KY, Li HM (2007) Precursor binding to an 880-kDa Toc complex as an early step during active import of protein into chloroplasts. Plant J 49:149–158
21. Chen KM, Holmstrom M, Raksajit W, Suorsa M, Piippo M, Aro EM (2010) Small chloroplast-targeted DnaJ proteins are involved in optimization of photosynthetic reactions in *Arabidopsis thaliana*. BMC Plant Biol 10:43
22. Chen KM, Piippo M, Holmstrom M, Nurmi M, Pakula E, Suorsa M, Aro EM (2011) A chloroplast-targeted DnaJ protein AtJ8 is negatively regulated by light and has rapid turnover in darkness. J Plant Physiol 168:1780–1783
23. Chiu CC, Chen LJ, Li HM (2010) Pea chloroplast DnaJ-J8 and Toc12 are encoded by the same gene and localized in the stroma. Plant Physiol 154:1172–1182
24. Chou ML, Fitzpatrick LM, Tu SL, Budziszewski G, Potter-Lewis S, Akita M, Levin JZ, Keegstra K, Li HM (2003) Tic40, a membrane-anchored co-chaperone homolog in the chloroplast protein translocon. EMBO J 22:2970–2980
25. Chou ML, Chu CC, Chen LJ, Akita M, Li HM (2006) Stimulation of transit-peptide release and ATP hydrolysis by a cochaperone during protein import into chloroplasts. J Cell Biol 175:893–900
26. Chow IT, Barnett ME, Zolkiewski M, Baneyx F (2005) The N-terminal domain of *Escherichia coli* ClpB enhances chaperone function. FEBS Letters 579:4242–4248
27. Chu CC, Li HM (2012) The amino-terminal domain of chloroplast Hsp93 is important for its membrane association and functions in vivo. Plant Physiol 158:1656–1665
28. Clarke AK, Eriksson MJ (1996) The cyanobacterium *Synechococcus sp.* PCC 7942 possesses a close homologue to the chloroplast ClpC protein of higher plants. Plant Mol Biol 31:721–730
29. Cloney LP, Wu HB, Hemmingsen SM (1992) Expression of plant chaperonin-60 genes in *Escherichia coli*. J Biol Chem 267:23327–23332
30. Constan D, Froehlich JE, Rangarajan S, Keegstra K (2004) A stromal Hsp100 protein is required for normal chloroplast development and function in *Arabidopsis*. Plant Physiol 136:3605–3615
31. Craig EA, Huang P, Aron R, Andrew A (2006) The diverse roles of J-proteins, the obligate Hsp70 co-chaperone. Rev Physiol Biochem Pharmacol 156:1–21
32. da Costa e Silva O et al (2004) The Etched1 gene of *Zea mays (L.)* encodes a zinc ribbon protein that belongs to the transcriptionally active chromosome (TAC) of plastids and is similar to the transcription factor TFIIS. Plant J 38:923–939
33. Dickson R, Weiss C, Howard RJ, Alldrick SP, Ellis RJ, Lorimer G, Azem A, Viitanen PV (2000) Reconstitution of higher plant chloroplast chaperonin 60 tetradecamers active in protein folding. J Biol Chem 275:11829–11835

34. Dorn KV et al (2010) Chloroplast DnaJ-like proteins 3 and 4 (CDJ3/4) from *Chlamydomonas reinhardtii* contain redox-active Fe-S clusters and interact with stromal HSP70B. Biochem J 427:205–215
35. Doyle SM, Wickner S (2009) Hsp104 and ClpB: protein disaggregating machines. Trends Biochem Sci 34:40–48
36. Doyle SM, Hoskins JR, Wickner S (2007) Collaboration between the ClpB AAA+ remodeling protein and the DnaK chaperone system. Proc Natl Acad Sci U S A 104:11138–11144
37. Drzymalla C, Schroda M, Beck CF (1996) Light-inducible gene HSP70B encodes a chloroplast-localized heat shock protein in *Chlamydomonas reinhardtii*. Plant Mol Biol 31:1185–1194
38. Duppre E, Rupprecht E, Schneider D (2011) Specific and promiscuous functions of multiple DnaJ proteins in *Synechocystis sp.* PCC 6803. Microbiology 157:1269–1278
39. Emelyanov VV (2002) Phylogenetic relationships of organellar Hsp90 homologs reveal fundamental differences to organellar Hsp70 and Hsp60 evolution. Gene 299:125–133
40. Felts SJ, Owen BA, Nguyen P, Trepel J, Donner DB, Toft DO (2000) The hsp90-related protein TRAP1 is a mitochondrial protein with distinct functional properties. J Biol Chem 275:3305–3312
41. Ferreira S, Hjerno K, Larsen M, Wingsle G, Larsen P, Fey S, Roepstorff P, Salome Pais M (2006) Proteome profiling of *Populus euphratica* Oliv. upon heat stress. Ann Bot 98:361–377
42. Flores-Perez U, Jarvis P (2012) Molecular chaperone involvement in chloroplast protein import. Biochim Biophys Acta 1833(2):332–340
43. Friso G, Majeran W, Huang M, Sun Q, van Wijk KJ (2010) Reconstruction of metabolic pathways, protein expression, and homeostasis machineries across maize bundle sheath and mesophyll chloroplasts: large-scale quantitative proteomics using the first maize genome assembly. Plant Physiol 152:1219–1250
44. Frydman J (2001) Folding of newly translated proteins in vivo: the role of molecular chaperones. Annu Rev Biochem 70:603–647
45. Fuhrmann E, Bultema JB, Kahmann U, Rupprecht E, Boekema EJ, Schneider D (2009) The vesicle-inducing protein 1 from *Synechocystis sp.* PCC 6803 organizes into diverse higher-ordered ring structures. Mol Biol Cell 20:4620–4628
46. Georgopoulos C (2006) Toothpicks, serendipity and the emergence of the *Escherichia coli* DnaK (Hsp70) and GroEL (Hsp60) chaperone machines. Genetics 174:1699–1707
47. Glover JR, Lindquist S (1998) Hsp104, Hsp70, and Hsp40: a novel chaperone system that rescues previously aggregated proteins. Cell 94:73–82
48. Glover JR, Tkach JM (2001) Crowbars and ratchets: hsp100 chaperones as tools in reversing protein aggregation. Biochem Cell Biol 79:557–568
49. Göhre V, Ossenbuhl F, Crevecoeur M, Eichacker LA, Rochaix JD (2006) One of two alb3 proteins is essential for the assembly of the photosystems and for cell survival in *Chlamydomonas*. Plant Cell 18:1454–1466
50. Gurley WB (2000) HSP101: a key component for the acquisition of thermotolerance in plants. Plant Cell 12:457–460
51. Halperin T, Adam Z (1996) Degradation of mistargeted OEE33 in the chloroplast stroma. Plant Mol Biol 30:925–933
52. Halperin T, Ostersetzer O, Adam Z (2001) ATP-dependent association between subunits of Clp protease in pea chloroplasts. Planta 213:614–619
53. Han W, Christen P (2003) Mechanism of the targeting action of DnaJ in the DnaK molecular chaperone system. J Biol Chem 278:19038–19043
54. Harrison C (2003) GrpE, a nucleotide exchange factor for DnaK. Cell Stress Chaperones 8:218–224
55. Harrison CJ, Hayer-Hartl M, Di Liberto M, Hartl F, Kuriyan J (1997) Crystal structure of the nucleotide exchange factor GrpE bound to the ATPase domain of the molecular chaperone DnaK. Science 276:431–435
56. Hartl FU (1996) Molecular chaperones in cellular protein folding. Nature 381:571–579

57. Hartl FU (2011) Chaperone-assisted protein folding: the path to discovery from a personal perspective. Nat Med 17:1206–1210
58. Hartl FU, Hayer-Hartl M (2009) Converging concepts of protein folding in vitro and in vivo. Nat Struct Mol Biol 16:574–581
59. Hartl FU, Bracher A, Hayer-Hartl M (2011) Molecular chaperones in protein folding and proteostasis. Nature 475:324–332
60. Haslberger T, Weibezahn J, Zahn R, Lee S, Tsai FT, Bukau B, Mogk A (2007) M domains couple the ClpB threading motor with the DnaK chaperone activity. Mol Cell 25:247–260
61. Haslberger T, Zdanowicz A, Brand I, Kirstein J, Turgay K, Mogk A, Bukau B (2008) Protein disaggregation by the AAA+chaperone ClpB involves partial threading of looped polypeptide segments. Nat Struct Mol Biol 15:641–650
62. Heide H, Nordhues A, Drepper F, Nick S, Schulz-Raffelt M, Haehnel W, Schroda M (2009) Application of quantitative immunoprecipitation combined with knockdown and cross-linking to *Chlamydomonas* reveals the presence of vesicle-inducing protein in plastids 1 in a common complex with chloroplast HSP90C. Proteomics 9:3079–3089
63. Heins L, Mehrle A, Hemmler R, Wagner R, Kuchler M, Hormann F, Sveshnikov D, Soll J (2002) The preprotein conducting channel at the inner envelope membrane of plastids. EMBO J 21:2616–2625
64. Hill JE, Hemmingsen SM (2001) *Arabidopsis thaliana* type I and II chaperonins. Cell Stress Chaperon 6:190–200
65. Hoppe G, Chai YC, Crabb JW, Sears J (2004) Protein s-glutathionylation in retinal pigment epithelium converts heat shock protein 70 to an active chaperone. Exp Eye Res 78:1085–1092
66. Horwich AL, Weber-Ban EU, Finley D (1999) Chaperone rings in protein folding and degradation. Proc Natl Acad Sci U S A 96:11033–11040
67. Horwich AL, Fenton WA, Chapman E, Farr GW (2007) Two families of chaperonin: physiology and mechanism. Annu Rev Cell Dev Biol 23:115–145
68. Hung GC, Masison DC (2006) N-terminal domain of yeast Hsp104 chaperone is dispensable for thermotolerance and prion propagation but necessary for curing prions by Hsp104 overexpression. Genetics 173:611–620
69. Hunt JF, Weaver AJ, Landry SJ, Gierasch L, Deisenhofer J (1996) The crystal structure of the GroES co-chaperonin at 2.8 A resolution. Nature 379:37–45
70. Ivey RA, 3rd, Bruce BD (2000) In vivo and in vitro interaction of DnaK and a chloroplast transit peptide. Cell Stress Chaperon 5:62–71
71. Ivey RA, 3rd, Subramanian C, Bruce BD (2000) Identification of a Hsp70 recognition domain within the rubisco small subunit transit peptide. Plant Physiol 122:1289–1299
72. Johnson JL (2012) Evolution and function of diverse Hsp90 homologs and cochaperone proteins. Biochim Biophys Acta 1823:607–613
73. Kamal A, Thao L, Sensintaffar J, Zhang L, Boehm MF, Fritz LC, Burrows FJ (2003) A high-affinity conformation of Hsp90 confers tumour selectivity on Hsp90 inhibitors. Nature 425:407–410
74. Kang PJ, Ostermann J, Shilling J, Neupert W, Craig EA, Pfanner N (1990) Requirement for hsp70 in the mitochondrial matrix for translocation and folding of precursor proteins. Nature 348:137–143
75. Kessler F, Blobel G (1996) Interaction of the protein import and folding machineries of the chloroplast. Proc Natl Acad Sci U S A 93:7684–7689
76. Kim YI, Levchenko I, Fraczkowska K, Woodruff RV, Sauer RT, Baker TA (2001) Molecular determinants of complex formation between Clp/Hsp100 ATPases and the ClpP peptidase. Nat Struct Biol 8:230–233
77. Kiyosue T, Yamaguchi-Shinozaki K, Shinozaki K (1993) Characterization of cDNA for a dehydration-inducible gene that encodes a CLP A, B-like protein in *Arabidopsis thaliana* L. Biochem Biophys Res Commun 196:1214–1220
78. Kluth J, Schmidt A, Marz M, Krupinska K, Lorbiecke R (2012) *Arabidopsis* Zinc Ribbon 3 is the ortholog of yeast mitochondrial HSP70 escort protein HEP1 and belongs to an ancient protein family in mitochondria and plastids. FEBS Lett 586:3071–3076

79. Ko K, Bornemisza O, Kourtz L, Ko ZW, Plaxton WC, Cashmore AR (1992) Isolation and characterization of a cDNA clone encoding a cognate 70-kDa heat shock protein of the chloroplast envelope. J Biol Chem 267:2986–2993
80. Koumoto Y, Shimada T, Kondo M, Hara-Nishimura I, Nishimura M (2001) Chloroplasts have a novel Cpn10 in addition to Cpn20 as co-chaperonins in *Arabidopsis thaliana*. J Biol Chem 276:29688–29694
81. Kourtz L, Ko K (1997) The early stage of chloroplast protein import involves Com70. J Biol Chem 272:2808–2813
82. Kovacheva S, Bedard J, Patel R, Dudley P, Twell D, Rios G, Koncz C, Jarvis P (2005) In vivo studies on the roles of Tic110, Tic40 and Hsp93 during chloroplast protein import. Plant J 41:412–428
83. Kovacheva S, Bedard J, Wardle A, Patel R, Jarvis P (2007) Further in vivo studies on the role of the molecular chaperone, Hsp93, in plastid protein import. Plant J 50:364–379
84. Krieger-Liszkay A, Rutherford AW (1998) Influence of herbicide binding on the redox potential of the quinone acceptor in photosystem II: relevance to photodamage and phytotoxicity. Biochemistry 37:17339–17344
85. Krishna P, Gloor G (2001) The Hsp90 family of proteins in *Arabidopsis thaliana*. Cell Stress Chaperones 6:238–246
86. Kunkel DD (1982) Thylakoid centers: structures associated with the cyanobacterial photosynthetic membrane system. Arch Microbiol 133:97–99
87. Langer T, Lu C, Echols H, Flanagan J, Hayer MK, Hartl FU (1992) Successive action of DnaK, DnaJ and GroEL along the pathway of chaperone-mediated protein folding. Nature 356:683–689
88. Latijnhouwers M, Xu XM, Moller SG (2010) *Arabidopsis* stromal 70-kDa heat shock proteins are essential for chloroplast development. Planta 232:567–578
89. Lee YR, Nagao RT, Key JL (1994) A soybean 101-kD heat shock protein complements a yeast HSP104 deletion mutant in acquiring thermotolerance. Plant Cell 6: 1889–1897
90. Lee DW, Lee S, Lee GJ, Lee KH, Kim S, Cheong GW, Hwang I (2006) Functional characterization of sequence motifs in the transit peptide of *Arabidopsis* small subunit of rubisco. Plant Physiol 140:466–483
91. Lee S, Choi JM, Tsai FT (2007) Visualizing the ATPase cycle in a protein disaggregating machine: structural basis for substrate binding by ClpB. Mol Cell 25:261–271
92. Lee U, Rioflorido I, Hong SW, Larkindale J, Waters ER, Vierling E (2007) The *Arabidopsis* ClpB/Hsp100 family of proteins: chaperones for stress and chloroplast development. Plant J 49:115–127
93. Lee S, Lee DW, Lee Y, Mayer U, Stierhof YD, Lee S, Jurgens G, Hwang I (2009) Heat shock protein cognate 70-4 and an E3 ubiquitin ligase, CHIP, mediate plastid-destined precursor degradation through the ubiquitin-26S proteasome system in *Arabidopsis*. Plant Cell 21:3984–4001
94. Lemaire SD, Guillon B, Le Marechal P, Keryer E, Miginiac-Maslow M, Decottignies P (2004) New thioredoxin targets in the unicellular photosynthetic eukaryote *Chlamydomonas reinhardtii*. Proc Natl Acad Sci U S A 101:7475–7480
95. Liberek K, Marszalek J, Ang D, Georgopoulos C, Zylicz M (1991) *Escherichia coli* DnaJ and GrpE heat shock proteins jointly stimulate ATPase activity of DnaK. Proc Natl Acad Sci U S A 88:2874–2878
96. Lin Y, Cheng CL (1997) A chlorate-resistant mutant defective in the regulation of nitrate reductase gene expression in *Arabidopsis* defines a new *HY* locus. Plant Cell 9:21–35
97. Liu C, Willmund F, Whitelegge JP, Hawat S, Knapp B, Lodha M, Schroda M (2005) J-domain protein CDJ2 and HSP70B are a plastidic chaperone pair that interacts with vesicle-inducing protein in plastids 1. Mol Biol Cell 16:1165–1177
98. Liu C, Willmund F, Golecki JR, Cacace S, Hess B, Markert C, Schroda M (2007) The chloroplast HSP70B-CDJ2-CGE1 chaperones catalyse assembly and disassembly of VIPP1 oligomers in *Chlamydomonas*. Plant J 50:265–277
99. Liu B et al (2010) Folding of Toll-like receptors by the HSP90 paralogue gp96 requires a substrate-specific cochaperone. Nat Commun 1:79

100. Lubben TH, Donaldson GK, Viitanen PV, Gatenby AA (1989) Several proteins imported into chloroplasts form stable complexes with the GroEL-related chloroplast molecular chaperone. Plant Cell 1:1223–1230
101. Madueno F, Napier JA, Gray JC (1993) Newly imported Rieske iron-sulfur protein associates with both Cpn60 and Hsp70 in the chloroplast stroma. Plant Cell 5:1865–1876
102. Marín-Navarro J, Manuell A, Wu J, Mayfield S (2007) Chloroplast translation regulation Photosynth Res 94:359–374
103. Marshall JS, DeRocher AE, Keegstra K, Vierling E (1990) Identification of heat shock protein hsp70 homologues in chloroplasts. Proc Natl Acad Sci U S A 87:374–378
104. May T, Soll J (2000) 14-3-3 proteins form a guidance complex with chloroplast precursor proteins in plants. Plant Cell 12:53–64
105. Mayer MP, Bukau B (2005) Hsp70 chaperones: cellular functions and molecular mechanism. Cell Mol Life Sci 62:670–684
106. Michelet L, Zaffagnini M, Vanacker H, Le Marechal P, Marchand C, Schroda M, Lemaire SD, Decottignies P (2008) *In vivo* targets of S-thiolation in *Chlamydomonas reinhardtii*. J Biol Chem 283:21571–21578
107. Miernyk JA (2001) The J-domain proteins of *Arabidopsis thaliana*: an unexpectedly large and diverse family of chaperones. Cell Stress Chaperon 6:209–218
108. Mogk A, Schlieker C, Strub C, Rist W, Weibezahn J, Bukau B (2003) Roles of individual domains and conserved motifs of the AAA+chaperone ClpB in oligomerization, ATP hydrolysis, and chaperone activity. J Biol Chem 278:17615–17624
109. Molik S, Karnauchov I, Weidlich C, Herrmann RG, Klosgen RB (2001) The Rieske Fe/S protein of the cytochrome b6/f complex in chloroplasts: missing link in the evolution of protein transport pathways in chloroplasts? J Biol Chem 276:42761–42766
110. Momose T, Ohshima C, Maeda M, Endo T (2007) Structural basis of functional cooperation of Tim15/Zim17 with yeast mitochondrial Hsp70. EMBO Rep 8:664–670
111. Mühlhaus T, Weiss J, Hemme D, Sommer F, Schroda M (2011) Quantitative shotgun proteomics using a uniform 15N-labeled standard to monitor proteome dynamics in time course experiments reveals new insights into the heat stress response of Chlamydomonas reinhardtii. Mol Cell Proteomics 10:M110 004739
112. Musgrove JE, Johnson RA, Ellis RJ (1987) Dissociation of the ribulosebisphosphate-carboxylase large-subunit binding protein into dissimilar subunits. Eur J Biochem 163:529–534
113. Myouga F, Motohashi R, Kuromori T, Nagata N, Shinozaki K (2006) An *Arabidopsis* chloroplast-targeted Hsp101 homologue, APG6, has an essential role in chloroplast development as well as heat-stress response. Plant J 48:249–260
114. Nakabayashi K, Ito M, Kiyosue T, Shinozaki K, Watanabe A (1999) Identification of clp genes expressed in senescing *Arabidopsis* leaves. Plant Cell Physiol 40:504–514
115. Nakashima K, Kiyosue T, Yamaguchi-Shinozaki K, Shinozaki K (1997) A nuclear gene, erd1, encoding a chloroplast-targeted Clp protease regulatory subunit homolog is not only induced by water stress but also developmentally up-regulated during senescence in *Arabidopsis thaliana*. Plant J 12:851–861
116. Nakrieko KA, Mould RM, Smith AG (2004) Fidelity of targeting to chloroplasts is not affected by removal of the phosphorylation site from the transit peptide. Eur J Biochem/FEBS 271:509–516
117. Nickelsen J, Rengstl B, Stengel A, Schottkowski M, Soll J, Ankele E (2011) Biogenesis of the cyanobacterial thylakoid membrane system-an update. FEMS Microbiol Lett 315:1–5
118. Nielsen E, Akita M, Davila-Aponte J, Keegstra K (1997) Stable association of chloroplastic precursors with protein translocation complexes that contain proteins from both envelope membranes and a stromal Hsp100 molecular chaperone. EMBO J 16:935–946
119. Nimura K, Yoshikawa H, Takahashi H (1994) Identification of dnaK multigene family in *Synechococcus sp.* PCC7942. Biochem Biophys Res Commun 201:466–471
120. Nishio K, Hirohashi T, Nakai M (1999) Chloroplast chaperonins: evidence for heterogeneous assembly of alpha and beta Cpn60 polypeptides into a chaperonin oligomer. Biochem Biophys Res Commun 266:584–587

121. Nordhues A, Miller SM, Muhlhaus T, Schroda M (2010) New insights into the roles of molecular chaperones in *Chlamydomonas* and *Volvox*. Int Rev Cell Mol Biol 285:75–113
122. Nordhues A et al (2012) Evidence for a role of VIPP1 in the structural organization of the photosynthetic apparatus in *Chlamydomonas*. Plant Cell 24:637–659
123. Olsen LJ, Keegstra K (1992) The binding of precursor proteins to chloroplasts requires nucleoside triphosphates in the intermembrane space. J Biol Chem 267:433–439
124. Orme W, Walker AR, Gupta R, Gray JC (2001) A novel plastid-targeted J-domain protein in *Arabidopsis thaliana*. Plant Mol Biol 46:615–626
125. Osorio H, Carvalho E, del Valle M, Gunther Sillero MA, Moradas-Ferreira P, Sillero A (2003) H2O2, but not menadione, provokes a decrease in the ATP and an increase in the inosine levels in *Saccharomyces cerevisiae*. An experimental and theoretical approach. Eur J Biochem 270:1578–1589
126. Ostersetzer O, Adam Z (1996) Effects of light and temperature on expression of ClpC, the regulatory subunit of chloroplastic Clp protease, in pea seedlings. Plant Mol Biol 31:673–676
127. Panaretou B, Prodromou C, Roe SM, O'Brien R, Ladbury JE, Piper PW, Pearl LH (1998) ATP binding and hydrolysis are essential to the function of the Hsp90 molecular chaperone in vivo. EMBO J 17:4829–4836
128. Park S, Rodermel SR (2004) Mutations in ClpC2/Hsp100 suppress the requirement for FtsH in thylakoid membrane biogenesis. Proc Natl Acad Sci U S A 101:12765–12770
129. Parsell DA, Kowal AS, Singer MA, Lindquist S (1994) Protein disaggregation mediated by heat-shock protein Hsp104. Nature 372:475–478
130. Pearl LH, Prodromou C (2006) Structure and mechanism of the Hsp90 molecular chaperone machinery. Annu Rev Biochem 75:271–294
131. Pearl LH, Prodromou C, Workman P (2008) The Hsp90 molecular chaperone: an open and shut case for treatment. Biochem J 410:439–453
132. Peltier JB, Friso G, Kalume DE, Roepstorff P, Nilsson F, Adamska I, van Wijk KJ (2000) Proteomics of the chloroplast: systematic identification and targeting analysis of lumenal and peripheral thylakoid proteins. Plant Cell 12:319–341
133. Peltier JB et al (2002) Central functions of the lumenal and peripheral thylakoid proteome of *Arabidopsis* determined by experimentation and genome-wide prediction. Plant Cell 14:211–236
134. Peltier JB, Ripoll DR, Friso G, Rudella A, Cai Y, Ytterberg J, Giacomelli L, Pillardy J, van Wijk KJ (2004) Clp protease complexes from photosynthetic and non-photosynthetic plastids and mitochondria of plants, their predicted three-dimensional structures, and functional implications. J Biol Chem 279:4768–4781
135. Peng L, Fukao Y, Myouga F, Motohashi R, Shinozaki K, Shikanai T (2011) A chaperonin subunit with unique structures is essential for folding of a specific substrate. PLoS Biol 9:e1001040
136. Pratt WB, Toft DO (2003) Regulation of signaling protein function and trafficking by the hsp90/hsp70-based chaperone machinery. Exp Biol Med (Maywood) 228:111–133
137. Prodromou C (2012) The 'active life' of Hsp90 complexes. Biochim Biophys Acta 1823:614–623
138. Queitsch C, Sangster TA, Lindquist S (2002) Hsp90 as a capacitor of phenotypic variation. Nature 417:618–624
139. Ratnayake RM, Inoue H, Nonami H, Akita M (2008) Alternative processing of *Arabidopsis* Hsp70 precursors during protein import into chloroplasts. Biosci Biotechnol Biochem 72:2926–2935
140. Rial DV, Ottado J, Ceccarelli EA (2003) Precursors with altered affinity for Hsp70 in their transit peptides are efficiently imported into chloroplasts. J Biol Chem 278:46473–46481
141. Rial DV, Arakaki AK, Almará AM, Orellano EG, Ceccarelli EA (2006) Chloroplast Hsp70s are not involved in the import of ferredoxin-NADP+ reductase precursor. Physiologia Plantarum 128:618–632
142. Rosano GL, Bruch EM, Ceccarelli EA (2011) Insights into the Clp/HSP100 chaperone system from chloroplasts of *Arabidopsis thaliana*. J Biol Chem 286:29671–29680

143. Rudiger S, Germeroth L, Schneider-Mergener J, Bukau B (1997) Substrate specificity of the DnaK chaperone determined by screening cellulose-bound peptide libraries. EMBO J 16:1501–1507
144. Rupprecht E, Gathmann S, Fuhrmann E, Schneider D (2007) Three different DnaK proteins are functionally expressed in the cyanobacterium *Synechocystis sp.* PCC 6803. Microbiology 153:1828–1841
145. Rutherford SL, Lindquist S (1998) Hsp90 as a capacitor for morphological evolution. Nature 396:336–342
146. Rutschow H, Ytterberg AJ, Friso G, Nilsson R, van Wijk KJ (2008) Quantitative proteomics of a chloroplast SRP54 sorting mutant and its genetic interactions with CLPC1 in *Arabidopsis*. Plant Physiol 148:156–175
147. Salvucci ME (2008) Association of Rubisco activase with chaperonin-60β: a possible mechanism for protecting photosynthesis during heat stress. J Exp Bot 59:1923–1933
148. Sangster TA, Salathia N, Undurraga S, Milo R, Schellenberg K, Lindquist S, Queitsch C (2008) HSP90 affects the expression of genetic variation and developmental stability in quantitative traits. Proc Natl Acad Sci U S A 105:2963–2968
149. Sanjuan Szklarz LK et al (2005) Inactivation of the mitochondrial heat shock protein zim17 leads to aggregation of matrix hsp70s followed by pleiotropic effects on morphology and protein biogenesis. J Mol Biol 351:206–218
150. Sato T, Minagawa S, Kojima E, Okamoto N, Nakamoto H (2010) HtpG, the prokaryotic homologue of Hsp90, stabilizes a phycobilisome protein in the cyanobacterium *Synechococcus* elongatus PCC 7942. Mol Microbiol 76:576–589
151. Schirmer EC, Glover JR, Singer MA, Lindquist S (1996) HSP100/Clp proteins: a common mechanism explains diverse functions. Trends Biochem Sci 21:289–296
152. Schirmer EC, Homann OR, Kowal AS, Lindquist S (2004) Dominant gain-of-function mutations in Hsp104p reveal crucial roles for the middle region. Mol Biol Cell 15:2061–2072
153. Schlicher T, Soll J (1996) Molecular chaperones are present in the thylakoid lumen of pea chloroplasts. FEBS Lett 379:302–304
154. Schlicher T, Soll J (1997) Chloroplastic isoforms of DnaJ and GrpE in pea. Plant Mol Biol 33:181–185
155. Schmid AB et al (2012) The architecture of functional modules in the Hsp90 co-chaperone Sti1/Hop. EMBO J 31:1506–1517
156. Schmitz G, Schmidt M, Feierabend J (1996) Characterization of a plastid-specific HSP90 homologue: identification of a cDNA sequence, phylogenetic descendence and analysis of its mRNA and protein expression. Plant Mol Biol 30:479–492
157. Schmitz G, Schmidt M, Feierabend J (1996) Comparison of the expression of a plastidic chaperonin 60 in different plant tissues and under photosynthetic and non-photosynthetic conditions. Planta 200:326–336
158. Schmollinger S, Strenkert D, Offeddu V, Nordhues A, Sommer F, Schroda M (2012) A Protocol for the identification of protein-protein interactions based on 15N metabolic labeling, immunoprecipitation, quantitative mass spectrometry and affinity modulation. J Vis Exp 67:e4083
159. Schnell DJ, Kessler F, Blobel G (1994) Isolation of components of the chloroplast protein import machinery. Science 266:1007–1012
160. Schroda M (2004) The *Chlamydomonas* genome reveals its secrets: chaperone genes and the potential roles of their gene products in the chloroplast. Photosynth Res 82:221–240
161. Schroda M, Mühlhaus T (2009) A 'foldosome' in the chloroplast? Plant Signal Behav 4:301–303
162. Schroda M, Vallon O, Wollman FA, Beck CF (1999) A chloroplast-targeted heat shock protein 70 (HSP70) contributes to the photoprotection and repair of photosystem II during and after photoinhibition. Plant Cell 11:1165–1178
163. Schroda M, Kropat J, Oster U, Rudiger W, Vallon O, Wollman FA, Beck CF (2001) Possible role for molecular chaperones in assembly and repair of photosystem II. Biochem Soc Trans 29: 413–418

164. Schroda M, Vallon O, Whitelegge JP, Beck CF, Wollman FA (2001) The chloroplastic GrpE homolog of *Chlamydomonas*: two isoforms generated by differential splicing. Plant Cell 13:2823–2839
165. Schroda M, Vallon O (2008) Chaperones and proteases. In: Stern DB (ed) The *Chlamydomonas* sourcebook. Organellar and metabolic processes, 2nd edn, vol 2. Elsevier, San Diego, pp 671–729
166. Schubert M, Petersson UA, Haas BJ, Funk C, Schroder WP, Kieselbach T (2002) Proteome map of the chloroplast lumen of *Arabidopsis thaliana.* J Biol Chem 277:8354–8365
167. Shanklin J, DeWitt ND, Flanagan JM (1995) The stroma of higher plant plastids contain ClpP and ClpC, functional homologs of *Escherichia coli* ClpP and ClpA: an archetypal two-component ATP-dependent protease. Plant Cell 7:1713–1722
168. Sharkia R, Bonshtien AL, Mizrahi I, Weiss C, Niv A, Lustig A, Viitanen PV, Azem A (2003) On the oligomeric state of chloroplast chaperonin 10 and chaperonin 20. Biochim Biophys Acta 1651:76–84
169. Sharma SK, Christen P, Goloubinoff P (2009) Disaggregating chaperones: an unfolding story. Curr Protein Pept Sci 10:432–446
170. Shi LX, Theg SM (2010) A stromal heat shock protein 70 system functions in protein import into chloroplasts in the moss Physcomitrella patens. Plant Cell 22:205–220
171. Shorter J, Lindquist S (2004) Hsp104 catalyzes formation and elimination of self-replicating Sup35 prion conformers. Science 304:1793–1797
172. Sichting M, Mokranjac D, Azem A, Neupert W, Hell K (2005) Maintenance of structure and function of mitochondrial Hsp70 chaperones requires the chaperone Hep1. EMBO J 24:1046–1056
173. Simpson SD, Nakashima K, Narusaka Y, Seki M, Shinozaki K, Yamaguchi-Shinozaki K (2003) Two different novel cis-acting elements of erd1, a clpA homologous *Arabidopsis* gene function in induction by dehydration stress and dark-induced senescence. Plant J 33:259–270
174. Singh A, Grover A (2010) Plant Hsp100/ClpB-like proteins: poorly-analyzed cousins of yeast ClpB machine. Plant Mol Biol 74:395–404
175. Singh A, Singh U, Mittal D, Grover A (2010) Genome-wide analysis of rice ClpB/HSP100, ClpC and ClpD genes. BMC Genomics 11:95
176. Sjogren LL, MacDonald TM, Sutinen S, Clarke AK (2004) Inactivation of the clpC1 gene encoding a chloroplast Hsp100 molecular chaperone causes growth retardation, leaf chlorosis, lower photosynthetic activity, and a specific reduction in photosystem content. Plant Physiol 136:4114–4126
177. Song H, Zhao R, Fan P, Wang X, Chen X, Li Y (2009) Overexpression of AtHsp90.2, AtHsp90.5 and AtHsp90.7 in *Arabidopsis thaliana* enhances plant sensitivity to salt and drought stresses. Planta 229:955–964
178. Spiess C, Meyer AS, Reissmann S, Frydman J (2004) Mechanism of the eukaryotic chaperonin: protein folding in the chamber of secrets. Trends Cell Biol 14:598–604
179. Su PH, Li HM (2008) *Arabidopsis* stromal 70-kD heat shock proteins are essential for plant development and important for thermotolerance of germinating seeds. Plant Physiol 146:1231–1241
180. Su PH, Li HM (2010) Stromal Hsp70 is important for protein translocation into pea and *Arabidopsis* chloroplasts. Plant Cell 22:1516–1531
181. Sung DY, Vierling E, Guy CL (2001) Comprehensive expression profile analysis of the *Arabidopsis* Hsp70 gene family. Plant Physiol 126:789–800
182. Suzuki K, Nakanishi H, Bower J, Yoder DW, Osteryoung KW, Miyagishima SY (2009) Plastid chaperonin proteins Cpn60 alpha and Cpn60 beta are required for plastid division in *Arabidopsis thaliana.* BMC Plant Biol 9:38
183. Taipale M, Jarosz DF, Lindquist S (2010) HSP90 at the hub of protein homeostasis: emerging mechanistic insights. Nat Rev Mol Cell Biol 11:515–528
184. Tanaka Y, Nishiyama Y, Murata N (2000) Acclimation of the photosynthetic machinery to high temperature in *Chlamydomonas reinhardtii* requires synthesis de novo of proteins encoded by the nuclear and chloroplast genomes. Plant Physiol 124:441–449

185. Tessarz P, Mogk A, Bukau B (2008) Substrate threading through the central pore of the Hsp104 chaperone as a common mechanism for protein disaggregation and prion propagation. Mol Microbiol 68:87–97
186. Thompson MD, Paavola CD, Lenvik TR, Gantt JS (1995) *Chlamydomonas* transcripts encoding three divergent plastid chaperonins are heat-inducible. Plant Mol Biol 27:1031–5
187. Tomoyasu T, Ogura T, Tatsuta T, Bukau B (1998) Levels of DnaK and DnaJ provide tight control of heat shock gene expression and protein repair in *Escherichia coli*. Mol Mircrobiol 30:567–581
188. Tran LS et al (2004) Isolation and functional analysis of *Arabidopsis* stress-inducible NAC transcription factors that bind to a drought-responsive cis-element in the early responsive to dehydration stress 1 promoter. Plant Cell 16:2481–2498
189. Tran LS et al (2007) Co-expression of the stress-inducible zinc finger homeodomain ZFHD1 and NAC transcription factors enhances expression of the ERD1 gene in *Arabidopsis*. Plant J 49:46–63
190. Tsai YC, Mueller-Cajar O, Saschenbrecker S, Hartl FU, Hayer-Hartl M (2012) Chaperonin cofactors, Cpn10 and Cpn20, of green algae and plants function as hetero-oligomeric ring complexes. J Biol Chem 287:20471–20481
191. Tsugeki R, Nishimura M (1993) Interaction of homologues of Hsp70 and Cpn60 with ferredoxin-NADP+ reductase upon its import into chloroplasts. FEBS Lett 320:198–202
192. Tyedmers J, Mogk A, Bukau B (2010) Cellular strategies for controlling protein aggregation. Nat Rev Mol Cell Biol 11:777–788
193. Ungewickell E, Ungewickell H, Holstein SE, Lindner R, Prasad K, Barouch W, Martin B, Greene LE, Eisenberg E (1995) Role of auxilin in uncoating clathrin-coated vesicles. Nature 378:632–635
194. Uniacke J, Zerges W (2009) Chloroplast protein targeting involves localized translation in *Chlamydomonas*. Proc Natl Acad Sci U S A 106:1439–1444
195. van de Meene AM, Hohmann-Marriott MF, Vermaas WF, Roberson RW (2006) The three-dimensional structure of the cyanobacterium *Synechocystis sp.* PCC 6803. Arch Microbiol 184:259–270
196. VanBogelen RA, Kelley PM, Neidhardt FC (1987) Differential induction of heat shock, SOS, and oxidation stress regulons and accumulation of nucleotides in *Escherichia coli*. J Bacteriol 169:26–32
197. Viitanen PV, Bacot K, Dickson R, Webb T (1998) Purification of recombinant plant and animal GroES homologs: chloroplast and mitochondrial chaperonin 10. Methods Enzymol 290:218–230
198. Vitha S, Froehlich JE, Koksharova O, Pyke KA, van Erp H, Osteryoung KW (2003) ARC6 is a J-domain plastid division protein and an evolutionary descendant of the cyanobacterial cell division protein Ftn2. Plant Cell 15:1918–1933
199. von Gromoff ED, Treier U, Beck CF (1989) Three light-inducible heat shock genes of *Chlamydomonas reinhardtii*. Mol Cell Biol 9:3911–3918
200. Wall D, Zylicz M, Georgopoulos C (1994) The NH2-terminal 108 amino acids of the *Escherichia coli* DnaJ protein stimulate the ATPase activity of DnaK and are sufficient for lambda replication. J Biol Chem 269:5446–5451
201. Wang H, Goffreda M, Leustek T (1993) Characteristics of an Hsp70 homolog localized in higher plant chloroplasts that is similar to DnaK, the Hsp70 of prokaryotes. Plant Physiol 102:843–850
202. Watanabe S, Kobayashi T, Saito M, Sato M, Nimura-Matsune K, Chibazakura T, Taketani S, Nakamoto H, Yoshikawa H (2007) Studies on the role of HtpG in the tetrapyrrole biosynthesis pathway of the cyanobacterium *Synechococcus* elongatus PCC 7942. Biochem Biophys Res Commun 352:36–41
203. Weaver LM, Froehlich JE, Amasino RM (1999) Chloroplast-targeted ERD1 protein declines but its mRNA increases during senescence in *Arabidopsis*. Plant Physiol 119: 1209–1216

204. Wegele H, Muller L, Buchner J (2004) Hsp70 and Hsp90-a relay team for protein folding. Rev Physiol Biochem Pharmacol 151:1–44
205. Wegele H, Wandinger SK, Schmid AB, Reinstein J, Buchner J (2006) Substrate transfer from the chaperone Hsp70 to Hsp90. J Mol Biol 356:802–811
206. Weibezahn J et al (2004) Thermotolerance requires refolding of aggregated proteins by substrate translocation through the central pore of ClpB. Cell 119:653–665
207. Weiss C, Bonshtien A, Farchi-Pisanty O, Vitlin A, Azem A (2009) Cpn20: siamese twins of the chaperonin world. Plant Mol Biol 69:227–238
208. Wendler P, Shorter J, Plisson C, Cashikar AG, Lindquist S, Saibil HR (2007) Atypical AAA+ subunit packing creates an expanded cavity for disaggregation by the protein-remodeling factor Hsp104. Cell 131:1366–1377
209. Wickner S, Hoskins J, McKenney K (1991) Monomerization of RepA dimers by heat shock proteins activates binding to DNA replication origin. Proc Natl Acad Sci U S A 88:7903–7907
210. Wickner S, Maurizi MR, Gottesman S (1999) Posttranslational quality control: folding, refolding, and degrading proteins. Science 286:1888–1893
211. Willmund F, Schroda M (2005) HEAT SHOCK PROTEIN 90C is a bona fide Hsp90 that interacts with plastidic HSP70B in *Chlamydomonas reinhardtii*. Plant Physiol 138:2310–2322
212. Willmund F, Muhlhaus T, Wojciechowska M, Schroda M (2007) The NH_2-terminal domain of the chloroplast GrpE homolog CGE1 is required for dimerization and cochaperone function in vivo. J Biol Chem 282:11317–11328
213. Willmund F, Dorn KV, Schulz-Raffelt M, Schroda M (2008) The chloroplast DnaJ homolog CDJ1 of *Chlamydomonas reinhardtii* is part of a multichaperone complex containing HSP70B, CGE1, and HSP90C. Plant Physiol 148:2070–2082
214. Willmund F, Hinnenberger M, Nick S, Schulz-Raffelt M, Muhlhaus T, Schroda M (2008) Assistance for a chaperone: *Chlamydomonas* HEP2 activates plastidic HSP70B for cochaperone binding. J Biol Chem 283:16363–16373
215. Winter J, Linke K, Jatzek A, Jakob U (2005) Severe oxidative stress causes inactivation of DnaK and activation of the redox-regulated chaperone Hsp33. Mol Cell 17:381–392
216. Wu C, Seibert FS, Ko K (1994) Identification of chloroplast envelope proteins in close physical proximity to a partially translocated chimeric precursor protein. J Biol Chem 269:32264–32271
217. Yang JY, Sun Y, Sun AQ, Yi SY, Qin J, Li MH, Liu J (2006) The involvement of chloroplast HSP100/ClpB in the acquired thermotolerance in tomato. Plant Mol Biol 62:385–395
218. Yeyati PL, Bancewicz RM, Maule J, van Heyningen V (2007) Hsp90 selectively modulates phenotype in vertebrate development. PLoS Genet 3:e43
219. Yokthongwattana K, Chrost B, Behrman S, Casper-Lindley C, Melis A (2001) Photosystem II damage and repair cycle in the green alga *Dunaliella salina*: involvement of a chloroplast-localized HSP70. Plant Cell Physiol 42:1389–1397
220. Zaltsman A, Ori N, Adam Z (2005) Two types of FtsH protease subunits are required for chloroplast biogenesis and Photosystem II repair in *Arabidopsis*. Plant Cell 17:2782–2790
221. Zheng B, Halperin T, Hruskova-Heidingsfeldova O, Adam Z, Clarke AK (2002) Characterization of Chloroplast Clp proteins in *Arabidopsis*: localization, tissue specificity and stress responses. Physiol Plant 114:92–101
222. Zietkiewicz S, Krzewska J, Liberek K (2004) Successive and synergistic action of the Hsp70 and Hsp100 chaperones in protein disaggregation. J Biol Chem 279:44376–44383
223. Zybailov B, Friso G, Kim J, Rudella A, Rodriguez VR, Asakura Y, Sun Q, van Wijk KJ (2009) Large scale comparative proteomics of a chloroplast Clp protease mutant reveals folding stress, altered protein homeostasis, and feedback regulation of metabolism. Mol Cell Proteomics 8:1789–1810

Chapter 14
Plastid Proteases

Zach Adam and Wataru Sakamoto

Abstract Steady-state levels of chloroplast proteins rely on the balance between synthesis and degradation rates. Thus, the importance of protein-degradation processes in shaping the chloroplast proteome, and hence proper organellar functioning, cannot be overestimated. Chloroplast proteases and peptidases participate in chloroplast biogenesis through maturation or activation of pre-proteins, adaptation to changing environmental conditions through degradation of certain proteins, and maintenance of protein quality through degradation of unassembled or damaged proteins. These activities are mediated by ATP-dependent and—independent proteases, many of which are encoded by multigene families. Newly imported proteins are processed by stroma- and thylakoid-localized peptidases that remove signal sequences, which are then further degraded. The multisubunit ATP-dependent Clp and FtsH complexes degrade housekeeping and oxidatively damaged proteins in the stroma and thylakoid membranes, respectively. A number of other chloroplast proteases have been identified, but their function and substrates are still unknown, as are the nature of degradation signals and determinants of protein instability. Future research is expected to focus on these questions.

Keywords Chaperones · Chloroplasts · Development · Proteases · Senescence · Thylakoids

Abbreviations

BARD	Binational agricultural research and development fund
BSF	Binational science foundation
CAO	Chlorophyllide *a* oxygenase
ELIP	Early light-inducible protein
ISF	Israel science foundation
JSPS	Japan society for the promotion of science

Z. Adam (✉)
The Robert H. Smith Institute of Plant Sciences and Genetics in Agriculture,
The Hebrew University, Rehovot 76100, Israel
e-mail: zach.adam@mail.huji.ac.il; zach@agri.huji.ac.il

W. Sakamoto
Institute of Plant Science and Resources, Okayama University,
Kurashiki, Okayama 710-0046, Japan
e-mail: saka@rib.okayama-u.ac.jp

S.M. Theg, F.-A. Wollman (eds.), *Plastid Biology,* Advances in Plant Biology 5,
DOI 10.1007/978-1-4939-1136-3_14,

JST	Japan science and technology agency
LHC	Light-harvesting complex
MAP	Methionine aminopeptidase
PDF	Peptide deformylase
PSII	Photosystem II
SsrA	Small stable RNA A

14.1 Introduction

The chloroplast proteome comprises more than 2000 nuclear- and chloroplast-encoded proteins. Steady-state levels of these proteins are determined by the balance between transcription and translation rates on one hand, and degradation rates on the other. Thus, the importance of protein-degradation processes in shaping the chloroplast proteome, and hence proper functioning of the organelle, cannot be overestimated. Proteolytic activities, defined as cleavage of peptide bonds, are carried out by proteases or peptidases, which differ in a number of aspects. Some activities are limited to the hydrolysis of a single bond in a given substrate, whereas others function processively. Products of such activities can be either peptides of different lengths, from di- and tri-peptides to much longer ones, or free amino acids. The hydrolysis itself can be catalyzed by different mechanisms, depending on the chemistry of their catalytic centers, giving rise to the categorization of proteases into seven different families: serine, cysteine, aspartic, metalloproteases, threonine, glutamic and peptidases of unknown catalytic mechanisms. Although cleavage of a peptide bond does not require metabolic energy, some proteases couple the hydrolysis of ATP to the unfolding of their substrates as a prerequisite for the actual cleavage. The in vivo contexts of proteolytic reactions are also highly variable: maturation or activation of pre-proteins require either N- or C-terminal processing by specific peptidases; proteolytic enzymes participate in some cases of signal transduction by releasing factors from membranes into the soluble phase; rapid turnover rates of certain regulatory proteins allow their function as 'timing proteins' in the control of gene expression; protein quality control is maintained by the degradation of unassembled or damaged proteins. Thus, proteolytic processes are intimately involved in almost every aspect of the cell's life cycle. Organelles such as chloroplasts are no exception. Although examples have been documented for the involvement of only some of the above proteolytic processes in chloroplasts, it is already clear that proteases play an essential role in this organelle's biogenesis and function.

In retrospective, research in the field of chloroplast proteolysis can be roughly divided into three periods. During the 1980s and early 1990s, a number of proteolytic processes were documented and characterized. However, attempts to identify the proteases involved in these processes, primarily through biochemical approaches, were largely unsuccessful. In the mid-1990s, the identities of chloroplast proteases began to be revealed. The first successes in this respect resulted from searching for homologues of known bacterial proteases, and identification of genes responsible for mutations in chloroplast development [19, 51, 86, 127, 142]. Completion of the Arabidopsis genome project enabled comprehensive homology searches, and in conjunction with the use of programs for predicting the intracellular location of proteins, a list of putative components of the proteolytic machinery of chloroplasts was compiled [134]. Research in the field in recent years has been characterized by attempts to link identified proteases with the previously described proteolytic processes, and to reveal their physiological roles, primarily through a reverse-genetics approach.

This chapter reviews the different components of the chloroplast proteolytic machinery, their function in different proteolytic processes delineated to date, and the limited information on substrate specificity and determinants of protein stability and instability in chloroplasts. Where possible, proteolytic enzymes are referred to according to their names and classification in the peptidase database MEROPS [112] (http://merops.sanger.ac.uk/index.htm) and its corresponding handbook [13].

14.2 Major Chloroplast Proteases

Given the prokaryotic evolutionary origin of chloroplasts, it is not surprising that all chloroplast proteases are homologues of known bacterial ones. In fact, this relationship facilitated the initial identification of some chloroplast proteases. Proteases involved in intracellular proteolysis in any biological system can be categorized, based on their energy requirement, into ATP-dependent and -independent ones. Hydrolysis of a peptide bond does not require metabolic energy. Thus, the requirement for ATP in certain enzymes is limited to unfolding the substrate and feeding it into a catalytic chamber, which is secluded from the cellular environment, a paradigm that led to classifying these enzymes as self- compartmentalizing proteases [14]. Similar to all bacteria, chloroplasts contain both ATP-dependent and independent proteases. However, whereas *Escherichia coli* and most other bacteria contain single genes encoding these enzymes, higher plants have evolved multiple genes for most of them [1, 3, 57, 116]. These enzymes (Fig. 14.1) are described below.

14.2.1 The Clp Protease

The Clp protease in *E. coli* is a multisubunit complex, composed of two main components, proteolytic and regulatory (for review see [74, 123]). The proteolytic chamber is made of two heptameric rings of the serine peptidase ClpP. Together the two rings form a barrel-like structure with two narrow inlets and an internal cavity where the active sites, composed of the catalytic triad of Ser-His-Asp, are located. The openings of the ClpP subcomplex are capped by hexameric rings of specific ATP-dependent chaperones of the AAA+ superfamily [100], either ClpA or ClpX, which recognize potential substrates, unfold them, and feed them into the catalytic chamber [66]. ClpAP and ClpXP specifically degrade different regulatory proteins,

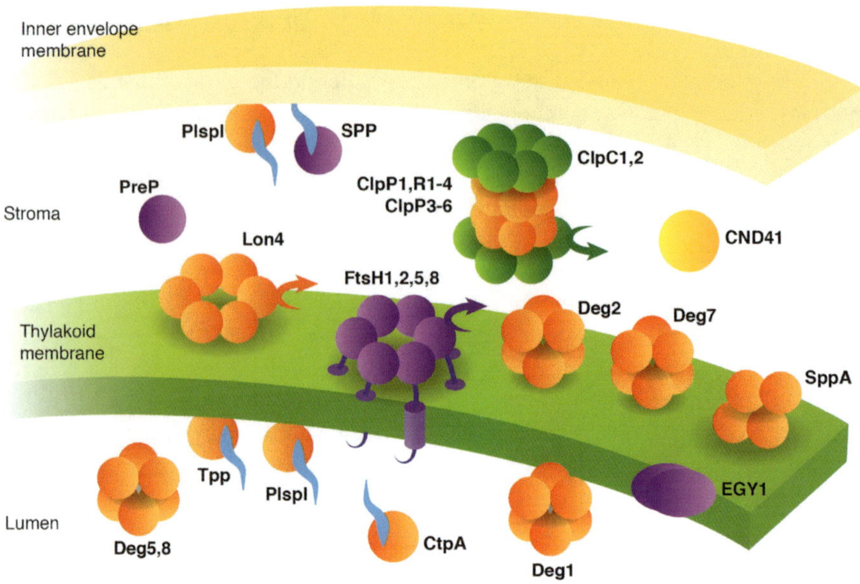

Fig. 14.1 The proteolytic machinery in chloroplasts. Different chloroplast proteases are depicted in their respective sub-organellar location. Serine proteases are colored in *orange*, metalloproteases in *purple* and an aspartic protease in *yellow*. ATPases are marked with *arrows*, and processing peptidases (described in Chap. 12) are marked by *blue* ribbons

and participate in protein quality control by degrading aggregated, misfolded and otherwise abnormal proteins [123].

The chloroplast Clp protease is much more complex (for reviews see [3, 22, 23, 102]. In Arabidopsis, there are six different genes encoding ClpP (peptidase S14.002), giving rise to proteins of 20–29 kDa, five of which are targeted to chloroplasts. The sixth, ClpP2, is targeted to mitochondria where, together with ClpX, it forms the mitochondrial Clp complex [44, 79, 109]. One of the ClpPs, ClpP1, is the only component of the chloroplast proteolytic machinery that is encoded in the organelle's genome. The Arabidopsis nuclear genome encodes four ClpP-like proteins, designated ClpR. These are similar in size and sequence to ClpP and located exclusively in chloroplasts, but lack the conserved residues of the catalytic triad, and are thus not expected to perform a proteolytic function. The ClpP cognate chaperones in Arabidopsis include two copies of ClpC (belonging to the Hsp100 family), the plant homologue of ClpA, and another related protein designated ClpD, which are found in the chloroplast, and three ClpX proteins that are located in the mitochondria. At the protein level, expression of all Clp proteins, with the exception of ClpD, appears to be constitutive under different short- and long-term stress conditions [167]. Nevertheless, abundance of their transcripts differs in a tissue specific manner [102]. Additional Clp proteins include one ClpS and two copies of ClpT. ClpS is homologous to the *E. coli* ClpS, a substrate modulator of the bacterial ClpAP complex [28], which is essential for the operation of the N-end rule pathway (see Sect. 14.6) in bacteria [29]. The function of the chloroplast ClpS is currently unclear. ClpT is unique to land plants, while absent from algae and cyanobacteria, and shares homology with the N terminus of ClpC [106]. ClpT appears to modulate the assembly of ClpP core complexes (see below).

Native isoelectric focusing followed by mass spectrometry revealed that the core of the chloroplast Clp protease is a complex of 325–350 kDa, composed of one to three copies of ClpP (ClpP1, ClpP3-ClpP6), four copies of ClpR (ClpR1-ClpR4), and one copy of ClpT [107] (it should be noted that the names of ClpS and ClpT were swapped in this paper). Interestingly, the same core Clp complex is found in the stroma of chloroplasts and in non-green plastids isolated from roots and flowers [109]. Another work, using native polyacrylamide gel electrophoresis followed by immunoblot analysis with specific antibodies for each of the Clp isomers, shed more light on the structure of the core Clp complex. Two sub-core complexes were observed, probably corresponding to the two different rings. One, a 335 kDa core contained all chloroplastic ClpP and ClpR subunits, while two smaller sub-complexes were identified: a 230 kDa complex containing ClpP1 and ClpR1-ClpR4, and a 180 kDa complex containing ClpP3-ClpP6 [133].

Presence of the two different core heptamers, designated P-ring (containing only P subunits) and R-ring (containing P and R subunits), has been recently confirmed. This was achieved by complementing Arabidopsis mutants lacking ClpP3 or ClpR4 with the corresponding proteins fused to a StrepII tag, followed by affinity purification and mass spectrometry analysis [101]. The complex purified by the tagged ClpP3 corresponded to the P-ring and contained ClpP4, ClpP5, and ClpP6 at a stoichiometry of P3:P4:P5:P6 = 1:2:3:1. The complex purified by the tagged ClpR4 corresponded to the R-ring and contained ClpP1, ClpR1, ClpR2 and ClpR3, at a stoichiometry of P1:R1:R2:R3:R4 = 3:1:1:1:1. Notably, composition of the R-ring (P:R = 3:4) is reminiscent of the essential R- ring in the cyanobacterium Synechococcus, in which subunit composition is much simpler and the R-ring consists of three ClpP3 and four ClpR in an alternating order [8]. Similarity of the R-ring composition in these two organisms suggests that diversification of P and R subunits coincided with the evolution of photosynthesis (reviewed in [22, 102]). Phylogenetic analysis suggests that Arabidopsis ClpR2 originated from the chloroplast-encoded ClpP1, which is closely related to the cyanobacterial ClpP3, whereas all other R subunits probably originated from the cyanobacterial ClpR.

The aforementioned diversity in the composition of PR subunits appears to coincide also with the presence of the land plant-specific ClpT proteins [22, 102, 106]. Function of ClpT is essential, as the lack of both ClpT1 and T2 results in seedling lethality in Arabidopsis [131]. Both of these proteins are found in the stroma as homodimers, where they appear to participate in the assembly of the Clp proteolytic core complex. They first monomerize by an unknown mechanism, ClpT1 then binds to the P-ring, followed by binding of ClpT2. This then facilitates the association of the P-ring with the R-ring. Thus, ClpT proteins apparently serve as regulators of the core assembly [131]. However, the mechanistic details of this regulation, as well as why ClpT proteins are essential in land plants but are missing from cyanobacteria, are still unknown.

Knockout and knockdown studies of Clp genes provided additional insight into both the structure of the complex and its physiological functions. Disruption of the chloroplast ClpP1 gene in tobacco resulted in loss of shoot development [78, 129]. Inactivation or downregulation of several ClpP and ClpR genes in Arabidopsis led to phenotypes of variable severity. Viable ClpP4 and ClpP6 knockout mutants could not be obtained, but repression of their expression by antisense constructs resulted in slow growth and a variegated 'yellow-heart' phenotype [133, 168]. Yellow variegated leaves were also observed in rice as a result of disrupting the ClpP5 gene [144]. A T-DNA-knockdown line of ClpR2 demonstrated a decrease in the level of all other Clp core-complex subunits, suggesting that their accumulation is coordinated and that they are all essential for the assembly and stability of the complex [114].

More systematic analysis of mutants lacking either one of the ClpP or ClpR subunits revealed that most of them displayed severe phenotypes, such as embryo or seedling lethality.

This strongly supports the notion that the multitudes of ClpP and ClpR subunits are not redundant, but are required for the formation of a functional ClpPR core [65, 102]. The only exception is *clpr1* knockout, which gives rise to viable plants despite its pale-green leaf phenotype [65, 71]. This mild phenotype of *clpr1* is partially explained by functional exchangeability between ClpR1 and ClpR3 [65].

Mutations in the regulatory ATPase have somewhat less severe consequences than the loss of ClpP and ClpR proteins. ClpC1 mutants can grow autotrophically, but are smaller and paler than wild-type plants [26, 72, 132]. In contrast, a ClpC2 mutant is indistinguishable from the wild type [105], suggesting that the two copies of ClpC are redundant. It is not clear why these two mutants have different phenotypes; one possibility is that the two isomers accumulate to different levels, and loss of the abundant one has more severe effects. A mutant lacking both ClpC1 and ClpC2 results in embryo lethality, indicating that the function of ClpC is essential [73]. In contrast, the function of ClpS remains unclear, as its knockout mutants have no obvious phenotype.

14.2.2 The FtsH Protease

The *E. coli* FtsH is a membrane-bound ATP-dependent metalloprotease [50]. Of all the ATP-dependent proteases in this organism, FtsH is the only essential one. Unlike the Clp protease, the proteolytic and ATPase domains of FtsH are found on the same polypeptide. The N terminus of the protein contains two trans-membrane helices, which anchor it to the plasma membrane. This region is followed by the ATPase domain, classifying FtsH as a member of the AAA+ superfamily [100]. The proteolytic domain is found at the C terminus and contains the zinc-binding motif His-Glu-X-X-His, which serves as the catalytic site. Similar to other ATP-dependent proteases, FtsH forms a hexameric ring- like structure, in which access to the proteolytic site is controlled by the ATPase domain. Details of these structural features were revealed with the determination of the three- dimensional structure of the soluble portion of bacterial FtsH [15, 141].

The Arabidopsis FtsH gene family contains 12 members, all found in the nuclear genome [2, 3, 116, 118, 148]. Products of three of these (FtsH3, FtsH4 and FtsH10) are targeted to mitochondria, whereas the other nine (FtsH1, FtsH2, FtsH5-FtsH9, FtsH11 and FtsH12) are targeted to chloroplasts, as revealed by transient-expression assays with GFP fusions [118]. In addition to these, there are four genes encoding FtsH-like proteins that show significant similarity to authentic FtsH, but that lack a histidine residue in the zinc- binding motif, and are thus expected to be proteolytically inactive [134]. Unlike ClpR in the core complex of the Clp protease, so far there is no evidence for the association of these FtsH-like proteins with any proteolytically active FtsH complex. It is worth mentioning that unimported copies of FtsH are degraded in the cytosol by the ubiquitin- dependent pathway [128].

Mass spectrometry analyses confirmed the presence of FtsH1, FtsH2, FtsH5 and FtsH8 in thylakoids [33, 130, 159]. Immunoblot analysis of isolated organelles suggested that FtsH4 is located exclusively in the mitochondria, whereas FtsH11 is dually targeted to both mitochondria and chloroplasts [145]. Within the chloroplast, FtsH11 is localized to the envelope membranes, as revealed by proteomic analysis of this sub-compartment [31].

Among those exclusively targeted to chloroplasts, FtsH1 and FtsH5, FtsH2 and FtsH8, and FtsH7 and FtsH9 comprise three pairs of duplicated genes (see phylogenetic trees in [2, 118, 159]. The two pairs, FtsH1/5 and FtsH2/8, seem to represent the major isoforms, with FtsH2 being the most abundant, followed by FtsH5, FtsH8 and FtsH1, in decreasing order [130]. The differential abundance of these four FtsHs is positively correlated with the severity of phenotypes associated with mutations in their corresponding genes. FtsH2 mutants have variegated leaves, containing distinct green and yellow/white sectors with viable cells [19, 60, 97, 120, 142](see Fig. 14.2). Mutants in FtsH5 have only slightly variegated leaves [117], whereas mutants in FtsH1 and FtsH8 are indistinguishable from wild type plants [118]. This mutant phenotype suggests that FtsH might be involved in chloroplast biogenesis. More importantly, dispensability of all FtsH isoforms, despite leaf variegation, suggests their redundancy (see below), unlike ClpP and ClpR subunits.

The size of the chloroplast FtsH monomer (peptidase M41.005) is ~74 kDa. The variants located in the thylakoid membrane have their ATPase and proteolytic domains facing towards the stroma [86]. Owing to their similarity to the *E. coli* FtsH, thylakoid FtsHs have been considered to have two transmembrane domains. A recent study involving protein import, however, revealed that they have a single membrane-spanning region, and that they are targeted to the thylakoid membranes through canonical transport pathways [113] (see Chap. 10 in this volume). Interestingly, FtsH2 (and also the related FtsH8) uses the Tat pathway that is dependent on a proton gradient, whereas FtsH5 (and FtsH1) uses the NTP-dependent Sec pathway. Consequently, the N-termini of these FtsHs are exposed to the thylakoid lumen.

In the thylakoid membranes, FtsH forms a complex of 400–450 kDa, which is most likely a hexamer [118, 159]. Several lines of evidence suggest that FtsH complexes are heteromeric: FtsH2 and FtsH5, identified by either specific antibodies or mass spectrometry, co-migrate on native gels, sucrose gradients and size- exclusion chromatography. Moreover, in mutants lacking one of these proteins, the level of the other is also reduced [118, 159], and double mutants lacking a pair of duplicated genes do not accumulate the products of the existing pair [164].

Fig. 14.2 Leaf-variegated phenotype caused by the lack of the thylakoid FtsH complex. *Upper panel*: FtsH2 is a major isoform comprising the FtsH hetero-complex. An Arabidopsis mutant lacking FtsH2 (*var2–6*) is viable due to its functional redundancy with FtsH8, but shows leaf variegation that does not appear in the wild type (Columbia). Coexsistence of another mutation (*fug1–1* in this figure) has been shown to suppress leaf variegation [96], providing an intriguing aspect of chloroplast biogenesis related to FtsH. Bar in each panel, 10 mm. *Lower panel*: Expression of an FtsH RNAi construct in tobacco led to suppression of overall FtsH protein. (A), transgenic tobacco transformed by a vector control; (B) to (D), three independent transgenic lines in which FtsH levels are reduced to 90, 70 and 50 % of the control, respectively (see also [62]). Photographs provided by the Sakamoto laboratory

The membrane-association of FtsH has prevented its purification to homogeneity in its native form. Nevertheless, insights into its composition in Arabidopsis could be obtained from overexpression experiments and analysis of single and double knockout mutants. Overexpression of FtsH8 compensates for the loss of its duplicated gene FtsH2 [159], and FtsH1 can compensate for the loss of its close homologue FtsH5 [160]. However, attempts to restore the wild type phenotype by overexpressing FtsH5 in the FtsH2- mutant background were unsuccessful. Furthermore, double mutants of duplicated genes, either FtsH1 and FtsH5, or FtsH2 and FtsH8, are completely albino, and can only grow on agar plates supplemented with sucrose [164]. In each of these double mutants, the presumably remaining FtsHs do not accumulate. Taken together, these results suggested that the chloroplast FtsH complex is a hetero-oligomer composed of two types of subunits, 'type A'—FtsH1 and FtsH5 and 'type B'—FtsH2 and FtsH8. Whereas subunits within a type are redundant, the presence of subunits from both types is essential for accumulation of the complex [2, 3, 164]. Although genetic and transgenic approaches allowed making these conclusions, the issue of stoichiometry between type A and type B subunits awaits the purification of the authentic complex.

The variegated phenotype in the mutants lacking FtsH2 or FtsH5 suggests that FtsH2 and FtsH5 are the major isoforms in the heterocomplex, whereas FtsH1 and FtsH8 might act as backup. Fitness test of mutants lacking one of the FtsHs reinforced the importance of FtsH2 and FtsH5 over other FtsHs [148]. Presence of Type A and Type B isoforms is well conserved not only in higher plants, but also in algae and cyanobacteria, demonstrating its evolutionary conservation in photosynthetic organisms. In contrast to these four thylakoid-membrane FtsHs, most other chloroplast FtsHs have been poorly characterized (see Sect. 14.4.5). Mutants in FtsH have not been characterized in plant species other than Arabidopsis, and only very recently the knockdown of FtsH (to ~50 %) in tobacco was shown to also cause leaf variegation [62] (Fig. 14.2).

The multiplication of FtsH genes had initially suggested that this could be reflected in spatial or temporal variation in their expression. Transcript levels of all FtsH genes increased in response to increase in light intensity [130]. However, results of a proteomic study on the response to high light revealed that the levels of chloroplast proteases or chaperones, including FtsH, are kept more or less constant [37]. In light of the transient decrease in the level of the FtsH protein itself in response to high light [163], it appears that changes in the transcript level only ensure a constant level of the protein. Similarly, differential spatial expression of different FtsHs could also be ruled out, as GUS-fusion experiments revealed similar patterns for FtsH1, FtsH2, FtsH5 and FtsH8 [159, 160].

As mutants lacking FtsH2 show a remarkable leaf-variegation phenotype, genetic studies have been extensively performed to unravel important aspects of FtsH structure- function relationship. Identification of many *ftsh2* alleles (also termed *var2*), which contain single amino-acid substitutions, showed that variegation is always related to mutations in the ATPase domain [119]. In contrast, none of the variegated mutants had a mutation in the C-terminal proteolytic domain, raising the possibility that not all proteolytic activities of FtsH heterocomplexes are required for its function. Complementation analysis of an *ftsh2*/*ftsh8* double mutant by expression of a proteolytically-inactive version of FtsH2 was indeed shown to recover leaf variegation [166]. These results suggest that the FtsH heterocomplex without protease activity in Type-B isomers is functional, and protease activity provided by Type-A isomers alone is sufficient for proper function of the heterocomplex. This situation is reminiscent of the chloroplast Clp protease described above, where the core proteolytic complex is composed of essential proteolytically active subunits and their inactive homologs [22, 102], and also of the eukaryotic 20 S proteasome, where most subunits of the proteolytic chamber lack proteolytic activity (e.g. [34]).

Suppressor analysis of the *ftsh2* mutant variegated phenotype has also provided interesting insight into the relations between the thylakoid FtsH protease and other chloroplast proteins [89, 115]. A number of suppressors have been identified to date (e.g.,

Fig. 14.2), whose mutations were found in the genes encoding Clp core and chaperone subunits [105, 161] factors involved in translation [88, 96] and ribosomal RNA maturation [161], and proteins with unknown function [90, 162]. These results highlighted the importance of FtsH threshold levels and the proper balance between protein synthesis and degradation during chloroplast biogenesis. Consistent with this was the recent demonstration that a cross between *ftsh2* mutant and a mutant defective in the chloroplast protein deformylase (whose activity is a prerequisite for removal of the N-terminal Met), active during protein synthesis within the chloroplast, also resulted in the abrogation of the variegated phenotype [4]. Nevertheless, none of the factors identified by suppressor screening appeared to physically interact with FtsH. Several proteins that interact with FtsH have been identified in bacteria and mitochondria, however, no such FtsH- interacting proteins have been found in chloroplasts so far.

14.2.3 The Lon Protease

Another important ATP-dependent protease in *E. coli* is Lon. Similar to FtsH protease, its ATPase and proteolytic domains are found on the same polypeptide. It is a hexameric serine protease that uses a Ser-Lys dyad in its active site [17], which is required for the degradation of abnormal as well as several short-lived regulatory proteins (for review see [40]). Plant homologues of Lon protease (peptidase S16.003) had been first identified in mitochondria [11, 122], but are also found in chloroplasts. Transient-expression assays of GFP fusions revealed that of the four genes found in Arabidopsis, the products of Lon1 and Lon2 are targeted to mitochondria and peroxisomes, respectively, whereas Lon4 is dually targeted to mitochondria and chloroplasts [104, 116]. Proteomic analysis of mitochondria has also revealed the presence of Lon3 in this organelle [46, 47]. Moreover, immunoblot analysis of purified chloroplasts with an antibody against Lon1 revealed a cross-reacting protein of the correct size that was peripherally associated with the stromal side of the thylakoid membrane [104]. This association is reminiscent of the previous localization of plant Lon to the inner membrane of the mt [122], and the archaeal Lon to the plasma membrane [35]. However, how Lon proteases are anchored to membranes is still unknown. Expression of Lon4 appears to be constitutive, as its transcript level does not change upon exposure to high light, or to low or high temperatures [130]. The oligomeric structure of the plant Lon protease, in chloroplasts or mitochondria, is not known. Although our current knowledge of the chloroplast Lon is very limited, it is expected that given its importance in prokaryotes, it should play an important role in chloroplast biology as well.

14.2.4 Deg Proteases

The *E. coli* DegP (also known as HtrA) is an ATP-independent serine protease, peripherally attached to the periplasmic side of the plasma membrane, which is essential for survival at elevated temperatures (for review see [24, 25]. Two related proteases, DegQ and DegS, are also located in the periplasm. DegP forms a hexameric complex, made up of two staggered trimmers, or even larger assemblies of these. Its monomer size is 48 kDa, composed of two distinct domains. The proteolytic domain, with a typical catalytic triad of Ser-His-Asp, is found at the N terminus. Two PDZ domains in tandem, implicated in protein-protein interactions, are located at the C terminus. In addition to its proteolytic activity, DegP demonstrates chaperone activity. Whereas the chaperone activity dominates at low temperatures (below 22 °C), the proteolytic one is manifested at elevated temperatures [135]. The transition between the two activities can be explained by the structure of the protein. At normal optimal growth temperature, the active site of the protease is blocked by segments of the protein itself, while at elevated temperature a conformational change is induced, making the active site accessible to substrates [75, 76].

Similar to the case of Clp and FtsH, a substantial number of DegP homologs are found in higher plants, which are now called 'Deg proteases' [57, 126]. The Arabidopsis genome encodes 16 homologues of DegP and a similar number is present in rice and populus. Of these, Deg1, 2, 5, 7, and 8 have been experimentally localized to chloroplasts, whereas others appear to be targeted to different organelles [126]. The chloroplast enzymes are variable in their domain content and arrangement: Deg5 has one protease domain and lacks a PDZ one; Deg1 and 8 have one protease domain followed by one PDZ domain; Deg2 is most similar to bacterial DegP, having one protease and two PDZ domains; Deg7 appears as a duplication of the latter arrangement in tandem, possessing a total of two protease domains and four PDZ ones [126].

Deg1, Deg5, and Deg8 are located in the lumen and attached to the thylakoid membrane [51, 108, 125], whereas Deg2 and Deg7 are located in the stroma and peripherally associated with the stromal side of this membrane [45, 138]. The association of Deg proteins with membranes is somewhat surprising because they lack any obvious membrane anchor. Interestingly, a significant portion of Deg7 is recruited to the thylakoid membranes in response to exposure to high-light irradiation, probably through interaction with photosystem II (PSII) [138], suggesting a role for this and perhaps other Deg proteins in the repair of photodamaged proteins (see below).

The proteolytic activity of bacterial Deg proteases is dependent on their oligomerization [24, 25]. Size-exclusion chromatography demonstrated that recombinant Deg1 forms a homohexamer [18, 67]. Deg5 and Deg8 were shown (in an in vitro pull-down experiment) to interact with each other, and likely form a heterohexamer with an equal representation of the two variants [137]. Recently, the crystal structure of hexameric Deg1 has been determined at 2.5 Å resolution, and together with the analysis of site-directed mutants, the mechanistic details of its hexamerization were deciphered [67]. This process is triggered by protonation of a specific His residue, His 244, which in turn stabilizes an N-terminal α-helix, now free to interact with a neighboring monomer. Such conformational changes in three adjacent monomers trigger the formation of trimeric intermediates that readily dimerize to form stable and active hexamers (see Fig. 14.3). The proteolytic sites are located within this cage-like structure, access into which is achieved through three side openings. The dimensions of these should allow cleavage of unfolded protein termini or loops. This novel pH-dependent activation mechanism suggests that Deg1 should be more active during the day, when photosynthetic electron transport operates and the lumen is acidic [67].

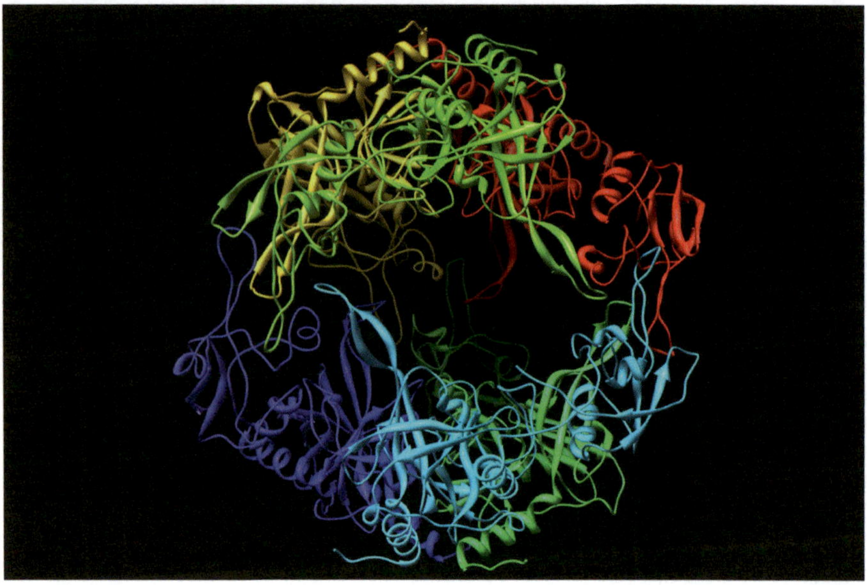

Fig. 14.3 Crystal structure of Deg1. Side view of the Deg1 hexamer is presented. Each one of the monomers is colored differently. The free ends at the top and bottom of the picture are the N-termini. Interaction between two trimers facing each other forms the active hexamer. Access into the active sites, found within the cage-like structure, is through three side windows. (Adopted from [67])

14.3 Other Proteases

The term 'intramembrane proteolysis' refers to the cleavage of peptide bonds within trans-membranes helices. Such cleavage events are catalyzed by four groups of proteases: S2P, Presenilin, SPP and Rhomboid (for reviews see [150, 152]. Initially, the idea of a hydrolytic reaction within the hydrophobic core of a membrane seemed paradoxical, but elucidation of the crystal structure of Rhomboid proteases demonstrated the presence of water molecules at the active site and unraveled a gating mechanism for substrate accessibility (e.g., [81, 153]).

Two studies identified homologues of S2P in chloroplasts. A genetic screen for Arabidopsis mutants deficient in both chlorophyll accumulation and ethylene-induced gravitropism revealed EGY1, a 59 kDa membrane-bound metalloprotease located in the chloroplast [20]. Although the intraorganellar location of EGY1 was not determined, mutant plants had reduced levels of grana stacking and light-harvesting complex (LHC) proteins, suggesting that this protease is required for proper chloroplast development. Another protease related to S2P, designated AraSP, was localized to the chloroplast inner-envelope membrane [16]. Antisense and T-DNA insertion lines of this protease demonstrated severely impaired chloroplast biogenesis. However, how these proteases are involved in chloroplast biogenesis is not clear.

Rhomboid proteases are widely distributed intramembrane serine proteases, integrated into the membrane by six or seven hydrophobic α-helices. They have their Ser-His catalytic dyad located within the hydrophobic core of the membrane [80, 146, 149]. Cleavage of transmembrane substrates by Rhomboids enables the release of soluble domains from the membrane. This regulated intramembrane proteolysis plays a critical role in signal transduction cascade in Drosophila and maturation of certain membrane proteins in mitochondria. To date, several reports have described the occurrence of plant Rhomboids [36, 52, 69, 70, 82]. In the Arabidopsis genome, at least 16 Rhomboid-like sequences exist, with three of them lacking the conserved catalytic dyad in their sequence [69]. Among the 13 possibly functional homologues, two (AtRBL1 and AtRBL2) are localized to the Golgi apparatus [52] and one (previously named AtRBL12, and currently AtPARL) is targeted to mitochondria [68]. Results obtained using the yeast mitochondria system suggest a potential link between a Rhomboid-like protease and the plastid translocon component Tic40 [54]. Nevertheless, to date, only AtRBL9 is known to localize to chloroplasts, as demonstrated by the transient expression of GFP fusion proteins [68]. The physiological functions of the Rhomboids in the chloroplast are still unclear.

SppA is a homologue of the *E. coli* SppA serine protease (peptidase S49.001), which functions as a signal-peptide peptidase in bacteria and contains a Ser-Lys catalytic dyad [64]. In chloroplasts, it is bound to the stromal side of the thylakoid membrane, and its expression is induced by light [83]. Interestingly, this is the only thylakoid protease whose level increases in response to high light [37].

14.4 Functions of Chloroplast Proteases

As briefly mentioned in the Introduction, proteases participate in a wide range of cellular and extracellular activities. In the following section, we will review characterized proteolytic processes in the chloroplast and the proteases responsible for them. Maturation of chloroplast pre-proteins is a common proteolytic event that many proteins are subjected to at the beginning of their life cycle. This includes removal of the initiating Met residue of many proteins synthesized within the organelle, cleavage of N-terminal targeting sequences of proteins synthesized in the cytosol, and a few examples of C-terminal processing. However, as processing peptidases are the subject of Chap. 12 in this volume, they will be excluded from this chapter.

14.4.1 Adaptation to Changing Light Intensities

Although light is essential to plants, it also has detrimental effects on them, a phenomenon known as 'photoinhibition' [6, 12]. Several strategies that prevent the harmful effects of light on the photosynthetic machinery have evolved, some of them involving proteolysis. Long-term adaptation to an increase in light intensity is accompanied by a decrease in the antenna size of PSII, leading to a decrease in the amount of excitation energy being funneled to the reaction center, and hence a decrease in the probability for damage. This modulation of antenna size is achieved by proteolytic degradation of a subset of LHCII subunits [85, 156]. Analysis of several Arabidopsis FtsH mutants suggested the involvement FtsH6 in this process [165]. This was criticized based on the relatively insensitive degradation assay used in the aforementioned work [3], and indeed, a recent study, re-examining the role of FtsH6 in LHCII degradation during senescence and acclimation to high light, found no differences between WT and FtsH6 knockout mutants [148]. In this respect, it is also worth mentioning another recent report, in which a similar degradation assay was used, concluding that the Deg2 protease was involved in the degradation of Lhcb6 in response to exposure to short-term stresses, including high light [92].

Another proposed candidate for LHCII degradation during adaptation to high light is the SppA protease. As mentioned above, this is the only thylakoid protease whose level increases in response to high light [37]. An interesting observation in this respect is the involvement of a cyanobacterial homologous enzyme in the degradation of phycobiliproteins, which are the antenna complexes in these organisms, during acclimation to increased light intensity [111]. In the corresponding Arabidopsis knockout lines, acclimation to high light is indeed altered [151]. Nevertheless, direct experimental support for the involvement of SppA in LHCII degradation during exposure to high light is still lacking.

The transition from high to low light is also accompanied by protein degradation. The best example in this context is the 'early light-inducible protein' (ELIP). This protein, which is structurally related to LHCs, is rapidly degraded upon such a transition [5]. In the aforementioned Arabidopsis mutants lacking SppA, ELIP1 is indeed stabilized during recovery from the high light treatment, supporting a role for this protease in ELIP degradation as well [151].

14.4.2 Protein Quality Control

Accumulation of all major photosynthetic complexes requires coordination between the chloroplast and nuclear genomes. Although in recent years advances in understanding how the two genomes communicate with each other have been made (see Sect. 14.4.6 and Chap. 3 of this volume), little is known about the mechanisms involved in regulating the correct stoichiometry between the different subunits of a given complex. It is assumed that fine-tuning of their levels is achieved by proteolytic degradation of super-stoichiometric subunits. The first support for this assumption came from a work published almost 30 years ago. Inhibition of protein synthesis in the chloroplast of *Chlamydomonas*, including that of the large subunit of Ribulose 1,5- bisphosphate carboxylase/oxygenase, resulted in degradation of the nuclear-encoded small subunit within the chloroplast [124]. These results suggested that unassembled subunits of multiprotein complexes are rapidly degraded, and indeed, similar observations have been made for other photosynthetic complexes. For instance, in *Chlamydomonas,* when cytochrome *b6*, subunit IV and the Rieske protein of the cytochrome *b6-f* complex cannot assemble with cytochrome *f,* they are rapidly degraded [77]. Similarly, a point mutation in the Rieske protein leads to a significant decrease in its level, as well as to the levels of other subunits of the cytochrome *b6-f* complex. Crossing this mutant with one possessing reduced levels of ClpP1 resulted in stabilization of these proteins, suggesting a role for Clp protease in the degradation of some unassembled proteins [93].

In vitro studies have hinted at a role for the Clp protease in the degradation of unassembled or abnormal proteins in the stroma as well. Mistargeting of the lumenal protein OE33 to the stroma resulted in its rapid degradation, with characteristics reminiscent of those of the Clp protease [42, 43]. A similar function may be fulfilled by other proteases as well. Experiments with wild type or mutant forms of the Rieske protein have demonstrated that molecules that fail to translocate across the thylakoid membrane are rapidly degraded by a membrane-bound metalloprotease. This in vitro degradation reaction could be inhibited by increasing amounts of an antibody against native FtsH [103], suggesting that this protease may be involved in protein quality control in the chloroplast as well.

Many mutants have been shown to contain decreased levels of subunits of a complex when one other subunit is missing. This observation is often interpreted as degradation of the not-fully-assembled complex. However, such conclusions should be viewed with caution when no direct evidence for degradation is provided, for instance, by pulse-chase experiments. In some cases, translation of a complex's subunits is regulated by another component of the complex (a regulatory mechanism known as 'control by epistasy', see [95] and references therein). Thus, lower levels of the subunits of a complex may not necessarily be due to degradation of unassembled components, but also a result of reduced rates of translation in the chloroplast.

It has long been known that chloroplast proteins are unstable when their cofactors are missing. For example, in the absence of chlorophyll, due to either inhibition of synthesis or mutation, chlorophyll-binding proteins are rapidly degraded (e.g., [63] and references therein). Similarly, the lack of a single copper ion is sufficient to destabilize the electron carrier plastocyanin [84]. These observations suggest that minor structural changes, induced by a lack of minor components of a protein, may render it susceptible to proteolysis. Nevertheless, although the above examples have been documented for some time, the proteases involved in degrading the protein substrates remain a mystery.

14.4.3 Removal of Oxidatively Damaged Proteins

Not all light-energy absorbed by the photosynthetic antenna is converted into chemical energy. Depending on environmental conditions, free radicals are generated in chloroplasts, and despite the presence of free-radical scavengers, chloroplast proteins are highly prone to oxidation processes, which may impair their structure and function. The best characterized oxidatively damaged protein in the chloroplast is the D1 protein of the PSII reaction center. Oxidative damage leads to PSII inactivation, and hence, to photoinhibition (for reviews, see [6, 7, 155]). A prerequisite for the repair of photoinhibited PSII is degradation of the D1 protein (for a model, see Fig. 14.4), and numerous attempts have been made to identify the protease(s) involved. Biochemical approaches have been largely unsuccessful. However, the identification of chloroplast proteases and availability of protease mutants have enabled testing the possible involvement of specific proteases in D1 degradation.

Attempts to test the potential involvement of FtsH in the degradation of D1 were first made using recombinant GST-FtsH1 [87]. These experiments demonstrated weak, albeit significant activity of the recombinant enzyme against the 23 kDa degradation product of the D1 protein, but not against the full-length protein. The weak activity and partial specificity may result from the homomeric nature of the recombinant enzyme used in vitro, as opposed to the heteromeric nature of the enzyme found in vivo, as described in Sect. 14.2.2. The variegated mutants of FtsH2 and FtsH5 provided an opportunity to test the possible involvement of the FtsH protease in the repair cycle of PSII in vivo as well. These mutants demonstrated increased sensitivity to photoinhibition compared to the wild type, as revealed by PSII-activity measurements [10, 117]. Consistent with this proposed role was the finding that the D1 protein is stabilized, probably in an inactive form, in FtsH mutant plants after exposing them to different light conditions [10, 56, 61]. It is also interesting to note that although no association between the thylakoid FtsH and PSII was observed in the initial characterization of this enzyme [86], a recent report demonstrated a close proximity between the two [158]. Similar to the role of FtsH in the repair of PSII from photoinhibition, its involvement in D1 degradation in response to heat stress was also demonstrated [157].

Deg2, associated with the stromal side of the thylakoid membrane, has also been implicated in D1 degradation. An in vitro study demonstrated that recombinant Deg2 could cleave the D1 protein at the stromal loop (DE loop) that connects its fourth and fifth transmembrane helices, yielding an N-terminal 23 kDa product and a C-terminal 10 kDa product, suggesting that this protease participates in the initial stages of D1 degradation [45]. However, an in vivo study with Arabidopsis mutants lacking Deg2 showed that the rate of D1 degradation under light stress that was comparable to the wild type, indicating that Deg2 is not essential for D1 degradation [49]. In a more recent study, another variant of the Deg protease, Deg7, was identified attached to the stromal side of Arabidopsis thylakoids. In its absence, sensitivity to photoinhibition was enhanced. Consistent with this, degradation of the D1 protein, as well as of other subunits of PSII, was inhibited [138]. These results foster the notion that, in addition to the FtsH complex, other proteases that face the stromal side of the thylakoid membrane contribute to the degradation of D1, and likely of other PSII subunits.

Deg proteases located on the lumenal side of the thylakoid membrane also participate in D1 degradation. Transgenic plants in which Deg1 was knocked down were smaller than wild type and more sensitive to photoinhibition. These plants

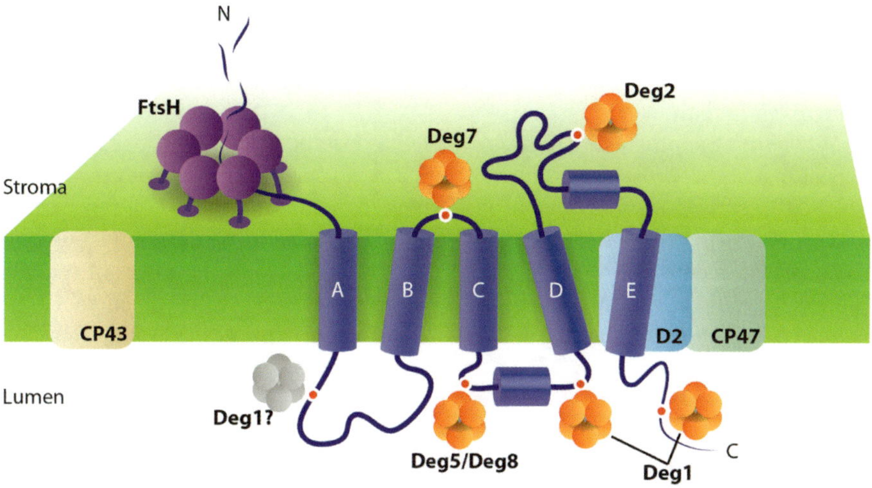

Fig. 14.4 A model for the degradation of the D1 protein of PSII reaction center. The D1 protein with its five trans-membrane helices is depicted in *blue*. Degradation presumably requires partial disassembly of the PSII complex, illustrated here by the detachment of the CP43 subunit. Cleavage of soluble domains of D1 in the stroma by Deg2 and Deg7 generates additional termini that can be recognized by the FtsH complex. Cleavage of lumenal domains by Deg1 and the Deg5/Deg8 complex shorten the hydrophobic segments, and thus facilitate their extraction from the membrane and their degradation by the ATP-dependent FtsH. Cleavage sites of Deg proteases are only schematic, as their exact locations have not been experimentally determined yet. We hypothesize that this model of degradation is not limited to D1, and should be applicable to other highly hydrophobic thylakoid membrane proteins as well

accumulated more of the D1 protein, probably in an inactive form, but less of its 16 and 5.2 kDa degradation products [53]. Moreover, addition of recombinant Deg1 to inside-out thylakoid membranes could induce the formation of the 5.2 kDa C-terminal D1fragment in vitro. As no homogyzous knockout lines could be obtained so far, Deg1 seems to be an essential protein [53]. In addition to Deg1, a lumenal complex of Deg5 and Deg8, also takes part in D1 degradation. Arabidopsis single and double mutants of these proteins demonstrate increased sensitivity to photoinhibition, as well as stabilization of the D1 protein. Taken together, the results suggest that lumenal Deg proteases cooperate with proteases found on the stromal side of the membrane in the degradation of D1 protein during repair from photoinhibition [53] (Fig. 14.4).

Further insight into the regulation of proteolytic activities involved in D1 protein degradation was recently obtained from the Deg1 structure and from characterization of the activity of its site-directed mutants [67]. These have unraveled a novel activation mechanism that relates generation of substrates to the activation of the Deg1 protease. In the dark, when the pH in the lumen is neutral or slightly basic, Deg1 is found in its monomeric, resting inactive conformation. Once light comes on, photosynthetic electron transport commences and components of PSII, especially the D1 protein, are oxidized. Coupled to electron transport, the lumen becomes acidified. Under these conditions, which contribute to accumulation of oxidative damage, Deg1 oligomerizes into the active hexamer. Thus, the same conditions that lead to the formation of oxidatively damaged proteins, also activate a protease that is involved in their degradation.

A *Chlamydomonas* ATP synthase mutant has also been shown to lose PSII upon exposure to light. Crossing this mutant with a strain containing lower levels of ClpP resulted in stabilization of several PSII subunits, including the D2 protein, CP43 and CP47 [94]. It is not known whether this degradation process is identical to the one that occurs during photoinhibition, or even whether the effect of the Clp protease is a direct one. Nevertheless, these results suggest the involvement of the soluble Clp protease in the degradation of membrane substrates as well.

14.4.4 Nutrient Stress and Senescence

Nutrient stress and senescence are both characterized by the need to remobilize internal cellular resources, some of which can be provided by the building blocks of existing proteins. Thus, massive protein degradation is expected to accompany the plants' attempts to deal with nutrient stress or their final developmental stage, senescence. However, only little is known about the involvement of specific proteases in these processes. Downregulation of ClpP1 in *Chlamydomonas* suggests involvement of the Clp complex in the degradation of thylakoid membrane proteins upon exposure to nutrient stress [93]. Nitrogen starvation results in degradation of subunits of the cytochrome *b6-f* complex. In cells containing reduced levels of ClpP1, this degradation process is retarded, suggesting that Clp protease may be involved in the adaptation to nitrogen starvation via the degradation of existing abundant proteins.

Iron homeostasis was recently linked to the Clp protease [154]. Either a point mutation or knockout of ClpC1 in Arabidopsis resulted in a chlorotic phenotype that could be overcome by an excess of iron. Nevertheless, it is not clear whether this function of ClpC1 is related to its role as the regulatory subunit of the Clp protease or simply to its chaperone activity.

Protein degradation in senescing leaves followed by nitrogen mobilization to younger ones is a well-documented phenomenon [48]. To date, one specific protease has been linked to the degradation of the most abundant protein in chloroplasts, Rubisco. CND41 (peptidase A01.050) is a 41 kDa aspartic protease that is associated with chloroplast nucleoids. It exhibits proteolytic activity against denatured Rubisco [98], and its level increases in senescing leaves [27]. Moreover, antisense plants demonstrated delayed senescence, along with stabilization of Rubisco as well as other chloroplast proteins [58]. Interestingly, CND41 itself must undergo a protelytic processing step for its activation [59]. The significance of CND41 binding to DNA is not known yet, but it could be a means of sequestering it from other chloroplast proteins. Degradation of plastid DNA during early stages of senescence may release CND41 to the stroma, allowing the initiation of massive protein degradation.

14.4.5 Thermotolerance

A study on a thermosensitive Arabidopsis mutant suggests the involvement of FtsH11 in thermotolerance [21]. This mutant was more sensitive to moderate high temperature (30 °C), and thus had lower photosynthetic capability, than the wild type. On the other hand, unlike the FtsH2 and FtsH5 mutants, the FtsH11 mutant was not sensitive to high light [21]. These results suggest that the physiological functions of the thylakoid FtsH complex and FtsH11 differ.

14.4.6 Retrograde Signaling

Communication between the chloroplast and the nucleus is essential for the biogenesis and function of the chloroplast. This has been the subject of intensive studies over the past two decades; however, the nature of the molecular relay has remained elusive (for review see [30, 110] and Chap. 3 of this volume). A major breakthrough in the field was the recent identification of a chloroplast envelope-bound transcription factor that possesses transmembrane domains. Upon proteolytic cleavage, the soluble N-terminal domain of this protein is released from the membrane and relocates to the nucleus, where it activates the transcription of the ABA response gene ABI4 [140]. The protease involved in this process is still unknown, but it is most likely one of the intramembrane proteases capable of cleaving within transmembrane α-helices.

14.4.7 Other Functions

Somewhat less characterized and not so well understood functions of chloroplast proteases include a recently proposed chaperone role for Deg1 in PSII assembly [139], degradation of the PsbF apoprotein by Deg5 in response to wounding [91], and a glutamyl endopeptidase capable of degrading the N-terminal domain of LHCII [32]. Further work will hopefully lead to better understanding of these processes in the near future.

14.5 Identification of Specific Substrates

The availability of specific protease mutants lends itself to the identification of their substrates, when these are unknown. Specific substrates of a protease are expected to be stabilized in a mutant background, and thus comparative proteomics has the potential to yield their unbiased identification. This approach has been successfully utilized to compare between stromal proteins in wild type and ClpP6, ClpR1 and

ClpR2 mutants [133, 136, 169]. Potential substrates of Clp protease found in these works include a nuclear exchange factor for the elongation factor Tu, the molecular chaperone HSP90, an RNA helicase, polynucleotide phosphorylase, the folding catalyst PPIase, the UPRT and NDP kinase proteins involved in nucleic acid synthesis, as well as other constitutive enzymes involved in different metabolic pathways. From these results it seems that Clp substrates are more involved in chloroplast homeostasis rather than stress responses [133, 136]. A similar approach should prove useful in the identification of substrates of other proteases as well. Nevertheless, results of such analyses should be viewed with care, as some of the up-regulated proteins may result from increase in their expression in response to protein folding stress, incurred by the reduction in protease level [169], rather than a decrease in their degradation.

Another promising approach is the use of epitope-tagged proteases such as the one reported for Clp [101]. Together with mutations in the corresponding proteolytic active sites, this will allow trapping of substrates within proteases and their identification following affinity purification and mass spectrometry analysis. Along with the comparative proteomics approach described above, they should yield more reliable lists of specific substrates.

More substrates will likely be revealed not only by large-scale approaches, but also by studies on specific substrates. One such example is the case of chlorophyllide *a* oxygenase (CAO), the enzyme responsible for chlorophyll *b* synthesis. The N-terminal domain of this enzyme controls CAO level in response to chlorophyll *b* accumulation [99]. A genetic screen that was based on fusion of this domain to GFP identified a mutation in ClpC1 that resulted in a decreased rate of degradation of the reporter protein, and hence established CAO as a specific substrate of the chloroplast Clp protease.

14.6 Determinants of Protein Instability

Similar to all proteins synthesized in prokaryotic organisms, the 80 or so proteins synthesized within the chloroplast contain an N-formyl Met residue at their N terminus. Most of these proteins undergo maturation that involves two hydrolytic reactions: the N-formyl group is removed by peptide deformylase (PDF), and in most cases, this is followed by the activity of methionine aminopeptidase (MAP) (peptidase M24.001), which removes the N-terminal Met residue [38, 39]. Thus, the activity of MAP has implications for the identity of the N-terminal residue of mature proteins encoded and synthesized in chloroplasts. Similarly, the activity of processing peptidases results in the exposure of different residues at the N-termini of different proteins. This may affect the stability of these proteins through the N-end rule pathway (see below).

Although progress has been made in identifying components of the chloroplast proteolytic machinery, and proteolytic processes have been documented, determinants of instability within the protein substrates themselves are still obscure. The

N-end rule, discovered and characterized in eukaryotic cells, relates the half-life of a protein to the identity of its N-terminal residue. Proteins carrying a destabilizing residue at their N terminus are ubiquitinated and degraded by the 26S proteasome [147]. The N-end rule was shown to operate also in *E. coli*, where degradation of substrates is mediated by the ClpAP protease [143]. The *E. coli* ClpS adaptor protein has been described as a modulator of ClpAP activity [28], and shown to be essential for operation of the N-end rule pathway in bacteria [29]. As plastids are descendants of a prokaryotic progenitor and many of their characteristics, including their proteolytic machinery, are prokaryote-like, it is highly likely that an N-end rule-like mechanism governs protein stability in plastids as well. As described in Sect. 14.2.1, plastids contain homologues of the bacterial ClpA, designated ClpC, and ClpS [109]. However, ClpS was not identified in proteomic studies of the Clp core complex [3], and thus may not be a component of the Clp protease core. Nevertheless, the presence of chloroplast homologues to components of the bacterial N-end rule pathway suggests that it may govern protein stability/instability in this organelle as well.

One comprehensive attempt for determination of stability determinants was recently reported [9]. GFP constructs, in which the penultimate N-terminal residues were systematically engineered, were expressed in tobacco chloroplasts. Assuming removal of the initiating Met, these constructs were expected to yield GFP with different N-termini within the chloroplast. Monitoring the levels of GFP in the 20 different lines revealed that Cys and His were highly destabilizing residues, whereas lines with Glu, Met and Val accumulated the highest level of the fusion protein [9]. Asp and Ile were also destabilizing and Asn, Arg and Gln gave intermediate levels. However, since this study was performed with a single substrate protein, it is still not clear whether general conclusions can be drawn from it.

The small stable RNA A (SsrA) system in *E. coli* tags proteins translated from incomplete mRNAs for degradation [55]. The ssrA RNA is a small molecule that acts both as a tRNA and an mRNA molecule. When a ribosome stalls on an incomplete mRNA, the ssrA molecule binds the ribosome, which then reads through to add 11 amino acids to the protein. This tag, containing a sequence of small nonpolar amino acid residues, is recognized in the cytoplasm by the ClpAP, ClpXP or FtsH proteases, or in the periplasm by the DegP protease. To date, there is no evidence for the presence of ssrA RNA in plastids of higher plants [41]. However, the presence of homologues of the bacterial proteases suggests that plastid proteins with C termini homologous to the SsrA tag would be short-lived. Moreover, even in the absence of the SsrA system, the identity of C- terminal residues may confer stability or instability on a protein. However, this has not been explored to date.

The case of CAO, the specific substrate of Clp protease described above, also provided an insight into the destabilizing sequence found in the N-terminal domain of this enzyme. Serial deletions from this domain identified the sequence QDLLTIMILH as essential for regulating the stability of CAO [121]. Moreover, destabilization of GFP by fusion to this sequence further corroborated its destabilizing nature.

14.7 Regulation of Proteolysis in Chloroplasts

The coexistence of proteases and their substrates within the same compartment raises the question how proteolysis is regulated, and how premature degradation is prevented. Earlier studies in the field, especially those looking at proteolytic processes occurring in response to high light, assumed regulation at the level of protease availability, and hence pursued the identification of proteases whose expression is induced in response to exposure to such conditions. However, this mode of regulation looks unlikely as it is now well established that many if not all chloroplast proteases are expressed in a constitutive manner.

One existing mode of regulation is the aforementioned 'self-compartmentalization' [14] (see Sect. 14.2), a characteristic common to all ATP-dependent proteases, including the chloroplast Clp, Lon and FtsH proteases. Access into the proteolytic chamber of these enzymes is restricted only to proteins that are recognized by the ATPase component of the protease and are unfolded by them. As mentioned above, the recognition signals within chloroplast substrates are largely unknown. Nevertheless, it appears that at least for inherently stable proteins, such as those of the photosynthetic machinery, a certain degree of conformational change is needed to allow their recognition by proteases.

A second mode of regulation is represented by the newly discovered pH-dependent oligomerization of Deg1 [67]. Here again, the active sites are within an internal sphere of the oligomer and not exposed at its surface. Access into this sphere is through side pores whose dimensions restrict entry only to unfolded structures. It will be interesting to see whether oligomerization and hence activation states of other chloroplast proteases change in response to different clues, be it pH, substrate availability, redox state, or any other clue.

Apart from the two modes mentioned above, which are strictly related to the proteases themselves, substrates may represent another level of regulation. Short-lived proteins such as 'timing proteins', are continuously degraded, probably due to having degradation signals or unstable structure. However, inherently stable proteins are also degraded at some point of their life cycle. What could trigger their degradation? Most likely it is conformational changes. These could be induced by posttranslational modifications such as oxidation, which do occur in chloroplasts at high rate, especially during exposure to high light, and/or other modifications. However, this scenario remains to be demonstrated.

14.8 Future Prospects

Plant sequence data that has accumulated over the past 15 years, and completion of genome sequencing projects in many species, suggest that the identity of most, if not all, chloroplast proteases and peptidases is now known. The major challenge ahead is to assign a function to each one of them, and relate the proteases and peptidases to known proteolytic processes. A striking feature of several chloroplast proteases is the relatively large number of genes encoding them, compared with their prokaryotic progenitors. It is now clear that at least the Clp and FtsH proteases are heteromeric complexes, with little redundancy between their components. As the presence of the P- and R-rings of the Clp complex is now established, it will be important to determine whether ClpRs have only a structural role, or perhaps other functions, such as substrate binding or recognition. The heteromeric nature of the thylakoid FtsH complex is also established, with redundancy between duplicated genes, but essentiality of the type A and type B isoforms. However, the structure of this heterocomplex, and especially the interaction between the two isoforms, await deciphering. The recent structural analysis of Deg1 provided important insights into the regulation of its activity. However, the relations between the three lumenal Deg proteases, and also between the stromal ones, need to be sorted out. Recruitment of Deg proteases to thylakoid membranes in response to light is also an intriguing issue that will have to be tackled.

Major insights into the functions of different chloroplast proteases have been gained using specific mutants. There are now a number of publicly available mutant collections, not only in Arabidopsis but also in many other species. Thus, these will probably continue to serve as main tools in deciphering the physiological functions of specific proteases. In vitro approaches will probably continue to complement these efforts. However, special attention should be paid to possible pleiotropic effects. Many mutants lacking different chloroplast proteins demonstrate a similar phenotype: slow growth, reduced pigmentation, altered chloroplast morphology and reduced levels of thylakoids. Thus, efforts should be made to distinguish between these general effects and the specific function of a given protease leading to these effects. This requires more specific assays for particular proteolytic processes, better linkage to substrate proteins, and attempts to understand their involvement in a given physiological response. Nevertheless, a combination of these mutant lines with advanced techniques such as modern mass spectrometry for proteomics and next-generation sequencing for transcriptomics should provide more insights into the function of these proteases.

Identification of specific substrates for each of the proteases will have to be accompanied by attempts to reveal recognition determinants. To date, understanding of the recognition mechanisms between chloroplast proteases and their substrates is almost totally lacking. This applies to both partners—the proteases and their substrates. Efforts will need to be made to identify subunits within a proteolytic complex, or domains within a given protease, that are responsible for substrate recognition and binding, and determinants on the substrates themselves that allow this recognition. All these questions will keep the growing community of scientists interested in chloroplast proteases busy for years to come.

Acknowledgements Work in Adam's lab was supported by grants from The Israel Science Foundation (ISF), The U.S.-Israel Binational Science Foundation (BSF), The U.S.-Israel Binational Agricultural Research and Development Fund (BARD) and The Sixth Framework Program of the EU. Work in Sakamoto's lab was supported by Japan Society for the Promotion of Science (JSPS), Japan Science and Technology Agency (JST), and Oohara Foundation. The critical reading and editorial work done by Dr. Dana Charuvi and graphic work by Gil Adam are greatly appreciated.

References

1. Adam Z et al (2001) Chloroplast and mitochondrial proteases in Arabidopsis. A proposed nomenclature. Plant Physiol 125:1912–1918
2. Adam Z, Zaltsman A, Sinvany-Villalobo G, Sakamoto W (2005) FtsH proteases in chloroplasts and cyanobacteria. Physiol Plant 123:386–390
3. Adam Z, Rudella A, van Wijk KJ (2006) Recent advances in the study of Clp, FtsH and other proteases located in chloroplasts. Curr Opin Plant Biol 9:234–240
4. Adam Z, Frottin F, Espagne C, Meinnel T, Giglione C (2011) Interplay between N-terminal methionine excision and FtsH protease is essential for normal chloroplast development and function in Arabidopsis. Plant Cell 23:3745–3760
5. Adamska I, Kloppstech K, Ohad I (1993) Early light-inducible protein in pea is stable during light stress but is degraded during recovery at low light intensity. J Biol Chem 268:5438–5444
6. Adir N, Zer H, Shochat S, Ohad I (2003) Photoinhibition-a historical perspective. Photosynth Res 76:343–370
7. Andersson B, Aro E-M (2001) Photodamage and D1 turnover in photosystem II. In: Andersson B, Aro E-M (eds) Regulation of photosynthesis. Kluwer, Dordrecht, pp 377–393
8. Andersson FI et al (2009) Structure and function of a novel type of ATP-dependent Clp protease. J Biol Chem 284:13519–13532
9. Apel W, Schulze WX, Bock R (2010) Identification of protein stability determinants in chloroplasts. Plant J 63:636–650
10. Bailey S, Thompson E, Nixon PJ, Horton P, Mullineaux CW, Robinson C, Mann NH (2002) A critical role for the Var2 FtsH homologue of *Arabidopsis thaliana* in the photosystem II repair cycle in vivo. J Biol Chem 277:2006–2011
11. Barakat S, Pearce DA, Sherman F, Rapp WD (1998) Maize contains a *Lon protease* gene that can partially complement a yeast pim1-deletion mutant. Plant Mol Biol 37:141–154
12. Barber J, Andersson B (1992) Too much of a good thing: light can be bad for photosynthesis. Trends Biochem Sci 17:61–66
13. Barrett AJ, Rawlings ND, Woessner JF (2004) Handbook of proteolytic enzymes, 2nd edn. Elsevier, Amsterdam
14. Baumeister W, Walz J, Zuhl F, Seemuller E (1998) The proteasome: paradigm of a self-compartmentalizing protease. Cell 92:367–380
15. Bieniossek C, Schalch T, Bumann M, Meister M, Meier R, Baumann U (2006) The molecular architecture of the metalloprotease FtsH. Proc Natl Acad Sci U S A 103:3066–3071
16. Bolter B, Nada A, Fulgosi H, Soll J (2006) A chloroplastic inner envelope membrane protease is essential for plant development. FEBS Lett 580:789–794
17. Botos I et al (2004) Crystal structure of the AAA+ alpha domain of *E. coli* Lon protease at 1.9 A resolution. J Struct Biol 146:113–122
18. Chassin Y, Kapri-Pardes E, Sinvany G, Arad T, Adam Z (2002) Expression and characterization of the thylakoid lumen protease DegP1 from *Arabidopsis*. Plant Physiol 130:857–864
19. Chen M, Choi Y, Voytas DF, Rodermel S (2000) Mutations in the *Arabidopsis* VAR2 locus cause leaf variegation due to the loss of a chloroplast FtsH protease. Plant J 22:303–313
20. Chen G, Bi YR, Li N (2005) EGY1 encodes a membrane-associated and ATP-independent metalloprotease that is required for chloroplast development. Plant J 41:364–375
21. Chen J, Burke JJ, Velten J, Xin Z (2006) FtsH11 protease plays a critical role in *Arabidopsis* thermotolerance. Plant J 48:73–84
22. Clarke AK (2012) The chloroplast ATP-dependent Clp protease in vascular plants-new dimensions and future challenges. Physiol Plant 145:235–244
23. Clarke AK, MacDonald TM, Sojogren LLE (2005) The ATP-dependent Clp protease in chloroplasts of higher plants. Physiol Plant 123:406–412
24. Clausen T, Kaiser M, Huber R, Ehrmann M (2011) HTRA proteases: regulated proteolysis in protein quality control. Nat Rev Mol Cell Biol 12:152–162

25. Clausen T, Southan C, Ehrmann M (2002) The HtrA family of proteases: implications for protein composition and cell fate. Mol Cell 10:443–455
26. Constan D, Froehlich JE, Rangarajan S, Keegstra K (2004) A stromal hsp100 protein is required for normal chloroplast development and function in *Arabidopsis*. Plant Physiol 136:3605–3615
27. Diaz C, Lemaitre T, Christ A, Azzopardi M, Kato Y, Sato F, Morot-Gaudry JF, Le Dily F, Masclaux-Daubresse C (2008) Nitrogen recycling and remobilization are differentially controlled by leaf senescence and development stage in *Arabidopsis* under low nitrogen nutrition. Plant Physiol 147:1437–1449
28. Dougan DA, Reid BG, Horwich AL, Bukau B (2002) ClpS, a substrate modulator of the ClpAP machine. Mol Cell 9:673–683
29. Erbse A, Schmidt R, Bornemann T, Schneider-Mergener J, Mogk A, Zahn R, Dougan DA, Bukau B (2006) ClpS is an essential component of the N-end rule pathway in *Escherichia coli*. Nature 439: 753–756
30. Fernandez AP, Strand A (2008) Retrograde signaling and plant stress: plastid signals initiate cellular stress responses. Curr Opin Plant Biol 11:509–513
31. Ferro M et al (2010) AT_CHLORO, a comprehensive chloroplast proteome database with subplastidial localization and curated information on envelope proteins. Mol Cell Proteomics 9:1063–1084
32. Forsberg J, Strom J, Kielsbach T, Larsson H, Aleciev K, Engstrom A, Akerlund H-E (2005) Protease activities in the chloroplast capable of cleaving an LCHPII N-terminal peptide. Physiol Plant 123:21–29
33. Friso G, Giacomelli L, Ytterberg AJ, Peltier JB, Rudella A, Sun Q, Wijk KJ (2004) In-depth analysis of the thylakoid membrane proteome of *Arabidopsis thaliana* chloroplasts: new proteins, new functions, and a plastid proteome database. Plant Cell 16:478–499
34. Fu H, Doelling JH, Arendt CS, Hochstrasser M, Vierstra RD (1998) Molecular organization of the 20S proteasome gene family from *Arabidopsis thaliana*. Genetics 149:677–692
35. Fukui T, Eguchi T, Atomi H, Imanaka T (2002) A membrane-bound archaeal Lon protease displays ATP-independent proteolytic activity towards unfolded proteins and ATP-dependent activity for folded proteins. J Bacteriol 184:3689–3698
36. Garcia-Lorenzo M, Sjodin A, Jansson S, Funk C (2006) Protease gene families in populus and *Arabidopsis*. BMC Plant Biol 6:30
37. Giacomelli L, Rudella A, van Wijk KJ (2006) High light response of the thylakoid proteome in *Arabidopsis* wild type and the ascorbate-deficient mutant vtc2-2. A comparative proteomics study. Plant Physiol 141:685–701
38. Giglione C, Meinnel T (2001) Organellar peptide deformylases: universality of the N-terminal methionine cleavage mechanism. Trends Plant Sci 6:566–572
39. Giglione C, Vallon O, Meinnel T (2003) Control of protein life-span by N-terminal methionine excision. EMBO J 22:13–23
40. Gottesman S (1996) Proteases and their targets in *Escherichia coli*. Annu Rev Genet 30:465–506
41. Gueneau de Novoa P, Williams KP (2004) The tmRNA website: reductive evolution of tm-RNA in plastids and other endosymbionts. Nucleic Acids Res 32:D104–D108
42. Halperin T, Adam Z (1996) Degradation of mistargeted OEE33 in the chloroplast stroma. Plant Mol Biol 30:925–933
43. Halperin T, Ostersetzer O, Adam Z (2001) ATP-dependent association between subunits of Clp protease in pea chloroplasts. Planta 213:614–619
44. Halperin T, Zheng B, Itzhaki H, Clarke AK, Adam Z (2001) Plant mitochondria contain proteolytic and regulatory subunits of the ATP-dependent Clp protease. Plant Mol Biol 45:461–468
45. Haussuhl K, Andersson B, Adamska I (2001) A chloroplast DegP2 protease performs the primary cleavage of the photodamaged D1 protein in plant photosystem II. EMBO J 20:713–722
46. Heazlewood JL, Millar AH (2005) AMPDB: the *Arabidopsis* mitochondrial protein database. Nucleic Acids Res 33:D605–D610

47. Heazlewood JL, Tonti-Filippini JS, Gout AM, Day DA, Whelan J, Millar AH (2004) Experimental analysis of the *Arabidopsis* mitochondrial proteome highlights signaling and regulatory components, provides assessment of targeting prediction programs, and indicates plant-specific mitochondrial proteins. Plant Cell 16:241–256
48. Hortensteiner S, Feller U (2002) Nitrogen metabolism and remobilization during senescence. J Exp Bot 53:927–937
49. Huesgen PF, Schuhmann H, Adamska I (2006) Photodamaged D1 protein is degraded in *Arabidopsis* mutants lacking the Deg2 protease. FEBS Lett 580:6929–6932
50. Ito K, Akiyama Y (2005) Cellular functions, mechanism of action, and regulation of FtsH protease. Annu Rev Microbiol 59:211–231
51. Itzhaki H, Naveh L, Lindahl M, Cook M, Adam Z (1998) Identification and characterization of DegP, a serine protease associated with the luminal side of the thylakoid membrane. J Biol Chem 273:7094–7098
52. Kanaoka MM, Urban S, Freeman M, Okada K (2005) An Arabidopsis Rhomboid homolog is an intramembrane protease in plants. FEBS Lett 579:5723–5728
53. Kapri-Pardes E, Naveh L, Adam Z (2007) The thylakoid lumen protease Deg1 is involved in the repair of photosystem II from photoinhibition in *Arabidopsis*. Plant Cell 19:1039–1047
54. Karakasis K, Taylor D, Ko K (2007) Uncovering a link between a plastid translocon component and rhomboid proteases using yeast mitochondria-based assays. Plant Cell Physiol 48:655–661
55. Karzai AW, Roche ED, Sauer RT (2000) The SsrA-SmpB system for protein tagging, directed degradation and ribosome rescue. Nat Struct Biol 7:449–455
56. Kato Y, Sakamoto W (2009) Protein quality control in chloroplasts: a current model of D1 protein degradation in the photosystem II repair cycle. J Biochem 146:463–469
57. Kato Y, Sakamoto W (2010) New insights into the types and function of proteases in plastids. Int Rev Cell Mol Biol 280:185–218
58. Kato Y, Murakami S, Yamamoto Y, Chatani H, Kondo Y, Nakano T, Yokota A, Sato F (2004) The DNA-binding protease, CND41, and the degradation of ribulose-1,5-bisphosphate carboxylase/oxygenase in senescent leaves of tobacco. Planta 220:97–104
59. Kato Y, Yamamoto Y, Murakami S, Sato F (2005) Post-translational regulation of CND41 protease activity in senescent tobacco leaves. Planta 222:643–651
60. Kato Y, Miura E, Matsushima R, Sakamoto W (2007) White leaf sectors in yellow variegated2 are formed by viable cells with undifferentiated plastids. Plant Physiol 144:952–960
61. Kato Y, Miura E, Ido K, Ifuku K, Sakamoto W (2009) The variegated mutants lacking chloroplastic FtsHs are defective in D1 degradation and accumulate reactive oxygen species. Plant Physiol 151:1790–1801
62. Kato Y, Kouso T, Sakamoto W (2012) Variegated tobacco leaves generated by chloroplast FtsH suppression: implication of FtsH function in the maintenance of thylakoid membranes. Plant Cell Physiol 53:391–404
63. Kim J, Klein PG, Mullet JE (1994) Vir-115 gene product is required to stabilize D1 translation intermediates in chloroplasts. Plant Mol Biol 25:459–467
64. Kim AC, Oliver DC, Paetzel M (2008) Crystal structure of a bacterial signal peptide peptidase. J Mol Biol 376:352–366
65. Kim J, Rudella A, Ramirez Rodriguez V, Zybailov B, Olinares PD, van Wijk KJ (2009) Subunits of the plastid ClpPR protease complex have differential contributions to embryogenesis, plastid biogenesis, and plant development in *Arabidopsis*. Plant Cell 21:1669–1692
66. Kirstein J, Moliere N, Dougan DA, Turgay K (2009) Adapting the machine: adaptor proteins for Hsp100/Clp and AAA+ proteases. Nat Rev Microbiol 7:589–599
67. Kley J et al (2011) Structural adaptation of the plant protease Deg1 to repair photosystem II during light exposure. Nat Struct Mol Biol 18:728–731
68. Kmiec-Wisniewska B, Krumpe K, Urantowka A, Sakamoto W, Pratje E, Janska H (2008) Plant mitochondrial rhomboid, AtRBL12, has different substrate specificity from its yeast counterpart. Plant Mol Biol 68:159–171
69. Knopf RR, Adam Z (2012) Rhomboid proteases in plants-still in square one? Physiol Plant 145:41–51

70. Koonin EV, Makarova KS, Rogozin IB, Davidovic L, Letellier MC, Pellegrini L (2003) The rhomboids: a nearly ubiquitous family of intramembrane serine proteases that probably evolved by multiple ancient horizontal gene transfers. Genome Biol 4:R19
71. Koussevitzky S, Stanne TM, Peto CA, Giap T, Sjogren LL, Zhao Y, Clarke AK, Chory J (2007) An *Arabidopsis thaliana* virescent mutant reveals a role for ClpR1 in plastid development. Plant Mol Biol 63:85–96
72. Kovacheva S, Bedard J, Patel R, Dudley P, Twell D, Rios G, Koncz C, Jarvis P (2005) In vivo studies on the roles of Tic110, Tic40 and Hsp93 during chloroplast protein import. Plant J 41:412–428
73. Kovacheva S, Bedard J, Wardle A, Patel R, Jarvis P (2007) Further in vivo studies on the role of the molecular chaperone, Hsp93, in plastid protein import. Plant J 50:364–379
74. Kress W, Maglica Z, Weber-Ban E (2009) Clp chaperone-proteases: structure and function. Res Microbiol 160:618–628
75. Krojer T, Garrido-Franco M, Huber R, Ehrmann M, Clausen T (2002) Crystal structure of DegP (HtrA) reveals a new protease-chaperone machine. Nature 416:455–459
76. Krojer T, Sawa J, Schafer E, Saibil HR, Ehrmann M, Clausen T (2008) Structural basis for the regulated protease and chaperone function of DegP. Nature 453:885–890
77. Kuras R, Wollman FA (1994) The assembly of cytochrome b6/f complexes: an approach using genetic transformation of the green alga *Chlamydomonas reinhardtii*. EMBO J 13:1019–1027
78. Kuroda H, Maliga P (2003) The plastid clpP1 protease gene is essential for plant development. Nature 425:86–89
79. Kwasniak M, Pogorzelec L, Migdal I, Smakowska E, Janska H (2012) Proteolytic system of plant mitochondria. Physiol Plant 145:187–195
80. Lee JR, Urban S, Garvey CF, Freeman M (2001) Regulated intracellular ligand transport and proteolysis control EGF signal activation in Drosophila. Cell 107:161–171
81. Lemberg MK, Freeman M (2007) Cutting proteins within lipid bilayers: rhomboid structure and mechanism. Mol Cell 28:930–940
82. Lemberg MK, Freeman M (2007) Functional and evolutionary implications of enhanced genomic analysis of rhomboid intramembrane proteases. Genome Res 17:1634–1646
83. Lensch M, Herrmann RG, Sokolenko A (2001) Identification and characterization of SppA, a novel light-inducible chloroplast protease complex associated with thylakoid membranes. J Biol Chem 276:33645–33651
84. Li HH, Merchant S (1995) Degradation of plastocyanin in copper-deficient *chlamydomonas reinhardtii* -evidence for a protease-susceptible conformation of the apoprotein and regulated proteolysis. J Biol Chem 270:23504–23510
85. Lindahl M, Yang DH, Andersson B (1995) Regulatory proteolysis of the major light-harvesting chlorophyll a/b protein of photosystem II by a light-induced membrane-associated enzymic system. Eur J Biochem 231:503–509
86. Lindahl M, Tabak S, Cseke L, Pichersky E, Andersson B, Adam Z (1996) Identification, characterization, and molecular cloning of a homologue of the bacterial FtsH protease in chloroplasts of higher plants. J Biol Chem 271:29329–29334
87. Lindahl M, Spetea C, Hundal T, Oppenheim AB, Adam Z, Andersson B (2000) The thylakoid FtsH protease plays a role in the light-induced turnover of the photosystem II D1 protein. Plant Cell 12:419–431
88. Liu X, Rodermel SR, Yu F (2010) A var2 leaf suppressor locus, SUPPRESSOR OF VARIAGATION3, encodes a putative chloroplast translation elongation factor that is important for chloroplast development in the cold. BMC Plant Biol 10:287
89. Liu X, Yu F, Rodermel S (2010) *Arabidopsis* chloroplast FtsH, var2 and suppressors of var2 leaf variegation: a review. J Integr Plant Biol 52:750–761
90. Liu X, Yu F, Rodermel S (2010) An *Arabidopsis* pentatricopeptide repeat protein, SUPPRESSOR OF VARIEGATION7, is required for FtsH-mediated chloroplast biogenesis. Plant Physiol 154:1588–1601
91. Lucinski R, Misztal L, Samardakiewicz S, Jackowski G (2011) Involvement of Deg5 protease in wounding-related disposal of PsbF apoprotein. Plant Physiol Biochem 49:311–320

92. Lucinski R, Misztal L, Samardakiewicz S, Jackowski G (2011) The thylakoid protease Deg2 is involved in stress-related degradation of the photosystem II light-harvesting protein Lhcb6 in *Arabidopsis thaliana*. New Phytol 192:74–86
93. Majeran W, Wollman FA, Vallon O (2000) Evidence for a role of ClpP in the degradation of the chloroplast cytochrome b(6)f complex. Plant Cell 12:137–150
94. Majeran W, Olive J, Drapier D, Vallon O, Wollman FA (2001) The light sensitivity of ATP synthase mutants of *Chlamydomonas reinhardtii*. Plant Physiol 126:421–433
95. Minai L, Wostrikoff K, Wollman FA, Choquet Y (2006) Chloroplast biogenesis of photosystem II cores involves a series of assembly-controlled steps that regulate translation. Plant Cell 18:159–175
96. Miura E, Kato Y, Matsushima R, Albrecht V, Laalami S, Sakamoto W (2007) The balance between protein synthesis and degradation in chloroplasts determines leaf variegation in *Arabidopsis* yellow variegated mutants. Plant Cell 19:1313–1328
97. Miura E, Kato Y, Sakamoto W (2010) Comparative transcriptome analysis of green/white variegated sectors in *Arabidopsis* yellow variegated2: responses to oxidative and other stresses in white sectors. J Exp Bot 61:2433–2445
98. Murakami S, Kondo Y, Nakano T, Sato F (2000) Protease activity of CND41, a chloroplast nucleoid DNA-binding protein, isolated from cultured tobacco cells. FEBS Lett 468:15–18
99. Nakagawara E, Sakuraba Y, Yamasato A, Tanaka R, Tanaka A (2007) Clp protease controls chlorophyll b synthesis by regulating the level of chlorophyllide a oxygenase. Plant J 49:800–809
100. Neuwald AF, Aravind L, Spouge JL, Koonin EV (1999) AAA+: a class of chaperone-like ATPases associated with the assembly, operation, and disassembly of protein complexes. Genome Res 9:27–43
101. Olinares PD, Kim J, Davis JI, van Wijk KJ (2011) Subunit stoichiometry, evolution, and functional implications of an asymmetric plant plastid ClpP/R protease complex in *Arabidopsis*. Plant Cell 23:2348–2361
102. Olinares PD, Kim J, van Wijk KJ (2011) The Clp protease system; a central component of the chloroplast protease network. Biochim Biophys Acta 1807:999–1011
103. Ostersetzer O, Adam Z (1997) Light-stimulated degradation of an unassembled Rieske FeS protein by a thylakoid-bound protease: the possible role of the FtsH protease. Plant Cell 9:957–965
104. Ostersetzer O, Kato Y, Adam Z, Sakamoto W (2007) Multiple intracellular locations of Lon protease in *Arabidopsis*: evidence for the localization of AtLon4 to chloroplasts. Plant Cell Physiol 48:881–885
105. Park S, Rodermel SR (2004) Mutations in ClpC2/Hsp100 suppress the requirement for FtsH in thylakoid membrane biogenesis. Proc Natl Acad Sci U S A 101:12765–12770
106. Peltier JB, Ytterberg J, Liberles DA, Roepstorff P, van Wijk KJ (2001) Identification of a 350 kDa ClpP protease complex with 10 different Clp isoforms in chloroplasts of *Arabidopsis thaliana*. J Biol Chem 276:16318–16327
107. Peltier JB, Ytterberg J, Liberles DA, Roepstorff P, van WKJ (2001) Identification of a 350 kDa ClpP protease complex with 10 different Clp isoforms in chloroplasts of *Arabidopsis thaliana*. J Biol Chem 276:16318–16327
108. Peltier JB et al (2002) Central functions of the lumenal and peripheral thylakoid proteome of *Arabidopsis* determined by experimentation and genome-wide prediction. Plant Cell 14:211–236
109. Peltier JB, Ripoll DR, Friso G, Rudella A, Cai Y, Ytterberg J, Giacomelli L, Pillardy J, van Wijk KJ (2004) Clp protease complexes from photosynthetic and non-photosynthetic plastids and mitochondria of plants, their predicted three-dimensional structures, and functional implications. J Biol Chem 279:4768–4781
110. Pogson BJ, Woo NS, Forster B, Small ID (2008) Plastid signalling to the nucleus and beyond. Trends Plant Sci 13:602–609
111. Pojidaeva E, Zinchenko V, Shestakov SV, Sokolenko A (2004) Involvement of the SppA1 peptidase in acclimation to saturating light intensities in *Synechocystis* sp. strain PCC 6803. J Bacteriol 186:3991–3999

112. Rawlings ND, Morton FR, Barrett AJ (2006) MEROPS: the peptidase database. Nucleic Acids Res 34:D270–D272
113. Rodrigues RA, Silva-Filho MC, Cline K (2011) FtsH2 and FtsH5: two homologous subunits use different integration mechanisms leading to the same thylakoid multimeric complex. Plant J 65:600–619
114. Rudella A, Friso G, Alonso JM, Ecker JR, van Wijk KJ (2006) Downregulation of ClpR2 leads to reduced accumulation of the ClpPRS protease complex and defects in chloroplast biogenesis in *Arabidopsis*. Plant Cell 18:1704–1721
115. Sakamoto W (2003) Leaf-variegated mutations and their responsible genes in *Arabidopsis thaliana*. Genes Genet Syst 78:1–9
116. Sakamoto W (2006) Protein degradation machineries in plastids. Annu Rev Plant Biol 57:599–621
117. Sakamoto W, Tamura T, Hanba-Tomita Y, Murata M (2002) The VAR1 locus of *Arabidopsis* encodes a chloroplastic FtsH and is responsible for leaf variegation in the mutant alleles. Genes Cells 7:769–780
118. Sakamoto W, Zaltsman A, Adam Z, Takahashi Y (2003) Coordinated regulation and complex formation of yellow variegated1 and yellow variegated2, chloroplastic FtsH metalloproteases involved in the repair cycle of photosystem II in *Arabidopsis* thylakoid membranes. Plant Cell 15:2843–2855
119. Sakamoto W, Miura E, Kaji Y, Okuno T, Nishizono M, Ogura T (2004) Allelic characterization of the leaf-variegated mutation var2 identifies the conserved amino acid residues of FtsH that are important for ATP hydrolysis and proteolysis. Plant Mol Biol 56:705–716
120. Sakamoto W, Uno Y, Zhang Q, Miura E, Kato Y, Sodmergen (2009) Arrested differentiation of proplastids into chloroplasts in variegated leaves characterized by plastid ultrastructure and nucleoid morphology. Plant Cell Physiol 50:2069–2083
121. Sakuraba Y, Tanaka R, Yamasato A, Tanaka A (2009) Determination of a chloroplast degron in the regulatory domain of chlorophyllide a oxygenase. J Biol Chem 284:36689–36699
122. Sarria R, Lyznik A, Vallejos CE, Mackenzie SA (1998) A cytoplasmic male sterility-associated mitochondrial peptide in common bean is post-translationally regulated. Plant Cell 10:1217–1228
123. Sauer RT et al (2004) Sculpting the proteome with AAA(+) proteases and disassembly machines. Cell 119:9–18
124. Schmidt GW, Mishkind ML (1983) Rapid degradation of unassembled ribulose 1,5-bisphosphate carboxylase small subunits in chloroplasts. Proc Natl Acad Sci U S A 80:2632–2636
125. Schubert M, Petersson UA, Haas BJ, Funk C, Schroder WP, Kieselbach T (2002) Proteome map of the chloroplast lumen of *Arabidopsis thaliana*. J Biol Chem 277:8354–8365
126. Schuhmann H, Adamska I (2012) Deg proteases and their role in protein quality control and processing in different subcellular compartments of the plant cell. Physiol Plant 145:224–234
127. Shanklin J, DeWitt ND, Flanagan JM (1995) The stroma of higher plant plastids contain ClpP and ClpC, functional homologs of *Escherichia coli* ClpP and ClpA: an archetypal two-component ATP-dependent protease. Plant Cell 7:1713–1722
128. Shen G, Adam Z, Zhang H (2007) The E3 ligase AtCHIP ubiquitylates FtsH1, a component of the chloroplast FtsH protease, and affects protein degradation in chloroplasts. Plant J 52:309–321
129. Shikanai T, Shimizu K, Ueda K, Nishimura Y, Kuroiwa T, Hashimoto T (2001) The chloroplast clpP gene, encoding a proteolytic subunit of ATP-dependent protease, is indispensable for chloroplast development in tobacco. Plant Cell Physiol 42:264–273
130. Sinvany-Villalobo G, Davydov O, Ben-Ari G, Zaltsman A, Raskind A, Adam Z (2004) Expression in multigene families. Analysis of chloroplast and mitochondrial proteases. Plant Physiol 135:1336–1345
131. Sjogren LL, Clarke AK (2011) Assembly of the chloroplast ATP-dependent Clp protease in *Arabidopsis* is regulated by the ClpT accessory proteins. Plant Cell 23:322–332

132. Sjogren LL, MacDonald TM, Sutinen S, Clarke AK (2004) Inactivation of the clpC1 gene encoding a chloroplast Hsp100 molecular chaperone causes growth retardation, leaf chlorosis, lower photosynthetic activity, and a specific reduction in photosystem content. Plant Physiol 136:4114–4126
133. Sjogren LL, Stanne TM, Zheng B, Sutinen S, Clarke AK (2006) Structural and functional insights into the chloroplast ATP-dependent Clp protease in *Arabidopsis*. Plant Cell 18:2635–2649
134. Sokolenko A, Pojidaeva E, Zinchenko V, Panichkin V, Glaser VM, Herrmann RG, Shestakov SV (2002) The gene complement for proteolysis in the cyanobacterium *Synechocystis* sp. PCC 6803 and *Arabidopsis thaliana* chloroplasts. Curr Genet 41:291–310
135. Spiess C, Beil A, Ehrmann M (1999) A temperature-dependent switch from chaperone to protease in a widely conserved heat shock protein. Cell 97:339–347
136. Stanne TM, Sjogren LL, Koussevitzky S, Clarke AK (2009) Identification of new protein substrates for the chloroplast ATP-dependent Clp protease supports its constitutive role in *Arabidopsis*. Biochem J 417:257–268
137. Sun X, Peng L, Guo J, Chi W, Ma J, Lu C, Zhang L (2007) Formation of DEG5 and DEG8 complexes and their involvement in the degradation of photodamaged photosystem II reaction center D1 protein in *Arabidopsis*. Plant Cell 19:1347–1361
138. Sun X, Fu T, Chen N, Guo J, Ma J, Zou M, Lu C, Zhang L (2010) The stromal chloroplast Deg7 protease participates in the repair of photosystem II after photoinhibition in *Arabidopsis*. Plant Physiol 152:1263–1273
139. Sun X, Ouyang M, Guo J, Ma J, Lu C, Adam Z, Zhang L (2010) The thylakoid protease Deg1 is involved in photosystem-II assembly in *Arabidopsis thaliana*. Plant J 62:240–249
140. Sun X, Feng P, Xu X, Guo H, Ma J, Chi W, Lin R, Lu C, Zhang L (2011) A chloroplast envelope-bound PHD transcription factor mediates chloroplast signals to the nucleus. Nat Commun 2:477
141. Suno R, Niwa H, Tsuchiya D, Zhang X, Yoshida M, Morikawa K (2006) Structure of the whole cytosolic region of ATP-dependent protease FtsH. Mol Cell 22:575–585
142. Takechi K, Sodmergen Murata M, Motoyoshi F, Sakamoto W (2000) The YELLOW VARIEGATED (VAR2) locus encodes a homologue of FtsH, an ATP-dependent protease in *Arabidopsis*. Plant Cell Physiol 41:1334–1346
143. Tobias JW, Shrader TE, Rocap G, Varshavsky A (1991) The N-end rule in bacteria. Science 254:1374–1377
144. Tsugane K, Maekawa M, Takagi K, Takahara H, Qian Q, Eun CH, Iida S (2006) An active DNA transposon nDart causing leaf variegation and mutable dwarfism and its related elements in rice. Plant J 45:46–57
145. Urantowka A, Knorpp C, Olczak T, Kolodziejczak M, Janska H (2005) Plant mitochondria contain at least two i-AAA-like complexes. Plant Mol Biol 59:239–252
146. Urban S, Lee JR, Freeman M (2001) Drosophila rhomboid-1 defines a family of putative intramembrane serine proteases. Cell 107:173–182
147. Varshavsky A (1992) The N-end rule. Cell 69:725–735
148. Wagner R, Aigner H, Pruzinska A, Jankanpaa HJ, Jansson S, Funk C (2011) Fitness analyses of *Arabidopsis thaliana* mutants depleted of FtsH metalloproteases and characterization of three FtsH6 deletion mutants exposed to high light stress, senescence and chilling. New Phytol 191:449–458
149. Wang Y, Zhang Y, Ha Y (2006) Crystal structure of a rhomboid family intramembrane protease. Nature 444:179–180
150. Weihofen A, Martoglio B (2003) Intramembrane-cleaving proteases: controlled liberation of proteins and bioactive peptides. Trends Cell Biol 13:71–78
151. Wetzel CM, Harmacek LD, Yuan LH, Wopereis JL, Chubb R, Turini P (2009) Loss of chloroplast protease SPPA function alters high light acclimation processes in *Arabidopsis thaliana* L. (Heynh.). J Exp Bot 60:1715–1727
152. Wolfe MS, Kopan R (2004) Intramembrane proteolysis: theme and variations. Science 305:1119–1123

153. Wu Z et al (2006) Structural analysis of a rhomboid family intramembrane protease reveals a gating mechanism for substrate entry. Nat Struct Mol Biol 13: 1084–1091
154. Wu H, Ji Y, Du J, Kong D, Liang H, Ling HQ (2010) ClpC1, an ATP-dependent Clp protease in plastids, is involved in iron homeostasis in *Arabidopsis* leaves. Ann Bot 105:823–833
155. Yamamoto Y (2001) Quality control of photosystem II. Plant Cell Physiol 42:121–128
156. Yang DH, Webster J, Adam Z, Lindahl M, Andersson B (1998) Induction of acclimative proteolysis of the light-harvesting chlorophyll a/b protein of photosystem II in response to elevated light intensities. Plant Physiol 118:827–834
157. Yoshioka M, Uchida S, Mori H, Komayama K, Ohira S, Morita N, Nakanishi T, Yamamoto Y (2006) Quality control of photosystem II. Cleavage of reaction center D1 protein in spinach thylakoids by FtsH protease under moderate heat stress. J Biol Chem 281:21660–21669
158. Yoshioka M, Nakayama Y, Yoshida M, Ohashi K, Morita N, Kobayashi H, Yamamoto Y (2010) Quality control of photosystem II: FtsH hexamers are localized near photosystem II at grana for the swift repair of damage. J Biol Chem 285:41972–41981
159. Yu F, Park S, Rodermel SR (2004) The *Arabidopsis* FtsH metalloprotease gene family: interchangeability of subunits in chloroplast oligomeric complexes. Plant J 37:864–876
160. Yu F, Park S, Rodermel SR (2005) Functional redundancy of AtFtsH metalloproteases in thylakoid membrane complexes. Plant Physiol 138:1957–1966
161. Yu F, Liu X, Alsheikh M, Park S, Rodermel S (2008) Mutations in SUPPRESSOR OF VARIEGATION1, a factor required for normal chloroplast translation, suppress var2-mediated leaf variegation in *Arabidopsis*. Plant Cell 20:1786–1804
162. Yu F et al (2011) SUPPRESSOR OF VARIEGATION4, a new var2 suppressor locus, encodes a pioneer protein that is required for chloroplast biogenesis. Mol Plant 4:229–240
163. Zaltsman A, Feder A, Adam Z (2005) Developmental and light effects on the accumulation of FtsH protease in *Arabidopsis* chloroplasts-implications for thylakoid formation and photosystem II maintenance. Plant J 42:609–17
164. Zaltsman A, Ori N, Adam Z (2005) Two types of FtsH protease subunits are required for chloroplast biogenesis and photosystem II repair in *Arabidopsis*. Plant Cell 17:2782–2790
165. Zelisko A, Garcia-Lorenzo M, Jackowski G, Jansson S, Funk C (2005) AtFtsH6 is involved in the degradation of the light-harvesting complex II during high-light acclimation and senescence. Proc Natl Acad Sci U S A 102:13699–13704
166. Zhang D, Kato Y, Zhang L, Fujimoto M, Tsutsumi N, Sodmergen, Sakamoto W (2010) The FtsH protease heterocomplex in *Arabidopsis*: dispensability of type-B protease activity for proper chloroplast development. Plant Cell 22:3710–3725
167. Zheng B, Halperin T, Hruskova-Heidingsfeldova O, Adam Z, Clarke AK (2002) Characterization of chloroplast clp proteins in *Arabidopsis*: localization, tissue specificity and stress responses. Physiol Plant 114:92–101
168. Zheng B, MacDonald TM, Sutinen S, Hurry V, Clarke AK (2006) A nuclear-encoded ClpP subunit of the chloroplast ATP-dependent Clp protease is essential for early development in *Arabidopsis thaliana*. Planta 224:1103–1115
169. Zybailov B, Friso G, Kim J, Rudella A, Rodriguez VR, Asakura Y, Sun Q, van Wijk KJ (2009) Large scale comparative proteomics of a chloroplast Clp protease mutant reveals folding stress, altered protein homeostasis, and feedback regulation of metabolism. Mol Cell Proteomics 8:1789–1810

Part IV
Chloroplast Photophysiology

Chapter 15
Photoprotective Mechanisms: Carotenoids

Luca Dall'Osto, Roberto Bassi and Alexander Ruban

Abstract Environmental light can frequently be detrimental to the photosynthetic machinery of plants. This chapter provides an up-to-date overview of the targets of the photo-oxidative damage caused by light and the multiple functions of photosynthetic carotenoids that minimize it. Recently acquired knowledge on the localisation and distribution of carotenoids in the photosynthetic apparatus of plants is presented. Mechanisms that control the light harvesting process in the photosynthetic antenna of higher plants, via protective energy dissipation, are compared and discussed. The role of functional genomics approaches to the study of the multiple functions of carotenoids are highlighted. The significance of carotenoid structure and the physico-chemical properties that enable fine control over the photosynthetic light harvesting processes are analysed and discussed in order to explain the variety of their types.

Keywords Photoinhibition · Carotenoids · Non-photochemical quenching · Functional genomics

Abbreviations

ABA	Abscisic acid
CT	*Charge-transfer*
DTT	Dithiothreitol
EL	Excess light
LHC	Light harvesting complex
NPQ	Non-photochemical quenching
PAM	Pulse amplitude modulated
PQ	Plastoquinone
PSII	Photosystem II

R. Bassi (✉) · L. Dall'Osto
Dipartimento di Biotecnologie, Università di Verona, Strada Le Grazie 15, Verona 37134, Italy
e-mail: roberto.bassi@univr.it

A. Ruban
School of Biological and Chemical Sciences, Queen Mary University of London,
Mile End Road, London E1 4NS, UK

S.M. Theg, F.-A. Wollman (eds.), *Plastid Biology*, Advances in Plant Biology 5,
DOI 10.1007/978-1-4939-1136-3_15,

qE	Energy-dependent quenching
Ql	Photoinhibitory quenching
qP	*Photochemical quenching*
qZ	Zea-dependent quenching
RC	Reaction center
ROS	Reactive oxygen species
SOD	Superoxide dismutase
szl1	*Suppressor of zeaxanthinless1*
VDE	Violaxanthin de-epoxidase
ZE	Zeaxanthin epoxidase

15.1 Damage of Photosynthetic Apparatus in Excess Light: Targets of Photoxidative Stress

Although light is essential for photosynthesis, its excess could be damaging. Significant reduction in photosynthetic efficiency, called "photo-inhibition", is observed upon exposure to light intensity exceeding plant capacity for electron transport (reflected in the photochemical chlorophyll fluorescence quenching). The extent of the photoinhibitory effect is not only dependent upon the photon flux density, environmental and metabolic conditions are important too. For example, prolonged restriction of CO_2 influx in leaves during drought stress or low temperatures during winter makes even moderate irradiances to exceed the maximal capacity for photosynthetic electron transport. The time-course of illumination is also crucial: rapid fluctuations in light intensity (sunflecks) frequently overload photosystems with excess light energy. These events could be very detrimental to both plant growth and crop yield [4, 162].

In the context of light phase reactions, *photooxidative stress* is a consequence of the accumulation of chlorophyll excited states resulting in the formation of strongly oxidizing intermediates in different steps of the process that inevitably lead to the release of reactive oxygen species (ROS). Excess light (EL) leads to multiple effects detrimental to many components of the photosynthetic apparatus and primarily Photosystem II (PSII). The PSII reaction center (RC) catalyzes photochemical oxidation of H_2O and reduction of plastoquinone (PQ). It binds the "special" Chl pair, P680 [132]. It becomes positively charged ($P680^+$) and step-oxidizes the Mn-containing OEC, thus generating the strongest oxidizing potential in biological systems (>1.0 Volts) that enables water splitting, resulting in O_2 release. The PSII reaction center is the most vulnerable component of the whole photosynthetic electron chain: due to its high oxidizing potential it can easily extract electrons and damage its own molecular environment. Moreover, due to EL intensities, electron flow may not keep the pace with charge separation, leading to increased lifetime of $P680^+$. When the PQ pool is over-reduced, charge recombination will occur within $P680^+$. As a consequence, the formed P680 excited triplet state can rapidly react with O_2, yielding singlet oxygen ($^1O_2$), a toxic species that preferentially destruct the chlorophylls of P680 and the critical pigment cofactors leading to D1 protein degradation. In addition, PSII was shown to produce both superoxide anion (O_2^-) and hydroxyl radical ($OH^\bullet$) during strong illumination. O_2^- is proposed to originate from the reduction of molecular oxygen on the electron acceptor side of PSII [33], while the reduction of peroxides bound to the PSII metal center (Fe^{2+}…OOH) was proposed to yield into the formation of $OH^\bullet$ [157]. Most of these problems also concern the PSI reaction center, but to a lesser extent. Indeed, $P700^+$ is far less oxidizing than $P680^+$ and acts as a very efficient quencher of the excitation energy collected by its LHCI antenna. Nevertheless, photoinhibition of the PSI reaction center can be observed, especially under chilling stress [178]. When light exceeds the capacity to use reducing equivalents for CO_2 fixation, a marked over-reduction of the $NADP^+$ pool leads to O_2^- production at the PSI donor side, in the so-called Mehler's reaction [9]. O_2^- can be metabolized to hydrogen peroxide (H_2O_2) or $OH^\bullet$. While O_2^- and H_2O_2 are relatively unreactive, $OH^\bullet$ is highly reactive and dangerous for photosynthetic components. It should be noted that, despite the spatial separation between the stroma-exposed membranes hosting PSI and the grana partitions where the PSII reaction centre is located, the major target for O_2^- damage is the PSII reaction centre. This is because the Q_B donor side is situated in the lipid layer where O_2^- is accumulated and easily diffuses [48].

Photoxidative stress does not exclusively originate in reaction centres. Obviously, chlorophylls bound to the light-harvesting complexes may well become dangerous photosensitizers. Below saturating light intensity, the continuous charge separation and scavenging reactions are sufficient for counteracting the low level of ROS produced. Indeed, efficient consumption of excitation energy captured by the photosynthetic antennae keeps the level of $^1Chl^*$ low. Thus rapid transfer of excitons between neighbor Chls towards the reaction center is an effective photoprotection system under open RC only. Gradual closure of RCII traps increases the lifetime of $^1Chl^*$, as well as the probability of Chl *a* triplet formation ($^3Chl^*$) by intersystem crossing. $^3Chl^*$ formation is an intrinsic property of excited antenna Chls. It is a long-living state able to promptly react with molecular oxygen to form $^1O_2$ [188].

$$\begin{array}{l} h\upsilon \ \mathrm{ISC} \\ {}^1\mathrm{Chl} \rightarrow {}^1\mathrm{Chl}^* \rightarrow {}^3\mathrm{Chl}^* \\ {}^3\mathrm{Chl}^* + {}^3\mathrm{O}_2 \rightarrow {}^1\mathrm{Chl} + {}^1\mathrm{O}_2 \end{array}$$

Although it was long assumed that $^3Chl^*$ in the antenna was efficiently quenched by xanthophylls bound to Lhc proteins [149], later studies showed that quenching was indeed incomplete [174]. A long-living (ms) $^3Chl^*$ state was identified in native LHCII complexes [126], implying that some antenna Chls are not efficiently coupled to potential quenchers, e.g. xanthophylls. Since the average lifetime of $^1Chl^*$ in PSII is far longer than in the PSI-LHCI complex, both PSII and its LHC antenna moiety become, when overexcited, the most important source of $^1O_2$ in the chloroplast. This ROS compromises membrane integrity, leading to peroxidation of unsaturated fatty acid chains abundant in the thylakoids [189], as well as oxidation of proteins and pigments in its immediate vicinity.

15.1.1 The Major Photoprotective Strategies in Higher Plants

The photoxidative events described above inevitably lead to chloroplast damage and decrease the overall rate of photosynthesis if the repair rate is overcome. During evolution plants have developed a network of adaptation mechanisms to cope with damaging excess light (EL) exposure. These can be of two major types: adaptations to control light absorption capacity and adaptations to deal with the captured light energy. To optimize light absorption, plants respond on different levels of organization. On the level of the whole organism, the EL adaptation involves adjustment of the leaf orientation [199]. This adaptation helps many land plants to cope with excess irradiation particularly during the midday. Leaf movements can be of developmental, passive (drought related) and active nature (reversible). The latter employs a blue-light absorbing pigment system that has been identified as the phototropins *phot1* and *phot2* (reviewed in [181]) as central regulators of phototropism, leaf flattening and leaf positioning [55]. This adaptive system can be very effective in some shade plants with low photosynthetic capacity during occasional exposure to light bursts. Some desert plants have also developed a number of adaptations to increase leaf reflectance and therefore reduce the amount of absorbed light. Building up inorganic deposits on the leaf surface (for example, salt crystals) or developing air-filled hairs are the typical examples. As a rule, the efficiency of these protective measures is good, but as with the developmental leaf movements, these adaptations occur on rather slow time scale.

On the cellular level, light absorption can be regulated by chloroplast movements [94]. These are relatively fast adaptations, occurring within minutes, but are only able to reduce light absorption by 10–20 % in the environment of excess light [200]. Generally, adaptations on the cell level or at the level of the whole plant have limited capacity. In the natural environment light tends to be scattered in all directions by clouds, fog or simply high air humidity, reducing the effectiveness of cellular as well as leaf orientation adaptations.

A most fundamental and efficient type of plant adaptation to light occurs at the molecular level. The regulation of light absorption occurs by long-term control of chlorophyll content in leaves. Other responses involve short-term adaptations of the photosynthetic membrane consisting of dynamic changes in light harvesting antenna efficiency and mechanisms of detoxification of ROS. Long-term response consists of modification of plant architecture and composition away from the energy conserving organization typical of light-limiting conditions and towards a different organization favoring photoprotection. These acclimatory responses need a time lag of days to be completed, since the response to photoxidative conditions results in dramatic changes in gene expression [99, 131, 161]. In *Arabidopsis*, this regulation is mediated directly by photoreceptors such as cryptochrome [172], or through biochemical and metabolic signals such as the plastoquinone redox state, the release of ROS, the redox state of the glutathione pool and the ATP/ADP ratio [64, 138, 150, 192]. Typically, the size of photosynthetic antenna systems, particularly PSII, is reduced, the PSII/PSI ratio decreased [32] and the stoichiometry of electron transport components and ATPase is up-regulated with respect to photosynthetic reaction centres, while the abundance of the enzymes of Calvin cycle is increased during acclimation to EL [5, 11–13, 118].

Diversion of energy from the normal assimilative pathway to dissipative ones can also be performed at the level of the electron transport chain, including direct reduction of O_2 by PSI, cyclic electron transport around PSI, and pseudocyclic electron transport around PSII which uses PTOX in order to reduce O_2 with electrons derived from H_2O, thus releasing over-reduction of PQ pool [106]. All together, these strategies control dissipative disposal in a safe mode of reducing equivalents produced by photosynthesis in EL, and avoid their direct univalent reduction of molecular oxygen.

Despite the excellent protective role fulfilled by the above mechanisms, ROS are produced in the chloroplast in many ways, particularly under unfavorable conditions. These can be deactivated by antioxidant molecules, including α-tocopherol, glutathione and ascorbate [78, 128, 159], or enzymes such as superoxide dismutase and ascorbate peroxidase for detoxification of H_2O_2/O_2^- [9, 100]. Yet this can be not enough and plants face the irreversible photodamage of PSII reaction centers by removing and degrading the damaged D1 proteins, followed by its replacement with a newly synthesized D1 polypeptide. PSII turn-over is a critical mechanism for photoprotection: indeed, inhibitors blocking protein synthesis in the chloroplast, result in selective loss of the D1 protein in EL, and to increased photo-inhibition, implying this mechanism is needed to maintain the photosynthetic efficiency [8].

A dramatic short-term response to EL exposure involves the dissipation of excitation energy in the PSII antenna system which down-regulates the concentration of ^{1}Chl* and thus the probability of excitons to be transferred to PSII RC whenever the lumen is acidified because of the insufficient activity of ATP synthetase [93]. This mechanism is called Non-Photochemical Quenching (NPQ); the reason for this name is that the energy dissipation process is measured indirectly by the quenching of antenna chlorophyll fluorescence. NPQ measures the extent of the excess energy dissipation in PSII, when reaction centers are closed, i.e., not receiving excitation energy from antenna. NPQ was found to be dependent not only upon the levels of lumen acidification (transthylakoid proton gradient) but, most importantly, upon the oxygenated carotenoids (i.e., xanthophylls) of the light harvesting antenna.

15.2 Carotenoid Location and Distribution in the Photosynthetic Apparatus of Plants

15.2.1 Arrangement of Carotenoids in the Photosynthetic Membrane

Many factors contribute to chloroplast photoprotection. One can argue that the crucial contribution comes from carotenoids (Cars) (Fig. 15.1). Carotenoids of the photosynthetic apparatus are classified as *accessory pigments*; indeed, they contribute to increase the light-harvesting capacity of chlorophyll-binding proteins in the blue

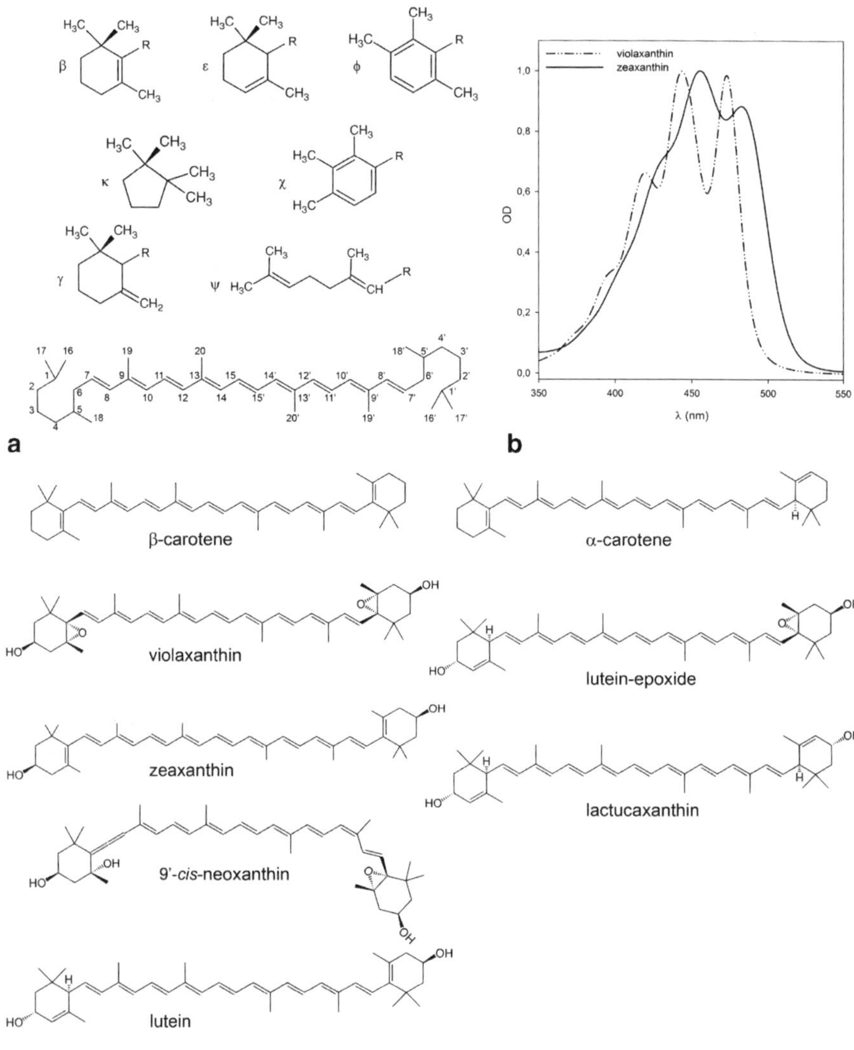

Fig. 15.1 Carotenoid structure and classification. **a** Characteristic end groups of carotenoids and structure of a generic carotenoid with common numbering system. **b** Visible absorption spectra of violaxanthin (—···—···—) and zeaxanthin (———) in acetone. **c** Structure of the main carotenoids species commonly found in higher plants

spectral region (enhancing antenna cross-section, hence the light harvesting function of carotenoids). Clearly, maximizing light harvesting efficiency implies that Cars are in close proximity and properly positioned with respect to Chls in light harvesting complexes of photosystems. Thus, the majority of carotenoids of the thylakoid membrane are bound to specific binding sites in pigment-protein complexes where the protein scaffold determines both the distance and orientation with respect to Chl molecules in the same complex. Actually, in the case of the LHC protein family, xanthophylls are structural determinants for the whole complex so that the holocomplex cannot fold in their absence. However, a minor fraction of unbound carotenoids was reported to be dispersed free in the lipid phase of the photosynthetic membranes where it plays an antioxidant function [76] and modulates both the structure and the fluidity of the lipid bilayer [73]. The size of the unbound carotenoid pool was recently reported by quantifying Cars distribution on pigment-protein complexes upon mild thylakoid solubilization and fractionation by sucrose gradient ultracentrifugation [49]. The lipid-free pool was found to be only ~15% of the total carotenoid content in *Arabidopsis*, and the distribution along density gradients upon ultracentrifugation reveals no major quantitative changes upon EL treatment. The earlier report from work on spinach thylakoid membranes showed that the free fraction of xanthophylls observed in experiments using even mild detergent treatment could originate from the weakly-bound xanthophyll cycle Cars into V1 site of LHCII [168]. Hence, caution is required for an accurate assessment of the protein-unbound lipid-soluble pigment fraction described above.

15.2.2 Pigment Content of the Various Photosynthetic Complexes

Photosynthesis is accomplished by the activity of four protein supercomplexes embedded into the thylakoid membrane of the chloroplast: two photosystems (PSI and PSII), the cytochrome b_6f complex and the ATP synthase. Light absorption by both photosystems fuels the electron transport from O_2 to $NADP^+$ into two energy steps, while the cytochrome b_6f complex converts part of the redox energy stored in the intermediate PQ electron acceptor into a transthylakoid ΔpH gradient, finally exploited by the ATP synthase to produce ATP [132]. PSI and PSII have each a complex array of xanthophyll-rich pigment-binding (antenna) proteins surrounding the core complex, which increases light absorption (Fig. 15.2). In PSII the light harvesting function is fulfilled by two classes of proteins: PsbB (CP47) and PsbC (CP43) are plastid-encoded components of the core complex and bind Chl *a* and *β*-carotene, while a peripheral layer of pigment-protein subunits is made by the nuclear-encoded LHC proteins, which bind Chls *a*, *b* and xanthophylls. Further components of PSII core include the D1/D2/cytochrome b_{559} RC complex, hosting primary charge separation and early steps of electron transport, and a number of small trans-membrane proteins. The outer antenna system is composed of two copies each of three monomeric LHCs, called Lhcb4 (CP29), Lhcb5 (CP26) and Lhcb6 (CP24), and, more peripherally, by two to four copies of the major antenna complex LHCII, an heterotrimer of the Lhcb1/2/3 subunits. Trimeric LHCIIs are distinct in S (strongly-bound), M (medium-bound) or L (loosely-bound) types in relation to their binding strength to the photosystems [101]. In PSI, the RC complex and the inner antenna are fused into a single complex where the Chl *a* and β-carotene chromophores are essentially bound to the PsaA and PsaB major subunits, together with most of the cofactors of the electron transfer chain. The structure of the PSI core with its outer antenna (PSI-LHCI supercomplex) from *P. sativum*, as resolved by x-ray crystallography, revealed a peripheral antenna system composed of one copy each of the Lhca1–4 polypeptides [18].

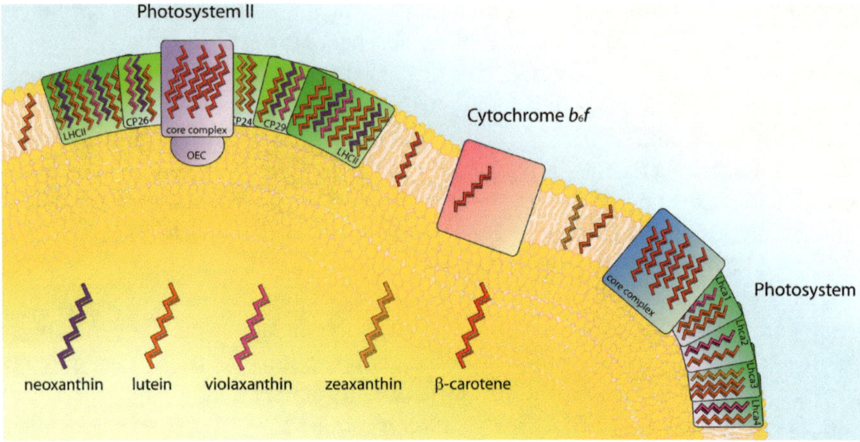

Fig. 15.2 Model for distribution of carotenoids in the thylakoid membrane and photosynthetic pigment-binding complexes. *OEC* oxygen-evolving complex

It is important to stress that the whole complement of protein-bound xanthophyll pools of thylakoids is bound to the PS outer antenna complexes, LHCs. There xanthophylls not only play a role of light harvesting pigments [190] but, as was mentioned above, are vital in the proper assembly and stabilization of LHC structure [145] and have been proposed to be involved in protective energy dissipation [2, 170]. The major LHCII from spinach was first reported to bind 13 chlorophylls [42] and four xanthophyll molecules: two luteins, one neoxanthin and one violaxanthin/zeaxanthin per monomer [168]. Owing to their cross-brace construction, the two luteins in all-*trans* configuration bind to the inner sites L1 and L2, and stabilize the whole complex [113]. In addition, the L1 site-bond lutein was discovered to be an effective excess energy quencher [170]. The β-cyclohexane rings of both lutein molecules are located towards the lumen, while the ε-cyclohexane rings are close to the stromal membrane surface. Binding sites for neoxanthin and violaxanthin are respectively indicated as N1 and V1. The cyclohexane ring of neoxanthin protrudes out of the protein into the lipid bilayer; the high selectivity of the N1 binding site for neoxanthin resides in the high specificity of the pocket for a 9′-*cis* stereoisomer [30]. The fourth xanthophyll ligand is located at the monomer-monomer interface, in a peripheral site called V1. Its polyene chain has an all-*trans* configuration and forms a small angle with the membrane normal [113]. Composition of the V1 site differs with species; in LHCII from *Arabidopsis*, it was found to bind also lutein in addition to violaxanthin or zeaxanthin [45]. The V1 site was proposed by Horton and co-workers to play a role in allosteric control of NPQ process [86]. In addition, this role has been attributed to site L2 by Bassi and co-workers [29, 122, 123]. Protonation of LHCII and replacement of violaxanthin by zeaxanthin were proposed to work co-operatively for induction of a conformational quenching state in the complex associated with its aggregation (c.f. the following sections).

Despite that each xanthophyll-binding site shows peculiar composition in wild-type LHCII, plasticity is observed *in vivo* with respect to the xanthophyll composition. This is evident from the study of *Arabidopsis* mutants with altered carotenoid composition: sites L1, L2 and V1 can bind lutein, violaxanthin or zeaxanthin when available as the only xanthophyll species [45, 76, 152]. The only binding site that showed higher selectivity was site N1 [34, 48, 63].

The structure of the monomeric antenna Lhcb4 (CP29) from spinach was recently made available [141]. It revealed binding sites for the three xanthophylls—lutein in L1, violaxanthin in L2 and neoxanthin in N1—and the absence of a peripheral V1-type site. In agreement with previous suggestions [16], site L2 binds violaxanthin rather than lutein. High resolution structural information is still missing for Lhcb5 and Lhcb6 and information on their xanthophyll composition and site occupancy can be obtained from either samples purified from chloroplasts using ion-exchange chromatography or isoelectrofocusing purification steps [43] which, however, may cause removal of xanthophyll bound to labile sites [29]. Alternatively, holoproteins can be obtained by *in vitro* refolding of recombinant proteins with purified pigments [16, 144]. Analysis of native and recombinant Lhcb5 (CP26) and Lhcb6 (CP24) [39, 45, 140, 143] was performed by combining reconstitution with different pigment complements and deconvolution of absorption and fluorescence excitation spectra. Since each binding site has a specific environment, which affects refraction index and shifts absorption of xanthophylls to a different extent, the presence of different binding sites can be recognized by the deconvolution of spectral forms [37, 65]. Site L1 was confirmed to bind lutein in both Lhcb5 and Lhcb6, while site L2 was proposed to bind violaxanthin as in the case of Lhcb4. Binding of violaxanthin into the L2 site in the minor antenna complexes is an interesting and much discussed phenomenon [168, 194]. It is possible that the replacement of lutein in this site affects the structure of the complex and enhances its stability in the monomeric form, as suggested by monomerization of LHCII in *lut2* [46, 114] and, to a lesser extent, in *npq2* [34]. Another role of L2 occupation by violaxanthin was proposed by Bassi and co-workers, that is, the participation of it in the xanthophyll cycle [122]. This prompted a series of experiments eventually leading Fleming's group to propose a role of zeaxanthin bound into L2 site of CP29 as another direct quencher of excess energy in PSII in addition to lutein bound to L1 site of the major LHCII [2]. In addition, Bassi and co-workers proposed that the L2 site has, in fact, an allosteric nature and binding of these two xanthophylls controls transition between two distinct protein conformations [45, 125]. These conformations possess different fluorescence lifetimes, thus controlling the efficiency of excitation energy transfer to the PSII core complex (see in the following sections).

Interestingly, neoxanthin in the minor antenna was found to be present only in Lhcb4 and Lhcb5. Lhcb6 was proposed to lack neoxanthin due to the absence of a tyrosine residue stabilizing site N1 [30, 141]. Native Lhca1–4 complexes are even harder to purify than Lhcbs, due to their high sequence homology: a partial purification of Lhca1/4 dimer from Lhca2/3 was obtained by isoelectrofocusing [40] or purification of LHCI from mutants depleted in individual Lhca proteins [124]. However, the approach of *in vitro* reconstitution of recombinant proteins allowed biochemical and spectroscopic characterization of all these complexes in purified form [175]. According to these results, Lhca proteins can be grouped into pairs with respect to their pigment binding properties: Lhca1 and Lhca3 bind three xanthophylls per polypeptide, mainly lutein and violaxanthin, while Lhca2 and Lhca4 bind only two xanthophylls, again lutein and violaxanthin. A small amount of β-carotene is also bound to LHCI [40, 107].

15.2.3 Changes in Composition and Distribution of Xanthophylls upon De-epoxidation

The carotenoid composition of thylakoids is not constant: it undergoes changes during rapid fluctuations of light intensity as well as during long-term acclimation of plants to stress conditions. Indeed, the xanthophyll zeaxanthin is hardly detectable in low light or dark-adapted leaves, while it accumulates in EL [57]. Light in excess with respect to the capacity for ATP hydrolysis by dark reactions or export to the cytoplasm leads to P_i depletion and block of the ATP synthase activity, and thus increases lumen acidification. Violaxanthin de-epoxidase (VDE), located in the thylakoid lumen compartment, is activated at low pH and becomes associated with the thylakoid membrane in a dimeric form [7] which can accommodate violaxanthin and catalyze the simultaneous de-epoxidation of the two β-rings, yielding zeaxanthin. The newly formed zeaxanthin is proposed to bind to either the V1 site of LHCII [29, 89] or to the L2 site of monomeric Lhcbs [45]. The extent and dynamics of zeaxanthin binding to Lhc proteins *in vivo* strongly differ among antenna proteins, and reflect different binding affinities, rather than the minor effects of protein steric hindrance in thylakoid membranes [122]. The monomeric antennae CP26 and CP24 showed the highest zeaxanthin content [22, 194]. The case of CP29 is somehow different, since isolation of this complex from EL-treated plants reveals much lower levels of bound zeaxanthin [61, 167, 168]. Recent theoretical assessment of the binding affinity of violaxanthin to V1 *vs.* L2 sites confirmed these experiments and provided a reason for the slow/inefficient de-epoxidation in the L2 site [60]. Interestingly, evidence *in vivo* comparing Lhcb4 knock-out mutants carrying or not the additional *npq1* mutation shows that zeaxanthin has a strong effect on the photoprotective activity associated to Lhcb4 that may not be associated with NPQ *per se* [54]. Finally, zeaxanthin was also found in the PSI-LHCI complex upon EL treatment [191].

15.3 Multiple Functions of Carotenoids *In Vivo*

15.3.1 Light Harvesting Function. Carotenoid-Mediated Energy Transfer in Light-Harvesting and Core Complexes: Excited States and Ultrafast Dynamics

Chlorophylls and carotenoids are the pigments that absorb and efficiently transduce sunlight energy in the photosynthetic membrane. The universal distribution of these molecules in photosynthetic organisms, ranging from bacteria to higher plants, suggests a vital role in the photosynthetic process. Participation in light-harvesting requires close proximity of chromophores as well as the reciprocal correct orientation in photosystems, where light-harvesting and photochemical reactions take place. Indeed, the pigments there are arranged in a highly ordered fashion in antennae and reaction center complexes.

The first event in photosynthesis involves the absorption of a photon by a pigment in the antenna system that eventually transfers the excitation to the reaction center [182]. Chl *b*→Chl *a* and Car→Chl *a* transfer efficiencies in isolated pigment-protein complexes can be evaluated by deconvolution of absorption and fluorescence excitation spectra of antennae complexes into their elementary components. The energy transfer efficiency from each pigment pool is calculated as the ratio of fluorescence excitation to absorption spectrum areas of individual components obtained as a result of deconvolution (Fig. 15.3).

The photophysical reactions taking place within light harvesting pigments must be extremely fast, occurring on the pico- and even femtosecond timescales. This is required by the competition between the energy transfer with other de-excitation mechanisms like internal conversion, inter-system crossing and fluorescence. Thus, the extremely rapid, early events of light harvesting and excitation energy transfer involving carotenoids can be followed only by means of time-resolved ultrafast laser spectroscopy. For example, in transient absorption spectroscopy, a "pump" laser pulse promotes an electronically excited state in a population of pigment molecules of the sample. A low intensity "probe" pulse directed to the sample with time delay τ with respect to the pump pulse is used to record a difference absorption spectrum ΔA ($A_{\text{excited state}} - A_{\text{ground state}}$). Its profile as a function of τ and λ contains information on the dynamic processes of excited energy transfer that occur in the system under investigation [19]. Several time-resolved spectroscopic studies allowed a better understanding of the structure-function organization of the light-harvesting process in photosystems.

As for all carotenoids in the photosynthetic apparatus, even β-carotene bound to the core complex of both photosystems showed a one-photon forbidden transition $S_0 \rightarrow S_1$. However, the S_1 excited state can be populated by internal conversion via the S_2 state and possesses a lifetime of ~10 ps. The $S_0 \rightarrow S_2$ transition is allowed and is responsible for the characteristic visible absorption. The S_2 state has a lifetime of 120–150 fs [115]. The excitation energy transfer from the Car singlet excited states to Chl is normally fast enough to compete with rapid internal conversion. Transient absorption spectroscopy work performed on isolated CP43, CP47 and PSII reaction center complexes of higher plants [56] revealed that the β-carotene S_2 state is the main donor in the energy transfer between Cars and Chls, while S_1 state does not participate. In all complexes, the yield of this transfer does not exceed 35 %, due to competing $S_2 \rightarrow S_1$ internal conversion (Fig. 15.4).

A number of ultrafast transient absorption studies on CP29 and on LHCII, both purified from leaves [41, 72] or reconstituted *in vitro* [38], gave information about the excited energy transfer in these complexes. All these studies concluded that excited energy transfer from the cross-branched, central xanthophylls to chlorophylls occurs with high efficiency (e.g., >80 % for both lutein of LHCII) from the S_2 excited state within 100 fs. However, Croce and co-workers [41] reported that

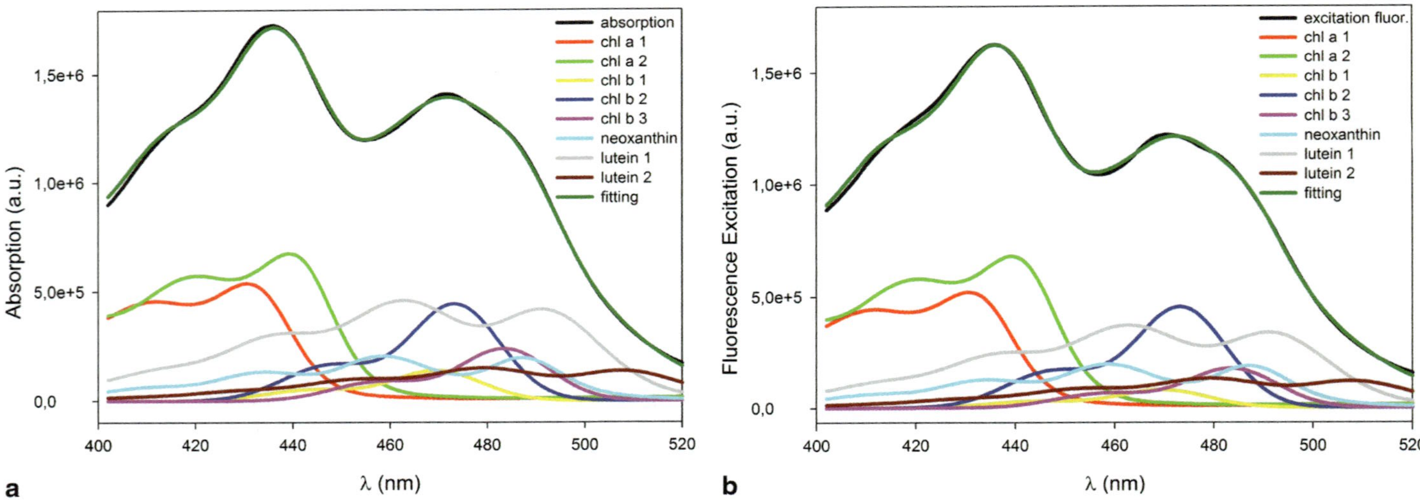

Fig. 15.3 Spectral deconvolution of absorption and fluorescence excitation spectra of the major trimeric LHCII complex in the Soret region. Absorption spectra (**a**) and fluorescence excitation spectra (for emission at 683 nm) (**b**) of purified LHC are described in terms of the absorption of individual pigments. The *experimental* spectra (*black*) are compared with those *reconstituted* as a sum of the absorption spectra of the individual pigments (*dark green*). The contribution of individual pigments is also shown: experimental spectra were fitted using 2 Chl *a*, 3 Chl *b* and 3 xanthophyll spectral forms [37]. Energy transfer efficiency can be calculated from the ratio of Chls and Cars amplitude in absorption *vs.* fluorescence excitation spectra

efficiency of excited energy transfer from the S_2 state dropped to 50% in the case of violaxanthin bound to site L2 in CP29. In LHCII, the contribution of excited energy transfer from carotenoid S_1 state to Chls is about 10–20% of the total energy absorbed by the chromophore [72]. Similar results were obtained by femtosecond fluorescence upconversion on LHCII from *Arabidopsis* [79]. No energy transfer from the carotenoid S_1 state to chlorophylls was observed in both CP26 and CP24 complexes, suggesting that this state is energetically below the chlorophyll Q_y energy level [116], consistent with measurements on LHCII [156].

Xanthophyll bound to site V1 was shown to be ineffective in energy transfer to Chls, thus it cannot function in light harvesting. As for β-carotene bound to the core complex of PSI, the efficiency of excited energy transfer to Chl *a* is high, reaching ~80%. Energy transfer from β-carotene to Chl *a* was found to proceed though both S_2 and S_1 excited states [96], with the majority of excitation transfer involving the S_2 state. The peripheral LHCI antenna carries chlorophylls and carotenoids and therefore provides an additional cross-section for the PSI reaction center. The transfer of excitation energy from carotenoids to Chl *a* molecules involves mainly the $S_2 \rightarrow Q_y$ channel and is completed within ~100 fs as measured in native dimeric LHCI [70] and recombinant Lhca4 [66].

A structure of the Cyt b_6f complex has been available for some time [179] and it confirmed that one chlorophyll *a* molecule and one carotenoid (9-*cis* β-carotene) is bound to the complex, as previously suggested [151, 198]. The role of this carotenoid still remains poorly understood: it was originally suggested to prevent the generation of $^1O_2$ by photoexcited Chl *a* [198]; however, neither triplet energy transfer from Chl to Car nor singlet energy transfer from Car to Chl did occur [149]. These observations are in agreement with structural data that show a distance of 14 Å between both pigments, thus too far for singlet/triplet excited energy transfer. Since the β-carotene molecule faces out of the complex, it cannot be excluded that functional interactions with other components of the photosynthetic apparatus take place. However, this hypothesis remains unproven.

15.3.2 ROS Scavenging and Chlorophyll Triplet Quenching by Carotenoids

Carotenoids play an important role in acclimation: the Car/Chl ratio increases in sun as compared to shade leaves [58]. Photoacclimation of *Arabidopsis* at low temperature induces the accumulation of antioxidants, including carotenoids [74]. The total xanthophyll content increases under strong light and the ratio between lutein and the xanthophyll-cycle components zeaxanthin, antheraxanthin and violaxanthin decreases [13]. Transcriptional analysis of an *Arabidopsis* mutant that accumulates $^1O_2$ to higher levels than the wild-type [3] showed that the carotenogenic genes in this mutant were strongly upregulated. This acclimation pattern strongly suggests that carotenoids are likely to have a crucial photoprotective function in plants under conditions that lead to $^1O_2$ production.

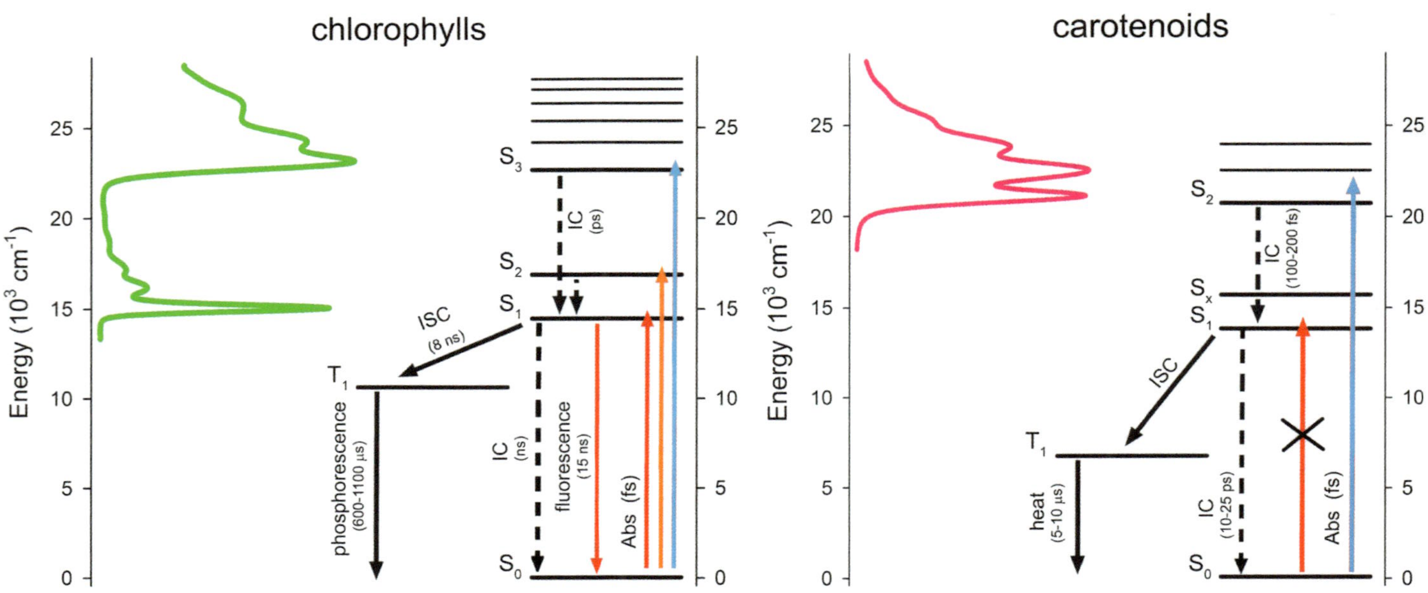

Fig. 15.4 Chlorophyll (*left*) and carotenoid (*right*) transition energy diagrams. Excitation and relaxation processes are indicated by *arrows* and corresponding time constants. The range of energies of pigment transitions are also related to the corresponding visible absorption spectrum of the pigment (*green*, chlorophyll; *purple*, carotenoid). *ISC* intersystem crossing, *IC* internal conversion, *Abs* absorption

Carotenoids contribute to the antioxidant network of the chloroplast, aimed to the detoxification of ROS generated by photosynthesis. The pool of carotenoids of the thylakoid membrane increases the resistance to both EL and treatment with photosensitizers that artificially increase the release of ROS into the chloroplast [14]. Carotenoids act by scavenging both O_2^- and $OH^\bullet$ [173] and by quenching $^1O_2$ in thylakoids, thus preventing lipid peroxidation [75]. Recently Havaux and coworkers [78] studied the *in vivo* antioxidant activity of carotenoids unbound from the photosynthetic complexes. They found that carotenoids provide protection against thylakoid membrane peroxidation. In particular, the antioxidant activity of zeaxanthin was noticeably higher with respect to all other carotenoids of *Arabidopsis* leaves.

As previously described, PSII and LHCs can become intense sources of ROS upon overexcitation. Carotenoids present in thylakoid membranes play their photoprotective role in two distinct forms that determines their mode of action. The observed free fraction of carotenoids (~15% of the total carotenoid pool) was suggested to perform their antioxidant function on ROS species released from LHC and RC complexes. Most carotenoids, however, are bound to photosynthetic machinery where they are in close contact with chlorophyll molecules—the major photosynthesizers in plant systems. The van-de-Waals contact with chlorophylls allows quenching of the potentially harmful $^3Chl^*$. Indeed, the population of carotenoid triplet excited states increases proportionally to the light intensity in leaves [195], thylakoids [87], isolated photosystems [117] and LHC complexes [126, 148] due to triplet-triplet energy transfer from the $^3Chl^*$. The triplet yield of chlorophylls is ~30% *in vitro* [103]. A comparable intersystem crossing yield has been estimated for chlorophyll bound to photosynthetic complexes [176]. The efficiency of $^3Chl^*$ quenching by carotenoid bound to LHC was estimated to be between 80 and 100% by [148]. More recent results showed that $^3Chl^* \rightarrow {}^3Car^*$ has a 95% efficiency in LHCII, thus leaving 5% of the chlorophyll triplets unquenched [126]. This effect appears to be intrinsically related to the molecular organization of the Lhcb proteins.

In LHC, triplets are expected to appear only on Chls *a* due to the very fast energy transfer from Chl *b* to Chl *a*. Triplet transfer between chlorophylls and carotenoids requires van der Waals contact distance between these molecules (<3.6 Å). The structure of LHCII [113] shows that at least one Chl *a* molecule is in close proximity to each xanthophyll ligand. In the PSII core most of the β-carotene molecules are in close contact with Chl head groups, as required for facilitating energy transfer to Chl and for quenching $^3Chl^*$. However, the two β-carotene ligands in the PSII reaction center localized at the D1/CP43 and the dimer interfaces are away from the special pair of chlorophylls. Indeed, their relative distance is longer than 3.6 Å, implying they cannot quench $^3P680^*$ by triplet-triplet transfer. The topology of β-carotene molecules within PSII reaction center reflects their functioning in a harshly oxidizing environment. Indeed, direct quenching of $^3P680^*$ seems to be impossible because carotenoids would be oxidized by the strong oxidizing potential formed during charge separation [185]. β-carotene molecules of the reaction center complex are likely to act as quenchers of $^1O_2$ produced during charge recombination (see above) and are in fact efficient in reducing chlorophyll bleaching in isolated PSII [187].

15.3.3 Involvement of Carotenoids in Quenching of Singlet Chlorophyll Excited States. Regulation of Light Harvesting by Non-photochemical Quenching

As shown above, the reaction $^1Chl^* \rightarrow {}^3Chl^*$ ($\rightarrow {}^1O_2$) is a constitutive and unavoidable process occurring in pigment-protein complexes upon illumination. While ROS scavenging by carotenoids and other anti-oxidants ensures prevention of photodamage under low or moderate light conditions, excess light for long periods leads to exhaustion of ROS scavenging compounds. Thus the capacity to control ROS formation, specifically $^1O_2$, is essential for maintaining the functional integrity of the photosynthetic apparatus. Photosynthetic processes that decrease the overall chlorophyll fluorescence yield (Φ_{Chl}) include *photochemical quenching* (qP), which is associated with the event of charge separation in the PSII reaction center, and *non-photochemical quenching* (NPQ), namely a fluorescence quenching event that is not dependent upon charge separation [85]. NPQ is a regulatory mechanism preventing formation of excess $^1Chl^*$ through its dissipation as heat in PSII reaching 90% [108] of the absorbed light energy. NPQ activity can be easily detected by the decrease in the PSII yield (Φ_{PSII}) due to down-regulation of the average chlorophyll excited state lifetime measured by fluorescence [69, 82]. Thus, a decrease in the fluorescence yield indicates that a dissipation channel for $^1Chl^*$ has been activated, which can harmlessly dissipate the excess absorbed energy (Fig. 15.5a). Besides decreasing the $^1Chl^*$ concentration, and thereby minimizing the generation of $^1O_2$ in the LHC and PSII reaction centers, NPQ prevents both over-acidification of the chloroplast lumen and generation of $^3P680^*$ [67, 84].

The early finding that NPQ depends upon the amplitude of the trans-thylakoid ΔpH qualifies this process as a feedback-regulatory mechanism for excitation energy transfer to PSII reaction centres [67]. The proton concentration in the lumen is determined by the rate of photosynthetic electron transport while the dissipation of the pH gradient is essentially determined by the activity of the ATP synthase complex. In EL conditions, the limiting factor for ATP synthase activity is the availability of the substrates, P_i and ADP, caused by ATP utilization by chloroplast dark reactions, such as the Calvin-Benson cycle for CO_2 assimilation, photorespiration, NO_3^- reduction or by the exchange of triose phosphates for P_i with the cytoplasm [93]. This regulation machinery ensures that quenching only applies to the fraction of Chl excited states exceeding the capacity for use by the cell metabolism or export of photosynthetic products to sink organs of the plant. The ability of plants to modulate the efficiency of light-utilization is of particular importance under fluctuating light intensity. Indeed, NPQ was shown to be crucial for plant fitness in natural environment [105].

The term NPQ refers a set of inducible fluorescence quenching mechanisms with different properties and likely based on distinct, although related, molecular processes (Fig. 15.5b). The predominant component is named qE, after *E*nergy *q*uenching. qE, in fact, depends upon the *energization* of the chloroplast, namely by the light-dependent acidification of the lumen. It is normally rapidly induced

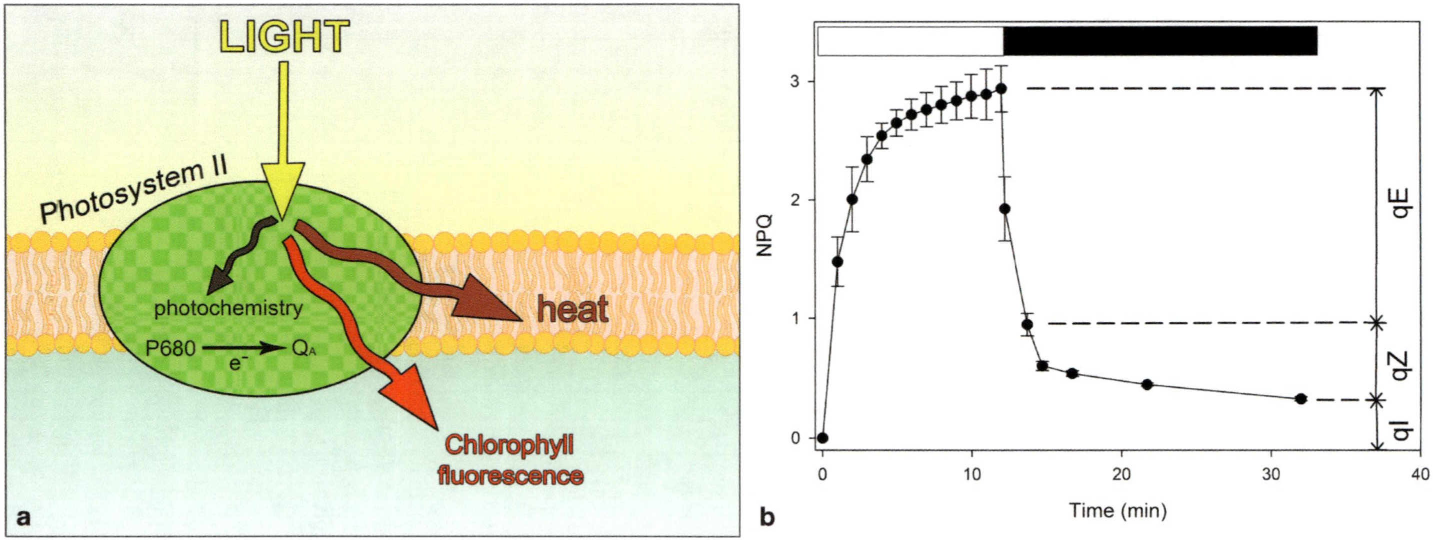

Fig. 15.5 The non-photochemical quenching of chlorophyll fluorescence. **a** Light absorbed by the PSII can be used for driving photosynthetic electron transport, or can be lost as fluorescence or heat. Since these processes are in competition with each other, an increased efficiency of a process leads to a decreased efficiency of the two others. **b** The amplitude of NPQ (non-photochemical quenching) accounts for the rate of light-induced excitation decay by heat loss. NPQ induction and relaxation time points are recorded with a pulse amplitude modulated fluorometer, during the time-course of illumination of dark-adapted *Arabidopsis* wild-type leaves, during 12 min of illumination at 1200 µmol photons m^{-2} s^{-1} (*white bar*). The NPQ dark relaxation kinetic (*black bar*) reveals that three different phases contribute to NPQ the energy-dependent quenching (qE), the Zea-dependent quenching (qZ) and the photoinhibitory quenching (qI)

upon transition from dark to light and is rapidly reversible upon return to dark [85]. Treatment with the ionophore nigericin collapses ΔpH and prevents the activation of qE [26] (Fig. 15.6a). Under rapid fluctuations of light intensity, qE is the major factor for prompt modulation of photosynthesis in the order of seconds to minutes. The slowest component of NPQ is named qI, after its early identification as photoinhibitory. This definition is based upon qI relaxation dependence of D1 protein re-synthesis [104, 139] with a half-time of several hours. Although in extreme stress conditions photoinhibition can be involved in slowly relaxing quenching, this is unlikely to be the major NPQ component in most conditions [92, 165, 171]. The analysis of the NPQ relaxation dynamics identified a third quenching component with intermediate lifetime, called qT. It was shown to relax within several minutes and was originally attributed to the quenching of a phosphorylated LHCII population upon its migration from grana to stroma membranes during the establishment of State 2 [85]. However, *stn7* mutant of *Arabidopsis* that lacks the state transition mechanism still displays unchanged NPQ kinetics and amplitude with respect to wild-type plants [17]. More recent results suggest involvement of zeaxanthin in the modulation of both the qI and qT [45, 133].

Despite the recent proposal that the actual quencher is a Chl dimer [129], a large body of evidence supports the view that xanthophylls present in Lhcb family of proteins play a key role in the modulation of the processes that gives rise to quenching of ^{1}Chl* in plants and green algae. Hence, the *ch1* mutant of *Arabidopsis* which is depleted from xanthophyll-binding complexes have a very low NPQ [78], while mutants in the xanthophyll biosynthetic pathway exhibited strong changes in both the maximal amplitude and the kinetics of the quenching [47, 152].

15.4 Mechanism of qE

15.4.1 Key Elements Involved in qE

qE, the NPQ component with normally the highest amplitude and fastest kinetics is activated by low lumenal pH. Lumen acidification is an activator of a few mechanistic steps of photoprotection mechanisms. Interestingly, *N*, *N*-dicyclohexylcarbodiimide (DCCD), a chemical that reacts with carboxyl groups in hydrophobic environments, can bind to protonatable residues of many protein subunits involved in photoprotection, such as Lhcb4 [193], Lhcb5 [147] and PsbS [59, 112], was discovered to inhibit qE *in vivo* [84]. The maximum amplitude of NPQ induced by a saturating light treatment depends upon the activation of the xanthophyll cycle, which synthesize zeaxanthin from pre-existing violaxanthin [58]. Both the triggering of qE and the synthesis of zeaxanthin are dependent upon low lumenal pH. The PSII subunit PsbS [110] bears two lumen-exposed glutamate residues essential for the transduction of acidic pH signal into activation of qE [111, 112]. Concomitantly, lumen acidification activates the VDE, a soluble protein at neutral pH which

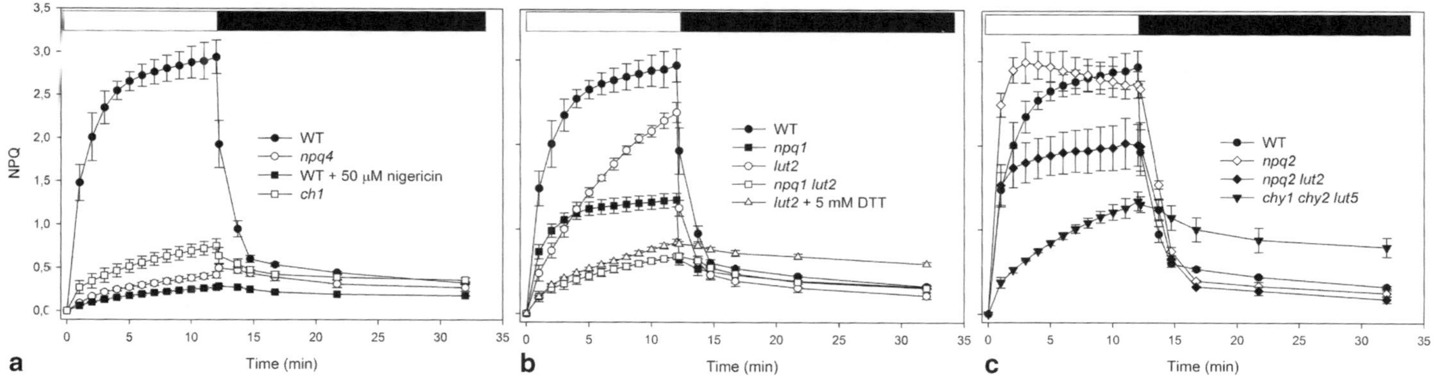

Fig. 15.6 NPQ analysis of *Arabidopsis* wild-type and mutant genotypes. Kinetics of NPQ upon transition from the dark to EL and subsequent relaxation in the dark are recorded as described in the previous figure. **a** NPQ kinetic of wild-type leaves is compared to those of mutants lacking PsbS (*npq4*) or LHC proteins (*ch1*), or with wild-type leaves treated with the ionophore nigericin, that hamper ΔpH build-up. **b**, **c** NPQ kinetics of wild-type and mutants with modified xanthophyll composition. *npq1*, *lut2*, and *npq1lut2* lack respectively the xanthophylls zeaxanthin, lutein or both. Note that residual qE in *lut2* is due to zeaxanthin, whose synthesis is inhibited by leaf infiltration with 5 mM DTT, a VDE inhibitor (panel **b**). *npq2* leaves lacks violaxanthin and neoxanthin, while *npq2lut2* and *chy1chy2lut5* accumulates respectively zeaxanthin and lutein as the only xanthophyll (panel **c**)

becomes a membrane-associated dimer [7] with a pH optimum of 5.2. The enzyme converts violaxanthin to zeaxanthin as part of the xanthophyll cycle. Experiments in leaves infiltrated with the VDE inhibitor dithiothreitol (DTT) have shown that most qE depends upon formation of zeaxanthin [197], while a correlation between zeaxanthin and high qE was established [23, 57]. The analysis of mutants affected in xanthophyll biosynthesis [134, 135] has confirmed the need for zeaxanthin and lutein for qE activity *in vivo* (see next section).

Although there is a general consensus that quenching events are catalyzed in pigment-binding proteins [2, 10, 170] an important role in the qE mechanism is played by the pigment-less subunit PsbS [25, 59, 110] (Fig. 15.6a). Indeed, mutation at the psbS locus is epistatic over mutations affecting Lhcb protein and xanthophylls biosynthesis enzyme in abolishing NPQ [25, 45]. Further work using site-directed mutant analysis *in vivo* highlighted the role of a conserved pair of lumen-facing glutamic residues (E122 and E226) required for qE induction, implying that PsbS role was the transduction of low lumenal pH signal into a conformational change of pigment-binding proteins where quenching occurs [112]. How the PsbS protonation event is coupled to the conformational change within Lhcb proteins is matter of debate. Recent work with the use of diaminodurene, an enhancer of proton gradient across the thylakoid membrane, demonstrated the full restoration of qE in chloroplasts of *npq4* mutants lacking the PsbS protein. It was concluded that PsbS simply enhances sensitivity of the LHCII antenna to protons, and its response leading to qE [88]. Further, it was found that PsbS greatly enhances mobility of PSII complexes in granal membranes, hence accelerating its transition into the qE state [71]. A key step towards elucidation of the nature of this PSII dynamic rearrangement leading to qE was made by the finding that the PSII-LHCII supercomplex dissociates into two moieties during NPQ induction, forming two separated domains, one enriched in the central components of PSII (PSII core dimer, CP29, CP26, LHCII-S) and the other built of more peripheral subunits (CP24, LHCII-M and LHCII-L) [21, 91]. Mutations that produced a constitutive dissociation of the PSII-LHCII supercomplex were found to have partial inhibiting effects on NPQ [53, 54, 102] or no effect at all [51]. These results are consistent with stabilization of Lhcb proteins in their energy-conserving, unquenched conformation when cooperatively assembled into a PSII-LHCII supercomplex while they readily assume a quenched conformation whilst dissociated from the supercomplex. Since a PsbS knock-out mutation is epistatic over CP24-depleted plants, it appears that PsbS, besides its effect in dissociating PSII supercomplex, might have a direct effect in promoting the quenching conformation of Lhcb proteins, possibly by direct interactions [20, 183].

15.4.2 Xanthophyll Biosynthesis Mutants and Modulation of qE

The role of specific gene products leading to a NPQ phenotype has been evaluated through reverse genetics in *Arabidopsis thaliana* falling into three classes: (i) mutants altered in the ΔpH formation [44, 130]; (ii) mutants altered in luminal pH sensing [112]; (iii) mutants affecting either Lhcb proteins [6, 53, 54] or the synthesis of their chromophores [45, 78, 134, 135, 154]. Among class (iii) mutants, those targeting genes of the carotenoid biosynthesis pathway have been produced and characterized, yielding plants with altered xanthophyll composition. Interestingly, most of these mutations affect both qE amplitude and kinetics, thus confirming that xanthophylls have important roles in qE.

The *npq1* mutant lacks VDE activity and is thus unable to convert violaxanthin into zeaxanthin upon exposure to EL [134]. The mutant displays a 70 % reduction in the maximum qE amplitude with respect to wild-type, implying that zeaxanthin accumulation is needed for full establishment of qE in plants. Despite the absence of a functional xanthophyll cycle, a residual qE activity is detected in *npq1* plants (Fig. 15.6b). It has been proposed to originate from the operation of xanthophylls constitutively present in LHC. Since the absence of neoxanthin does not affect NPQ [46], such xanthophyll is likely to be lutein bound to the L1 site of all LHCs and to the site L2 of the trimeric LHCII complex. This proposal was confirmed by the complete loss of qE in the *npq1lut2* double mutant that lacks both lutein and zeaxanthin [135]. Lutein-less plants showed a lower qE and a slower kinetic of qE induction than wild-type plants [46, 114, 154]. This residual quenching was attributed to zeaxanthin. Indeed, treatment of *lut2* leaves with the inhibitor of xanthophyll cycle DTT abolishes qE. A positive proof for lutein activity in qE was further provided by the analysis of the *szl1* (*suppressor of zeaxanthinless1*) mutant which is characterized by an increased lutein/β-xanthophyll ratio with respect to wild-type. In an *npq1* background, *szl1* is effective in releasing the qE restriction owing to the lack of zeaxanthin, implying that lutein replaces violaxanthin from its binding site(s) and acts in qE [109].

The phenotypes of mutants described above suggest that lutein and zeaxanthin have both a role in the full expression of qE. However, these molecules were found to modulate distinct kinetic components. The most rapid phase in qE induction (0–1 min) is retained in *npq1* plants, suggesting that lutein catalyzes quenching in the first minute upon transition from dark to EL. Consistently, the onset of quenching is slower in the *lut2* mutant with respect to the wild-type, implying that zeaxanthin, initially absent, needs to be synthesized by VDE and be incorporated in Lhc proteins, thus delaying the onset of quenching. That the slow quenching component in *lut2* leaves is due to zeaxanthin is actually confirmed by the observation that DTT treatment blocks qE, thus phenocopying the *npq1lut2* phenotype (Fig. 15.6b). Conversely, when zeaxanthin is constitutively accumulated as in the *npq2* mutant lacking zeaxanthin epoxidase (ZE) [134, 161], thus abolishing the lag phase due to zeaxanthin synthesis and binding to Lhcb proteins, the kinetics of NPQ is much faster and reaches its maximal amplitude in 1 min rather than 8–10 min (Fig. 15.6c). This led Bassi and co-workers to propose that lutein and zeaxanthin act in the two distinct quenching sites of LHCII antenna. It should be noted that *chy1chy2lut5* plants that accumulate lutein only [47] and *npq2lut2* with zeaxanthin only [76] have both reduced levels of NPQ in comparison to the wild type.

More recent work proposed a new description of the xanthophyll molecular properties by the differences in their *hydrophobicity* expressed as the so-called H-parameter. Variation in this parameter was hypothesized to reflect the variations in the conformational energy of xanthophyll-binding domains, and thus to affect the efficiency of the transition from a *conservative* (F_{max}) to a *dissipative* (NPQ) state [164]. Xanthophyll H-parameter was proposed to affect the allosteric character of NPQ regulation [164], namely the hysteretic relationship between ΔpH and NPQ amplitude [146] (more detail to follow below).

15.4.3 Proposed Molecular Mechanisms of Excess Energy Dissipation

Despite the study of many *npq* mutants and those performed in isolated LHC proteins, fundamental questions about Chl fluorescence quenching, namely the identity of the quenching specie(s) and the mechanism by which quenching is catalyzed, still do not have a single answer. It is possible that this is to the existence of more than one qE quenching mechanism/site.

As for the physical mechanism of quenching, at least two models have been proposed:

Model 1: Aggregation-Dependent LHCII Quenching This model was proposed based on the early evidence that aggregation of isolated LHC, induced by low detergent concentration and/or low pH, yields a decrease in Chl fluorescence lifetime [83]. Aggregation was later shown to be instrumental in catalyzing conformational change(s) within the LHCII protein, and the spectral signatures associated to this event were interpreted to indicate the formation of a tight interaction between lutein bound into the site L1 and terminal emitter Chl *a* [170]. Extrapolation to NPQ *in vivo* relies on the detection of similar various spectral changes in leaves upon NPQ induction [142, 170].

Model 2: CT (Charge-Transfer) Quenching Mechanism The formation of a charge-transfer state between Chl *a* and zeaxanthin was proposed based on theoretical work and ultrafast spectroscopy on isolated thylakoids of *Arabidopsis* [81]. In this model, qE activation involves a charge separation between a Chl-zeaxanthin heterodimer that produce a transient zeaxanthin radical cation ($Zea^{\cdot+}$) with a short relaxation time (50–200 ps). The spectroscopic signal of $Zea^{\cdot+}$ was found in isolated monomeric Lhcbs [10]. Further mutation analysis of Chl-binding sites in Lhcb4 led to the proposal that a Chl pair (Chl A5 and Chl B5) located in close proximity to the L2 xanthophyll binding site is involved in this charge-transfer event [2].

The charge transfer model is based upon the work performed by Bassi and coworkers on the xanthophyll cycle activity in the monomeric Lhcb antenna complexes assuming that *Arabidopsis* violaxanthin could be replaced by zeaxanthin under EL conditions [122]. However, the somewhat lower de-epoxidation efficiency of violaxanthin bound into the L2 site of Lhcb4 mentioned above [168], that may be explained by the higher binding affinity of this xanthophyll [60], is likely to require a so far unknown structural mechanism of undocking for its de-epoxidation. Further work is in progress to shed more light on this issue.

Recent analysis of ultrafast Chl fluorescence kinetics in intact leaves of *Arabidopsis* [120, 121] provided some evidence that at least two different quenching sites contribute to qE *in vivo*: a PsbS-dependent site located in LHCII that becomes detached from PSII and aggregated upon illumination, and a PsbS-independent site located within the minor antennae attached to PSII. A red-shifted fluorescence lifetime component was detected in both quenched wild-type leaves and aggregated LHCII trimers binding zeaxanthin. Furthermore, Car S_1–Chl excited state coupling was recently measured in isolated LHCII and correlated with qE amplitude *in vivo* in mutants such as *npq1*, *npq2* and *lut2* [24]; thus it was proposed that a short-lived low excitonic Car–Chl states, formed upon lumen acidification, may also function *in vivo* as excess energy dissipation valves.

15.5 Functional Genomics of Carotenoids

15.5.1 Significance of Conserved Xanthophyll Composition of the Land Plants

Despite the great diversity generated by evolution, the xanthophyll content of land plants is extremely well conserved with respect to both the overall composition and localization in chloroplast structures. The large majority of carotenoids is bound to the photosynthetic complexes and shows a constant distribution among their different components: β-carotene is bound to the reaction centres while light-harvesting complexes bind xanthophylls: lutein, violaxanthin, neoxanthin and, upon its accumulation under EL, zeaxanthin (see Sect. 15.2). The conservation of carotenoid composition across a wide range of plant taxa suggests a unique role for each molecular species. Yet, all these pigments have similar spectroscopic properties in terms of absorption wavelengths and extinction coefficients, and all are able to prevent photooxidation of chlorophyll-binding proteins and membrane lipids by either quenching ^{3}Chl* or detoxifying ROS. Furthermore, the energy level of the S_1 excited state, critical for energy exchange with Chls has been reported to be very similar for all xanthophylls [155, 156]. The reason for the strong conservation of this diversity has long been poorly understood, while a comprehensive understanding of functions for each molecular species has been a major challenge. Early work has been based on the use of recombinant antenna proteins for model studies in isolated pigment-binding proteins. This work suggested that the function of individual xanthophyll species can be understood within the framework of their binding to LHC proteins [35, 36, 63]. Later, reverse genetics of xanthophyll biosynthesis enzymes in *Arabidopsis thaliana* has been instrumental for dissection of the function of each carotenoid species in plants. Based on homology with bacteria most plant carotenoid biosynthesis genes could be identified and knock-out lines were isolated (Table I). Additional genes involved in the pathway were identified by a forward genetic approach based on phenotypes e.g. the *aba4* gene whose deletion leads to lack of neoxanthin [136]. Combination of these mutations allowed construction of a library of mutants, including plants deficient in each one of the carotenoids species accumulated in wild-type plants as well as plants retaining only one carotenoid species. The physiological characterization of these plants has shed light on uniqueness of each type of carotenoid in the higher plant photosynthetic membrane.

15.5.2 Lutein

Lutein is the most abundant carotenoid in the photosynthetic apparatus of plants and has been found to be the exclusive ligand for the conserved binding site L1 in all light-harvesting complexes. Occupancy of site L1 was found to be essential for the folding of Lhc proteins [63]. By measuring the kinetics of chlorophyll photobleaching in recombinant proteins refolded *in vitro* with different xanthophylls species, the best photoprotection was obtained using complexes binding both lutein and one or more β,β-xanthophylls, implying a synergic activity in quenching of harmful ^{3}Chl* states and scavenging of ROS produced by the reaction of ^{3}Chl* with molecular oxygen. Since lutein was known to be less efficient that β,β-xanthophylls in scavenging in model systems [196], it was proposed that lutein-dependent enhancement of photoprotection was due to a special role in triplet quenching [63]. The first isolation of two viable lutein-deficient mutants of *Arabidopsis*, *lut1* and *lut2,* by Pogson and coworkers [153], showed that lutein could be substituted by violaxanthin in *lut2* plants [114, 153]. More detailed studies showed that the *lut2* mutation caused biochemical and physiological alterations, including changes in kinetic and amplitude of NPQ (see Sect. 15.6) and monomerization of LHCII trimers [46, 114, 154].

Functional alterations due to the *lut2* mutation were also revealed *in vivo*, consisting of a lower capacity for state1-state 2 transitions and a lower accumulation of Lhcb1 and Lhcb2 gene products, slower growth rate, and higher photoinhibition with respect to wild-type in EL [46, 95]. One of the reasons for all these phenotypes was suggested by the use of *triplet-minus-singlet* spectroscopy that showed a slower rate of ^{3}Chl* quenching in complexes with violaxanthin *vs.* lutein, and a consequent increased production of $^1O_2$. ROS stress, in fact, induces defense responses typical of EL acclimation [3, 12, 13], with reduction of the PSII antenna size. It was concluded that lutein is the most suitable molecule for ^{3}Chl* quenching in antenna (Fig. 15.7a). It should be noted that the photosensitive phenotype of the *lut2* mutant is difficult to reveal in moderate light stress conditions because of the rescuing effect of zeaxanthin, which is synthesized more promptly in *lut2* with respect to WT and is an excellent ROS scavenger [78]. When the xanthophyll cycle is switched off, as in the double mutant *npq1lut2* lacking both zeaxanthin and lutein, the photosensitivity phenotype becomes strikingly evident; when compared to the *npq1* control, *npq1lut2* is much more sensitive to EL and low temperature [46, 135].

Since ^{3}Chl* quenching prevents reaction with oxygen and $^1O_2$ formation, it was expected that mutants enriched in lutein would be more effective in preventing oxidative stress and would increase their resistance to EL. However, this did not fit

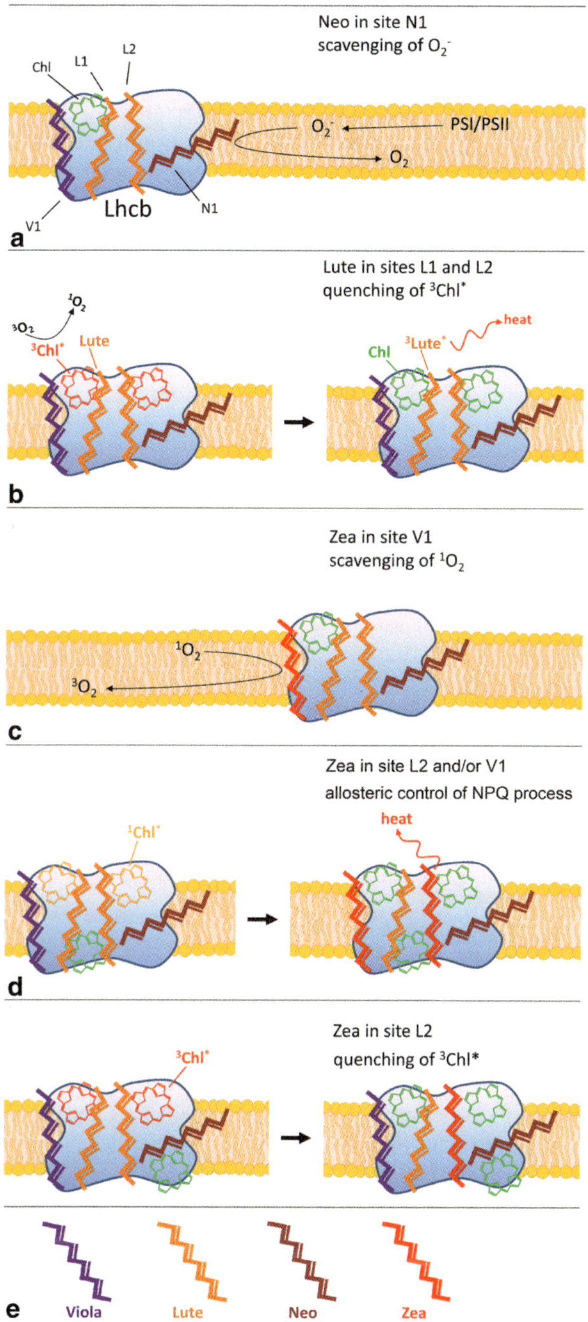

Fig. 15.7 Functional genomics of higher plant carotenoids. Reverse genetic of xanthophylls biosynthesis enzymes in *Arabidopsis thaliana* has been instrumental in order to dissect the function of each carotenoid species in plants. **a** Lutein has the specific property of quenching harmful $^3Chl^*$ by binding at sites L1 and L2 of LHCII, thus preventing ROS formation. **b** Neo preserves PSII from photoinactivation and protects membrane lipids from photooxidation, being particularly active against O_2^-. **c** The protein/lipid interface is the active site for the antioxidant activity of Zea; when bound to site V1 of LHCII, Zea mediates stress tolerance by protecting against lipid peroxidation. **d** Binding of Zea to both sites L2 of monomeric Lhcs and/or V1 of the major trimeric LHCII complex enhances NPQ amplitude *in vivo*. **e** Binding of Zea to site L2 of monomeric antennae modulates $^3Chl^*$ formation *in vivo*: by lowering the yield of potentially dangerous chlorophyll excited states on the complexes, it prevents the release of $^1O_2$

with the observation that recombinant Lhc proteins reconstituted with lutein only underwent faster photobleaching than control complexes binding both β,β- and β,ε-xanthophylls [36, 39, 63]. Consistently, acclimation to EL leads to a strong increase of the xanthophyll cycle pigments violaxanthin and zeaxanthin [58], while over-accumulation of the xanthophyll cycle pigment pool by over-expressing carotene β-hydroxylase in *Arabidopsis* enhances EL tolerance [52, 89]. Overall, these results strongly suggested that lutein alone was not enough for effective photoprotection, while β,β-xanthophylls have a distinct, although complementary, role. These conclusions were strengthened by the analysis of the triple *Arabidopsis* mutant *chy-1chy2lut5,* lacking three carotene hydroxylase enzymes and accumulating lutein as the only xanthophylls species [62]. Lutein-only plants could grow in low light only (<50 μmol photons m^{-2} sec^{-1}) and were extremely sensitive to even moderate light, showing rapid photoinhibition at 150 μmol photons m^{-2} sec^{-1} as well as higher rates of $^1O_2$ release with respect to the wild-type, as measured from leaves, thylakoids and purified LHC proteins [47]. It should be noted that, in contrast, purified PSII and PSI core complexes exhibited the same level of resistance to EL, consistent with their binding of carotenes only [47].

The important conclusion from the above work is that light harvesting antenna system of plants is far from being the oxygen-tight protected system predicted from earlier studies. Even at low to moderate light intensity, long-lived (ms) $^3Chl^*$ are accumulated that cannot be quenched by lutein in the μs time range, thus allowing for constitutive $^1O_2$ production. $^1O_2$ must be continuously scavenged. Impairing either the $^3Chl^*$ quenching or $^1O_2$ scavenging functions leads to photodamage. Indeed, subsequent results [126] showed that violaxanthin and neoxanthin, bound respectively in sites V1 and N1 of LHCII, do not directly contribute to the quenching of $^3Chl^*$ that originate in the complex, such as lutein molecules do; neoxanthin bound to the N1, however, acts as a barrier for oxygen entering the core of LHC domains, resulting into a limited level of $^1O_2$ production.

15.5.3 Neoxanthin

Apart from a few parasitic plants [27], neoxanthin is always present in the light harvesting system, accounting for ~15% of total carotenoids. Its major binding sites are the N1 sites of LHCII [35, 1113]. The two monomeric complexes Lhcb4 and Lhcb5 also bind this pigment. Neoxanthin is not present in Lhcb6 in any Lhca antenna complexes. Neoxanthin was found not to be essential for folding of recombinant LHC proteins and is unable to promote the process on its own when supplied with Chl *a* and Chl *b* [63]. Besides, its efficiency in transferring excitation energy to Chl *a* is one of the lowest among xanthophylls. These properties make neoxanthin different from lutein, leading to the suggestion that its major role is in a metabolic pathway as the precursor of the plant growth regulator abscisic acid (ABA) [119] and in regulation of its synthesis. Although neoxanthin-depleted mutants were available relatively early [160], identification of phenotypes specifically associated to the absence of neoxanthin has been hampered by the coupling with accumulation of zeaxanthin at the expense of violaxanthin [134, 161] in *Arabidopsis* mutants such as *aba1* and *npq2*. Indeed, the genetic lesion was located at the ZE reaction, a few step upstream in the β,β-xanthophylls pathway, thus implying pleiotropic effects. Further screening among *Arabidopsis* mutants affected in ABA accumulation led to the identification of the *aba4* mutant [136], which was only partially deficient in ABA, and was depleted in neoxanthin but not in other pigment species, thus avoiding most pleiotropic effects. Further characterization of the *aba4* mutant of *Arabidopsis* showed that, in the N1 site of LHCII, Lhcb5 and Lhcb6 proteins, neoxanthin was replaced by violaxanthin in its 9-*cis* configuration, similar to the earlier reported observations performed on *Cuscuta reflexa* LHCII [177]. Because of its specificity, this mutant has been a useful tool for the determination of the physiological role of neoxanthin and how its function differs from the replacing violaxanthin. Photosynthesis in the *aba4* mutant was not significantly affected under mild environmental conditions. However, it appeared to be more sensitive to photoxidative stress in EL, suggesting a highly specific function for neoxanthin. This was identified by challenging leaves with artificial sources of the different ROS species produced in plants during stress, namely $^1O_2$, O_2^- or H_2O_2. *Aba4* plants were selectively more sensitive to damage from O_2^- with respect to wild-type leaves [48]. Consistently, photodamage was increased by both inhibition of superoxide dismutase (SOD) and chilling stress [48], two conditions leading to the accumulation of O_2^-. An intriguing characteristic of neoxanthin is that, in spite of being localized within PSII in grana partitions, it was found to be active against a ROS species produced by PSI in the stroma membranes. A possible explanation for this contradiction could be the high solubility of O_2^- in the lipid phase, allowing its diffusion into membranes towards the PSII reaction centre where the Q_B site is exposed to lipid-soluble agents. Neoxanthin binding Lhcb components forms a shell around PSII, with neoxanthin molecules protruding with their allene groups [113] that could potentially act in O_2^- scavenging (Fig. 15.7b). While enzymatic superoxide detoxification based on the SOD system [9] appears to be active on the soluble phase of the chloroplast, neoxanthin would act within the thylakoid lipid phase, complementing the action of antioxidant molecules in the chloroplast [77, 127].

15.5.4 Xanthophyll Cycle Carotenoids

Zeaxanthin function in photoprotection is self-evident from its synthesis in EL conditions only, while it is absent in dark or low light adapted leaves. The zeaxanthin synthesizing enzyme, VDE, is located in the chloroplast lumen in its inactive, water soluble form at pH above 5.2. When light intensity exceeds the rate of energy utilization by downstream metabolic processes [58], which refuel ATP synthase with ADP and Pi, acidification occurs because of decreased proton consumption by ATP synthase. In these conditions, VDE dimerizes and exposes a hydrophobic surface to the chloroplast lipid phase, where the violaxanthin substrate is released from LHCII V1 site and becomes available [7]. A maximal de-epoxidation state of 70 % can be obtained in O_2^- within 15–30 min under saturating light intensity. The kinetics of zeaxanthin accumulation in leaves is clearly biphasic: a fast initial phase reflects rapid de-epoxidation of pre-existing, lipid-free violaxanthin, while a subsequent slow kinetic component accounts for the equilibrium of exchange of newly synthesized zeaxanthin with violaxanthin bound to sites L2 and/or V1 in Lhcb proteins. ZE catalyzes the zeaxanthin → violaxanthin conversion. Its activity is modulated by plant stress experience, thus acclimation to light stress or overwintering in evergreen plants lead to ZE retardation or complete blockage [1, 158]. Indeed, a stable zeaxanthin-dependent feature, called 'cold hard band', is found in evergreen plants, and allows photosynthetic machinery to withstand freezing conditions by maintaining a permanently photoprotected LHCII in its quenched, aggregated state [68]. Once synthesized from pre-existing violaxanthin, zeaxanthin appears to up-regulate several protection mechanisms of plants.

Zeaxanthin was shown to increase resistance to both EL and exogenously provided photosensitizers [14] by scavenging $^1O_2$ and thus preventing lipid peroxidation [75, 78] (Fig. 15.7c). Indeed, comparison of photosensitivity in mutants either affected in zeaxanthin synthesis (*npq1*) or qE (*npq4*) showed a peculiar photoporetective function for zeaxanthin, independent from its role in the modulation of qE [75]. The relative size of the VAZ pool free in the lipid phase is indeed modulated by environmental conditions; it accounts for 15 % of total VAZ of thylakoids under mid light intensity, while higher values are measured in plants acclimated to EL and low temperature [29].

Demmig-Adams and coworkers early described the correlation between de-epoxidation state of the xanthophyll cycle pigments (VAZ) and NPQ amplitude in a number of plant species [58]. Zeaxanthin has peculiar roles in the enhancement of both the PsbS-dependent energy quenching mechanism qE [80, 134] and the PsbS-independent $^1Chl^*$ quenching state qI [45] (Fig. 15.7d). Present knowledge on the functional role of zeaxanthin in NPQ was obtained by the identification and characterization of the xanthophyll cycle mutants *npq1* and *npq2* in *Arabidopsis* [134] (Fig. 15.6). The physiological characterization of these mutants led to the conclusion that zeaxanthin provides the most important contribution among xanthophylls in the photoprotection of thylakoids, and the amplitude of its effect is enhanced upon its binding to antenna proteins [49, 89].

Recent results by Dall'Osto et al. [50] showed that, in addition to the previously described effects in quenching $^1Chl^*$ and scavenging $^1O_2$, an additional photoprotection mechanism is elicited by zeaxanthin binding to Lhc proteins, namely, the modulation of $^3Chl^*$ (Fig. 15.7d). Indeed, binding of zeaxanthin to monomeric Lhcbs causes a reduced yield of $^1O_2$ production from both purified Lhcbs and PSII supercomplexes containing these subunits. Moreover, analysis by fluorescence-detected magnetic resonance showed decreased amplitude of $^3Chl^*$ in thylakoids as well as in monomeric antennae upon zeaxanthin binding, implying this xanthophyll has a specific effect in decreasing the yield of harmful triplet excited states. Furthermore, changes in the carotenoid composition in thylakoid membranes upon activation of the xanthophyll cycle leads to modification of the fluidity of these membranes. Thus, the physic-chemical features of zeaxanthin affect structural and dynamic properties of the lipid bilayer, contribute to the membrane stabilization, and decrease the lipid membrane susceptibility to oxidative degradation [73, 180].

Artificial over accumulation of zeaxanthin induces monomerization and degradation of the major light-harvesting antenna complex, thus decreasing over-excitation of PSII. The *Arabidopsis* mutant *npq2lut2*, that accumulates zeaxanthin as the only xanthophyll, showed a reduced photochemical efficiency in low light and a photosynthesis rate saturated at higher light intensities with respect to wild-type leaves. Therefore, even when grown in low light conditions, this mutant showed a photosynthetic phenotype resembling that of EL-acclimated plants [76].

In conclusion, zeaxanthin synthesis appears as a general rescuing mechanism under extreme EL stress, working as an "allosteric effector" aimed to amplify the photoprotective mechanisms already present in the photosynthetic apparatus [171]. However, its release needs to be controlled in order to avoid constitutive energy dissipation, and thus lower light harvesting efficiency and growth under limiting light [45].

15.5.5 β-carotene

While carotenoid biosynthesis mutants of *Arabidopsis* have been instrumental in revealing the function of xanthophyll components *in vivo*, the role of β-carotene has been more difficult to dissect due to the highly pleiotropic effects shown by mutations up-stream of lycopene β-cycles reaction [28, 137] leading to synthesis of β-carotene. Yet, β-carotene is a component of both PSI and PSII reaction centre core complexes (see Sect. 15.4), suggesting it has a role in mitigating oxidative damage under EL conditions, especially in PSII, in agreement with the report that β-carotene acts in quenching of $^1O_2$ generated by the triplet state of the primary electron donor in isolated PSII core complexes [184, 186]. No plant mutants lacking carotene molecules in photosystem core complexes have been hitherto reported, supporting their vital role, not only in photosynthesis but also for the survival of the plant cell. Despite the lack of mutants exhibiting a complete β-carotene-less phenotype, cases of alterations have recently been described. *lut5* lacks the cytochrome P450 hydroxylase CYP97A3, primarily responsible for catalyzing hydroxylation of the β-ring of both α-carotene and, to a minor extent, β-carotene [97]. The *lut5* mutation causes an accumulation of α-carotene, normally present in trace amounts, to a level similar to that of β-carotene. α-carotene efficiently replaces β-carotene in photosynthetic reaction centre complexes, however photoprotection capacity of *cyp97a3* (*lut5*) plants was affected and photoinhibition was higher than in wild-type under EL conditions [98]. The suggestion that α-carotene accumulation impairs photoprotection, however, is not consistent with the unchanged resistance to photobleaching as well as light-dependent $^1O_2$ yield of PSII core complexes isolated from *lut5* leaves with respect to wild-type complexes [47]. It should be noted, however, that the *lut5* mutant has a 30% decrease of xanthophyll per chlorophyll content with respect to wild-type plants, a condition that leads to a reduced content in antenna proteins [47] which have a strong photoprotective effect [49]. Although the results from *lut5* mutant are difficult to interpret, the mutation *szl1* recently identified in *Arabidopsis* [109] appears to offer better perspectives. The s*zl1* mutant originates from a point mutation of the lycopene β-cyclase gene, and causes a higher lutein/*β*-xanthophyll ratio in plants. Mutant leaves express a still functional but less active β-cyclase relative to the wild-type, thus yielding the same Chl/Car ratio but lower total carotenes (−30 %) and correspondingly higher total xanthophylls (+6 %) than in wild-type. It leads to a strong depletion of β-carotene content in core complexes of both photosystems, which, however, appears to be normally assembled, although with a reduced carotene complement. Despite both PSI and PSII from *szl1* plants being similarly depleted in carotenes, PSI activity was far more sensitive to photoxidative stress than PSII activity, as shown by the stronger photoinhibition of PSI when plants were exposed to EL at chilling temperature, and by the higher rate of $^1O_2$ release from isolated PSI-LHCI complexes of *szl1* with respect to the wild-type [31]. This implies that β-carotene ligands of PSI have a crucial role in the photoprotection of the complex, especially in low-temperature conditions.

15.6 Structural Properties of Antenna Xanthophylls and the Control of Light Harvesting

15.6.1 Hydrophobicity Parameter

Another property of xanthophylls that was used to explain their conserved nature and variety in higher plants is their hydrophobicity. The hydrophobicity of a xanthophyll molecule is a property that is traditionally exploited in the HPLC separation technique. The more oxygenated and polar xanthophylls like neoxanthin and violaxanthin elute much faster than the less polar lutein and zeaxanthin. In spite of the identical molecular mass, the latter two have slightly different mobility because of configurationally differences in their end-group orientations. Hence, not only the number of polar atoms but also molecular configuration and conformation are important determinants of the molecular interactions with environment. This can be explained by the strong sensitivity of the dipole moment of xanthophyll molecule to the molecular geometry, i.e. whether the molecule adopts a twisted or planar conformation of the cyclic head groups relative to the C=C backbone. Our dipole moment calculations indicate that zeaxanthin dipole moment is ~10 % smaller than that of lutein. The stronger dipole moment of lutein makes it less hydrophobic then zeaxanthin. To further explore and quantify hydrophobic properties, work on dissolving isolated xanthophylls in a range of water/ethanol mixtures at different ratios to modulate the solvent polarity was undertaken [166]. Ethanol–water mixtures proved to be a good system in which to test hydrophobicity of xanthophylls using the solvent ratio at which solute molecules became insoluble and formed dimers and higher order aggregates. Dependency of the ratio between soluble and insoluble (aggregated) molecules of four major LHCII xanthophylls has been obtained and the point of 50 % transition from dissolved to insoluble molecular form estimated. The percentage of ethanol in water/ethanol mixture at this point was taken as an empirical measure of the xanthophyll hydrophobicity and called hydrophobicity parameter or H-parameter, by analogy to the amino acid polarity scale crucial for defining and predicting hydrophobic membrane-spanning a-helixes on membrane proteins. The H-parameter values for neoxanthin, violaxanthin, lutein and zeaxanthin were found to be approximately 32, 43, 50 and 64 % respectively.

15.6.2 H-parameter and Quantum Efficiency of Photosystem II

The work on xanthophyll biosynthesis mutants described in the previous paragraph enabled the testing of key photosystem II and antenna efficiency characteristics against the xanthophylls H-parameter. It has been observed that PSII efficiency in xanthophyll biosynthesis mutants is different from the wild-type plants [76, 154]. The causes of this were not apparent. The quantum efficiency of electron transport in photosystem II is determined as the ratio between number of electrons produced into the electron transfer chain by PSII and number of photons absorbed by the pigments of the light harvesting antenna. This is expressed using the pulse amplitude modulated chlorophyll fluorescence analysis technique (PAM) as (Fm-Fo)/Fm, where Fm is the relative fluorescence yield when all PSII reaction centers are closed and Fo is the yield when they are all open. Fm expressed in terms of fluorescence lifetime is a powerful parameter reflecting the absolute fluorescence yield and hence alone reflects the efficiency of the LHCII antenna: a longer lifetime corresponds to a more efficient antenna and, therefore in essence, PSII. Fluorescence lifetime experiments performed on leaves revealed a clear relationship between the quantum efficiency of PSII and Fm lifetime for the range of xanthophyll compositions, varying from zeaxanthin-only to neoxanthin- and violaxanthin-only LHCII antennae [92]. The yield of PSII in xanthophyll mutants was found to be modulated by the antenna xanthophylls via variation in the chlorophyll fluorescence/excited state lifetime. Remarkably, while some xanthophylls decreased the PSII efficiency, noticeably in mutants containing only zeaxanthin, others increased it, as in the *npq1lut2* mutant possessing NVVV composition—the two least hydrophobic xanthophylls. This fact suggests that the PSII efficiency is likely to be determined by the LHCII antenna, controlled by its xanthophyll composition and is not limited to 80 %. Indeed, it is possible that PSII yield can be increased simply by the extension of the antenna ^{1}Chl* lifetime [92]. The work on isolated LHCII complexes revealed a very clear reciprocal relationship between the chlorophyll fluorescence lifetime and H-index, so that the complexes that contain more polar xanthophylls, neoxanthin and violaxanthin, had the longest lifetime whilst those containing zeaxanthin only possessed the shortest fluorescence lifetime. Extrapolation suggests that PSII can potentially work at 100 % quantum efficiency when the excited state lifetime at Fm reaches about 4 ns, a realistically achievable value. Thus, it seems the quantum efficiency of PSII is sensitive to an average ‘ensemble’ effect of the total number and structure of the xanthophylls bound to LHCII. Variations in xanthophyll hydrophobicity suggest the crucial differences between the protein and chlorophyll domains they are bound to, hence implying their strong heterogeneity.

15.6.3 H-parameter and qE

The relationship between chlorophyll fluorescence lifetime in NPQ/quenched LHCII aggregates and H-parameter was found to be almost identical, implying the common nature of energy dissipation in vivo and in vitro [92]. The shorter fluorescence lifetime corresponds to the stronger NPQ and therefore to the better protection of PSII. Interestingly, the shortest lifetime for both, NPQ and LHCII aggregates, corresponding to the H-parameter 48–54 %, was found in leaves and LHCII aggregates with the wild-type xanthophyll composition, NLLZ, or the *npq2* mutant lacking neoxanthin, _LLZ [90]. The fluorescence lifetime for _ZZZ composition (*npq2lut2* mutant) was found to be slightly longer than that of the wild-type for both NPQ and aggregates of LHCII, suggesting that the zeaxanthin-only composition is not the most effective in NPQ-type protection of PSII. These observations provided evidence that it is the ensemble structural effect of the xanthophylls, quantified by the H-parameter, rather than simply the number of conjugated C=C bonds or the mere presence of zeaxanthin, that is crucial for the maximum level of photoprotection. Previously, it was proposed that xanthophylls could directly quench ^{1}Chl* by energy transfer between their low-lying S_1 excited states, followed by rapid internal conversion. However, since the energies of the lowest lying S_1 excited state of all of the xanthophylls bound to LHCII (even neoxanthin and violaxanthin) lie below that of chlorophyll *a* [156], it seems that if xanthophylls do indeed act as quenchers that this property must depend largely upon other factors. One possibility is that for LHCII proteins to adopt the maximum dissipative state it is necessary for structural reasons to have lutein not zeaxanthin bound to the intrinsic L1 and L2 binding sites, at least in the trimeric LHCII. One possible explanation is that a precise orientation of pigments is required in the chlorophyll terminal emitter domain where most excitation energy is concentrated in order to allow either energy transfer or excitonic mixing between the dipole forbidden xanthophyll S_1 state and the S_1 state of chlorophyll. The configuration and interactions of zeaxanthin, and indeed violaxanthin, with chlorophylls in the L1 and L2 sites could be less optimal than those of lutein in creating such a quenching complex. Thus, while the more hydrophobic H-parameters generally favor a shorter ^{1}Chl* lifetime by favoring condensation of LHCII, there is a clear structural optimum for this effect. It is interesting to note that nature has exploited a xanthophyll composition that offers the optimal flexibility, in terms of the difference in excited state lifetime, between the non-condensed light harvesting (Fm) and condensed aggregated (NPQ) states of LHCII.

ΔpH versus qE titrations revealed an additive nature of the effect of zeaxanthin and PsbS on the p*Ka* of protonatable residues associated with qE. Furthermore, the xanthophyll composition of LHC was shown to affect the p*Ka* of lumenal amino acids, likely as consequence of LHCs structure modulation [164]. These results imply that, even in absence of PsbS and irrespective of xanthophyll composition, LHC can become a good quencher by giving a sufficiently high level of ΔpH. Thus, the structural differences between xanthophyll species and their effects on LHC conformation determine how closely they can interact with Chls in order to form efficient quenchers; alternatively, both lutein and zeaxanthin could not be directly involved in the quenching mechanism [129], rather their effects on qE could result by effect (allosteric) on the tertiary structure of the LHC proteins [45].

15.6.4 "Molecular Memory" of Light Exposure

In addition to the investigation of NPQ strength, it was insightful to study its kinetic properties, the rates of formation and relaxation. These kinetic parameters are important in order to characterize the transition to and from the NPQ state as an enzymatic reaction [169]. Previously, data was obtained in isolated wild-type LHCII complexes in which the V1 site contained either violaxanthin or zeaxanthin, showing that the latter increased the tendency and rate of formation of the condensed aggregated state while the former resisted it [163]. NPQ formation and relaxation was found to be strongly dependent upon the xanthophyll hydrophobicity parameter of various mutants *in vivo*. The formation and relaxation rate plots were found to be almost symmetric showing that the increase in the rate of NPQ formation is followed by the decrease in the relaxation rate [164]. This is a typical behavior of transitions with memory, like the ones occurring in magnetic or elastic materials. More hydrophobic xanthophylls cause faster development of NPQ but drastically decrease its relaxation.

Crucially, not only zeaxanthin bound to the external site (V1), but intrinsically bound xanthophylls seem to control the NPQ rate. Thus, while the *npq2* (_LLZ) and *npq2lut2* (_ZZZ) mutants with constitutively high levels of zeaxanthin are able to form NPQ more rapidly than the zeaxanthin-enriched wild-type plants (NLLZ), which in turn form NPQ more rapidly than the *lut2* mutant (NVVZ). These data support the proposal by Bassi and co-workers on the activity of the L2 site in the minor antenna for the allosteric control of its transition into a dissipative state and its role in NPQ [15, 125]. Thus, the xanthophyll H-parameter controls not only the absolute ^{1}Chl* lifetime of LHCII but also the kinetics of the transition from the light harvesting to NPQ state. Therefore xanthophyll H-parameter is an important factor enabling the memory of the NPQ state. In the wild-type a natural flexibility in H-parameter is governed by light in the form of a gradual replacement of violaxanthin by zeaxanthin via the action of the xanthophyll cycle during illumination of plants, as was mentioned above. This is a way to provide a light exposure memory or counter for plants living in the frequently changing environments. Increased light exposure will cause an increase in zeaxanthin concentration and make NPQ more sensitive/responsive to illumination while concomitantly slowing its relaxation. This light conditioning of photoprotection is a remarkable achievement of the molecular evolution of LHCII antenna components, in particular, xanthophylls. Finally, genetic manipulation of xanthophyll biosynthesis created an array of plants with various light adaptation strategies, the understanding of which could be utilised for the future crop development and creating plants that are more resistant to the ever increasingly changing environment on our planet.

References

1. Adams Iii WW, Demmig-Adams B, Rosenstiel TN, Brightwell AK, Ebbert V (2002) Photosynthesis and photoprotection in overwintering plants. Plant Biol 4:545–557
2. Ahn TK, Avenson TJ, Ballottari M, Cheng YC, Niyogi KK, Bassi R, Fleming GR (2008) Architecture of a charge-transfer state regulating light harvesting in a plant antenna protein. Science 320:794–797
3. Alboresi A, Dall'Osto L, Aprile A, Carillo P, Roncaglia E, Cattivelli L, Bassi R (2011) Reactive oxygen species and transcript analysis upon excess light treatment in wild-type *Arabidopsis thaliana* vs a photosensitive mutant lacking zeaxanthin and lutein. BMC Plant Biol 11:62
4. Alter P, Dreissen A, Luo FL, Matsubara S (2012) Acclimatory responses of *Arabidopsis* to fluctuating light environment: comparison of different sunfleck regimes and accessions. Photosynth Res 113:221–237
5. Anderson MD, Prasad TK, Stewart CR (1995) Changes in isozyme profiles of catalase, peroxidase, and glutathione reductase during acclimation to chilling in mesocotyls of maize seedlings. Plant Physiol 109:1247–1257
6. Andersson J, Wentworth M, Walters RG, Howard CA, Ruban AV, Horton P, Jansson S (2003) Absence of the Lhcb1 and Lhcb2 proteins of the light-harvesting complex of photosystem II—effects on photosynthesis, grana stacking and fitness. Plant J 35:350–361
7. Arnoux P, Morosinotto T, Saga G, Bassi R, Pignol D (2009) A structural basis for the pH-dependent xanthophyll cycle in *Arabidopsis thaliana*. Plant Cell 21:2036–2044
8. Aro EM, McCaffery S, Anderson JM (1994) Recovery from photoinhibition in peas (*Pisum sativum* L.) acclimated to varying growth irradiances (role of D1 protein turnover). Plant Physiol 104:1033–1041
9. Asada K (1999) The water-water cycle in chloroplasts: scavenging of active oxygens and dissipation of excess photons. Annu Rev Plant Physiol Plant Mol Biol 50:601–639
10. Avenson TJ, Ahn TK, Zigmantas D, Niyogi KK, Li Z, Ballottari M, Bassi R, Fleming GR (2008) Zeaxanthin radical cation formation in minor light-harvesting complexes of higher plant antenna. J Biol Chem 283:3550–3558
11. Bailey S, Walters RG, Jansson S, Horton P (2001) Acclimation of *Arabidopsis thaliana* to the light environment: the existence of separate low light and high light responses. Planta 213:794–801
12. Bailey S, Horton P, Walters RG (2004) Acclimation of *Arabidopsis thaliana* to the light environment: the relationship between photosynthetic function and chloroplast composition. Planta 218:793–802
13. Ballottari M, Dall'Osto L, Morosinotto T, Bassi R (2007) Contrasting behavior of higher plant photosystem I and II antenna systems during acclimation. J Biol Chem 282:8947–8958
14. Baroli I, Gutman BL, Ledford HK, Shin JW, Chin BL, Havaux M, Niyogi KK (2004) Photooxidative stress in a xanthophyll-deficient mutant of *Chlamydomonas*. J Biol Chem 279: 6337–6344
15. Bassi R, Caffarri S (2000) Lhc proteins and the regulation of photosynthetic light harvesting function by xanthophylls. Photosynth Res 64:243–256
16. Bassi R, Croce R, Cugini D, Sandona D (1999) Mutational analysis of a higher plant antenna protein provides identification of chromophores bound into multiple sites. Proc Natl Acad Sci U S A 96:10056–10061
17. Bellafiore S, Bameche F, Peltier G, Rochaix JD (2005) State transitions and light adaptation require chloroplast thylakoid protein kinase STN7. Nature 433:892–895
18. Ben Shem A, Frolow F, Nelson N (2003) Crystal structure of plant photosystem I. Nature 426:630–635
19. Berera R, van Grondelle R, Kennis JT (2009) Ultrafast transient absorption spectroscopy: principles and application to photosynthetic systems. Photosynth Res 101:105–118
20. Bergantino E, Segalla A, Brunetta A, Teardo E, Rigoni F, Giacometti GM, Szabo I (2003) Light- and pH-dependent structural changes in the PsbS subunit of photosystem II. Proc Natl Acad Sci U S A 100:15265–15270

21. Betterle N, Ballottari M, Zorzan S, de Bianchi S, Cazzaniga S, Dall'Osto L, Morosinotto T, Bassi R (2009) Light-induced dissociation of an antenna hetero-oligomer is needed for non-photochemical quenching induction J Biol Chem 284:15255–15266
22. Betterle N, Ballottari M, Hienerwadel R, Dall'Osto L, Bassi R (2010) Dynamics of zeaxanthin binding to the Photosystem II monomeric antenna protein Lhcb6 (CP24) and modulation of its photoprotection properties. Arch Biochem Biophys 504:67–77
23. Bilger W, Bjoerkman O (1990) Role of the xanthophyll cycle in photoprotection elucidated by measurements of light-induced absorbance changes, fluorescence and photosynthesis in leaves of *Hedera canariensis*. Photosynth Res 25:173–185
24. Bode S, Quentmeier CC, Liao PN, Hafi N, Barros T, Wilk L, Bittner F, Walla PJ (2009) On the regulation of photosynthesis by excitonic interactions between carotenoids and chlorophylls. Proc Natl Acad Sci U S A 106:12311–12316
25. Bonente G, Howes BD, Caffarri S, Smulevich G, Bassi R (2007) Interactions between the photosystem II subunit psbs and xanthophylls studied in vivo and in vitro. J Biol Chem 283:8434–8445
26. Briantais JM, Vernotte C, Picaud M, Krause GH (1979) A quantitative study of the slow decline of chlorophyll a fluorescence in isolated chloroplasts. Biochim Biophys Acta 548:128–138
27. Bungard RA, Ruban AV, Hibberd JM, Press MC, Horton P, Scholes JD (1999) Unusual carotenoid composition and a new type of xanthophyll cycle in plants. Proc Natl Acad Sci U S A 96:1135–1139
28. Busch M, Seuter A, Hain R (2002) Functional analysis of the early steps of carotenoid biosynthesis in tobacco. Plant Physiol 128:439–453
29. Caffarri S, Croce R, Breton J, Bassi R (2001) The major antenna complex of photosystem II has a xanthophyll binding site not involved in light harvesting. J Biol Chem 276:35924–35933
30. Caffarri S, Passarini F, Bassi R, Croce R (2007) A specific binding site for neoxanthin in the monomeric antenna proteins CP26 and CP29 of photosystem II. FEBS Lett 581:4704–4710
31. Cazzaniga S, Li Z, Niyogi KK, Bassi R, Dall'Osto L (2012) The *Arabidopsis* szl1 mutant reveals a critical role of?-carotene in photosystem I photoprotection. Plant Physiol 159:1745–1758
32. Chow WS, Melis A, Anderson JM (1990) Adjustments of photosystem stoichiometry in chloroplasts improve the quantum efficiency of photosynthesis. Proc Natl Acad Sci U S A 87:7502–7506
33. Cleland RE, Grace SC (1999) Voltammetric detection of superoxide production by photosystem II. FEBS Lett 457:348–352
34. Connelly JP, Muller MG, Bassi R, Croce R, Holzwarth AR (1997) Femtosecond transient absorption study of carotenoid to chlorophyll energy transfer in the light-harvesting complex II of photosystem II. Biochemistry 36:281–287
35. Croce R, Remelli R, Varotto C, Breton J, Bassi R (1999) The neoxanthin binding site of the major light harvesting complex (LHC II) from higher plants. FEBS Lett 456:1–6
36. Croce R, Weiss S, Bassi R (1999) Carotenoid-binding sites of the major light-harvesting complex II of higher plants. J Biol Chem 274:29613–29623
37. Croce R, Cinque G, Holzwarth AR, Bassi R (2000) The soret absorption properties of carotenoids and chlorophylls in antenna complexes of higher plants. Photosynth Res 64:221–231
38. Croce R, Muller MG, Bassi R, Holzwarth AR (2001) Carotenoid-to-chlorophyll energy transfer in recombinant major light- harvesting complex (LHCII) of higher plants. I. femtosecond transient absorption measurements. Biophys J 80:901–915
39. Croce R, Canino g, Ros F, Bassi R (2002) Chromophore organization in the higher-plant photosystem II antenna protein CP26. Biochemistry 41:7334–7343
40. Croce R, Morosinotto T, Castelletti S, Breton J, Bassi R (2002) The Lhca antenna complexes of higher plants photosystem I. Biochim Biophys Acta (Bioenergetics) 1556:29–40
41. Croce R, Muller MG, Caffarri S, Bassi R, Holzwarth AR (2003) Energy transfer pathways in the minor antenna complex CP29 of photosystem II: a femtosecond study of carotenoid to chlorophyll transfer on mutant and WT complexes. Biophys J 84:2517–2532

42. Dainese P, Bassi R (1991) Subunit stoichiometry of the chloroplast photosystem- II antenna system and aggregation state of the component chlorophyll-a/b binding proteins. J Biol Chem 266:8136–8142
43. Dainese P, Hoyer-Hansen G, Bassi R (1990) The resolution of chlorophyll a/b binding proteins by a preparative method based on flat bed isoelectric focusing. Photochem Photobiol 51:693–703
44. Dalcorso G, Pesaresi P, Masiero S, Aseeva E, Schunemann D, Finazzi G, Joliot P, Barbato R, Leister D (2008) A complex containing PGRL1 and PGR5 is involved in the switch between linear and cyclic electron flow in *Arabidopsis*. Cell 132:273–285
45. Dall'Osto L, Caffarri S, Bassi R (2005) A mechanism of nonphotochemical energy dissipation, independent from PsbS, revealed by a conformational change in the antenna protein CP26. Plant Cell 17:1217–1232
46. Dall'Osto L, Lico C, Alric J, Giuliano G, Havaux M, Bassi R (2006) Lutein is needed for efficient chlorophyll triplet quenching in the major LHCII antenna complex of higher plants and effective photoprotection in vivo under strong light. BMC Plant Biol 6:32
47. Dall'Osto L, Fiore A, Cazzaniga S, Giuliano G, Bassi R (2007) Different roles of a- and b-branch xanthophylls in photosystem assembly and photoprotection. J Biol Chem 282:35056–35068
48. Dall'Osto L, Cazzaniga S, North H, Marion-Poll A, Bassi R (2007) The *Arabidopsis* aba4-1 mutant reveals a specific function for neoxanthin in protection against photoxidative stress. Plant Cell 19:1048–1064
49. Dall'Osto L, Cazzaniga S, Havaux M, Bassi R (2010) Enhanced photoprotection by protein-bound vs free xanthophyll pools: a comparative analysis of chlorophyll b and xanthophyll biosynthesis mutants. Mol Plant 3:576–593
50. Dall'Osto L, Holt NE, Kaligotla S, Fuciman M, Cazzaniga S, Carbonera D, Frank HA, Alric J, Bassi R (2012) Zeaxanthin protects plant photosynthesis by modulating chlorophyll triplet yield in specific light-harvesting antenna subunits. J Biol Chem 287:41820–41834
51. Damkjaer J, Kereiche S, Johnson MP, Kovacs L, kiss az, Boekema EJ, Ruban AV, Horton P, Jansson S (2009) The photosystem II light-harvesting protein Lhcb3 affects the macrostructure of photosystem II and the rate of state transitions in *Arabidopsis*. Plant Cell 21:3245–3256
52. Davison PA, Hunter CN, Horton P (2002) Overexpression of beta-carotene hydroxylase enhances stress tolerance in *Arabidopsis*. Nature 418:203–206
53. de Bianchi S, Dall'Osto L, Tognon G, Morosinotto T, Bassi R (2008) Minor antenna proteins CP24 and CP26 affect the interactions between photosystem II subunits and the electron transport rate in grana membranes of *Arabidopsis*. Plant Cell 20:1012–1028
54. de Bianchi S, Betterle N, Kouril R, Cazzaniga S, Boekema E, Bassi R, Dall'Osto L (2011) *Arabidopsis* mutants deleted in the light-harvesting protein Lhcb4 have a disrupted photosystem II macrostructure and are defective in photoprotection. Plant Cell 23:2659–2679
55. de Carbonnel M et al (2010) The *Arabidopsis* PHYTOCHROME KINASE SUBSTRATE2 protein is a phototropin signaling element that regulates leaf flattening and leaf positioning. Plant Physiol 152:1391–1405
56. de Weerd FL, van Stokkum IHM, Van Amerongen H, Dekker JP, van Grondelle R (2002) Pathways for energy transfer in the core light-harvesting complexes CP43 and CP47 of photosystem II. Biophys J 82:1586–1597
57. Demmig-Adams B (1990) Carotenoids and photoprotection in plants: a role for the xanthophyll zeaxanthin. Biochim Biophys Acta 1020:1–24
58. Demmig-Adams B, Adams WW (1992) Photoprotection and other responses of plants to high light stress. Ann Rev Plant Physiol Plant Mol Biol 43:599–626
59. Dominici P, Caffarri S, Armenante F, Ceoldo S, Crimi M, Bassi R (2002) Biochemical properties of the PsbS subunit of photosystem II either purified from chloroplast or recombinant. J Biol Chem 277:22750–22758
60. Duffy CD, Ruban AV (2012) A theoretical investigation of xanthophyll-protein hydrogen bonding in the photosystem II antenna. J Phys Chem B 116:4310–4318

61. Färber A, Young AJ, Ruban AV, Horton P, Jahns P (1997) Dynamics of xanthophyll-cycle activity in different antenna subcomplexes in the photosynthetic membranes of higher plants. Plant Physiol 115:1609–1618
62. Fiore A, Dall'Osto L, Fraser PD, Bassi R, Giuliano G (2006) Elucidation of the beta-carotene hydroxylation pathway in *Arabidopsis thaliana*. FEBS Lett 580:4718–4722
63. Formaggio E, Cinque G, Bassi R (2001) Functional architecture of the major light-harvesting complex from higher plants. J Mol Biol 314:1157–1166
64. Frigerio S et al (2007) Photosynthetic antenna size in higher plants is controlled by the plastoquinone redox state at the post-transcriptional rather than transcriptional level. J Biol Chem 282:29457–29469
65. Gastaldelli M, Canino g, Croce R, Bassi R (2003) Xanthophyll binding sites of the CP29 (Lhcb4) subunit of higher plant photosystem II investigated by domain swapping and mutation analysis. J Biol Chem 278:19190–19198
66. Gibasiewicz K, Croce R, Morosinotto T, Ihalainen JA, van Stokkum IH, Dekker JP, Bassi R, van Grondelle R (2005) Excitation energy transfer pathways in Lhca4. Biophys J 88:1959–1969
67. Gilmore AM (1997) Mechanistic aspects of xanthophyll cycle-dependent photoprotection in higher plant chloroplasts and leaves. Physiol Plant 99:197–209
68. Gilmore AM, Ball MC (2000) Protection and storage of chlorophyll in overwintering evergreens. Proc Natl Acad Sci U S A 97:11098–11101
69. Gilmore AM, Hazlett TL, Govindjee (1995) Xanthophyll cycle-dependent quenching of photosystem II chlorophyll a fluorescence: formation of a quenching complex with a short fluorescence lifetime. Proc Natl Acad Sci U S A 92:2273–2277
70. Gobets B et al (2001) Excitation energy transfer in dimeric light harvesting complex I: a combined streak-camera/fluorescence upconversion study. J Phys Chem B 105:10132–10139
71. Goral TK, Johnson MP, Duffy CD, Brain AP, Ruban AV, Mullineaux CW (2012) Light-harvesting antenna composition controls the macrostructure and dynamics of thylakoid membranes in *Arabidopsis*. Plant J 69:289–301
72. Gradinaru CC, van Stokkum IHM, Pascal AA, van Grondelle R, Van Amerongen H (2000) Identifying the pathways of energy transfer between carotenoids and chlorophylls in LHCII and CP29. A multicolor, femtosecond pump—probe study. J Phys Chem B 104:9330–9342
73. Gruszecki WI, Strzalka K (2005) Carotenoids as modulators of lipid membrane physical properties. Biochim Biophys Acta 1740:108–115
74. Havaux M, Kloppstech K (2001) The protective functions of carotenoid and flavonoid pigments against excess visible radiations at chilling temperature investigated in *Arabidopsis* npq and tt mutants. Planta 213:953–966
75. Havaux M, Niyogi KK (1999) The violaxanthin cycle protects plants from photooxidative damage by more than one mechanism. Proc Natl Acad Sci U S A 96:8762–8767
76. Havaux M, Dall'Osto L, Cuine S, Giuliano G, Bassi R (2004) The effect of zeaxanthin as the only xanthophyll on the structure and function of the photosynthetic apparatus in *Arabidopsis thaliana*. J Biol Chem 279:13878–13888
77. Havaux M, Eymery F, Porfirova S, Rey P, Dormann P (2005) Vitamin E protects against photoinhibition and photooxidative stress in *Arabidopsis thaliana* Plant Cell 17:3451–3469
78. Havaux M, Dall'Osto L, Bassi R (2007) Zeaxanthin has enhanced antioxidant capacity with respect to all other xanthophylls in *Arabidopsis* leaves and functions independent of binding to PSII antennae. Plant Physiol 145:1506–1520
79. Holt NE, Kennis JTM, Dall'Osto L, Bassi R, Fleming GR (2003) Carotenoid to chlorophyll energy transfer in light harvesting complex II from *Arabidopsis thaliana* probed by femtosecond fluorescence upconversion. Chem Phys Lett 379:305–313
80. Holt NE, Fleming GR, Niyogi KK (2004) Toward an understanding of the mechanism of nonphotochemical quenching in green plants. Biochemistry 43:8281–8289
81. Holt NE, Zigmantas D, Valkunas L, Li XP, Niyogi KK, Fleming GR (2005) Carotenoid cation formation and the regulation of photosynthetic light harvesting. Science 307:433–436
82. Holzwarth AR (1986) Fluorescence lifetimes in photosynthetic systems. Photochem Photobiol 43:707–725

83. Horton P, Ruban AV, Rees D, Pascal AA, Noctor G, Young AJ (1991) Control of the light-harvesting function of chloroplast membranes by aggregation of the LHCII chlorophyll-protein complex. FEBS Lett 292:1–4
84. Horton P, Ruban AV, Walters RG (1994) Regulation of light harvesting in green plants. Indication by nonphotochemical quenching of chlorophyll fluorescence. Plant Physiol 106:415–420
85. Horton P, Ruban AV, Walters RG (1996) Regulation of light harvesting in green plants. Annu Rev Plant Physiol Plant Mol Biol 47:655–684
86. Horton P, Ruban A, Wentworth M (2000) Allosteric regulation of the light-harvesting system of photosystem II. Philos Trans R Soc Lond B Biol Sci 355:1361–1370
87. Javorfi T, Garab G, Naqvi KR (2000) Reinvestigation of the triplet-minus-singlet spectrum of chloroplasts. Spectrochim Acta A Mol Biomol Spectrosc 56A:211–214
88. Johnson MP, Ruban AV (2011) Restoration of rapidly-reversible photoprotective energy dissipation in the absence of PsbS protein by enhanced DeltapH. J Biol Chem 286:19973–19981
89. Johnson MP, Havaux M, Triantaphylides C, Ksas B, Pascal AA, Robert B, Davison PA, Ruban AV, Horton P (2007) Elevated zeaxanthin bound to oligomeric LHCII enhances the resistance of *Arabidopsis* to photooxidative stress by a lipid-protective, antioxidant mechanism. J Biol Chem 282:22605–22618
90. Johnson MP, Zia A, Horton P, Ruban AV (2010) Effect of xanthophyll composition on the chlorophyll excited state lifetime in plant leaves and isolated LHCII. J Chem Phys 373:23–32
91. Johnson MP, Goral TK, Duffy CD, Brain AP, Mullineaux CW, Ruban AV (2011) Photoprotective energy dissipation involves the reorganization of photosystem II light-harvesting complexes in the grana membranes of spinach chloroplasts. Plant Cell 23:1468–1479
92. Johnson MP, Zia A, Ruban AV (2012) Elevated DpH restores rapidly reversible photoprotective energy dissipation in *Arabidopsis* chloroplasts deficient in lutein and xanthophyll cycle activity. Planta 235:193–204
93. Joliot PA, Finazzi G (2010) Proton equilibration in the chloroplast modulates multiphasic kinetics of nonphotochemical quenching of fluorescence in plants. Proc Natl Acad Sci U S A 107:12728–12733
94. Kagawa T, Sakai T, Suetsugu N, Oikawa K, Ishiguro S, Kato T, Tabata S, Okada K, Wada M (2001) *Arabidopsis* NPL1: a phototropin homolog controlling the chloroplast high-light avoidance response. Science 291:2138–2141
95. Kalituho L, Rech J, Jahns P (2007) The roles of specific xanthophylls in light utilization. Planta 225:423–439
96. Kennis JTM, Gobets B, van Stokkum IHM, Dekker JP, van Grondelle R, Fleming GR (2001) Light harvesting by chlorophylls and carotenoids in the photosystem I core complex of *Synechococcus elongatus*: a fluorescence upconversion study. J Phys Chem B 105:4485–4494
97. Kim J, DellaPenna D (2006) Defining the primary route for lutein synthesis in plants: the role of *Arabidopsis* carotenoid beta-ring hydroxylase CYP97A3. Proc Natl Acad Sci U S A 103:3474–3479
98. Kim J, Smith JJ, Tian L, DellaPenna D (2009) The evolution and function of carotenoid hydroxylases in *Arabidopsis*. Plant Cell Physiol 50:463–479
99. Kimura M et al (2003) Identification of *Arabidopsis* genes regulated by high light-stress using cDNA microarray. Photochem Photobiol 77:226–233
100. Kliebenstein DJ, Monde RA, Last RL (1998) Superoxide dismutase in *Arabidopsis*: an eclectic enzyme family with disparate regulation and protein localization. Plant Physiol 118:637–650
101. Kouril R, Dekker JP, Boekema EJ (2011) Supramolecular organization of photosystem II in green plants. Biochim Biophys Acta doi:10.1016/j.bbabio.2011.05.024
102. Kovacs L, Damkjaer J, Kereiche S, Ilioaia C, Ruban AV, Boekema EJ, Jansson S, Horton P (2006) Lack of the light-harvesting complex CP24 affects the structure and function of the grana membranes of higher plant chloroplasts. Plant Cell 18:3106–3120

103. Krasnovsky AA (1982) Delayed fluorescence and phosphorescence of plant pigments. Photochem Photobiol 36:733–741
104. Krause GH (1988) Photoinhibiton of photosynthesis. An evaluation of damaging and protective mechanisms. Physiol Plant 74:566–574
105. Kulheim C, Agren J, Jansson S (2002) Rapid regulation of light harvesting and plant fitness in the field. Science 297:91–93
106. Kuntz M (2004) Plastid terminal oxidase and its biological significance. Planta 218:896–899
107. Lam E, Ortiz W, Malkin R (1984) Chlorophyll a/b proteins of photosystem I. FEBS Lett 168:10–14
108. Lavaud J, Rousseau B, van Gorkom HJ, Etienne AL (2002) Influence of the diadinoxanthin pool size on photoprotection in the marine planktonic diatom *Phaeodactylum tricornutum*. Plant Physiol 129:1398–1406
109. Li Z et al (2009) Lutein accumulation in the absence of zeaxanthin restores nonphotochemical quenching in the *Arabidopsis thaliana* npq1 mutant. Plant Cell 21:1798–1812
110. Li XP, Bjorkman O, Shih C, Grossman AR, Rosenquist M, Jansson S, Niyogi KK (2000) A pigment-binding protein essential for regulation of photosynthetic light harvesting. Nature 403:391–395
111. Li XP, Gilmore AM, Niyogi KK (2002) Molecular and global time-resolved analysis of a *psbS* gene dosage effect on pH- and xanthophyll cycle-dependent nonphotochemical quenching in photosystem II. J Biol Chem 277:33590–33597
112. Li XP, Gilmore AM, Caffarri S, Bassi R, Golan T, Kramer D, Niyogi KK (2004) Regulation of photosynthetic light harvesting involves intrathylakoid lumen pH sensing by the PsbS protein. J Biol Chem 279:22866–22874
113. Liu Z, Yan H, Wang K, Kuang T, Zhang J, Gui L, An X, Chang W (2004) Crystal structure of spinach major light-harvesting complex at 2.72 A resolution. Nature 428:287–292
114. Lokstein H, Tian L, Polle JE, DellaPenna D (2002) Xanthophyll biosynthetic mutants of *Arabidopsis thaliana*: altered nonphotochemical quenching of chlorophyll fluorescence is due to changes in photosystem II antenna size and stability. Biochim Biophys Acta 1553:309–319
115. MacPherson AN, Gillbro T (1998) Solvent dependence of the ultrafast S-2-S-1 internal conversion rate of beta-carotene. J Phys Chem A 102:5049–5058
116. Marin A, Passarini F, van Stokkum IH, van Grondelle R, Croce R (2011) Minor complexes at work: light-harvesting by carotenoids in the photosystem II antenna complexes CP24 and CP26. Biophys J 100:2829–2838
117. Mathis P, Butler WL, Satoh K (1979) Carotenoid triplet state and chlorophyll fluorescence quenching in chloroplasts and sub chloroplasts particles. Photochem Photobiol 30:603–614
118. Melis A (1991) Dynamics of photosynthetic membrane composition and function. Biochim Biophys Acta 1058:87–106
119. Milborrow BV (2001) The pathway of biosynthesis of abscisic acid in vascular plants: a review of the present state of knowledge of ABA biosynthesis. J Exp Bot 52:1145–1164
120. Miloslavina Y, Wehner A, Lambrev PH, Wientjes E, Reus M, Garab G, Croce R, Holzwarth AR (2008) Far-red fluorescence: a direct spectroscopic marker for LHCII oligomer formation in non-photochemical quenching. FEBS Lett 582:3625–3631
121. Miloslavina Y, de Bianchi S, Dall'Osto L, Bassi R, Holzwarth AR (2011) Quenching in *Arabidopsis thaliana* mutants lacking monomeric antenna proteins of photosystem II. J Biol Chem 286:36830–36840
122. Morosinotto T, Baronio R, Bassi R (2002) Dynamics of chromophore binding to Lhc proteins in vivo and in vitro during operation of the xanthophyll cycle. J Biol Chem 277:36913–36920
123. Morosinotto T, Caffarri S, Dall'Osto L, Bassi R (2003) Mechanistic aspects of the xanthophyll dynamics in higher plant thylakoids. Physiologia Plantarum 119:347–354
124. Morosinotto T, Mozzo M, Bassi R, Croce R (2005) Pigment-pigment interactions in the higher plants photosystem I antenna complex Lhca4. A mutagenesis study (submitted)

125. Moya I, Silvestri M, Vallon O, Cinque G, Bassi R (2001) Time-resolved fluorescence analysis of the photosystem II antenna proteins in detergent micelles and liposomes. Biochemistry 40:12552–12561
126. Mozzo M, Dall'Osto L, Hienerwadel R, Bassi R, Croce R (2008) Photoprotection in the antenna complexes of photosystem II: role of individual xanthophylls in chlorophyll triplet quenching. J Biol Chem 283:6184–6192
127. Muller-Moule P, Conklin PL, Niyogi KK (2002) Ascorbate deficiency can limit violaxanthin de-epoxidase activity in vivo. Plant Physiol 128:970–7
128. Muller-Moule P, Havaux M, Niyogi KK (2003) Zeaxanthin deficiency enhances the high light sensitivity of an ascorbate-deficient mutant of *Arabidopsis*. Plant Physiol 133:748–760
129. Muller MG, Lambrev P, Reus M, Wientjes E, Croce R, Holzwarth AR (2010) Singlet energy dissipation in the photosystem II light-harvesting complex does not involve energy transfer to carotenoids. Chemphyschem 11:1289–1296
130. Munekage Y, Hojo M, Meurer J, Endo T, Tasaka M, Shikanai T (2002) PGR5 is involved in cyclic electron flow around photosystem I and is essential for photoprotection in *Arabidopsis*. Cell 110:361–371
131. Murchie EH, Hubbart S, Peng S, Horton P (2005) Acclimation of photosynthesis to high irradiance in rice: gene expression and interactions with leaf development. J Exp Bot 56:449–460
132. Nelson N, Ben Shem A (2004) The complex architecture of oxygenic photosynthesis. Nature 5:1–12
133. Nilkens M, Kress E, Lambrev P, Miloslavina Y, Muller M, Holzwarth AR, Jahns P (2010) Identification of a slowly inducible zeaxanthin-dependent component of non-photochemical quenching of chlorophyll fluorescence generated under steady-state conditions in *Arabidopsis*. Biochim Biophys Acta 1797:466–475
134. Niyogi KK, Grossman AR, Bjorkman O (1998) *Arabidopsis* mutants define a central role for the xanthophyll cycle in the regulation of photosynthetic energy conversion. Plant Cell 10:1121–1134
135. Niyogi KK, Shih C, Soon Chow W, Pogson BJ, Dellapenna D, Bjorkman O (2001) Photoprotection in a zeaxanthin- and lutein-deficient double mutant of *Arabidopsis*. Photosynth Res 67:139–145
136. North HM, De Almeida A, Boutin JP, Frey A, To A, Botran L, Sotta B, Marion-Poll A (2007) The *Arabidopsis* ABA-deficient mutant aba4 demonstrate that the major route for stress-induced ABA accumulation is via neoxanthin isomers. Plant J 50:810–824
137. Okada K, Kawaide H, Kuzuyama T, Seto H, Curtis IS, Kamiya Y (2002) Antisense and chemical suppression of the nonmevalonate pathway affects ent-kaurene biosynthesis in *Arabidopsis*. Planta 215:339–344
138. op den Camp RG et al (2003) Rapid induction of distinct stress responses after the release of singlet oxygen in *Arabidopsis*. Plant Cell 15:2320–2332
139. Öquist G, Chow WS, Anderson JM (1992) Photoinhibition of photosynthesis represents a mechanism for the long-term regulation of photosystem-II. Planta 186:450–460
140. Pagano A, Cinque G, Bassi R (1998) In vitro reconstitution of the recombinant photosystem II light-harvesting complex CP24 and its spectroscopic characterization. J Biol Chem 273:17154–17165
141. Pan X, Li M, Wan T, Wang L, Jia C, Hou Z, Zhao X, Zhang J, Chang W (2011) Structural insights into energy regulation of light-harvesting complex CP29 from spinach. Nat Struct Mol Biol 18:309–315
142. Pascal AA et al (2005) Molecular basis of photoprotection and control of photosynthetic light-harvesting. Nature 436:134–137
143. Passarini F, Wientjes E, Hienerwadel R, Croce R (2009) Molecular basis of light harvesting and photoprotection in CP24: unique features of the most recent antenna complex. J Biol Chem 284:29536–29546
144. Paulsen H (1995) Chlorophyll a/b-binding proteins. Photochem Photobiol 62:367–382

145. Paulsen H, Hobe S (1992) Pigment-binding properties of mutant light-harvesting chlorophyll-a/b-binding protein. Eur J Biochem 205:71–76
146. Perez-Bueno ML, Johnson MP, Zia A, Ruban AV, Horton P (2008) The Lhcb protein and xanthophyll composition of the light harvesting antenna controls the DeltapH-dependency of non-photochemical quenching in *Arabidopsis thaliana*. FEBS Lett 582:1477–1482
147. Pesaresi P, Sandona D, Giuffra E, Bassi R (1997) A single point mutation (E166Q) prevents dicyclohexylcarbodiimide binding to the photosystem II subunit CP29. FEBS Lett 402:151–156
148. Peterman EJ, Dukker FM, van Grondelle R, Van Amerongen H (1995) Chlorophyll a and carotenoid triplet states in light-harvesting complex II of higher plants. Biophys J 69:2670–2678
149. Peterman EJ, Gradinaru CC, Calkoen F, Borst JC, van Grondelle R, van Amerongen H (1997) Xanthophylls in light-harvesting complex II of higher plants: light harvesting and triplet quenching. Biochemistry 36:12208–12215
150. Pfannschmidt T, Nilsson A, Allen JF (1999) Photosynthetic control of chloroplast gene expression. Nature 397:625–628
151. Pierre Y, Breyton C, Lemoine Y, Robert B, Vernotte C, Popot JL (1997) On the presence and role of a molecule of chlorophyll a in cytochrome b6 f. complex. J Biol Chem 272:21901–21908
152. Pogson B, Rissler HM (2000) Genetic manipulation of carotenoid biosynthesis and photoprotection. Phil Trans R Soc Lond B 355:1395–1403
153. Pogson B, McDonald KA, Truong M, Britton G, DellaPenna D (1996) *Arabidopsis* carotenoid mutants demonstrate that lutein is not essential for photosynthesis in higher plants. Plant Cell 8:1627–1639
154. Pogson BJ, Niyogi KK, Bjorkman O, DellaPenna D (1998) Altered xanthophyll compositions adversely affect chlorophyll accumulation and nonphotochemical quenching in *Arabidopsis* mutants. Proc Natl Acad Sci U S A 95: 13324–13329
155. Polivka T, Herek JL, Zigmantas D, Akerlund HE, Sundstrom V (1999) Direct observation of the (forbidden) S1 state in carotenoids. Proc Natl Acad Sci U S A 96:4914–4917
156. Polivka T, Zigmantas D, Sundstrom V, Formaggio E, Cinque G, Bassi R (2002) Carotenoid S(1) state in a recombinant light-harvesting complex of photosystem II. Biochemistry 41:439–450
157. Pospisil P, Arato A, Krieger-Liszkay A, Rutherford AW (2004) Hydroxyl radical generation by photosystem II. Biochemistry 43:6783–6792
158. Reinhold C, Niczyporuk S, Beran KC, Jahns P (2008) Short-term down-regulation of zeaxanthin epoxidation in *Arabidopsis thaliana* in response to photo-oxidative stress conditions. Biochim Biophys Acta 1777:462–469
159. Rey P, Becuwe N, Barrault MB, Rumeau D, Havaux M, Biteau B, Toledano MB (2007) The *Arabidopsis thaliana* sulfiredoxin is a plastidic cysteine-sulfinic acid reductase involved in the photooxidative stress response. Plant J 49:505–514
160. Rock CD, Zeevaart JA (1991) The aba mutant of *Arabidopsis thaliana* is impaired in epoxy-carotenoid biosynthesis. Proc Natl Acad Sci U S A 88:7496–7499
161. Rossel JB, Wilson IW, Pogson BJ (2002) Global changes in gene expression in response to high light in *Arabidopsis*. Plant Physiol 130:1109–1120
162. Roy SJ, Tucker EJ, Tester M (2011) Genetic analysis of abiotic stress tolerance in crops. Curr Opin Plant Biol 14:232–239
163. Ruban AV, Horton P (1999) The xanthophyll cycle modulates the kinetics of nonphotochemical energy dissipation in isolated light-harvesting complexes, intact chloroplasts, and leaves of spinach. Plant Physiol 119: 531–542
164. Ruban AV, Johnson MP (2010) Xanthophylls as modulators of membrane protein function. Arch Biochem Biophys 504:78–85
165. Ruban AV, Murchie EH (2012) Assessing the photoprotective effectiveness of non-photochemical chlorophyll fluorescence quenching: a new approach. Biochim Biophys Acta 1817:977–982

166. Ruban AV, Horton P, Young AJ (1993) Aggregation of higher plant xanthophylls: difference in absorption spectra and in the dependency on solvent polarity. J Photochem Photobiol B 21:229–234
167. Ruban AV, Young AJ, Pascal AA, Horton P (1994) The effects of illumination on the xanthophyll composition of the photosystem II light-harvesting complexes of spinach thylakoid membranes. Plant Physiol 104:227–234
168. Ruban AV, Lee PJ, Wentworth M, Young AJ, Horton P (1999) Determination of the stoichiometry and strength of binding of xanthophylls to the photosystem II light harvesting complexes. J Biol Chem 274:10458–65
169. Ruban AV, Wentworth M, Horton P (2001) Kinetic analysis of nonphotochemical quenching of chlorophyll fluorescence. 1. Isolated chloroplasts. Biochemistry 40:9896–9901
170. Ruban AV et al (2007) Identification of a mechanism of photoprotective energy dissipation in higher plants. Nature 450:575–578
171. Ruban AV, Johnson MP, Duffy CD (2012) The photoprotective molecular switch in the photosystem II antenna. Biochim Biophys Acta 1817:167–181
172. Ruckle ME, DeMarco SM, Larkin RM (2007) Plastid signals remodel light signaling networks and are essential for efficient chloroplast biogenesis in *Arabidopsis*. Plant Cell 19:3944–3960
173. Sachindra NM, Sato E, Maeda H, Hosokawa M, Niwano Y, Kohno M, Miyashita K (2007) Radical scavenging and singlet oxygen quenching activity of marine carotenoid fucoxanthin and its metabolites. J Agric Food Chem 55:8516–8522
174. Santabarbara S, Bordignon E, Jennings RC, Carbonera D (2002) Chlorophyll triplet states associated with photosystem II of thylakoids. Biochemistry 41:8184–8194
175. Schmid VHR, Cammarata KV, Bruns BU, Schmidt GW (1997) In vitro reconstitution of the photosystem I light-harvesting complex LHCI-730: heterodimerization is required for antenna pigment organization. Proc Natl Acad Sci U S A 94:7667–7672
176. Schodel R, Irrgang KD, Voigt J, Renger G (1999) Quenching of chlorophyll fluorescence by triplets in solubilized light-harvesting complex II (LHCII). Biophys J 76:2238–2248
177. Snyder AM, Clark BM, Robert B, Ruban AV, Bungard RA (2004) Carotenoid specificity of light-harvesting complex II binding sites. J Biol Chem 279:5162–5168
178. Sonoike K (1996) Photoinhibition of photosystem I: its physiological significance in the chilling sensitivity of plants. Plant & Cell Physiol 37:239–247
179. Stroebel D, Choquet Y, Popot JL, Picot D (2003) An atypical haem in the cytochrome b(6) f complex. Nature 426:413–418
180. Strzalka K, Gruszecki WI (1997) Modulation of thylakoid membrane fluidity by exogenously added carotenoids. J Biochem Mol Biol Biophys 1:103–108
181. Suetsugu N, Wada M (2007) Chloroplast photorelocation movement mediated by phototropin family proteins in green plants. Biol Chem 388:927–935
182. Sundstrom V (2008) Femtobiology. Annu Rev Phys Chem 59:53–77
183. Teardo E, De Laureto PP, Bergantino E, Dalla VF, Rigoni F, Szabo I, Giacometti GM (2007) Evidences for interaction of PsbS with photosynthetic complexes in maize thylakoids. Biochim Biophys Acta 1767:703–711
184. Telfer A (2005) Too much light? How b-carotene protects the photosystem II reaction centre. Photochem Photobiol Sci 4:950–956
185. Telfer A, De Las Rivas J, Barber J (1991) Beta-Carotene within the isolated photosystem II reaction centre: photooxidation and irreversible bleaching of this chromophore by oxidised P680. Biochim Biophys Acta 1060:106–114
186. Telfer A, Bishop SM, Phillips D, Barber J (1994) Isolated photosynthetic reaction center of photosystem II as a sensitizer for the formation of singlet oxygen—detection and quantum yield determination using a chemical trapping technique. J Biol Chem 269:13244–13253
187. Telfer A, Dhami S, Bishop SM, Phillips D, Barber J (1994) Beta-carotene quenches singlet oxygen formed by isolated photosystem II reaction centers. Biochemistry 33:14469–14474
188. Triantaphylides C, Havaux M (2009) Singlet oxygen in plants: production, detoxification and signaling. Trends Plant Sci 14:219–228

189. Triantaphylides C, Krischke M, Hoeberichts FA, Ksas B, Gresser G, Havaux M, Van Breusegem F, Mueller MJ (2008) Singlet oxygen is the major reactive oxygen species involved in photooxidative damage to plants. Plant Physiol 148:960–968
190. Van Amerongen H, van Grondelle R (2001) Understanding the energy transfer function of LHCII, the major light-harvesting complex of green plants. J Phys Chem B 105:604–617
191. Verhoeven AS, Adams WW, Demmig-Adams B, Croce R, Bassi R (1999) Xanthophyll cycle pigment localization and dynamics during exposure to low temperatures and light stress in Vinca major. Plant Physiol 120:727–737
192. Walters RG (2005) Towards an understanding of photosynthetic acclimation. J Exp Bot 56:435–447
193. Walters RG, Ruban AV, Horton P (1996) Identification of proton-active residues in a higher plant light- harvesting complex. Proc Natl Acad Sci U S A 93:14204–14209
194. Wehner A, Grasses T, Jahns P (2006) De-epoxidation of violaxanthin in the minor antenna proteins of photosystem II, LHCB4, LHCB5, and LHCB6. J Biol Chem 281:21924–21933
195. Witt HT (1971) Coupling of quanta, electrons, fields, ions and phosphorylation in the functional membrane of photosynthesis. Q Rev Biophys 4:365–477
196. Woodall AA, Lee SW, Weesie RJ, Jackson MJ, Britton G (1997) Oxidation of carotenoids by free radicals: relationship between structure and reactivity. Biochim Biophys Acta 1336:33–42
197. Yamamoto HY, Kamite L (1972) The effects of dithiothreitol on violaxanthin deepoxidation and absorbance changes in the 500 nm region. Biochim Biophys Acta 267:538–543
198. Zhang H, Huang D, Cramer WA (1999) Stoichiometrically bound beta-carotene in the cytochrome b6 f. complex of oxygenic photosynthesis protects against oxygen damage. J Biol Chem 274:1581–1587
199. Powles S.B., Bjorkman O. (1981) Leaf movement in the shade species *Oxalis oregano*. II. Role in protection against injury by intense light. Carnegie Institute Washington Yearbook 80, 63–66
200. Brugnoli E, Björkman O. (1992) Chloroplast movements in leaves: Influence on chlorophyll fluorescence and measurements of light-induced absorbance changes related to ΔpH and zeaxanthin formation.Photosynth Res. 32(1):23–35

Chapter 16
Regulation of Electron Transport in Photosynthesis

Giles N. Johnson, Pierre Cardol, Jun Minagawa and Giovanni Finazzi

Abstract Photosynthetic organisms display a remarkable flexibility in their capacity to adjust photosynthetic performances in response to changes in their environment. This flexibility arises from the interplay of a range of different responses, including fast changes in light harvesting, changes in the pathways of electron flow and slower changes in the protein composition of the photosynthetic machinery. An array of possible adaptative responses is available to most photosynthetic organisms, which have in general selected from amongst these during their evolution to cope with the environmental circumstances of their specific environment. In this chapter we describe some representative strategies employed by eukaryotic photosynthetic organisms to adapt electron transfer capacity. We discuss processes in well characterised organisms from the green lineage (*Arabidopsis thaliana* and *Chlamydomonas reinhardtii*), and then focus on some peculiar strategies that have emerged in other organisms, in particular in marine phytoplankton.

Keywords Photosynthesis · Linear electron flow · Cyclic electron flow · Water-water cycle · Mitochondria-chloroplast metabolic interactions

---

G. Finazzi (✉)
Centre National Recherche Scientifique, Unité Mixte Recherche 5168, Laboratoire Physiologie Cellulaire et Végétale, 38054 Grenoble, France
e-mail: giovanni.finazzi@ceaf.fr

Commissariat à l'Energie Atomique et Energies Alternatives, l'Institut de Recherches en Technologies et Sciences pour le Vivant, 38054 Grenoble, France

Université Grenoble Alpes, 38041 Grenoble, France

Institut National Recherche Agronomique, UMR1200, 38054 Grenoble, France

G. N. Johnson
Faculty of Life Sciences, University of Manchester, Oxford Road, Manchester M13 9PT, UK

P. Cardol
Laboratoire de Génétique des Microorganismes, Institut de Botanique, Université de Liège, 27 bld du rectorat, 4000 Liège, Belgium

J. Minagawa
National Institute for Basic Biology (NIBB), Nishigonaka 38, Myodaiji, Okazaki 444-8585, Japan

S.M. Theg, F.-A. Wollman (eds.), *Plastid Biology,* Advances in Plant Biology 5,
DOI 10.1007/978-1-4939-1136-3_16,

Abbreviations

AOX	Mitochondrial alternative oxidase
APX	Ascorbate peroxidase
CBB	Calvin Benson Bassham cycle
CEF	Cyclic electron flow
Cyt	Cytochrome
Cyt b_6f	The cytochrome b_6f complex
Fd	Ferredoxin
FNR	Ferredoxin-$NADP^+$ oxidoreductase
FQR	Ferredoxin quinone reductase
GAPDH	Glyceraldehyde-3-phosphate dehydrogenase
LEF	Linear electron flow
MDA	Monodehydroascorbate radical
NDH	NAD(P)H dehydrogenase
PC	Plastocyanin
PQ	Plastoquinone
PQH_2	Plastoquinol
PS	Photosystem
PTOX	Plastoquinone terminal oxidase
P_{700}	Primary electron donor to PSI
SOD	Superoxide dismutase

16.1 Introduction

Eukaryotic photosynthesis first evolved probably more than 1.8 billion years ago [49] and revolutionised life on this planet. It allowed different groups of photosynthetic eukaryotes to inhabit freshwater and marine environments, and ultimately led to the colonisation of land. For this purpose, organisms developed different chloroplast structures, light-harvesting apparatuses and photosynthetic metabolism. However, the basic mechanisms of photosynthesis, as described by the Z scheme proposed by Hill and Bendall [62] are practically unchanged in all photosynthetic eukaryotes so far studied. This process starts with light absorption by dedicated pigment-containing complexes (the light harvesting or "antenna" complexes), which provide energy for charge separation by two photosystems (PSII and PSI). These photosystems are linked by a series of electron carriers, eventually leading to NADPH synthesis and the net movement of protons into the thylakoid lumen. The electrochemical proton gradient (the $\Delta\mu_H{}^+$) generated in this way is consumed for ATP synthesis by a CF_0-F_1 ATP synthase (ATPase) complex. ATP and NADPH fuel the fixation of CO_2 in the Calvin Benson Bassham (CBB) cycle.

In PSII, photochemical conversion of absorbed light leads to water oxidation by a Mn_4-Ca cluster by a sequential process known as the Joliot-Kok clock (or the "S states" mechanism, Joliot and Kok [74]), the molecular principles of which are now being elucidated thanks to the high resolution structures of this complex [101, 141]. Altough the H^+/e^- are variable during the operation of the clock [82], on the average, 1 proton is released into the lumen per electron subtracted from water. Electrons withdrawn from water are injected into the plastoquinone (PQ) pool, producing plastoquinol (PQH_2), which is the substrate of the cytochrome b_6f complex (Cyt b_6f) (as well as of PQ terminal oxidase (PTOX), see Sect. 16.4). As a member of the *bc*-type proteins family, Cyt b_6f couples proton translocation across the membrane to electron transfer from this lipophilic quinone to a hydrophilic one-electron acceptor protein (plastocyanin (PC) or a *c*-type cytochrome (cyt c_6)). During this reaction, electrons originating from PQH_2 are injected into a high potential chain (also called the linear path) formed by the Rieske protein and cyt *f* and into a cyclic route that comprises the *b* hemes (Q cycle; reviewed in [30]). Owing to the less positive E_m of the cyt composing this path (when compared to cyt *f* and the Rieske) this route is referred to as the low potential chain. According to the Q cycle mechanism proposed by Mitchell [94] and modified by Crofts et al. [31], quinones are oxidized and reduced at two distinct sites in the protein, the Qo and Qi (or Qp and Qn) sites respectively, which are located on the opposite side of the membrane [30]. PQH_2 oxidation on the lumenal side is associated with the reduction of both cyt *f* and b_L [31] and the release of protons into the lumen. Oxidation of the b_6 hemes occurs through a two step reduction of a PQ molecule at the Qi site, possibly involving the recently discovered c' heme [30]. The electron transfer sequence is the reduction of cyt b_L by PQH_2, electron transfer to b_H, and then a double electron transfer from these hemes to a PQ molecule located on the stromal side, a process that is coupled to proton uptake from the stroma. Overall, the Q cycle increases the H^+/e^- ratio of photosynthetic electron transfer. Previous work has suggested that this cycle is active under physiological conditions [120], and can be bypassed only upon drastic modifications of the redox features of the low potential chain [88]. On average, each electron passes twice through Cyt b_6f, once through the low and once the high potential pathways, and releases 2 protons into the lumen. The ultimate product of this complex, reduced PC or cyt c_6 (depending on species and growth conditions), is released from the b_6f complex and binds to PSI, where it reduces the primary electron donor of this complex, P_{700}. In contrast to the PQ pool, PC and cyt c_6 are extremely mobile in plant chloroplasts in the light [78], thus probably allowing the functional connection between the grana stacks and the stroma lamellae, where PSII and PSI are mainly concentrated.

Electrons arriving on the donor side of PSI are passed to the acceptor side via a charge separation reaction and then either continue their journey towards carbon assimilation, or provide reducing power for other cellular metabolic processes (nitrogen and sulphur metabolism, lipid, amino acid, pigment biosynthesis etc). In oxygenic photosynthesis, carbon assimilation is mainly driven by linear electron flow (LEF), which requires the in-series activity of the two photosystems. ATP and NADPH are produced in this process, although probably in a ratio not sufficient to support the formation of glyceraldehyde-3-phosphate, the export product of the CBB cycle [see 3 for a discussion]. Certainly, the ATP/NADPH ratio coupled to LEF does not exceed a value of 1.5, i.e. the stoichiometry required for CO_2 fixation. This means that LEF at best produces enough ATP to match the production of NADPH used in CO_2 fixation. Alternative electron consuming reactions, such as nitrogen metabolism, synthesis of lipids, amino acids, pigments, proteins and gene expression (all of which require reducing equivalents and ATP in variable stoichiometries) will alter the relative demands for ATP and NADPH, meaning that the synthesis of these products of the electron transport chain cannot be directly coupled in a simple way—the 'energy balance' issue (reviewed by [12], see Sect. 16.4). Besides this, various types of stress reduce the capacity of plants and algae to produce ATP and NADPH in the light, further exacerbating the difficulty of supplying different processes with the appropriate "energy stocks". It follows that other mechanisms must operate *in vivo* to control the relative production of NADPH and ATP. In plants and algae, alternative electron flow pathways exist, which can compete with the CBB cycle for reducing equivalents generated by the electron transport chain, thereby reducing the overall quantum yield of CO_2 fixation (reviewed by [105]). Amongst these, molecular oxygen can act as an acceptor for electrons, either from PSII (e.g. via PQH_2 and the so-called plastoquinone terminal oxidase PTOX), or during the Mehler reaction at the PSI reducing side (see review by [105]). Respiration can also act as a sink of photosynthetic electron flow [77] with electrons being exported from the chloroplast via the oxalate-malate shunt. Another process able to produce ATP without net NADPH generation is cyclic electron flow (CEF) around PSI [70, 124], the extent of which is extremely important in the green alga *Chlamydomonas reinhardtii* under specific metabolic conditions (see Sect. 16.3). Our understanding of the importance of all these different pathways is growing and it is becoming clear that the efficiency with which electron flow is diverted to these alternative pathways could be extremely high. Owing to their different evolutionary origin, different organisms seem to have chosen a limited number of alternative electron flow processes to cope with the energy balance issue. In this chapter, we will describe the core specifics of regulation of photosynthetic electron flow in plants and microalgae, summarising knowledge obtained developed the two best characterised model systems, flowering plants such as *Arabidopsis thaliana* and the green alga *Chlamydomonas reinhardtii*. Then, we will focus on some details on the particular strategies developed in the marine environments.

16.2 Electron Transport in Flowering Plants

Flowering plants growing in natural conditions face the challenge that they have little or no control over the concentration of the principle substrates required for photosynthesis. The light intensity varies from approx. zero to up to 2000 μmol m^{-2} s^{-1} during the course of a day but can also vary by as much as two orders of magnitude on a second to second basis as cloud cover changes or the sun moves relative to shading objects. The CO_2 concentration in the atmosphere does not fluctuate significantly but the availability of CO_2 to the chloroplast does, due both to changing demand and also changes in supply caused by stomata opening and closing.

Effects of variations in light and CO_2 may be further complicated by changes in leaf temperature from day to day and through the day, which will differentially affect different parts of the photosynthetic apparatus [28]. The result of all this is that the balance between energy supply to and demand for photosynthetic electron transport is unpredictable and highly variable. Variations in supply and demand place the leaf under stress. Specifically, excess light at any moment is liable to give rise to the production of reactive oxygen species, either through the photoreduction of oxygen to form superoxide, or through the generation of singlet excited oxygen [6]. It is increasingly recognised that such fluctuations play a major role in determining the fitness of plants [80] and that most of the regulatory processes that control the electron transport chain have evolved to mitigate against that stress.

16.2.1 Regulation of NADP(H) Redox Poise

The end product of LEF is NADPH, which is mainly used to drive sugar synthesis by the CBB cycle. When conditions arise to limit CO_2 fixation, for example when drought causes stomatal closure leading to a low leaf CO_2 concentration, we would expect that the NADPH concentration should rise. This should cause Fd and the electron acceptors in PSI to become reduced. Reduced FeS centres react readily with oxygen, producing superoxide [91]. Reduction of PSI acceptors blocks PSI charge separation. PSII turnover will then continue to produce reducing equivalents which should lead to the reduction all intermediates in the electron transport chain. In fact, apart from transiently, this does not occur. The events that prevent this from occurring are complex and intricate and still only partly understood.

Exposure of a leaf to a step change from high to low CO_2 concentration results, after a short response time, in a decrease in CO_2 fixation and a reduction of the electron acceptors of PSII (Q_A and the PQ pool). Electron carriers in Cyt b_6f, PC and P700 typically all become more oxidised. This effect results from a slowing down of the turnover of PQH_2 oxidation by the Qo site of Cyt b_6f. From this we can conclude that there must be a mechanism of feedback regulation sensing the demands of CO_2 fixation and regulation of Cyt b_6f. Two non-mutually exclusive models have been put forward to explain this regulation—pH regulation and redox control.

It has long been known, from early classic experiments on both mitochondria and thylakoid membranes, that the oxidation of quinones by Cyt *bc*-type complexes is sensitive to pH. There is evidence that protonation of the Rieske FeS subunit of Cyt b_6f is known to play an important role in the kinetics of PQH_2 oxidation [42, 67]. It is clear that the lumen pH influences the oxidation kinetics of PQH_2 and so the overall flow of electrons through the electron transport chain.

The physiological importance of lumen pH in controlling electron transport has been debated. When plants, under conditions of high CO_2 are exposed to light the pH of the thylakoid lumen drops, becoming more acidic with increasing irradiance. We cannot directly measure the ΔpH *in vivo*, although we can get an indication of its extent indirectly. Perhaps the most employed indicator is the extent of pH dependent non-photochemical quenching (qE). This term describes the enhancement of thermal dissipation of absorbed energy that occurs in the pigment-containing proteins of PSII, whenever light absorption exceeds the maximum rate of CO_2 assimilation. Based on an extensive body of work, we know that the pH dependence of qE is itself variable, in particular depending on the concentration of zeaxanthin in the light harvesting complexes [63]. In the presence of zeaxanthin, qE is induced below about pH 7.5 (pK ~ 7) [115]. The wild type Cyt b_6f shows an *in vitro* pH inhibition with a pK of approx. 6–6.5 [67]. In wild type plants in high CO_2, it is possible, by increasing light, to induce qE to a high level, suggesting the induction of a substantial ΔpH. At the same time, P_{700} (the primary electron donor to PSI) becomes progressively more oxidised. Measurments of the kinetics of P_{700} reduction following a light-dark transition under such experimental conditions, an indicator of flux through Cyt b_6f, indicate that no inhibition of PQH_2 oxidation is occurring [58, 106]. This implies that the *in vivo* lumen pH is too high to inhibit Cyt b_6f. Interestingly a mutant of *Arabidopsis*, pgr1, has been shown to have a mutation in the Rieske protein that results in an altered pH sensitivity, with a pK of 6.5–7 [67]. This shift results in plants which are deficient in qE—the flux through Cyt b_6f is inhibited at a pH which is too high to induce significant quenching.

When a leaf is exposed to changes in internal CO_2 concentration, either due to drought or as a result of altering the external CO_2 supply, regulation of Cyt b_6f can clearly be seen. Ott et al. [106] observed that when leaves of red campion were exposed to ambient, compared to elevated, CO_2, there was a clear inhibition of P_{700} reduction kinetics, across a wide range of irradiances. Except at the lowest irradiances, there was no change in reduction kinetics of P_{700} with irradiance however. Over the same irradiance range, reversible non-photochemical quenching (an indicator of qE) varied substantially but did not differ between ambient and high CO_2. In other words, the variation seen in P_{700} reduction kinetics cannot simply be explained by lumen pH. Similar non-correlations between qE and flux through Cyt b_6f were observed by Clarke and Johnson [28] and Golding and Johnson [47] in experiments examining responses to temperature and drought, respectively.

The alternative hypothesis for regulation of Cyt b_6f was proposed by Johnson [69]. In isolated spinach thylakoids, the flow through Cyt b_6f was shown to be sensitive to the presence of the thiol reducing agent dithiothreitol. Titration of this inhibition indicated the reduction with a pH sensitive midpoint potential in the region of −300 to −400 mV depending on the pH, i.e. between the midpoint potentials of Fd and NADPH. This led to the suggestion that thioredoxin might be responsible for feedback regulation of Cyt b_6f. This would give the redox poise of the PSI acceptor pool a direct role in regulating electron transport. To test this hypothesis, Hald et al. [54] examined plants in which electron flux away from PSI was inhibited due to reduced levels of either Fd-$NADP^+$ reductase (FNR) or glyceraldehyde-3-phosphate dehydrogenase (GAPDH), respectively the last enzyme in electron transfer and the first in the CBB cycle. A lack of FNR is predicted to result in reduction of Fd but in oxidation of NADP. Lack of GAPDH might be expected to give rise to reduction of both Fd and NADP. Plants lacking FNR were found to be very stressed, with elevated lipid peroxidation and loss of chlorophyll. The electron transport chain in such plants was substantially reduced, indicating a failure of feedback regulation. In contrast, plants lacking GAPDH, whilst inhibited in growth, showed no signs of stress and had an electron transport chain that was more oxidised under any set of conditions. These results were taken as evidence that it is the redox poise of the NADP/NADPH pool, rather than Fd and thioredoxin, that regulates the electron transport chain.

The above experiments provide evidence that control of Cyt b_6f is not a simple function of lumen pH. This does not exclude a role for pH in that control however. The *in vitro* redox sensitivity observed by Johnson [69] was seen to be pH sensitive in a way that would increase the sensitivity of PQH_2 oxidation to redox control at low pH. More recently, Joliot and Johnson [71] demonstrated that partial uncoupling of the thylakoid membrane by infiltrating leaves with nigericin, resulted in an inhibition of NPQ, as expected if the ΔpH is inhibited, and also in a net reduction of the electron transport chain. This is consistent with the lumen pH playing a role in regulating the activity of Cyt b_6f. In summary, both redox and pH signals can be shown *in vitro* to affect the flow through electron transport and it is likely that both play a role, possibly through a common mechanism, in the regulation of flow *in vivo*.

16.2.2 Regulation of the Trans-Thylakoid pH Gradient

The formation of a ΔpH across the thylakoid membrane is essential for the synthesis of ATP. It also plays an essential role in the regulation of light harvesting, through the process of high energy state quenching [63], and also in the regulation of LEF [71]. The generation and the regulation of this ΔpH is therefore essential to the proper function of the photosynthetic apparatus. ΔpH is generated during electron flux, however the balance between generation and utilisation of this gradient is a topic that has long been debated (see 2]). This discussion centred to a large extent on two points: is the Q-cycle in Cyt b_6f an obligate reaction; and what is the stoichiometry of ATP synthesis? Assuming that the Q-cycle is obligate [120], then each electron that is removed from water results in 3 protons being released into the thylakoid lumen (or 12 protons per O_2 evolved). The stoichiometry of ATP synthesis is related to the stoichiometry of subunits in the ATP synthase and specifically of the c subunit of the Fo part of the enzyme. In flowering plants, there are 14 c subunits, so 14 protons are required to synthesise 3 ATP molecules (discussed in [36]). If there were a strict coupling of LEF and ATP synthesis to CO_2 fixation in the CBB cycle, this would mean that there was a small deficit in ATP production.

To some extent this discussion was futile. Alternative pathways for consumption of reducing power and of ATP mean that there is no simple fixed requirement for a particular ratio of ATP to NADPH. Furthermore it cannot be assumed that the coupling of proton transfer to ATP synthesis is perfect—almost certainly, a proportion of protons will leak through the membrane. Nevertheless, what is certain is that the cell needs to maintain the balance of ATP and NADPH to ensure that the fixation of CO_2 and other metabolic processes can proceed in a balanced way and that the chloroplast does not become over-reduced. The latter is of particular concern, as reducing conditions can result in damage to the photosynthetic apparatus itself and also to cell in general, through the excess production of reactive oxygen species [6]. Increasingly therefore we must recognise the need for regulation of different electron transport pathways in a way that ensures that ATP and NADPH concentrations are controlled.

If we accept that LEF from water to CO_2 does not give rise to a net generation of ΔpH, then how are plants able to generate, and more importantly control, the pH gradient? In terms of generation of pH gradient, it follows that there must be some flexibility in the coupling between electron flow and ATP consumption. This could result from alternative electron flow, other than to CO_2 fixation, occurring from the import of ATP of respiratory origin into the chloroplast or possibly from changes in the ATP/H^+ coupling ratio of the ATPase. Owing to the complex regulation of the pH gradient, especially given the simultaneous need for ΔpH to generate ATP and regulate light harvesting, all these hypotheses have been previously considered as possible (e.g. [79]). In a perfectly coupled system (i.e. where the movement of protons out of the lumen is strictly linked to ATP synthesis) only a small number of protons need to be pumped into the lumen over and above those used for ATP synthesis. Once there, the ΔpH would be maintained. Kramer and co-workers have discussed that regulation of the pH gradient could then be achieved through regulation of the conductance of the ATPase, such that the potential gradient required to drive ATP synthesis might vary. This is suggested to occur through changes in chloroplast phosphate concentration [75], which in turn will reflect changes in the metabolic status of the ATP pool and CBB cycle intermediates. In this way, the pH gradient required to give a particular rate of ATP synthesis might vary. An alternative view is that different electron flows, especially CEF, generate additional ΔpH and it is the relative extent of these flows compared to LEF that controls the steady state ΔpH. It is now widely accepted that CEF occurs in flowering plants and that the rate of this can be substantial (see Sect. 16.2.4).

Obviously the above models are not mutually exclusive. However, there is growing experimental evidence for the occurrence of alternative electron flows, in addition to a simple LEF to CO_2 fixation *in vivo*. Different alternative flows are discussed in the following sections.

16.2.3 Sinks for Electron Flow Not Linked to ATP Consumption

A number of pathways exist that accept reducing equivalents from PSI but do not consume ATP. Reduction of inorganic nitrogen, via nitrate reductase and nitrite reductase, and assimilation of sulphate all consume reducing equivalents from photosynthetic electron transport [57, 59]. Reducing equivalents can also be exported from the chloroplast via the malate-oxalate cycle [77]. The latter pathway in particular is known to be regulated in direct response to chloroplast stroma redox poise, via thioredoxin-linked activation of the enzyme malate dehydrogenase [93]. Activation of malate dehydrogenase occurs at a substantially more negative redox potential than that required for activation of enzymes involved in CO_2 fixation, similar to the potential seen to inhibit Cyt b_6f [69]. Recent data have provided a molecular platform to understand the dynamics of the malate shuttle and suggest that the exchange of ATP and reducing power between the chloroplast and the mitochondrion is important for the optimization of carbon assimilation in vascular plants [77].

Electron flux to oxygen in the Mehler reaction will also generate ΔpH. Oxygen is reduced by iron sulphur centres on the acceptor side of PSI, generating superoxide which is detoxified by a series of reactions referred to collectively as the Mehler-Ascorbate Peroxidase pathway (see Sect. 16.3.4). There is probably always a basal flux through this pathway, possibly as high as 5–10 % of total electron flux, and this will generate ΔpH. It has been suggested that this pathway might play an important role in regulating the ΔpH [105, 114]. However, when leaves are illuminated at low O_2 concentrations, expected to lower the rate of Mehler reaction, they are still able to generate a ΔpH to drive high energy state quenching (see e.g. [28]). Since superoxide detoxification and oxidative stress impose a significant metabolic load on the leaf, it seems unlikely that flux to the Mehler reaction is actively used to regulate ΔpH. Nevertheless, it is probable that this flux plays an important role in overall chloroplast behaviour.

16.2.4 Cyclic Electron Flow in Flowering Plants

Fluxes through the electron transport chain to these alternative electron sinks will all support proton pumping into the thylakoid lumen without being strictly coupled to ATP consumption. Hence such fluxes will generate net ΔpH. It is not however clear that they can be regulated in a way that would allow for a responsive regulation of ΔpH. Rather, regulation of these flows probably occurs in ways that are intended to limit over-reduction of the chloroplast stroma, rather than specifically to generate ΔpH. CEF, involving just PSI, almost certainly acts as the major process that allows plants to control their ΔpH independently of redox potential. CEF in flowering plants has been the subject of a number of reviews in recent years (e.g. [70, 124]) and we do not intend here to review in detail the evidence for its occurrence. Rather the focus will be on our growing understanding of how this pathway is regulated and the role it plays in flowering plants.

The detailed functioning of CEF in plants is almost certainly somewhat different to that in green algae (see Sect. 16.3.1), although the basic pathways involved are almost certainly the same. Reduced Fd, produced by PSI, is oxidised in a way that results in the reduction of PQ, feeding electrons back into the electron transport chain between the two photosystems. That reaction may be catalysed by either an NAD(P)H dehydrogenase (NDH) complex homologous to Complex I in mitochondrial electron transport [118, 124], or via an alternative reaction presumed to involve Cyt b_6f. Our understanding of the structure and function of the former complex has grown significantly in recent years [111], although the precise role of this reaction remains to be elucidated. Plants deficient in the NDH complex appear to perform well across a wide range of conditions, though there is some evidence that they have increased sensitivity to drought stress [83, 118]. It seems however, that NDH is not essential for plant growth.

In addition to the NDH pathway, electron transfer can take place via a pathway often referred to as the FQR (ferredoxin quinone reductase) pathway [13]. In early studies, the FQR pathway was proposed to involve Cyt b_6f [132]. However, later studies hypothesized a distinct enzyme for FQR that bypasses Cyt b_6f and directly interfaces with PQ pool reduction by Fd [13]. No one, however, was able to identify or purify the FQR specific enzyme. On the other hand, physical association of PSI, Cyt b_6f, and FNR has long been suggested as a platform for the electron transfer in the FQR pathway [5, 26]. Mathematical modelling of the electron transfer [81] and *in vivo* observation of its high efficiency [73] suggest that FQR activity might operate in a complex. Indeed, Cramer and his co-workers reported that Cyt b_6f was co-purified with FNR and was reduced by Fd [149]. More recently, physical interactions between an integral membrane protein PGRL1, which was essential for FQR activity in *Arabidopsis*, and PsaD (subunit of PSI), PetB (cyt b_6), FNR, and PGR5 were shown by a yeast two-hybrid assay using the corresponding genes from *Arabidopsis* [32]. A *c*-type cyt found close to the stromal face of Cyt b_6f is suggested to be required to mediate the transfer of electrons from Fd/FNR to the Qi (quinone reducing) pocket in Cyt b_6f. [128] This model of CEF has in recent years gained an acceptance that probably exceeds the experimental evidence that supports it and several groups worldwide are working to establish a more solid mechanistic understanding of the events that are involved.

A key point in our understanding of CEF is the question of how the relative fluxes through LEF and CEF are regulated. Important points will be the steps where the two pathways converge or diverge—i.e. what is the fate of reduced Fd and how do PSII and CEF compete for PQ reduction. The need to regulate relative CEF and LEF fluxes is central to controlling ΔpH and it is widely thought that there needs to be some form of separation of the two pathways to achieve this regulation. For example, under conditions of high light or low CO_2, the PQ pool associated with PSII is known to be highly reduced. Under such conditions, CEF needs to be able to complete effectively with electron flow from PSII; indeed these are exactly the conditions that promote CEF [47, 95]. So how is it possible to ensure that the cyclic pathway has access to oxidised PQ?

In green algae there is strong evidence, both functional and structural, that supercomplexes of PSI and Cyt b_6f exist that are necessary for CEF to occur (see Sect. 16.3.2). The evidence for such complexes in flowering plants is however absent [21], and indeed, it has been argued that no such complexes are required [70, 71]. In particular, the presence in flowering plants of distinct membrane regions—the granal stacks and the stromal lamellae—provide the potential for regulating CEF and LEF in a way that is less possible in, for example, *Chlamydomonas*, where membrane stacking is much less prominent.

Key to our understanding of the structure of flowering plant chloroplasts is the observation that photosynthetic complexes are not evenly distributed in the thylakoid membrane [1] (c.f. Chap. 5 in this Volume). Functional PSII is localised in the membrane stacks. PSI is distributed between the grana margins (~70%) and the stromal lamellae. Cyt b_6f is evenly distributed in the membrane. It has been shown that the diffusion of PQ/PQH_2 in the thylakoid membrane is highly restricted, probably due to the very high protein concentrations that characterise this membrane [139]. This means that PQ localised in the granal stacks will primarily act in mediating electron transport between PSII and those Cyt b_6f localised in the stacked regions. For reasons of access, it is unlikely that electrons will be transferred from Fd to Cyt b_6f complexes localised in the grana regions. Rather, a separate pool of Cyt b_6f, with associated PQ will be found in the stromal membranes. The physical distance between the different domains and the limitations on PQ diffusion mean that these different pools could be quite separate.

Beyond Cyt b_6f, electrons are passed in flowering plants to PC. PC is soluble in the thylakoid lumen and, in flowering plants at least, much more mobile than PQ [78]. Therefore, there is the potential for the LEF and CEF pathways to mix at this point. This is however not such a problem. Once electrons have been injected from Fd or PSII into the high potential chain, their fate, as either "cyclic" or "linear" is determined. Thus, the presence of distinct Cyt b_6f and PQ pools is sufficient to allow for the co-existence of CEF and LEF pathways.

According to this model, the point at which the relative fluxes through CEF and LEF flows is regulated will be through competition for the oxidation of reduced Fd. Insights into the regulation of this step can be obtained from examination of various mutants altered in proteins required for CEF and LEF flows. Plants deficient in PSI complex have been shown to be impaired in LEF but are still able to maintain a significant proportion of CEF, giving a higher steady state pH gradient (as indicated by non-photochemical fluorescence quenching) [55]. In other words, CO_2 fixation does not simply extract reducing equivalents at a rate determined by its own capacity. Tobacco plants over-expressing Fd from *Arabidopsis* were shown to have increased CEF, even though their capacity for CO_2 assimilation was unaltered [147]. This led to the suggestion that Fd represents a limitation on CEF. This is perhaps a surprising conclusion, since the observation that the acceptor side of PSI is maintained in an oxidised state over most conditions suggests that Fd is not functionally limiting for electron flow, but it does suggest that the behaviour of Fd plays a major role in controlling electron partitioning. Plants deficient in Fd do show an inability to generate a high ΔpH, implying that at low Fd concentrations, CEF is limited [144].

The predominant reaction oxidising reduced Fd is via FNR. There is evidence that FNR exists in multiple forms [20, 53, 96]. In *Arabidopsis* there are two genes encoding this protein. Maize has three, including one that is thought to be specific to bundle sheath cells, where CEF is thought to be the predominant form of electron transfer [104]. These genes encode proteins with different pK values and with different tendencies to attach to the thylakoid membrane. These different isoforms can undergo different post translational modifications, including N-terminal truncation and possibly phosphorylation. FNR is known to be free in solution but also to bind to the acceptor side of PSI and the stromal side of Cyt b_6f. As such, it makes a good candidate as a point of regulation of CEF. Plants lacking FNR are found to be significantly impaired in the regulation of electron transport but are still able to perform CEF and to generate a significant ΔpH, at least under some conditions, although this is not greater than in wild type plants [54, 71]. This is consistent with a role for FNR in CEF, but there is little direct evidence for control of this step. Measurements of the distribution of FNR between different membrane-bound pools under different conditions provided no evidence of a regulatory re-distribution of FNR [21].

A protein-complex that has captured the imagination in terms of its role in CEF is the one formed by PGR5 and PRGL1 [32]. The nature of that involvement remains somewhat unclear. It was first suggested that it may form a Fd quinone oxidoreductase (FQR) [98], however there is no direct evidence for a catalytic role of either PGR5 or PGRL1 and no evidence of redox active cofactors. Nandha et al. [100] presented evidence that plants lacking PGR5 are in fact still capable of CEF at rates similar that those of wild type plants, but that they are impaired in CEF under most conditions, due to impairment of the redox poising of the chloroplast electron transport chain. Notably, the high potential portion of the electron transport chain is maintained in a reduced state in the light in PGR5 mutants. PGR5 mutants are capable of generating a significant ΔpH under conditions where photosynthesis is sink-limited, due to removal of CO_2 from the atmosphere, however the high potential is still reduced. This supports a model where PGR5 is required for feedback regulation of Cyt b_6f and it is that failure that prevents CEF from competing with LEF under most conditions. Whatever the precise role of PGR5, it is clear that it plays a crucial role in the regulation of photosynthesis. Plants lacking PGR5 are unable to survive in fluctuating growth conditions [136].

16.2.5 Alternative Electron Flow from PSII—the Plastid Terminal Oxidase

In addition to LEF through both PSII and PSI and CEF only involving PSI, there is strong evidence for electron transport reactions in flowering plants that involve PSII alone. There have been various indications that oxygen may act as an electron acceptor, from PSII directly [19] or from PQH_2 either at PSII, free in the membrane or at Cyt b_6f. [29, 76] These reactions are thermodynamically plausible. Direct reaction in the membrane between O_2 and PQH_2 is likely to be kinetically limited but reactions at quinone binding sites in either protein complex may occur, if a semiquinone exists for any time. Such reactions may occur significantly in some plants under some conditions, especially under stress, however their significance remains to be established.

Less controversial is the observation of a plastid terminal oxidase (Ptox) as a significant sink for electron transport from PSII. The PTOX protein was first identified in plants showing a phenotype where leaves developed with mottled white patches—the IMMUTANS phenotype seen in *Arabidopsis* and tobacco [25]. The protein is a di-iron non-haem protein showing homology with the alternative oxidase found in mitochondria. PTOX is thought to act as a PQH_2 water oxidoreductase. The IMMUTANS phenotype is explained as being due to an involvement of Ptox in carotenoid biosynthesis, providing a mechanism for the oxidation of phytoene. This role is probably however only required at crucial stages in development and, given the right conditions, IMMUTANS mutants can be grown that develop normal green leaves [117]. Recently a link was established between PGR5 and leaf variegation in immutans plants [103], leading to the hypothesis that when leaves are exposed to an excitation pressure that overcomes a threshold level (e.g. in some leaf patches of *immutans* plants), pigment bleacing occurs [90].

The existence of PTOX led to the suggestion that this may act as a sink for electron transport from PSII and therefore may have a direct role in photosynthesis. Phenotypes relating to small transient changes in chlorophyll fluorescence following light-dark transitions have been used as evidence that PTOX may be able to oxidise the PQ pool associated with PSII [85]. However there is no evidence in *Arabidopsis* tobacco or tomato plants that PTOX acts as a significant electron sink [61, 116, 140]

Such evidence has however been found in some less widely studied plant species. Streb et al. [127] studied electron transport in the alpine species *Ranunculus glacialis* and found evidence both for significant levels of Ptox and for significant electron transport from PSII to oxygen. Perez Torres et al. [112] found evidence for electron flow to oxygen in the arctic grass *Deschampsia Antarctica*, although there was no direct evidence found for PTOX protein. Stepien and Johnson [126] presented evidence for a substantial level of PSII to PTOX electron flow in the salt tolerant model species *Thellungiella halophilla.* This flow accounted for up to ~30 % of total PSII electron flow and was especially prominent at high light. There was no evidence that PTOX was competing with LEF at low light, suggesting that this flux is somehow regulated, such that it only occurs when LEF is saturated. Electron flow to PTOX resulted in the PQ pool being significantly more oxidised, consistent with a role in protecting PSII from stress. Surprisingly, inhibition of PTOX activity by lowering O_2 concentration did not affect the generation of non-photochemical quenching, suggesting that PTOX is not important in generating ΔpH, in contrast to the conclusion drawn in other systems (see Sect. 16.4).

The idea that Ptox might act as a significant sink for electron transport has encouraged the notion that it might be over-expressed to increase stress tolerance. To date, attempts to induce PTOX activity have met with little success. *Arabidopsis* plants over-expressing PTOX do not show any significant increase in O_2-sensitive electron transport [116], and tobacco over-expressers show elevated reactive oxygen production and increased stress sensitivity [61]. It is possible that there are differences between Ptox in *Arabidopsis* and in species where PTOX acts as a substantial electron sink, however the recent publication of the *Thellungiella* genome sequence provides no evidence for significant differences in sequence. Rather, it is likely that the known PTOX polypeptide is not acting alone but needs to be expressed alongside other peptides.

16.3 Chlamydomonas reinhardtii

Microalgae represent another ideal system to study photosynthetic electron transfer because, while they share most of the basic features of this process with plants, their growth rates are in general higher than those of plants. It is easy to get uniform cell cultures by controlling the microenvironment around the cells and to prepare a large number of samples for biochemical characterizations. Moreover, microalgae are extremely amenable to molecular and genetic manipulation. Their study has allowed us to elucidate the mechanisms of water oxidation (the Joliot-Kok clock, [72]), state transitions [18], and chlororespiration [14, 64, 68]. Recently there has been a large growth in interest in microalgae in general, and the freshwater alga *Chlamydomonas reinhardtii* in particular, as possible sources of biofuels, and an understanding of their photosynthesisis is central to this.

Chlamydomonas has been widely used as a model system to study photosynthesis in eukaryotic algae due to the fact that it can grow in the complete absence of photosynthesis (allowing mutants devoid in some of the major functions of this process to be easily characterized), has a well characterized sexual cycle (allowing mutations to be dissected and double mutants to be generated), and is suitable for molecular manipulation. This organism has recently been employed to study the functional organization of the photosynthetic apparatus (e.g. the epistatic control of the assembly of different complexes within the thylakoid membranes [27]), and the molecular mechanisms of photoprotection (state transitions and NPQ [84, 86] respectively). More recently, *Chlamydomonas* has allowed the establishment, for the first time, of a molecular basis for the regulation of alternative electron flow in photosynthesis [66] (see below).

16.3.1 Cyclic Electron Flow around PSI

CEF in *Chlamydomonas* is subjected to the same general rules that govern this process in plants, however the conditions giving rise to this flow may be quite different. Conditions leading to the reduction of the soluble electron carriers (e.g. anaerobic conditions) trigger the appearance of CEF [43], but in contrast to plants, it is the redox state of the PQ pool (instead of the redox state of PSI soluble acceptors) which primarly affects the partitioning between LEF and CEF. The mechanism linking changes in the chloroplast redox poise to the appearance of CEF is also different in this alga. In anaerobic conditions, reduction of the PQ promotes a transition to State 2, a condition in which most of the PSII antenna complexes are functionally connected to PSI, as a consequence of their phosphorylation by a specific kinase, Stt7 [34]. Enhancement of CEF is observed during the State 1 to State 2 transition, as shown by the diminished sensitivity of electron flow through Cyt b_6f [43] and PSI [24, 66] to the addition of the PSII inhibitor DCMU. Still, it is not known whether a true causal relation exists between the transition to State 2 and the onset of CEF. Indeed PSI absorption is increased to ~4 times that of PSII in *Chlamydomonas* cells acclimated to State 2, thus largely limiting the capacity of this complex to drive electron flow. A recent study however has reported that mutants with a largely impaired CEF capacity can undergo state transitions with the same efficiency as the wild type strain [135]. This clearly challenges the existence of a strict cause and effect relationship between state transtions and cyclic flow capacity

16.3.2 Mechanisms of CEF in Chlamydomonas

As discussed above, two major routes for CEF have been proposed in oxygenic photosynthesis. The first one involves the activity of a chloroplast NDH complex. In *Chlamydomonas*, as well as in most of the microalgae studied so far, this complex is not a homologue to the respiratory Complex I, but rather is represented by a monomeric complex [35, 68]. The Nad2 protein of *Chlamydomonas*, which is the only one studied in details, is extremely active in reducing the PQ pool at the expenses of stromal reducing power [35, 68]. However, the real contribution of the NDH pathway to CEF has not yet been tested experimentally. As an alternative to the NDH pathway, the FQR is defined as the complex that catalyzes the reduction of the PQ pool using Fd as a substrate (see Sect. 16.2.4).

While the nature of any complexes involved in FQR activity is still under debate in plants, recent analysis in *Chlamydomonas* has likely provided the first molecular information concerning the machinery in charge of the FQR pathway in this alga. Iwai et al. [66] used solubilized thylakoid membranes from *Chlamydomonas* cells under State 2 conditions. Using sucrose density gradient, they purified a super-supercomplex composed of the PSI-LHCI supercomplex with LHCIIs, Cyt b_6f, FNR, and PGRL1 in a fraction heavier than the PSI-LHCI supercomplex. Spectroscopic analyses of this super-supercomplex indicated that, upon illumination, reducing equivalents downstream of PSI were transferred to Cyt b_6f, while the oxidized PSI was re-reduced by reducing equivalents from Cyt b_6f [66]. When *Chalmdyomonas* cells are in State 2, where more LHCIIs are associated with PSI, CEF can operate in this supercomplex involving PSI, Cyt b_6f, and FNR. Since CEF and LEF share several redox carriers (e.g., PQ, Cyt b_6f, PC, PSI, Fd, and FNR), they are potentially in competition with one another. Furthermore, the redox poise of the CEF components could be disturbed if reduced components for LEF coexist [2]. By localizing the mobile electron carriers (PQ, Fd, and PC) within a restricted space, the super-supercomplex could compartmentalise the CEF components, generating a functional pool. In doing so, super-supercomplex formation would play a similar role to that suggested for the segregation of PSI and PSII in the grana and stroma lamellae in plants (see Sect. 16.2.4), although providing a total physical compartmentation of the two pathways. In line with this conclusion, limitation of the overall rate of electron flow by PC diffusion between PSI and Cyt b_6f has been observed in *Arabidopsis* under some conditions [122], while, no such limitation is seen in *Chlamydomonas*, where PC release from the PSI is the limiting step of the electron flow between Cyt b_6f and PSI [44]. This could reflect the fact that, while

PC is freely diffusing (in an almost bidimensional space) in plants in the light [48, 77], the presence of a supercomplex of supercomplexes in the alga would reduce the distance to be travelled by PC to being essentially negligible, making this process non-limiting for photosynthesis. It is of note that the association of LHCII and Cyt b_6f with PSI is in line with their behaviors in the membrane domains during a State 1-to-2 transition. During this transition, LHCII and Cyt b_6f migrate from the appressed region in the thylakoid membranes, where PSII resides, to the non-appressed region, where PSI resides [142].

16.3.3 Physiological Consequences of CEF in *Chlamydomonas*

In contrast to plants, large state transitions are induced in *Chlamydomonas* upon oxygen deprivation. This treatment rapidly leads to a decrease in the cellular ATP content, which in turn triggers an increase in the cellular NADH/NADPH concentration, due to the well known enhancement of glycolysis by a reduced ATP content (the reverse Pasteur effect [22]. PSI absorption cross section is substantially increased in State 2 [33] due to the extensive association of LHCII proteins (including the minor monomeric LHCIIs) with PSI [133, 137]. Moreover, the generation of tightly bound super-supercomplexes allows a complete thermodynamic segregation of the soluble electron carriers from the PSII-driven LEF pathway, a process that is not observed in plants, where CEF requires a small electron input from PSII [73], to compensate for the losses of electrons towards the CBB cycle (see Sect. 16.2.4).

What then is the rationale for such major rearrangements of the photosynthetic machinery in *Chlamydomonas* upon a State 2 transition? In general, a switch to State 2 is observed upon nutrient starvation in *Chlamydomonas* (see e.g. [60, 146]). Phosphorous and sulfur deficiencies, which decrease the rate of oxygen evolution induce a systematic transition to State 2 and a loss of the ability to perform LEF [146], thus giving rise to an increased CEF capacity. Nitrogen starvation also induces a systematic transition to State 2 [110], which correlates with overreduction of the PQ pool and a loss of LEF activity [60]. Overall, the rationale for these observations would be to maintain a good capacity for ATP synthesis via CEF for housekeeping purposes every time that photosynthetic performance is limited by nutrient availability (see also [36]). Indeed, from an energetic point of view, state transitions in *Chlamydomonas* mimic a shift from an oxygenic type of photosynthesis (that generates both reducing power and ATP, State 1) to an anoxygenic bacterial one, where only ATP is synthesised (State 2). The generation of the PSI-Cyt b_6f supercomplex probably provides the molecular platform for this major metabolic change in *Chlamydomonas*. This switch may provide an advantage in different environmental conditions. First of all it would increase the capacity to survive under oxygen deprivation. This situation is probably often encountered by this alga, which is normally found in eutrophic shallow ponds, rich in biomass and therefore possibly subjected to periods of anoxia. By maintaining a high quantum yield of ATP synthesis in State 2, cells might be able to maintain vital processes and therefore to cope successfully with these unfavorable conditions, and to rapidly recover photosynthetic carbon fixation once O_2 is resupplied. Even under oxygen replete conditions, CEF would provide a benefit, but providing ATP for carbon assimilation. However, in contrast to plants [99], no growth phenotype is observed in mutants with an altered CEF capacity in *Chlamydomonas* [24, 44, 138]. This is only observed in mutants where both respiration (dums, for dark uniparental minus) and CEF (via state transitions) are down regulated [24]. In these lines, a drastic decrease in growth is seen [24], due to a diminished photosynthetic activity. It appears that in *Chlamydomonas* (and possibly in other green algae) lack of CEF is compensated by an efficient exchange of reducing equivalents (and/or of ATP) between the mitochondrion and the chloroplast. The two metabolisms are intimately linked in this alga, owing to the interplay between respiration, reduction of the PQ pool, state transitions and CEF [36], and it is therefore reasonable to assume that they may both contribute in maintaining the cellular energy charge in this alga through optimum light utilization. This hypothesis is consistent with the finding that mutants with decreased PGRL1 (and therefore possibly impaired in their CEF capacity) avoid degrading PSI in Fe-starved cultures of *Chlamydomonas* to maintain a given CEF capacity [113], despite the very high cellular pressure to mobilise Fe for other metabolic processes. Moreover, these strains do not reduce their respiratory capacity under Fe starvation despite the high Fe requirement of the mitochondrial electron transport chain, possibly to maintain a high energetic metabolism. Recent (unpublished) data, indicate that exposure to high light in the absence of an external carbon source (as required to promote the induction of the LHCSR protein, and therefore NPQ onset in *Chlamydomonas*) also results in a transtion to State 2. Again, this could reflect a change in the cellular energy metabolism to enhance PSII protection from photoinhibition. Indeed, reducing the PSII antenna size (via state transtions) and increasing the capacity to generate a ΔpH (via CEF) could reduce the risk of photodamage in this complex.

16.3.4 Other Alternative Electron Flow Processes in Chlamydomonas

The observation that O_2 depletion triggers a reduction of the PQ pool, leading to state transitions, suggests that O_2 may act as a significant sink for electron transport in Chlamydomonas, either via a PTOX-type activity or through the Mehler reaction. This process would allow oxidation of PQH_2 under conditions where CO_2 assimilation is impaired, and by so doing, maintain the cell in State 1. As discussed above, the Mehler reaction results in the production of superoxide, which is rapidly converted into H_2O_2 by the activity of the superoxide dismutase (SOD) enzyme. H_2O_2can be efficiently scavenged by a chloroplast-associated ascorbate peroxidase (APX), leading to the production of a monodehydroascorbate radical (MDA) from ascorbate and H_2O_2 [6]. In *Chlamydomonas*, the existence of an ascorbate peroxidase similar to that of plants has been reported [134]. In this alga, the ascorbate concentration is very low and the ascorbate peroxidase enzyme has an extremely high affinity for this metabolite [134].

In *Chlamydomonas*, an additional role for these ROS has been recently proposed. Expression of a nuclear-encoded reporter gene coupled to an H_2O_2 sensitive promoter was shown to respond not only to the levels of exogenously added H_2O_2 but also to light. The higher induction of the reporter gene seen in light-H_2O_2 treated cells was correlated with a lower H_2O_2 scavenging activity [123]. Therefore, the authors concluded that the enhanced H_2O_2 concentration observed upon light exposure would represent a molecular switch to activate a specific ROS signalling pathway within the cell.

16.4 Photosynthesis in the Oceans

Photosynthesis by marine algae has been the subject of studies for about 80 years (e.g. [9, 89, 119]). The ocean is the place where about half of the global carbon assimilation occurs [11, 41] thanks to the photosynthetic activity of phytoplankton communities composed of prokaryotes (mainly cyanobacteria *Prochlorococcus* and *Synechococcus*) and eukaryotes (mainly diatoms, dinoflagellates chlorophyta and haptophyta). This performance is astonishing *per se*, since marine phytoplankton probably contribute ~1% of the total photosynthetic biomass on Earth [38, 41]. This high efficiency of carbon assimilation in the oceans arises from the very fast growth rates and the high photosynthetic performances shown by phytoplankton in some regions of the oceans. Far from being homogeneous, the marine world offers a variety of environments ranging from warm nutrient-rich coastal areas to vast cold and oligotrophic oceanic areas. The oligotrophic sections of the oceans represents about 70% of the marine environment, are generally distant from costal zones, and are thus characterized by very low iron (Fe) and nitrogen (N) content. These environments have provided microalgae with strong selection pressures leading to the emergence of multiple adaptation or acclimation strategies in the photosynthetic apparatus.

The analysis of the response of marine microalgae to changes in the light intensity has revealed two opposite and paradigmatic strategies for light acclimation [39]: the modification of number of reaction centers, a strategy termed n-type photoacclimation, or of the antenna cross-section, a strategy termed σ-type acclimation. These strategies are exemplified in the diatom *Skeletonema costatum* that responds to decreasing irradiance by increasing its antenna size, and in the chlorophyte *Dunaliella tertiolecta* which withstands changes in irradiance by modifying the number of reaction centers [40, 131].

Analysis of the chlorophyll distribution between the nutrient-rich coastal regions and the vast oligotrophic parts of the oceans has also shown the cost of nutrient limitation on photosynthetic activity *in situ* [10]. In a laboratory environment, nutrient limitation usually leads to a decrease in photosynthesis, due to changes in the activity and stoichiometry of the photosynthetic complexes [92]. Acclimation to nutrient accessibility is particularly well characterized upon iron starvation, which consists of modifications of both the electron flow and light absorption capacities of the cells. Iron is of primary importance in biological systems, notably because it is a central constituent of prosthetic groups such as hemes and iron-sulfur clusters bound to enzyme complexes involved in energy-conversion processes. Iron mobilization is thus critical for microalgae in the iron-limited oceanic areas, as it is for terrestrial plants growing on neutral and alkaline soils, where iron availability is limited, as a result of the prominent form of iron in the presence of oxygen being the poorly soluble Fe^{3+}.

Despite the fact that several mechanisms exist to ensure iron homeostasis in plants and algae [55, 102, 107], when deprivation of this metal occurs, the most prominent effects on photosynthetic apparatus are at the level of PSI. Cyanobacteria respond to iron-deprivation by lowering the relative abundance of PSI and by forming an additional light harvesting antenna around the remaining PSI. *Prochlorococcus sp.*, which dominates certain regions of the oligotrophic oceans, synthesizes Pcb proteins, while fresh water species express a chlorophyll binding protein similar to CP43 (IsiA) [15–17, 125]. Modifications in PSI antenna are also observed in various eukaryotic algae. In *Chlamydomonas reinhardtii*, the LHCI-PSI association and stoichiometry are altered [97] and in the halotolerant eukaryotic alga *Dunalliela salina* a light-harvesting chlorophyll a/b-binding protein, Tidi, accumulates [143]. But the most dramatic effect of iron limitation is a marked drop in the amount of PSI centers relative to other molecular constituents of the photosynthesic apparatus. PSI has the highest iron content (12 Fe per reaction centre) among the photosynthetic complexes and requires Fe for stable assembly [46]. Up to a four-fold decrease in the PSI/PSII ratio has been observed in Fe-depleted cyanobacteria [e.g. 52, 121] or eukaryotic algae, including *Chlamydomonas reinhardtii* [97, 113] and the diatom *Phaeodactylum tricornutum* [3]. In the cyanobacterium *Synechococcus* WH8102, the prasinophyte *Ostreococcus* RCC809 and the central diatom *Thalasiossira oceanica,* three species isolated from oligotrophic regions, a markedly lower content of PSI and Cyt b_6f (which contains 6 Fe atoms) relative to PSII has been measured, even if algae are cultivated in Fe-replete medium ([8, 23], Cardol and Finazzi, unpublished, [129]). The constitutively low Cyt b_6f and PSI contents seen in oligotrophic species contrasts with the almost equimolar ratios between major photosynthetic complexes observed in closely related coastal species (*Ostreococcus tauri* and *Thalassiosira weissflogii*) [23, 129]. The PSI decrease relative to PSII also raises the question of how cells respond to a reduced capacity to reoxidize the PQ pool in the light. In *Synechococcus* WH8102 and in *Ostreococcus RCC809*, it has been shown that an enhanced PTOX activity allows rerouting of PSII-generated electrons into a water-to-water cycle to an extent of about 50 % [7, 23]. A similar observation has been made *in* situ on open ocean picophytoplankton communities dominated by the *Prochlorococcus* genus [87]. Though this efficient electron flow to oxygen might take place at the expense of CO_2 fixation, it alleviates the redox pressure on the PSII acceptor side and allows maintainance of an electrochemical proton gradient in the light, despite the very strong limitation of electron flow by Cyt b_6f and PSI (in contrast to plants, see Sect. 16.2.4). This proton gradient may in turn serve the purpose of maintaining efficient ATP synthesis and/or developing photoprotective responses (NPQ), which are triggered by lumen acidification and protect PSII from photoinhibition. This strategy could be relevant in the overall economy of photosynthesis in iron-limited marine environments [50, 148].

Interestingly, light-stimulated oxygen uptake by phytoplankton has been described for several other species (e.g. the diatom *T. weissflogii* [145], the cryptophyte *Storeatula major*, the prasinophyte *Pycnococcus provasolii*, and the dinophyte *Prorocentrum minimum* [130]. In the case of *Nannochloropsis sp.* and *Emiliania huxleyi* (Chromalveolates), the light-stimulated O_2 uptake probably occurs close to PSII [37], but we don't know yet if it involves a PTOX-dependent mechanism, the PSI-associated Mehler reaction or even the cytochromes and alternative oxidases in the mitochondrion.

In the diatom *Phaeodactylum tricornutum,* under Fe starvation, electrons could be redirected to the respiratory chain through the alternative oxidase [3]. Recent proteomic analysis [51] has shown that the chlororespiratory complex NDH, which is an essential component of the photosynthetic machinery (see Sect. 16.2) is absent in this organism. Following the model established in *Chlamydomonas* where both CEF and respiration in the light contribute to the generation of "extra" ATP for carbon assimilation [24], it has been postulated that the strong interaction between the two energetic metabolisms observed upon Fe starvation would stem from the necessity to compensate for the diminished CEF capacity [45].

Beside water-to-water cycle strategies, another strategy to adapt the photosynthetic apparatus under iron-limitation has been described in marine algae. Surprisingly, oceanic diatom species have a greater need for copper (Cu) compared to coastal strains [4, 108]. In photosynthetic organisms, a common strategy to partially deal with Cu availability consists of employing the Fe-containing cyt c_6 as the soluble electron carrier between the Cyt b_6f and PSI complexes, instead of the Cu-containing PC [92]. In the case of the oceanic diatom *Thalassiosira oceanica*, it has been shown that this Cu requirement is due to the single Cu-containing protein, PC, which is absolutely needed for photosynthetic electron transport [109]. In contrast, the costal species *T. weissflogii* expresses a classical cyt c_6 as electron carrier [65]. These observations suggested that the selection pressure imposed by Fe limitation in oligotrophic marine areas has resulted in the use of PC which reduces the need for Fe, Cu being relatively more abundant in the open sea [109].

The analysis of photosynthesis in the ocean has provided an experimental basis for the notion that photosynthesis cannot fuel carbon assimilation and the cellular anabolic processes at the same time, (the so called 'energetic management' issue [12]). In *Prochlorococcus* PCC 9511 cells, whilst a constant photosynthetic activity can be measured during a day (i.e. the entire life span of this prokaryote), the light generated ATP and reducing power are employed to supply different metabolic pathways during this period (carbohydrate vs aminoacid synthesis) as required to complete all the steps of the life cycle of this organism. So far, no such information has been provided in the case of photosynthetic eukaryotes, but it is tempting to propose that a similar phenomenon may take place, because the same rules governing the efficiency of ATP and NADPH synthesis and consumption exist in both kingdoms.

16.5 Conclusion

Regulating electron flow in photosynthetic organisms has two purposes: adjusting the generation of ATP and reducing power, and allowing a proper response to environmental changes. Because it is well established that the ATP/NADPH ratio coupled to photosynthetic electron transport *in vivo* (although probably variable in different conditions) is not going to exceed 1.5, clearly photosynthesis cannot supply carbon assimilation and other metabolic pathways simultaneously. Alternative electron transport pathways could overcome this limitation, although they have to be tightly regulated to avoid an excessive reduction of the quantum yield of CO_2 assimilation. A survey of the literature indicates that CEF, a water-water cycle (via either PTOX or the ascorbate-Mehler reaction), and the malate shunt have the capacity to solve the "energy balance" and "ATP shortage" issues, and can operate with high efficicncies (reviewed in [36]). Obviously, if they were all operating at the same time and at their maximum capacity, the overall yield of carbon assimilation would certainly be too low to allow photosynthetic growth in a natural environment. Evolution has therefore provided plants and microalgae with the capacity to choose the most appropriate alternative elctron flow pathway among the array of processes available. Although some relevant progress has been made in the elucidation of the molecular mechanisms allowing plants and algae to regulate their electron flow capacity, future effort is still required to explore the mechanisms allowing photosynthetic organisms to cope with the energy requirements of carbon assimilation.

References

1. Albertsson P-A (2001) A quantitative model of the domain structure of the photosynthetic membrane. Trends Plant Sci 6:349–354
2. Allen JF (2003) Cyclic, pseudocyclic and noncyclic photophosphorylation: new links in the chain. Trends Plant Sci 8:15–19
3. Allen AE, LaRoche J, Maheswari U, Lommer M, Schauer N, Lopez PJ, Finazzi G, Fernie AR, Bowler C (2008) Whole-cell response of the pennate diatom Phaeodactylum tricornutum to iron starvation. Proc Natl Acad Sci U S A 105:10438–10443
4. Annett AL, Lapi S, Ruth TJ, Maldonado MT (2008) The effects of Cu and Fe availability on the growth and Cu: C ratios of marine diatoms. Limnol Oceanogr 53:2451–2461
5. Arnon DI (1995) Divergent pathways of photosynthetic electron transfer: the autonomous oxygenic and anoxygenic photosystems. Photosynth Res 46:47–71
6. Asada K (2006) Production and scavenging of reactive oxygen species in chloroplasts and their functions. Plant Physiol 141:391–396
7. Bailey S, Grossman A (2008) Photoprotection in cyanobacteria: regulation of light harvesting. Photochem Photobiol 84:1410–1420
8. Bailey S et al (2008) Alternative photosynthetic electron flow to oxygen in marine Synechococcus. Biochim Biophys Acta 1777:269–276
9. Barker HA (1935) Photosynthesis in diatoms. Arch Microbiol 6:141–156
10. Behrenfeld MJ, Falkowski PG (1997) Photosynthetic rates derived from satellite-based chlorophyll concentration. Limnol Oceanogr 42:1–20

11. Behrenfeld MJ, Worthington K, Sherrell RM, Chavez FP, Strutton P, McPhaden M, Shea DM (2006) Controls on tropical Pacific Ocean productivity revealed through nutrient stress diagnostics. Nature 442:1025–1028
12. Behrenfeld MJ, Halsey KH, Milligan AJ (2008) Evolved physiological responses of phytoplankton to their integrated growth environment. Philos Trans R Soc Lond B Biol Sci 363:2687–2703
13. Bendall DS, Manasse RS (1995) Cyclic photophosphorylation and electron-transport. Biochim Biophys Acta 1229:23–38
14. Bennoun P (1982) Evidence for a respiratory chain in the chloroplast. Proc Natl Acad Sci U S A 79:4352–4356
15. Bibby TS, Nield J, Barber J (2001) Iron deficiency induces the formation of an antenna ring around trimeric photosystem I in cyanobacteria. Nature 412:743–745
16. Bibby TS, Nield J, Partensky F, Barber J (2001) Oxyphotobacteria-antenna ring around photosystem I. Nature 413:590–590
17. Boekema EJ, Hifney A, Yakushevska AE, Piotrowski M, Keegstra W, Berry S, Michel KP, Pistorius EK, Kruip J (2001) A giant chlorophyll-protein complex induced by iron defciency in cyanobacteria. Nature 412:745–748
18. Bonaventura C, Myers J (1969) Fluorescence and oxygen evolution from *Chlorella pyrenoidosa*. Biochim Biophys Acta 189:366–383
19. Bondarava N, De Pascalis L, Al-Babili S, Goussias C, Golecki JR, Beyer P, Bock R, Krieger-Liszkay A (2003) Evidence that cytochrome b(559) mediates the oxidation of reduced plastoquinone in the dark. J Biol Chem 278:13554–13560
20. Bowsher CG et al (2011) Identification of N-terminal regions of wheat leaf ferredoxin NADP(+) oxidoreductase important for interactions with ferredoxin. Biochemistry 50:1778–1787
21. Breyton C, Nandha B, Johnson GN, Joliot P, Finazzi G (2006) Redox modulation of cyclic electron flow around photosystem I in C3 plants. Biochemistry 45:13465–13475
22. Bulté L, Gans P, Rebéillé F, Wollman F-A (1990) ATP control on state transitions *in vivo* in *Chlamydomonas reinhardtii*. Biochim Biophys Acta 1020:72–80
23. Cardol P et al (2008) An original adaptation of photosynthesis in the marine green alga *Ostreococcus*. Proc Natl Acad Sci U S A 105:7881–7886
24. Cardol P, Alric J, Girard-Bascou J, Franck F, Wollman F-A, Finazzi G (2009) Impaired respiration discloses the physiological significance of state transitions in *Chlamydomonas*. Proc Natl Acad Sci U S A 106:15979–15984
25. Carol P, Stevenson D, Bisanz C, Breitenbach J, Sandmann G, Mache R, Coupland G, Kuntz M (1999) Mutations in the *Arabidopsis* gene immutans cause a variegated phenotype by inactivating a chloroplast terminal oxidase associated with phytoene desaturation. Plant Cell 11:57–68
26. Carrillo N, Vallejos RH (1983) The light-dependent modulation of photosynthetic electron-transport. Trends Biochem Sci 8:52–56
27. Choquet Y, Wollman FA (2002) Translational regulations as specific traits of chloroplast gene expression. FEBS Lett 529:39–42
28. Clarke JE, Johnson GN (2001) *In vivo* temperature dependence of cyclic and pseudocyclic electron transport in barley. Planta 212:808–816
29. Cleland RE, Grace SC (1999) Voltametric detection of supreoxide production by photosystem II. FEBS Lett 457:348–352
30. Cramer WA, Hasan SS, Yamashita E (2011) The Q cycle of cytochrome bc complexes: a structure perspective. Biochim Biophys Acta 1807:788–802
31. Crofts AR, Meinahrdt SW, Jones KR, Snozzi M (1983) The role of the quinone pool in the cyclic electron-transfer chain of *Rhodopseudomonas spaeroides*. A modified Q-cycle mechanism. Biochim Biophys Acta 723:202–218
32. DalCorso G, Pesaresi P, Masiero S, Aseeva E, Nemann DS, Finazzi G, Joliot P, Barbato R, Leister D (2008) A complex containing PGRL1 and PGR5 is involved in the switch between linear and cyclic electron flow in *Arabidopsis*. Cell 132:273–285

33. Delosme R, Olive J, Wollman FA (1996) Changes in light energy distribution upon state transitions: an *in vivo* photoacoustic study of the wild type and photosynthesis mutants from *Chlamydomonas reinhardtii*. Biochim Biophys Acta 1273:150–158
34. Depege N, Bellafiore S, Rochaix JD (2003) Rote of chloroplast protein kinase Stt7 in LHCII phosphorylation and state transition in *Chlamydomonas*. Science 299:1572–1575
35. Desplats C, Mus F, Cuine S, Billon E, Cournac L, Peltier G (2009) Characterization of Nda2, a plastoquinone-reducing type II NAD(P)H dehydrogenase in *Chlamydomonas* chloroplasts. J Biol Chem 284:4148–4157
36. Eberhard S, Finazzi G, Wollman F-A (2008) The dynamics of photosynthesis. Annu Rev Genet 42:463–515
37. Eisenstadt D, Barkan E, Luz B, Kaplan A (2010) Enrichment of oxygen heavy isotopes during photosynthesis in phytoplankton. Photosynth Res 103:97–103
38. Falkowski PG (1994) The role of phytoplankton photosynthesis in global biogeochemical cycles. Photosynth Res 39:235–258
39. Falkowski PG, Owens TG (1980) Light-shade adaptation −2 strategies in marine phytoplankton. Plant Physiol 66:592–595
40. Falkowski PG, Owens TG, Ley AC, Mauzerall DC (1981) Effects of growth irradiance levels on the ratio of reaction centres in 2 species of marine phytoplankton. Plant Physiol 68:969–973
41. Field CB, Behrenfeld MJ, Randerson JT, Falkowski P (1998) Primary production of the biosphere: integrating terrestrial and oceanic components. Science 281:237–240
42. Finazzi G (2002) Redox-coupled proton pumping activity in cytochrome b(6)f, as evidenced by the pH dependence of electron transfer in whole cells of *Chlamydomonas reinhardtii*. Biochemistry 41:7475–7482
43. Finazzi G, Furia A, Barbagallo RP, Forti G (1999) State transitions, cyclic and linear electron transport and photophosphorylation in *Chlamydomonas reinhardtii*. Biochim Biophys Acta 1413:117–129
44. Finazzi G, Sommer F, Hippler M (2005) Release of oxidized plastocyanin from photosystem I limits electron transfer between photosystem I and cytochrome b(6)f complex *in vivo*. Proc Natl Acad Sci U S A 102:7031–7036
45. Finazzi G, Moreau H, Bowler C (2010) Genomic insights into photosynthesis in eukaryotic phytoplankton. Trends Plant Sci 15:565–572
46. Golbeck JH (2003) The binding of cofactors to photosystem I analyzed by spectroscopic and mutagenic methods. Annu Rev Biophys Biomol Struct 32:237–256
47. Golding AJ, Johnson GN (2003) Down-regulation of linear and activation of cyclic electron transport during drought. Planta 218:107–114
48. Golding AJ, Joliot P, Johnson GN (2005) Equilibration between cytochrome f and P700 in intact leaves. Biochim Biophys Acta 1706:105–109
49. Gould SB, Waller RF, McFadden GI (2008) Plastid evolution. Annu Rev Plant Biol 59: 491–517
50. Grossman AR, Mackey KRM, Bailey S (2010) A perspective on photosynthesis in oligotrophic oceans: hypothesis concerning alternate routes of electron flow. J Phycol 46:629–634
51. Grouneva I, Rokka A, Aro E-M (2011) The thylakoid membrane proteome of two marine diatoms outlines both diatom-specific and species-specific features of the photosynthetic machinery. J Proteome Res 10:5338–5353
52. Guikema JA, Sherman LA (1984) Influence of iron deprivation on the membrane-composition of anacystis-nidulans. Plant Physiol 74:90–95
53. Gummadova JO, Fletcher GJ, Moolna A, Hanke GT, Hase T, Bowsher CG (2007) Expression of multiple forms of ferredoxin NADP(+) oxidoreductase in wheat leaves. J Exp Bot 58:3971–3985
54. Hald S, Nandha B, Gallois P, Johnson GN (2008) Feedback regulation of photosynthetic electron transport by NADP(H) redox poise. Biochim Biophys Acta 1777:433–440
55. Hald S, Pribil M, Leister D, Gallois P, Johnson GN (2008) Competition between linear and cyclic electron flow in plants deficient in Photosystem I. Biochim Biophys Acta 1777:1173–1183

56. Hanikenne M, Merchant S, Hamel P (2009) Metal nutrition: a balance between deficiency and toxicity. In: Stern D, Harris E (eds) The chlamydomonas sourcebook. Organellar and metabolic processes, vol 2. Academic, New York, pp 333–339
57. Hanke GT, Okutani S, Satomi Y, Takao T, Suzuki A, Hase T (2005) Multiple iso-proteins of FNR in *Arabidopsis*: evidence for different contributions to chloroplast function and nitrogen assimilation. Plant Cell Environ 28:1146–1157
58. Harbinson J, Hedley CL (1989) The kinetics of P-700$^+$ reduction in leaves-a novel *in situ* probe of thylakoid functioning. Plant Cell Environ 12:357–369
59. Hawkesford MJ, De Kok LJ (2006) Managing sulphur metabolism in plants. Plant Cell Environ 29:382–395
60. Herzig R, Falkowski PG (1989) Nitrogen limitation in *Isochrysis galbana* (Haptophyceae) 1. Photosynthetic energy conversion and growth efficiencies. J Phycol 25:462–471
61. Heyno E, Gross CM, Laureau C, Culcasi M, Pietri S, Krieger-Liszkay A (2009) Plastid alternative oxidase (PTOX) promotes oxidative stress when overexpressed in tobacco. J Biol Chem 284:31174–31180
62. Hill R, Bendall F (1960) Function of the two cytochrome components in chloroplasts: a working hypothesis. Nature 186:136–137
63. Horton P, Ruban AV, Walters RG (1996) Regulation of light harvesting in green plants. Annu Rev Plant Physiol Plant Mol Biol 47:655–684
64. Houille-Vernes L, Rappaport F, Wollman F-A, Alric J, Johnson X (2011) Plastid terminal oxidase 2 (PTOX2) is the major oxidase involved in chlororespiration in *Chlamydomonas*. Proc Natl Acad Sci U S A 108:20820–20825
65. Inda LA, Erdner DL, Peleato ML, Anderson DM (1999) Cytochrome c(6) isolated from the marine diatom *Thalassiosira weissflogii*. Phytochemistry 51:1–4
66. Iwai M, Takizawa K, Tokutsu R, Okamuro A, Takahashi Y, Minagawa J (2010) Isolation of the elusive supercomplex that drives cyclic electron flow in photosynthesis. Nature 464:1210–1213
67. Jahns P, Graf M, Munekage Y, Shikanai T (2002) Single point mutation in the Rieske iron-sulfur subunit of cytochrome b(6)/f leads to an altered pH dependence of plastoquinol oxidation in *Arabidopsis*. FEBS Lett 519:99–102
68. Jans F et al (2008) A type II NAD(P) H dehydrogenase mediates light-independent plastoquinone reduction in the chloroplast of *Chlamydomonas*. Proc Natl Acad Sci U S A 105:20546–20551
69. Johnson GN (2003) Thiol regulation of the thylakoid electron transport chain-a missing link in the regulation of photosynthesis? Biochemistry 42:3040–3044
70. Johnson GN (2011) Physiology of PSI cyclic electron transport in higher plants. Biochim Biophys Acta 1807:384–389
71. Joliot P, Johnson GN (2011) Regulation of cyclic and linear electron flow in higher plants. Proc Natl Acad Sci U S A 108:13317–13322
72. Joliot A, Joliot P (1964) Etude Cinetique De La Reaction Photochimique Liberant Loxygene Au Cours De La Photosynthese. C R Hebd Seances Acad Sci 258:4622–4625
73. Joliot P, Joliot A (2002) Cyclic electron transport in plant leaf. Proc Natl Acad Sci U S A 99:10209–10214
74. Joliot P, Kok B (1975) Oxygen evolution in photosynthesis. In Govindjee (ed) Bioenergetics of photosynthesis. Academic, New York, pp 387–412
75. Kanazawa A, Kramer DM (2002) *In vivo* modulation of nonphotochemical exciton quenching (NPQ) by regulation of the chloroplast ATP synthase. Proc Natl Acad Sci U S A 99:12789–12794
76. Khorobrykh SA, Ivanov BN (2002) Oxygen reduction in a plastoquinone pool of isolated pea thylakoids. Photosynth Res 71:209–219
77. Kinoshita H et al (2011) The chloroplastic 2-oxoglutarate/malate transporter has dual function as the malate valve and in carbon/nitrogen metabolism. Plant J 65:15–26
78. Kirchhoff H, Hall C, Wood M, Herbstová M, Tsabari O, Nevo R, Charuvi D, Shimoni E, Reich Z (2011) Dynamic control of protein diffusion within the granal thylakoid lumen. Proc Natl Acad Sci U S A 108:20248–20253

79. Kramer DM, Cruz JA, Kanazawa A (2003) Balancing the central roles of the thylakoid proton gradient. Trends Plant Sci 8:27–32
80. Kulheim C, Agren J, Jansson S (2002) Rapid regulation of light harvesting and plant fitness in the field. Science 297:91–93
81. Laisk A (1993) Mathematical modelling of free-pool and channelled electron transport in photosynthesis: evidence for a functional supercomplex around photosystem 1. Proc R Soc Lond Ser B Biol Sci 251:243–251
82. Lavergne J, Junge W (1993) Proton release during the redox cycle of the water oxidase. Photosynth Res 38:279–296
83. Lehtimaki N, Lintala M, Allahverdiyeva Y, Aro EM, Mulo P (2010) Drought stress-induced upregulation of components involved in ferredoxin-dependent cyclic electron transfer. J Plant Physiol 167:1018–1022
84. Lemeille S, Rochaix J-D (2010) State transitions at the crossroad of thylakoid signalling pathways. Photosynth Res 106:33–46
85. Lennon AM, Prommeenate P, Nixon PJ (2003) Location, expression and orientation of the putative chlororespiratory enzymes, Ndh and IMMUTANS, in higher-plant plastids. Planta 218:254–260
86. Li Z, Wakao S, Fischer BB, Niyogi KK (2009) Sensing and responding to excess light. Annu Rev Plant Biol 60:239–260
87. Mackey KRM, Paytan A, Grossman AR, Bailey S (2008) A photosynthetic strategy for coping in a high-light, low-nutrient environment. Limnol Oceanogr 53:900–913
88. Malnoe A, Wollman F-A, de Vitry C, Rappaport F (2011) Photosynthetic growth despite a broken Q-cycle. Nat Commun 2:301
89. Marshall S, Orr A (1928) The photosynthesis of diatom cultures in the sea. J Mar Biol Ass UK 15:321–360
90. McDonald AE, Ivanov AG, Bode R, Maxwell DP, Rodermel SR, Hüner NPA (2011) Flexibility in photosynthetic electron transport: the physiological role of plastoquinol terminal oxidase (PTOX). Biochim Biophys Acta 1807:954–967
91. Mehler AH (1951) Studies on reactions of illuminated chloroplasts. I. Mechanism of the reduction of oxygen and other hill reagents. Arch Biochem Biophys 33:65–77
92. Merchant SS, Allen MD, Kropat J, Moseley JL, Long JC, Tottey S, Terauchi AM (2006) Between a rock and a hard place: trace element nutrition in *Chlamydomonas*. Biochim Biophys Acta 1763:578–594
93. Miginiac-Maslow M, Lancelin JM (2002) Intrasteric inhibition in redox signalling: light activation of NADP-malate dehydrogenase. Photosynth Res 72:1–12
94. Mitchell P (1975) The protonmotive Q cycle: a general formulation. FEBS Lett 59:137–199
95. Miyake C, Horiguchi S, Makino A, Shinzaki Y, Yamamoto H, Tomizawa K (2005) Effects of light intensity on cyclic electron flow around PSI and its relationship to non-photochemical quenching of Chl fluorescence in tobacco leaves. Plant Cell Physiol 46:1819–1830
96. Moolna A, Bowsher CG (2010) The physiological importance of photosynthetic ferredoxin NADP(+) oxidoreductase (FNR) isoforms in wheat. J Exp Bot 61:2669–2681
97. Moseley JL, Allinger T, Herzog S, Hoerth P, Wehinger E, Merchant S, Hippler M (2002) Adaptation to Fe-deficiency requires remodeling of the photosynthetic apparatus. EMBO J 21:6709–6720
98. Munekage Y, Hojo M, Meurer J, Endo T, Tasaka M, Shikanai T (2002) PGR5 is involved in cyclic electron flow around photosystem I and is essential for photoprotection in *Arabidopsis*. Cell 110:361–371
99. Munekage Y, Hashimoto M, Miyake C, Tomizawa K-I, Endo T, Tasaka M, Shikanai T (2004) Cyclic electron flow around photosystem I is essential for photosynthesis. Nature 429:579–582
100. Nandha B, Finazzi G, Joliot P, Hald S, Johnson GN (2007) The role of PGR5 in the redox poising of photosynthetic electron transport. Biochim Biophys Acta 1762:1252–1259
101. Nelson N, Yocum CF (2006) Structure and function of photosystems I and II. Annu Rev Plant Biol 57:521–565

102. Nouet C, Motte P, Hanikenne M (2011) Chloroplastic and mitochondrial metal homeostasis. Trends Plant Sci 16:395–404
103. Okegawa Y, Kobayashi Y, Shikanai T (2010) Physiological links among alternative electron transport pathways that reduce and oxidize plastoquinone in *Arabidopsis*. Plant J 63:458–468
104. Okutani S, Hanke GT, Satomi Y, Takao T, Kurisu G, Suzuki A, Hase T (2005) Three maize leaf ferredoxin: NADPH oxidoreductases vary in subchloroplast location, expression, and interaction with ferredoxin. Plant Physiol 139:1451–1459
105. Ort DR, Baker NR (2002) A photoprotective role for O(2) as an alternative electron sink in photosynthesis? Curr Opin Plant Biol 5:193–198
106. Ott T, Clarke J, Birks K, Johnson G (1999) Regulation of the photosynthetic electron transport chain. Planta 209:250–258
107. Palmer CM, Guerinot ML (2009) Facing the challenges of Cu, Fe and Zn homeostasis in plants. Nat Chem Biol 5:333–340
108. Peers G, Price NM (2005) A role for copper in the photosynthetic apparatus of a diatom from the open ocean. Plant Biol (Rockville) 2005:145
109. Peers G, Price NM (2006) Copper-containing plastocyanin used for electron transport by an oceanic diatom. Nature 441:341–344
110. Peltier G, Schmidt GW (1991) Chlororespiration-an adaptation to nitrogen deficiency in *chlamydomonas-reinhardtii*. Proc Natl Acad Sci U S A 88:4791–4795
111. Peng L, Yamamoto H, Shikanai T (2011) Structure and biogenesis of the chloroplast NAD(P)H dehydrogenase complex. Biochim Biophys Acta 1807:945–953
112. Perez-Torres E, Bravo LA, Corcuera LJ, Johnson GN (2007) Is electron transport to oxygen an important mechanism in photoprotection? Contrasting responses from Antarctic vascular plants. Physiol Plant 130:185–194
113. Petroutsos D, Terauchi AM, Busch A, Hirschmann I, Merchant SS, Finazzi G, Hippler M (2009) PGRL1 participates in iron-induced remodeling of the photosynthetic apparatus and in energy metabolism in *Chlamydomonas reinhardtii*. J Biol Chem 284:32770–32781
114. Polle A (1996) Mehler reaction: friend or foe in photosynthesis? Botanica Acta 109:84–89
115. Rees D, Noctor G, Ruban AV, Crofts J, Young A, Horton P (1992) Ph dependent chlorophyll fluorescence quenching in spinach thylakoids from light treated or dark adapted leaves. Photosynth Res 31:11–19
116. Rosso D et al (2006) IMMUTANS does not act as a stress-induced safety valve in the protection of the photosynthetic apparatus of *Arabidopsis* during steady-state photosynthesis. Plant Physiol 142:574–585
117. Rosso D et al (2009) Photosynthetic redox imbalance governs leaf sectoring in the *Arabidopsis thaliana* variegation mutants immutans, spotty, var1, and var2. Plant Cell 21:3473–3492
118. Rumeau D, Peltier G, Cournac L (2007) Chlororespiration and cyclic electron flow around PSI during photosynthesis and plant stress response. Plant Cell Environ 30:1041–1051
119. Ryther JH (1956) Photosynthesis in the ocean as a function of light intensity. Limnol Oceanogr 1:61–70
120. Sacksteder CA, Kanazawa A, Jacoby ME, Kramer DM (2000) The proton to electron stoichiometry of steady-state photosynthesis in living plants: a proton -pumping Q cycle is continuously engaged. Proc Natl Acad Sci U S A 97:14283–14288
121. Sandstrom S, Ivanov AG, Park YI, Oquist G, Gustafsson P (2002) Iron stress responses in the cyanobacterium *Synechococcus* sp PCC7942. Physiol Plant 116:255–263
122. Schottler MA, Kirchhoff H, Weis E (2004) The role of plastocyanin in the adjustment of the photosynthetic electron transport to the carbon metabolism in tobacco. Plant Physiol 136:4265–4274
123. Shao N, Beck CF, Lemaire SD, Krieger-Liszkay A (2008) Photosynthetic electron flow affects H2O2 signaling by inactivation of catalase in *Chlamydomonas reinhardtii*. Planta 228:1055–1066

124. Shikanai T (2007) Cyclic electron transport around photosystem I: genetic approaches. Annu Rev Plant Biol 58:199
125. Singh AK, Sherman LA (2007) Reflections on the function of IsiA, a cyanobacterial stress-inducible, Chl-binding protein. Photosynth Res 93:17–25
126. Stepien P, Johnson GN (2009) Contrasting responses of photosynthesis to salt stress in the glycophyte *Arabidopsis thaliana* and the halophyte *Thellungiella halophila*. Role of the plastid terminal oxidase as an alternative electron sink. Plant Physiol 149:1154–1165
127. Streb P, Josse EM, Gallouet E, Baptist F, Kuntz M, Cornic G (2005) Evidence for alternative electron sinks to photosynthetic carbon assimilation in the high mountain plant species *Ranunculus glacialis*. Plant Cell Environ 28:1123–1135
128. Stroebel D, Choquet Y, Popot JL, Picot D (2003) An atypical haem in the cytochrome b(6) f complex. Nature 426:413–418
129. Strzepek RF, Harrison PJ (2004) Photosynthetic architecture differs in coastal and oceanic diatoms. Nature 431:689–692
130. Suggett DJ, MacIntyre HL, Kana TM, Geider RJ (2009) Comparing electron transport with gas exchange: parameterising exchange rates between alternative photosynthetic currencies for eukaryotic phytoplankton. Aquat Microb Ecol 56:147–162
131. Sukenik A, Bennett J, Mortainbertrand A, Falkowski PG (1990) Adaptation of the photosynthetic apparatus to irradiance in dunaliella-tertiolecta-a kinetic-study. Plant Physiol 92:891–898
132. Tagawa K, Tsujimoto HY, Arnon DI (1963) Role of chloroplast ferredoxin in the energy conversion process of photosynthesis. Proc Natl Acad Sci U S A 49:567–572
133. Takahashi H, Iwai M, Takahashi Y, Minagawa J (2006) Identification of the mobile light-harvesting complex II polypeptides for state transitions in *Chlamydomonas reinhardtii*. Proc Natl Acad Sci U S A 103:477–482
134. Takeda T, Ishikawa T, Shigeoka S (1997) Metabolism of hydrogen peroxide by the scavenging system in *Chlamydomonas reinhardtii*. Physiol Plant 99:49–55
135. Terashima M et al (2012) Calcium-dependent regulation of cyclic photosynthetic electron transfer by a CAS, ANR1, and PGRL1 complex. Proc Natl Acad Sci U S A 109:17717–17722
136. Tikkanen M, Grieco M, Kangasjarvi S, Aro EM (2010) Thylakoid protein phosphorylation in higher plant chloroplasts optimises electron transfer under fluctuating light. Plant Physiol 152:723–735
137. Tokutsu R, Iwai M, Minagawa J (2009) CP29, a monomeric light-harvesting complex II protein, is essential for state transitions in *Chlamydomonas reinhardtii*. J Biol Chem 284:7777–7782
138. Tolleter D et al (2011) Control of hydrogen photoproduction by the proton gradient generated by cyclic electron flow in *Chlamydomonas reinhardtii*. Plant Cell 23:2619–2630
139. Tremmel IG, Kirchhoff H, Weis E, Farquhar GD (2003) Dependence of plastoquinol diffusion on the shape, size, and density of integral thylakoid proteins. Biochim Biophys Acta 1607:97–109
140. Trouillard M, Shahbazi M, Moyet L, Rappaport F, Joliot P, Kuntz M, Finazzi G (2012) Kinetic properties and physiological role of the plastoquinone terminal oxidase (PTOX) in a vascular plant. Biochim Biophys Acta 1817:2140–2148
141. Umena Y, Kawakami K, Shen J-R, Kamiya N (2011) Crystal structure of oxygen-evolving photosystem II at a resolution of 1.9 angstrom. Nature 473:U55–U65
142. Vallon O, Bulte L, Dainese P, Olive J, Bassi R, Wollman FA (1991) Lateral redistribution of cytochrome B6/F complexes along thylakoid membranes upon state transitions. Proc Natl Acad Sci U S A 88:8262–8266
143. Varsano T, Wolf SG, Pick U (2006) A chlorophyll a/b-binding protein homolog that is induced by iron deficiency is associated with enlarged photosystem I units in the eucaryotic alga *Dunaliella salina*. J Biol Chem 281:10305–10315
144. Voss I et al (2011) FdC1, a novel ferredoxin protein capable of alternative electron partitioning, increases in conditions of acceptor limitation at photosystem I. J Biol Chem 286:50–59

145. Weger HG, Herzig R, Falkowski PG, Turpin DH (1989) Respiratory losses in the light in a marine diatom-measurements by short-term mass-spectrometry. Limnol Oceanogr 34:1153–1161
146. Wykoff DD, Davies JP, Melis A, Grossman AR (1998) The regulation of photosynthetic electron transport during nutrient deprivation in *Chlamydomonas reinhardtii*. Plant Physiol 117:129–139
147. Yamamoto H et al (2006) Ferredoxin limits cyclic electron flow around PSI (CEF-PSI) in higher plants-stimulation of CEF-PSI enhances non-photochemical quenching of Chl fluorescence in transplastomic tobacco. Plant Cell Physiol 47:1355–1371
148. Zehr JP, Kudela RM (2009) Photosynthesis in the open ocean. Science 326:945–946
149. Zhang HM, Whitelegge JP, Cramer WA (2001) Ferredoxin: NADP(+) oxidoreductase is a subunit of the chloroplast cytochrome b(6)f complex. J Biol Chem 276:38159–38165

Chapter 17
Ion homeostasis in the Chloroplast

Marc Hanikenne, María Bernal and Eugen-Ioan Urzica

Abstract The chloroplast is an organelle of high demand for macro- and micronutrient ions, which are required for the maintenance of the photosynthetic process. To avoid deficiency while preventing excess, homeostasis mechanisms must be tightly regulated. Here, we describe the needs for nutrient ions in the chloroplast and briefly highlight their functions in the chloroplastidial metabolism. We further discuss the impact of nutrient deficiency on chloroplasts and the acclimation mechanisms that evolved to preserve the photosynthetic apparatus. We finally present what is known about import and export mechanisms for these ions. Whenever possible, a comparison between cyanobacteria, algae and plants is provided to add an evolutionary perspective to the description of ion homeostasis mechanisms in photosynthesis.

Keywords Mineral nutrition · Macronutrients · Micronutrients · Metal · Cation · Anion · Transport

Abbreviations

Arabidopsis	*Arabidopsis thaliana*
ABC	ATP-binding-cassette
APS	Adenosine 5′-phosphosulfate
ATP	Adénosine-5′-triphosphate
ATPS	ATP Sulfurylase
CaM	Calmodulin

M. Hanikenne (✉)
Functional Genomics and Plant Molecular Imaging, Center for Protein Engineering (CIP), PhytoSystems, B22, Department of Life Sciences, University of Liège, Bld du Rectorat, 27, Sart Tilman, 4000 Liège, Belgium
e-mail: marc.hanikenne@ulg.ac.be

M. Bernal
Plant Nutrition Department, Estación Experimental De Aula Dei, Consejo Superior de Investigaciones Científicas (CSIC), Avda. Montañana 1005, 50059 Zaragoza, Spain

Department of Plant Physiology, Ruhr University Bochum, 44801 Bochum, Germany

E.-I. Urzica
Department of Chemistry and Biochemistry, UCLA, Los Angeles, CA 90095, USA

S.M. Theg, F.-A. Wollman (eds.), *Plastid Biology,* Advances in Plant Biology 5,
DOI 10.1007/978-1-4939-1136-3_17,

CAS	Calcium-sensing protein
Chl	Chlorophyll
CPA2	Cation/proton antiporters 2
CuRE	Copper responsive element
Cytc_6	Cytochrome c_6
CytOx	Cytochrome *c* oxidase
Chlamydomonas	*Chlamydomonas reinhardtii*
Ery-4-P	Erythrose 4-phosphate
Fe/S	Iron/Sulfur cluster
FD	Ferredoxin
Glu-6-P	Glucose 6-phosphate
GOGAT	Glutamate synthase
GPT	Glucose 6-phosphate/phosphate translocator
GS	Glutamine synthetase
GSH	Glutathione
GUN	Genomes uncoupled
ISC	Iron-sulfur cluster
LHC	Light harvesting complex
NADK	NAD kinase
NAP	Non intrinsic ABC protein
NIF	NItrogenase fixation
NiR	Nitrite reductase
NR	Nitrate reductase
OEC	Oxygen-evolving complex
OEP	Outer envelop proteins
PC	Plastocyanin
PEP	Phosphoenolpyruvate
Pi	Inorganic orthophosphate
PNPase	Ribonuclease polynucleotide phosphorylase
PPT	Phosphoenolpyruvate/phosphate translocator
PSI	Photosystem I
PSII	Photosystem II
PTOX	Plastoquinol terminal oxidase
Pyr	Pyruvate
Rib-5-P	Ribulose 5-phosphate
ROS	Reactive oxygen species
RuBisCO	Ribulose-1,5-bisphosphate carboxylase
SOD	Superoxide dismutase
SUF	Sulfur fixation
TIC	Translocon at the Inner envelope membrane of Chloroplast
TOC	Translocon at the Outer envelope membrane of Chloroplast
TP	Triose phosphate
TPT	Triose phosphate translocator (TPT)
XPT	Xylulose 5-phosphate/phosphate translocator
Xyl-5-P	Xylulose 5-phosphate

17.1 Introduction

Chloroplasts are central in the metabolism of photosynthetic organisms, and more generally for global primary productivity in waters and soils. Photosynthesis provides energy for many major metabolic pathways, e.g. CO_2 assimilation and starch, fatty acid, amino acid, nucleic acid synthesis, as well as reductive assimilation of inorganic ions like nitrate and sulfate. These numerous processes heavily rely on nutrient ions, a term which in this chapter is used for all essential macro- and micronutrients, i.e. cation and anion nutrients that are required for proper plant growth and development. Hence, nutrient deficiency often results in leaf chlorosis, i.e. in chlorophyll (Chl) loss, and in impaired photosynthesis [207, 208]. Chloroplasts are thus organelles with high demand for nutrient ions, and that tightly regulate ion homeostasis mechanisms to ensure the maintenance of the photosynthetic function.

In this chapter, we will describe the requirements of chloroplasts for each macro- (Mg, Ca, K, Na, P, S, N, Cl) and micro- (Fe, Cu, Zn, Mn, Co, Ni) nutrients and detail their key functions in the chloroplast metabolism (Table 17.1). We will review the impact of nutrient deficiency on chloroplasts and our growing understanding of the acclimation mechanisms acting at the organelle, cellular and organism levels to preserve the photosynthetic function under nutrient deficiency.

Chloroplasts have three membrane systems, the outer- and inner-envelope membranes and the thylakoid membrane, stemming from the endosymbiotic origin of the organelle [90, 320]. Several import and export systems are required for nutrient ion movements across these membrane systems. Although several solute channels (or OEP for Outer Envelop Proteins) for a variety of substrates are found in the outer envelope membrane of chloroplasts (for reviews, see [34, 86, 353, 362]), the transport of nutrient ions through this membrane is generally viewed as non-selective. In contrast, their transport through the inner envelope and thylakoid membranes requires specific transporters. Several proteomic studies have been conducted to identify chloroplastidial proteins and to determine their sub-plastidial localization [85, 98–100, 104, 160, 291, 324, 362]. The outcome of these efforts is available at the AT_Chloro database (http://www.grenoble.prabi.fr/at_chloro/) [100] and the Plant Proteomics database (PPDB; http://ppdb.tc.cornell.edu/) [324]. Around 1300 proteins are present in the databases and sub-plastidial localizations are provided for >800 proteins. Among those, several hundreds of proteins are located in the envelope or the thylakoid membranes, including transporters for most macro- and micro-nutrients (Table 17.2, Figs. 17.1 and 17.2). Genome sequence analyses in cyanobacteria, algae and plants also allowed identification of a number of putative transporters in chloroplasts. However, in many cases, the actual proteins responsible for specific transport activities that have been measured biochemically remain to be identified. We will nevertheless detail our current knowledge on proteins involved in import and export of nutrient ions across chloroplast membranes.

Finally, we will briefly discuss the role of chelation and metallochaperones in the homeostasis of metal micronutrients. So-called metallochaperones are shuttle proteins that ensure proper metal delivery to target apoproteins and also prevent metal

Table 17.1 Essential functions of macro- and micronutrient ions in plastids

Nutrients		Functions[a]
A. Macronutrients		
1. Magnesium	Mg^{2+}	Chlorophyll synthesis
		Activation of metabolic enzymes
		Arrangement and stacking of thylakoid membranes
		Nucleic acid metabolism
2. Calcium	Ca^{2+}	Oxygen evolution, photoassembly of photosystem II–manganese cluster
		CO_2 fixation, Calvin Benson Cycle
		Stomatal closure
		Photoacclimation
		Redox reactions (ferredoxin)
		Metabolism (NADK)
		Protein import
3. Potassium	K^+	Counter-ion for light-dependent proton movement (electroneutrality and maintenance of pH)
		Chloroplast development
		Control of stomatal resistance
4. Sodium	Na^+	Regulation of stromal acidification and alkalinization
		Counter-ion for pyruvate uptake
		Osmotic balance in response to high salinity
5. Phosphorus	PO_4^{2-} (Pi)	ATP synthesis
		Organellar DNA synthesis
		Organellar RNA synthesis and degradation
		Phospholipid synthesis
		Regulator of protein activity in the photosynthetic apparatus
		Transport of carbon compounds between stroma and cytoplasm
6. Sulfur	SO_4^{2-}	Assimilation of sulfate (SO_4^{2-}) into sulfide (S^{2-})
		Cystein synthesis
		Fe/S cluster synthesis
		Sulfolipid synthesis
7. Nitrogen	NO_2^-/NH_4^+	Assimilation of nitrite (NO_2^-) into ammonium (NH_4^+) (nitrite reductase)
		Assimilation of ammonium (NH_4^+) into glutamate (Glu) (GS/GOGAT)
		Re-assimilation of the ammonium produced in the mitochondria (GS/GOGAT)
8. Chloride	Cl^-	pH homeostasis across the envelope and thylakoid membranes
B. Micronutrients		
1. Iron	Fe^{3+}/Fe^{2+}	Photosynthetic electron transport chain (photosystems II and I, cytochrome b_6f, ferredoxins, plastoquinol terminal oxidase)
		Fe/S cluster synthesis
		Heme synthesis
		Redox control (FeSOD, ferritin)
2. Copper	Cu^{2+}/Cu^+	Photosynthetic electron transport chain (plastocyanin)
		Redox control (Cu/ZnSOD)

Table 17.1 (continued)

Nutrients		Functions[a]
3. Manganese	Mn^{2+}	Catalytic center of the water–splitting complex in photosystem II
4. Zinc	Zn^{2+}	Cofactor for the RNA polymerase and zinc finger nucleic acid–binding proteins
		Cofactor of several enzymes (e.g. carbonic anhydrase, D-ribulose-5-phosphate 3-epimerase)
		Cofactor for proteolytic activities
		Repair of photosystem II
5. Cobalt	Co^{2+}	?
6. Nickel	Ni^{2+}	?

[a] See main text for a detailed description and references

toxicity by maintaining virtually no free metal ions in cells [24, 54, 101, 256, 287, 363]. Only a few chaperones have been identified so far in plant chloroplasts [255].

17.2 Macronutrients

17.2.1 Magnesium

Magnesium is involved in numerous physiological and biochemical processes affecting growth and development in plants [47, 211, 307]. A few examples of functions of magnesium in chloroplasts include the activation of metabolic enzymes, the arrangement and stacking of thylakoid membranes, and nucleic acid metabolism [16, 110, 212, 295]. The role of magnesium as the central atom of Chl molecules is its best-known function in photosynthetic organisms. Chl *a* and *b* are the most abundant tetrapyrrole molecules in plants and, depending on the magnesium status of the plant, up to 35 % of the total magnesium can be found in chloroplasts, mostly associated to Chl [47, 307, 366]. In the tetrapyrrole biosynthetic pathway, protoporphyrin IX, a closed macrocycle lacking a chelated ion, represents the branch point of the Chl and heme biosynthesis (see also Sect. 17.3.1). Insertion of magnesium (as Mg^{2+}) into protoporphyrin IX by the magnesium chelatase (Mg-chelatase) to form Mg-protoporphyrin IX is the first step of the so-called Mg branch, the pathway specifically committed to Chl synthesis [211]. Chl biosynthesis is tightly coordinated with photosynthetic activity. Hence, Mg-chelatase activity is regulated by a diurnal cycle and photosynthetic electron transport. Moreover, as 95 % of the ~3000 chloroplastidial proteins are encoded in the nuclear genome, the expression of organellar and nuclear genes is tightly coordinated through communication between the two compartments to respond to environmental and developmental cues (see also Chap. 3 in this volume). Nuclear gene expression is regulated by so-called retrograde signaling pathways coming from damaged or malfunctioning chloroplasts [116, 211, 254, 369]. One retrograde signal was initially thought to be the accumu-

Table 17.2 Ion transporters in Arabidopsis plastids

Name	Alternative name	AGI number	Putative substrate		Localization							References
			Predicted[a]	Experimental evidence	ARAMEMNON[b]	Experimental evidence: GFP	Experimental evidence: Fractionation[c]	Experimental evidence: Immunoloc[d]	Proteomics: AT_Chloro[e]	Proteomics: PPDB[f]	Proteomics: Other[g]	
A. Macronutrient ions												
MRS2-11	MGT10	At5g22830	Mg^{2+}	Yeast, *Salmonella*	CHLORO	ENV	IM	–	IM	IM	–	[24, 108, 192, 194]
ACA1	PEA1	At1g27770	Ca^{2+}	–	ENV	–	IM	–	–	IM	ER	[85, 146]
ALB3	–	At2g28800	Ca^{2+}	Mutant phenotype[h]	ENV	–	–	ENV	THYL	THYL	–	[193, 325, 357]
KT12	KUP12	At1g60160	K^+	–	–	–	–	–	–	–	CHLORO/TONO	[160, 367]
KEA1	–	At1g01790	K^+/H^+	–	ENV	–	–	–	ENV	IM	–	–
KEA2	–	At4g00630	K^+/H^+	–	–	–	–	–	ENV	ENV	–	–
TPK3	KCO6	At4g18160	K^+	–	THYL	–	THYL	–	–	–	–	[384]
CHX23	–	At1g05580	$K^+(Na^+)/H^+$	Mutant phenotype	ENV	ENV	–	–	–	ENV	–	[318]
BASS2	–	At2g26900	$Pyr/Na^+–H^+$	Mutant phenotype	ENV	Plastid[i]	–	CHLORO[i]	IM	IM	–	[106]
PHT2;1	–	At3g26570	$Pi/H^+(Na^+)$	Yeast, Mutant phenotype	–	ENV	–	–	IM	IM	–	[348]
PHT4;1	ANTR1	At2g29650	$Pi/H^+(Na^+)$	Yeast, *E. coli*	CHLORO	Plastid/IM	THYL	–	IM	IM/THYL	–	[120, 292, 294]
PHT4;2	–	At2g38060	Pi/H^+	Yeast	CHLORO	IM	Plastid (root only)	–	–	IM	–	[120, 149]
PHT4;3	–	At3g46980	Pi/H^+	Yeast	–	–	–	–	IM	Plastid	–	–
PHT4;4	ANTR2	At4g00370	Pi/H^+	Yeast	IM	IM	IM	–	IM	IM	–	[120, 292, 294]
PHT4;5		At5g20380	Pi/H^+	Yeast	CHLORO	IM			–	Plastid	–	[120]

Table 17.2 (continued)

Name	Alternative name	AGI number	Putative substrate		Localization							References
			Predicted[a]	Experimental evidence	ARAMEMNON[b]	Experimental evidence			Proteomics			
						GFP	Fractionation[c]	Immunoloc[d]	AT_Chloro[e]	PPDB[f]	Other[g]	
–	–	At1g68570	NO_2^-	Yeast, Mutant phenotype	–	–	ENV	–	–	IM	–	[323]
CLCe	–	At4g35440	Cl^-/NO_2^-	Yeast, Mutant phenotype	THYL	CHLORO	THYL	–	–	THYL	–	[206, 230]
B. Micronutrient ions												
PIC1	TIC21	At2g15290	Fe^{3+}/Fe^{2+}?	Yeast, Mutant phenotype	CHLORO	IM	IM	–	IM	IM	–	[87]
IREG3	MAR1/ FPN3	At5g26820	Fe^{2+}(–NA?) NA?	Mutant phenotype	CHLORO	ENV	–	–	–	ENV	–	[72]
PAA1	HMA6	At4g33520	Cu^+	Mutant phenotype	CHLORO	IM	CHLORO[j]	–	ENV	IM	–	[3, 312]
PAA2	HMA8	At5g21930	Cu^+	Mutant phenotype	CHLORO	THYL[k]	THYL[j]	–	–	THYL	–	[3]
HMA1	–	At4g37270	Cu^{2+}, Zn^{2+}, Ca^{2+}, Cd^{2+}?	Yeast, Mutant phenotype	ENV	ENV	ENV	–	IM	ENV	–	[158, 234, 305]
NiCoT	–	At2g16800	Ni^{2+}/Co^{2+}	–	CHLORO	ENV	–	–	–	–	–	[89]
NiCoT	–	At4g35080	Ni^{2+}/Co^{2+}	–	CHLORO	ENV	–	–	–	–	–	[89]

CHLORO chloroplast, *ENV* chloroplast envelope, *ER* endoplasmic reticulum, *IM* inner envelope membrane, *NA* nicotianamine, *OM* outer envelope membrane, *Plastid* heterotrophic plastid and chloroplast, *Pyr* pyruvate, *THYL* thylakoid membrane, *TONO* tonoplast; [a] Substrate prediction based on sequence homology; [b] ARAMEMNON database, http://aramemnon.uni-koeln.de/, [304]; [c] Purification of chloroplast membranes followed by a Western blot analysis; [d] Detection of the protein by immunocytochemistry; [e] AT_Chloro database, http://www.grenoble.prabi.fr/at_chloro/, [100]; [f] PPDB, the Plant Proteomics database, http://ppdb.tc.cornell.edu/, [324]; [g] Conflicts in protein localization by different proteomics studies are indicated; [h] Functional analysis was conducted for the pea homolog (PPF1) of ALB3 [193, 357]; [i] Detection of the BASS2 protein of *Flaveria trinervia* by immolocalization or in fusion with GFP expressed in tobacco root epidermal cells [106]; [j] Determined by analysis of transit peptide functionality [3, 312]; [k] A truncated PAA2 protein (residues 1 to 306 including the first four predicted transmembrane domains) was fused to GFP [3]

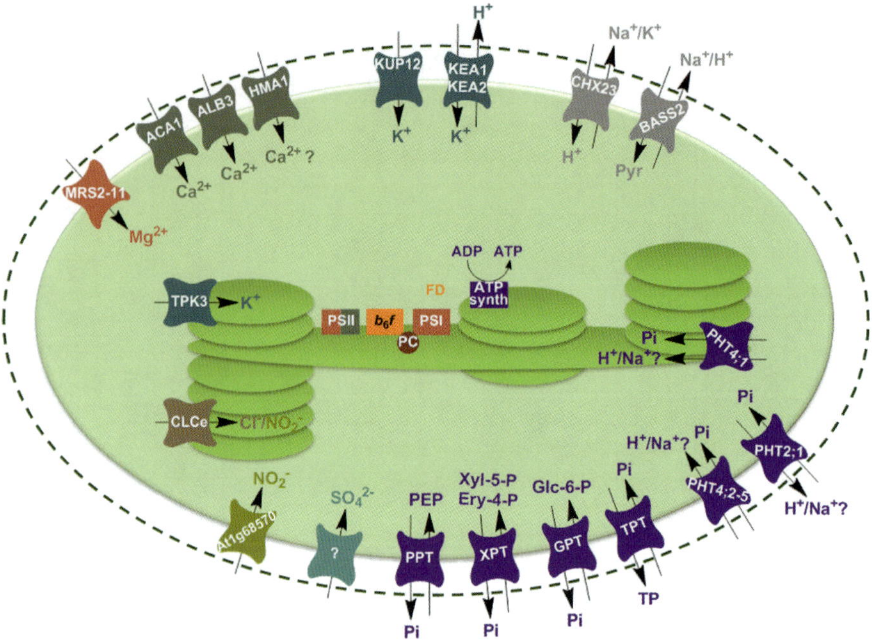

Fig. 17.1 Macronutrient ion homeostasis in plastids. Transmembrane transporters are represented by boxes and arrows, together with their putative substrates. Proteins and processes associated to each macronutrient are differentiated by colors. *Questions marks* (?) are pointing to uncertain functions or unidentified components. Note that some functions are only present in chloroplasts or in heterotrophic plastids, respectively. See main text for details

lation of Mg-protoporphyrin IX in damaged chloroplasts, which led to the down-regulation of hundreds of nuclear genes [269, 321, 369]. Indeed, four of the five *gun* (*genomes uncoupled*) mutants identified in *Arabidopsis thaliana* (Arabidopsis) are affected in tetrapyrrole biosynthesis and are involved in modulating the level of Mg-protoporphyrin IX. *gun2* and *gun3* are mutants of the genes encoding heme oxygenase and phytochromobilin synthase, respectively, whereas *gun5* is a mutant of the H subunit of Mg-chelatase [226]. GUN4 is a porphyrin-binding protein that activates Mg-chelatase [6, 7, 183]. However, more recent studies have established that protoporphyrin IX does not act as a direct signaling molecule [227, 241]. It is likely that the perturbation of tetrapyrrole synthesis in the *gun* mutants leads to localized production of reactive oxygen species (ROS) or changes in the redox state of the chloroplast, which could mediate retrograde signaling (see below in this section) [161, 227, 241]. In red and green algae, Mg-protoporphyrin IX appears to play a role in the signaling between chloroplasts and nucleus [165, 172, 173, 352]. Hence, Mg-protoporphyrin IX alters significantly gene expression in *Chlamydomonas reinhardtii* (Chlamydomonas) [352]. Other retrograde pathways respond to inhibition of plastid gene expression and to the redox state of the photosynthetic electron transport chain (which itself notably depends on redox state of the plastoquinone

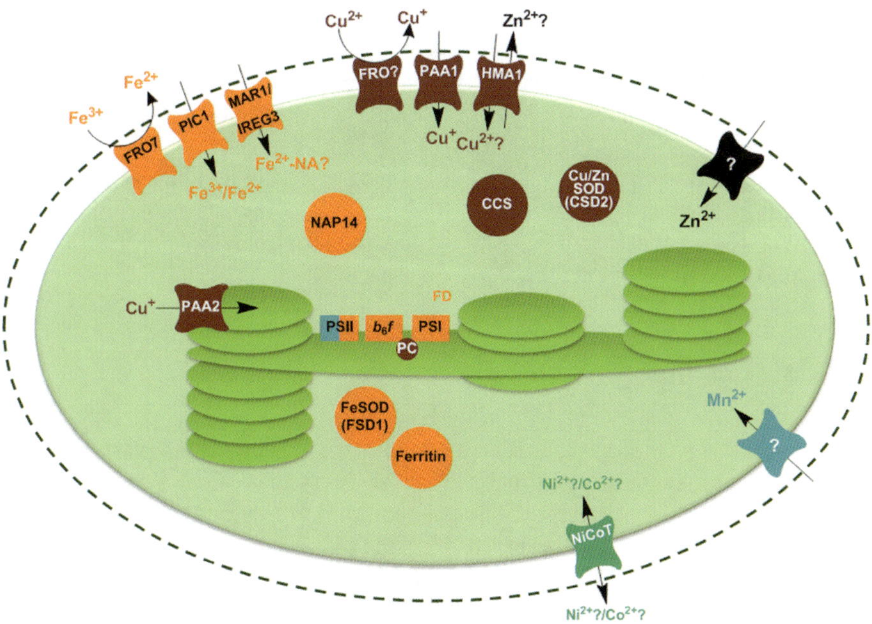

Fig. 17.2 Micronutrient ion homeostasis in plastids. Transmembrane transporters are represented by boxes and arrows, together with their putative substrates. Proteins and processes associated to each micronutrient are differentiated by colors. *Questions marks* (?) are pointing to uncertain functions or unidentified components. See main text for details

pool and ROS production), respectively. All pathways are integrated via GUN1, a chloroplastidial pentatricopeptide-repeat protein, and the nuclear transcription factor ABI4 (abscisic acid–insensitive 4, an APETALA 2–type transcription factor) to adapt nuclear transcriptional activity to the chloroplast functional state [166].

Magnesium deficiency is classically associated with the development of interveinal leaf chlorosis. However, increased sugar (sucrose and starch) concentration and altered sucrose export from young source leaves precedes noticeable effects of magnesium deficiency on photosynthetic activity. Magnesium deficiency impairs loading of sucrose into the phloem, possibly by affecting the Mg-ATP-dependent activity of proton pumps. The early sugar accumulation in young leaves accounts for the decrease in Chl content rather than a lack of magnesium for Chl biosynthesis [47, 48, 135, 136]. In addition, another early response to magnesium deficiency is an increase of anti-oxidative mechanisms. Indeed, the production of ROS is potentiated by the magnesium deficiency-triggered over-reduction in the photosynthetic electron transport chain, especially at high light intensity [47]. In agreement, about 50 genes involved in oxidative stress defence and in photoprotection of the photosynthetic apparatus are up-regulated in response to long-term magnesium deficiency in Arabidopsis [138].

Contrary to other mineral deficiencies, magnesium deficiency does not induce higher expression of magnesium uptake genes [137, 138]. Among the ten magnesium transporters (named MSR2 or MGT) of the CorA family identified in Arabidopsis [163, 192, 286], only AtMRS2-11/MGT10 localizes in the chloroplast (Table 17.2, Fig. 17.1) [84]. AtMRS2-11 transports Mg^{2+} in both bacteria and yeast [84, 108, 192, 194]. The *AtMRS2-11* gene displays a diurnal cycling of expression, with higher expression in the light than in the dark. The gene is highly expressed in the mesophyll of cotyledons and of developing and adult leaves. It is also highly expressed in stomata guard cells [84, 108]. A *mrs2-11* mutant has no observable phenotype [84], suggesting that unknown additional mechanisms are involved in chloroplastidial magnesium uptake.

17.2.2 Calcium

Calcium (Ca^{2+}) is important for several key processes inside chloroplasts. Chloroplasts play an essential role in maintaining low cytosolic calcium concentrations and it has been proposed that the free calcium ion in the stroma regulates key enzymes involved in photosynthesis [36, 169] and that it is essential for O_2 evolution by Photosystem II (PSII) [118]. In particular, calcium ions are essential for the function of the oxygen-evolving complex (OEC), a complex responsible for the light-dependent oxygen evolution in plants. PSII requires, besides several polypeptide subunits, a cluster of three inorganic ions (manganese, calcium and chloride, see also Sect. 17.3.3) [223, 224]. Additionally, it has been demonstrated that calcium plays an essential role in photo-assembly of PSII-manganese cluster [57]. Crystal structures of PSII from *Thermosynechococcus elongatus* indicated that calcium and manganese are integral components of the OEC [97]. The PSBP and PSBO subunits of PSII are capable of calcium binding, but molecular mechanisms about calcium delivery into PSII are unknown [326].

A connection between redox processes and calcium regulation in chloroplasts was found in the ability of ferredoxin (FD) to bind calcium with highest affinity in the reduced state and in the light when the stromal calcium concentrations are low [327]. NAD kinase (NADK) which catalyzes the phosphorylation of NAD(H) in the presence of ATP, was the first calmodulin (CaM)-regulated enzyme identified in plants [13]. One of the three NADK isoforms in Arabidopsis (NADK2) is responsible for CaM-dependent NADK activity in chloroplasts [347, 355]. Other examples of chloroplast proteins that bind to CaM are AtPSAN (subunit of Photosystem I, PSI), the chaperonin AtCPN10 and two AAA^+-ATPases, AtCIP111 and AtFIG1L1 [40, 46, 282, 379]. Although, these proteins might play a role in calcium regulation network, the functional relevance of these interactions is still unclear.

CaM can promote chloroplast protein import via TOC (Translocon at the Outer envelope membrane of Chloroplast) and TIC (Translocon at the Inner envelope membrane of Chloroplast) [63]. Regulation of protein import by calcium occurs at the TIC complex and is most likely due to the interaction of CaM with Tic32 [64]. In a more recent study [22], electrophysiological measurements indicated that calcium specifically affects the channel activity of Tic110.

Several studies demonstrated that calcium plays also an important role in the regulation of key enzymes of the Calvin-Benson cycle. Kreimer et al. [169] showed that calcium has a positive effect in the activation of fructose-1,6-biphosphatase and this calcium-dependent activation is related to the redox state of the enzyme [55]. High calcium concentrations result in inhibition of the catalytic activities of fructose-1,6-biphosphatase and sedoheptulose-1,7-biphosphatase [56]. Calcium fluxes in chloroplasts have been shown to be dependent on daily rhythms after the transition from light to dark [153, 298]. This calcium flux into the chloroplast stroma after transition to dark is proposed to be responsible for inhibiting the CO_2 fixation during the night. All these characteristics of calcium fluxes suggest that chloroplastidial import of calcium from the cytosol occurs during the light period and most likely that calcium is stored in the thylakoid membrane.

Recent studies have also investigated the effect of calcium on processes not directly linked to photosynthesis. Calcium-sensing protein (CAS) was initially characterized in Arabidopsis as a plasma membrane protein that mediates extracellular calcium sensing in guard cells [123]. However, later studies demonstrated that CAS is a thylakoid membrane protein in Arabidopsis [264] and Chlamydomonas [9, 334] and binds calcium with low affinity [123]. The Arabidopsis CAS plays a crucial role for proper stomatal regulation in response to elevations of external calcium concentrations through modulation of the cytoplasmic calcium levels [253, 365]. In recent studies, CAS has been suggested to regulate stomatal closure in guard cells through elevated levels of nitric oxide and hydrogen peroxide, suggesting a signaling interaction between calcium and the antioxidant enzymatic system in stomatal movement [358]. In Chlamydomonas, CAS has been shown to be an essential component of acclimation and adaptation to high-light stress and CAS together with calcium is essential for dissipating excess energy by efficient qE quenching (i.e. energy-dependent quenching, the main component of Non Photochemical Quenching, NPQ) [266].

Although the role of calcium in photosynthesis and signal transduction has been studied in detail, there have been relatively few investigations on calcium uptake by chloroplasts. Pioneering studies indicated that chloroplasts accumulate calcium to concentrations as high as 13–25 mM [184, 252, 376]. While cytosolic calcium concentrations are in the nanomolar range [153], free stromal calcium content in the dark is between 2–6 μM [169]. Early experiments demonstrated that light induces calcium import into isolated chloroplasts [168]. All these observations suggested a mechanism for calcium transport into the chloroplast that is facilitated by an energy-dependent calcium pump [168] or by a H^+/Ca^{2+}antiporter [244]. The presence of a H^+/Ca^{2+}antiporter was supported by later experiments which demonstrated that calcium is translocated across the thylakoidal membrane in a light- and energy-dependent manner. The transport process of calcium is sensitive to proton-translocating uncouplers like nigericin/K^+ [92]. An alternative calcium import mechanism has been proposed by Kreimer et al. [168] that would involve an electrogenic uniport-type carrier. The light-dependent calcium uptake by isolated chloroplasts is stimulated by a negative membrane potential and inhibited by ruthenium red, a known inhibitor of electrogenic calcium influx mediated by uniport-type carriers

[168, 290]. To date, the molecular identity of the calcium transporter/pump has not been described. However, there is evidence for the existence of two potential Ca^{2+}-ATPases in the chloroplast envelope (Table 17.2, Fig. 17.1). The first described one is ACA1 (PEA1) from Arabidopsis, which is a member of the auto-inhibited Ca^{2+}-ATPases family [146]. AtACA1 has been shown to be localized in the inner chloroplast envelope and is more abundant in root plastids. Nevertheless, in two independent studies no calcium-dependent ATPase activity or ATP-dependent calcium uptake could be detected in the chloroplast envelope [146, 290]. Moreover, the chloroplast localization of ACA1 was further questioned by studies which indicated that ACA1 is present in the Arabidopsis ER [85]. The second potential chloroplast Ca^{2+}-ATPase is HMA1, which is a member of P_{IB} ATPase family and was first identified as a chloroplast protein in a proteomic study [99]. Arabidopsis HMA1 shares a sequence signature common to sarcoplasmic/endoplasmic reticulum Ca^{2+}-ATPase (SERCA)-type pumps and it has been shown to complement the mutant phenotype displayed by two calcium transport-deficient yeast strains [234]. Moreover, HMA1-dependent calcium uptake had a high affinity with a K_m of 370 nM, an activity that was strongly inhibited by thapsigargin [234], a known specific inhibitor of SERCA-type Ca^{2+}-pumps [386]. HMA1 was also described to transport transition metals (see below, Sect. 17.3.2) [158, 234, 305]. The translocase ALBINO3 (ALB3) which plays a role in chloroplast biogenesis [188, 325] might be involved in calcium transport into chloroplast. The pea (*Pisum sativum*) homolog of ALB3, PPF1 (*Pisum*-post-floral-specific gene 1) has been shown to be involved in calcium influx into chloroplast. PPF1 has a major impact on chloroplast calcium stores, since isolated chloroplasts from Arabidopsis plants over-expressing PPF1 showed high levels of calcium, whereas suppression of the *PPF1* gene resulted in low calcium levels inside chloroplasts [193, 357]. Although several transporters were proposed to function in chloroplast calcium influx, a more comprehensive work is needed to link these proteins to chloroplast calcium homeostasis.

17.2.3 Potassium

Potassium (K^+) is an essential micronutrient for plant growth and development. It is the most abundant inorganic cation in plant cells (2–10 % of dry weight) and cytosolic concentration of potassium is around 100 mM. As a univalent cation with poor antagonistic affinity for sites requiring divalent cations, potassium plays several major functions in the cell: it ensures electrical neutrality by neutralizing the negative charges of organic acid and inorganic anions (e.g. carboxylic groups, phosphate esters of biological macromolecules) and is responsible for cell turgor by maintaining osmotic equilibrium. It is also involved in several steps of mRNA translation into proteins and plays a key role in enzyme activation by inducing stabilization of the proteins through conformational changes [18, 66, 186, 207, 289]. Potassium deficiency thus has serious deleterious effects for the organisms. When intracellular potassium levels drop because of insufficient external supply, electroneutrality is maintained by protons (H^+), which in turn decreases internal pH and has further detrimental effects.

Potassium homeostasis is key for the photosynthetic function and potassium deficiency strongly impacts both photosynthesis and photorespiration [207]. Hence, the massive light-dependent transport of protons into the thylakoid lumen is compensated by an efflux of potassium and magnesium to equilibrate electrical charges [93, 271, 306, 336]. Potassium is also the counter-ion for protons establishing the pH gradient for photophosphorylation [207]. In addition, the synthesis of ribulose-1,5-bisphosphate carboxylase (RuBisCO), the most abundant protein in chloroplasts, is strongly impaired by potassium deficiency [207]. The alkaline stromal pH (~8) required for proper CO_2 fixation by RuBisCO is maintained by a proton flux from the stroma to the cytosol compensated by potassium [152, 272]. Excess magnesium results in stromal acidification and photosynthesis inhibition: binding of excess magnesium to the negative surface charges of the chloroplast membrane inhibits potassium conductance across the membrane and thus impairs stromal alkalinisation and photosynthesis [27, 307]. Moreover, potassium channels in the plasma membrane of guard cells control stomatal aperture, and thus gas exchanges, essential for photosynthesis. Consequently, potassium deficiency increases stomatal resistance to CO_2 [130, 152, 159].

Although activities of potassium channels conducting potassium for counter-exchange with protons between cytoplasm and stroma and between stroma and thylakoid lumen (see above in this section) have been measured in the chloroplast inner envelope membrane [132, 221, 272, 356] and in thylakoid membrane [93, 271, 336], respectively, the molecular identity of these channels remain unknown. A large number of potassium transporters and channels have been identified in plant genomes, e.g. 34 in Arabidopsis [74, 210], and a few of them are predicted to localize in the chloroplast based on protein sequences and/or proteomic studies (Table 17.2, Fig. 17.1). With a few exceptions, these proteins have not been functionally characterized yet. The localization of the KT/KUP12 potassium transporter [210] remains unclear as it was identified in proteomic studies of both tonoplast and chloroplast membranes in Arabidopsis [160, 367]. The KEA1 and KEA2 proteins of Arabidopsis belong to a family of K^+/H^+ antiporters (Cation/Proton Antiporters or CPA2) [210] and have been identified in the chloroplast envelope [100, 324]. AtCHX23 is another member of the large CPA2 family [210]. It was localized in the chloroplast envelope using a GFP fusion. An Arabidopsis *chx23* mutant has a smaller size, is chlorotic and displays a defect in chloroplast development. Leaves from the mutant have less and smaller chloroplast compared to wild-type. The *chx23* mutant shows a higher cytoplasmic pH suggesting a role of AtCHX23 in a movement of cations (K^+ or Na^+) from cytosol into the stroma that would lead to proton loss [318]. The mutant phenotype is partially rescued by increased potassium external supply [318]. Altogether, these observations suggest that AtCHX23 is a K^+/H^+ antiporter involved in the control of stromal pH [262, 318]. It is also involved in salt tolerance (see Sect. 17.2.4). Recently, the potassium channel SynK was shown to localize in the plasma membrane and the thylakoid membrane in the cyanobacteria *Synechocystis* [384]. The closest homolog of SynK in Arabidopsis is the protein KCO6/TPK3 [384], a calcium-activated outward rectifying potassium channel [210, 350]. Similar to the six members of the family, TPK3 was initially localized in the tonoplast using a YFP fusion [350], and later in the chloroplast by immunolocalization [384].

17.2.4 Sodium

Although sodium (Na^+) is not an essential nutrient but rather considered as beneficial for plants [207], we will discuss its function in chloroplasts in macronutrient section together with other cations found abundantly in the environment. A controlled balance between sodium influxes and effluxes must be maintained since high salinity leads to osmotic and oxidative stress [387, 388]. Na^+/H^+ antiporters are membrane proteins involved in the regulation of ion homeostasis and pH balance. The Arabidopsis genome encodes more than 40 putative Na^+/H^+ antiporters [210]. In plants, the most characterized Na^+/H^+ antiporters belong to two different classes, NHX and SOS1 [128, 337]. NHX1 from Arabidopsis has been shown to be localized to the tonoplast [15] and is involved in maintaining the vacuolar pH homeostasis [105, 171] whereas SOS1-type proteins are localized at the plasma membrane [309, 310] where they control sodium efflux and sodium transport from root to shoot [310]. Several studies provided biochemical and physiological evidence for an yet uncharacterized Na^+/H^+ antiporter localized in the chloroplast envelope that regulates stromal acidification and alkalinization [147, 213, 265]. Only recently, Song et al. [318] showed that AtCHX23 is a Na^+-K^+/H^+ exchanger on the chloroplast membrane (Table 17.2, Fig. 17.1). In addition to the growth and chloroplastic defects described above (see Sect. 17.2.3), *chx23* mutants are highly sensitive to salt (NaCl) indicating that in the absence of the antiporter, the plants suffer from high cytosolic sodium toxicity [318]. AtCHX23 shows similarities to a Na^+/K^+ antiporter from Synechocystis, NhaS3. In *Synechocystis* sp. PCC 6803, NhaS3 is localized to the thylakoid membrane and essential for growth [346]. A *nhaS3* mutant shows enhanced sensitivity to high sodium concentrations. NhaS3 may function to sequester sodium into the thylakoid lumen by utilizing the transmembrane pH gradient and may serve as a putative uncoupler of the electrochemical proton gradient generated by photosynthesis [346].

First identified in bacteria [259], members of the NhaD family of transporters have been characterized as Na^+ (Li^+)/H^+ antiporters. NhaD from *Physcomitrella patens* [25] and *Mesembryanthemum crystallinum* [75] have been recently characterized and shown to mediate sodium transport into the chloroplast. Putative transporters of the NhaD family have been identified in many other photosynthetic organisms, including land plants, red and green algae and photosynthetic stramenopiles [25]. Maintaining a high cytosolic K^+:Na^+ ratio by extruding sodium ions outside the cell or compartmentalizing them into the vacuole plays an important role in plants exposed to high salinity. In *Mesembryanthemum crystallinum*, three sodium antiporters (*McSOS1*, *McNHX1* and *McNhaD*) are induced at transcript level in leaves in response to salt stress indicating an important role of these antiporters in sodium compartmentation in *M. crystallinum* under salt stress [75]. Up-regulation of *McSOS1* at transcript level in leaves but not in roots indicates that this antiporter functions in maintaining low sodium concentrations in the cytosol and most likely is localized at the plasma membrane as the Arabidopsis homolog. The NhaD antiporter from *P. patens* [25] and *M. crystallinum* [75] are both localized to the chloroplast envelope. In *M. crystallinum* exposure to salt stress results in the induction of the transcript levels of McNhaD and sodium accumulation inside the chloroplast. The physiological role of McNhaD might be to ensure an osmotic balance in the cytosol in response to high salinity.

Previous studies indicated that, in C_4 plants, pyruvate transport across the mesophyll chloroplasts is dependent on two distinct mechanisms: either proton-dependent [258] or sodium-dependent [257]. Transcriptome comparison between a C3 plant (*Flaveria pringlei*) and C_4 plants (*Flaveria trinervia* and *Flaveria bidentis*) identified a novel C_4-abundant gene related to bile acid:sodium symporter family of proteins (BASS2) [106]. The BASS2 protein is localized at the chloroplast envelope membrane and is highly abundant in C_4 plants that have the sodium-dependent pyruvate transporter. Sodium influx is balanced by a Na^+/H^+ antiporter (NHD1), which was mimicked in recombinant *E. coli* cells expressing both BASS2 and NHD1. Arabidopsis mutants lacking BASS2 show no pyruvate uptake into chloroplasts, which affects plastid-localized isopentenyl diphosphate synthesis. Orthologues of BASS2 can be detected in all the genomes of land plants that have been characterized so far, thus indicating the widespread importance of sodium-coupled pyruvate import into plastids.

17.2.5 Phosphorus

Chloroplasts generate ATP through photosynthetic activity and hence determine, with mitochondria, the energy status of green cells. In addition to its many roles in chloroplasts, e.g. in ATP synthesis, DNA and RNA synthesis, phospholipid synthesis and as regulator of the activity of proteins of the photosynthetic apparatus via post-translational modifications, phosphorus has also an important function in metabolic exchanges between chloroplasts and cytoplasm. A number of phosphate translocators located in the plastid inner envelope catalyze the antiport exchange of inorganic orthophosphate (Pi) and phosphorylated carbon molecules ensuring that the transport of carbon compounds between stroma and cytoplasm is neutral for phosphate homeostasis in both compartments (Fig. 17.1). This topic has been extensively reviewed and the reader is referred to recent reviews for details [102, 103, 162, 360, 362]. In short, these translocators include: (i) the triose phosphate translocator (TPT), which is mostly active in green tissues and translocate triose phosphate (TP) synthesized during the day out of the chloroplast maintaining Pi delivery into the chloroplasts to support ATP synthesis. Triose phosphate is then available for sucrose and cell-wall synthesis; (ii) the glucose 6-phosphate/phosphate translocator (GPT), which is active in non-green tissues and imports glucose 6-phosphate (Glu-6-P) from the cytoplasm into the stroma where it is used for starch biosynthesis and in the oxidative pentose phosphate pathway generating reducing power (i.e. NADPH) necessary for e.g. fatty acid biosynthesis or nitrite reduction (see Sect. 17.2.7). The dephosphorylation of sugar compounds during starch synthesis provides the Pi for exchange with Glu-6-P; (iii) the xylulose 5-phosphate/phosphate translocator (XPT), which is active in both green and non-green tissues. It imports xylulose 5-phosphate (Xyl-4-P), ribulose 5-phosphate (Rib-5-P) and erythrose 4-phosphate (Ery-4-P) in exchange for Pi into plastids, providing intermediates for the oxidative pentose phosphate pathway and other biosynthetic pathways; (iv) the phosphoenolpyruvate/phosphate translocators (PPT), which is active in both green and non-green tissues. It imports phosphoenolpyruvate (PEP) in exchange for Pi into plastids. PEP is a precursor for aromatic amino acids, fatty acids and secondary metabolites (e.g. anthocyanins and flavonoids) synthesis in plastids.

The importance of the establishment of a controlled metabolic connection between endosymbiont and host during (primary and secondary) endosymbiosis has been discussed extensively by Weber and Linka [360] in a comparison of envelope transporters in green plants, red algae, glaucophytes and apicomplexans.

The response to phosphate deprivation in plants has been studied in details. Several mechanisms are activated to increase phosphate mobilization and uptake from soils: secretion of organic acids and hydrolytic enzymes, induction of uptake systems and morphological changes of the root system and, in many species, symbiosis with mycorrhizal fungi to increase soil/uptake system interface. It also includes mobilization of phosphate from internal storage (mostly from vacuoles) and metabolic adaptations to reduce phosphate use [41, 250, 278, 300]. Complex signaling cascades control the plant response to phosphate deficiency and regulate a large set of genes in both root and shoot [122, 225, 233, 300, 371]. Hence, several genes encoding proteins involved in photosynthesis, e.g. components of PSII, PSI, the Calvin cycle, Chl biosynthesis and photorespiration, are down-regulated in response to phosphate starvation [233, 371]. However, this transcriptional repression of photosynthesis possibly represents a secondary response to phosphate deprivation, resulting from a lower need for photosynthetic assimilates caused by high sugar levels [233]. Indeed, the induction of several genes would support the higher carbohydrate synthesis observed under phosphate limitation, which allows recycling of Pi from phosphorylated carbon compounds and increasing starch synthesis [233]. Moreover, phosphate starvation induces a decrease in phospholipids. This loss is compensated by increased synthesis of galactolipids and sulfolipids (see Sect. 17.2.6), hence saving phosphate for other use. In addition, this would allow maintaining photosynthesis by adjusting lipid composition in chloroplast envelopes [225, 233].

Equally complex acclimation responses have been described in Chlamydomonas [117, 150, 239, 240, 373] and also includes reduced expression of photosynthesis genes [239]. The putative chloroplastidial LPB1 protein possesses a domain with similarity to nucleotide-diphospho-sugar transferases, suggesting a role in sugar metabolism. A *lpb1* Chlamydomonas mutant is hypersensitive to phosphate starvation [53]. Although the function of LPB1, and its plant homologs, remains to be determined [53], this indicates further the need to adapt carbon metabolism in chloroplasts in response to phosphate starvation. Moreover, a reduction of ribonuclease polynucleotide phosphorylase (PNPase) activity, an enzyme involved in chloroplastidial RNA degradation, triggers accumulation of RNA in chloroplasts of Chlamydomonas under limited phosphate supply. Both PNPase transcript and protein are repressed in these conditions and this repression is dependent on the major phosphate deprivation regulator PSR1. The degradation of RNA by PNPase requires Pi and reduced PNPase activity would save Pi to preserve other functions in chloroplasts under phosphate limitation [381]. The PNPase is also part of the phosphate starvation response in Arabidopsis and the phenotype associated to a loss of PNPase function is more complex [203]. The precise function of the chloroplastidial PNPase in phosphate starvation acclimation and signaling remains to be determined. Finally, the amount of chloroplastidial DNA decreases in Chlamydomonas cells under limited phosphate supply, which is also part of a phosphate sparing strategy [381].

In Arabidopsis, four families of phosphate transporters are found with different affinity for phosphate and intracellular localization. The PHT1 (PHT1;1 to 1;9) and PHT3 (PHT3;1 to 3;3) transporters have high affinity for phosphate and localize in the plasma membrane and mitochondrial membrane, respectively [300]. The PHT2 (PHT2;1) and PHT4 (PHT4;1 to 4;6) proteins are low and high affinity transporters, respectively and are both located in the chloroplast (Table 17.2, Fig. 17.1), with the exception of PTH4;6 that is located in the Golgi apparatus [119, 120, 149, 292, 294, 348]. The *PHT2;1* gene is expressed in shoots and is induced by light. A *pht2;1* mutant display alteration of Pi allocation in leaves and de-regulation of genes induced by Pi starvation [76, 348]. The five *PHT4* genes encoding chloroplast localized transporters have distinct expression patterns and functions. *PHT4;1* and *PHT4;4* are mostly expressed in green tissues and display a circadian expression and an induction by light, respectively [119, 120]. PHT4;1 is localized in the thylakoid membrane and would export Pi from the thylakoid lumen into the stroma, whereas PHT4;4 localizes in the inner envelope and would import Pi from the cytosol into the stroma [120, 294]. *PHT4;3* and *PHT4;5* are expressed mainly in leaf phloem [119] and *PHT4;2* is only expressed in roots [119, 120]. PHT4;2 exports Pi from the stroma in heterotrophic plastids in roots and the reduced level of starch in roots of a *pht4;2* mutant is consistent with an accumulation of Pi in the stroma. The mutant also displays increased leaf size resulting from increased cellular proliferation, suggesting a role of the transporter in signaling throughout the plant [149]. PHT4 proteins are symporters. Transport of Pi by PHT4;1 required either H^+ or Na^+ depending on the heterologous system used for transport analyses [120, 294]. Directed mutagenesis of a serine residue that is conserved in all PHT4 proteins revealed its key role in Na^+-dependent Pi transport by the Arabidopsis PHT4;1 [294]. Pi export by PHT4;2 is also dependent on Na^+ when tested on isolated root plastids [149]. Further functional analyses *in planta* will help determining the nature of the ion co-transported with Pi.

17.2.6 Sulfur

Sulfur is an essential macronutrient for all organisms. It is taken up by plants from soils as sulfate (SO_4^{2-}). The uptake and distribution of sulfate in plant tissues is a highly coordinated process (for reviews, see [129, 134]). Chloroplasts play a key role in sulfur metabolism: the reduction pathway leading to the assimilation of sulfate into sulfide (S^{2-}) takes place exclusively in plastids of both photoautotrophic and heterotrophic tissues. This pathway includes three steps: (i) ATP-dependent activation of sulfate into adenosine 5′-phosphosulfate (APS) by ATP sulfurylase (ATPS) occurring in both plastids and cytoplasm, (ii) reduction of APS into sulfite (SO_3^{2-}) by APS reductase in plastids and (iii) reduction of sulfite into sulfide by sulfite reductase in plastids [134, 157]. Sulfide is then readily available for incorporation into cysteine, which synthesis takes place in plastids, mitochondria and cytoplasm [133]. Once incorporated in numerous proteins (e.g. thioredoxins, glutaredoxins) and peptides (glutathione, GSH), cysteines are pivotal in controlling the redox status in chloroplasts. Moreover, cysteine is the source of sulfur used in the first step of iron/sulfur (Fe/S) cluster biosynthesis catalyzed by cysteine desulfurase [21]. Fe/S clusters are essential cofactors involved in numerous cellular processes, including photosynthetic electron transport (see Sect. 17.3.1). In addition to incorporation into proteins through cysteines, sulfates are components of sulfolipids, i.e. sulfoquinosovyldiacylgycerides that constitute the largest part of polar lipids in plastids [26].

Long-term sulfur deficiency has a characteristic phenotype: plants develop chlorosis of interveinal sections of young leaves. As sulfate release from vacuoles of mature leaves is a slow process, young leaves suffer more strongly from deficiency. The chlorosis observed under sulfate deficiency illustrates the need for reduced sulfur for photosynthesis [70, 134, 372]. More generally, sulfur deficiency reduces *de novo* protein biosynthesis, protein turnover and ultimately growth. In Chlamydomonas, sulfur deprivation inhibits protein synthesis in chloroplasts, which prevents the repair of the D1 subunit of PSII after photodamage [372]. Note that the resulting inhibition of photosynthetic oxygen evolution leads to anaerobiosis which allows hydrogen production by oxygen-sensitive iron-hydrogenase in chloroplasts [214, 216]. The decline of photosynthetic activity is an essential component of acclimation mechanisms to sulfur deprivation in Chlamydomonas. It is essential for survival and is coordinated by two actors of the sulfur deprivation response, i.e. SAC1, a protein related to sulfate transporters possibly acting as a sensor of extracellular sulfate, and SNRK2.1, a plant-specific SNF-1 related kinase [77, 113, 114, 240]. A large set of transcripts encoding components of the photosynthetic apparatus are down-regulated under sulfur starvation in Chlamydomonas. This includes proteins of PSII and PSI, light-harvesting complexes (LHC) of PSII and PSI, subunits of the cytochrome b_6f complex, and proteins involved in photosynthetic electron transport and Chl biosynthesis. The transcripts for proteins constituting peripheral PSII antennae are among the most highly sensitive. A large part of these gene expression alterations are dependent on SAC1 and/or SNRK2.1

[114, 385]. The re-arrangement of PSII antennae might participate in sulfur sparing and in maintaining the integrity of PSII by limiting the photoproduction of singlet oxygen ($^1O_2$). The *sac1* and *snrk2.1* mutants photobleach and die much faster than wild-type cells under sulfur deprivation. $^1O_2$ causes massive photooxidative damages in chloroplasts. Induction of several genes encoding antioxidant enzymes (GSH peroxidase, thioredoxin) is observed in the *snrk2.1* mutants under sulfur deprivation, highlighting the importance of SNRK2.1-dependent re-organization of PSII [114].

In plants, the response to short term sulfur deficiency mostly involves induction of sulfate uptake systems (see below in this section) and reduced sulfate assimilation into cysteine and GSH. A longer exposure to sulfur deprivation (>24–72 h) triggers more drastic physiological and morphological changes directed at energy saving and seed production. In particular, reduced S-adenosyl-methionine, which is required for Chl biosynthesis, results in lower Chl content accompanied by reduced photosynthesis and increased photorespiration. Several genes encoding LHC of PSII are down-regulated in these conditions [11, 134, 142, 143, 248, 249].

In cyanobacteria, and more generally in prokaryotes, sulfate uptake is driven by an ATP-Binding-Cassette (ABC) multi-unit transport system which consists of a periplasmic sulfate-binding subunit, two membrane proteins forming a channel and a cytoplasmic ATP-binding subunit [129]. A similar sulfate permease system is found at the envelope of the chloroplast of Chlamydomonas. Four nuclear genes encode the subunits of the complex: a cytoplasmic sulfate-binding protein (SBP), two transmembrane units (SULP and SULP2) and a stromal ATP-binding subunit (SABC) [58–60, 196, 215]. Transcripts and proteins of all four subunits accumulate in response to sulfur starvation [58, 196]. Knock-down strains with reduced *SULP* levels display typical sulfur deficiency phenotypes (see above in this section), including diminished *de novo* protein biosynthesis in chloroplast, reduced ability to repair PSII after photodamage and impaired photosynthesis [58]. Accordingly, these strains sustain anaerobic hydrogen production in the presence of sulfate concentrations in the medium, contrary to the wild-type [60].

No ABC-type system for sulfate import into chloroplasts is found in vascular plants [59, 129]. In contrast, sulfate transport in vascular plants is mediated by proton/sulfate co-transporters (Sultr) of the SulP family. Note that despite they have the same name, SulP proteins of vascular plants and Chlamydomonas are unrelated. Among the 14 Sultr proteins found in Arabidopsis [41, 129], Sultr4;1 was initially shown to be localized in chloroplasts [329], but it was later established that it is a vacuolar transporter [154]. Biochemical experiments have measured sulfate uptake into isolated chloroplasts. This uptake was inhibited by phosphate, which suggested that the chloroplastidial triose-phosphate/phosphate translocator was responsible for sulfate uptake [242]. However, no functional evidence allowed substantiating this hypothesis. Although sulfate transport across the plastid envelope is absolutely required as it allows sulfate assimilation into cysteine, the nature of the sulfate transporter in chloroplasts of higher plants remains unknown (Fig. 17.1).

17.2.7 Nitrogen

Nitrogen (N) is a primary constituent of nucleotides, amino acids and proteins and therefore is one of the most essential nutrient for plants [80, 375]. Plants take up mineral nitrogen (as nitrate, NO_3^-, or ammonium, NH_4^+) from the soil, although some plant species (e.g. legumes) are able to use gaseous nitrogen due to their capacity of a symbiotic relationship with specific microorganisms [115]. Nitrate represents the major source of nitrogen for plants in aerobic soils, whereas ammonium is the second most abundant nitrogen source in soils and represents the primary nitrogen source in anoxic soil conditions [115]. In the green alga Chlamydomonas, ammonium is the preferred nitrogen source, although several uptake systems for nitrate have also been described [94].

The impact of varying nitrogen availability and type of nitrogen source in soils on growth and development is studied extensively as it is key in determining yield. These efforts revealed complex mechanisms of nitrate and ammonium uptake and distribution, as well as nitrogen sensing and signaling, in plants and Chlamydomonas [10, 80, 94, 349, 375].

Environmental variations in nitrogen source, nitrogen limitation or deprivation profoundly affect the metabolism of photosynthetic cells. Transcriptomic studies in Arabidopsis revealed that several genes encoding proteins involved in photosynthesis (both in PSII and PSI), in Chl biosynthesis and in plastid protein synthesis are coordinately down-regulated at low nitrogen supply and rapidly up-regulated when nitrogen supply increases [30, 303, 349].

Nitrogen assimilation partly takes place in chloroplasts, which are therefore essential in nitrogen metabolism. Indeed, after its transport to the cytosol of mesophyll cells of leaves, nitrate is reduced to nitrite (NO_2^-) by nitrate reductase (NR) which is then imported into the chloroplast (see below in this section) for further reduction to ammonium (NH_4^+) by nitrite reductase (NiR) [80, 375]. The reducing equivalents and ATP required for this chloroplastidial reaction are provided by carbon skeletons produced by photosynthesis. Another major source of ammonium that is assimilated in chloroplasts is the oxidation of glycine by glycine decarboxylase in mitochondria during the photorespiration cycle [179, 359]. Several models have been proposed for the shuttling of nitrogen (as ammonium, glutamine or citrulline) between mitochondria and chloroplasts [197].

The ammonium released from nitrite reduction or by photorespiration is assimilated into glutamate via the glutamine synthetase/glutamate synthase (GS/GOGAT) cycle in the stroma of the chloroplasts [359]. In leaves of most plant species, the predominant GS/GOGAT isoenzymes, GS2 and Ferredoxin-GOGAT (FD-GOGAT), respectively, are found in the chloroplast and more minor isoforms, GS1 and NADH-GOGAT, are located in the cytosol. GS2 is also found in mitochondria [328, 359]. The integration of ammonium into its organic form further requires a malate-coupled two-translocator system located at the chloroplast inner envelope [197, 284, 331, 361]. The precursor of ammonium assimilation, 2-oxoglutarate which is synthesized in the mitochondria and in the cytosol [179], is imported into chloroplasts by the 2-oxoglutarate/malate transporter DiT1. Glutamate, the end product of ammonium assimilation, is then exported out of the chloroplast into the cytosol by the glutamate/malate translocator DiT2 [197, 284, 331, 361]. In the cytosol, glutamate is used as an universal organic nitrogen donor for the biosynthesis of amino acids, nucleic acids, and other nitrogen-containing compounds via several transamination reactions catalyzed by a number of different aminotransferases [177].

As mentioned above, after reduction of nitrate into nitrite in the cytosol, nitrite needs to enter the chloroplasts for further reduction. However, the transport of nitrite from the cytosol into chloroplasts is still poorly understood in plants. It was initially thought that nitrite movement occurred by rapid diffusion across the inner chloroplast envelope membrane as nitrous acid and, consequently, that a chloroplast transporter was not necessary [314]. However, the physiological nitrite concentration is too low to allow this free diffusion [156]. Moreover, nitrite uptake measurements within intact chloroplast from pea showed saturation kinetics which suggested the presence of a nitrite transporter [38, 39]. In Chlamydomonas, at least two membrane proteins (NAR1.1 and NAR1.2) located in the inner envelop are involved in nitrite transport into the chloroplast. NAR1.1 is a nitrite transporter, whereas NAR1.2 co-transports nitrite and bicarbonate [204, 205, 285]. No NAR1 orthologs have been identified in plants. However, two transporters belonging to the proton-dependent oligopeptide transporter (POT) family, Nitr1-L and Nirt1-S, were proposed to be involved in nitrite transport in cucumber (*Cucumis sativus*) [323]. In contrast to Nitr1-S, Nitr1-L has a putative transit peptide and its predicted chloroplast localization was confirmed using GFP fusions [323]. This transporter is also able to transport nitrite when expressed in yeast [323]. The Nitr proteins are member of the NRT1 nitrate transporter family [345]. Mutants of the Arabidopsis Nitr1-L homolog (At1g68570) show nitrite accumulation in leaves compared to wild-type, suggesting a role of At1g68570 in nitrite transport [323].

In Arabidopsis, a member of the CLC family (chloride channel), AtCLCe, is suggested to be involved in nitrite transport from the stroma into the thylakoid lumen (see also Sect. 17.2.8). Co-localization of a GFP fusion with Chl and Western blot analysis indicated the localization of AtCLCe in the thylakoid membrane [206]. Knock-out mutants of AtCLCe have an altered photosynthetic activity and reduced level of nitrate associated with an over-accumulation of nitrite [230]. The *Atclce* mutants also show reduced expression of several genes involved in nitrate uptake (NRT1.1 or NRT2.1), indicating interaction between nitrate uptake and nitrate assimilation [230].

The PII protein is a sensor of carbon/nitrogen balance and energy status. It is present in bacteria, cyanobacteria, and plants [195]. An array of data suggests that PII is involved in the down-regulation of nitrite uptake into chloroplasts in Arabidopsis. Indeed, mutants of the single PII homolog (named GLB1) identified in Arabidopsis display higher sensitivity to nitrite toxicity [95, 145]. In addition, radiolabelled nitrite uptake experiments with intact chloroplasts of wild-type and *glb1* mutant plants demonstrated an increase of light-dependent nitrite uptake into chloroplast in the *glb1* mutant plants compared to wild-type [96]. Finally, the expression of the nitrite transporter At1g68570 is not altered in the *glb1* mutants. The mode of action of PII and its interaction partners remain unknown.

In Chlamydomonas, two putative ammonium transporters of the AMT1 family (AMT1;2 and AMT1;4) are predicted to localize in the chloroplast inner envelope [94, 112]. These transporters would enable the direct uptake into chloroplasts of the ammonium taken up in the cell at the plasma membrane. The five AMT1 ammonium transporters found in Arabidopsis are mostly expressed in roots and none of them are localized in plastids [111, 382]. The molecular identity of an ammonium transporter in plant chloroplasts remains unknown.

17.2.8 Chloride

Anion channels represent important players in the maintenance of electrochemical gradients and in signaling pathways that are essential for plant adaptation to biotic and abiotic stresses. The presence of anion transporters in plants cells has been documented for all types of membranes, including plasma membrane, tonoplast, endoplasmic reticulum, mitochondria and chloroplasts [79]. In chloroplasts, pH homeostasis across the envelope and thylakoid membranes requires the participation of many ion channels, including anion transporters. In the chloroplast stroma, the concentration of chloride (Cl^-) is in the 50 mM range [81]. The inner envelope membrane has been previously shown to be permeable to chloride [370]. Using the patch-clamp technique and ion-selective microelectrodes, Pottosin [270] and Heiber et al. [132] showed that chloride channels are also present in the chloroplast envelope of green algae *Nitellopsis* sp. and *Eremosphaera viridis*. The only genes encoding for anion channels that have been described so far in plants belong to the CLC (chloride channel) family [131, 200]. CLC-type transporters are ubiquitously present in prokaryotes and eukaryotes and mechanistically are divided into two classes: Cl^- channels and Cl^-/H^+ exchangers (also called antiporters) [222]. Seven members of the CLC-family have been identified in Arabidopsis (*AtCLCa-AtCLCg*) [131, 206] and in rice (*OsCLC1-OsCLC7*) [83]. Phylogenetic analysis revealed that most plant CLCs, including Arabidopsis members, are related to an eukaryotic branch, except for AtCLCe which is part of a distinct subfamily together with AtCLCf and some other CLCs from tomato and rice, all closely related to the bacterial CLC proteins [206]. Characterization of AtCLCa [109] and AtCLCc [127] mutants suggested a role of these transporters in the regulation of nitrate concentrations in Arabidopsis. *AtCLCa* transcript levels are induced by nitrate in both roots and shoots [109] and subcellular localization studies using AtCLCa fused to GFP indicated vacuolar membrane localization [78]. Whereas AtClCf and AtCLCd were shown to be targeted to the Golgi membranes [206] and *trans*-Golgi network [351], respectively, AtCLCe is targeted to the thylakoid membranes of the chloroplast (see also Sect. 17.2.7) [206]. The subcellular localization of AtCLCb, AtCLCc and AtCLCg is unknown. In rice, OsCLC1 and OsCLC2, which are closely related to AtCLCc, were localized at the vacuolar membrane [245] and their transcripts were found to be regulated by salt stress [83].

Transcriptomics data available at Genevestigator [144] indicate that *AtClCe* expression is higher in green tissues compared to roots. The mutant gene-chip data sets indicate induction of *AtCLCe* in *lec1* (a leaf developmental mutant) and in a mutant with altered plastid signaling pathways (*gun4gun5*, see Sect. 17.2.1). The loss-of-function mutants of AtCLCa and AtCLCe showed no developmental or growth phenotype, whereas OsCLC1 and OsCLC2 knock-down mutants showed reduce growth at all developmental stages [245]. However, *clce* mutant plants showed altered photosynthetic activity, most likely due to changes in the ionic strength or osmotic properties of the lumen resulting from an impaired anionic permeability of the thylakoid membrane [206]. On the other hand, *clce* mutants have reduced nitrate content as observed for the Arabidopsis *clca* mutants [78, 79, 230]. Analysis of structure and transport mechanisms of bacterial CLCec1 protein indicated an essential role of two glutamate residues (Glu148 and Glu203) in a Cl^--binding region [5, 222]. The presence of a glutamate residue in a position equivalent to Glu203 in CLCec1 represents a feature specific only for CLC antiporters [5]. In Arabidopsis, AtClCa has been shown to function as a nitrate/proton exchanger and its protein sequence contains the two glutamate residues like the AtCLCb, AtCLCc, AtCLCd and AtCLCg. In contrast, AtCLCe and AtCLCf possess only the first glutamate residue suggesting a different transport mechanism. To date, the transport activity and substrate of the only chloroplast localized CLC protein (AtCLCe) remains unknown.

17.3 Micronutrients

17.3.1 Iron

Iron is an essential micronutrient for all living cells because it is a component of a number of important macromolecules, including those involved in respiration, photosynthesis, DNA synthesis and metabolism [164]. In plants, the largest iron pool is found in chloroplasts, which accumulate about 80–90% of the iron of leaf cells [255, 335]. Indeed, iron is required for photosynthesis, Chl and heme biosynthesis, and Fe/S cluster assembly, all of which take place in chloroplasts. There are three different groups of iron proteins in chloroplasts: (i) heme proteins, e.g. cytochromes (b_{559}, b_6f and c_6) and P450 proteins; (ii) soluble or membrane proteins containing Fe/S clusters, e.g. ferredoxins (FD); (iii) proteins that bind iron ions directly, e.g. ferritins and proteins of PSII and PSI [255]. As a cofactor of several electron carriers, iron is thus a key nutrient for the photosynthetic transport chain.

Four types of tetrapyrrole molecules are found in plants: Chl, heme, siroheme (prosthetic group of sulfite and nitrite reductase, see Sects. 17.2.6 and 17.2.7) and phytochromobilin (chromophore of phytochromes). The tetrapyrrole biosynthesis pathway thus represents an essential process for plant metabolism, which mainly takes place in chloroplasts. However, note that the two last steps of heme synthesis have also been observed in mitochondria (see the following reviews for details about the tetrapyrrole biosynthetic and degradation pathways: [73, 228, 330]). Iron is a key component of this pathway because the heme precursor, protoporphyrin IX, needs to chelate ferrous iron (Fe^{2+}) to produce functional heme. Additionally, iron is also needed in the biosynthesis of Chl, the major tetrapyrrole molecule in chloroplasts, because the enzyme that catalyzes the synthesis of protochlorophyllide, a Chl precursor, binds two iron atoms (see also Sect. 17.3.2) [236, 343].

Chloroplasts synthesize Fe/S clusters via the SUF (SUlfur Fixation) pathway which stems from its endosymbiotic origin and is related to the system found in cyanobacteria. The sulfur is acquired from cysteine (see Sect. 17.2.6), whereas the source of iron for Fe/S cluster synthesis in chloroplasts remains unknown. An evolutionary distinct Fe/S cluster assembly pathway, named ISC (Iron-Sulfur Cluster) is present in mitochondria, and is required for proper assembly of Fe/S proteins in the cytosol [21, 374]. Plants also possess members of a third Fe/S synthesis pathway, the NIF (NItrogenase Fixation) system, in both chloroplasts and mitochondria [187, 344]. The respective contribution and interaction of the SUF and NIF pathways in chloroplasts remain to be established [21]. The most abundant Fe/S proteins are the Rieske protein of the cytochrome b_6f complex, PSI and FD that localize in the thylakoid membranes and are part of the photosynthetic electron transport chain. In the stroma, the most abundant Fe/S requiring enzymes are GOGAT, NiR, and the sulfur assimilation enzymes, sulfite reductase and APS-reductase (see sulfur and nitrogen Sects. 17.2.6 and 17.2.7, respectively) [21].

Owing to the many essential functions of iron in chloroplasts, iron deficiency has a strong impact on the photosynthetic apparatus. Hence, iron deficiency results in chlorosis due to decreased light harvesting pigments, such as Chls and carotenoids [231, 232], and in reduced photosynthetic rates and efficiency [319]. In addition, iron deficiency causes a reduction of granal and stromal lamellae per chloroplast [319]. Iron deficiency also reduces dramatically both the light harvesting and core complexes of PSI and PSII. PSI is the complex most affected by iron deficiency, probably because it contains 12 iron atoms per monomer, followed by cytochrome b_6f and PSII whereas the ATP synthase is unaffected [251].

Photosynthetic organisms have evolved acclimation mechanisms to preserve the photosynthetic function under iron deficiency. Hence, cyanobacteria respond to iron deficiency by (i) replacing FD by the iron-free flavodoxin; (ii) reducing the PSI/PSII ratio from 4:1 to 1:1, thus reducing iron use; and (iii) inducing a new antenna for PSI named CP43′ that binds Chl and surrounds the PSI trimeric reaction center forming a ring of 18 molecules and may be functioning as a light-harvesting system or as a dissipater of light energy [31, 33, 185, 301]. In Chlamydomonas, the response to iron deficiency involves a progressive remodeling of the antenna complexes depending on the iron nutritional status that ends with a complete disconnection of the LHCI of PSI and degradation of LHCI under severe iron deficiency conditions [125, 217, 237, 246]. This response avoids photo-oxidative damage induced by the loss of Fe/S clusters of PSI. In iron-deficient plants, the electron transfer between PSI and PSII is also impaired leading to photooxidative damage. It has been recently shown that remodeling the major light-harvesting antenna protein of PSII protects barley (*Hordeum vulgare*) leaves from photoinhibition under prolonged iron deficiency [299]. Proteomic studies in Chlamydomonas [247] and plants [12, 176, 340] showed a reduction of the electron transfer complexes under iron deficiency compared to control conditions.

Iron is also the most analyzed transition metal in oceans due to its low availability [35]. Indeed, iron is highly insoluble in oceans and usually is the limiting nutrient for phytoplankton growth. It has been observed that oceanic cyanobacteria, diatoms and algae have been subjected to a long-term adaptation to iron deficiency via decreased levels of the major iron users PSI and cytochrome b_6f complexes and re-routing of the electrons flow to oxygen directly downstream of PSII through a plastoquinol terminal oxidase (PTOX) [19, 49, 198, 322].

As described above, iron is an essential micronutrient for all living organisms. However, excess free iron is highly toxic due to its ability to react with oxygen and produce hydroxyl radicals via the Fenton reaction. Thus, iron homeostasis, as for other nutrients, must be tightly controlled. Ferritins are a superfamily of iron storage proteins, found in all living organisms, with the exception of yeast, that protect cells against oxidative stress [14, 17, 37, 280]. Plant ferritins are mostly located in chloroplasts [37] but can also be found in mitochondria [333, 383]. In Arabidopsis, there are four ferritin genes (*FER*) that are expressed in seeds (*FER2*), shoots or flowers (*FER1, FER3, FER4*), respectively. Analysis of several mutants lacking seed (*fer2*) or leaf ferritins (*fer1fer3fer4*) revealed that ferritins are part of the defense machinery against oxidative damage, but are not an essential iron pool for seedling development or proper functioning of the photosynthetic apparatus [280]. Similarly, two ferritins, encoded by *FER1* and *FER2*, are located in the chloroplast in Chlamydomonas and prevent photo-oxidative stress, in particular when PSI is degraded under iron limitation [45, 175, 199]. Another iron protein located in the chloroplast is the iron superoxide dismutase (FeSOD or FSD1) [8, 268]. FeSOD, as ferritins, might also be involved in the oxidative damage protection by converting the superoxide anion, produced during the water-water cycle in PSII, into hydrogen peroxide that is subsequently removed by ascorbate oxidase [8, 268].

Despite the importance of iron in chloroplasts, knowledge about iron transport and homeostasis control within the organelle is still limited. Physiological studies with chloroplasts from barley showed that the ferric iron (Fe^{3+}) uptake within chloroplast is light-dependent [42]. Further, biochemical *in vitro* assays using vesicles prepared from the chloroplast inner envelope showed that ferrous iron is transported across the chloroplast inner membrane [315, 316]. This is in agreement with the identification of a chloroplast ferric chelate reductase, encoded by *FRO7*, in Arabidopsis (Table 17.2, Fig. 17.2) [151]. FRO7 is a member of the ferric chelate reductase FRO family [243]. Chloroplasts of *fro7* mutants possess 4-fold reduced ferric chelate reductase activity compared to wild-type, which results in restricted iron uptake and thus iron content (33 % reduction) in *fro7* chloroplasts. *fro7* mutants also present impaired photosynthesis and severe chlorosis in alkaline soils suggesting that the FRO7 function is essential for iron import into the chloroplast under iron-limiting conditions. However, the loss of FRO7 is not lethal under normal conditions suggesting that chloroplasts likely transport both ferrous and ferric iron [151].

Both ferrous and ferric iron uptake pathways have been characterized in cyanobacteria: the Fut (ferric iron uptake) transporter, an ABC transporter, is responsible for ferric iron uptake whereas ferrous iron uptake depends on the transporter FeoB, a member of the bacterial G protein family. FeoB functions only when the ferric uptake system is disabled [155]. In Arabidopsis, a chloroplast homolog of the Fut protein, named NAP14 (non-intrinsic ABC protein 14), has been identified [313]. *nap14* mutant plants overaccumulate iron compare to wild-type plants suggesting a role of this protein in the iron homeostasis mechanism within chloroplasts [255].

PIC1, an ortholog of the cyanobacterial COGs (cluster of orthologous groups), has been identified as an iron transporter which localizes to the chloroplast inner envelope but it is still unclear whether it transports ferric or ferrous iron [87, 88]. Expression of PIC1 complements the phenotype of a yeast mutant defective in iron uptake. *pic1* mutants are severely chlorotic and dwarf with altered iron homeostasis, accumulation of ferritin and impaired chloroplast development. These phenotypes are consistent with a role of PIC1 in iron compartmentalization into the chloroplast [87, 88]. Recently, MAR1/IREG3 (for Multiple Antibiotic REsistance 1/Iron-REGulated protein 3) was defined as a plastid member of the ferroportin/IREG transporter family [71, 72]. MAR1/IREG3 specifically transports aminoglycoside antibiotics, however, due to sequence homology to the metal transporters IREG1 and IREG2 [235, 302] and the fact that MAR1 overexpression produces leaf chlorosis that can be rescued by exogenous iron, it has been proposed that MAR1/IREG3 may function in the uptake of the iron-chelator polyamine nicotianamine (NA) or of Fe^{2+}-NA chelates into chloroplasts (Table 17.2, Fig. 17.2) [71, 72].

Surprisingly, despite the high number of iron-containing proteins observed in all living organisms, a single iron chaperone has been identified so far. It was shown that the iron chaperone PCBP1 (Human Poly r(C)-Binding Protein 1) delivers iron to human ferritins [311].

17.3.2 Copper

In plants, chloroplasts contain more than half of the copper found in cells. In this organelle, copper is mostly associated to the electron carrier plastocyanin (PC) and the copper/zinc superoxide dismutase (Cu/ZnSOD or CSD2) [44, 126]. In Chlamydomonas, a deficiency response allows maintenance of photosynthesis under copper starvation. It includes the replacement of PC by the heme-containing cytochrome c_6 (Cytc_6) through degradation of PC and transcriptional activation of the *CYC6* gene, which allows maintaining electron transfer from cytochrome b_6f to PSI at low copper supply [190, 218, 219]. The induction of Cytc_6 synthesis increases the need for heme, which is supported by the up-regulation of *CPX1*, encoding a soluble plastid coproporphyrinogen III oxidase [140, 274–276]. This exchange of electron carrier might be facilitated by a modification of thylakoid membrane physical properties in copper-deficient cells where an increased expression of plastid-acting desaturases results in enhanced galactolipid desaturation. This might allow a better mobility of Cyt$c_6$between the cytochrome b_6f and the PSI complexes [50]. In addition to *CPX1*, *CRD1* and *CTH1* are other copper-regulated genes encoding enzymes in the tetrapyrrole pathway. These di-iron enzymes catalyze the aerobic oxidative cyclization of Mg-protoporphyrin IX monomethylester in Chl biosynthesis [236, 238,

343]. The balance of *CRD1* and *CTH1* expression allows adjusting the interactions between photosystems and LHCs in response to copper status [125, 236, 238].

The CRR1 transcription factor, a Squamosa promoter Binding-like Protein, is a master regulator of the copper deficiency response in Chlamydomonas. Through binding to GATC Copper Responsive Elements (CuREs), it controls about half of the genes whose expression is altered by copper deficiency, including *CYC6*, *CPX1*, *CRD1*, *CTH1* and plastidial lipid desaturase-encoding genes [50, 91, 174, 238]. Moreover, CRR1 is also responsible for the increased expression of *CTR1* and *CTR2* genes, encoding the main plasma membrane copper transporters involved in cellular copper uptake [261]. Similar to the PC/Cytc_6 switch, several additional proteins (e.g. involved in nitrogen and iron assimilation pathways) whose expression is induced by CRR1 could serve as backups for copper-requiring functions under copper deficiency [50]. This would allow maintenance of essential functions and/or spare copper for other functions. Indeed, prioritization has been suggested for copper [24] which would be reallocated from chloroplastidial PC to mitochondrial cytochrome *c* oxidase (CytOx) under copper deficiency [217]. The observation at limited copper supply of the CRR1-dependent decreased expression of *PCC1*, a putative plastidial copper chaperone, and the CRR1-independent increased expression of *COX17*, which is required for copper assembly into CytOx [20, 283, 287] supports this hypothesis [50]. The down-regulation of PCC1 might allow changing the intracellular copper flux away from PC to CytOx [50]. Finally, note that many genes induced by copper deficiency in a CRR1-depedent (e.g. *CYC6*, *CPX1*, *CRD1*) or independent manner (e.g. the hydrogenase-encoding genes *HYDEF*, *HYD1*, *HYDG*) are also activated by oxygen deprivation [50, 91, 174, 236]. For the latter, it is suggested that the upregulation of the anaerobic metabolic pathway is a consequence of a possible cofactor role of copper in an oxygen sensor [50]. For the former, although no physiological significance has been established so far, several oxygen-dependent steps (e.g., *CPX1*, *CRD1*/*CTH1*) in the tetrapyrrole pathway are impacted by copper deficiency. CRR1 is directly involved—through binding to GATC motifs—in the regulation of these genes, and others (e.g. the ferredoxin encoding *FDX5*), by hypoxia [50, 91, 174, 178, 236].

In plant chloroplasts, Cytc_6 is absent and PC is indispensable [202, 229, 364]. It is therefore suggested that PC has priority for copper delivery [1, 3, 217]. Moreover, one PC isoform acts as a copper buffering system in chloroplasts under copper excess in Arabidopsis [1]. The closest homolog of CRR1 in Arabidopsis, SPL7, is also involved in the coordination of the copper deficiency response, mainly by increasing the expression of copper uptake systems, reducing the expression of non-essential copper-requiring systems and replacing those by copper-independent backups, which ultimately spares copper for essential functions [2, 29, 267, 377, 378]. Hence, in chloroplasts, SPL7 controls a decrease in Cu/ZnSOD (CSD2) via the action of microRNAs (miR398), and the concomitant increase of FeSOD (FSD1) to compensate for loss of Cu/ZnSOD (CSD2) [268, 377, 378]. A recent study in poplar (*Populus trichocarpa*) further supports the hypothesis that a prioritization of copper use is mediated by copper microRNAs. The photosynthetic function is strongly impacted by copper deficiency. However, deficiency/resupply experiments revealed that the first functions to recover after copper resupply are PC and photosynthetic electron transport, whereas recovery of the other copper-dependent activities is delayed [281].

CUTA is a protein possibly involved in chloroplast divalent copper homeostasis. It is suggested to localize in the envelope intermembrane space with its N-terminal end inserted in the inner envelope membrane. The analysis of several *cuta* mutant lines showed that CUTA is not essential for copper homeostasis *per se* in the chloroplast but could be involved in copper signaling [43].

Several transporters have been involved in copper transport into chloroplasts (Table 17.2, Fig. 17.2). The PAA1 and PAA2 proteins belong to a subgroup of I_B P-type ATPases transporting monovalent cations (mostly copper as Cu^+), which also includes proteins involved in copper delivery to the secretory pathway in a wide range of organisms [124, 201, 368]. In plants, PAA1 and PAA2 are responsible for copper transport across the inner envelope and thylakoid membranes, respectively [3, 28, 312]. The PAA1 protein is activated by Cu^+ and has high affinity for copper [51]. A *paa1* mutant is defective both in stromal Cu/ZnSOD and PC, whereas the *paa2* mutant only lacks PC. A *paa1paa2* double mutant is seedling lethal, highlighting the crucial role of copper in chloroplasts [3, 312]. This uptake system is evolutionary conserved: the CtaA and PacS proteins found in the plasma membrane and thylakoid membrane of cyanobacteria are homologs of PAA1 and PAA2 [308, 341]. Similarly, two putative copper-transporting P-type ATPases (CTP2 and CTP3) are predicted to be localized in the chloroplast in Chlamydomonas. Although it is likely that CTP2 and CTP3 play roles similar to PAA1 and PAA2, their subcellular localizations and copper delivery activities remain to be established [124, 220]. At the transcript level, the expression of these copper transporters is not induced by copper limitation. They are thus not part of the CRR1/SPL7-dependent deficiency responses in Chlamydomonas and Arabidopsis, respectively [29, 50, 332, 378]. Moreover, PAA2 protein stability is controlled by copper chloroplast levels in Arabidopsis, whereas PAA1 levels are unaffected. This control is independent of SPL7, but is determined by PC levels [332].

PAA1 and PAA2 transport monovalent copper (Cu^+), which suggests the need for a copper reductase in the chloroplast membrane as it has been shown for iron (Fig. 17.2). Whether a member of the ferric chelate reductase family is conducting this function remains to be determined (see Sect. 17.3.1) [29, 151]. Moreover, biochemical studies showed that Cu^{2+} is transported across pea thylakoid membranes [317], which suggests the existence of a divalent copper transporter in the thylakoid membrane.

HMA1 is another I_B P-type ATPase which has been involved in chloroplastidial copper import. HMA1 belongs to a subgroup of I_B P-type ATPase transporting divalent cations (e.g. zinc, cadmium, lead) that is only found in prokaryotes and plants [124, 368]. HMA1 proteins found in red and green algae and in plants, have an uncharacteristic Ser/Pro/Cys motif in the sixth predicted transmembrane domain instead of the common Cys-Pro-Cys/His/Ser motif, suggesting that they might have a different metal specificity than other members of the group [68, 124, 368]. In addition to its possible role in calcium uptake (see Sect. 17.2.2), the Arabidopsis HMA1 protein transports both zinc and copper in yeast and localizes to the chloroplast envelope. A *hma1* mutant accumulates less copper in plastids, displays reduced SOD activity, but normal PC level, and has a defect in the photosynthetic water-water cycle, an alternative pathway to dissipate electrons [139, 234, 305]. However, another study suggests that HMA1 is involved in zinc export from the chloroplast, based on zinc hypersensitivity and increased zinc accumulation in the chloroplast of the mutant [158]. AtHMA1 also complements a cadmium-hypersensitive yeast mutant and confers high cadmium tolerance to wild-type yeast. Cadmium-stimulated ATPase activity of AtHMA1 confirms that this pump plays a role in cadmium transport [234].

Copper has two main targets in chloroplasts: PC and Cu/ZnSOD (see above in this section) [3, 44, 126]. Apo- and holoforms of PC have a very similar structure, suggesting that PC could spontaneously acquire copper in the thylakoid lumen after import by PAA2 [44, 191]. In cyanobacteria, the ATX1 protein interacts with both CtaA and PacS, and was shown to directly deliver copper to a PacS homolog [23, 52, 342]. In plants, two cytosolic ATX1-like proteins have been involved in copper delivery to P_{1B}-ATPases, but no homolog has been localized to chloroplasts or described to interact with PAA1 or PAA2 [141, 273]. In contrast, a copper chaperone for SOD (CCS) that is localized in both cytosolic and plastidial compartments has been identified in Arabidopsis. CCS is responsible for copper delivery and activation of cytosolic, peroxisomal and chloroplastidial Cu/ZnSODs [4, 65, 69, 268, 297]. In Arabidopsis, a *ccs* mutant has less than 2% of chloroplastidial SOD compared to the wild-type in copper sufficient conditions [69]. The *CCS* gene is another target of SPL7-induced microRNAs (miR398) and is down-regulated under copper deficiency (see above in this section) [29, 378].

17.3.3 Manganese

Manganese atoms are part of the catalytic center of the water-splitting complex in PSII [217]. Manganese deficiency strongly affects the photosynthetic apparatus, resulting in PSII photoinhibition, loss of the D1 PSII subunit that binds the manganese cluster, and the production of ROS [8, 170, 380]. In Arabidopsis, the NRAMP3 and NRAMP4 proteins are transporting iron and manganese out of the vacuole [181, 182, 338, 339]. An *nramp3nramp4* double mutant is unable to remobilize vacuolar manganese in adult leaves and contains less functional PSII. In contrast, it displays normal levels of mitochondrial MnSOD, which suggests prioritization for manganese or a limited role of the vacuolar compartment in manganese delivery to the mitochondria [182]. In barley, differential ability to grow at low manganese supply was recently linked to differences in PSII damages and state transitions under manganese deficiency [148]. No transporter similar to the cyanobacterial high affinity manganese uptake system (MntABC) is present in plants [308] and the uptake mechanisms for manganese in chloroplasts remain unknown (Fig. 17.2).

17.3.4 Zinc

Zinc plays multiple roles in chloroplasts. It is a cofactor for the RNA polymerase and zinc fingers-containing nucleic acid-binding proteins and thus is crucial for plastidial transcription. Numerous enzymes (e.g. carbonic anhydrase, D-ribulose-5-phosphate 3-epimerase) use zinc as cofactor. Moreover, many zinc-dependent proteolytic activities take place inside chloroplasts (e.g. repair of PSII after photodamage of the D1 protein, cleavage and processing of the signal peptide) [126, 354]. Despite these multiple functions, little is known about the impact of zinc deficiency [277] or excess on the photosynthetic function [62, 82, 296]. A better maintenance of the photochemical capacity under zinc limitation has been linked to higher zinc efficiency of rice (*Oryza sativa*) cultivars [61]. While zinc excess has little effect on photochemistry, it results in a major decrease of stomatal and mesophyll conductances to CO_2, which impairs photosynthesis, in sugar beet (*Beta vulgaris*) [296]. So far, no transport system for zinc import into chloroplasts has been identified in plants. The system called ZnuABC involved in zinc uptake in cyanobacteria is not conserved in plants [32, 52]. Recently, putative ubiquist zinc chaperones have been identified, which may play a role in cellular zinc homeostasis and possibly in chloroplasts [121].

17.3.5 Other Transition Metals

In addition to iron, copper, zinc and manganese, other transition metals have a physiological role in plants. Hence, cobalt is considered a beneficial element for plants, but no precise function has been determined [167, 207]. Putative cobalt-transporting ABC transporters related to bacterial systems are found in cyanobacteria, algae and plants (Table 17.2, Fig. 17.2) [52, 89, 125, 288]. The plant and algal proteins are predicted to localize either to chloroplasts or mitochondria, but their function and localization has not been determined. Whether cobalt may substitute for another metal in plants as it does in marine organisms [180] remains to be established.

Nickel is an essential micronutrient in plants, where the only known nickel-requiring enzyme is urease [67, 89, 167]. Nickel deficiency induces the accumulation of toxic urea concentrations in several plant species [167]. Putative high-affinity nickel transporters are localized in the chloroplastidial envelope (Table 17.2, Fig. 17.2) [89, 125], but their function remains to be established.

17.4 Conclusions and Perspectives

Both macro- and micronutrient ions fulfil multiple key functions in chloroplasts, which thus have important requirements for those ions. Recent work revealed that complex acclimation mechanisms evolved to allow maintenance of the photosynthetic function under ion deficiency. Several transcription factors that coordinate the deficiency response have been identified. However, further work will be required to determine how the ion status is sensed in the chloroplast and how it is signaled to other compartments to adapt uptake and trafficking in the cell. Hence, electrophysiology experiments have established that chloroplast activity controls ion transport at the plasma membrane of photosynthetic cells, possibly through changes in the redox status [209].

Although uptake has been measured by biochemical methods, the uptake systems for many ions (e.g. sulfur, zinc, manganese) in chloroplasts remain to be identified. A question of interest is how ions might compete for uptake into chloroplasts. Hence, copper in excess competes with both iron and manganese for uptake, impacting the photosynthetic apparatus [260, 263].

Recent studies started to shed light on the interactions between ion homeostasis of chloroplasts, mitochondria and/or vacuoles [182, 189, 281]. Remobilization of ions from vacuolar stores (by AtNRAMP3 and 4 for manganese and iron or by COPT5 for copper) is part of the deficiency response and is required to preserve the photosynthetic function [107, 182]. Similarly, reduced iron storage in the vacuole results in higher ferritin, and thus iron, content in Arabidopsis seeds [279]. How chloroplasts and mitochondria interact with each other and with the cytosol for proper heme and Fe/S cluster biosynthesis and distribution to their respective targets remain unclear [21, 228].

So far, little attention has been paid to the examination of possible cross-talks between the homeostatic network of macro- or micronutrients, and how this influences the photosynthetic activity. Integration of the regulatory circuits of sulfate and phosphate deficiencies is starting to be revealing in both Chlamydomonas and Arabidopsis [240, 293].

In conclusion, many open questions remain regarding ion homeostasis in chloroplasts, which represent topical avenues for future investigations.

Acknowledgments Research in the M.H.'s laboratory is supported by the "Fonds de la Recherche Scientifique—FNRS" (grant nos. 2.4638.05, 2.4540.06, 2.4583.08 and 2.4581.10) and the "Fonds Spéciaux du Conseil de la Recherche" from the University of Liège. M.H. is a Research Associate of the FNRS. M.B. acknowledges the support of the CSIC JAE-DOC Program and Juan de la Cierva Program of the Spanish Ministry of Science and Innovation. M.B. research is supported by a grant from MCINN (AGL2011-23574) and a grant for young scientist from ARAID. The work of E.I.U. at UCLA is funded by the Division of Chemical Sciences, Geosciences and Biosciences, Office of Basic Energy Sciences of the US Department of Energy (DE-FD02-04ER15529).

References

1. Abdel-Ghany SE (2009) Contribution of plastocyanin isoforms to photosynthesis and copper homeostasis in *Arabidopsis thaliana* grown at different copper regimes. Planta 229:767–779
2. Abdel-Ghany SE, Pilon M (2008) MicroRNA-mediated systemic down-regulation of copper protein expression in response to low copper availability in *Arabidopsis*. J Biol Chem 283:15932–15945
3. Abdel-Ghany SE, Müller-Moulé P, Niyogi KK, Pilon M, Shikanai T (2005) Two P-type ATPases are required for copper delivery in *Arabidopsis thaliana* chloroplasts. Plant Cell 17:1233–1251

4. Abdel-Ghany SE, Burkhead JL, Gogolin KA, Andrés-Colás N, Bodecker JR, Puig S, Peñarrubia L, Pilon M (2005) AtCCS is a functional homolog of the yeast copper chaperone Ccs1/Lys7. FEBS Lett 579:2307–2312
5. Accardi A, Walden M, Nguitragool W, Jayaram H, Williams C, Miller C (2005) Separate ion pathways in a Cl^-/H^+ exchanger. J Gen Physiol 126:563–570
6. Adhikari ND, Orler R, Chory J, Froehlich JE, Larkin RM (2009) Porphyrins promote the association of GENOMES UNCOUPLED 4 and a Mg-chelatase subunit with chloroplast membranes. J Biol Chem 284:24783–24796
7. Adhikari ND, Froehlich JE, Strand DD, Buck SM, Kramer DM, Larkin RM (2011) GUN4-porphyrin complexes bind the ChlH/GUN5 subunit of Mg-Chelatase and promote chlorophyll biosynthesis in Arabidopsis. Plant Cell 23:1449–1467
8. Allen MD, Kropat J, Tottey S, Del Campo JA, Merchant SS (2007) Manganese deficiency in chlamydomonas results in loss of photosystem II and MnSOD function, sensitivity to peroxides, and secondary phosphorus and iron deficiency. Plant Physiol 143:263–277
9. Allmer J, Naumann B, Markert C, Zhang M, Hippler M (2006) Mass spectrometric genomic data mining: novel insights into bioenergetic pathways in Chlamydomonas reinhardtii. Proteomics 6:6207–6220
10. Alvarez JM, Vidal EA, Gutiérrez RA (2012) Integration of local and systemic signaling pathways for plant N responses. Curr Opin Plant Biol 15:185–191
11. Amtmann A, Armengaud P (2009) Effects of N, P, K and S on metabolism: new knowledge gained from multi-level analysis. Curr Opin Plant Biol 12:275–283
12. Andaluz S, López-Millán AF, De Las Rivas J, Aro EM, Abadía J, Abadía A (2006) Proteomic profiles of thylakoid membranes and changes in response to iron deficiency. Photosynth Res 89:141–155
13. Anderson JM, Charbonneau H, Jones HP, McCann RO, Cormier MJ (1980) Characterization of the plant nicotinamide adenine dinucleotide kinase activator protein and its identification as calmodulin. Biochemistry 19:3113–3120
14. Andrews SC, Robinson AK, Rodríguez-Quiñones F (2003) Bacterial iron homeostasis. FEMS Microbiol Rev 27:215–237
15. Apse MP, Sottosanto JB, Blumwald E (2003) Vacuolar cation/H^+ exchange, ion homeostasis, and leaf development are altered in a T-DNA insertional mutant of AtNHX1, the *Arabidopsis* vacuolar Na^+/H^+ antiporter. Plant J 36:229–239
16. Aran M, Ferrero D, Wolosiuk A, Mora-García S, Wolosiuk RA (2011) ATP and Mg^{2+} promote the reversible oligomerization and aggregation of chloroplast 2-Cys peroxiredoxin. J Biol Chem 286:23441–23451
17. Arosio P, Ingrassia R, Cavadini P (2009) Ferritins: a family of molecules for iron storage, antioxidation and more. Biochim Biophys Acta 1790:589–599
18. Ashley MK, Grant M, Grabov A (2006) Plant responses to potassium deficiencies: a role for potassium transport proteins. J Exp Bot 57:425–436
19. Bailey S et al (2008) Alternative photosynthetic electron flow to oxygen in marine *Synechococcus*. Biochim Biophys Acta 1777:269–276
20. Balandin T, Castresana C (2002) AtCOX17, an *Arabidopsis* homolog of the yeast copper chaperone COX17. Plant Physiol 129:1852–1857
21. Balk J, Pilon M (2011) Ancient and essential: the assembly of iron-sulfur clusters in plants. Trends Plant Sci 16:218–226
22. Balsera M, Goetze TA, Kovács-Bogdán E, Schürmann P, Wagner R, Buchanan BB, Soll J, Bölter B (2009) Characterization of Tic110, a channel-forming protein at the inner envelope membrane of chloroplasts, unveils a response to Ca^{2+} and a stromal regulatory disulfide bridge. J Biol Chem 284:2603–2616
23. Banci L, Bertini I, Ciofi-Baffoni S, Kandias NG, Robinson NJ, Spyroulias GA, Su XC, Tottey S, Vanarotti M (2006) The delivery of copper for thylakoid import observed by NMR. Proc Natl Acad Sci U S A 103:8320–8325
24. Banci L, Bertini I, Ciofi-Baffoni S, Kozyreva T, Zovo K, Palumaa P (2010) Affinity gradients drive copper to cellular destinations. Nature 465:645–648

25. Barrero-Gil J, Rodríguez-Navarro A, Benito B (2007) Cloning of the PpNHAD1 transporter of *Physcomitrella patens*, a chloroplast transporter highly conserved in photosynthetic eukaryotic organisms. J Exp Bot 58:2839–2849
26. Benning C (2010) The anionic chloroplast membrane lipids: phosphatidylglycerol and sulfoquinovosyldiacylglycerol. In: Rebeiz CA et al (eds) The chloroplast. Springer, Netherlands, pp 171–184
27. Berkowitz GA, Wu W (1993) Magnesium, potassium flux and photosynthesis. Magnes Res 6:257–265
28. Bernal M, Testillano PS, Alfonso M, Carmen Risueño Md, Picorel R, Yruela I (2007) Identification and subcellular localization of the soybean copper $P1_B$-ATPase GmHMA8 transporter. J Struct Biol 158:46–58
29. Bernal M et al (2012) Transcriptome sequencing identifies *SPL7*-regulated copper acquisition genes *FRO4*/*FRO5* and the copper dependence of iron homeostasis in *Arabidopsis*. Plant Cell 24:738–761
30. Bi Y-M, Wang R-L, Zhu T, Rothstein S (2007) Global transcription profiling reveals differential responses to chronic nitrogen stress and putative nitrogen regulatory components in *Arabidopsis*. BMC Genomics 8:281
31. Bibby TS, Nield J, Barber J (2001) Iron deficiency induces the formation of an antenna ring around trimeric photosystem I in cyanobacteria. Nature 412:743–745
32. Blindauer CA (2008) Zinc-handling in cyanobacteria: an update. Chem Biodivers 5:1990–2013
33. Boekema EJ, Hifney A, Yakushevska AE, Piotrowski M, Keegstra W, Berry S, Michel KP, Pistorius EK, Kruip J (2001) A giant chlorophyll-protein complex induced by iron deficiency in cyanobacteria. Nature 412:745–748
34. Bölter B, Soll J (2001) Ion channels in the outer membranes of chloroplasts and mitochondria: open doors or regulated gates? EMBO J 20:935–940
35. Boyd PW, Ellwood MJ (2010) The biogeochemical cycle of iron in the ocean. Nat Geosci 3:675–682
36. Brand JJ, Becker DW (1984) Evidence for direct roles of calcium in photosynthesis. J Bioenerg Biomembr 16:239–249
37. Briat JF, Ravet K, Arnaud N, Duc C, Boucherez J, Touraine B, Cellier F, Gaymard F (2010) New insights into ferritin synthesis and function highlight a link between iron homeostasis and oxidative stress in plants. Ann Bot 105:811–822
38. Brunswick P, Cresswell CF (1988) Nitrite uptake into intact pea chloroplasts. 1. Kinetics and relationship with nitrite assimilation. Plant Physiol 86:378–383
39. Brunswick P, Cresswell CF (1988) Nitrite uptake into intact pea chloroplasts. 2. Influence of electron-transport regulators, uncouplers, ATPase and anion uptake inhibitors and protein binding reagents. Plant Physiol 86:384–389
40. Buaboocha T, Liao B, Zielinski RE (2001) Isolation of cDNA and genomic DNA clones encoding a calmodulin-binding protein related to a family of ATPases involved in cell division and vesicle fusion. Planta 212:774–781
41. Buchner P, Takahashi H, Hawkesford MJ (2004) Plant sulphate transporters: co-ordination of uptake, intracellular and long-distance transport. J Exp Bot 55:1765–1773
42. Bughio N, Takahashi M, Yoshimura E, Nishizawa NK, Mori S (1997) Light-dependent iron transport into isolated barley chloroplasts. Plant Cell Physiol 38:101–105
43. Burkhead JL, Abdel-Ghany SE, Morrill JM, Pilon-Smits EAH, Pilon M (2003) The *Arabidopsis thaliana* CUTA gene encodes an evolutionarily conserved copper binding chloroplast protein. Plant J 34:856–867
44. Burkhead JL, Reynolds KA, Abdel-Ghany SE, Cohu CM, Pilon M (2009) Copper homeostasis. New Phytol 182:799–816
45. Busch A, Rimbauld B, Naumann B, Rensch S, Hippler M (2008) Ferritin is required for rapid remodeling of the photosynthetic apparatus and minimizes photo-oxidative stress in response to iron availability in *Chlamydomonas reinhardtii*. Plant J 55:201–211

46. Bussemer J, Chigri F, Vothknecht UC (2009) *Arabidopsis* ATPase family gene 1-like protein 1 is a calmodulin-binding AAA+-ATPase with a dual localization in chloroplasts and mitochondria. FEBS J 276:3870–3880
47. Cakmak I, Kirkby EA (2008) Role of magnesium in carbon partitioning and alleviating photooxidative damage. Physiol Plant 133:692–704
48. Cakmak I, Hengeler C, Marschner H (1994) Changes in phloem export of sucrose in leaves in response to phosphorus, potassium and magnesium deficiency in bean plants. J Exp Bot 45:1251–1257
49. Cardol P et al (2008) An original adaptation of photosynthesis in the marine green alga *Ostreococcus*. Proc Natl Acad Sci U S A 105:7881–7886
50. Castruita M et al (2011) Systems biology approach in *chlamydomonas* reveals connections between copper nutrition and multiple metabolic steps. Plant Cell 23:1273–1292
51. Catty P, Boutigny S, Miras R, Joyard J, Rolland N, Seigneurin-Berny D (2011) Biochemical characterization of AtHMA6/PAA1, a chloroplast envelope Cu(I)-ATPase. J Biol Chem 286:36188–36197
52. Cavet JS, Borrelly GP, Robinson NJ (2003) Zn, Cu and Co in cyanobacteria: selective control of metal availability. FEMS Microbiol Rev 27:165–181
53. Chang C-W, Moseley JL, Wykoff D, Grossman AR (2005) The *LPB1* gene is important for acclimation of *Chlamydomonas reinhardtii* to phosphorus and sulfur deprivation. Plant Physiol 138:319–329
54. Changela A, Chen K, Xue Y, Holschen J, Outten CE, O'Halloran TV, Mondragón A (2003) Molecular basis of metal-ion selectivity and zeptomolar sensitivity by CueR. Science 301:1383–1387
55. Chardot T, Meunier J-C (1990) Fructose-1,6-bisphosphate and calcium activate oxidized spinach (*Spinacia oleracea*) chloroplast fructose-1,6-bisphosphatase. Plant Sci 70:1–9
56. Charles SA, Halliwell B (1980) Action of calcium ions on spinach (*Spinacia oleracea*) chloroplast fructose bisphosphatase and other enzymes of the Calvin cycle. Biochem J 188:775–779
57. Chen C, Kazimir J, Cheniae GM (1995) Calcium modulates the photoassembly of photosystem II $(Mn)_4$-clusters by preventing ligation of nonfunctional high-valency states of manganese. Biochemistry 34:13511–13526
58. Chen H-C, Melis A (2004) Localization and function of SulP, a nuclear-encoded chloroplast sulfate permease in *Chlamydomonas reinhardtii*. Planta 220:198–210
59. Chen H-C, Yokthongwattana K, Newton AJ, Melis A (2003) *SulP*, a nuclear gene encoding a putative chloroplast-targeted sulfate permease in *Chlamydomonas reinhardtii*. Planta 218:98–106
60. Chen H-C, Newton AJ, Melis A (2005) Role of SulP, a nuclear-encoded chloroplast sulfate permease, in sulfate transport and H_2 evolution in *Chlamydomonas reinhardtii*. Photosynth Res 84:289–296
61. Chen W, Yang X, He Z, Feng Y, Hu F (2008) Differential changes in photosynthetic capacity, 77 K chlorophyll fluorescence and chloroplast ultrastructure between Zn-efficient and Zn-inefficient rice genotypes (*Oryza sativa*) under low zinc stress. Physiol Plant 132:89–101
62. Cherif J, Derbel N, Nakkach M, Bergmann H, Jemal F, Lakhdar ZB (2010) Analysis of in vivo chlorophyll fluorescence spectra to monitor physiological state of tomato plants growing under zinc stress. J Photochem Photobiol B 101:332–339
63. Chigri F, Soll J, Vothknecht UC (2005) Calcium regulation of chloroplast protein import. Plant J 42:821–831
64. Chigri F, Hörmann F, Stamp A, Stammers DK, Bölter B, Soll J, Vothknecht UC (2006) Calcium regulation of chloroplast protein translocation is mediated by calmodulin binding to Tic32. Proc Natl Acad Sci U S A 103:16051–16056
65. Chu CC, Lee WC, Guo WY, Pan SM, Chen LJ, Li HM, Jinn TL (2005) A copper chaperone for superoxide dismutase that confers three types of copper/zinc superoxide dismutase activity in *Arabidopsis*. Plant Physiol 139:425–436

66. Clarkson DT, Hanson JB (1980) The mineral nutrition of higher plants. Annu Rev Plant Biol 31:239–298
67. Claus-Peter W (2011) Urea metabolism in plants. Plant Sci 180:431–438
68. Cobbett CS, Hussain D, Haydon MJ (2003) Structural and functional relationships between type 1B heavy metal-transporting P-type ATPases in *Arabidopsis*. New Phytol 159:315–321
69. Cohu CM, Abdel-Ghany SE, Gogolin Reynolds KA, Onofrio AM, Bodecker JR, Kimbrel JA, Niyogi KK, Pilon M (2009) Copper delivery by the copper chaperone for chloroplast and cytosolic copper/zinc-superoxide dismutases: regulation and unexpected phenotypes in an *Arabidopsis* mutant. Mol Plant 2:1336–1350
70. Collier JL, Grossman AR (1992) Chlorosis induced by nutrient deprivation in *Synechococcus* sp. strain PCC 7942: not all bleaching is the same. J Bacteriol 174:4718–4726
71. Conte SS, Lloyd AM (2010) The MAR1 transporter is an opportunistic entry point for antibiotics. Plant Signal Behav 5:49–52
72. Conte S, Stevenson D, Furner I, Lloyd A (2009) Multiple antibiotic resistance in *Arabidopsis* is conferred by mutations in a chloroplast-localized transport protein. Plant Physiol 151:559–573
73. Cornah JE, Terry MJ, Smith AG (2003) Green or red: what stops the traffic in the tetrapyrrole pathway? Trends Plant Sci 8:224–230
74. Corratgé-Faillie C, Jabnoune M, Zimmermann S, Véry AA, Fizames C, Sentenac H (2010) Potassium and sodium transport in non-animal cells: the Trk/Ktr/HKT transporter family. Cell Mol Life Sci 67:2511–2532
75. Cosentino C, Fischer-Schliebs E, Bertl A, Thiel G, Homann U (2010) Na^+/H^+ antiporters are differentially regulated in response to NaCl stress in leaves and roots of *Mesembryanthemum crystallinum*. New Phytol 186:669–680
76. Daram P, Brunner S, Rausch C, Steiner C, Amrhein N, Bucher M (1999) Pht2;1 encodes a low-affinity phosphate transporter from *Arabidopsis*. Plant Cell 11:2153–2166
77. Davies JP, Yildiz FH, Grossman A (1996) Sac1, a putative regulator that is critical for survival of *Chlamydomonas reinhardtii* during sulfur deprivation. EMBO J 15:2150–2159
78. De Angeli A, Monachello D, Ephritikhine G, Frachisse JM, Thomine S, Gambale F, Barbier-Brygoo H (2006) The nitrate/proton antiporter AtCLCa mediates nitrate accumulation in plant vacuoles. Nature 442:939–942
79. De Angeli A, Monachello D, Ephritikhine G, Frachisse JM, Thomine S, Gambale F, Barbier-Brygoo H (2009) CLC-mediated anion transport in plant cells. Philos Trans R Soc Lond B Biol Sci 364:195–201
80. Dechorgnat J, Nguyen CT, Armengaud P, Jossier M, Diatloff E, Filleur S, Daniel-Vedele F (2011) From the soil to the seeds: the long journey of nitrate in plants. J Exp Bot 62:1349–1359
81. Demmig B, Gimmler H (1983) Properties of the isolated intact chloroplast at cytoplasmic K concentrations: I. light-induced cation uptake into intact chloroplasts is driven by an electrical potential difference. Plant Physiol 73:169–174
82. Di Baccio D, Galla G, Bracci T, Andreucci A, Barcaccia G, Tognetti R, Sebastiani L (2011) Transcriptome analyses of *Populus x euramericana* clone I-214 leaves exposed to excess zinc. Tree Physiol 31:1293–1308
83. Diédhiou CJ, Golldack D (2006) Salt-dependent regulation of chloride channel transcripts in rice. Plant Sci 170:793–800
84. Drummond RSM, Tutone A, Li YC, Gardner RC (2006) A putative magnesium transporter AtMRS2-11 is localized to the plant chloroplast envelope membrane system. Plant Sci 170:78–89
85. Dunkley TP et al (2006) Mapping the *Arabidopsis* organelle proteome. Proc Natl Acad Sci U S A 103:6518–6523
86. Duy D, Soll J, Philippar K (2007) Solute channels of the outer membrane: from bacteria to chloroplasts. Biol Chem 388:879–889
87. Duy D, Wanner G, Meda AR, von Wirén N, Soll J, Philippar K (2007) PIC1, an ancient permease in *Arabidopsis* chloroplasts, mediates iron transport. Plant Cell 19:986–1006

88. Duy D, Stübe R, Wanner G, Philippar K (2011) The chloroplast permease PIC1 regulates plant growth and development by directing homeostasis and transport of iron. Plant Physiol 155:1709–1722
89. Eitinger T, Suhr J, Moore L, Smith JA (2005) Secondary transporters for nickel and cobalt ions: theme and variations. Biometals 18:399–405
90. Elias M, Archibald JM (2009) Sizing up the genomic footprint of endosymbiosis. Bioessays 31:1273–1279
91. Eriksson M, Moseley JL, Tottey S, Del Campo JA, Quinn J, Kim Y, Merchant S (2004) Genetic dissection of nutritional copper signaling in Chlamydomonas distinguishes regulatory and target genes. Genetics 168:795–807
92. Ettinger WF, Clear AM, Fanning KJ, Peck ML (1999) Identification of a $Ca2^{+}/H^{+}$ antiport in the plant chloroplast thylakoid membrane. Plant Physiol 119:1379–1386
93. Fang Z, Mi F, Berkowitz GA (1995) Molecular and physiological analysis of a thylakoid K+ channel protein. Plant Physiol 108:1725–1734
94. Fernández E, Galván A (2007) Inorganic nitrogen assimilation in Chlamydomonas. J Exp Bot 58:2279–2287
95. Ferrario-Méry S, Bouvet M, Leleu O, Savino G, Hodges M, Meyer C (2005) Physiological characterisation of Arabidopsis mutants affected in the expression of the putative regulatory protein PII. Planta 223:28–39
96. Ferrario-Méry S, Meyer C, Hodges M (2008) Chloroplast nitrite uptake is enhanced in Arabidopsis PII mutants. FEBS Lett 582:1061–1066
97. Ferreira KN, Iverson TM, Maghlaoui K, Barber J, Iwata S (2004) Architecture of the photosynthetic oxygen-evolving center. Science 303:1831–1838
98. Ferro M, Salvi D, Riviere-Rolland H, Vermat T, Seigneurin-Berny D, Grunwald D, Garin J, Joyard J, Rolland N (2002) Integral membrane proteins of the chloroplast envelope: identification and subcellular localization of new transporters. Proc Natl Acad Sci U S A 99:11487–11492
99. Ferro M, Salvi D, Brugière S, Miras S, Kowalski S, Louwagie M, Garin J, Joyard J, Rolland N (2003) Proteomics of the chloroplast envelope membranes from *Arabidopsis thaliana*. Mol Cell Proteomics 2:325–345
100. Ferro M et al (2010) AT_CHLORO, a comprehensive chloroplast proteome database with subplastidial localization and curated information on envelope proteins. Mol Cell Proteomics 9:1063–1084
101. Finney LA, O'Halloran TV (2003) Transition metal speciation in the cell: insights from the chemistry of metal ion receptors. Science 300:931–936
102. Flügge U-I, Häusler RE, Ludewig F, Fischer K (2003) Functional genomics of phosphate antiport systems of plastids. Physiol Plant 118:475–482
103. Flügge UI, Häusler RE, Ludewig F, Gierth M (2011) The role of transporters in supplying energy to plant plastids. J Exp Bot 62:2381–2392
104. Froehlich JE, Wilkerson CG, Ray WK, McAndrew RS, Osteryoung KW, Gage DA, Phinney BS (2003) Proteomic study of the *Arabidopsis thaliana* chloroplastic envelope membrane utilizing alternatives to traditional two-dimensional electrophoresis. J Proteome Res 2:413–425
105. Fukada-Tanaka S, Inagaki Y, Yamaguchi T, Saito N, Iida S (2000) Colour-enhancing protein in blue petals. Nature 407:581
106. Furumoto T et al (2011) A plastidial sodium-dependent pyruvate transporter. Nature 476:472–475
107. Garcia-Molina A, Andrés-Colás N, Perea-García A, Del Valle-Tascón S, Peñarrubia L, Puig S (2011) The intracellular *Arabidopsis* COPT5 transport protein is required for photosynthetic electron transport under severe copper deficiency. Plant J 65:848–860
108. Gebert M, Meschenmoser K, Svidová S, Weghuber J, Schweyen R, Eifler K, Lenz H, Weyand K, Knoop V (2009) A root-expressed magnesium transporter of the MRS2/MGT gene family in *Arabidopsis thaliana* allows for growth in low-Mg^{2+} environments. Plant Cell 21:4018–4030

109. Geelen D, Lurin C, Bouchez D, Frachisse JM, Lelièvre F, Courtial B, Barbier-Brygoo H, Maurel C (2000) Disruption of putative anion channel gene AtCLC-a in *Arabidopsis* suggests a role in the regulation of nitrate content. Plant J 21:259–267
110. Gertz M, Seelert H, Dencher NA, Poetsch A (2007) Interactions of rotor subunits in the chloroplast ATP synthase modulated by nucleotides and by Mg^{2+}. Biochim Biophys Acta 1774:566–574
111. Glass ADM et al (2002) The regulation of nitrate and ammonium transport systems in plants. J Exp Bot 53:855–864
112. González-Ballester D, Camargo A, Fernández E (2004) Ammonium transporter genes in *Chlamydomonas*: the nitrate-specific regulatory gene *Nit2* is involved in *Amt1*;1 expression. Plant Mol Biol 56:863–878
113. González-Ballester D, Pollock SV, Pootakham W, Grossman AR (2008) The central role of a SNRK2 kinase in sulfur deprivation responses. Plant Physiol 147:216–227
114. González-Ballester D, Casero D, Cokus S, Pellegrini M, Merchant SS, Grossman AR (2010) RNA-seq analysis of sulfur-deprived *Chlamydomonas* cells reveals aspects of acclimation critical for cell survival. Plant Cell 22:2058–2084
115. Gordon T, Lea PJ, Rosenberg C, Trinchnat JC (2001) Nodule formation and function. In: Lea PJ, Morot-Gaudry J-F (eds) Plant nitrogen. Springer, Berlin, pp 101–146
116. Gray JC (2003) Chloroplast-to-nucleus signalling: a role for Mg-protoporphyrin. Trends Genet 19:526–529
117. Grossman AR, González-Ballester D, Shibagaki N, Pootakham W, Moseley J (2010) Responses to macronutrient deprivation. In: Pareek A, Sopory SK, Bohnert HJ (eds) Abiotic stress adaptation in plants: physiological, molecular and genomic foundation. Springer, Netherlands, pp 307–348
118. Grove GN, Brudvig GW (1998) Calcium binding studies of photosystem II using a calcium-selective electrode. Biochemistry 37:1532–1539
119. Guo B, Irigoyen S, Fowler TB, Versaw WK (2008) Differential expression and phylogenetic analysis suggest specialization of plastid-localized members of the PHT4 phosphate transporter family for photosynthetic and heterotrophic tissues. Plant Signal Behav 3:784–790
120. Guo B, Jin Y, Wussler C, Blancaflor EB, Motes CM, Versaw WK (2008) Functional analysis of the *Arabidopsis* PHT4 family of intracellular phosphate transporters. New Phytol 177:889–898
121. Haas CE, Rodionov DA, Kropat J, Malasarn D, Merchant SS, de Crecy-Lagard V (2009) A subset of the diverse COG0523 family of putative metal chaperones is linked to zinc homeostasis in all kingdoms of life. BMC Genomics 10:470
122. Hammond JP et al (2003) Changes in gene expression in *Arabidopsis* shoots during phosphate starvation and the potential for developing smart plants. Plant Physiol 132:578–596
123. Han S, Tang R, Anderson LK, Woerner TE, Pei ZM (2003) A cell surface receptor mediates extracellular Ca^{2+} sensing in guard cells. Nature 425:196–200
124. Hanikenne M, Krämer U, Demoulin V, Baurain D (2005) A comparative inventory of metal transporters in the green alga *Chlamydomonas reinhardtii* and the red alga *Cyanidioschizon merolae*. Plant Physiol 137:428–446
125. Hanikenne M, Merchant SS, Hamel P (2009) Transition metal nutrition: a balance between deficiency and toxicity. In: Stern D, Harris EH (eds) The *Chlamydomonas*. Organellar and metabolic processes, vol 2. Elsevier, Amsterdam, pp 333–399
126. Hänsch R, Mendel RR (2009) Physiological functions of mineral micronutrients (Cu, Zn, Mn, Fe, Ni, Mo, B, Cl). Curr Opin Plant Biol 12:259–266
127. Harada H, Kuromori T, Hirayama T, Shinozaki K, Leigh RA (2004) Quantitative trait loci analysis of nitrate storage in *Arabidopsis* leading to an investigation of the contribution of the anion channel gene, AtCLC-c, to variation in nitrate levels. J Exp Bot 55:2005–2014
128. Hauser F, Horie T (2010) A conserved primary salt tolerance mechanism mediated by HKT transporters: a mechanism for sodium exclusion and maintenance of high K^+/Na^+ ratio in leaves during salinity stress. Plant Cell Environ 33:552–565

129. Hawkesford MJ (2008) Uptake, distribution and subcellular transport of sulfate. In: Hell R et al (eds) Sulfur metabolism in phototrophic organisms. Springer, Netherlands, pp 15–30
130. Haydon MJ, Bell LJ, Webb AAR (2011) Interactions between plant circadian clocks and solute transport. J Exp Bot 62:2333–2348
131. Hechenberger M, Schwappach B, Fischer WN, Frommer WB, Jentsch TJ, Steinmeyer K (1996) A family of putative chloride channels from *Arabidopsis* and functional complementation of a yeast strain with a CLC gene disruption. J Biol Chem 271:33632–33638
132. Heiber T, Steinkamp T, Hinnah S, Schwarz M, Flügge UI, Weber A, Wagner R (1995) Ion channels in the chloroplast envelope membrane. Biochemistry 34:15906–15917
133. Hell R, Wirtz M (2008) Metabolism of cysteine in plants and phototrophic bacteria. In: Hell R et al (eds) Sulfur metabolism in phototrophic organisms. Springer, Netherlands, pp 59–91
134. Hell R, Khan M, Wirtz M (2010) Cellular biology of sulfur and its functions in plants. In: Hell R, Mendel R-R (eds) Cell biology of metals and nutrients. Springer, Berlin/Heidelberg, pp 243–279
135. Hermans C, Verbruggen N (2005) Physiological characterization of Mg deficiency in *Arabidopsis thaliana*. J Exp Bot 56:2153–2161
136. Hermans C, Bourgis F, Faucher M, Strasser RJ, Delrot S, Verbruggen N (2005) Magnesium deficiency in sugar beets alters sugar partitioning and phloem loading in young mature leaves. Planta 220:541–549
137. Hermans C, Vuylsteke M, Coppens F, Craciun A, Inzé D, Verbruggen N (2010) Early transcriptomic changes induced by magnesium deficiency in *Arabidopsis thaliana* reveal the alteration of circadian clock gene expression in roots and the triggering of abscisic acid-responsive genes. New Phytol 187:119–131
138. Hermans C, Vuylsteke M, Coppens F, Cristescu SM, Harren FJ, Inzé D, Verbruggen N (2010) Systems analysis of the responses to long-term magnesium deficiency and restoration in *Arabidopsis thaliana*. New Phytol 187:132–144
139. Higuchi M, Ozaki H, Matsui M, Sonoike K (2009) A T-DNA insertion mutant of AtHMA1 gene encoding a Cu transporting ATPase in *Arabidopsis thaliana* has a defect in the water-water cycle of photosynthesis. J Photochem Photobiol B 94:205–213
140. Hill KL, Merchant S (1995) Coordinate expression of coproporphyrinogen oxidase and cytochrome c_6 in the green alga *Chlamydomonas reinhardtii* in response to changes in copper availability. EMBO J 14:857–865
141. Himelblau E, Mira H, Lin SJ, Culotta VC, Penarrubia L, Amasino RM (1998) Identification of a functional homolog of the yeast copper homeostasis gene ATX1 from *Arabidopsis*. Plant Physiol 117:1227–1234.
142. Hirai MY et al (2005) Elucidation of gene-to-gene and metabolite-to-gene networks in *Arabidopsis* by integration of metabolomics and transcriptomics. J Biol Chem 280:25590–25595
143. Hoefgen R, Nikiforova VJ (2008) Metabolomics integrated with transcriptomics: assessing systems response to sulfur-deficiency stress. Physiol Plant 132:190–198
144. Hruz T, Laule O, Szabo G, Wessendorp F, Bleuler S, Oertle L, Widmayer P, Gruissem W, Zimmermann P (2008) Genevestigator v3: a reference expression database for the meta-analysis of transcriptomes. Adv Bioinformatics 2008:420747
145. Hsieh MH, Lam HM, van de Loo FJ, Coruzzi G (1998) A PII-like protein in *Arabidopsis*, putative role in nitrogen sensing. Proc Natl Acad Sci U S A 95:13965–13970
146. Huang L, Berkelman T, Franklin AE, Hoffman NE (1993) Characterization of a gene encoding a Ca^{2+}-ATPase-like protein in the plastid envelope. Proc Natl Acad Sci U S A 90:10066–10070
147. Huber SC, Maury W (1980) Effects of magnesium on intact chloroplasts: I. EVIDENCE FOR ACTIVATION OF (SODIUM) POTASSIUM/PROTON EXCHANGE ACROSS THE CHLOROPLAST ENVELOPE. Plant Physiol 65:350–354
148. Husted S, Laursen KH, Hebbern CA, Schmidt SB, Pedas P, Haldrup A, Jensen PE (2009) Manganese deficiency leads to genotype-specific changes in fluorescence induction kinetics and state transitions. Plant Physiol 150:825–833

149. Irigoyen S, Karlsson PM, Kuruvilla J, Spetea C, Versaw WK (2011) The sink-specific plastidic phosphate transporter PHT4;2 influences starch accumulation and leaf size in *Arabidopsis*. Plant Physiol 157:1765–1777
150. Irihimovitch V, Yehudai-Resheff S (2008) Phosphate and sulfur limitation responses in the chloroplast of *Chlamydomonas reinhardtii*. FEMS Microbiol Lett 283:1–8
151. Jeong J, Cohu C, Kerkeb L, Pilon M, Connolly EL, Guerinot ML (2008) Chloroplast Fe(III) chelate reductase activity is essential for seedling viability under iron limiting conditions. Proc Natl Acad Sci U S A 105:10619–10624
152. Jin SH, Huang JQ, Li XQ, Zheng BS, Wu JS, Wang ZJ, Liu GH, Chen M (2011) Effects of potassium supply on limitations of photosynthesis by mesophyll diffusion conductance in *Carya cathayensis*. Tree Physiol 31:1142–1151
153. Johnson CH, Knight MR, Kondo T, Masson P, Sedbrook J, Haley A, Trewavas A (1995) Circadian oscillations of cytosolic and chloroplastic free calcium in plants. Science 269:1863–1865
154. Kataoka T, Watanabe-Takahashi A, Hayashi N, Ohnishi M, Mimura T, Buchner P, Hawkesford MJ, Yamaya T, Takahashi H (2004) Vacuolar sulfate transporters are essential determinants controlling internal distribution of sulfate in *Arabidopsis*. Plant Cell 16:2693–2704
155. Katoh H, Hagino N, Grossman AR, Ogawa T (2001) Genes essential to iron transport in the cyanobacterium *Synechocystis* sp. strain PCC 6803. J Bacteriol 183:2779–2784
156. Kawamura Y, Takahashi M, Arimura G, Isayama T, Irifune K, Goshima N, Morikawa H (1996) Determination of levels of NO_3^-, NO_2^-, and NO_4^- ions in leaves of various plants by capillary electrophoresis. Plant Cell Physiol 37:878–880
157. Khan MS et al (2010) Sulfite reductase defines a newly discovered bottleneck for assimilatory sulfate reduction and is essential for growth and development in *Arabidopsis thaliana*. Plant Cell 22:1216–1231
158. Kim YY, Choi H, Segami S, Cho HT, Martinoia E, Maeshima M, Lee Y (2009) AtHMA1 contributes to the detoxification of excess Zn(II) in *Arabidopsis*. Plant J 58:737–753
159. Kim T-H, Böhmer M, Hu H, Nishimura N, Schroeder JI (2010) Guard cell signal transduction network: advances in understanding abscisic acid, CO_2, and Ca^{2+} signaling. Annu Rev Plant Biol 61:561–591
160. Kleffmann T, Russenberger D, von Zychlinski A, Christopher W, Sjölander K, Gruissem W, Baginsky S (2004) The *Arabidopsis thaliana* chloroplast proteome reveals pathway abundance and novel protein functions. Curr Biol 14:354–362
161. Kleine T, Voigt C, Leister D (2009) Plastid signalling to the nucleus: messengers still lost in the mists? Trends Genet 25:185–192
162. Knappe S, Flugge UI, Fischer K (2003) Analysis of the plastidic phosphate translocator gene family in *Arabidopsis* and identification of new phosphate translocator-homologous transporters, classified by their putative substrate-binding site. Plant Physiol 131:1178–1190
163. Knoop V, Groth-Malonek M, Gebert M, Eifler K, Weyand K (2005) Transport of magnesium and other divalent cations: evolution of the 2-TM-GxN proteins in the MIT superfamily. Mol Genet Genomics 274:205–216
164. Kobayashi T, Nishizawa NK (2012) Iron uptake, translocation, and regulation in higher plants. Annu Rev Plant Biol 63:131–152
165. Kobayashi Y, Kanesaki Y, Tanaka A, Kuroiwa H, Kuroiwa T, Tanaka K (2009) Tetrapyrrole signal as a cell-cycle coordinator from organelle to nuclear DNA replication in plant cells. Proc Natl Acad Sci U S A 106:803–807
166. Koussevitzky S, Nott A, Mockler TC, Hong F, Sachetto-Martins G, Surpin M, Lim J, Mittler R, Chory J (2007) Signals from chloroplasts converge to regulate nuclear gene expression. Science 316:715–719
167. Krämer U, Clemens S (2005) Functions and homeostasis of zinc, copper, and nickel in plants. In: Tamás M, Martinoia E (eds) Molecular biology of metal homeostasis and detoxification. Springer, Berlin, pp 216–271

168. Kreimer G, Melkonian M, Latzko E (1985) An electrogenic uniport mediates light-dependent Ca^{2+} influx into intact spinach chloroplasts. FEBS Lett 180:253–258
169. Kreimer G, Melkonian M, Holtum JA, Latzko E (1988) Stromal free calcium concentration and light-mediated activation of chloroplast fructose-1,6-bisphosphatase. Plant Physiol 86:423–428
170. Krieger A, Rutherford AW, Vass I, Hideg E (1998) Relationship between activity, D1 loss, and Mn binding in photoinhibition of photosystem II. Biochemistry 37:16262–16269
171. Kronzucker HJ, Britto DT (2010) Sodium transport in plants: a critical review. New Phytol 189:54–81
172. Kropat J, Oster U, Rüdiger W, Beck CF (1997) Chlorophyll precursors are signals of chloroplast origin involved in light induction of nuclear heat-shock genes. Proc Natl Acad Sci U S A 94:14168–14172
173. Kropat J, Oster U, Rüdiger W, Beck CF (2000) Chloroplast signalling in the light induction of nuclear HSP70 genes requires the accumulation of chlorophyll precursors and their accessibility to cytoplasm/nucleus. Plant J 24:523–531
174. Kropat J, Tottey S, Birkenbihl RP, Depège N, Huijser P, Merchant S (2005) A regulator of nutritional copper signaling in *Chlamydomonas* is an SBP domain protein that recognizes the GTAC core of copper response element. Proc Natl Acad Sci U S A 102:18730–18735
175. La Fontaine S, Quinn JM, Nakamoto SS, Page MD, Göhre V, Moseley JL, Kropat J, Merchant S (2002) Copper-dependent iron assimilation pathway in the model photosynthetic eukaryote *Chlamydomonas reinhardtii*. Eukaryot Cell 1:736–757.
176. Laganowsky A, Gómez SM, Whitelegge JP, Nishio JN (2009) Hydroponics on a chip: analysis of the Fe deficient *Arabidopsis* thylakoid membrane proteome. J Proteomics 72:397–415
177. Lam HM, Coschigano K, Oliveira IC, M-O R, Coruzzi G (1996) The molecular-genetics of nitrogen assimilation into amino acids in higher plants. Annu Rev Plant Physiol Plant Mol Biol 47:569–593
178. Lambertz C, Hemschemeier A, Happe T (2010) Anaerobic expression of the ferredoxin-encoding *FDX5* gene of *Chlamydomonas reinhardtii* is regulated by the Crr1 transcription factor. Eukaryotic Cell 9:1747–1754
179. Lancien M, Gadal P, Hodges M (2000) Enzyme redundancy and the importance of 2-oxoglutarate in higher plant ammonium assimilation. Plant Physiol 123:817–824
180. Lane TW, Morel FM (2000) Regulation of carbonic anhydrase expression by zinc, cobalt, and carbon dioxide in the marine diatom *Thalassiosira weissflogii*. Plant Physiol 123:345–352
181. Lanquar V et al (2005) Mobilization of vacuolar iron by AtNRAMP3 and AtNRAMP4 is essential for seed germination on low iron. EMBO J 24:4041–4051
182. Lanquar V, Ramos MS, Lelièvre F, Barbier-Brygoo H, Krieger-Liszkay A, Krämer U, Thomine S (2010) Export of vacuolar manganese by AtNRAMP3 and AtNRAMP4 is required for optimal photosynthesis and growth under manganese deficiency. Plant Physiol 152:1986–1999
183. Larkin RM, Alonso JM, Ecker JR, Chory J (2003) GUN4, a regulator of chlorophyll synthesis and intracellular signaling. Science 299:902–906
184. Larkum AWD (1968) Ionic relations of chloroplasts in vivo. Nature 218:447
185. Laudenbach DE, Reith ME, Straus NA (1988) Isolation, sequence analysis, and transcriptional studies of the flavodoxin gene from *Anacystis nidulans* R2. J Bacteriol 170:258–265
186. Leigh RA, Wyn Jones RG (1984) A hypothesis relating critical potassium concentrations for growth to the distribution and functions of this ion in the plant cell. New Phytol 97:1–13
187. Léon S, Touraine B, Ribot C, Briat JF, Lobréaux S (2003) Iron-sulphur cluster assembly in plants: distinct NFU proteins in mitochondria and plastids from *Arabidopsis thaliana*. Biochem J 371:823–830
188. Lewis NE, Marty NJ, Kathir KM, Rajalingam D, Kight AD, Daily A, Kumar TK, Henry RL, Goforth RL (2010) A dynamic cpSRP43-Albino3 interaction mediates translocase

regulation of chloroplast signal recognition particle (cpSRP)-targeting components. J Biol Chem 285:34220–34230
189. Li L, Kaplan J (2004) A mitochondrial-vacuolar signaling pathway in yeast that affects iron and copper metabolism. J Biol Chem 279:33653–33661
190. Li HH, Merchant S (1995) Degradation of plastocyanin in copper-deficient *Chlamydomonas reinhardtii*. J Biol Chem 270:23504–23510
191. Li HM, Theg SM, Bauerle CM, Keegstra K (1990) Metal-ion-center assembly of ferredoxin and plastocyanin in isolated chloroplasts. Proc Natl Acad Sci U S A 87:6748–6752
192. Li L, Tutone AF, Drummond RS, Gardner RC, Luan S (2001) A novel family of magnesium transport genes in *Arabidopsis*. Plant Cell 13:2761–2775
193. Li J, Wang DY, Li Q, Xu YJ, Cui KM, Zhu YX (2004) PPF1 inhibits programmed cell death in apical meristems of both G2 pea and transgenic *Arabidopsis* plants possibly by delaying cytosolic Ca^{2+} elevation. Cell Calcium 35:71–77
194. Li L-G, Sokolov LN, Yang Y-H, Li D-P, Ting J, Pandy GK, Luan S (2008) A mitochondrial magnesium transporter functions in *Arabidopsis* pollen development. Mol Plant 1:675–685
195. Lillo C (2008) Signalling cascades integrating light-enhanced nitrate metabolism. Biochem J 415:11–19
196. Lindberg P, Melis A (2008) The chloroplast sulfate transport system in the green alga *Chlamydomonas reinhardtii*. Planta 228:951–961
197. Linka M, Weber APM (2005) Shuffling ammonia between mitochondria and plastids during photorespiration. Trends Plant Sci 10:461–465
198. Lommer M, Roy AS, Schilhabel M, Schreiber S, Rosenstiel P, LaRoche J (2010) Recent transfer of an iron-regulated gene from the plastid to the nuclear genome in an oceanic diatom adapted to chronic iron limitation. BMC genomics 11:718
199. Long JC, Sommer F, Allen MD, Lu SF, Merchant SS (2008) *FER1* and *FER2* encoding two ferritin complexes in *Chlamydomonas reinhardtii* chloroplasts are regulated by iron. Genetics 179:137–147
200. Lurin C, Geelen D, Barbier-Brygoo H, Guern J, Maurel C (1996) Cloning and functional expression of a plant voltage-dependent chloride channel. Plant Cell 8:701–711
201. Lutsenko S, Petris MJ (2003) Function and regulation of the mammalian copper-transporting ATPases: insights from biochemical and cell biological approaches. J Membr Biol 191:1–12
202. Marcaida MJ, Schlarb-Ridley BG, Worrall JA, Wastl J, Evans TJ, Bendall DS, Luisi BF, Howe CJ (2006) Structure of cytochrome c6A, a novel dithio-cytochrome of *Arabidopsis thaliana*, and its reactivity with plastocyanin: implications for function. J Mol Biol 360:968–977
203. Marchive C et al (2009) Abnormal physiological and molecular mutant phenotypes link chloroplast polynucleotide phosphorylase to the phosphorus deprivation response in *Arabidopsis*. Plant Physiol 151:905–924
204. Mariscal V, Rexach J, Fernández E, Galván A (2004) The plastidic nitrite transporter NAR1;1 improves nitrate use efficiency for growth in *Chlamydomonas*. Plant Cell Environ 27:1321–1328
205. Mariscal V, Moulin P, Orsel M, Miller AJ, Fernández E, Galván A (2006) Differential regulation of the *Chlamydomonas Nar1* gene family by carbon and nitrogen. Protist 157:421–433
206. Marmagne A et al (2007) Two members of the *Arabidopsis* CLC (chloride channel) family, AtCLCe and AtCLCf, are associated with thylakoid and golgi membranes, respectively. J Exp Bot 58:3385–3393
207. Marschner H (1995) Mineral nutrition of higher plants, 2nd edn. Academic, London
208. Marschner P (2012) Marschner's mineral nutrition of higher plants, 3rd edn. Academic, London
209. Marten I, Deeken R, Hedrich R, Roelfsema MRG (2010) Light-induced modification of plant plasma membrane ion transport. Plant Biol (Stuttg) 12:64–79

210. Mäser P et al (2001) Phylogenetic relationships within cation transporter families of *Arabidopsis*. Plant Physiol 126:1646–1667
211. Masuda T (2008) Recent overview of the Mg branch of the tetrapyrrole biosynthesis leading to chlorophylls. Photosynth Res 96:121–143
212. Matsushima R, Tang LY, Zhang L, Yamada H, Twell D, Sakamoto W (2011) A conserved, Mg^{2+}-dependent exonuclease degrades organelle DNA during *Arabidopsis* pollen development. Plant Cell 23:1608–1624
213. Maury WJ, Huber SC, Moreland DE (1981) Effects of magnesium on intact chloroplasts: II. CATION SPECIFICITY AND INVOLVEMENT OF THE ENVELOPE ATPase IN (SODIUM) POTASSIUM/PROTON EXCHANGE ACROSS THE ENVELOPE. Plant Physiol 68:1257–1263
214. Melis A (2007) Photosynthetic H_2 metabolism in *Chlamydomonas reinhardtii* (unicellular green algae). Planta 226:1075–1086
215. Melis A, Chen H-C (2005) Chloroplast sulfate transport in green algae-genes, proteins and effects. Photosynth Res 86:299–307
216. Melis A, Zhang L, Forestier M, Ghirardi ML, Seibert M (2000) Sustained photobiological hydrogen gas production upon reversible inactivation of oxygen evolution in the green alga *Chlamydomonas reinhardtii*. Plant Physiol 122:127–136
217. Merchant SS (2006) Trace metal utilization in chloroplasts. In: Wise RR, Hoober JK (eds) The structure and function of plastids. Springer, Netherlands, pp 199–218
218. Merchant S, Bogorad L (1986) Regulation by copper of the expression of plastocyanin and cytochrome c552 in *Chlamydomonas reinhardtii*. Mol Cell Biol 6:462–469
219. Merchant S, Bogorad L (1987) Metal ion regulated gene expression: use of a plastocyanin-less mutant of *Chlamydomonas reinhardtii* to study the Cu(II)-dependent expression of cytochrome *c*-552. EMBO J 6:2531–2535
220. Merchant S, Allen MD, Kropat J, Moseley JL, Long JC, Tottey S, Terauchi AM (2006) Between a rock and a hard place: trace element nutrition in *Chlamydomonas*. Biochim Biophys Acta 1763:578–594
221. Mi F, Peters JS, Berkowitz GA (1994) Characterization of a chloroplast inner envelope K^+ channel. Plant Physiol 105:955–964
222. Miller C (2006) ClC chloride channels viewed through a transporter lens. Nature 440:484
223. Miller AF, Brudvig GW (1989) Manganese and calcium requirements for reconstitution of oxygen-evolution activity in manganese-depleted photosystem II membranes. Biochemistry 28:8181–8190
224. Miqyass M, van Gorkom HJ, Yocum CF (2007) The PSII calcium site revisited. Photosynth Res 92:275–287
225. Misson J et al (2005) A genome-wide transcriptional analysis using *Arabidopsis thaliana* Affymetrix gene chips determined plant responses to phosphate deprivation. Proc Natl Acad Sci U S A 102:11934–11939
226. Mochizuki N, Brusslan JA, Larkin R, Nagatani A, Chory J (2001) *Arabidopsis* genomes uncoupled 5 (GUN5) mutant reveals the involvement of Mg-chelatase H subunit in plastid-to-nucleus signal transduction. Proc Natl Acad Sci U S A 98:2053–2058
227. Mochizuki N, Tanaka R, Tanaka A, Masuda T, Nagatani A (2008) The steady-state level of Mg-protoporphyrin IX is not a determinant of plastid-to-nucleus signaling in *Arabidopsis*. Proc Natl Acad Sci U S A 105:15184–15189
228. Mochizuki N, Tanaka R, Grimm B, Masuda T, Moulin M, Smith AG, Tanaka A, Terry MJ (2010) The cell biology of tetrapyrroles: a life and death struggle. Trends Plant Sci 15:488–498
229. Molina-Heredia FP, Wastl J, Navarro JA, Bendall DS, Hervas M, Howe CJ, De La Rosa MA (2003) Photosynthesis: a new function for an old cytochrome? Nature 424:33–34
230. Monachello D, Allot M, Oliva S, Krapp A, Daniel-Vedele F, Barbier-Brygoo H, Ephritikhine G (2009) Two anion transporters AtClCa and AtClCe fulfil interconnecting but not redundant roles in nitrate assimilation pathways. New Phytol 183:88–94

231. Morales F, Abadía A, Abadía J (1991) Chlorophyll fluorescence and photon yield of oxygen evolution in iron-deficient sugar beet (*Beta vulgaris* L.) leaves. Plant Physiol 97:886–893
232. Morales F, Cerovic ZG, Moya I (1994) Characterization of blue-green fluorescence in the mesophyll of sugar beet (*Beta vulgaris* L.) leaves affected by iron deficiency. Plant Physiol 106:127–133
233. Morcuende R et al (2007) Genome-wide reprogramming of metabolism and regulatory networks of *Arabidopsis* in response to phosphorus. Plant Cell Environ 30:85–112
234. Moreno I, Norambuena L, Maturana D, Toro M, Vergara C, Orellana A, Zurita-Silva A, Ordenes VR (2008) AtHMA1 is a thapsigargin sensitive Ca^{2+}/heavy metal pump. J Biol Chem 283:9633–9641
235. Morrissey J, Baxter IR, Lee J, Li L, Lahner B, Grotz N, Kaplan J, Salt DE, Guerinot ML (2009) The ferroportin metal efflux proteins function in iron and cobalt homeostasis in *Arabidopsis*. Plant Cell 21:3326–3338
236. Moseley J, Quinn J, Eriksson M, Merchant S (2000) The *Crd1* gene encodes a putative di-iron enzyme required for photosystem I accumulation in copper deficiency and hypoxia in *Chlamydomonas reinhardtii*. EMBO J 19:2139–2151.
237. Moseley JL, Allinger T, Herzog S, Hoerth P, Wehinger E, Merchant S, Hippler M (2002) Adaptation to Fe-deficiency requires remodeling of the photosynthetic apparatus. EMBO J 21:6709–6720.
238. Moseley JL, Page MD, Alder NP, Eriksson M, Quinn J, Soto F, Theg SM, Hippler M, Merchant S (2002) Reciprocal expression of two candidate di-iron enzymes affecting photosystem I and light-harvesting complex accumulation. Plant Cell 14:673–688
239. Moseley JL, Chang C-W, Grossman AR (2006) Genome-based approaches to understanding phosphorus deprivation responses and PSR1 control in *Chlamydomonas reinhardtii*. Eukaryot Cell 5:26–44
240. Moseley JL, González-Ballester D, Pootakham W, Bailey S, Grossman AR (2009) Genetic interactions between regulators of *Chlamydomonas* phosphorus and sulfur deprivation responses. Genetics 181:889–905
241. Moulin M, McCormac AC, Terry MJ, Smith AG (2008) Tetrapyrrole profiling in *Arabidopsis* seedlings reveals that retrograde plastid nuclear signaling is not due to Mg-protoporphyrin IX accumulation. Proc Natl Acad Sci U S A 105:15178–15183
242. Mourioux G, Douce R (1979) Transport du sulfate à travers la double membrane limitante, ou enveloppe, des chloroplastes d'épinard. Biochimie 61:1283–1292
243. Mukherjee I, Campbell NH, Ash JS, Connolly EL (2006) Expression profiling of the *Arabidopsis* ferric chelate reductase (FRO) gene family reveals differential regulation by iron and copper. Planta 223:1178–1190
244. Muto S, Izawa S, Miyachi S (1982) Light-induced Ca^{2+} uptake by intact chloroplasts. FEBS Lett 139:250–254
245. Nakamura A, Fukuda A, Sakai S, Tanaka Y (2006) Molecular cloning, functional expression and subcellular localization of two putative vacuolar voltage-gated chloride channels in rice (*Oryza sativa* L.). Plant Cell Physiol 47:32–42
246. Naumann B, Stauber EJ, Busch A, Sommer F, Hippler M (2005) N-terminal processing of Lhca3 Is a key step in remodeling of the photosystem I-light-harvesting complex under iron deficiency in *Chlamydomonas reinhardtii*. J Biol Chem 280:20431–20441
247. Naumann B, Busch A, Allmer J, Ostendorf E, Zeller M, Kirchhoff H, Hippler M (2007) Comparative quantitative proteomics to investigate the remodeling of bioenergetic pathways under iron deficiency in *Chlamydomonas reinhardtii*. Proteomics 7:3964–3979
248. Nikiforova V, Freitag J, Kempa S, Adamik M, Hesse H, Hoefgen R (2003) Transcriptome analysis of sulfur depletion in *Arabidopsis thaliana*: interlacing of biosynthetic pathways provides response specificity. Plant J 33:633–650
249. Nikiforova VJ, Kopka J, Tolstikov V, Fiehn O, Hopkins L, Hawkesford MJ, Hesse H, Hoefgen R (2005) Systems rebalancing of metabolism in response to sulfur deprivation, as revealed by metabolome analysis of *Arabidopsis* plants. Plant Physiol 138:304–318

250. Nilsson L, Müller R, Nielsen TH (2010) Dissecting the plant transcriptome and the regulatory responses to phosphate deprivation. Physiol Plant 139:129–143
251. Nishio JN, Abadía J, Terry N (1985) Chlorophyll-proteins and electron transport during iron nutrition-mediated chloroplast development. Plant Physiol 78:296–299
252. Nobel PS (1969) Light-induced changes in the ionic content of chloroplasts in *Pisum sativum*. Biochim Biophys Acta 172:134–143
253. Nomura H, Komori T, Kobori M, Nakahira Y, Shiina T (2008) Evidence for chloroplast control of external Ca^{2+}-induced cytosolic Ca^{2+} transients and stomatal closure. Plant J 53:988–998
254. Nott A, Jung HS, Koussevitzky S, Chory J (2006) Plastid-to-nucleus retrograde signaling. Annu Rev Plant Biol 57:739–759
255. Nouet C, Motte P, Hanikenne M (2011) Chloroplastic and mitochondrial metal homeostasis. Trends Plant Sci 16:395–404
256. O'Halloran TV, Culotta VC (2000) Metallochaperones, an intracellular shuttle service for metal ions. J Biol Chem 275:25057–25060
257. Ohnishi J-I, Kanai R (1987) Na^+-induced uptake of pyruvate into mesophyll chloroplasts of a C4 plant, *Panicum miliaceum*. FEBS Lett 219:347–350
258. Ohnishi J, Kanai R (1990) Pyruvate uptake induced by a pH jump in mesophyll chloroplasts of maize and sorghum, NADP-malic enzyme type C4 species. FEBS Lett 269:122–124
259. Padan E, Venturi M, Gerchman Y, Dover N (2001) Na^+/H^+ antiporters. Biochim Biophys Acta 1505:144–157
260. Pádua M, Cavaco AM, Aubert S, Bligny R, Casimiro A (2010) Effects of copper on the photosynthesis of intact chloroplasts: interaction with manganese. Physiol Plant 138:301–311
261. Page MD, Kropat J, Hamel PP, Merchant SS (2009) Two *Chlamydomonas* CTR copper transporters with a novel cys-met motif are localized to the plasma membrane and function in copper assimilation. Plant Cell 21:928–943
262. Pardo JM, Cubero B, Leidi EO, Quintero FJ (2006) Alkali cation exchangers: roles in cellular homeostasis and stress tolerance. J Exp Bot 57:1181–1199
263. Pätsikkä E, Kairavuo M, Šeršen F, Aro EM, Tyystjärvi E (2002) Excess copper predisposes photosystem II to photoinhibition in vivo by outcompeting iron and causing decrease in leaf chlorophyll. Plant Physiol 129:1359–1367
264. Peltier JB, Ytterberg AJ, Sun Q, van Wijk KJ (2004) New functions of the thylakoid membrane proteome of *Arabidopsis thaliana* revealed by a simple, fast, and versatile fractionation strategy. J Biol Chem 279:49367–49383
265. Peters JS, Berkowitz GA (1991) Studies on the system regulating proton movement across the chloroplast envelope: effects of ATPase inhibitors, Mg, and an amine anesthetic on stromal pH and photosynthesis. Plant Physiol 95:1229–1236
266. Petroutsos D et al (2011) The chloroplast calcium sensor CAS is required for photoacclimation in *Chlamydomonas reinhardtii*. Plant Cell 23:2950–2963
267. Pilon M, Cohu CM, Ravet K, Abdel-Ghany SE, Gaymard F (2009) Essential transition metal homeostasis in plants. Curr Opin Plant Biol 12:347–357
268. Pilon M, Ravet K, Tapken W (2011) The biogenesis and physiological function of chloroplast superoxide dismutases. Biochim Biophys Acta 1807:989–998
269. Pontier D, Albrieux C, Joyard J, Lagrange T, Block MA (2007) Knock-out of the magnesium protoporphyrin IX methyltransferase gene in *Arabidopsis*. J Biol Chem 282:2297–2304
270. Pottosin II (1992) Single channel recording in the chloroplast envelope. FEBS Lett 308:87–90
271. Pottosin II, Schönknecht G (1996) Ion channel permeable for divalent and monovalent cations in native spinach thylakoid membranes. J Membr Biol 152:223–233
272. Pottosin II, Muñiz J, Shabala S (2005) Fast-activating channel controls cation fluxes across the native chloroplast envelope. J Membr Biol 204:145–156
273. Puig S et al (2007) Higher plants possess two different types of ATX1-like copper chaperones. Biochem Biophys Res Commun 354:385–390

274. Quinn JM, Merchant S (1995) Two copper-responsive elements associated with the *Chlamydomonas Cyc6* gene function as targets for transcriptional activators. Plant Cell 7:623–638
275. Quinn JM, Nakamoto SS, Merchant S (1999) Induction of coproporphyrinogen oxidase in *Chlamydomonas* chloroplasts occurs via transcriptional regulation of *Cpx1* mediated by copper response elements and increased translation from a copper deficiency- specific form of the transcript. J Biol Chem 274:14444–14454
276. Quinn JM, Barraco P, Eriksson M, Merchant S (2000) Coordinate copper- and oxygen-responsive *Cyc6* and *Cpx1* expression in *Chlamydomonas* is mediated by the same element. J Biol Chem 275:6080–6089
277. Randall PJ, Bouma D (1973) Zinc deficiency, carbonic anhydrase, and photosynthesis in leaves of spinach. Plant Physiol 52:229–232
278. Rausch C, Bucher M (2002) Molecular mechanisms of phosphate transport in plants. Planta 216:23–37
279. Ravet K, Touraine B, Kim SA, Cellier F, Thomine S, Guerinot ML, Briat JF, Gaymard F (2009) Post-translational regulation of AtFER2 ferritin in response to intracellular iron trafficking during fruit development in *Arabidopsis*. Mol Plant 2:1095–1106
280. Ravet K, Touraine B, Boucherez J, Briat JF, Gaymard F, Cellier F (2009) Ferritins control interaction between iron homeostasis and oxidative stress in *Arabidopsis*. Plant J 57:400–412
281. Ravet K, Danford FL, Dihle A, Pittarello M, Pilon M (2011) Spatiotemporal analysis of copper homeostasis in *Populus trichocarpa* reveals an integrated molecular remodeling for a preferential allocation of copper to plastocyanin in the chloroplasts of developing leaves. Plant Physiol 157:1300–1312
282. Reddy VS, Ali GS, Reddy AS (2002) Genes encoding calmodulin-binding proteins in the *Arabidopsis* genome. J Biol Chem 277:9840–9852
283. Remacle C, Coosemans N, Jans F, Hanikenne M, Motte P, Cardol P (2010) Knock-down of the *COX3* and *COX17* gene expression of cytochrome c oxidase in the unicellular green alga *Chlamydomonas reinhardtii*. Plant Mol Biol 74:223–233
284. Renné P, Dressen U, Hebbeker U, Hille D, Flügge UI, Westhoff P, Weber APM (2003) The *Arabidopsis* mutant *dct* is deficient in the plastidic glutamate/malate translocator DiT2. Plant J 35:316–331
285. Rexach J, Fernández E, Galván A (2000) The *Chlamydomonas reinhardtii Nar1* gene encodes a chloroplast membrane protein involved in nitrite transport. Plant Cell 12:1441–1453
286. Richard CG (2003) Genes for magnesium transport. Curr Opin Plant Biol 6:263–267
287. Robinson NJ, Winge DR (2010) Copper metallochaperones. Annu Rev Biochem 79:537–562
288. Rodionov DA, Hebbeln P, Gelfand MS, Eitinger T (2006) Comparative and functional genomic analysis of prokaryotic nickel and cobalt uptake transporters: evidence for a novel group of ATP-binding cassette transporters. J Bacteriol 188:317–327
289. Rodríguez-Navarro A, Rubio F (2006) High-affinity potassium and sodium transport systems in plants. J Exp Bot 57:1149–1160
290. Roh MH, Shingles R, Cleveland MJ, McCarty RE (1998) Direct measurement of calcium transport across chloroplast inner-envelope vesicles. Plant Physiol 118:1447–1454
291. Rolland N, Ferro M, Seigneurin-Berny D, Garin J, Douce R, Joyard J (2003) Proteomics of chloroplast envelope membranes. Photosynth Res 78:205–230
292. Roth C, Menzel G, Petetot JM, Rochat-Hacker S, Poirier Y (2004) Characterization of a protein of the plastid inner envelope having homology to animal inorganic phosphate, chloride and organic-anion transporters. Planta 218:406–416
293. Rouached H, Secco D, Arpat B, Poirier Y (2011) The transcription factor PHR1 plays a key role in the regulation of sulfate shoot-to-root flux upon phosphate starvation in *Arabidopsis*. BMC Plant Biol 11:19

294. Ruiz-Pavón L, Lundh F, Lundin B, Mishra A, Persson BL, Spetea C (2008) *Arabidopsis* ANTR1 is a thylakoid Na^+-dependent phosphate transporter: functional characterization in *Escherichia coli*. J Biol Chem 283:13520–13527
295. Rumak I, Gieczewska K, Kierdaszuk B, Gruszecki WI, Mostowska A, Mazur R, Garstka M (2010) 3-D modelling of chloroplast structure under (Mg^{2+}) magnesium ion treatment. Relationship between thylakoid membrane arrangement and stacking. Biochim Biophys Acta 1797:1736–1748
296. Sagardoy R, Vazquez S, Florez-Sarasa ID, Albacete A, Ribas-Carbo M, Flexas J, Abadia J, Morales F (2010) Stomatal and mesophyll conductances to CO_2 are the main limitations to photosynthesis in sugar beet (*Beta vulgaris*) plants grown with excess zinc. New Phytol 187:145–158
297. Sagasti S, Yruela I, Bernal M, Lujan MA, Frago S, Medina M, Picorel R (2011) Characterization of the recombinant copper chaperone (CCS) from the plant *Glycine* (*G.*) *max*. Metallomics 3:169–175
298. Sai J, Johnson CH (2002) Dark-stimulated calcium ion fluxes in the chloroplast stroma and cytosol. Plant Cell 14:1279–1291
299. Saito A, Lino T, Sonoike K, Miwa E, Higuchi K (2010) Remodeling of the major light-harvesting antenna protein of PSII protects the young leaves of barley (*Hordeum vulgare* L.) from photoinhibition under prolonged iron deficiency. Plant Cell Physiol 51:2013–2030
300. Sánchez-Calderón L, Chacon-López A, Pérez-Torres C-A, Herrera-Estrella L (2010) Phosphorus: plant strategies to cope with its scarcity. In: Hell R, Mendel R-R (eds) Cell biology of metals and nutrients. Springer, Berlin, pp 173–198
301. Sandstrom S, Park YI, Oquist G, Gustafsson P (2001) CP43', the *isiA* gene product, functions as an excitation energy dissipater in the cyanobacterium *Synechococcus* sp. PCC 7942. Photochem Photobiol 74:431–437
302. Schaaf G, Honsbein A, Meda AR, Kirchner S, Wipf D, von Wirén N (2006) AtIREG2 encodes a tonoplast transport protein involved in iron-dependent nickel detoxification in *Arabidopsis thaliana* roots. J Biol Chem 281:25532–25540
303. Scheible W-R et al (2004) Genome-wide reprogramming of primary and secondary metabolism, protein synthesis, cellular growth processes, and the regulatory infrastructure of *Arabidopsis* in response to nitrogen. Plant Physiol 136:2483–2499
304. Schwacke R, Schneider A, van der Graaff E, Fischer K, Catoni E, Desimone M, Frommer WB, Flügge U-I, Kunze R (2003) ARAMEMNON, a novel database for *Arabidopsis* integral membrane proteins. Plant Physiol 131:16–26
305. Seigneurin-Berny D et al (2006) HMA1, a new Cu-ATPase of the chloroplast envelope, is essential for growth under adverse light conditions. J Biol Chem 281:2882–2892
306. Shabala S (2003) Regulation of potassium transport in leaves: from molecular to tissue level. Ann Bot 92:627–634
307. Shaul O (2002) Magnesium transport and function in plants: the tip of the iceberg. Biometals 15:309–323
308. Shcolnick S, Keren N (2006) Metal homeostasis in cyanobacteria and chloroplasts. Balancing benefits and risks to the photosynthetic apparatus. Plant Physiol 141:805–810
309. Shi H, Ishitani M, Kim C, Zhu JK (2000) The *Arabidopsis thaliana* salt tolerance gene SOS1 encodes a putative Na^+/H^+ antiporter. Proc Natl Acad Sci U S A 97:6896–6901
310. Shi H, Lee BH, Wu SJ, Zhu JK (2003) Overexpression of a plasma membrane Na^+/H^+ antiporter gene improves salt tolerance in *Arabidopsis thaliana*. Nat Biotechnol 21:81–85
311. Shi H, Bencze KZ, Stemmler TL, Philpott CC (2008) A cytosolic iron chaperone that delivers iron to ferritin. Science 320:1207–1210
312. Shikanai T, Muller-Moule P, Munekage Y, Niyogi KK, Pilon M (2003) PAA1, a P-type ATPase of *Arabidopsis*, functions in copper transport in chloroplasts. Plant Cell 15:1333–1346
313. Shimoni-Shor E, Hassidim M, Yuval-Naeh N, Keren N (2010) Disruption of Nap14, a plastid-localized non-intrinsic ABC protein in *Arabidopsis thaliana* results in the over-

accumulation of transition metals and in aberrant chloroplast structures. Plant Cell Environ 33:1029–1038
314. Shingles R, Roh MH, McCarty RE (1996) Nitrite transport in chloroplast inner envelope vesicles. 1. Direct measurement of proton-linked transport. Plant Physiol 112:1375–1381
315. Shingles R, North M, McCarty RE (2001) Direct measurement of ferrous ion transport across membranes using a sensitive fluorometric assay. Anal Biochem 296:106–113
316. Shingles R, North M, McCarty RE (2002) Ferrous ion transport across chloroplast inner envelope membranes. Plant Physiol 128:1022–1030
317. Shingles R, Wimmers LE, McCarty RE (2004) Copper transport across pea thylakoid membranes. Plant Physiol 135:145–151
318. Song C-P, Guo Y, Qiu Q, Lambert G, Galbraith DW, Jagendorf A, Zhu J-K (2004) A probable $Na^+(K^+)/H^+$ exchanger on the chloroplast envelope functions in pH homeostasis and chloroplast development in *Arabidopsis thaliana*. Proc Natl Acad Sci U S A 101:10211–10216
319. Spiller S, Terry N (1980) Limiting factors in photosynthesis: II. IRON STRESS DIMINISHES PHOTOCHEMICAL CAPACITY BY REDUCING THE NUMBER OF PHOTOSYNTHETIC UNITS. Plant Physiol 65:121–125
320. Stiller J (2007) Plastid endosymbiosis, genome evolution and the origin of green plants. Trends Plant Sci 12:391–396
321. Strand A, Asami T, Alonso J, Ecker JR, Chory J (2003) Chloroplast to nucleus communication triggered by accumulation of Mg-protoporphyrinIX. Nature 421:79–83
322. Strzepek RF, Harrison PJ (2004) Photosynthetic architecture differs in coastal and oceanic diatoms. Nature 431:689–692
323. Sugiura M, Georgescu MN, Takahashi M (2007) A nitrite transporter associated with nitrite uptake by higher plant chloroplasts. Plant Cell Physiol 48:1022–1035
324. Sun Q, Zybailov B, Majeran W, Friso G, Olinares PDB, van Wijk KJ (2009) PPDB, the plant proteomics database at Cornell. Nucl Acids Res 37:D969–D974
325. Sundberg E, Slagter JG, Fridborg I, Cleary SP, Robinson C, Coupland G (1997) ALBINO3, an *Arabidopsis* nuclear gene essential for chloroplast differentiation, encodes a chloroplast protein that shows homology to proteins present in bacterial membranes and yeast mitochondria. Plant Cell 9:717–730
326. Suorsa M, Aro EM (2007) Expression, assembly and auxiliary functions of photosystem II oxygen-evolving proteins in higher plants. Photosynth Res 93:89–100
327. Surek B, Kreimer G, Melkonian M, Latzko E (1987) Spinach ferredoxin is a calcium-binding protein. Planta 171:565–568
328. Taira M, Valtersson U, Burkhardt B, Ludwig RA (2004) *Arabidopsis thaliana* GLN2-encoded glutamine synthetase is dual targeted to leaf mitochondria and chloroplasts. Plant Cell 16:2048–2058
329. Takahashi H, Asanuma W, Saito K (1999) Cloning of an *Arabidopsis* cDNA encoding a chloroplast localizing sulphate transporter isoform. J Exp Bot 50:1713–1714
330. Tanaka R, Kobayashi K, Masuda T (2011) Tetrapyrrole metabolism in *Arabidopsis thaliana*. Arabidopsis Book 9:e0145
331. Taniguchi M, Taniguchi Y, Kawasaki M, Takeda S, Kato T, Sato S, Tabata S, Miyake H, Sugiyama T (2002) Identifying and characterizing plastidic 2-oxoglutarate/malate and dicarboxylate transporters in *Arabidopsis thaliana*. Plant Cell Physiol 43:706–717
332. Tapken W, Ravet K, Pilon M (2012) Plastocyanin controls the stabilization of the thylakoid Cu-transporting P-type ATPase PAA2/HMA8 in response to low copper in *Arabidopsis*. J Biol Chem 287:18544–18550
333. Tarantino D, Casagrande F, Soave C, Murgia I (2010) Knocking out of the mitochondrial AtFer4 ferritin does not alter response of *Arabidopsis* plants to abiotic stresses. J Plant Physiol 167:453–460
334. Terashima M, Specht M, Naumann B, Hippler M (2010) Characterizing the anaerobic response of *Chlamydomonas reinhardtii* by quantitative proteomics. Mol Cell Proteomics 9:1514–1532

335. Terry N, Abadia J (1986) Function of iron in chloroplasts. J Plant Nutr 9:609–646
336. Tester M, Blatt MR (1989) Direct measurement of K^+ channels in thylakoid membranes by incorporation of vesicles into planar lipid bilayers. Plant Physiol 91:249–252
337. Tester M, Davenport R (2003) Na^+ tolerance and Na^+ transport in higher plants. Ann Bot 91:503–527
338. Thomine S, Wang R, Ward JM, Crawford NM, Schroeder JI (2000) Cadmium and iron transport by members of a plant metal transporter family in *Arabidopsis* with homology to Nramp genes. Proc Natl Acad Sci U S A 97:4991–4996
339. Thomine S, Lelièvre F, Debarbieux E, Schroeder JI, Barbier-Brygoo H (2003) AtNRAMP3, a multispecific vacuolar metal transporter involved in plant responses to iron deficiency. Plant J 34:685–695
340. Timperio AM, D'Amici GM, Barta C, Loreto F, Zolla L (2007) Proteomics, pigment composition, and organization of thylakoid membranes in iron-deficient spinach leaves. J Exp Bot 58:3695–3710
341. Tottey S, Rich PR, Rondet SAM, Robinson NJ (2001) Two menkes-type ATPases supply copper for photosynthesis in *Synechocystis* sp. 6803. J Biol Chem 276:19999–20004
342. Tottey S, Rondet SAM, Borrelly GPM, Robinson PJ, Rich PR, Robinson NJ (2002) A copper metallochaperone for photosynthesis and respiration reveals metal-specific targets, interaction with an importer, and alternative sites for copper acquisition. J Biol Chem 277:5490–5497
343. Tottey S, Block MA, Allen M, Westergren T, Albrieux C, Scheller HV, Merchant S, Jensen PE (2003) *Arabidopsis* CHL27, located in both envelope and thylakoid membranes, is required for the synthesis of protochlorophyllide. Proc Natl Acad Sci U S A 100:16119–16124
344. Touraine B, Boutin JP, Marion-Poll A, Briat JF, Peltier G, Lobréaux S (2004) Nfu2: a scaffold protein required for [4Fe-4S] and ferredoxin iron-sulphur cluster assembly in *Arabidopsis* chloroplasts. Plant J 40:101–111
345. Tsay YF, Chiu CC, Tsai CB, Ho CH, Hsu PK (2007) Nitrate transporters and peptide transporters. FEBS Lett 58:2290–2300
346. Tsunekawa K et al (2009) Identification and characterization of the Na^+/H^+ antiporter Nhas3 from the thylakoid membrane of *Synechocystis* sp. PCC 6803. J Biol Chem 284:16513–16521
347. Turner WL, Waller JC, Vanderbeld B, Snedden WA (2004) Cloning and characterization of two NAD kinases from *Arabidopsis*. Identification of a calmodulin binding isoform. Plant Physiol 135:1243–1255
348. Versaw WK, Harrison MJ (2002) A chloroplast phosphate transporter, PHT2;1, influences allocation of phosphate within the plant and phosphate-starvation responses. Plant Cell 14:1751–1766
349. Vidal EA, Gutiérrez RA (2008) A systems view of nitrogen nutrient and metabolite responses in *Arabidopsis*. Curr Opin Plant Biol 11:521–529
350. Voelker C, Schmidt D, Mueller-Roeber B, Czempinski K (2006) Members of the *Arabidopsis* AtTPK/KCO family form homomeric vacuolar channels in planta. Plant J 48:296–306
351. von der Fecht-Bartenbach J, Bogner M, Krebs M, Stierhof YD, Schumacher K, Ludewig U (2007) Function of the anion transporter AtCLC-d in the trans-Golgi network. Plant J 50:466–474
352. Voß B, Meinecke L, Kurz T, Al-Babili S, Beck CF, Hess WR (2011) Hemin and magnesium-protoporphyrin IX induce global changes in gene expression in *Chlamydomonas reinhardtii*. Plant Physiol 155:892–905
353. Vothknecht UC, Soll J (2005) Chloroplast membrane transport: interplay of prokaryotic and eukaryotic traits. Gene 354:99–109
354. Wagner R, Aigner H, Funk C (2011) FtsH proteases located in the plant chloroplast. Physiol Plant 145:203–214

355. Waller JC, Dhanoa PK, Schumann U, Mullen RT, Snedden WA (2010) Subcellular and tissue localization of NAD kinases from *Arabidopsis*: compartmentalization of de novo NADP biosynthesis. Planta 231:305–317
356. Wang X, Berkowitz GA, Peters JS (1993) K^+-conducting ion channel of the chloroplast inner envelope: functional reconstitution into liposomes. Proc Natl Acad Sci U S A 90:4981–4985
357. Wang D, Xu Y, Li Q, Hao X, Cui K, Sun F, Zhu Y (2003) Transgenic expression of a putative calcium transporter affects the time of *Arabidopsis* flowering. Plant J 33:285–292
358. Wang WH et al (2012) Calcium-sensing receptor regulates stomatal closure through hydrogen peroxide and nitric oxide in response to extracellular calcium in *Arabidopsis*. J Exp Bot 63:177–190
359. Weber A, Flügge UI (2002) Interaction of cytosolic and plastidic nitrogen metabolism in plants. J Exp Bot 53:865–874
360. Weber APM, Linka N (2011) Connecting the plastid: transporters of the plastid envelope and their role in linking plastidial with cytosolic metabolism. Annu Rev Plant Biol 62: 53–77
361. Weber A, Menzlaff E, Arbinger B, Gutensohn M, Eckerskorn C, Flügge UI (1995) The 2-oxoglutarate/malate translocator of chloroplast envelope membranes: molecular cloning of a transporter containing a 12-helix motif and expression of the functional protein in yeast cells. Biochemistry 34:2621–2627
362. Weber APM, Schwacke R, Flügge U-I (2005) Solute transporters of the plastid envelope membrane. Annu Rev Plant Biol 56:133–164
363. Wegner SV, Sun F, Hernandez N, He C (2011) The tightly regulated copper window in yeast. Chem Commun 47:2571–2573
364. Weigel M, Varotto C, Pesaresi P, Finazzi G, Rappaport F, Salamini F, Leister D (2003) Plastocyanin is indispensable for photosynthetic electron flow in *Arabidopsis thaliana*. J Biol Chem 278:31286–31289
365. Weinl S, Held K, Schlucking K, Steinhorst L, Kuhlgert S, Hippler M, Kudla J (2008) A plastid protein crucial for Ca^{2+}-regulated stomatal responses. New Phytol 179:675–686
366. White PJ, Broadley MR (2009) Biofortification of crops with seven mineral elements often lacking in human diets-iron, zinc, copper, calcium, magnesium, selenium and iodine. New Phytol 182:49–84
367. Whiteman S-A, Serazetdinova L, Jones AME, Sanders D, Rathjen J, Peck SC, Maathuis FJM (2008) Identification of novel proteins and phosphorylation sites in a tonoplast enriched membrane fraction of *Arabidopsis thaliana*. Proteomics 8:3536–3547
368. Williams LE, Mills RF (2005) P(1B)-ATPases-an ancient family of transition metal pumps with diverse functions in plants. Trends Plant Sci 10:491–502
369. Woodson JD, Chory J (2008) Coordination of gene expression between organellar and nuclear genomes. Nat Rev Genet 9:383–395
370. Wu W, Berkowitz GA (1991) Lidocaine and ATPase inhibitor interaction with the chloroplast envelope. Plant Physiol 97:1551–1557
371. Wu P, Ma L, Hou X, Wang M, Wu Y, Liu F, Deng XW (2003) Phosphate starvation triggers distinct alterations of genome expression in *Arabidopsis* roots and leaves. Plant Physiol 132:1260–1271
372. Wykoff DD, Davies JP, Melis A, Grossman AR (1998) The regulation of photosynthetic electron transport during nutrient deprivation in *Chlamydomonas reinhardtii*. Plant Physiol 117:129–139
373. Wykoff DD, Grossman AR, Weeks DP, Usuda H, Shimogawara K (1999) Psr1, a nuclear localized protein that regulates phosphorus metabolism in *Chlamydomonas*. Proc Natl Acad Sci U S A 96:15336–15341
374. Xu XM, Møller SG (2011) Iron-sulfur clusters: biogenesis, molecular mechanisms and their functional significance. Antioxid Redox Signal 15:271–307
375. Xu G, Fan X, Miller AJ (2012) Plant nitrogen assimilation and use efficiency. Annu Rev Plant Biol 63:153–182

376. Yamagishi A, Satoh K, Katoh S (1981) The concentrations and thermodynamic activities of cations in intact *Bryopsis* chloroplasts. Biochim Biophys Acta 637:252–263
377. Yamasaki H, Abdel-Ghany SE, Cohu CM, Kobayashi Y, Shikanai T, Pilon M (2007) Regulation of copper homeostasis by micro-RNA in *Arabidopsis*. J Biol Chem 282:16369–16378
378. Yamasaki H, Hayashi M, Fukazawa M, Kobayashi Y, Shikanai T (2009) SQUAMOSA promoter binding protein-like7 is a central regulator for copper homeostasis in *Arabidopsis*. Plant Cell 21:347–361
379. Yang T, Poovaiah BW (2000) *Arabidopsis* chloroplast chaperonin 10 is a calmodulin-binding protein. Biochem Biophys Res Commun 275:601–607
380. Yanykin DV, Khorobrykh AA, Khorobrykh SA, Klimov VV (2010) Photoconsumption of molecular oxygen on both donor and acceptor sides of photosystem II in Mn-depleted subchloroplast membrane fragments. Biochim Biophys Acta 1797:516–523
381. Yehudai-Resheff S, Zimmer SL, Komine Y, Stern DB (2007) Integration of chloroplast nucleic acid metabolism into the phosphate deprivation response in *Chlamydomonas reinhardtii*. Plant Cell 19:1023–1038
382. Yuan L, Loqué D, Kojima S, Rauch S, Ishiyama K, Inoue E, Takahashi H, von Wirén N (2007) The organization of high-affinity ammonium uptake in *Arabidopsis* roots depends on the spatial arrangement and biochemical properties of AMT1-type transporters. Plant Cell 19:2636–2652
383. Zancani M et al (2004) Evidence for the presence of ferritin in plant mitochondria. Eur J Biochem 271:3657–3664
384. Zanetti M et al (2010) A novel potassium channel in photosynthetic cyanobacteria. PLoS ONE 5:e10118
385. Zhang Z, Shrager J, Jain M, Chang CW, Vallon O, Grossman AR (2004) Insights into the survival of *Chlamydomonas reinhardtii* during sulfur starvation based on microarray analysis of gene expression. Eukaryot Cell 3:1331–1348
386. Zhong L, Inesi G (1998) Role of the S3 stalk segment in the thapsigargin concentration dependence of sarco-endoplasmic reticulum Ca^{2+} ATPase inhibition. J Biol Chem 273:12994–12998
387. Zhu JK (2001) Plant salt tolerance. Trends Plant Sci 6:66–71
388. Zhu JK (2003) Regulation of ion homeostasis under salt stress. Curr Opin Plant Biol 6: 441–445

Part V
Chloroplast Biotechnology

Chapter 18
Synthesis of Recombinant Products in the Chloroplast

Ghislaine Tissot-Lecuelle, Saul Purton, Manuel Dubald and Michel Goldschmidt-Clermont

Abstract For the engineering of new traits in plants and algae, modification of the plastid genome offers significant advantages. These include transgene integration by homologous recombination, high levels of expression, multigene engineering in polycistronic units, and reduced transmission through pollen. Numerous tools are available for DNA delivery, selection of transformants, removal of undesirable markers and efficient expression of single or multiple transgenes. The list of plants and algae with successful plastid transformation is increasing and includes some crop plants. Its further extension to agronomically relevant species and appropriate cultivars, together with advances in synthetic biology, will be key to the future development of plastids as green factories for the production of high-value metabolites or proteins.

Keywords Transformation · Plastome · Chloroplast genome · Homologous recombination · Herbicide tolerance · Biolistic transformation · Homoplasmic · Heteroplasmic · Uniparental inheritance · Maternal inheritance · Transcription · RNA polymerase · Promoter · 5′ untranslated region · Shine-Dalgarno sequence · 3′ untranslated region · Inducible expression · Polycistronic unit · Oral vaccines · Metabolic engineering · Chlamydomonas

Abbreviations

3′UTR 3′-Untranslated region
5′UTR 5′-Untranslated region

M. Goldschmidt-Clermont (✉)
University of Geneva, 30 quai E. Ansermet, 1211 Geneva 4, Switzerland
e-mail: michel.goldschmidt-clermont@unige.ch

G. Tissot-Lecuelle
Alganelle, 813 Chemin de Pingon, 73290 La Motte-Servolex, France
e-mail: ghislaine.tissot@gmail.com

S. Purton
Institute of Structural and Molecular Biology, University College London, Gower Street, WC1E 6BT London, UK
e-mail: s.purton@ucl.ac.uk

M. Dubald
Bayer CropScience, 3500 Paramount Parkway, Morrisville, NC 27560, USA
e-mail: manuel.dubald@bayer.com

S.M. Theg, F.-A. Wollman (eds.), *Plastid Biology,* Advances in Plant Biology 5, DOI 10.1007/978-1-4939-1136-3_18,

AHAS	Acetohydroxyacid synthase
Bt	*Bacillus thuringiensis*
CES	Control by epistasy of synthesis
CTB	Cholera toxin B
DB	Downstream box
GOI	Gene of interest
GRAS	Generally recognized as safe
HPPD	Hydroxyphenyl pyruvate dioxygenase
HSA	Human serum albumin
IEE	Intercistronic expression element
IFT	Isoxaflutole
IR	Inverse repeat regions of the plastome
LHRR	Left homologous recombination region
NEP	Nucleus-encoded RNA polymerase
NTRC	NADPH-dependent thioredoxin reductase C
PEG	Polyethylene glycol
PEP	Plastid-encoded RNA polymerase
RBPs	RNA-Binding-Proteins
RHRR	Right homologous recombination region
TSP	Total soluble protein

18.1 Introduction

The majority of commercial recombinant proteins are synthesized in industrial fermenters using simple microbial expression systems such as *Escherichia coli* and *Saccharomyces cerevisiae*. More complex therapeutic proteins, such as vaccines and hormones that require glycosylation, are typically produced in cultured mammalian cells. Each of these expression systems has its advantages and drawbacks, which include production cost, scalability, safety, or ability to perform specific post-translational modifications. The rising global demand for recombinant proteins has prompted research into alternative production platforms [150, 158]. Plants and algae offer some of the most attractive and diverse new expression systems, because of their potential for mass production at low anticipated cost [150, 193]. The photosynthetic plastids (*i.e.* chloroplasts) of these organisms are particularly attractive platforms for production of foreign proteins and other recombinant products as they are the sites of very active metabolism and protein synthesis, and contain multiple copies of a small genome that can be genetically engineered in a precise and predictable way. Furthermore, chloroplasts possess some of the sophisticated machinery lacking in bacteria which is required to fold complex human proteins [58]. Plants and green algae are generally regarded as safe production platforms, since they avoid the risk of potential contamination with human pathogens associated with animal cell-culture systems, or the harmful endotoxins present in *E. coli*. Edible species of plants or green algae can therefore be engineered into low-cost and low-tech delivery vectors for oral vaccines [158]. Although the scale of production of recombinant proteins in crop plant is much greater than that of algae—building as it does on well-established technologies of intensive farming, harvesting and bulk downstream processing—single-cell microalgae such as *Chlamydomonas reinhardtii* have some key advantages over plant systems. Microalgae can readily be grown as clonal cultures in enclosed bioreactors, reducing the risk of airborne or soil contaminants, and avoiding the containment issues that arise when genetically modified crops are released in the field. Containment and precise control of growth conditions are also critical in ensuring reproducibility in both the yield and quality of the desired product. The high growth rates of algae and the shorter time required to generate stable transgenic lines means that the initial evaluation of protein production is considerably shortened—a matter of months compared to years with higher plants [57, 58].

Transformation of the plastid genome was first achieved in the unicellular green alga *C. reinhardtii* [19], followed two years later by tobacco (*Nicotiana tabacum*), a flowering plant [202]. Engineering of the plastid genome (or 'transplastomics') differs from nuclear genome transformation in many ways, as will be discussed below in more detail. Plastid transformation and its biotechnology potential have been demonstrated extensively in tobacco and in chlamydomonas, with numerous novel products and traits reported over the last 25 years. However, the extension to other plant and algal species has been slow and is still technically challenging. A growing number of proteins of medical interest, in particular therapeutics such as protein vaccines, hormones and antibodies have been expressed in the plastid—often at high level, illustrating the potential of this technology for products beyond the farm gate. However, we have yet to see any commercial product on the market derived from transgenic plastids.

In this review, we will present an update on the technology of plastid transformation and cover issues that are important from the product perspective such as DNA delivery into the plastid genome, gene expression, protein accumulation and product stability.

18.2 The Chloroplast Genome: A Target for Engineering

As detailed in other chapters of this book, the chloroplast is the photosynthetic form of the group of semi-autonomous organelles termed plastids that are derived from a cyanobacterial ancestor and are found in the cells of plants, algae and some protists. The number of plastids per cell is highly variable, ranging from none in pollen grains of some higher plants, to one single chloroplast in unicellular green algae such as chlamydomonas, and up to hundreds in wheat leaf cells [225]. Although all plastid types contain the same small genome (the 'plastome'), the copy number and gene expression is highest in chloroplasts; reflecting the fact that the primary role of the genome is to encode core components of the photosynthetic apparatus, as discussed below. The plastome is generally described as a circular double stranded DNA molecule made up of two inverted repeated regions (IR) separating two single copy regions (a large LSC and a small SSC). However, the reality is more complex since linear, branched or circular multimers of the plastome have been described, for instance in maize [154, 222] and in *Medicago truncatula* [183]. Moreover some plastomes, such as those of *Pisum sativum* or *M. truncatula*, do not contain an IR [183]. The size of plant plastomes ranges from 108 to 165 kb (with some exceptions: 171 kb for *N. accuminati*, 180 kb for the duckweed *Lemna gibba* and 217 kb for *Pelargonium hortorum* [26, 68]; the size of algal plastomes is also variable [30]. In plants, plastome copy number depends on the cell type, physiological state and species. It can vary from about 1000 per mesophyll cell in *Arabidopsis thaliana* (containing about 120 chloroplasts) to up to more than 10,000 in wheat and barley [225] or in tobacco (containing up to 100 chloroplasts) [182]. In comparison, the plastome of *C. reinhardtii* contains approximately 50–80 copies within the single chloroplast. Thus, ptDNA represents a significant fraction of total cellular DNA in both plants and algae (up to 10–20 %).

Plastids present many features that reflect their eubacterial ancestry [136]. Their genomes contain clusters of genes organized in polycistronic units [13, 186], and their transcription and translation machineries are prokaryotic in nature. The plastid genomes harbor between 110 and 130 genes (reviewed in [15, 167]. The largest set of genes is involved in transcription and translation, and encodes ribosomal RNAs and proteins, RNA polymerase subunits, and probably all required tRNAs. A second set of genes encodes products involved in the photosynthetic apparatus, such as proteins found in photosystems I and II, the cytochrome b_6f complex, the large subunit of Rubisco, and subunits of the ATP synthase and NADH dehydrogenase. The last and very limited set of genes is diverse, and ranges from genes involved in lipid metabolism (*accD*) to genes of still unknown function. The other thousands of proteins implicated in plastid metabolism and its regulation are nuclear-encoded and generally have a cleavable N-terminal transit peptide which directs their import into the plastid. Complex mechanisms have evolved to coordinate the expression of plastid and nuclear genes with the changing developmental and functional requirements of the plant cell.

Plastid genetic engineering differs from nuclear transformation in many ways, as recently reviewed in [3, 38, 132, 180, 193]: (i) the incorporation of transforming DNA by homologous recombination in the plastome, (ii) the apparent absence of epigenetic regulation, (iii) the reduced probability of gene transfer to other cultivars or species *via* pollen because in many plant species the maternal inheritance of plastids offers natural containment of the transgenes, (iv) the potentially very high level of expression of the recombinant protein in the chloroplasts of green tissues or photosynthetic algae, and (v) the possibility to envisage engineering of complex multigene pathways using expression cassettes organized in polycistronic units.

18.2.1 Precise Engineering by Homologous Recombination

In all species studied to-date, it is found that the principal mechanism by which exogenous DNA delivered into the chloroplast integrates into the genome involves recombination between homologous sequences shared between the DNA and the genome. Consequently, any foreign DNA can be integrated at a defined locus by flanking it with plastid targeting sequences on the transformation vector such that homologous recombination events occurring either side ensures integration at a precise locus on the host plastome (Fig. 18.1). This precise mechanism, mediated by the organelle machinery, avoids the variability due to random integration and the resulting 'position effects' observed with nuclear transformants. The screening of the selected events is therefore considerably simplified because they are in principle predictable and genetically identical. A typical plastid transformation vector carries two plastid DNA fragments, a left homologous recombination region (LHRR) and a right homologous recombination region (RHRR) (Fig. 18.1). The plastid recombination sequences are generally about 1–2 kb in size [212]. The backbone of the transformation vectors derives from *E. coli* plasmids, which are not capable of replication in plastids.

Stable genetic transformation of plastids is a two-step process and starts with the production of an initial transformant. Delivery of the transforming DNA into the organelle compartment is achieved by particle bombardment (biolistics) [19, 202], PEG treatment of protoplasts [65, 149] or in the case of chlamydomonas, agitation in the presence of glass beads [95] (see Sect. 18.2.1). Delivery is followed by DNA integration through homologous recombination [19]. Integration of the transforming DNA in the plastid genome is a very rare event, presumably occurring initially in only one or a few plastids per transformed cell, and directly modifying only a small percentage of plastid genomes. The second step is the selection of transplastomes (transformed plastid genomes) by successive cycles of subculturing under strong selection pressure until the homoplasmic stage is reached. Remaining copies of wild-type plastomes in the selected events should be eliminated entirely since a heteroplasmic state is not genetically stable. In higher plants, this can require repeated cycles of *in vitro* tissue regeneration that involve plastid dedifferentiation to a proplastid stage and the associated strong reduction in plastid number and plastome polyploidy. This constitutes a bottleneck that contributes to the loss of the wild-type copies by genetic drift [38, 132]. Multiple cycles of regeneration are not always necessary because homoplasmic lines can be obtained after a single round, and it might not be crucial for analysis if the objective is not a gene knockout or replacement. A low level of heteroplasmy will not interfere with the evaluation of the phenotype of the selected events as long as this is monitored by DNA analysis, but could interfere with the transgene stability. Selection of plastid lines in tobacco does not take significantly more time than with nuclear transformation, and unlimited plant material (T1 generation) is available six months after the start of transformation. The evaluation of homoplasmy can sometimes be complicated by the presence in the nuclear genome of highly homologous stretches of plastid sequence. These can extend over more than 100 kb with more than 99 % identity to their plastid counterpart [81]. As a consequence, weak signals can be wrongly attributed to remaining wild-type plastome, whereas in fact they correspond to pre-existing nuclear plastid DNA insertions.

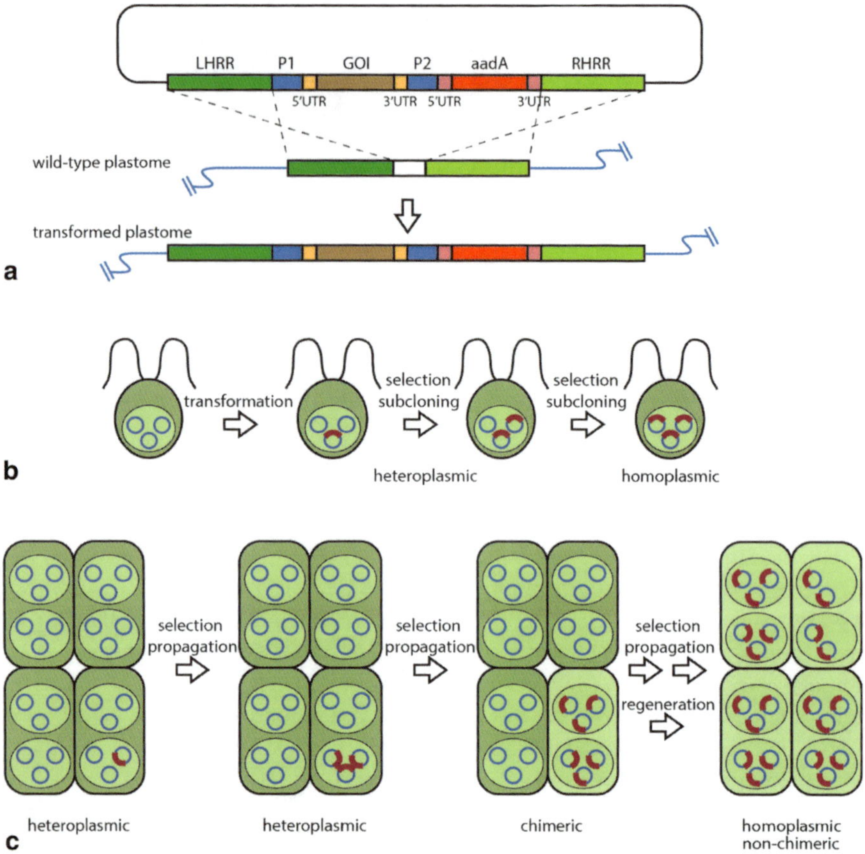

Fig. 18.1 Chloroplast transformation. **a** A generic plastid transformation vector contains a gene of interest (GOI) under the control of a promoter (P1), a 5′ (5′UTR) with cis-acting elements for mRNA stability and translation initiation, and a 3′ untranslated region (3′UTR) for mRNA processing and stability. The vector also contains a selectable marker such as *aadA*, under the control of a promoter (P2), 5′UTR and 3′UTR. Left and Right Homologous Recombination Regions (*LHRR* and *RHRR*) direct integration at a specific site in the host wild-type plastome. **b** In the single chloroplast of chlamydomonas the transformation vector presumably initially integrates in one of the chloroplast genomes which are present in approximately eighty copies. By repeated subculturing on selective medium, homoplasmic transformed clones can be obtained where all copies are transformed. A homoplasmic state can be reached only if the integration does not disrupt any essential function, otherwise a heteroplasmic mixture of wild-type and transformed genomes will be maintained. **c** In the multicellular higher plants, each cell contains many plastids, each of which in turn contains multiple copies of the plastome. Thus after the initial transformation event, strong selection is required to first obtain a homoplasmic plastid in a cell that may also harbor un-transformed plastids, then to select cells with only transformed homoplasmic plastids within a chimeric tissue, and finally to generate homoplasmic plants. Several rounds of propagation under selection are usually required; regeneration favors the derivation of homoplasmic plants. Reproduced from [38]

18.2.2 Post-translational Modifications and the Exploitation of Plastid Compartments

Plastids have complex structures with membrane systems separating three distinct aqueous phases. The chloroplast envelope, which separates the organelle from the cytosol, encloses an inter-membrane space between the outer and inner membranes. The second and also major compartment is the stroma, which is the site of most metabolic reactions and where transcription and translation occur. Finally, within the stroma, an internal membrane system constitutes the interconnected thylakoid network, where the photosynthetic machinery is located. This membrane network encloses a further soluble phase, the thylakoid lumen. These plastid compartments can be appropriate sites to accumulate certain proteins or their biosynthetic products that would be harmful if they were present in large amounts in other cell compartments [17]. This was illustrated with cell-wall degrading enzymes such as xylanases [112] and with trehalose production [109].

Various post-translational modifications can occur in proteins synthesized in chloroplasts. Protein phosphorylation [205, 210], N-acetylation and palmitoylation have all been detected [209]. Some very specific bacterial modifications such as the lipidation of the OspA protein have been demonstrated to occur in transgenic chloroplasts [64]. As in *E. coli,* post-translational removal of N-formylmethionine in proteins synthesized in plastids depends on the characteristics of the second residue [62]. Oxidation of proteins takes place in plastids as in other organisms or cell compartments, but could be more prevalent in chloroplasts due to the reactive oxygen species that are generated during photosynthesis. From an applied perspective, the most important feature differentiating plastids from prokaryotes is their ability to correctly fold proteins containing disulfide bonds in the stroma. This was demonstrated with human growth hormone [194], cholera toxin B [33], human serum albumin [54], a single chain antibody [227], human interferon alpha [34], a bacterial alkaline phosphatase [6], aprotinin [206], a human monoclonal antibody [207] and antimicrobial peptides [111]. The redox poise of the plastid stroma is sufficiently oxidizing for disulfide bond formation in proteins. Such conditions can only be achieved in double mutants of *E. coli,* and at a cost, since these strains grow more slowly than the wild type [14]. Redox signaling in chloroplasts in both stroma and thylakoids is the subject of very active research [5, 20]. A considerable number of plastid thioredoxins and glutaredoxins are involved in the regulation of various enzymatic activities [60, 171]. Transcription can be regulated by redox-sensitive factors [123]. Protein translation in the chlamydomonas chloroplast is also redox-controlled, as illustrated with the eukaryotic-type protein disulfide isomerase RB60 which regulates the synthesis of the D1 protein [93], or with the RB40-Nac2 complex which is regulated by the NADPH-dependent thioredoxin reductase C (NTRC) and regulates the synthesis of the D2 protein [179].

Some recombinant proteins require specific N-termini, lacking N-formylmethionine, for their function and stability. Several strategies for removing the N-terminal methionine were explored. The human growth hormone, somatotropin, was expressed in the stroma with ubiquitin as an N-terminal extension [194]. During extraction, this fusion protein is exposed to cytosolic ubiquitin hydrolases and mature somatotropin is released. The processing is however not complete, and predominantly occured at an unexpected position. A different strategy was used for human serum albumin (HSA), which was expressed from the tobacco plastome as a fusion with a transit peptide from the small subunit of Rubisco. The chimeric protein was apparently correctly processed within the organelle leading to mature HSA [54]. A third strategy was to target the recombinant protein to the lumen of thylakoids using signal peptides for the SEC or the TAT pathways. This was exemplified with the accumulation in the thylakoid lumen of an alkaline phosphatase from *E. coli* [6] and of aprotinin, a disulfide-bond containing protein [206]. In the latter case, the processing of the signal peptides took place precisely at the expected site, releasing the mature and active protease inhibitor, with an amino terminal arginine residue. The thylakoid lumen could be particularly appropriate for disulfide-bond containing recombinant proteins, since this compartment is expected to be more oxidizing than the stroma [20]. Finally, targeting to the thylakoid lumen should be considered for recombinant proteins when it is suspected that these are prone to protease degradation or are toxic when expressed in the stroma.

18.2.3 High Levels of Expression

The very active chloroplast transcription and translation machineries confer the potential for higher levels of recombinant protein expression compared to those of nuclear transgenes. Heterologous proteins have been expressed in transplastomic lines to remarkably high levels, up to 70 % of total soluble protein (TSP) in tobacco plastids [151]. Levels of 10–20 % of TSP have been claimed in the chlamydomonas chloroplast [165, 199], (reviewed in [16, 35, 132, 164, 180, 193]).

18.2.4 Reduced Transgene Transmission Through Pollen/mt(−)

The plastids of many Angiosperms are inherited solely from the maternal parent because they are eliminated at different stages of male gametophyte development, or excluded at fertilization [70]. This reduced transmission *via* pollen is one of the attractive features of chloroplast transformation. It provides a natural genetic containment system and greatly reduces the probability of transgene transfer from transgenic plants to wild species, weeds or to neighboring fields with non-genetically modified plants (see Sect. 18.5.2). In *C. reinhardtii*, a similar process of uniparental inheritance is observed with the chloroplast genome inherited from the mating type plus (mt+) parent in more than 95 % of the progeny. Thus, chloroplast transformation of a mt(−) strain also offers a degree of biological containment.

18.3 Chloroplast Transformation

18.3.1 DNA Delivery Methods

The most commonly used method for introducing exogenous DNA into the chloroplast is the biolistic particle delivery system [202]. This was the technique that first allowed the stable transformation of chloroplasts in chlamydomonas [19, 66] and in tobacco [202]. The method is relatively expensive since a gene-gun device, generally helium-driven, is needed. Transforming DNA is coated onto the surface of tungsten or gold microparticles, which are generally smaller than 1 μm in diameter (0.4–1.0 μm). The microparticles are propelled at high velocity under a partial vacuum to penetrate the target host cells. Biolistic transformation can be used with intact leaf tissue or with cultured cells spread as a lawn on agar plates (Tables 18.1 and 18.2). The process of foreign DNA delivery has been studied in plants [52]. During biolistic transformation, transforming DNA can be delivered to compartments other than the plastids, and some transformants might contain additional copies of the transgene inserted in the nuclear genome.

An alternative DNA delivery method involves treatment of plant protoplasts with polyethylene glycol (PEG) [44, 65, 149]. This transformation method can only be used for species that can be efficiently regenerated from protoplasts, but this may lead to the generation of polyploid lines as observed in lettuce [113]. PEG has also been used to transfer entire transgenic plastids of tobacco into another *Solanaceous* species by protoplast fusion [203]. Plastid transformation *via Agrobacterium* infection has never been confirmed, even with a plastid-targeted engineered virD2 protein [75], and there have been no reports of successful plant chloroplast transformation by electroporation. Another physical process for the introduction of the transforming DNA is microinjection [99]. However, this method has not yet yielded stable transplastomic plants. With chlamydomonas and other unicellular algae, the standard method of DNA delivery is microparticle bombardment of an algal lawn [161]. For chlamydomonas, agitation of a suspension of cell-wall-deficient cells in the presence of glass beads and DNA has been shown to also yield chloroplast transformants [94].

18.3.2 Target Tissues, Tissue Culture and Plant Regeneration

Plant chloroplasts derive directly or indirectly from the differentiation of proplastids during development, and therefore carry the same genetic information. As a consequence, tissue containing transgenic chloroplasts that is used for plant regeneration can generate new plants in which all plastids carry the foreign DNA. However, the most difficult step in generating these homoplasmic, non-chimeric plants is the tissue culture and plant regeneration. In higher plants, the first plastid transformants were generated in tobacco by organogenesis using young leaves as starting mate-

Table 18.1 Plastid transformation in plants

Organism	Species	Transformation method	Explant/ regeneration	Selectable marker/gene of interest	Resistance	Plastome insertion site	References	Comments
Alfalfa	*Medicago sativa*	Biolistic	Tissue culture-derived embryos and leaf explants	*aadA-gfp(dicistron)*	Spectinomycin	*trnI/trnA*	[214]	Homoplasmic, fertility and inheritance not reported
Arabidopsis	*Arabidopsis thaliana*	Biolistic	Leaves	*aadA*	Spectinomycin	*trnV/rps12/7*	[188]	Sterile and not stable plants
Cabbage	*Brassica oleracea*	Biolistic	Leaves	*aadA, uiadA*	Spectinomycin -Streptomycin	*rrn16S/trnI*	[120]	Fertile and homoplasmic plants
Carrot	*Daucus carota*	Biolistic	Fine cell suspension cultures	*aadA*, *badh*	Spectinomycin	*trnI/trnA*	[104]	Homoplasmic, fertility and inheritance not reported
Cauliflower	*Brassica oleracea*	PEG-treated protoplasts	Protoplasts from leaves	*aadA*	Spectinomycin	*accD/rbcL*	[148]	One homoplasmic event. Fertility and inheritance not reported
Cotton	*Gossypium hirsutum*	Biolistic	Embryogenic green calli from hypocotyl explants	*aphA6, nptII*	Kanamycin	*trnI/trnA*	[105]	Fertile and homoplasmic plants
Eggplant	*Solanum melongena* L.	Biolistic	Stem explants	*aadA*	Zeatin riboside and spectinomycin	*trnV/rps12/7*	[189]	Homoplasmic and fertile

Table 18.1 (continued)

Organism	Species	Transformation method	Explant/ regeneration	Selectable marker/gene of interest	Resistance	Plastome insertion site	References	Comments
Lettuce	*Lactuca sativa*	PEG-treated protoplasts	Protoplasts from leaves	*aadA*, *gfp*	Spectinomycin	*trnI*/*trnA*	[113]	Fertile and homoplasmic plants
		Biolistic	Young leaves	*aadA*, *gfp*	Spectinomycin	*accD*/*rbcL*	[86]	Fertile and heteroplasmic plants
		Biolistic	Leaves	*aadA*, *CTB-Pins (dicistron)*	Spectinomycin	*trnI*/*trnA*	[174]	Fertile and homoplasmic plants
Petunia	*Petunia hybrida*	Biolistic	Leaves	*aadA*, *gusA*	Spectinomycin	*accD*/*rbcL*	[226]	Fertile and homoplasmic plants
Poplar	*Populus alba*	Biolistic	Leaves	*aadA*, *gfp*	Spectinomycin	*accD*/*rbcL*	[153]	Homoplasmic plants
Potato	*Solanum tuberosum* breeding line FL1607	Biolistic	Leaves	*aadA*, *gfp*	Spectinomycin	*accD*/*rbcL* *trnV*/*rps12*/*7*	[187]	Homoplasmic plants Sterile
	Solanum tuberosum commercially important cultivar Désirée	Biolistic	Leaves	*aadA* *aadA*, *gfp*	Spectinomycin	*accD*/*rbcL* *trnV*/*rps12*/*7*	[146]	Homoplasmic plants
	Solanum tuberosum	Biolistic	Organogenesis Petiole explant	*aadA*, *gfp*	Spectinomycin	*accD*/*rbcL*	[208]	Expression in leaves and tubers increase of 15–18 fold efficinecy using homologous potatoe sequence

Table 18.1 (continued)

Organism	Species	Transformation method	Explant/ regeneration	Selectable marker/gene of interest	Resistance	Plastome insertion site	References	Comments
Rice	*Oryza sativa*	Biolistic	Embryogenic calli	*aadA-gfp*	Streptomycin	*trnV/rps12/7*	[91]	Heteroplasmic Plants not stable
		Biolistic	Partly embryogenic calli	*aadA*	Streptomycin	*trnI/trnA*	[110]	Heteroplasmic
Soybean	*Glycine max*	Biolistic	Embryogenic calli	*aadA*	Spectinomycin	*trnV/rps12/7*	[48]	Fertile homoplasmic plants
Sugar beet	*Beta vulgaris*	Biolistic	Petioles	*aadA, gfp*	Spectinomycin	*rrn16/rps12*	[40]	Homoplasmic, fertility and inheritance not reported
Tobacco	*Nicotiana tabacum*	Biolistic	Leaves	16S rDNA	Spectinomycin	*16S rDNA*	[202]	Fertile homoplasmic plants
		PEG-treated protoplasts	Leaves	16S rDNA	Spectinomycin	*16S rDNA*	[65]	Fertile homoplasmic plants
		Biolistic	Leaves	*aadA*	Spectinomycin	*accD/rbcL*	[200]	Fertile homoplasmic plants
		Biolistic	Leaves	*nptII*	Kanamycin	*accD/rbcL*	[21]	Fertile plants
		Biolistic	Leaves	*cat*	Chloramphenicol	*trnfM/trnG*	[117]	Fertile homoplasmic plants
	N. plumbaginifolia	PEG-treated protoplasts	Leaves	16S rDNA	Spectinomycin	*16S rDNA*	[149]	Fertile plants
	Nicotiana benthamiana	Biolistic	Leaves	*aadA-gfp (dicistronic)*	Spectinomycin	*trnI/trnA*	[36]	Fertile homoplasmic plants
Tomato	*Lycopersicon esculentum*	Biolistic	Leaves	*aadA*	Spectinomycin	*trnfM/trnG*	[172]	Fertile homoplasmic plants
		PEG-treated protoplasts	Protoplasts from leaves	16S rDNA	Spectinomycin	*rrn16/rps12*	[147]	Fertile plants
	Solanum lycopersicum	Biolistic	Leaves	*aadA*	Spectinomycin	*trnfM/trnG*	[2]	Fertile homoplasmic plants

Table 18.1 (continued)

Organism	Species	Transformation method	Explant/ regeneration	Selectable marker/gene of interest	Resistance	Plastome insertion site	References	Comments
Oilseed rape	*Brassica napus*	Biolistic	Organogenesis Cotyledon petiole	*aadA, cry1Aa10*	Spectinomycin	*rps7/ndhB*	[79]	Heteroplasmic
		Biolistic	Cotyledons	*aadA*	Spectinomycin	*trnI/trnA*	[23]	Heteroplasmic
Oilseed Brassicacea	*Lesquerella fendleri*	Biolistic	Leaves	*aadA, gfp*	Spectinomycin	*trnV/rps12/7*	[190]	Fertile plants
Wheat	*Triticum aestivum* L.	Biolistic	Scutella	*nptII, gfp*	G418	*atpB/rbcL*	[31]	Homoplasmic and heteroplasmic fertile plants

Table 18.2 Plastid transformation in algae

Species	Transformation method	Selectable marker	Selection	References
Chlamydomonas reinhardtii	Particle gun	Photosynthesis markers	Photoautotrophy	[19]
		aadA	Spectinomycin, streptomycin	[66]
		psbA	Herbicides such as metribuzin, DCMU	[145, 159]
		aphA-6	Kanamycin, amikacin	[11]
		16S or 23S rDNA	Spectinomycin, streptomycin, erythromycin	[144]
		ARG9	Arginine auxotrophy	[169]
	Glass beads	*tscA*	Photoautotrophy	[95]
Haematococcus pluvialis	Particle gun	*aadA*	Spectinomycin	[69]
Dunaliella tertiolecta	Particle gun	*ereB*	Erythromycin	[61]
Porphyridium sp.	Particle gun	*AHAS* (W492S)	Sulfometuron methyl	[108]
Euglena gracilis	Particle gun	*aadA*	Streptomycin and spectinomycin	[45]

rial and a shoot induction medium in the presence of the antibiotic selective agent. It was long believed that this was the only appropriate tissue because it contains chloroplasts which are relatively large (a few μm diameter), present in high numbers, and metabolically very active. As discussed above, the primary transformants are often heteroplasmic. Only one or a small number of copies of the plastome contain the transgene, and only one or a few chloroplasts are initially transformed (Fig. 18.1). The primary transformants thus can sometimes display a variegated phenotype with pale-green sectors. Repeated cycles of plant regeneration from somatic cells under selective conditions are routinely employed in most laboratories to obtain homoplastomic lines [132].

This reliance on somatic cell regeneration by organogenesis precluded the facile application of the technology to most crops of agronomic interest, which are often regenerated *via* somatic embryogenesis. Other obstacles could reside in low transformation efficiency, in low transgene expression, in the difficulty of achieving transgene expression in non-green plastids or in the lack of selectable markers. This view has largely changed with publications describing the transformation of tobacco suspension cell cultures [107], embryogenic cultures of various species [48, 104, 105] and improved plastid transformation efficacy in potato [208] (Table 18.1).

18.3.3 Target Loci in the Plastome

For most biotechnological applications, the choice of the insertion site within the plastome is generally made such as not to disrupt an essential gene or interrupt the expression of a polycistronic unit. To date, several sites corresponding to intergenic regions have been used to insert transgenes in the plastome in various species (Table 18.1). The *trnI–trnA* integration site is particular attractive because it is adjacent to the 16S rRNA promoter, which drives read-through transcription through this integration site, potentially allowing the use of promoterless genes of interest [212]. Moreover, a transgene targeted into the IR is present in two copies per genome because it is rapidly copied over into the second repeat region through a gene conversion mechanism.

The choice of the insertion site in the plastome may have some effect on the level of protein accumulation. The length of the two flanking regions is also an important parameter to consider, as length is positively correlated to the recombination rate within a certain range, as is the case for *E. coli* [184]. Flanking regions of 1–2 kb are most often used and there is no evidence that longer sequences are beneficial. The high homology of plastome sequences between plant species, especially in some conserved regions, has allowed various groups to transform other species using vectors containing tobacco plastid flanking regions [172, 187]. Nevertheless, this strategy leads to a drop in transformation efficiency [42], and it has been demonstrated that the transformation efficiency can be improved using species-specific vectors with flanking sequences highly homologous to the host plastome [175, 208]. This explains why most groups use species or even sometimes cultivar-specific homologous plastid sequences in their transforming vectors. This is facilitated by the growing list of published complete plastome sequences. The transformation efficiency could be further enhanced by the addition of a plastid origin of replication on the transforming plasmid allowing amplification and prolonged maintenance of free plasmid copies in the chloroplast [32], or by using site-specific recombination systems [92, 128, 129].

In chlamydomonas, many sites of integration have been used to transform the chloroplast genome. One strategy is to choose as the host a photosynthetic mutant and to link the transgene expression cassette to a wild-type copy of the affected gene in order to restore photosynthetic function [19]. For example, photosynthetic mutants deleted in the *atpB* gene [19, 114, 141, 160] or *psbH* [11, 166] have been used. Another strategy is to insert the transgene linked to a selectable marker in a silent intergenic region of the chlamydomonas plastome, for instance between the *psbA* and 5S *rRNA* genes [9, 59, 141], the *rbcL* and *psaB* genes [89, 160] or *psaA ex3* and *psbH* [166].

18.3.4 Transgene Expression: The Requirement for Endogenous cis Elements

Plastids exhibit several prokaryotic features reminiscent of their eubacterial origin. Plastids have their own transcription and translation machineries which combine components originated from a cyanobacterial ancestor and those acquired during their endosymbiotic evolution (see Chaps. 1–3 of this volume). In order to be expressed, the coding sequence of a transgene (encoding a gene of interest or a selectable marker) needs to be integrated into an expression cassette such that it is fused to appropriate *cis* elements from endogenous plastid genes, and is therefore under the control of the plastid regulatory elements that mediate transcriptional and post-transcriptional steps [178]; for a review [84, 195, 204, 208]. Thus, the transgene coding region is placed under the control of a promoter, 5′-untranslated region (5′UTR) and 3′-untranslated region (or 3′UTR) as illustrated in Fig. 18.1a.

In higher plants, transcription of plastid genes involves three distinct DNA-dependent RNA polymerases. The nucleus-encoded polymerases (NEP) RPOTp and RPOTmp (which also localizes to mitochondria) are monomeric and are homologous to the phage-type RNA polymerases. The third RNA polymerase, called plastid-encoded RNA polymerase (PEP), is encoded by the plastome [71]; reviewed by [43, 118]. The PEP is a multi-subunit enzyme, homologous to the cyanobacterial-type RNA polymerase [115]. The core subunits are encoded by the *rpoA*, *rpoB*, *rpoC1*, and *rpoC2* genes of the plastome. Moreover, the PEP holoenzyme is formed by the interaction of the core subunits with nuclear-encoded sigma factors for promoter recognition. Many plastid promoters contain variations of the −35 and −10 (TATAATAT) elements typical of σ^{70}-type *E. coli* promoters [118]. Although several studies demonstrated that NEP and PEP enzymes preferentially transcribe plastid genes encoding proteins involved in gene expression or photosynthesis related genes, respectively, it has been shown that some genes contain promoters for both polymerases (for reviews see [115, 118]). A simpler situation is found in algal plastids where there is no evidence of NEPs and all plastid genes are transcribed by the plastome-encoded PEP enzyme [191]. In the construction of a functional transgene, a primary requirement is an efficient promoter (Fig. 18.1a). In plants, there is only a short list of plastid promoters that have been tested for transgene expression. Most laboratories use either the strong promoters of *psbA* or of *16S* rRNA (*Prrn*), and derivatives thereof, which are very actively transcribed by the PEP polymerase. The modified *Prrn* promoters are strong and constitutively expressed in plants [200] as compared to the promoter of *psbA*, which is induced by light and thus weakly expressed in non-photosynthetic tissues such as roots [50].

In both plants and algae, plastid gene expression is further regulated at the post-transcriptional level through the processing and stabilization, and translation of the RNA transcript [8]. In higher plants, transcripts may also be subject to RNA editing [195]. Most of these post-transcriptional mechanisms require the formation of RNA-protein complexes with the 5′-UTR of the transcript often crucial for mRNA accumulation, transcript stability and message translation, and hence for the accumulation of proteins in large quantities [9, 165, 197]. Plastid translation shares various features with that of eubacteria, such as initiation factors, rRNAs, tRNAs and prokaryotic-type 70S ribosomes. Its regulation is mainly controlled in the initiation phase through an interaction of sequence elements in the 5′-UTR and nuclear encoded translation factors [8, 135, 137, 155]. Many 5′UTR of endogenous plastid mRNAs contain key elements for translational regulation, such as a ribosome-binding site (GGAGG or GGA SD-like sequence) with significant homology with the prokaryote Shine-Dalgarno sequences. This sequence is found upstream of the bacterial AUG translation initiation codon. The sequence is complementary to the 3′-end of the 16S ribosomal RNA and leads to the recruitment of the 70S ribosome. Nevertheless, in tobacco chloroplasts, 30 of the 79 protein-coding genes do not contain a SD-like sequence in their 5′UTR mRNA, but contain *cis*-acting elements often forming secondary structures that facilitate the interaction with specific nucleus-encoded RNA-Binding-Proteins (RBPs) [195]. Most chloroplast genes contain an AU-rich 3′UTR with an inverted repeat sequence that can potentially form a stem-loop structure in transcripts. Unlike in prokaryotes where they function as efficient transcription terminators, these inverted repeat sequences stabilize upstream sequences, mediate correct 3′-end processing and are involved in translation initiation [67, 170].

The role of these *cis*-acting elements in mediating efficient expression of foreign genes in the plastid has been studied in plants [76, 175, 218, 223], and in chlamydomonas [9, 67, 122, 166]; for a review see [195]. Several studies have compared the efficiency of endogenous regulatory elements with heterologous elements from plastid genes of related species [63, 175]. These studies have concluded that RNA regulatory elements are very much species-specific and therefore efficient post-translational steps in the expression of transgenes require the use of endogenous elements, particularly with respect to the 5′UTR. Several plastid 5′UTR sequences (such as those of *psbA* or *rbcL*), which include SD-like sequences [78], have been shown to drive efficient transgene translation in transplastomic plants [39, 49, 194, 212]. Strong synthetic or bacteriophage-derived SD sequences have also been employed in chloroplast biotechnology to maximize transgene expression. For instance the strongest known expression signals in plastids used in numerous species is a hybrid construct comprising the constitutive *Prrn* promoter followed by the gene *10* leader from phage T7 (*G10L* [46, 106, 152, 174, 224]). The highest reported expression of a foreign protein (70 % of total soluble protein) was obtained with this regulatory element in tobacco chloroplasts [151]. Similarly, several endogenous 3′UTR elements have been compared for their effect on transgene expression in plant chloroplasts: namely, the 3′UTRs from the *rbcL, psbA, petD, rps16, rpl32* and *rpoA* genes [188, 200, 204]. Interestingly, the heterologous terminator region from the *E. coli rrnB* operons seems to be more efficient for mRNA accumulation and stability in tobacco chloroplasts than those of the endogenous *rbcL, psbA, petD* or *rpoA* genes [113, 204].

In *C. reinhardtii,* several promoters and 5′UTRs were used to drive transgene expression in the chloroplast, such as those of the *rbcL, atpA, psbA, psbD,* and *psaA-exon1* genes [9, 83, 141]. To date, two promoters and 5′UTRs have been reported to drive the highest levels of heterologous protein accumulation, those from the *psbA* gene [134, 165, 199] and from the *psaA-exon1* gene [141]. The *psbA* gene encodes the D1 protein of the photosystem II reaction center, one of the highest expressed proteins in the plastid and for which the translation is light-activated. However, the levels of heterologous protein accumulation from the *psbA* promoter/5′UTR were high only in D1-deficient strains [134, 141, 142, 165, 199]. This feature is explained either by a reduced competition with endogenous *psbA* for limiting transcription or translation factors, or by a negative feedback of the D1 protein on *psbA* expression, a mechanism called control by epistasy of synthesis (CES) [25]. A similar enhancement of transgene expression from the *psaA exon1* promoter/5′UTR was observed in strains that do not accumulate the PsaA protein because of a nuclear mutation that affects *psaA* trans-splicing [141]. However, a drawback with these strategies is that the D1-deficient or PsaA-deficient host strains are non-photosynthetic, and therefore transplastomic lines need to be cultured on media containing a fixed carbon source. Alternatively, the *Prrn* promoter fused to the *atpA* 5′UTR can drive strong transgene expression and protein accumulation in photosynthetically competent strains [166]. Different 3′UTR have been tested for chlamydomonas chloroplast transformation. The 3′-UTRs of the *rbcL, atpA, psbA* or *trnR* genes showed only minor differences in transgene expression [9]. The 3′UTR of *rbcL* is often used in expression cassettes because it contains two redundant *cis*-acting elements involved in efficient RNA processing and stability [67].

As discussed above, post-transcriptional events are clearly the most important and limiting factors for expression in plastids, and the 5′UTR is key to this as it contains the binding sites for regulatory factors, possible SD-like sequences for ribosome binding and the translation start site. However, in some cases, another important element required for efficient translation is the amino acid sequence immediately downstream of the initiation codon, called the downstream box (DB) region [106]. In some cases, inclusion of an N-terminal stabilization sequence or fusion to another protein has been essential for the successful expression of the protein of interest in transplastomic lines [112, 174, 218]. Protein accumulation can vary over several orders of magnitude when the DB region is altered. For instance, human proinsulin expressed alone is rapidly degraded in transgenic chloroplasts. Fusion with non-toxic cholera toxin B (CTB) allowed its high-level expression (up to 16% TSP in tobacco and 2.5% TSP in lettuce, as compared to 0.1% TSP in nuclear transformants of potato). This has facilitated studies on oral delivery to achieve protection against the development of insulitis in non-obese diabetic mice [174]. In chlamydomonas, translational fusion to the N-terminal part of an endogenous chloroplast protein can enhance expression of the protein of interest [88], but there are also examples where this has no effect [141]. Fusion of the hexogen 10FN3 protein to bovine serum albumin A3 (M-SAA) allowed its accumulation to high levels (10% of TSP) in the chlamydomonas chloroplast [134, 165].

Differences in expression level of up to four orders of magnitude have been observed for the same recombinant protein, driven by the same promoter, but with various combinations of 5′-UTR and N-terminal coding region [51, 106, 218]. High-level expression is not always observed or predictable with plastid transformants and requires optimization. The secondary structure in the mRNA around the translation start site is important for the accessibility to ribosomes and for translation initiation. This can be modeled and optimized *in silico*. The pragmatic approach to high-level expression in plastids is to test a few different 5′-UTRs or ribosome-binding sites (*G10 L*, *psbA*, *rbcL*, etc.) and when possible combine them with downstream box variants [131, 165, 166]. The similarity of bacterial and chloroplast gene expression has prompted various groups to use expression in *E. coli* to test the functionality of plastid expression vectors and to predict and optimize results anticipated in chloroplasts [32, 176, 212], however in some cases no correlation was found [12, 130].

One aspect of transgene design that does not seem to be critical in plants is the optimization of the coding sequence itself [127, 218]. Both GC and AT rich sequences have been successfully expressed in higher plants [131]. This does not rule out the possibility that in some instances codon usage might be a limiting factor. The situation seems to be different in chlamydomonas, where codon usage has been shown to play a significant role in protein accumulation [59, 138, 217].

18.3.5 Inducible or Repressible Expression, and Multigene Expression

The ability to regulate recombinant protein expression in the chloroplast could be essential for the production of certain types of recombinant proteins, including those that would normally be lethal to the host cell or detrimental to its growth [130]. Inducible expression systems have been described in plants, based either on a chloroplast-targeted T7 RNA polymerase engineered in the nuclear genome [126, 130] or on the *lac* repressor from *E. coli* in the chloroplast genome [143]. The possibility of using riboswitches to regulate the expression of a reporter gene has also been demonstrated, although expression levels were low [211].

In the chlamydomonas chloroplast, the regulated transcription of a chimeric *uidA* reporter was demonstrated using the *lacI* repressor from *E. coli* [89]. An alternative approach for conditional expression is to engineer anterograde signaling from the nucleus to the chloroplast. The nuclear promoter for the cytochrome c_6 gene is induced in the absence of copper or in the presence of nickel [55, 77]. This promoter has been used to regulate the expression of the *NAC2* gene in a nuclear transformant line lacking an endogenous copy of *NAC2*. Since the Nac2 protein is normally imported in the chloroplast where it binds to the 5′UTR of *psbD* transcripts and protects them against exonucleolytic degradation, this nuclear transformant can be used as a host for controlled expression of transgenes in the chloroplast. Transgenes are fused to the *psbD* 5′UTR such that their translation is dependent on Nac2 availability, and therefore translation occurs only in the absence of copper or presence of nickel [198]. This system has been further improved by placing the *NAC2* gene under the control of the cobalamin-repressible *METE* promoter and the thiamine-regulated *THI4* riboswitch, allowing vitamin-mediated control of transgene expression [163]. In principle, any *trans*-acting factor involved in anterograde signaling to the chloroplast could be similarly used to control transgene expression in plastids.

For many applications, in particular for metabolic engineering of the chloroplast, it is desirable to express multiple transgenes in the chloroplast. There are two options when designing the transgene expression cassettes [131]. The first strategy consists of providing independent transcription units for each transgene such that each has its own promoter and UTRs. The second strategy takes advantage of the polycistronic organization of most plastid genes, allowing the expression of several transgenes from one transcription unit [186]. A polycistronic construct would contain a single promoter, 5′ and 3′UTR. Within this unit, each ORF is preceded by specific sequence for translation initiation included in the linker sequence [85], or as a plastid Intercistronic Expression Element (IEE) [223].

Several heterologous polycistronic units have been expressed in transgenic plant chloroplasts. The *aadA* gene was expressed together with the petunia 5-enol-pyruvate shikimate-3-phosphate synthase gene [32], the *Bacillus thuringiensis Cry2Aa2* operon [22, 39, 162], or the yeast trehalose phosphate synthase gene [109] under the control of a single promoter in the tobacco plastid. With the *Cry2Aa2* operon from *B. thuringiensis* (Bt), coding for the insecticidal protein delta-endotoxin, the recombinant protein was expressed to levels of 46 % of the total leaf protein [39]. The native *E. coli merAB* operon encoding mercuric ion reductase (*merA*) and organomercurial lyase (*merB*) was expressed from the tobacco plastome and conferred a high level of resistance to organomercurial compounds [176]. Other operons have been introduced in the tobacco plastid genome, such as the *R. eutropha* PHB operons for significant accumulation of the biodegradable polyester polyhydroxybutyrate [4, 18, 125, 126], the *luxCDABEG* operon from *Photobacterium leiognathi* [102], or a dicistronic construct designed for increased astaxanthin content [73]. Polycistronic expression is technically challenging because of the size of the transgenic fragment inserted into the plastome. To date the highest number of genes inserted is seven, the *aadA* marker and six genes (approximately 6.5 kb) from the *lux* operon, which were introduced in the *trnI-trnA* intergenic region of the tobacco plastome [102]. In the case of chlamydomonas, the only reported example of successful polycistronic expression involved a cyanobacterial dicistronic operon encoding the large and small subunits of allophycocyanin [196].

18.3.6 Selection Markers

Delivery of exogenous DNA into the plastid compartment and its integration in the plastid genome are rare events, presumably occurring initially in one or a few copies of the plastome in a single plastid. In order to recover homoplasmic lines, it is necessary to use a dominant marker gene that can provide a selectable advantage to the transformed cell even when expressed at low level in a transformed plastid. In plants continued propagation under selective conditions and regeneration are needed to progressively eliminate all remaining wild-type plastome copies and wild-type plastids (Fig. 18.1). In unicellular microalgae, repeated sub-culturing is required to obtain a homoplasmic state in the single chloroplast. Only a few selectable marker genes have allowed reliable selection of plastid transformants in higher plants and algae, as reviewed by [38] (Tables 18.1 and 18.2). The markers available for transformation include (a) chloroplast genes that rescue the photosynthetic activity of corresponding host mutants, (b) chloroplast genes with mutations conferring antibiotic or herbicide tolerance, (c) bacterial genes affording resistance to antibiotics and (d) metabolic enzymes that rescue the respective auxotrophic mutant. The first report of successful chloroplast transformation in chlamydomonas used a recipient host with a deletion of the *atpB* gene and the corresponding wild-type gene as a marker allowing selection of photautotrophy on minimal medium [19]. Mutations in the ribosomal RNA genes conferring tolerance to antibiotics were later used in the alga, and similar markers derived from a line resistant to spectinomycin and streptomycin were used for the first successful transformation of tobacco [202]. A major improvement was obtained with the dominant bacterial *aadA* marker gene, with the advantage that the portable cassette could be introduced at any site in the plastome [66]. The *aadA* gene encodes an aminoglycoside-3″-adenylyltransferase that inactivates spectinomycin and streptomycin. It was soon found to also be effective for plant chloroplast transformation [200]. When fused to GFP, this marker can be used to track the selection process [91]. The *aadA* gene is still the most commonly used selectable marker for high-frequency production of plastid transformants in many higher plant species (Table 18.1). Genes encoding resistance to other antibiotics, such as *cat* (chloramphenicol acetyl transferase) [116], *nptII* [21, 106] and *aphA-6* [11, 20] are also possible options, and could be more appropriate in some species [104]. The amino-acid analogs 4-methylindole and 7-methyl-DL-Trp have also been successfully employed as selective agents in tobacco using the feedback-insensitive α-subunit gene (*ASA2*) as the selectable marker [10].

In plants, herbicide tolerance genes are widely used to select nuclear transformants, but the direct selection of plastid transformants could not be achieved

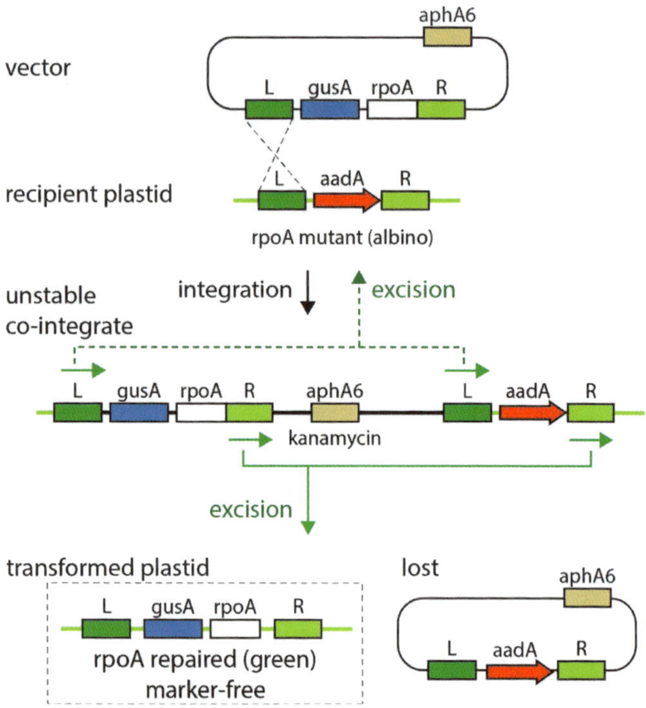

Fig. 18.2. Marker exchange. In this scheme [98], the host plant contains an insertion of the *aadA* marker that disrupts the *rpoA* locus, resulting in an albino mutant. Transformation with a vector containing the gene of interest (gusA) and the wild-type *rpoA* gene, as well as the aphA6 marker placed outside the regions of homology leads to an unstable co-integrate that is kanamycin resistant. If homologous recombination subsequently occurs between the repeats of the right flanking region (R), excision allows the loss of the *aadA* marker and rescues the *rpoA* mutation so that pigmentation is restored and the transplastomic tissue can be identified. Reproduced from [38]

with phosphinothricin [127] or glyphosate [219]. The herbicide tolerance provided by the few recombinant plastome copies at the start of selection is probably not strong enough to provide a selectable advantage. Nevertheless, herbicide tolerance genes can be successfully used for selection after a first amplification step of the recombinant plastomes using antibiotic selection [219]. Neither was it possible to isolate directly tobacco plastid transformants using a bacterial gene encoding 4-hydroxyphenyl pyruvate dioxygenase (HPPD), despite the fact that such transplastomic lines are highly tolerant to various herbicides acting on this enzyme [50, 53]. In contrast, selection based on mutant *psbA* genes conferring tolerance to the herbicides DCMU or metribuzin are effective in chlamydomonas [145]. Selection for herbicide resistance (sulfuron methyl) has also been reported for *Porphyridium*, a red unicellular alga [108].

Similar to the selection for photoautotrophy in mutants of photoautotrophy, an elegant strategy based on the restoration of photosynthetic activity function in albino tobacco lines has been described in plants (Fig. 18.2) [97]. Homoplasmic knockout lines were first generated by insertion of the *aadA* marker gene. The restoration of photosynthesis and pigmentation with the wild-type alleles could not be used directly on this material for the selection of transformants, because this would require the development of photoautotrophic cell culture systems. Nevertheless, complementation was accomplished using kanamycin selection with the *aphA-6* gene, with the simultaneous removal of the *aadA* gene. The advantages of this system are that no false positive transformants are generated, the transformation efficiency is increased, and homoplasmic plants are more rapidly and reliably produced.

18.4 Plants and Algae with Successful Plastid Transformation

Over the last 20 years, plastid transformation has been developed for a diverse group of plant species (Table 18.1). The technology, which for a long time was routine only in tobacco, has been progressively extended, albeit with lower transformation frequencies, to other solanaceous species such as *Nicotiana plumbaginifolia* [149], *Solanum tuberosum* (potato), [187], *Lycopersicon esculentum* (tomato), [2, 147, 172, 224], *Petunia hybrida* [226], and *Nicotiana benthamiana* [36]. Although in the initial work with potato the transformed plants were sterile [187], this issue has been resolved in a more recent report [146].

Transformation of *Arabidopsis thaliana* plastids has been achieved but the transformation frequency was low and the transformed plants were sterile [188]. *Lesquerella fendleri* (*Brassicaceae*) was transformed using a chimeric *Arabidopsis* and tobacco vector, which resulted in a low frequency of fertile, transformed plants [190]. Nonetheless, major achievements have been published concerning the production, at a reasonable frequency, of homoplasmic and fertile transplastomic plants in a wider range of species including soybean (*Glycine max*; [48]; species specific plastid vector), cotton (*Gossypium hirsutum*; [105]), carrot (*Daucus carota*; [104]) and lettuce (*Lactuca sativa*; [86, 113]), sugar beet (*Beta vulgaris*; [40]) and alfalfa [214]. For all these experiments, the plastid transformation vectors contained species-specific homologous recombination regions. Two *Brassica napus* (oilseed rape) vectors targeting different regions of the plastome were used to transform oilseed rape but only heteroplasmic plastid transformed lines were obtained [23, 79]. In monocots, since the first attempt of rice plastid transformation [91], only limited progress has been made in obtaining stable and fertile transplastomic lines [110]. A recent report of plastid transformation in wheat using a *gfp* reporter has been retracted.

The transfer of transplastomic technology to a wider range of algal species has also been disappointingly slow (Table 18.2). After the first successful chloroplast transformation of *Chlamydomonas reinhardtii* in 1988, chloroplast transformation was reported for *Euglena gracilis* using a homologous *aadA* transformation cassette [45]. In addition, in the unicellular red alga *Porphyridium sp.* chloroplasts were stably transformed by replacing the native plastid gene encoding acetohydroxyacid synthase (AHAS) with a mutant form resistant to the herbicide sulfometuron methyl [108]. However, there have not been any subsequent reports of transformation of these species. Transient chloroplast transformation of the halotolerant green alga *Dunaliella salina* was achieved using a di-cistronic construct of the *eGFP* and *hptII* gene, inserted in a rice chloroplast transformation vector [116], and this has been followed recently with a report of stable transformation of the related species *Dunaliella tertiolecta* using an erythromycin resistance marker, although homoplasmic transformants could not be recovered [61]. Finally, transformation of the green alga *Haematococcus pluvialis* has been established using the *aadA* marker [69].

18.5 Traits of Interest

Transplastomics offers the opportunity to produce not only high-value recombinant proteins, but to engineer a wide range of useful phenotypic traits into plant and algal species. A number of excellent recent reviews present exhaustive lists of genes or traits that have already been introduced [35, 132, 164, 180]. This list is growing rapidly. Chloroplast engineering should be considered seriously for a number of applications and reasons.

18.5.1 Input Traits for Higher Plants

A variety of genes have been expressed in higher plants plastids which have a direct impact on yield or quality. Two major traits have been addressed, (i) resistance to leaf chewing pests, expressing insecticidal toxins from *B. thuringiensis* [22, 39, 49, 79, 101, 103, 119, 121, 139, 168] and (ii) tolerance to herbicides, such as glyphosate [24, 32, 218], glufosinate [82, 87, 127]), 4-hydroxyphenyl pyruvate inhibitors [50, 53] and acetolactate synthase inhibitors [185]. Other traits such as resistance against microbial diseases [42, 111] or tolerance to drought [109, 221] and cold [29] have also been described in the literature. With unicellular algae, tolerance to herbicides such as metribuzin or DCMU was described in chlamydomonas [145, 159], and to sulfometuron methyl in a *Porphyridium* species [108].

These pilot studies have been performed mostly in tobacco and evaluated in the lab or the greenhouse, rarely in the field. It is really a drawback that efficient protocols for plastid transformation do not exist for cereals and most other major cultivated species. This largely explains the current absence of commercial transplastomic crops. Also, in most cases, similar phenotypes and agronomic performance can be achieved in relevant crops using nuclear transgenes, despite their more modest upper limit of expression in green tissues. It is also anticipated that expression levels in leaves above a certain threshold (a few percent of total soluble proteins) will negatively impact crop yield [151], even though high expressing tobacco lines seem to develop normally in greenhouse conditions [7]. Finally, the potential for expression in non-photosynthetic tissues, such as roots and seeds, is unclear and certainly much lower than in green leaves. Various groups are working on improving the performance of Rubisco in higher plants [215], the enzyme responsible for the first step in C_3 carbon fixation. Since the large subunit is plastid-encoded, it will be necessary to engineer the chloroplast genome of higher plants once promising variants are identified. This could become the main driver for the application of the technology to cereals and other recalcitrant species.

18.5.2 Massive Expression of High Value Proteins

Plant chloroplasts can accumulate recombinant proteins at extremely high levels, up to 70 % total soluble proteins in leaves in tobacco [151]. As mentioned above, such expression levels are not needed or even desired for input traits in higher plants, but this potential makes chloroplasts very attractive for the production of pharmaceutical proteins and for other types of high value proteins. Tobacco is an almost ideal higher plant platform for such molecular farming applications, since it can be easily genetically engineered, produces high leaf biomass and is not part of the food chain. Chloroplasts compare favorably with bacterial systems in the range of proteins that can be expressed. Even complex disulfide-bond containing full-length monoclonal antibodies can be produced and correctly folded, as shown in chlamydomonas [207]. In addition, recombinant proteins made in chloroplasts generally remain soluble and do not form inclusion bodies unlike *E. coli*. However, one limitation concerns the synthesis of glycosylated proteins since this post-translational modification is absent in chloroplasts, as in prokaryotic systems.

There is still almost no commercial application for proteins purified from transgenic plants, except a few proteins such as avidin produced from nuclear transgenic corn, and sold for research applications through Sigma-Aldrich. Extraction and purification of recombinant proteins represent a significant proportion of the production cost, and this cost is not reduced with plant expression platforms. Furthermore the production cost of a therapeutic protein is generally not the most critical parameter. Chloroplast technology may have a brighter future for proteins that are needed in massive amounts, possibly without purification, such as industrial enzymes. This is the case of plant cell wall degrading enzymes which can increase the amount of fermentable sugars in the process of biofuel production from biomass. A high number of those lytic enzymes have been recently produced at remarkably high levels in transplastomic tobacco [156, 213].

18.5.3 Edible Vaccines

A high number of publications have recently documented the potential of chloroplast-expressed antigens for low-cost immunization, exemplified in animal models. Major diseases such as hepatitis, AIDS, plague, anthrax, cholera or malaria have been targeted, and the oral delivery of corresponding antigens expressed in tobacco or lettuce has shown some promising efficacy [37, 124]. Lettuce, carrot or other edible plants could be eaten raw or be processed if formulated products are required. The concept of cheap oral vaccination using genetically engineered plants is also appealing for the immunization of animals, as shown for the treatment of post-weaning diarrhea in piglets [100]. As with therapeutic proteins, the possibility of easy up or down scaling of plant-based systems makes the production of antigens in plants very attractive, and could provide a quick response for potential pandemic agents [177]. In the same way, transplastomic algae also offer the potential for low cost production of edible vaccines with the added advantage of containment and controlled cultivation in industrial fermenters. Green algal species such as *C. reinhardtii, Dunaliella salina* and *Haematococcus pluvialis* are already classified as GRAS organisms (Generally Recognized As Safe), and therefore can be considered as edible without any downstream processing. One example of an oral vaccine produced in the chlamydomonas chloroplast is a vaccine against *Staphylococcus aureus* that was shown to confer protection from infection to mice [47]. A similar demonstration of the potential of oral delivery involved the synthesis of a bacterial phytase enzyme in the algal chloroplast. When a whole-cell lysate was fed to chickens there was a marked improvement in the uptake of phytates from the diet [220].

18.5.4 Metabolic Engineering

Chloroplasts are ideal bioreactors for metabolic engineering applications. These organelles host a great variety of primary and secondary metabolic pathways, which are much more complex and diverse than in most bacteria. As in other prokaryotes, multiple genes can be organized into polycistronic units and co-expressed under the control of a single promoter, as demonstrated with the expression in transplastomic tobacco of the entire bacterial luciferin operon encoding six distinct enzymes [102]. Efforts have also built upon existing plastid biosynthetic pathways, as exemplified with the engineering of the carotenoid pathway and increases in pro-vitamin A levels in tomato fruit [2]. The same pathway has been modified further downstream in transplastomic tobacco for the production of a high value carotenoid compound, astaxanthin [73], and upstream by overexpression of a cyanobacterial 1-deoxy-D-xylulose-5-phosphate reductoisomerase, leading to major changes in isoprenoid composition [74]. Metabolic engineering will become even more attractive when transformation of the plastid genome will become technically feasible in species that naturally already produce high value secondary metabolites, such as alkaloids, whose production could be optimized and tailored by genetic engineering.

18.6 Biosafety, Containment and Public Acceptance

18.6.1 Marker Removal

The removal of antibiotic resistance markers after the generation of transgenic plants is desirable to eliminate the risk of gene flow both to other plants and to pathogenic bacteria. The very high ploidy of the plastome in plant cells (the number of copies of the marker can reach 10,000 per cell, as compared to two marker copies for a diploid nuclear transformant) further exacerbates this risk, and raises serious public concern. Moreover, as only a few efficient selectable markers are available for plastid transformation, especially in algae, marker removal allows subsequent re-transformation when stacking of several transgenes is required. Finally, the maintenance and expression of the marker itself in the chloroplast must impose some (albeit, small) metabolic burden on the transplastomic line, so marker removal is desirable from a productivity perspective. One option to avoid undesired selection markers is simply to use an endogenous gene for selection, such as a chloroplast gene that encodes a protein involved in photosynthesis, however this typically restricts the transformation host to a mutant line that is defective in the corresponding gene (see Sect. 18.2.5).

Several methods, reviewed in [38], have been developed to remove the antibiotic resistance marker from transplastomic plants. These include the generation of deletion derivatives in which the marker gene is excised by homologous recombination between flanking direct repeats using the *Cre/loxP* and *Int/att* site-specific recombination systems from bacteriophages. The approach described by [98] is particularly attractive since marker-free lines can be routinely and rapidly generated in the T_0 generation, but the negative side is that it requires first the production of albino mutants by targeted gene disruption. In recombinase protocols, the marker genes are excised by the Cre recombinase from phage P1 or the recombinase from phage phiC31, which recognize *loxP* or *att* sites respectively, that are contained within direct repeats flanking the marker [28, 72, 96, 129]. Excision is induced by expressing the site-specific recombinase gene from the plant nuclear genome and targeting the enzyme to the chloroplast stroma.

Excision by homologous recombination between direct repeats placed on either side of the marker is also possible simply by exploiting the natural homologous recombination activity in the chloroplast. This was first demonstrated in chlamydomonas [56, 141]. The same strategy was also used successfully in plants [50, 82, 98]. An elegant variation of this approach was exemplified with a gene encoding 4-hydroxyphenyl pyruvate dioxygenase (HPPD) that provides tolerance to the herbicide isoxaflutole [50]. An *aadA* antibiotic resistance cassette was inserted in a non-functional herbicide tolerance gene. In this construct, approximately 400 bp of the coding sequence of *hppd* were duplicated in direct orientation on either side of the *aadA* cassette (Fig. 18.3). The construct was used to generate plastid transformants in tobacco, using selection for *aadA* on spectinomycin-containing media. Recombination by the plastid machinery, or slippage during replication, between the two repeats then led to the elimination of the antibiotic-resistance cassette, and the reconstitution of a functional herbicide tolerance *hppd* gene. This is precisely what was observed in some of the lines of generation T2 by selection/screening with isoxaflutole (IFT). This process produced lines that displayed a strong herbicide tolerance and that were susceptible to spectinomycin.

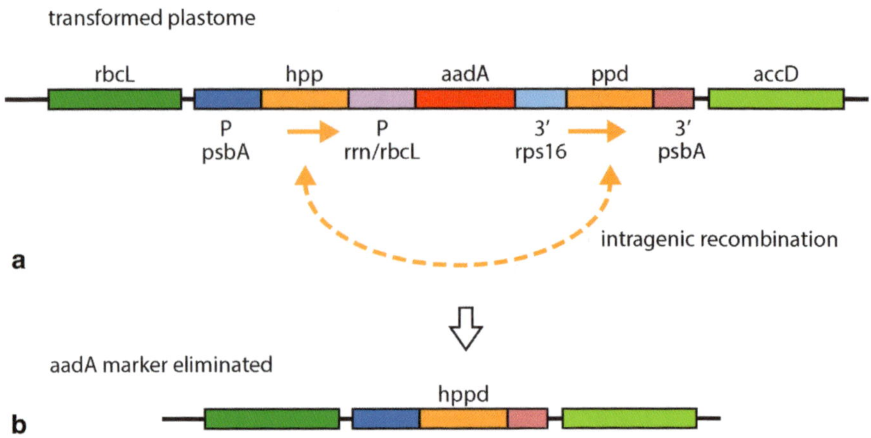

Fig. 18.3 Marker excision. **a** In the example of marker excision depicted here [50], the initial transformant contains the *aadA* selectable marker cassette (*Prrn/rbcL::aadA::3′rps16*) flanked by two overlapping subfragments of the *hppd* gene with appropriate promoter (*PpsbA*) and 3′UTR sequences. **b** Intragenic recombination between the two overlapping parts of *hppd* leads to the elimination of the *aadA* marker, and to the reconstitution of a functional *hppd* gene conferring tolerance to the herbicide diketonitrile

18.6.2 Pollen Transmission/Uniparental Inheritance

The plastids of a majority of flowering plant species are eliminated at different stages of male gametophyte development, and are therefore inherited maternally [70]. Some species like alfalfa have a biparental heredity [192], and others like kiwi plants transmit plastids predominantly by the pollen [27]. Uniparental inheritance can provide a strong level of natural genetic containment so that gene flow of the engineered traits through pollen is reduced. In *Nicotiana tabacum*, which shows strong maternal inheritance, paternal transmission of the plastid genome has been studied in detail, and is detected in less than 0.1 % of the progeny [140, 173, 201]. Thus, uniparental inheritance offers a degree of transgene containment and minimizes the possibility of outcrossing transgenes to related weeds or crops. In *Chlamydomonas reinhardtii*, plastid inheritance is largely from the mating-type plus (mt+) parent, although biparental inheritance is observed with frequencies of a few percent [181]. Introducing the transgene in the plastome of the mt− parent thus also offers a degree of containment. Since the algae are usually propagated vegetatively, it is also possible to envisage using strains with mutations that affect conjugation, zygote maturation, meiosis or spore germination.

18.6.3 Horizontal Gene Transfer

As mentioned above, horizontal transfer of antibiotic resistance genes from the chloroplast to other organisms is a significant concern. Since the genes are expressed from bacterial-type regulatory elements, in the event of horizontal transfer to a bacterium, they could possibly be expressed. Experimental results showing that horizontal gene transfer can take place between tobacco plastid transformants and naturally occurring bacteria have been published [41, 90, 157]. Nevertheless, this is a very rare event, which could be demonstrated only with genetically modified microorganisms containing sequences homologous to those of the recombinant tobacco plastids. Additional containment measures could be envisaged, such as the engineering of plastid introns in the coding regions of the transgenes, or RNA editing-dependent translation start sites [129].

18.7 Future Directions

While there are examples of remarkably high levels of transgenic protein accumulation in plant chloroplasts, with up to 70% of total soluble leaf protein being the highest claim, such amounts are not routinely achieved and a wide range of values have been reported. In microalgae, recombinant protein levels are altogether more modest, typically ranging from 1 to 5%, with many cases of undetectable levels. In spite of major advances in our understanding of plastid gene regulation, optimization of transgene expression still remains somewhat empirical. One factor that seems to limit transgene expression in chlamydomonas is the negative feedback regulation that unassembled subunits exert on the translation of their own messenger RNA (termed 'control by epistasy of synthesis,' or CES; cf. Chap. 3 of this volume). For transgenes placed under the control of the *psaAex1* promoter/5′UTR, it was shown that mutations that abolish *trans*-splicing of the endogenous *psaA* mRNA, and thus expression of PsaA, had a strong beneficial effect on transgene expression [141, 216]. However the strains were not photoautotrophic because PsaA is an essential subunit of PSI. A better understanding of the molecular mechanism of CES may allow engineering *Chlamydomonas* strains where this type of feedback inhibition is circumvented, but photoautotrophy is retained.

Proteolysis can also be a limiting factor for the expression of the desired transgenic protein [141]. Recent progress has allowed the identification of many plant chloroplast proteases [1] and the isolation of some corresponding mutants in chlamydomonas [133]. Although this knowledge has not yet been applied to transgenic plants or algae, it opens the possibility of using protease-deficient host strains for enhanced protein expression. While genetic modifications of the host are beginning to be explored along these lines in chlamydomonas, it will be a challenge in the future to translate the results to other species of algae or to plants that are of agronomic interest.

The prospect of large scale culture of microalgae in industrial systems such as raceways raises the question of how the genetically modified organisms will be contained if their release into the open is to be avoided. It may be technically difficult and costly to set up physical containment measures to ensure that the microalgae are not released into the environment. An alternative will be to implement biological containment with strains that are disabled and grow poorly in natural conditions. Host strains could be derived that have defects in motility, that are auxotrophic for specific nutrients or that are defective in sexual reproduction (see Sect. 18.4).

Synthetic biology offers the possibility of engineering complex metabolic pathways to derive strains that produce new generations of biofuels or other molecules of interest. Because the chloroplast harbors photosynthesis and many other important biochemical pathways, its metabolic engineering holds great promise. The field is still in its infancy, and its development will require many advances that may be facilitated by rapid parallel progress in systems biology.

Thousands of years of agricultural practice have allowed the domestication of many plants, modern breeding strategies have produced the crops that are widely cultivated, and genetic engineering has opened new opportunities which are now widely used, albeit for a very limited number of traits. The number of crops where chloroplast transformation has been achieved is steadily increasing, but the development of the appropriate methods remains the limiting step in many instances. To date, no transplastomic plants are exploited commercially. As reviewed in this chapter, chloroplast transformation has significant advantages and many traits have already been engineered, so it can be expected that transplastomic plants will eventually find practical applications. For algae the situation is more challenging, because the number of species that are in agronomic or industrial use is quite small, and few of these can be transformed. The growing interest in microalgae for many applications suggests that rapid advances can be expected both in the identification of appropriate species and in their domestication, a process where chloroplast transformation may play a decisive role.

Acknowledgements The authors thank Nicolas Roggli for preparing figures. SP acknowledges support for work on algal transplastomics from the UK Biotechnology and Biological Sciences Research Council (grant BB/I007660/1). Work in the laboratory of MGC was supported by the University of Geneva and by grants from the Swiss NRP59 programme (Benefits and Risks of the Deliberate Release of Genetically Modified Plants), from the European Union FP7 KBBE 2009-3 Sunbiopath (GA 245070) and from the Swiss National Science Foundation (3100A-117712).

References

1. Adam Z (2007) Protein stability and degradation in plastids. In: Bock R (ed) Cell and molecular biology of plastids. Springer, Heidelberg, pp 315–338
2. Apel W, Bock R (2009) Enhancement of carotenoid biosynthesis in transplastomic tomatoes by induced lycopene-to-provitamin A conversion. Plant Physiol 151:59–66
3. Apel W, Schulze WX, Bock R (2010) Identification of protein stability determinants in chloroplasts. Plant J 63:636–650

4. Arai Y, Shikanai T, Doi Y, Yoshida S, Yamaguchi I, Nakashita H (2004) Production of polyhydroxybutyrate by polycistronic expression of bacterial genes in tobacco plastid. Plant Cell Physiol 45:1176–1184
5. Baier M, Dietz KJ (2005) Chloroplasts as source and target of cellular redox regulation: a discussion on chloroplast redox signals in the context of plant physiology. J Exp Bot 56:1449–1462
6. Bally J, Paget E, Droux M, Job C, Job D, Dubald M (2008) Both the stroma and thylakoid lumen of tobacco chloroplasts are competent for the formation of disulphide bonds in recombinant proteins. Plant Biotechnol J 6:46–61
7. Bally J, Nadai M, Vitel M, Rolland A, Dumain R, Dubald M (2009) Plant physiological adaptations to the massive foreign protein synthesis occurring in recombinant chloroplasts. Plant Physiol 150:1474–1481
8. Barkan A (2011) Expression of plastid genes: organelle-specific elaborations on a prokaryotic scaffold. Plant Physiol 155:1520–1532
9. Barnes D, Franklin S, Schultz J, Henry R, Brown E, Coragliotti A, Mayfield SP (2005) Contribution of 5′- and 3′-untranslated regions of plastid mRNAs to the expression of *Chlamydomonas reinhardtii* chloroplast genes. Mol Genet Genomics 274:625–636
10. Barone P, Zhang XH, Widholm JM (2009) Tobacco plastid transformation using the feedback-insensitive anthranilate synthase [alpha]-subunit of tobacco (ASA2) as a new selectable marker. J Exp Bot 60:3195–3202
11. Bateman JM, Purton S (2000) Tools for chloroplast transformation in *Chlamydomonas*: expression vectors and a new dominant selectable marker. Mol Gen Genet 263:404–410
12. Bellucci M, De Marchis F, Mannucci R, Bock R, Arcioni S (2005) Cytoplasm and chloroplasts are not suitable subcellular locations for beta-zein accumulation in transgenic plants. J Exp Bot 56:1205–1212
13. Bendich AJ (1987) Why do chloroplasts and mitochondria contain so many copies of their genome? Bioessays 6:279–282
14. Bessette PH, Aslund F, Beckwith J, Georgiou G (1999) Efficient folding of proteins with multiple disulfide bonds in the *Escherichia coli* cytoplasm. Proc Natl Acad Sci U S A 96:13703–13708
15. Bock R (2007) Plastid biotechnology: prospects for herbicide and insect resistance, metabolic engineering and molecular farming. Curr Opin Biotechnol 18:100–106
16. Bock R, Warzecha H (2010) Solar-powered factories for new vaccines and antibiotics. Trends Biotechnol 28:246–252
17. Bogorad L (2000) Engineering chloroplasts: an alternative site for foreign genes, proteins, reactions and products. Trends Biotechnol 18:257–263
18. Bohmert-Tatarev K, McAvoy S, Daughtry S, Peoples OP, Snell KD (2011) High levels of bioplastic are produced in fertile transplastomic tobacco plants engineered with a synthetic operon for the production of polyhydroxybutyrate. Plant Physiol 155:1690–1708
19. Boynton JE et al (1988) Chloroplast transformation in Chlamydomonas with high velocity microprojectiles. Science 240:1534–1538
20. Buchanan BB, Luan S (2005) Redox regulation in the chloroplast thylakoid lumen: a new frontier in photosynthesis research. J Exp Bot 56:1439–1447
21. Carrer H, Hockenberry TN, Svab Z, Maliga P (1993) Kanamycin resistance as a selectable marker for plastid transformation in tobacco. Mol Gen Genet 241:49–56
22. Chakrabarti SK, Lutz KA, Lertwiriyawong B, Svab Z, Maliga P (2006) Expression of the *cry9Aa2 B.t.* gene in tobacco chloroplasts confers resistance to potato tuber moth. Transgenic Res 15:481–488
23. Cheng L, Li HP, Qu B, Huang T, Tu JX, Fu TD, Liao YC (2010) Chloroplast transformation of rapeseed (*Brassica napus*) by particle bombardment of cotyledons. Plant Cell Rep 29:371–381
24. Chin HG, Kim GD, Marin I, Mersha F, Evans TC Jr, Chen L, Xu MQ, Pradhan S (2003) Protein trans-splicing in transgenic plant chloroplast: reconstruction of herbicide resistance from split genes. Proc Natl Acad Sci U S A 100:4510–4515

25. Choquet Y, Wollman FA (2002) Translational regulations as specific traits of chloroplast gene expression. FEBS Lett 529:39–42
26. Chumley TW, Palmer JD, Mower JP, Fourcade HM, Calie PJ, Boore JL, Jansen RK (2006) The complete chloroplast genome sequence of Pelargonium x hortorum: organization and evolution of the largest and most highly rearranged chloroplast genome of land plants. Mol Biol Evol 23:2175–2190
27. Cipriani G, Testolin R, Morgante M (1995) Paternal inheritance of plastids in interspecific hybrids of the genus *Actinidia* revealed by PCR-amplification of chloroplast DNA fragments. Mol Gen Genet 247:693–697
28. Corneille S, Lutz K, Svab Z, Maliga P (2001) Efficient elimination of selectable marker genes from the plastid genome by the CRE-*lox* site-specific recombination system. Plant J 27:171–178
29. Craig W et al (2008) Transplastomic tobacco plants expressing a fatty acid desaturase gene exhibit altered fatty acid profiles and improved cold tolerance. Transgenic Res 17:769–782
30. Crosby K, Smith DR (2012) Does the mode of plastid inheritance influence plastid genome architecture? PLoS ONE 7:e46260
31. Cui C et al (2011) Stable chloroplast transformation of immature scutella and inflorescences in wheat (*Triticum aestivum* L.). Acta Biochim Biophys Sinica 43:284–291
32. Daniell H, Datta R, Varma S, Gray S, Lee SB (1998) Containment of herbicide resistance through genetic engineering of the chloroplast genome. Nat Biotechnol 16:345–348
33. Daniell H, Lee SB, Panchal T, Wiebe PO (2001) Expression of the native cholera toxin B subunit gene and assembly as functional oligomers in transgenic tobacco chloroplasts. J Mol Biol 311:1001–1009
34. Daniell H, Cohill PR, Kumar S, Dufourmantel N (2004) Chloroplast genetic engineering. In: Daniell H, Chase CD (eds) Molecular biology and biotechnology of plant organelles. Kluwer, The Netherlands, pp 443–490
35. Daniell H, Singh ND, Mason H, Streatfield SJ (2009) Plant-made vaccine antigens and biopharmaceuticals. Trends Plant Sci 14:669–679
36. Davarpanah SJ, Jung SH, Kim YJ, Park YI, Min SR, Liu JR, Jeong WJ (2009) Stable plastid transformation in *Nicotiana benthamiana*. J Plant Biol 52:244–250
37. Davoodi-Semiromi A, Samson N, Daniell H (2009) The green vaccine: a global strategy to combat infectious and autoimmune diseases. Hum Vaccin 5:488–493
38. Day A, Goldschmidt-Clermont M (2011) The chloroplast transformation toolbox: selectable markers and marker removal. Plant Biotechnol J 9:540–553
39. De Cosa B, Moar W, Lee SB, Miller M, Daniell H (2001) Overexpression of the Bt *cry2Aa2* operon in chloroplasts leads to formation of insecticidal crystals. Nature Biotechnol 19:71–74
40. De Marchis F, Wang Y, Stevanato P, Arcioni S, Bellucci M (2009) Genetic transformation of the sugar beet plastome. Transgenic Res 18:17–30
41. de Vries J, Herzfeld T, Wackernagel W (2004) Transfer of plastid DNA from tobacco to the soil bacterium *Acinetobacter* sp. by natural transformation. Mol Mircobiol 53:323–334
42. DeGray G, Rajasekaran K, Smith F, Sanford J, Daniell H (2001) Expression of an antimicrobial peptide via the chloroplast genome to control phytopathogenic bacteria and fungi. Plant Physiol 127:852–862
43. Demarsy E, Buhr F, Lambert E, Lerbs-Mache S (2012) Characterization of the plastid-specific germination and seedling establishment transcriptional programme. J Exp Bot 63:925–939
44. Dix PJ, Kanvangh TA (1995) Transforming the plastome: genetic markers and DNA delivery systems. Euphytica 85:29–34
45. Doetsch NA, Favreau MR, Kuscuoglu N, Thompson MD, Hallick RB (2001) Chloroplast transformation in *Euglena gracilis*: splicing of a group III twintron transcribed from a transgenic psbK operon. Curr Genet 39:49–60
46. Drechsel O, Bock R (2011) Selection of Shine-Dalgarno sequences in plastids. Nucleic Acids Res 39:1427–1438

47. Dreesen IA, Charpin-El Hamri G, Fussenegger M (2010) Heat-stable oral alga-based vaccine protects mice from *Staphylococcus aureus* infection. J Biotechnol 145:273–280
48. Dufourmantel N, Pelissier B, Garcon F, Peltier G, Ferullo JM, Tissot G (2004) Generation of fertile transplastomic soybean. Plant Mol Biol 55:479–489
49. Dufourmantel N, Tissot G, Goutorbe F, Garcon F, Muhr C, Jansens S, Pelissier B, Peltier G, Dubald M (2005) Generation and analysis of soybean plastid transformants expressing *Bacillus thuringiensis* Cry1Ab protoxin. Plant Mol Biol 58:659–668
50. Dufourmantel N et al (2007) Generation and characterization of soybean and marker-free tobacco plastid transformants over-expressing a bacterial 4-hydroxyphenylpyruvate dioxygenase which provides strong herbicide tolerance. Plant Biotechnol J 5:118–133
51. Eibl C, Zou Z, Beck A, Kim M, Mullet J, Koop HU (1999) *In vivo* analysis of plastid *psbA*, *rbcL* and *rpl32* UTR elements by chloroplast transformation: tobacco plastid gene expression is controlled by modulation of transcript levels and translation efficiency. Plant J 19:333–345
52. Elghabi Z, Ruf S, Bock R (2011) Biolistic co-transformation of the nuclear and plastid genomes. Plant J 67:941–948
53. Falk J, Brosch M, Schafer A, Braun S, Krupinska K (2005) Characterization of transplastomic tobacco plants with a plastid localized barley 4-hydroxyphenylpyruvate dioxygenase. J Plant Physiol 162:738–742
54. Fernandez-San Millan A, Farran I, Molina A, Mingo-Castel AM, Veramendi J (2007) Expression of recombinant proteins lacking methionine as N-terminal amino acid in plastids: human serum albumin as a case study. J Biotechnol 127:593–604
55. Ferrante P, Catalanotti C, Bonente G, Giuliano G (2008) An optimized, chemically regulated gene expression system for *Chlamydomonas*. PLoS ONE 3:e3200
56. Fischer N, Stampacchia O, Redding K, Rochaix JD (1996) Selectable marker recycling in the chloroplast. Mol Gen Genet 251:373–380
57. Fletcher SP, Muto M, Mayfield SP (2007) Optimization of recombinant protein expression in the chloroplasts of green algae. Adv Exp Med Biol 616:90–98
58. Franklin SE, Mayfield SP (2004) Prospects for molecular farming in the green alga *Chlamydomonas*. Curr Opin Plant Biol 7:159–165
59. Franklin S, Ngo B, Efuet E, Mayfield SP (2002) Development of a GFP reporter gene for *Chlamydomonas reinhardtii* chloroplast. Plant J 30:733–744
60. Gelhaye E, Rouhier N, Navrot N, Jacquot JP (2005) The plant thioredoxin system. Cell Mol Life Sci 62:24–35
61. Georgianna DR, Hannnon MJ, Marcuschi M, Wu S, Botsch K, Lewis AJ, Hyun J, Mendez M, Mayfield SP (2012) Production of recombinant enzymes in the marine alga *Dunaliella tertiolecta*. Algal Res 2:2–9
62. Giglione C, Meinnel T (2001) Organellar peptide deformylases: universality of the N-terminal methionine cleavage mechanism. Trends Plant Sci 6:566–572
63. Gimpel JA, Mayfield SP (2012) Analysis of heterologous regulatory and coding regions in algal chloroplasts. Appl Microbiol Biotechnol 97:4499–4510
64. Glenz K, Bouchon B, Stehle T, Wallich R, Simon MM, Warzecha H (2006) Production of a recombinant bacterial lipoprotein in higher plant chloroplasts. Nat Biotechnol 24:76–77
65. Golds T, Maliga P, Koop HU (1993) Stable plastid transformation in PEG-treated protoplasts of *Nicotiana tabacum*. Nat Biotechnol 11:95–97
66. Goldschmidt-Clermont M (1991) Transgenic expression of aminoglycoside adenine transferase in the chloroplast: a selectable marker of site-directed transformation of *Chlamydomonas*. Nucleic Acids Res 19:4083–4089
67. Goldschmidt-Clermont M, Rahire M, Rochaix JD (2008) Redundant cis-acting determinants of 3′ processing and RNA stability in the chloroplast *rbcL* mRNA of *Chlamydomonas*. Plant J 53:566–577
68. Guisinger MM, Kuehl JV, Boore JL, Jansen RK (2011) Extreme reconfiguration of plastid genomes in the angiosperm family *Geraniaceae*: rearrangements, repeats, and codon usage. Mol Biol Evol 28:583–600

69. Gutierrez CL, Gimpel J, Escobar C, Marshall SH, Henriquez V (2012) Chloroplast genetic tool for the green microalgae *Haematococcus Pluvialis* (*Chlorophyceae, Volvocales*). J Phycol 48:976–983
70. Hagemann R (2004) The sexual inheritance of plant organelles. In: Daniell H, Chase CD (eds) Molecular biology and biotechnology of plant organelles. Springer, The Netherlands, pp 93–113
71. Hajdukiewicz PT, Allison LA, Maliga P (1997) The two RNA polymerases encoded by the nuclear and the plastid compartments transcribe distinct groups of genes in tobacco plastids. EMBO J 16:4041–4048
72. Hajdukiewicz PT, Gilbertson L, Staub JM (2001) Multiple pathways for *Cre/lox*-mediated recombination in plastids. Plant J 27:161–170
73. Hasunuma T, Miyazawa S, Yoshimura S, Shinzaki Y, Tomizawa K, Shindo K, Choi SK, Misawa N, Miyake C (2008) Biosynthesis of astaxanthin in tobacco leaves by transplastomic engineering. Plant J 55:857–868
74. Hasunuma T, Takeno S, Hayashi S, Sendai M, Bamba T, Yoshimura S, Tomizawa K, Fukusaki E, Miyake C (2008) Overexpression of 1-deoxy-D-xylulose-5-phosphate reductoisomerase gene in chloroplast contributes to increment of isoprenoid production. J Biosci Bioeng 105:518–526
75. Heifetz PB (2000) Genetic engineering of the chloroplast. Biochimie 82:655–666
76. Herz S, Fussl M, Steiger S, Koop HU (2005) Development of novel types of plastid transformation vectors and evaluation of factors controlling expression. Transgenic Res 14:969–982
77. Hill KL, Li HH, Singer J, Merchant S (1991) Isolation and structural characterization of the *Chlamydomonas reinhardtii* gene for cytochrome c6. Analysis of the kinetics and metal specificity of its copper-responsive expression. J Biol Chem 266:15060–15067
78. Hirose T, Sugiura M (2004) Functional Shine-Dalgarno-like sequences for translational initiation of chloroplast mRNAs. Plant Cell Physiol 45:114–117
79. Hou BK, Zhou YH, Wan LH, Zhang ZL, Shen GF, Chen ZH, Hu ZM (2003) Chloroplast transformation in oilseed rape. Transgenic Res 12:111–114
80. Huang FC, Klaus SM, Herz S, Zou Z, Koop HU, Golds TJ (2002) Efficient plastid transformation in tobacco using the *aphA-6* gene and kanamycin selection. Mol Genet Genomics 268:19–27
81. Huang CY, Grunheit N, Ahmadinejad N, Timmis JN, Martin W (2005) Mutational decay and age of chloroplast and mitochondrial genomes transferred recently to angiosperm nuclear chromosomes. Plant Physiol 138:1723–1733
82. Iamtham S, Day A (2000) Removal of antibiotic resistance genes from transgenic tobacco plastids. Nat Biotechnol 18:1172–1176
83. Ishikura K, Takaoka Y, Kato K, Sekine M, Yoshida K, Shinmyo A (1999) Expression of a foreign gene in *Chlamydomonas reinhardtii* chloroplast. J Biosci Bioeng 87:307–314
84. Jacobs J, Kuck U (2011) Function of chloroplast RNA-binding proteins. Cell Mol Life Sci 68:735–748
85. Jeong SW, Jeong WJ, Woo JW, Choi DW, Park YI, Liu JR (2004) Dicistronic expression of the green fluorescent protein and antibiotic resistance genes in the plastid for selection and tracking of plastid-transformed cells in tobacco. Plant Cell Rep 22:747–751
86. Kanamoto H, Yamashita A, Asao H, Okumura S, Takase H, Hattori M, Yokota A, Tomizawa K (2006) Efficient and stable transformation of *Lactuca sativa* L. cv. Cisco (lettuce) plastids. Transgenic Res 15:205–217
87. Kang TJ, Seo JE, Loc NH, Yang MS (2003) Herbicide resistance of tobacco chloroplasts expressing the *bar* gene. Mol Cells 16:60–66
88. Kasai S, Yoshimura S, Ishikura K, Takaoka Y, Kobayashi K, Kato K, Shinmyo A (2003) Effect of coding regions on chloroplast gene expression in *Chlamydomonas reinhardtii*. J Biosci Bioeng 95:276–282
89. Kato K, Marui T, Kasai S, Shinmyo A (2007) Artificial control of transgene expression in *Chlamydomonas reinhardtii* chloroplast using the *lac* regulation system from *Escherichia coli*. J Biosci Bioeng 104:207–213

90. Kay E, Vogel TM, Bertolla F, Nalin R, Simonet P (2002) In situ transfer of antibiotic resistance genes from transgenic (transplastomic) tobacco plants to bacteria. Appl Environ Microbiol 68:3345–3351
91. Khan MS, Maliga P (1999) Fluorescent antibiotic resistance marker for tracking plastid transformation in higher plants. Nat Biotechnol 17:910–915
92. Khan MS, Khalid AM, Malik KA (2005) Phage phiC31 integrase: a new tool in plastid genome engineering. Trends Plant Sci 10:1–3
93. Kim J, Mayfield SP (2002) The active site of the thioredoxin-like domain of chloroplast protein disulfide isomerase, RB60, catalyzes the redox-regulated binding of chloroplast poly(A)-binding protein, RB47, to the 5′ untranslated region of psbA mRNA. Plant Cell Physiol 43:1238–1243
94. Kindle KL (1990) High-frequency nuclear transformation of *Chlamydomonas reinhardtii*. Proc Natl Acad Sci U S A 87:1228–1232
95. Kindle KL, Richards KL, Stern DB (1991) Engineering the chloroplast genome: techniques and capabilities for chloroplast transformation in *Chlamydomonas reinhardtii*. Proc Natl Acad Sci U S A 88:1721–1725
96. Kittiwongwattana C, Lutz K, Clark M, Maliga P (2007) Plastid marker gene excision by the phiC31 phage site-specific recombinase. Plant Mol Biol 64:137–143
97. Klaus SM, Huang FC, Eibl C, Koop HU, Golds TJ (2003) Rapid and proven production of transplastomic tobacco plants by restoration of pigmentation and photosynthesis. Plant J 35:811–821
98. Klaus SM, Huang FC, Golds TJ, Koop HU (2004) Generation of marker-free plastid transformants using a transiently cointegrated selection gene. Nat Biotechnol 22:225–229
99. Knoblauch M, Hibberd JM, Gray JC, van Bel AJ (1999) A galinstan expansion femtosyringe for microinjection of eukaryotic organelles and prokaryotes. Nat Biotechnol 17:906–909
100. Kolotilin I, Kaldis A, Devriendt B, Joensuu J, Cox E, Menassa R (2012) Production of a subunit vaccine candidate against porcine post-weaning diarrhea in high-biomass transplastomic tobacco. PLoS ONE 7:e42405
101. Kota M, Daniell H, Varma S, Garczynski SF, Gould F, Moar WJ (1999) Overexpression of the *Bacillus thuringiensis* (Bt) Cry2Aa2 protein in chloroplasts confers resistance to plants against susceptible and Bt-resistant insects. Proc Natl Acad Sci U S A 96:1840–1845
102. Krichevsky A, Meyers B, Vainstein A, Maliga P, Citovsky V (2010) Autoluminescent plants. PLoS ONE 5:e15461
103. Kumar H, Kumar V (2004) Tomato expressing Cry1A(b) insecticidal protein from *Bacillus thuringiensis* protected against tomato fruit borer, *Helicoverpa armigera* (Hübner) (Lepidoptera: Noctuidae) damage in the laboratory, greenhouse and field. Crop Prot 23:135–139
104. Kumar S, Dhingra A, Daniell H (2004) Plastid-expressed betaine aldehyde dehydrogenase gene in carrot cultured cells, roots, and leaves confers enhanced salt tolerance. Plant Physiol 136:2843–2854
105. Kumar S, Dhingra A, Daniell H (2004) Stable transformation of the cotton plastid genome and maternal inheritance of transgenes. Plant Mol Biol 56:203–216
106. Kuroda H, Maliga P (2001) Sequences downstream of the translation initiation codon are important determinants of translation efficiency in chloroplasts. Plant Physiol 125:430–436
107. Langbecker CL, Ye GN, Broyles DL, Duggan LL, Xu CW, Hajdukiewicz PT, Armstrong CL, Staub JM (2004) High-frequency transformation of undeveloped plastids in tobacco suspension cells. Plant Physiol 135:39–46
108. Lapidot M, Raveh D, Sivan A, Arad SM, Shapira M (2002) Stable chloroplast transformation of the unicellular red alga *Porphyridium* species. Plant Physiol 129:7–12
109. Lee SB, Kwon HB, Kwon SJ, Park SC, Jeong MJ, Han SE, Byun MO, Daniell H (2003) Accumulation of trehalose within transgenic chloroplasts confers drought tolerance. Mol Breed 11:1–13
110. Lee SM, Kang KS, Chung H, Yoo SH, Xu XM, Lee SB, Cheong JJ, Daniell H, Kim M (2006) Plastid transformation in the monocotyledonous cereal crop, rice (*Oryza sativa*) and transmission of transgenes to their progeny. Mol Cells 21:401–410

111. Lee SB, Li B, Jin S, Daniell H (2011) Expression and characterization of antimicrobial peptides Retrocyclin-101 and Protegrin-1 in chloroplasts to control viral and bacterial infections. Plant Biotechnol J 9:100–115
112. Leelavathi S, Gupta N, Maiti S, Ghosh A, Reddy VS (2003) Overproduction of an alkali- and thermostable xylanase in tobacco chloroplasts and efficient recovery of the enzyme. Mol Breed 11:59–67
113. Lelivelt CL, McCabe MS, Newell CA, Desnoo CB, van Dun KM, Birch-Machin I, Gray JC, Mills KH, Nugent JM (2005) Stable plastid transformation in lettuce (*Lactuca sativa* L.). Plant Mol Biol 58:763–774
114. Lemaire C, Wollman FA, Bennoun P (1988) Restoration of phototrophic growth in a mutant of *Chlamydomonas reinhardtii* in which the chloroplast *atpB* gene of the ATP synthase has a deletion: an example of mitochondria-dependent photosynthesis. Proc Natl Acad Sci U S A 85:1344–1348
115. Lerbs-Mache S (2011) Function of plastid sigma factors in higher plants: regulation of gene expression or just preservation of constitutive transcription? Plant Mol Biol 76:235–249
116. Li D, Han X, Zuo J, Xie L, He R, Gao J, Chang L, Yuan L, Cao M (2011) Construction of rice site-specific chloroplast transformation vector and transient expression of *EGFP* gene in *Dunaliella salina*. J Biomed Nanotechnol 7:801–806
117. Li WM, Ruf S, Bock R (2011) Chloramphenicol acetyltransferase as selectable marker for plastid transformation. Plant Mol Biol 76:443–451
118. Liere K, Weihe A, Borner T (2011) The transcription machineries of plant mitochondria and chloroplasts: composition, function, and regulation. J Plant Physiol 168:1345–1360
119. Lin CH, Chen YY, Tzeng CC, Tsay HS, Chen LJ (2003) Expression of a *Bacillus thuringiensis cry1C* gene in plastid confers high insecticidal efficacy against tobacco cutworm—a *Spodoptera* insect. Bot Bull Acad Sinica 44:199–210
120. Liu CW, Lin CC, Chen JJ, Tseng MJ (2007) Stable chloroplast transformation in cabbage (*Brassica oleracea* L. var. *capitata* L.) by particle bombardment. Plant Cell Rep 26:1733–1744
121. Liu CW, Lin CC, Yiu JC, Chen JJ, Tseng MJ (2008) Expression of a *Bacillus thuringiensis* toxin (cry1Ab) gene in cabbage (*Brassica oleracea* L. var. *capitata* L.) chloroplasts confers high insecticidal efficacy against *Plutella xylostella*. Theoret Appl Genet 117:75–88
122. Loiselay C, Gumpel NJ, Girard-Bascou J, Watson AT, Purton S, Wollman FA, Choquet Y (2008) Molecular identification and function of *cis*- and *trans*-acting determinants for *petA* transcript stability in *Chlamydomonas reinhardtii* chloroplasts. Mol Cell Biol 28:5529–5542
123. Loschelder H, Homann A, Ogrzewalla K, Link G (2004) Proteomics-based sequence analysis of plant gene expression—the chloroplast transcription apparatus. Phytochemistry 65:1785–1793
124. Lossl AG, Waheed MT (2011) Chloroplast-derived vaccines against human diseases: achievements, challenges and scopes. Plant Biotechnol J 9:527–539
125. Lossl A, Eibl C, Harloff HJ, Jung C, Koop HU (2003) Polyester synthesis in transplastomic tobacco (*Nicotiana tabacum* L.): significant contents of polyhydroxybutyrate are associated with growth reduction. Plant Cell Rep 21:891–899
126. Lossl A, Bohmert K, Harloff H, Eibl C, Muhlbauer S, Koop HU (2005) Inducible trans-activation of plastid transgenes: expression of the *R. eutropha phb* operon in transplastomic tobacco. Plant Cell Physiol 46:1462–1471
127. Lutz KA, Knapp JE, Maliga P (2001) Expression of *bar* in the plastid genome confers herbicide resistance. Plant Physiol 125:1585–1590
128. Lutz KA, Corneille S, Azhagiri AK, Svab Z, Maliga P (2004) A novel approach to plastid transformation utilizes the phiC31 phage integrase. Plant J 37:906–913
129. Lutz KA, Bosacchi MH, Maliga P (2006) Plastid marker-gene excision by transiently expressed CRE recombinase. Plant J 45:447–456
130. Magee AM, Coyne S, Murphy D, Horvath EM, Medgyesy P, Kavanagh TA (2004) T7 RNA polymerase-directed expression of an antibody fragment transgene in plastids causes a semi-lethal pale-green seedling phenotype. Transgenic Res 13:325–337

131. Maliga P (2003) Progress towards commercialization of plastid transformation technology. Trends Biotechnol 21:20–28
132. Maliga P, Bock R (2011) Plastid biotechnology: food, fuel, and medicine for the 21st century. Plant Physiol 155:1501–1510
133. Malnoe A, Wollman FA, de Vitry C, Rappaport F (2011) Photosynthetic growth despite a broken Q-cycle. Nat Commun 2:301
134. Manuell AL, Beligni MV, Elder JH, Siefker DT, Tran M, Weber A, McDonald TL, Mayfield SP (2007) Robust expression of a bioactive mammalian protein in *Chlamydomonas chloroplast*. Plant Biotechnol J 5:402–412
135. Manuell AL, Quispe J, Mayfield SP (2007) Structure of the chloroplast ribosome: novel domains for translation regulation. PLoS Biol 5:e209
136. Margulis L (1975) The microbes' contribution to evolution. Biosystems 7:266–292
137. Marin-Navarro J, Manuell AL, Wu J, S PM (2007) Chloroplast translation regulation. Photosyn Res 94:359–374
138. Mayfield SP, Schultz J (2004) Development of a luciferase reporter gene, *luxCt*, for *Chlamydomonas reinhardtii* chloroplast. Plant J 37:449–458
139. McBride KE, Svab Z, Schaaf DJ, Hogan PS, Stalker DM, Maliga P (1995) Amplification of a chimeric *Bacillus* gene in chloroplasts leads to an extraordinary level of an insecticidal protein in tobacco. Biotechnol 13:362–365
140. Medgyesy P, Pay A, Marton L (1986) Transmission of paternal chloroplasts in *Nicotiana*. Mol Gen Genet 204:195–198
141. Michelet L, Lefebvre-Legendre L, Burr SE, Rochaix JD, Goldschmidt-Clermont M (2011) Enhanced chloroplast transgene expression in a nuclear mutant of *Chlamydomonas*. Plant Biotechnol J 9:565–574
142. Minai L, Wostrikoff K, Wollman FA, Choquet Y (2006) Chloroplast biogenesis of photosystem II cores involves a series of assembly-controlled steps that regulate translation. Plant Cell 18:159–175
143. Muhlbauer SK, Koop HU (2005) External control of transgene expression in tobacco plastids using the bacterial *lac* repressor. Plant J 43:941–946
144. Newman SM, Boynton JE, Gillham NW, Randolph-Anderson BL, Johnson AM, Harris EH (1990) Transformation of chloroplast ribosomal RNA genes in *Chlamydomonas*: molecular and genetic characterization of integration events. Genetics 126:875–888
145. Newman SM, Harris EH, Johnson AM, Boynton JE, Gillham NW (1992) Nonrandom distribution of chloroplast recombination events in *Chlamydomonas reinhardtii*: evidence for a hotspot and an adjacent cold region. Genetics 132:413–429
146. Nguyen TT, Nugent G, Cardi T, Dix PJ (2005) Generation of homoplasmic plastid transformants of a commercial cultivar of potato (*Solanum tuberosum* L.). Plant Sci 168:1495–1500
147. Nugent GD, Ten HM, van der Gulik A, Dix PJ, Uijtewaal BA, Mordhorst AP (2005) Plastid transformants of tomato selected using mutations affecting ribosome structure. Plant Cell Rep 24:341–349
148. Nugent GD, Coyne S, Nguyen TT, Kavanagh TA, Dix PJ (2006) Nuclear and plastid transformation of *Brassica oleracea* var. botrytis (cauliflower) using PEG-mediated uptake of DNA into protoplasts. Plant Sci 170:135–142
149. O'Neill C, Horvath GV, Horvath E, Dix PJ, Medgyesy P (1993) Chloroplast transformation in plants: polyethylene glycol (PEG) treatment of protoplasts is an alternative to biolistic delivery systems. Plant J 3:729–738
150. Obembe OO, Popoola JO, Leelavathi S, Reddy SV (2011) Advances in plant molecular farming. Biotechnol Adv 29:210–222
151. Oey M, Lohse M, Kreikemeyer B, Bock R (2009) Exhaustion of the chloroplast protein synthesis capacity by massive expression of a highly stable protein antibiotic. Plant J 57:436–445
152. Oey M, Lohse M, Scharff LB, Kreikemeyer B, Bock R (2009) Plastid production of protein antibiotics against pneumonia via a new strategy for high-level expression of antimicrobial proteins. Proc Natl Acad Sci U S A 106:6579–6584

153. Okumura S, Sawada M, Park YW, Hayashi T, Shimamura M, Takase H, Tomizawa K (2006) Transformation of poplar (*Populus alba*) plastids and expression of foreign proteins in tree chloroplasts. Transgenic Res 15:637–646
154. Oldenburg DJ, Bendich AJ (2004) Changes in the structure of DNA molecules and the amount of DNA per plastid during chloroplast development in maize. J Mol Biol 344:1311–1330
155. Peled-Zehavi H, Danon A (2007) Translation and translational regulation in chloroplasts. In: Bock R (ed) Cell and molecular biology of plastids, Springer, Berlin, pp 249–281
156. Petersen K, Bock R (2011) High-level expression of a suite of thermostable cell wall-degrading enzymes from the chloroplast genome. Plant Mol Biol 76:311–321
157. Pontiroli A, Rizzi A, Simonet P, Daffonchio D, Vogel TM, Monier JM (2009) Visual evidence of horizontal gene transfer between plants and bacteria in the phytosphere of transplastomic tobacco. Appl Environ Microbiol 75:3314–3322
158. Potvin G, Zhang Z (2010) Strategies for high-level recombinant protein expression in transgenic microalgae: a review. Biotechnol Adv 28:910–918
159. Przibilla E, Heiss S, Johanningmeier U, Trebst A (1991) Site-specific mutagenesis of the D1 subunit of photosystem II in wild-type *Chlamydomonas*. The Plant Cell 3:169–174
160. Purton S (2007) Tools and techniques for chloroplast transformation of *Chlamydomonas*. Adv Exptl Med Biol 616:34–45
161. Purton S, Szaub JB, Wannathong T, Young RH, Economou CK (2013) Genetic engineering of algal chloroplasts: progress and prospects. Russ J Plant Physiol 60:491–499
162. Quesada-Vargas T, Ruiz ON, Daniell H (2005) Characterization of heterologous multigene operons in transgenic chloroplasts: transcription, processing, and translation. Plant Physiol 138:1746–1762
163. Ramundo S, Rahire M, Schaad O, Rochaix JD (2013) Repression of essential chloroplast genes reveals new signaling pathways and regulatory feedback loops in *Chlamydomonas*. Plant Cell 25:167–186
164. Rasala BA, Mayfield SP (2011) The microalga *Chlamydomonas reinhardtii* as a platform for the production of human protein therapeutics. Bioengineered Bugs 2:50–54
165. Rasala BA et al (2010) Production of therapeutic proteins in algae, analysis of expression of seven human proteins in the chloroplast of *Chlamydomonas reinhardtii*. Plant Biotechnol J 8:719–733
166. Rasala BA, Muto M, Sullivan J, Mayfield SP (2011) Improved heterologous protein expression in the chloroplast of *Chlamydomonas reinhardtii* through promoter and 5′ untranslated region optimization. Plant Biotechnol J 9:674–683
167. Raubeson LA, Jansen RK (2005) Chloroplast genomes of plants. In: Daniell H (ed), Plant diversity and evolution: genotypic and phenotypic variation in higher plants, CAB International, Wallingford, pp 45–68
168. Reddy VS, Leelavathi S, Selvapandiyan A, Raman R, Giovanni F, Shukla V, Bhatnagar RK (2002) Analysis of chloroplast transformed tobacco plants with *cry1Ia5* under rice *psbA* transcriptional elements reveal high level expression of Bt toxin without imposing yield penalty and stable inheritance of transplastome. Mol Breed 9:259–269
169. Remacle C, Cline S, Boutaffala L, Gabilly S, Larosa V, Barbieri MR, Coosemans N, Hamel PP (2009) The *ARG9* gene encodes the plastid-resident N-acetyl ornithine aminotransferase in the green alga *Chlamydomonas reinhardtii*. Eukaryot Cell 8:1460–1463
170. Rott R, Levy H, Drager RG, Stern DB, Schuster G (1998) 3′-Processed mRNA is preferentially translated in *Chlamydomonas reinhardtii* chloroplasts. Mol Cell Biol 18:4605–4611
171. Rouhier N, Gelhaye E, Jacquot JP (2004) Plant glutaredoxins: still mysterious reducing systems. Cell Mol Life Sci 61:1266–1277
172. Ruf S, Hermann M, Berger IJ, Carrer H, Bock R (2001) Stable genetic transformation of tomato plastids and expression of a foreign protein in fruit. Nat Biotechnol 19:870–875
173. Ruf S, Karcher D, Bock R (2007) Determining the transgene containment level provided by chloroplast transformation. Proc Natl Acad Sci U S A 104:6998–7002

174. Ruhlman T, Ahangari R, Devine A, Samsam M, Daniell H (2007) Expression of cholera toxin B-proinsulin fusion protein in lettuce and tobacco chloroplasts-oral administration protects against development of insulitis in non-obese diabetic mice. Plant Biotechnol J 5:495–510
175. Ruhlman T, Verma D, Samson N, Daniell H (2010) The role of heterologous chloroplast sequence elements in transgene integration and expression. Plant Physiol 152:2088–2104
176. Ruiz ON, Hussein HS, Terry N, Daniell H (2003) Phytoremediation of organomercurial compounds via chloroplast genetic engineering. Plant Physiol 132:1344–1352
177. Rybicki EP (2010) Plant-made vaccines for humans and animals. Plant Biotechnol J 8:620–637
178. Schmitz-Linneweber C, Small I (2008) Pentatricopeptide repeat proteins: a socket set for organelle gene expression. Trends Plant Sci 13:663–670
179. Schwarz C, Bohne AV, Wang F, Cejudo FJ, Nickelsen J (2012) An intermolecular disulfide-based light switch for chloroplast *psbD* gene expression in *Chlamydomonas reinhardtii*. Plant J 72:378–389
180. Scotti N, Rigano MM, Cardi T (2012) Production of foreign proteins using plastid transformation. Biotechnol Adv 30:387–397
181. Sears BB, Boynton JE, Gillham NW (1980) The effect of gametogenesis regimes on the chloroplast genetic system of *Chlamydomonas reinhardtii*. Genetics 96:95–114
182. Shaver JM, Oldenburg DJ, Bendich AJ (2006) Changes in chloroplast DNA during development in tobacco, *Medicago truncatula*, pea, and maize. Planta 224:72–82
183. Shaver JM, Oldenburg DJ, Bendich AJ (2008) The structure of chloroplast DNA molecules and the effects of light on the amount of chloroplast DNA during development in *Medicago truncatula*. Plant Physiol 146:1064–1074
184. Shen P, Huang HV (1986) Homologous recombination in *Escherichia coli*: dependence on substrate length and homology. Genetics 112:441–457
185. Shimizu M, Goto M, Hanai M, Shimizu T, Izawa N, Kanamoto H, Tomizawa K, Yokota A, Kobayashi H (2008) Selectable tolerance to herbicides by mutated acetolactate synthase genes integrated into the chloroplast genome of tobacco. Plant Physiol 147:1976–1983
186. Shinozaki K et al (1986) The complete nucleotide sequence of the tobacco chloroplast genome: its gene organization and expression. EMBO J 5:2043–2049
187. Sidorov VA, Kasten D, Pang SZ, Hajdukiewicz PT, Staub JM, Nehra NS (1999) Technical advance: stable chloroplast transformation in potato: use of green fluorescent protein as a plastid marker. Plant J 19:209–216
188. Sikdar SR, Serino G, Chaudhuri S, Maliga P (1998) Plastid transformation in *Arabidopsis thaliana*. Plant Cell Rep 18:20–24
189. Singh AK, Verma SS, Bansal KC (2010) Plastid transformation in eggplant (*Solanum melongena* L.). Transgenic Res 19:113–119
190. Skarjinskaia M, Svab Z, Maliga P (2003) Plastid transformation in *Lesquerella fendleri*, an oilseed *Brassicacea*. Transgenic Res 12:115–122
191. Smith AC, Purton S (2002) The transcriptional apparatus of algal plastids. Eur J Phycol 37:301–311
192. Smith SE, Bingham ET, Fulton RW (1986) Transmission of chlorophyll deficiencies in *Medicago sativa*—evidence for biparental inheritance of plastids. J Hered 77:35–38
193. Specht E, Miyake-Stoner S, Mayfield S (2010) Micro-algae come of age as a platform for recombinant protein production. Biotechnol Lett 32:1373–1183
194. Staub JM et al (2000) High-yield production of a human therapeutic protein in tobacco chloroplasts. Nat Biotechnol 18:333–338
195. Stern DB, Goldschmidt-Clermont M, Hanson MR (2010) Chloroplast RNA metabolism. Annu Rev Plant Biol 61:125–155
196. Su ZL, Qian KX, Tan CP, Meng CX, Qin S (2005) Recombination and heterologous expression of allophycocyanin gene in the chloroplast of *Chlamydomonas reinhardtii*. Acta Biochim Biophys Sinica 37:709–712

197. Suay L, Salvador ML, Abesha E, Klein U (2005) Specific roles of 5′ RNA secondary structures in stabilizing transcripts in chloroplasts. Nucleic Acids Res 33:4754–4761
198. Surzycki R, Cournac L, Peltier G, Rochaix JD (2007) Potential for hydrogen production with inducible chloroplast gene expression in *Chlamydomonas*. Proc Natl Acad Sci U S A 104:17548–17553
199. Surzycki R, Greenham K, Kitayama K, Dibal F, Wagner R, Rochaix JD, Ajam T, Surzycki S (2009) Factors effecting expression of vaccines in microalgae. Biologicals 37:133–138
200. Svab Z, Maliga P (1993) High-frequency plastid transformation in tobacco by selection for a chimeric *aadA* gene. Proc Natl Acad Sci U S A 90:913–917
201. Svab Z, Maliga P (2007) Exceptional transmission of plastids and mitochondria from the transplastomic pollen parent and its impact on transgene containment. Proc Natl Acad Sci U S A 104:7003–7008
202. Svab Z, Hajdukiewicz P, Maliga P (1990) Stable transformation of plastids in higher plants. Proc Natl Acad Sci U S A 87:8526–8530
203. Sytnik E, Komarnytsky I, Gleba Y, Kuchuk N (2005) Transfer of transformed chloroplasts from *Nicotiana tabacum* to the *Lycium barbarum* plants. Cell Biol Int 29:71–75
204. Tangphatsornruang S, Birch-Machin I, Newell CA, Gray JC (2011) The effect of different 3′ untranslated regions on the accumulation and stability of transcripts of a *gfp* transgene in chloroplasts of transplastomic tobacco. Plant Mol Biol 76:385–396
205. Tiller K, Link G (1993) Phosphorylation and dephosphorylation affect functional characteristics of chloroplast and etioplast transcription systems from mustard (*Sinapis alba* L.). EMBO J 12:1745–1753
206. Tissot G, Canard H, Nadai M, Martone A, Botterman J, Dubald M (2008) Translocation of aprotinin, a therapeutic protease inhibitor, into the thylakoid lumen of genetically engineered tobacco chloroplasts. Plant Biotechnol J 6:309–320
207. Tran M, Zhou B, Pettersson PL, Gonzalez MJ, Mayfield SP (2009) Synthesis and assembly of a full-length human monoclonal antibody in algal chloroplasts. Biotechnol Bioeng 104:663–673
208. Valkov VT, Gargano D, Manna C, Formisano G, Dix PJ, Gray JC, Scotti N, Cardi T (2011) High efficiency plastid transformation in potato and regulation of transgene expression in leaves and tubers by alternative 5′ and 3′ regulatory sequences. Transgenic Res 20:137–151
209. van Wijk KJ (2004) Plastid proteomics. Plant Physiol Biochem 42:963–977
210. Vener AV, Ohad I, Andersson B (1998) Protein phosphorylation and redox sensing in chloroplast thylakoids. Curr Opin Plant Biol 1:217–223
211. Verhounig A, Karcher D, Bock R (2010) Inducible gene expression from the plastid genome by a synthetic riboswitch. Proc Natl Acad Sci U S A 107:6204–6209
212. Verma D, Daniell H (2007) Chloroplast vector systems for biotechnology applications. Plant Physiol 145:1129–1143
213. Verma D, Kanagaraj A, Jin S, Singh ND, Kolattukudy PE, Daniell H (2010) Chloroplast-derived enzyme cocktails hydrolyse lignocellulosic biomass and release fermentable sugars. Plant Biotechnol J 8:332–350
214. Wei Z, Liu Y, Lin C, Wang Y, Cai Q, Dong Y, Xing S (2011) Transformation of alfalfa chloroplasts and expression of green fluorescent protein in a forage crop. Biotechnol Lett 33:2487–2494
215. Whitney SM, Houtz RL, Alonso H (2011) Advancing our understanding and capacity to engineer nature's CO_2-sequestering enzyme, Rubisco. Plant Physiol 155:27–35
216. Wostrikoff K, Girard-Bascou J, Wollman FA, Choquet Y (2004) Biogenesis of PSI involves a cascade of translational autoregulation in the chloroplast of *Chlamydomonas*. EMBO J 23:2696–2705
217. Wu S, Xu L, Huang R, Wang Q (2011) Improved biohydrogen production with an expression of codon-optimized hemH and lba genes in the chloroplast of *Chlamydomonas reinhardtii*. Bioresour Technol 102:2610–2616

218. Ye GN, Hajdukiewicz PT, Broyles D, Rodriguez D, Xu CW, Nehra N, Staub JM (2001) Plastid-expressed 5-enolpyruvylshikimate-3-phosphate synthase genes provide high level glyphosate tolerance in tobacco. Plant J 25:261–270
219. Ye GN, Colburn SM, Xu CW, Hajdukiewicz PT, Staub JM (2003) Persistence of unselected transgenic DNA during a plastid transformation and segregation approach to herbicide resistance. Plant Physiol 133:402–410
220. Yoon SM, Kim SY, Li KF, Yoon BH, Choe S, Kuo MM (2011) Transgenic microalgae expressing *Escherichia coli* AppA phytase as feed additive to reduce phytate excretion in the manure of young broiler chicks. Appl Microbiol Biotechnol 91:553–563
221. Zhang J, Tan W, Yang XH, Zhang HX (2008) Plastid-expressed choline monooxygenase gene improves salt and drought tolerance through accumulation of glycine betaine in tobacco. Plant Cell Rep 27:1113–1124
222. Zheng Q, Oldenburg DJ, Bendich AJ (2011) Independent effects of leaf growth and light on the development of the plastid and its DNA content in *Zea* species. J Exp Bot 62:2715–2730
223. Zhou F, Karcher D, Bock R (2007) Identification of a plastid intercistronic expression element (IEE) facilitating the expression of stable translatable monocistronic mRNAs from operons. Plant J 52:961–972
224. Zhou F et al (2008) High-level expression of human immunodeficiency virus antigens from the tobacco and tomato plastid genomes. Plant Biotechnol J 6:897–913
225. Zoschke R, Liere K, Borner T (2007) From seedling to mature plant: arabidopsis plastidial genome copy number, RNA accumulation and transcription are differentially regulated during leaf development. Plant J 50:710–722
226. Zubkot MK, Zubkot EI, van Zuilen K, Meyer P, Day A (2004) Stable transformation of petunia plastids. Transgenic Res 13:523–530
227. Mayfield SP, Franklin SE, Lerner RA (2003) Expression and assembly of a fully active antibody in algae. Proc Natl Acad Sci U S A 100:438–442

Chapter 19
Hydrogen and Biofuel Production in the Chloroplast

Yonghua Li-Beisson, Gilles Peltier, Philipp Knörzer, Thomas Happe and Anja Hemschemeier

Abstract Plastids are the compartments in which oxygenic photosynthesis of higher plants and eukaryotic microalgae converts the energy of sunlight into chemical energy. The latter is used by green cells to generate the whole palette of organic molecules needed to build a cell. Mankind has made use of the green powerhouses from the beginning of its existence on, but in the last decades, products other than food or fire wood have gained importance. Facing the deprivation of fossil fuels and climate changes due to anthropologically caused greenhouse effects, we want to use photosynthetically converted light energy as an energy source to generate renewable energy carriers. Above all, the synthesis of biodiesel made from plant or algal lipids is a promising strategy. As the cultivation of microalgae does not compete with food production, research focuses on understanding and engineering lipid biosynthesis in these unicellular organisms. Additionally, in contrast to higher plants, some green algae are capable of using the process of photosynthesis for the generation of another biofuel: molecular hydrogen. In this chapter, the pathways resulting in the generation of hydrogen and lipids in the plastid are reviewed. Additionally, proven and anticipated targets of biotechnological optimization are highlighted.

Keywords Biofuels · Chlamydomonas · Hydrogen · Hypoxia · Lipids · Microalgae · Nutrient starvation · Sustainable energy

Abbreviations

ACCase	Acetyl-CoA carboxylase
ACP	Acyl carrier protein
CoA	Coenzyme A
DAG	Diacylglycerol

T. Happe (✉) · P. Knörzer · A. Hemschemeier
Fakultät für Biologie und Biotechnologie, AG Photobiotechnologie, Ruhr-Universität Bochum, Universitätsstr. 150, 44801 Bochum, Germany
e-mail: thomas.happe@rub.de

Y. Li-Beisson · G. Peltier
Institut de Biologie Environnementale et Biotechnologie, CEA/CNRS/Aix Marseille Université, CEA Cadarache, 13108 Saint-Paul-lez-Durance, France

S.M. Theg, F.-A. Wollman (eds.), *Plastid Biology*, Advances in Plant Biology 5,
DOI 10.1007/978-1-4939-1136-3_19,

DGAT	Diacylglycerol acyltransferase
DGDG	Digalactosyldiacylglycerol
DGTS	Diacylglycerol-N,N,N-trimethylhomoserine
ER	Endoplasmic reticulum
FAS	Fatty acid synthase
FAT	Fatty acyl-ACP thioesterase
G3P	Glycerol-3-phosphate
GPAT	Glycerol-3-phosphate acyltransferase
LACS	Long chain acyl-CoA synthetase
LPA	Lysophosphatidic acid
LPAT	Lysophosphatidic acid acyltransferase
MGDG	Monogalactosyldiacylglycerol
MLDP	Major lipid droplet protein
PA	Phosphatidic acid
PAP	Phosphatidic acid phosphatase
PDAT	Phospholipid:diacylglycerol acyltransferase
PtdEtn	Phosphatidylethanolamine
PtdGro	Phosphatidylglycerol
PtdIns	Phosphatidylinositol
SQDG	Sulfoquinovosyldiacylglycerol
TAG	Triacylglycerol

19.1 Introduction

The inevitable decrease in fossil fuel reserves and the rising concerns about global warming urge us to search for alternative energy sources and new energy vectors for transportation. Photosynthesis, as the major entry port of energy into living matter, can be considered as a key technological module allowing storage of solar energy into carbohydrates and lipids. Optimized through millions of years of evolution, requiring only light, water and CO_2 as an input, photosynthesis has early been integrated into the survival strategy of our economies. The products of photosynthesis (biomass) were first used by mankind to cover its alimentation needs. Nowadays, crops have been domesticated through breeding and selection to optimize productivity. Biomass was subsequently used to cover energy needs through heating, and more recently it has been considered as a source of energy for transportation (biofuel). Facing the rising food demand of a growing population and the competition in land uses, considerable interest lies in producing biofuels from other feedstocks than crops. Microalgae have recently emerged as a promising alternative for the production of biofuels (also called algofuels). Unicellular microalgae have developed unique metabolic properties, such as the ability to produce molecular hydrogen (H_2) or to accumulate high amounts of storage lipids. They have a very high surface productivity and can be cultivated on non-arable land (therefore not competing with food production by agriculture). Additionally, algae can recycle industrial or urban wastes by using those as nutrients (CO_2, nitrogen or phosphate) for growth and biofuel production. However, these species have not been subjected to domestication through genetic improvement yet. H_2 production or lipid accumulation do not occur in conditions of optimal growth and generally correspond to an acclimatory response to specific environmental constraints, for instance stop in cell division resulting from nutrient deprivation. A major challenge for research will be to optimally utilize photosynthesis of these unicellular organisms to cover part of our future energy needs without competing with food production. In this chapter, we review the metabolic reactions involved in chloroplast production of H_2 and lipids that can be used as a valuable source of biofuels, and give an overview of experimental strategies currently developed to improve production of these energy rich compounds.

19.1.1 Bioenergetics and Efficiencies of Photosynthesis

A typical determinant of (bio-) technological applications is efficiency. In light-driven approaches, the efficiency value is usually defined as the ratio of light-energy input and the energy content of the output in terms of designated product. Individual steps of photosynthetic electron chemistry can display efficiencies higher than 95 %, such as the light harvesting efficiency resulting in the excited state of the photosystem 2 (PS2) reaction center, P680*. In contrast, the solar energy conversion efficiency resulting in dry biomass or molecular hydrogen was estimated to be 5 % in the best case, while reaching about 1 to 3 % in real instances [29, 51]. This contrasts with today's photovoltaic cells that can reach 18 %, and sun-light driven photolysis of water using current provided by present silicon photovoltaic cells was estimated to reach 11 % efficiency [11]. However, complete life-cycle assessments are necessary to really judge the "efficiency" of a system, and this assessment cannot solely be based on energy/cost input *versus* energy/profit output. Rather, the whole infrastructure of the system—risks or toxic by-products arising during manufacturing, competition to the food market (in case of biomass production), interactions with the environment and the climate—have to be considered [11]. Regarding the latter two points, plants and microalgae—or chloroplasts as the main players of this book—are probably superior. Moreover, when talking about biofuel production in the chloroplast, one usually thinks about the generation of lipids and H_2 as described in this chapter or maybe biomass in general. However, we should not neglect the fact that chloroplasts deliver the most important product in terms of "fuelling biology" independent from the final product of photosynthetic electron transport: molecular oxygen (O_2). Though still controversial in some points, most scientists agree that the oxygenation of the earth's atmosphere was mainly due to the invention of oxygenic photosynthesis [39]. The availability of the O_2 molecule allowed the evolution of today's intricate multicellular life forms due to its favorable thermodynamics and thus high energy yields via oxidative phosphorylation [65, 149]. Additionally, O_2

has been implemented in biochemical networks as one of the most frequently used substrates or reagents, resulting in more than 1000 metabolic pathways depending directly or indirectly on O_2 in contemporary aerobic organisms [121, 122]. Not the least, by-products of O_2 metabolism, reactive oxygen species (ROS), have been adopted to essential signaling pathways and pathogen defense reactions [60, 149]. To date, the constant removal of CO_2 from and the delivery of O_2 into the atmosphere are vital to contemporary ecosystems and survival of nearly all higher organisms. Thus, the cultivation of photosynthetic organisms for bioenergy purposes is integrated into the natural ecosystems.

19.2 Light-Dependent Hydrogen Production

19.2.1 Photosynthetic H_2-Production by Chlamydomonas—*The Principles*

Chlamydomonas reinhardtii [61, 62, 124] and several other microalgal species [42, 141, 150, 161] have in their chloroplast [FeFe]-hydrogenases that serve as additional electron sinks for photosynthetically provided high energy electrons [43]. Hydrogenases generate molecular hydrogen (H_2) from protons and low potential electrons and are wide-spread in prokaryotes [154, 155]. Three types of hydrogenases have been described, [FeFe]-, [NiFe]- and [Fe]-only hydrogenases [154]. As the latter are only found in methanogenic bacteria and generate H_2 from methylenetetrahydromethanopterin, they form an individual group [138]. [FeFe]- and [NiFe]-hydrogenases have in common that they have an unique active site in which [Fe]-atoms are coordinated by CO- and CN-ligands. In [NiFe]-hydrogenases, the active site cluster consists of a binuclear complex with one Ni- and one Fe-atom, while in [FeFe]-hydrogenases, the prosthetic group, the so-called H-cluster, is built from a classical [4Fe4S]-cluster bridged to the hydrogenase-specific di-iron cluster with its unique ligands [43]. [NiFe]- and [FeFe]-hydrogenases also show differences in their catalytic activity and their sensitivity towards O_2. [NiFe]-hydrogenases have a lower specific activity [43] and are metabolically involved in both H_2 generation and oxidation [154]. They are usually not irreversibly inhibited by O_2 [157]. The [NiFe]-hydrogenases of knallgas bacteria are even O_2-stable, and thus provide the metabolic backbone of the energy metabolism of these prokaryotes in the presence of both H_2 and O_2 [19]. In contrast, [FeFe]-hydrogenases have extraordinarily high specific activities, are mostly employed in electron disposal and H_2-production, respectively, and are very sensitive towards O_2 [54, 156]. Though the degree of O_2 intolerance varies, all [FeFe]-hydrogenases analyzed so far are irreversibly inactivated at low O_2 concentrations within minutes [54, 143].

Knowing these facts about [FeFe]-hydrogenases, the discovery of the enzymes in organisms carrying out oxygenic photosynthesis and, indeed, their electronic coupling to photosynthetic electron transport, was a surprise, and in earlier times was regarded as one of nature's curiosities or as an evolutionary left-over. However, much progress has been made in the last 15 years of algal H_2 research. Based on these findings, the common agreement today is that the "photosynthetic hydrogenases" of microalgae fulfill important physiological functions in these organisms, which dwell in diverse habitats and often find themselves in hypoxic conditions despite the presence of light (and "self-made" O_2). While photosynthetic H_2 evolution is a fascinating metabolism itself, it also holds the promise of being biotechnologically applicable. As noted in the introduction, photosynthesis is regarded as one solution to mankind's energy crisis and the atmospheric pollution created by burning of fossil fuels. In microalgae, this machinery could be used to go a step beyond biomass production and generate a highly efficient and "clean" energy carrier (H_2) directly.

H_2 uptake in green algae was first observed in *Scenedesmus obliquus* cells [46]. Later, light-dependent H_2 evolution was also observed in this species and in *Chlamydomonas* [47, 144]. While low H_2 evolution takes place in dark-anaerobic cell suspensions, relatively high H_2 production rates are only observed in the light [47]. Inhibitor studies revealed that PS2 contributes to electron supply, but is dispensable, while photosystem 1 (PS1) is essential for H_2 generation in illuminated algae [49, 144]. The [FeFe]-hydrogenase of *C. reinhardtii* is located in the chloroplast and accepts electrons from photosynthetic ferredoxin PETF [62, 63, 124]. Two hydrogenase-encoding genes, *HYDA1* and *HYDA2*, were isolated and characterized [44, 61].

Common to all observations and studies was that an anaerobic incubation of the cell suspensions was a prerequisite to H_2 conversion and the presence of (active) hydrogenases in the cells [61, 62]. The above-mentioned O_2-sensitivity of the [FeFe]-hydrogenases is one of the reasons for the dependence on anaerobiosis. However, the expression of hydrogenase encoding genes is also induced only in low O_2 concentrations [44, 61, 116]. Furthermore, it was noticed that H_2 is not the only product of the anaerobic or hypoxic metabolism of *C. reinhardtii*. Instead, a mixture of fermentative products, mainly formate, ethanol and acetate, is excreted by the cells, showing that H_2 generation is intimately connected to a complex fermentation metabolism [48, 68, 88, 112, 117]. Notably, the key enzyme of pyruvate fermentation in *C. reinhardtii* is pyruvate:formate lyase (PFL1) [5, 68, 117], an enzyme commonly found in bacteria [129], but rather rarely in eukaryotes [6, 59]. A further "bacterial" fermentation enzyme is pyruvate:ferredoxin oxidoreductase (PFR1) [117, 147]. This enzyme oxidatively decarboxylates pyruvate, yielding acetylCoA, CO_2 and reduced ferredoxin. The latter is proposed to serve as the reductant for the low rates of dark H_2 production in *C. reinhardtii* [58, 117]. Though both enzymes, PFL1 and PFR1, are probably key to dark anaerobic pathways in the alga, they are worth mentioning in a chapter about chloroplasts, as both were identified in chloroplast proteomes [147]. Together with the hydrogenases, these enzymes might provide the maintenance of redox and energy balance in the chloroplast compartment in the dark.

From a biotechnological point of view, early observations of H_2 generation and mixed acid fermentation in *C. reinhardtii* or other microalgae were not of interest. Dark H_2-generation by these organisms is very low, and photosynthetic H_2 production, despite of high initial rates, is rapidly inactivated [50]. The latter is due both to the O_2 sensitivity of [FeFe]-hydrogenases and the metabolic take-over of assimilatory electron sinks like the Calvin-cycle [25, 126]. A breakthrough regarding both novel physiological insights and biotechnological applicability was the discovery that illuminated *C. reinhardtii* cultures evolve relatively large amounts of H_2 over several days when deprived of the macro-nutrient sulfur (S) [101]. Following this report and its implications for using microalgae as H_2 generating factories, many studies were conducted on this complex metabolism, allowing us to draw a quite comprehensive picture on what is going on in the cells. In fact, these insights have already contributed substantially to finding strategies which enhance the efficiency and total yield of H_2 generation.

As anaerobiosis is indispensable to hydrogenase activity, one metabolic response of *C. reinhardtii* to S deficiency which is absolutely a prerequisite to H_2 metabolism is the diminution of PS2 activity. Within one or two days, O_2 evolution rates of S-deprived algal cultures drop below respiratory O_2-consumption rates, resulting in a net uptake of O_2 in sealed flasks [30, 101, 165]. Several factors contribute to the decrease of PS2-activity. One of these is simply a passive reaction to the limitation in S containing amino acids [22]. As one of the core-subunits of PS2, the D1 protein (encoded by the *psbA* gene), has a high turn-over due to the radicals naturally generated during PS2 photochemistry, its frequent replenishment is impaired by the lack of cysteine [170]. A second factor which influences the rate of D1 degradation is light-stress. Due to a cessation in cell division induced by nutrient deprivation [30, 101, 170] and thus to the diminution of assimilatory electron sinks [67, 170], photosynthetic electron carriers become over-reduced, resulting in a backing-up of electrons at the PS2 acceptor side. In the beginning of S deprivation, the decrease of PS2 quantum yield can be reversed by supplying O_2 [2]. However, as over-excitation of PS2 proceeds, PS2 complexes become irreversibly damaged and, subsequently, degraded [2, 101, 165].

In response to the reduced state of the photosynthetic electron transport chain and the plastoquinone (PQ) pool, respectively, the photosynthetic apparatus of S-deprived *Chlamydomonas* cells switches to state-2 conditions [165]. State-transitions, during which light-harvesting complexes (LHC) of PS2 migrate between association mainly with PS2 (state 1) or PS1 (state 2), can be very pronounced in *C. reinhardtii* [33]. A complete uncoupling of PS2 from the photosynthetic electron transport chain has been observed in full state 2 conditions [41]. Upon S deficiency, establishment of state 2 has been shown, further diminishing actual PS2 activity [165] (Fig. 19.1).

In addition to these passive reactions, the decrease of PS2 activity seems to be a regulated process rather specific to the absence of sulfur. Indeed, a mutant deficient for a major regulator of the response of *C. reinhardtii* to S deficiency, SAC1 [30], does not exhibit the drop of PS2 activity observable in wild type cells in the first 24 to 48 h [30, 165]. As *sac1* mutants rapidly die upon S deprivation and can be rescued both by adding the PS2 inhibitor DCMU or by darkness [30], regulated PS2 inactivation seems to be an essential response to S deprivation. The specificity to

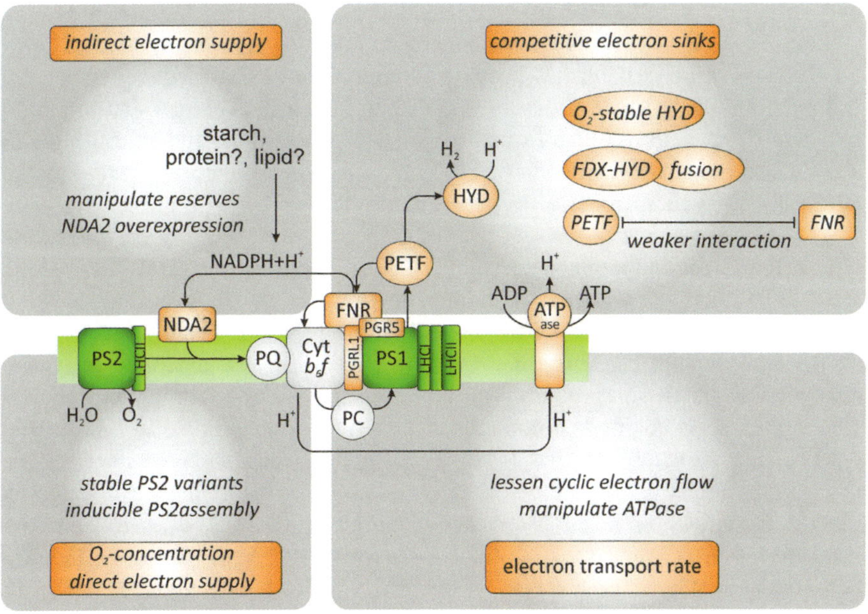

Fig. 19.1. Schematic of photosynthetic electron transport during H_2-production by S-deficient *C. reinhardtii* cells and main points of (possible) optimization. Upon S deprivation, PS2 activity decreases, but still contributes to electron supply to the [FeFe]-hydrogenase (*HYD*) via plastoquinone (*PQ*), the cytochrome b_6/f complex (Cytb_6f), plastocyanin (*PC*), PS1 and ferredoxin PETF. Light harvesting complexes (*LHC*) of PS2 (LHCII) are mainly associated with PS1. PS1 also donates electrons to $NADP^+$ via ferredoxin and ferredoxin-$NADP^+$ reductase (*FNR*). During cyclic electron flow, these electrons are transported back to the cytochrome b_6f complex. PGRL1 and PGR5 are involved in cyclic electron flow. A proton-gradient is built up during linear and cyclic electron transport and used by an ATP-synthase (ATPase) to generate ATP. A further source of reductant for the PQ-pool is starch. Electrons are transferred to PQ by NDA2. Several steps of this process might be or have already been optimized to enhance H_2 yields and are described in detail in the text

the nutrient sulfur is further supported by the fact that both phosphorous (P) [165] or nitrogen (N) [118] deprivation do not result in similar photosynthetic responses.

In order to reach high H_2 evolution rates, additional processes such as the provision of low-potential electrons to the hydrogenase are required. In response of S deprivation, growth stops, *C. reinhardtii* cells being unable to maintain the delivery of building blocks for the generation of new cells. CO_2 assimilation rates and the amount of ribulose bisphosphate carboxylase/oxygenase (Rubisco) decrease within one or two days of S deprivation [67, 170]. This results in the loss of the major electron sink of photosynthetic electron transport. The importance of the absence of competing electron sinks is emphasized in *Chlamydomonas* mutant strains that become hypoxic in replete medium due to higher respiratory *versus* photosynthetic rates [126]. Despite a high *in vitro* hydrogenase activity detectable in the cells (which is measured in cell lysates and in the presence of artificial electron donors), only an inhibition of the Calvin cycle results in significant light-dependent H_2 generation [126]. Similar observations have been made analyzing a mutant deficient for Rubisco, which establishes hypoxic conditions in the light when incubated in full medium and evolves H_2 under these conditions [67]. The fact that this light-sensitive *C. reinhardtii* mutant was fitter in sealed flasks, thereby establishing anaerobic conditions and hydrogenase activity, respectively, supported the theory that HYDA1 acts as an alternative electron sink for the algal cells [67] (Fig. 19.1).

Two pathways have been described for delivering electrons to the hydrogenase, termed the direct and the indirect pathways [24]. The direct pathway is fuelled by residual PS2 activity, which in the time periods studied so far never drops down to zero, but is kept at a low level for days [2, 101, 118, 170]. The indirect pathway, on the other hand, makes use of electrons derived from the oxidation of organic reserves like starch and proteins, which are then transferred to the photosynthetic electron transport chain via non-photochemical PQ reduction [24, 101, 118]. Both reserves accumulate in the first 24 to 48 h of S deficiency, when PS2 activity and CO_2 assimilation pathways are still active [45, 101, 170]. It has been suggested that the accumulation of starch serves to divert electrons away and to allow photosynthetic electron transport to continue [57]. Notably, triacylglycerols, which also accumulate to substantial levels in N- but also S-deficient algae [14, 36, 104, 151] (see below) do not seem to be degraded and used for H_2 generation [89, 151].

19.2.2 Molecular Aspects of H_2-Production—Keys to Optimization

In the last years, various molecular players of photosynthetic H_2 production by *C. reinhardtii* have been identified and their manipulation shown to increase H_2 yields (Fig. 19.1). Furthermore, the knowledge about individual steps gave rise to attempts achieving H_2 production by *Chlamydomonas* incubated in S-replete medium, as S deprivation is a severe stress condition for the cells. In principal, the main points supporting H_2 production by S-deficient *C. reinhardtii* cells, namely reduced O_2-evolution activity, the presence of large reserves for non-photochemical electron supply and the loss of competing electron sinks, have to be provided.

19.2.2.1 Manipulation of PS2 Activity

As mentioned above, despite its significantly reduced activity, PS2 contributes substantially to H_2 production in S-deprived green algae. This became especially clear when a study showed that starch-less *C. reinhardtii* cells produce similar amounts of H_2 upon S-deprivation, as long as PS2 is not inactivated by DCMU [24]. It was also noted that in algal strains possessing a D1 protein L159I-N230Y variant, PS2 was more stable and delivered more electrons for H_2-generation, indicating that the natural instability of PS2 might be engineered to yield algal strains with a higher H_2 evolution capacity [133].

One of the first attempts to provide hypoxic conditions in nutrient-replete algal cells was the inhibition of PS2 by DCMU or the use of PS2 mutants of *C. reinhardtii*. However, neither of these strategies resulted in significant or prolonged H_2 evolution [45, 67], and the reason is mainly that PS2 activity is required for the build-up of starch and other organic reserves. S-deprived algal cells treated with DCMU after starch has accumulated generate H_2, but less than cells in which PS2 is not inhibited [45, 67]. Still, as one can imagine a "design organism" in which all photosynthetic components involved in H_2 evolution are controllable, switching PS2 activity on and off is a desirable feature. Here, progress has been made by developing an inducible chloroplast expression system based on a factor, NAC2, essential for *psbD* mRNA stability [12]. By equipping the *NAC2* gene with a promoter inducible by copper deficiency, it was possible to control PS2 activity by the addition or removal of copper [146]. This was a first proof that a switch-on and -off of PS2 can be established genetically. Recently, a targeted triggering of PS2 or PS1 activity by wavelengths specific for PS2 and PS1 reaction centers has been reported, which would simplify the process and allow the use of wild type strains [71].

19.2.2.2 Non-photochemical PQ Reduction

Non-photochemical electron sources are important for H_2-generation by anaerobic or S-deficient *Chlamydomonas* cultures [45, 49, 67, 111]. The enzyme responsible for transferring electrons from NAD(P)H to the PQ pool in *C. reinhardtii* is the NDA2 protein [35, 78]. In contrast to the higher-plant NDH-1 complex composed of more than 15 subunits [145], NDA2 is monomeric. It is localized in the thylakoid membranes and contributes to PS2-independent H_2-generation in *Chlamydomonas* [78]. NDA2-deficient cells do not produce any H_2 at all upon PS2 inhibition [106], showing once more that both the direct and indirect pathway of electron supply have to work coordinately to result in sustained and optimal H_2-photoproduction. Therefore, homologous over-expression of *NDA2* is a target for future optimization of H_2 photo-production (Fig. 19.1).

19.2.2.3 Suppression of Competing Pathways

As the [FeFe]-hydrogenase accepts electrons from photosynthetic ferredoxin PETF [63, 162] and ferredoxin donates electrons to various reductive processes in the chloroplast [86, 148, 163], a competition between HYDA1 and other electron sinks seems obvious. In nutrient-depleted medium, the stop in assimilatory processes makes the hydrogenase the most important electron sink of photosynthetic electron flow. In replete medium, however, assimilatory electron sinks would compete with the hydrogenase as noted above [25, 67, 126]. Though *Chlamydomonas* has more than six ferredoxin-encoding genes [103, and recent genome annotation on phytozome.net, Chlamydomonas reinhardtii v4.3] and some of those, especially *FDX5*, are induced upon anaerobiosis [77, 90, 148], "good old" photosynthetic ferredoxin PETF is apparently the optimal electron donor for HYDA1 [162, 163].

One possibility to circumvent competition for PETF might be the use of hydrogenase-ferredoxin fusion proteins [166]. Using isolated PS1 and thylakoids, it was shown that FNR competes strongly for electrons delivered by ferredoxin. In contrast, a hydrogenase-ferredoxin fusion protein was able to scavenge more than 60 % of the electrons provided by PS1 [166]. Another approach might be engineering ferredoxin PETF, HYDA1 and FNR in a way that the affinity between FNR and PETF is weakened and the interaction of HYDA1 and PETF is optimized (Fig. 19.1). The interacting residues of HYDA1 and PETF are known [162] so that a starting point for targeted manipulation is provided.

A further constraint to photosynthetic H_2 generation is cyclic electron flow, which is activated in *C. reinhardtii* cells upon transition to state 2 [41]. This electron flow around PS1 builds up a proton gradient impairing electron flow. It was shown recently that cyclic electron flow is a major obstacle to H_2 generation, as a mutant deficient for one protein involved in this process, PROTON GRADIENT REGULATION LIKE 1 (PGRL1) has much higher H_2 evolution capacities both in anaerobiosis and S deprivation [152].

19.3 Semi-Artificial Applications for Photosynthetic H_2 Production

Despite recent progress in optimizing H_2 production by microalgae, the efficiency of the natural process is still relatively low. In contrast, the isolated enzymes, photosystems and [FeFe]-hydrogenases, might outperform artificial or electric catalysts. [FeFe]-hydrogenases are capable of generating 9000 molecules of H_2 per second when supplied with an efficient electron donor, and photosystems are the most efficient light-conversion machines known to date [29, 102]. Therefore, efforts have been made to create and analyze semi-artificial systems in which the natural catalysts, photosystems and hydrogenases, are combined with artificial electron donors, linkers or electron acceptors. Though the generation of large amounts of pure, stable and highly active proteins is also a challenge, their potential is significant.

The first semi-artificial system for photosynthetic H_2 production was already reported in 1961, when H_2 production in a mixture of spinach thylakoids and a [NiFe]-hydrogenase from *Chromatium vinosum* could be detected [4]. However, due to the O_2 sensitivity of the hydrogenase, H_2 formation could not be sustained by photosynthetic water oxidation, but relied on the sacrificial electron donor thiosulfate. Later, water could be established as electron source for H_2 production in an improved system, using a hydrogenase from *Clostridium kluyveri* [9]. However, O_2 generated at PS2 was always an obstacle to H_2 production in these systems. Today, experimental and theoretical investigations of how exactly O_2 attacks the active site H-cluster of [FeFe]-hydrogenases led to new insights which might result in the creation of O_2-stable [FeFe]-hydrogenases [17, 54, 70, 91, 142, 143].

One focus of research is the optimization of the electron transfer from PS1 towards hydrogenases. An important issue for successfully combining the two enzymes is the efficiency of the electron transfer between PS1 as the photoactive element and the hydrogenase as the proton reducing element. In the beginning, natural electron mediators such as ferredoxin or artificial electron carriers like methylviologen were applied [87]. Because of relatively slow diffusion of soluble mediators [164], however, researchers sought to establish a direct electron transfer. The first study used spinach PS1 and *Clostridium pasteurianum* [FeFe]-hydrogenase [100]. A direct electron transfer from PS1 to hydrogenase via the FeS-clusters of PS1 subunit PsaC was proposed, albeit in the presence of high protein concentration. Nevertheless, the idea of direct electron transfer between PSI and hydrogenase advanced and resulted in studies during which both units were physically linked. One strategy was a protein-based linker achieved by genetic techniques. A fusion protein was generated made of the PsaE subunit from *Thermosynechococcus elongatus* PS1 and the O_2-tolerant [NiFe]-hydrogenase MBH from *Ralstonia eutropha* [76, 132]. This protein was subsequently mixed with modified PS1 lacking PsaE, and a light-driven H_2 production rate of 0.58 μmol H_2 mg chlorophyll^{-1} (Chl) h^{-1} could be determined *in vitro*. However, in the presence of FNR and ferredoxin, simulating natural conditions in the cell, H_2 production was abolished. To improve electron transport between PS1 and hydrogenase, cytochrome c_3 was included in the system. Cytochrome c_3 serves as electron donor of hydrogenases from bacteria such as *Desulfovibrio* species [99]. Cross-linked to the PsaE subunit of PS1, cytochrome c_3 served as a mediator to the [NiFe]-hydrogenase and allowed a H_2 production rate of 0.3 μmol H_2 mg Chl^{-1} h^{-1} even in the presence of FNR and ferredoxin [75].

A very successful approach was the introduction of a chemical "electron wire" linking PS1 and a [FeFe]-hydrogenase. While this approach cannot serve as a blueprint for optimizing H_2 evolution in living cells, it demonstrated the high potential of semi-artificial system [96, 97]. Briefly, a dithiol linker was introduced by producing a variant of the PS1 subunit PsaC in which one cysteine residue coordinating the [4Fe4S]-cluster F_B [80] was exchanged by glycine. The [4Fe-4S]-cluster was then reconstituted *in vitro* and chemically rescued by thiol ligands [3]. A similar procedure generated a [FeFe]-hydrogenase in which a surface-exposed [4Fe4S]-cluster was reconstituted by a thiol. The modifications in PsaC and hydrogenase subsequently allowed a direct "wiring" of PS1 and hydrogenase via a dithiol linker [96]. The primary rate of light driven hydrogen production measured for this system was 30.3 μmol H_2 mg Chl^{-1} h^{-1} [96].

While the electron transfer between PS1 and hydrogenase could be improved substantially using this system, the electron supply to PS1 was still a bottleneck. One strategy to improve this transfer was the addition of natural electron donors (plastocyanin and cytochrome c_6) in the presence of the artificial electron donors [56]. Cross-linking cytochrome c_6 to PS1 turned out to be even more efficient, resulting in a light-driven H_2 production of 2832 μmol H_2 mg Chl^{-1} h^{-1}, using the dithiol-linked PS1 and [FeFe]-hydrogenase described above [97].

Despite the high rates of H_2 evolution that can be achieved using semi-artificial systems, their long-term stability remains a challenge. In this respect, utilization of living cells has some advantages, since a unique characteristic of living organisms is their ability to self-repair and self-replicate. However, improvements on photobioreactor costs and design are still needed [110], together with improvements of strain properties, before such technologies are applicable in an industrial scale. One might imagine a tailored organism for improved H_2 photoproduction in which an O_2-stable [FeFe]-hydrogenase is coupled to a photosynthetic electron chain modified to carry out the most efficient linear electron flow (Fig. 19.1). The capacity of alternative electron sinks such as CO_2 assimilation has to be reduced. The resulting organism would then, theoretically, generate H_2, while still being able to grow and self-replicate at a slow rate. Although some improvement in H_2 production has been possible thanks to the development of forward genetic approaches [89, 152], we believe that future improvements of H_2 production in *C. reinhardtii* will depend on the development of an efficient molecular toolbox allowing targeted gene modification or efficient transgene expression.

19.4 Photosynthetic Lipid Production

In photosynthetic organisms, the chloroplast is also the central compartment of lipid synthesis. Under normal growth conditions, fatty acids produced in the chloroplast are building blocks of all membrane lipids. Higher plants and microalgae have been found to synthesize a large variety of fatty acids and lipids [64, 72], the composition of which often reflects changes of environmental conditions [64]. In response to stress, many eukaryotic microalgae have the ability to accumulate significant amounts (20–50 % of dry biomass) of triacylglycerols (TAGs, i.e. oils) [72, 136]. This characteristic, combined with the fact that many microalgal species grow rapidly and can be cultivated in a wide range of environments, has led to the postulation that microalgae could be used as cell factories for production of oils and other lipids or fatty acids for biofuels and valuable biomaterials [8, 134, 136, 160].

Sustainable production of oils or other fatty acid-derived products from microalgae has not yet reached an industrial level. This is partly due to technological limitations (costs of growing, cell harvesting and lipid extraction) and to biological limitations [34]. Among the latter, the most widely recognized is the requirement of a stress condition (usually nitrogen (N-) starvation) to induce oil accumulation, which limits the overall productivity of the system [72]. Maximal lipid yields obtained so far in large scale cultivation systems are 10 to 20 times lower than the theoretical maximum (5000–15,000 gallons per acre per year) [136]. To circumvent the dependence on stress, a deeper knowledge of the underlying biochemistry, cell biology and genetics of lipid metabolism in this group of organisms is needed. Latest development in -omics technologies (genomics, transcriptomics, proteomics, metabolomics) together with transformation and molecular genetic toolboxes available have provided ample opportunities for lipid scientists to redesign algal metabolism toward production of oils or other chemical molecules useful for industrial applications.

Many laboratories have adopted *C. reinhardtii* as a reference organism for studying TAG accumulation [104]. Stressed *Chlamydomonas* cells accumulate oils in so-called oil bodies [107, 140, 159]. This process is dynamic and reversible [140]. Depending on the strains used and stress conditions, the oil content can reach 50% of dry biomass in starch-less mutants in response to N starvation [92]. With the versatile molecular genetic tools available, *Chlamydomonas* thus serves as an excellent model organism for addressing fundamental biological questions behind oil synthesis and degradation. In this section, we will summarize our current understanding of lipid metabolic pathways in this model alga, and point out key steps as potential targets for genetic engineering. Focus is given to the regulation and biosynthesis of TAGs.

19.4.1 Lipid Composition of Chlamydomonas

The fatty acid and lipid composition of *C. reinhardtii* is summarized in Fig. 19.2. Like in most land plants, thylakoid membranes of the alga contain monogalactosyldiacylglycerol (MGDG), digalactosyldiacylglycerol (DGDG), sulfoquionvosyldiacylglycerol (SQDG) and phosphatidylglycerol (PtdGro). Extraplastidial lipids include phosphatidylethanolamine (PtdEtn) and a betaine lipid, diacylglycerol-*N,N,N*-trimethylhomoserine (DGTS) [52, 53, 153]. *Chlamydomonas* contains no phosphatidylcholine (PtdCho), the major cytoplasmic lipid of higher plants. In the latter, PtdCho serves as key substrate for the modification of fatty acids such as desaturation or hydroxylation and acyl editing [93]. Due to the structural similarity of PtdCho to DGTS, it is generally assumed that in *Chlamydomonas,* DGTS plays a similar role as PC in higher plants [109, 123], but direct evidence for this is still lacking and is the subject of current investigation.

Fatty acids in *Chlamydomonas* usually have similar acyl chain lengths (C16, C18) as those in higher plants (Fig. 19.2b). However, contrary to plants where one or two double bonds are common, fatty acids with three and four double bonds are abundant in the alga. This higher level of desaturation makes the oil more fluid, but also more susceptible to oxidation. Contrary to most vascular plants where the *sn*-2 position of plastidial membrane lipids (MGDG, DGDG, PtdGro, SQDG) is generally esterified with a mixture of both C16 and C18 fatty acids or in some cases only with C18 fatty acids [15, 16, 108], in *Chlamydomonas* this position is almost exclusively occupied by a C16 fatty acid (Fig. 19.2c). This difference in stereochemical distribution of acyl-chains indicates major divergence in lipid synthetic pathways between the two green lineages [16].

19.4.2 Lipid Metabolism as Target for Genetic Engineering

Our current understanding of lipid metabolism in microalgae is very limited and mostly inferred from higher plant models where extensive knowledge on lipid

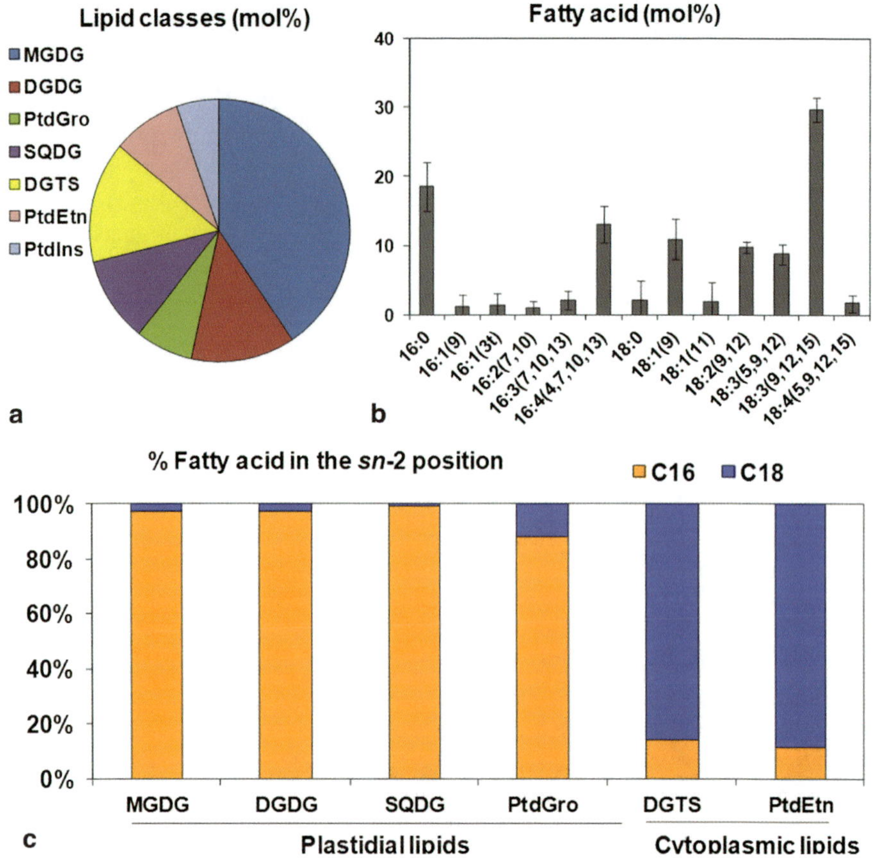

Fig. 19.2 Lipid classes (**a**), whole cell fatty acid composition (**b**) and the stereochemical distribution of fatty acids to major lipid classes (**c**) in *C. reinhardtii* cultivated under standard growth conditions [based on 52, 140]. *MGDG* monogalactosyldiacylglycerol, *DGDG* digalactosyldiacylglycerol, *DGTS* diacylglycerol-*N,N,N*-trimethylhomoserine, *SQDG* sulfoquinovosyldiacylglycerol, *PtdGro* phosphatidylglycerol, *PtdEtn* phosphatidylethanolamine, *PtdIns* phosphatidylinositol

synthetic pathways has been gained through the use of the model plant *Arabidopsis thaliana* [93]. For example, >600 proteins of the lipid metabolism in *Arabidopsis* have been annotated, among which >250 have been characterized [93]. This is in contrast to only about 10 proteins of lipid-related pathways characterized in *Chlamydomonas*. Based on genome comparisons and protein homology searches, it is generally thought that the basic pathways of fatty acid and lipid synthesis are conserved in the two green organisms [104, 123]. TAGs are made from acylation of a glycerol molecule with three fatty acids. Oil biosynthesis can be broken down into three independently regulated and spatially separated steps that are fatty acid synthesis in the chloroplast, glycerolipid assembly and its final packaging into oil bodies (Fig. 19.3). Below we summarize the key biochemical steps required for oil biosynthesis and highlight the potential targets for genetic engineering.

19.4.3 Plastidial Fatty Acid Synthesis

Lipid biosynthesis starts with the *de novo* synthesis of fatty acids, which, in photosynthetic cells, does not occur in the cytosol but in the plastids. Fatty acids are the building blocks of all cellular lipids including TAGs. Fatty acids are made by two major enzymatic complexes, the acetyl-CoA carboxylase (ACCase) and fatty acid synthase complex (FAS). ACCase catalyzes the first committed step of fatty acid synthesis and this reaction produces malonyl-CoA from acetyl-CoA and bicarbonate. Due to the essential role of ACCase, this enzyme has been intensively targeted for genetic engineering [37, 98, 125]. For example, overexpression of the plastid *accD* gene, encoding the β-carboxyl transferase subunit of ACCase, resulted in increased leaf lipid content in tobacco [98]. Overexpression of this enzyme has also been tested in the diatom *Cyclotella cryptica*, although no significant increase in oil content could be detected in the transgenic lines [125].

The FAS complexes catalyze a series of two-carbon chain elongating reactions which lead to the production of C16 or C18 acyl chains. This reaction requires stoichiometric amounts of ATP, acetyl-CoA and NADPH for each two-carbon moiety added to the growing acyl chain. In photosynthetic organisms, photochemical reactions are thus essential not only in providing the carbon source, but also in generating reducing (NADH and NADPH) and phosphorylating (ATP) equivalents to drive fatty acid synthesis [115, 120, 127].

One of the chain terminating reactions is catalyzed by fatty acyl-ACP thioesterases (FAT). The action of this enzyme produces free fatty acids via cleavage from the acyl carrier protein (ACP). The specificity of this enzyme usually determines the final chain length of the product emerging from the plastids. For most plant and algal species, this is a C16 or C18 fatty acid, however in some species such as California bay, the major fatty acid produced is lauric acid (C12:0) [158]. Biodiesel with medium chain fatty acids (C10 to C12) (MCFAs) has improved cold flow properties. Hence, isolating acyl-ACP thioesterase specific for this type of fatty acids is of great biotechnological interests [38]. Production of MCFAs has been achieved *via* transgenic expression of shorter-chain specific thioesterases in oilseed crops [32, 158] as well as in the diatom *Phaeodactylum tricornutum* [119]. This remains to be demonstrated for green microalgae.

The released free fatty acids are ultimately activated to form CoA esters by a long-chain acyl-CoA synthetase (LACS) and subsequently exported to the endoplasmic reticulum (ER) where they serve as substrates for various acyl-CoA dependent acyltransferases. Silencing of members of the acyl-CoA synthetase proteins led to lipid secretion in yeast [130] as well as in cyanobacteria [81]. Three acyl-activating enzyme homologues have been identified in the *Chlamydomonas* genome, and two of those have been found to be associated with oil bodies [107, 113]. These enzymes thus serve as potential targets for directing lipid secretion in eukaryotic microalgae.

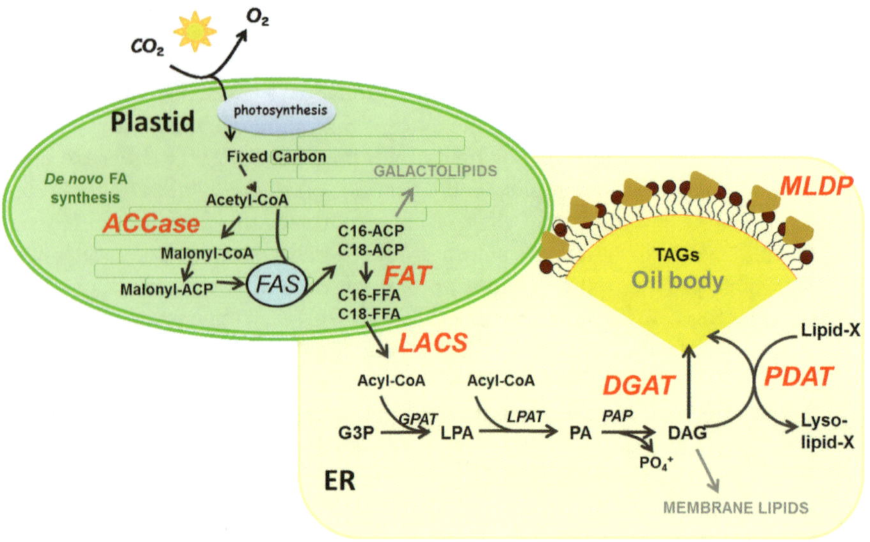

Fig. 19.3 Pathways of fatty acid synthesis and lipid assembly as targets for genetic engineering studies. The scheme of the subcellular organization of lipid metabolic pathways is based on that of plants, unless specific experimental evidence is provided for algal species. Names of enzymes are in italic, and those enzymes described in this chapter are highlighted in red. Lipid-X means that the exact substrate for this enzyme is unknown. *ACCase* acetyl-CoA carboxylase, *ACP* acyl carrier protein, *CoA* coenzyme A, *DAG* diacylglycerol, *DGAT* diacylglycerol acyltransferase, *FAS* fatty acid synthase, *ER* endoplasmic reticulum, *FAT* fatty acyl-ACP thioesterase, *G3P* glycerol-3-phosphate, *GPAT* glycerol-3-phosphate acyltransferase, *LACS* long chain acyl-CoA synthetase, *LPA* lysophosphatidic acid, *LPAT* lysophosphatidic acid acyltransferase, *MLDP* major lipid droplet protein, *PA* phosphatidic acid, *PDAT* phospholipid:diacylglycerol acyltransferase, *PAP* phosphatidic acid phosphatase, *TAG* triacylglycerol

19.4.4 Triacylglycerol Biosynthesis

The best known TAG biosynthetic pathway involves the sequential acylation of *sn*-glycerol-3-phosphate (G3P) with three acyl-CoAs catalyzed by three distinct acyltransferases (Fig. 19.3). It is initiated by G3P acyltransferase (GPAT) to produce lysophosphatidic acid (LPA), which is then further acylated by LPA acyltransferase (LPAT) to form phosphatidic acid (PA). Phosphatidic acid phosphatase (PAP) catalyzes the removal of the phosphate group from PA to generate *sn*-1,2-diacylglycerol (DAG), the central intermediate of all glycerolipids. The last and committed step to oil synthesis is catalyzed by acyl-CoA:diacylglycerol acyltransferase (DGAT). This enzyme has been subjected to intensive studies, including overexpression, directed evolution and quantitative trait loci mapping [18, 139, 172]. In *Chlamydomonas*, homology searches identified five type-2 DGATs (encoded by DGTT1–5) and one type-1 DGAT (DGAT-1). *DGTT1* exhibits increased transcript abundance in N-starvation conditions, and it has been demonstrated to be able to complement a yeast quadruple mutant deficient for TAG synthesis [14]. Engineering strategies involving overexpression of *DGTT1* alone or in combination with other enzymes might be a possible way to increase oil content.

An alternative reaction important for oil synthesis is catalyzed by phospholipid:diacylglycerol acyltransferase (PDAT) contributing to TAG synthesis using phosphatidylcholine as an acyl donor and *sn*-1,2-diacylglycerol as an acyl acceptor [28, 171]. Contrary to all three acyltransferases described above, PDAT does not require acyl-CoA as donor. Therefore, its reaction is often termed acyl-CoA independent pathway. PDAT has been well characterized in both plants and yeast [28, 171]. A homolog of this enzyme is present in *Chlamydomonas* and has lately been shown to be important for TAG accumulation as insertional null mutants (*pdat1–1* and *pdat1–2*) accumulate 25 % less TAG compared to the parent strain [14]. This evidence for a trans-acylation pathway in TAG synthesis in *Chlamydomonas* was corroborated by the observation that cell lines carrying *PDAT*-directed amiRNA silencing constructs accumulate up to 30 % less TAG compared to the wild-type strain [168]. However, as *pdat* mutants exhibit reduced, but not abolished TAG accumulation, DGTT1 must also contribute to oil synthesis [14]. The same overlapping function of PDAT and DGAT has been demonstrated in the model plant *Arabidopsis thaliana* [171] in which minor reductions in oil content could be observed in either of the single mutants, whereas the double mutation is embryo-lethal.

As well as their acylation to glycerol, fatty acyl chains are modified by fatty acid-modifying enzymes including desaturases, epoxidases, elongases, and hydroxylases. Desaturases catalyze the reduction of a C-C bond to form a C=C bond in an existing acyl chain [135]. The number of double bonds in a fatty acid molecule plays a determinant role in its final utility. For example, biodiesel containing a too high proportion of saturated fatty acids turns to gel even at ambient temperatures. On the other hand, when too many unsaturated fatty acids are present, the biodiesel will have a good cold flow but will be prone to oxidation. Desaturases have long been used as targets to engineer fatty acid compositions in higher plants [79, 95, 135]. Four desaturases have been characterized in *Chlamydomonas* [23, 82, 128, 169] and many more have been identified based on sequence homology searches. Molecular manipulation of these desaturases constitutes a promising way to engineer fatty acid composition in *Chlamydomonas*.

19.4.5 Accumulation of Oil Bodies

After a certain amount of TAGs has accumulated in specific domains of the ER or the plastid, oil bodies or lipid droplets bud off and form distinct subcellular organelles. Oil bodies are spherical organelles consisting of a neutral lipid core enclosed by a membrane lipid monolayer coated with proteins [74]. Oil body biogenesis and its associated proteins have been well studied in yeast [26, 27], as well as in plant oilseeds [73]. Only recently, compositions of lipid body-associated proteins have been analyzed in *Chlamydomonas* and >200 proteins have been identified [107, 113]. One protein of ~28 kDa is the most abundant of these and was thus named major lipid droplet protein (MLDP). MLDP has been postulated to play a similar structural role as oleosin in oilseeds. Besides MLDP, numerous metabolic enzymes (acyltransferases, lipases) or trafficking proteins are also present, indicating the dynamic nature of *Chlamydomonas* oil bodies. The knowledge about oil body-associated proteins provided by these studies represents a rich source for the exploration of oil accumulation mechanisms in general, and also elucidates biotechnological targets. For example, either N- or C- terminal fusion of a desired protein to MLDP could potentially direct it to oil bodies, as has been demonstrated for oleosins [10].

One unique feature of *Chlamydomonas* oil bodies is that they are not only present in the ER (as is true for most organisms studied), but also in the plastid [40, 55]. This has been shown by Transmission Electron Microscope (TEM) and is further supported by the strong enrichment in C16 fatty acids at the *sn*-2 position in both TAGs and chloroplast membrane lipids (but not in extra-plastidial lipids). This finding has implications for our overall understanding of the subcellular organization of glycerolipid metabolism and of the specificities of key lipid metabolic enzymes involved. A plastid TAG synthesis pathway could provide additional advantages because engineering of lipid metabolic pathways could be achieved *via* a synthetic biology approach based on manipulation of the plastid genome. Unlike the still-problematic transgene expression in the *C. reinhardtii* nuclear genome [131], it is a well-established technique in the plastid genome and transgene expression can reach very high levels (over 70% of total protein) [114]. Transgenes can be delivered to the plastid genome *via* biolistic bombardment and they are integrated by homologous recombination [31, 105]. Successful introduction of a 50 kb DNA fragment into the plastid genome of tobacco has been reported [1]. This opened up the possibility of introducing several genes simultaneously in the plastid genome using a synthetic biology approach.

19.4.6 Transcriptional Regulation of TAG Biosynthesis

WRINKLED1 (WRI1), belonging to the APETALA2-ethylene responsive element-binding protein (AP2-EREBP), family is the only transcription factor identified in regulation of fatty acid synthesis in *Arabidopsis* [7, 21] and maize [137]. It has also been implicated in regulating oil synthesis in other species such as oil palm [13]. Overexpression of WRI1 leads to a large increase in seed oil content in maize [137] and in tubers [69]. No WRI1 homolog could be identified in the genome of *Chlamydomonas*, but comparative transcriptomic studies have led to identification of two regulatory proteins, NRR1–1 (nitrogen responsive regulator) [14] and an stress-induced lipid trigger [167]. Overexpression or silencing of the genes encoding these proteins led to altered cellular oil content, but the exact mechanism and downstream target(s) of these proteins remain to be tested.

19.5 Closing Remarks

Plastids are the power house of all photosynthetic cells. Photosynthesis converts the abundant energy of the sun into high-energy electrons and chemical energy equivalents. Accordingly, chloroplasts of microalgae are sources of valuable compounds such as molecular hydrogen, starch and lipids. Recent studies of H_2 and lipid metabolic pathways in microalgal models have led to significant advances in our understanding of the molecular and biochemical mechanisms [66, 94, 104]. Currently, the vast majority of studies on microalgal biofuel are focused on understanding and boosting the generation of H_2 and the accumulation of TAGs [20, 66, 83, 84]. In our view, generation of oil is only the first step toward the engineering of algal cell factories. Production of value-added fatty acid-derived molecules such as alkanes, free fatty acids, wax esters and fatty alcohols will constitute the next major step. At the moment, significant effort has been put on analyzing the H_2 and lipid metabolism of the model microalga *C. reinhardtii*. However, the genomes of around ten microalgal species have been sequenced so far, and many more are currently being sequenced. Intensive efforts are underway to develop molecular genetic tools for *Chlamydomonas* and other algae. For example, the occurrence of homologous recombination in *Nannochloropsis* sp. has been reported [85]. This development, together with our knowledge gained through examining model systems, should aid in the master design of an ideal algal cell factory for the production of industrially desirable molecules.

References

1. Adachi T, Takase H, Tomizawa KI (2007) Introduction of a 50 kbp DNA fragment into the plastid genome. Biosci Biotechnol Biochem 71:2266–2273
2. Antal TK, Krendeleva TE, Laurinavichene TV, Makarova VV, Ghirardi ML, Rubin AB, Tsygankov AA, Seibert M (2003) The dependence of algal H_2 production on Photosystem II and O_2 consumption activities in sulfur-deprived *Chlamydomonas reinhardtii* cells. Biochim Biophys Acta 1607:153–160
3. Antonkine ML et al (2007) Chemical rescue of a site-modified ligand to a [4Fe-4S] cluster in PsaC, a bacterial-like dicluster ferredoxin bound to Photosystem I. Biochim Biophys Acta 1767:712–724
4. Arnon DI, Losada M, Nozaki M, Tagawa K (1961) Photoproduction of hydrogen, photofixation of nitrogen and a unified concept of photosynthesis. Nature 190:601–606
5. Atteia A, van Lis R, Gelius-Dietrich G, Adrait A, Garin J, Joyard J, Rolland N, Martin W (2006) Pyruvate formate-lyase and a novel route of eukaryotic ATP synthesis in *Chlamydomonas* mitochondria. J Biol Chem 281:9909–9918
6. Atteia A, van Lis R, Tielens AG, Martin WF (2013) Anaerobic energy metabolism in unicellular photosynthetic eukaryotes. Biochim Biophys Acta 1827:210–223
7. Baud S, Santos-Mendoza M, To A, Harscoet E, Lepiniec L, Dubreucq B (2007) WRINKLED1 specifies the regulatory action of LEAFY COTYLEDON2 towards fatty acid metabolism during seed maturation in Arabidopsis. Plant J 50:825–838
8. Beer LL, Boyd ES, Peters JW, Posewitz MC (2009) Engineering algae for biohydrogen and biofuel production. Curr Opin Biotechnol 20:264–271
9. Benemann JR, Berenson JA, Kaplan NO, Kamen MD (1973) Hydrogen evolution by a chloroplast-ferredoxin-hydrogenase system. Proc Natl Acad Sci USA 70:2317–2320
10. Bhatla SC, Kaushik V, Yadav MK (2010) Use of oil bodies and oleosins in recombinant protein production and other biotechnological applications. Biotechnol Adv 28:293–300
11. Blankenship RE et al (2011) Comparing photosynthetic and photovoltaic efficiencies and recognizing the potential for improvement. Science 332:805–809
12. Boudreau E, Nickelsen J, Lemaire SD, Ossenbuhl F, Rochaix JD (2000) The *Nac2* gene of *Chlamydomonas* encodes a chloroplast TPR-like protein involved in *psbD* mRNA stability. EMBO J 19:3366–3376

13. Bourgis F, Kilaru A, Cao X, Ngando-Ebongue GF, Drira N, Ohlrogge JB, Arondel V (2011) Comparative transcriptome and metabolite analysis of oil palm and date palm mesocarp that differ dramatically in carbon partitioning. Proc Natl Acad Sci U S A 108:12527–12532
14. Boyle NR et al (2012) Three acyltransferases and nitrogen-responsive regulator are implicated in nitrogen starvation-induced triacylglycerol accumulation in *Chlamydomonas*. J Biol Chem 287:15811–15825
15. Browse J, McCourt P, Somerville C (1986) A mutant of *Arabidopsis* deficient in C18–3 and C16–3 leaf lipids. Plant Physiol 81:859–864
16. Browse J, Warwick N, Somerville CR, Slack CR (1986) Fluxes through the prokaryotic and eukaryotic pathways of lipid-synthesis in the 16–3 plant *Arabidopsis-thaliana*. Biochem J 235:25–31
17. Bruska MK, Stiebritz MT, Reiher M (2011) Regioselectivity of H cluster oxidation. J Am Chem Soc 133:20588–20603
18. Burgal J, Shockey J, Lu CF, Dyer J, Larson T, Graham I, Browse J (2008) Metabolic engineering of hydroxy fatty acid production in plants: RcDGAT2 drives dramatic increases in ricinoleate levels in seed oil. Plant Biotechnol J 6:819–831
19. Burgdorf T, Lenz O, Buhrke T, van der Linden E, Jones AK, Albracht SP, Friedrich B (2005) [NiFe]-hydrogenases of *Ralstonia eutropha* H16: modular enzymes for oxygen-tolerant biological hydrogen oxidation. J Mol Microbiol Biotechnol 10:181–196
20. Burgess SJ, Tamburic B, Zemichael F, Hellgardt K, Nixon PJ (2011) Solar-driven hydrogen production in green algae. Adv Appl Microbiol 75:71–110
21. Cernac A, Benning C (2004) WRINKLED1 encodes an AP2/EREB domain protein involved in the control of storage compound biosynthesis in *Arabidopsis*. Plant J 40:575–585
22. Chen HC, Newton AJ, Melis A (2005) Role of SulP, a nuclear-encoded chloroplast sulfate permease, in sulfate transport and H_2 evolution in *Chlamydomonas reinhardtii*. Photosynthesis Res 84:289–296
23. Chi XY, Zhang XW, Guan XY, Ding L, Li YX, Wang MQ, Lin HZ, Qin S (2008) Fatty acid biosynthesis in eukaryotic photosynthetic microalgae: identification of a microsomal delta 12 desaturase in *Chlamydomonas reinhardtii*. J Microbiol 46:189–201
24. Chochois V, Dauvillee D, Beyly A, Tolleter D, Cuine S, Timpano H, Ball S, Cournac L, Peltier G (2009) Hydrogen production in *Chlamydomonas*: photosystem II-dependent and -independent pathways differ in their requirement for starch metabolism. Plant Physiol 151:631–640
25. Cinco RM, MacInnis JM, Greenbaum E (1993) The role of carbon dioxide in light-activated hydrogen production by *Chlamydomonas reinhardtii*. Photosynthesis Res 38:27–33
26. Coleman RA, Lee DP (2004) Enzymes of triacylglycerol synthesis and their regulation. Prog Lipid Res 43:134–176
27. Czabany T, Wagner A, Zweytick D, Lohner K, Leitner E, Ingolic E, Daum G (2008) Structural and biochemical properties of lipid particles from the yeast *Saccharomyces cerevisiae*. J Biol Chem 283:17065–17074
28. Dahlqvist A, Stahl U, Lenman M, Banas A, Lee M, Sandager L, Ronne H, Stymne H (2000) Phospholipid: diacylglycerol acyltransferase: an enzyme that catalyzes the acyl-CoA-independent formation of triacylglycerol in yeast and plants. Proc Natl Acad Sci U S A 97:6487–6492
29. Dau H, Zaharieva I (2009) Principles, efficiency, and blueprint character of solar-energy conversion in photosynthetic water oxidation. Acc Chem Res 42:1861–1870
30. Davies JP, Yildiz FH, Grossman A (1996) Sac1, a putative regulator that is critical for survival of *Chlamydomonas reinhardtii* during sulfur deprivation. EMBO J 15:2150–2159
31. Day A, Goldschmidt-Clermont M (2011) The chloroplast transformation toolbox: selectable markers and marker removal. Plant Biotechnol J 9:540–553
32. Dehesh K, Jones A, Knutzon DS, Voelker TA (1996) Production of high levels of 8:0 and 10:0 fatty acids in transgenic canola by overexpression of Ch FatB2, a thioesterase cDNA from *Cuphea hookeriana*. Plant J 9:167–172
33. Delosme R, Olive J, Wollman F-A (1996) Changes in light energy distribution upon state transitions: an in vivo photoacoustic study of the wild type and photosynthesis mutants from *Chlamydomonas reinhardtii*. Biochim Biophys Acta (BBA)-Bioenerg 1273:150–158

34. Delrue F, Setier PA, Sahut C, Cournac L, Roubaud A, Peltier G, Froment AK (2012) An economic, sustainability, and energetic model of biodiesel production from microalgae. Bioresour Technol 111:191–200
35. Desplats C, Mus F, Cuine S, Billon E, Cournac L, Peltier G (2009) Characterization of Nda2, a plastoquinone-reducing type II NAD(P)H dehydrogenase in *Chlamydomonas* chloroplasts. J Biol Chem 284:4148–4157
36. Doebbe A, Keck M, La Russa M, Mussgnug JH, Hankamer B, Tekce E, Niehaus K, Kruse O (2010) The interplay of proton, electron, and metabolite supply for photosynthetic H_2 production in *Chlamydomonas reinhardtii*. J Biol Chem 285:30247–30260
37. Dunahay TG, Jarvis EE, Dais SS, Roessler PG (1996) Manipulation of microalgal lipid production using genetic engineering. Appl Biochem Biotechnol 57–58:223–231
38. Durrett TP, Benning C, Ohlrogge J (2008) Plant triacylglycerols as feedstocks for the production of biofuels. Plant J 54:593–607
39. Falkowski PG (2006) Evolution. Tracing oxygen's imprint on earth's metabolic evolution. Science 311:1724–1725
40. Fan JL, Andre C, Xu CC (2011) A chloroplast pathway for the de novo biosynthesis of triacylglycerol in *Chlamydomonas reinhardtii*. FEBS Lett 585:1985–1991
41. Finazzi G, Furia A, Barbagallo RP, Forti G (1999) State transitions, cyclic and linear electron transport and photophosphorylation in *Chlamydomonas reinhardtii*. Biochim Biophys Acta 1413:117–129
42. Florin L, Tsokoglou A, Happe T (2001) A novel type of iron hydrogenase in the green alga *Scenedesmus obliquus* is linked to the photosynthetic electron transport chain. J Biol Chem 276:6125–6132
43. Fontecilla-Camps JC, Volbeda A, Cavazza C, Nicolet Y (2007) Structure/function relationships of [NiFe]- and [FeFe]-hydrogenases. Chem Rev 107:4273–4303
44. Forestier M, King P, Zhang LP, Posewitz M, Schwarzer S, Happe T, Ghirardi ML, Seibert M (2003) Expression of two [Fe]-hydrogenases in *Chlamydomonas reinhardtii* under anaerobic conditions. Eur J Biochem 270:2750–2758
45. Fouchard S, Hemschemeier A, Caruana A, Pruvost K, Legrand J, Happe T, Peltier G, Cournac L (2005) Autotrophic and mixotrophic hydrogen photoproduction in sulfur-deprived *Chlamydomonas* cells. Appl Environ Microbiol 71:6199–6205
46. Gaffron H (1939) Reduction of carbon dioxide with molecular hydrogen in green algae. Nature 143:204–205
47. Gaffron H, Rubin J (1942) Fermentative and photochemical production of hydrogen in algae. J Gen Physiol 26:219–240
48. Gfeller RP, Gibbs M (1984) Fermentative metabolism of *Chlamydomonas reinhardtii*: I. Analysis of fermentative products from starch in dark and light. Plant Physiol 75:212–218
49. Gfeller RP, Gibbs M (1985) Fermentative metabolism of *Chlamydomonas reinhardtii*: II. Role of plastoquinone. Plant Physiol 77:509–511
50. Ghirardi ML, Togasaki RK, Seibert M (1997) Oxygen sensitivity of algal H_2-production. Appl Biochem Biotechnol 63–65:141–151
51. Giannelli L, Scoma A, Torzillo G (2009) Interplay between light intensity, chlorophyll concentration and culture mixing on the hydrogen production in sulfur-deprived *Chlamydomonas reinhardtii* cultures grown in laboratory photobioreactors. Biotechnol Bioeng 104:76–90
52. Giroud C, Eichenberger W (1988) Fatty acids of *Chlamydomonas reinhardtii*-structure, positional distribution and biosynthesis. Biol Chem Hoppe Seyler 369:18–19
53. Giroud C, Gerber A, Eichenberger W (1988) Lipids of *Chlamydomonas reinhardtii*-analysis of molecular species and intracellular sites(s) of biosynthesis. Plant Cell Physiol 29:587–595
54. Goldet G, Brandmayr C, Stripp ST, Happe T, Cavazza C, Fontecilla-Camps JC, Armstrong FA (2009) Electrochemical kinetic investigations of the reactions of [FeFe]-hydrogenases with carbon monoxide and oxygen: comparing the importance of gas tunnels and active-site electronic/redox effects. J Am Chem Soc 131:14979–14989
55. Goodson C, Roth R, Wang ZT, Goodenough U (2011) Structural correlates of cytoplasmic and chloroplast lipid body synthesis in *Chlamydomonas reinhardtii* and stimulation of lipid body production with acetate boost. Eukaryot Cell 10:1592–1606

56. Grimme RA, Lubner CE, Bryant DA, Golbeck JH (2008) Photosystem I/molecular wire/metal nanoparticle bioconjugates for the photocatalytic production of H_2. J Am Chem Soc 130:6308
57. Grossman A (2000) Acclimation of *Chlamydomonas reinhardtii* to its nutrient environment. Protist 151:201–224
58. Grossman AR, Catalanotti C, Yang W, Dubini A, Magneschi L, Subramanian V, Posewitz MC, Seibert M (2011) Multiple facets of anoxic metabolism and hydrogen production in the unicellular green alga *Chlamydomonas reinhardtii*. New Phytol 190:279–288
59. Hackstein JH, Akhmanova A, Boxma B, Harhangi HR, Voncken FG (1999) Hydrogenosomes: eukaryotic adaptations to anaerobic environments. Trends Microbiol 7:441–447
60. Halliwell B (2006) Reactive species and antioxidants. Redox biology is a fundamental theme of aerobic life. Plant Physiol 141:312–322
61. Happe T, Kaminski A (2002) Differential regulation of the Fe-hydrogenase during anaerobic adaptation in the green alga *Chlamydomonas reinhardtii*. Eur J Biochem 269:1022–1032
62. Happe T, Naber JD (1993) Isolation, characterization and N-terminal amino acid sequence of hydrogenase from the green alga *Chlamydomonas reinhardtii*. Eur J Biochem 214:475–481
63. Happe T, Mosler B, Naber JD (1994) Induction, localization and metal content of hydrogenase in the green alga *Chlamydomonas reinhardtii*. Eur J Biochem 222:769–774
64. Harwood JL, Guschina IA (2009) The versatility of algae and their lipid metabolism. Biochimie 91:679–684
65. Hedges SB, Blair JE, Venturi ML, Shoe JL (2004) A molecular timescale of eukaryote evolution and the rise of complex multicellular life. BMC Evol Biol 4:2
66. Hemschemeier A, Happe T (2011) Alternative photosynthetic electron transport pathways during anaerobiosis in the green alga *Chlamydomonas reinhardtii*. Biochim Biophys Acta 1807:919–926
67. Hemschemeier A, Fouchard S, Cournac L, Peltier G, Happe T (2008) Hydrogen production by *Chlamydomonas reinhardtii*: an elaborate interplay of electron sources and sinks. Planta 227:397–407
68. Hemschemeier A, Jacobs J, Happe T (2008) Biochemical and physiological characterization of the pyruvate formate-lyase Pfl1 of *Chlamydomonas reinhardtii*, a typically bacterial enzyme in a eukaryotic alga. Eukaryot Cell 7:518–526
69. Hofvander P, Carlsson AS, A.S. F (2011) Arabidopsis WRL1 induces a large increase of TAG accumulation in potato microtubers. 5th European symposium on plant lipids: 88
70. Hong G, Pachter R (2012) Inhibition of biocatalysis in [Fe-Fe] hydrogenase by oxygen: molecular dynamics and density functional theory calculations. ACS Chem Biol 7:1268–1275
71. Hoshino T, Johnson DJ, Cuello JL (2012) Design of new strategy for green algal photohydrogen production: spectral-selective photosystem I activation and photosystem II deactivation. Bioresour Technol 120:233–240
72. Hu Q, Sommerfeld M, Jarvis E, Ghirardi M, Posewitz M, Seibert M, Darzins A (2008) Microalgal triacylglycerols as feedstocks for biofuel production: perspectives and advances. Plant J 54:621–639
73. Huang AHC (1992) Oil bodies and oleosins in seeds. Annu Rev Plant Physiol Plant Mol Biol 43:177–200
74. Huang AHC (1994) Structure of plant seed oil bodies. Curr Opin Struct Biol 4:493–498
75. Ihara M, Nakamoto H, Kamachi T, Okura I, Maeda M (2006) Photoinduced hydrogen production by direct electron transfer from photosystem I cross-linked with cytochrome c_3 to [NiFe]-hydrogenase. Photochem Photobiol 82:1677–1685
76. Ihara M et al (2006) Light-driven hydrogen production by a hybrid complex of a [NiFe]-hydrogenase and the cyanobacterial photosystem I. Photochem Photobiol 82:676–682
77. Jacobs J, Pudollek S, Hemschemeier A, Happe T (2009) A novel, anaerobically induced ferredoxin in *Chlamydomonas reinhardtii*. FEBS Lett 583:325–329
78. Jans F et al (2008) A type II NAD(P)H dehydrogenase mediates light-independent plastoquinone reduction in the chloroplast of *Chlamydomonas*. Proc Natl Acad Sci U S A 105:20546–20551

79. Jaworski J, Cahoon EB (2003) Industrial oils from transgenic plants. Curr Opin Plant Biol 6:178–184
80. Jensen PE, Bassi R, Boekema EJ, Dekker JP, Jansson S, Leister D, Robinson C, Scheller HV (2007) Structure, function and regulation of plant photosystem I. Biochim Biophys Acta Bioenerg 1767:335–352
81. Kaczmarzyk D, Fulda M (2010) Fatty acid activation in cyanobacteria mediated by acyl-acyl carrier protein synthetase enables fatty acid recycling. Plant Physiol 152:1598–1610
82. Kajikawa M, Yamato KT, Kohzu Y, Shoji S, Matsui K, Tanaka Y, Sakai Y, Fukuzawa H (2006) A front-end desaturase from *Chlamydomonas reinhardtii* produces pinolenic and coniferonic acids by omega13 desaturation in methylotrophic yeast and tobacco. Plant Cell Physiol 47:64–73
83. Khozin-Goldberg I, Cohen Z (2011) Unraveling algal lipid metabolism: recent advances in gene identification. Biochimie 93:91–100
84. Khozin-Goldberg I, Iskandarov U, Cohen Z (2011) LC-PUFA from photosynthetic microalgae: occurrence, biosynthesis, and prospects in biotechnology. Appl Microbiol Biotechnol 91:905–915
85. Kilian O, Benemann CSE, Niyogi KK, Vick B (2011) High-efficiency homologous recombination in the oil-producing alga *Nannochloropsis* sp. Proc Natl Acad Sci U S A 108:21265–21269
86. Knaff DB (2004) Ferredoxin and ferredoxin-dependent enzymes oxygenic photosynthesis: the light reactions. (Ort DR, Yocum CF, Heichel IF (eds)). Springer, Netherlands, pp 333–361
87. Krasnovsky AA, Van Ni C, Nikandrov VV, Brin GP (1980) Efficiency of hydrogen photoproduction by chloroplast-bacterial hydrogenase systems. Plant Physiol 66:925–930
88. Kreuzberg K (1984) Starch fermentation via a formate producing pathway in *Chlamydomonas reinhardii*, *Chlorogonium elongatum* and *Chlorella fusca*. Physiologia Plantarum 61:87–94
89. Kruse O, Rupprecht J, Bader KP, Thomas-Hall S, Schenk PM, Finazzi G, Hankamer B (2005) Improved photobiological H_2 production in engineered green algal cells. J Biol Chem 280:34170–34177
90. Lambertz C, Hemschemeier A, Happe T (2010) Anaerobic expression of the ferredoxin-encoding *FDX5* gene of *Chlamydomonas reinhardtii* is regulated by the Crr1 transcription factor. Eukaryot Cell 9:1747–1754
91. Lambertz C, Leidel N, Havelius KG, Noth J, Chernev P, Winkler M, Happe T, Haumann M (2011) O_2 reactions at the six-iron active site (H-cluster) in [FeFe]-hydrogenase. J Biol Chem 286:40614–40623
92. Li Y, Han D, Hu G, Dauvillee D, Sommerfeld M, Ball S, Hu Q (2010) *Chlamydomonas* starchless mutant defective in ADP-glucose pyrophosphorylase hyper-accumulates triacylglycerol. Metab Eng 12:387–391
93. Li-Beisson Y SB, Beisson F, Andersson M, Arondel V, Bates P, Baud S, Bird D, DeBono A, Durrett T, Franke R, Graham I, Katayama K, Kelly A, Larson T, Markham J, Miquel M, Molina I, Nishida I, Rowland O, Samuels L, Schmid K, Wada H, Welti R, Xu C, Zallot R, Ohlrogge J (2010) Acyl lipid metabolism in the Arabidopsis book. Last R (ed). American Society of Plant Biologists Rockville, MD
94. Liu B, Benning C (2013) Lipid metabolism in microalgae distinguishes itself. Curr Opin Biotechnol 24:300–309
95. Lu CF, Napier JA, Clemente TE, Cahoon EB (2011) New frontiers in oilseed biotechnology: meeting the global demand for vegetable oils for food, feed, biofuel, and industrial applications. Curr Opin Biotechnol 22:252–259
96. Lubner CE, Knörzer P, Silva PJN, Vincent KA, Happe T, Bryant DA, Golbeck JH (2010) Wiring an [Fe-Fe]-hydrogenase with photosystem I for light-induced hydrogen production. Biochemistry 49:10264–10266
97. Lubner CE, Applegate AM, Knörzer P, Ganago A, Bryant DA, Happe T, Golbeck JH (2011) Solar hydrogen-producing bionanodevice outperforms natural photosynthesis. Proc Natl Acad Sci U S A 108:20988–20991

98. Madoka Y, Tomizawa KI, Mizoi J, Nishida I, Nagano Y, Sasaki Y (2002) Chloroplast transformation with modified *accD* operon increases acetyl-CoA carboxylase and causes extension of leaf longevity and increase in seed yield in tobacco. Plant Cell Physiol 43:1518–1525
99. Matias PM, Soares CM, Saraiva LM, Coelho R, Morais J, Le Gall J, Carrondo MA (2001) [NiFe] hydrogenase from *Desulfovibrio desulfuricans* ATCC 27774: gene sequencing, three-dimensional structure determination and refinement at 1.8 A and modelling studies of its interaction with the tetrahaem cytochrome c_3. J Biol Inorg Chem 6:63–81
100. McTavish H (1998) Hydrogen evolution by direct electron transfer from photosystem I to hydrogenases. J Biochem 123:644–649
101. Melis A, Zhang L, Forestier M, Ghirardi ML, Seibert M (2000) Sustained photobiological hydrogen gas production upon reversible inactivation of oxygen evolution in the green alga *Chlamydomonas reinhardtii*. Plant Physiol 122:127–136
102. Merchant S, Sawaya MR (2005) The light reactions: a guide to recent acquisitions for the picture gallery. Plant Cell 17:648–663
103. Merchant SS, Allen MD, Kropat J, Moseley JL, Long JC, Tottey S, Terauchi AM (2006) Between a rock and a hard place: trace element nutrition in *Chlamydomonas*. Biochim Biophys Acta 1763:578–594
104. Merchant SS, Kropat J, Liu B, Shaw J, Warakanont J (2012) TAG, You're it! *Chlamydomonas* as a reference organism for understanding algal triacylglycerol accumulation. Curr Opin Biotechnol 23:352–363
105. Michelet L, Lefebvre-Legendre L, Burr SE, Rochaix JD, Goldschmidt-Clermont M (2011) Enhanced chloroplast transgene expression in a nuclear mutant of *Chlamydomonas*. Plant Biotechnol J 9:565–574
106. Mignolet E, Lecler R, Ghysels B, Remacle C, Franck F (2012) Function of the chloroplastic NAD(P)H dehydrogenase Nda2 for H_2 photoproduction in sulphur-deprived *Chlamydomonas reinhardtii*. J Biotechnol 162:81–88
107. Moellering ER, Benning C (2010) RNA interference silencing of a major lipid droplet protein affects lipid droplet size in *Chlamydomonas reinhardtii*. Eukaryot Cell 9:97–106
108. Mongrand S, Bessoule J-J, Cabantous F, Cassagne C (1998) The C16:3\C18:3 fatty acid balance in photosynthetic tissues from 468 plant species. Phytochemistry 49:1049–1064
109. Moore TS, Du Z, Chen Z (2001) Membrane lipid biosynthesis in *Chlamydomonas reinhardtii*. In vitro biosynthesis of diacylglyceryltrimethylhomoserine. Plant Physiol 125:423–429
110. Morweiser M, Kruse O, Hankamer B, Posten C (2010) Developments and perspectives of photobioreactors for biofuel production. Appl Microbiol Biotechnol 87:1291–1301
111. Mus F, Cournac L, Cardettini V, Caruana A, Peltier G (2005) Inhibitor studies on non-photochemical plastoquinone reduction and H_2 photoproduction in *Chlamydomonas reinhardtii*. Biochim Biophys Acta 1708:322–332
112. Mus F, Dubini A, Seibert M, Posewitz MC, Grossman AR (2007) Anaerobic acclimation in *Chlamydomonas reinhardtii*: anoxic gene expression, hydrogenase induction, and metabolic pathways. J Biol Chem 282:25475–25486
113. Nguyen HM et al (2011) Proteomic profiling of oil bodies isolated from the unicellular green microalga *Chlamydomonas reinhardtii*: With focus on proteins involved in lipid metabolism. Proteomics 11:4266–4273
114. Oey M, Lohse M, Kreikemeyer B, Bock R (2009) Exhaustion of the chloroplast protein synthesis capacity by massive expression of a highly stable protein antibiotic. Plant J 57:436–445
115. Ohlrogge J, Pollard M, Bao X, Focke M, Girke T, Ruuska S, Mekhedov S, Benning C (2000) Fatty acid synthesis: from CO_2 to functional genomics. Biochem Soc Trans 28:567–574
116. Pape M, Lambertz C, Happe T, Hemschemeier A (2012) Differential expression of the *Chlamydomonas* [FeFe]-hydrogenase-encoding *HYDA1* gene is regulated by the copper response regulator1. Plant Physiol 159:1700–1712
117. Philipps G, Krawietz D, Hemschemeier A, Happe T (2011) A pyruvate formate lyase-deficient *Chlamydomonas reinhardtii* strain provides evidence for a link between fermentation and hydrogen production in green algae. Plant J 66:330–340

118. Philipps G, Happe T, Hemschemeier A (2012) Nitrogen deprivation results in photosynthetic hydrogen production in *Chlamydomonas reinhardtii*. Planta 235:729–745
119. Radakovits R, Eduafo PM, Posewitz MC (2011) Genetic engineering of fatty acid chain length in *Phaeodactylum tricornutum*. Metab Eng 13:89–95
120. Rawsthorne S (2002) Carbon flux and fatty acid synthesis in plants. Prog Lipid Res 41:182–196
121. Raymond J, Blankenship RE (2004) Biosynthetic pathways, gene replacement and the antiquity of life. Geobiology 2:199–203
122. Raymond J, Segre D (2006) The effect of oxygen on biochemical networks and the evolution of complex life. Science 311:1764–1767
123. Riekhof WR, Sears BB, Benning C (2005) Annotation of genes involved in glycerolipid biosynthesis in *Chlamydomonas reinhardtii*: discovery of the betaine lipid synthase BTA1Cr. Eukaryot Cell 4:242–252
124. Roessler PG, Lien S (1984) Purification of hydrogenase from *Chlamydomonas reinhardtii*. Plant Physiol 75:705–709
125. Roessler PG, Ohlrogge JB (1993) Cloning and characterization of the gene that encodes acetyl-coenzyme-A carboxylase in the alga *Cyclotella cryptica*. J Biol Chem 268:19254–19259
126. Rühle T, Hemschemeier A, Melis A, Happe T (2008) A novel screening protocol for the isolation of hydrogen producing *Chlamydomonas reinhardtii* strains. BMC Plant Biol 8:107
127. Ruuska SA, Girke T, Benning C, Ohlrogge JB (2002) Contrapuntal networks of gene expression during *Arabidopsis* seed filling. Plant Cell 14:1191–1206
128. Sato N, Fujiwara S, Kawaguchi A, Tsuzuki M (1997) Cloning of a gene for chloroplast omega6 desaturase of a green alga, *Chlamydomonas reinhardtii*. J Biochem 122:1224–1232
129. Sawers RG (2005) Formate and its role in hydrogen production in *Escherichia coli*. Biochem Soc Trans 33:42–46
130. Scharnewski M, Pongdontri P, Mora G, Hoppert M, Fulda M (2008) Mutants of *Saccharomyces cerevisiae* deficient in acyl-CoA synthetases secrete fatty acids due to interrupted fatty acid recycling. FEBS J 275:2765–2778
131. Schroda M (2006) RNA silencing in *Chlamydomonas*: mechanisms and tools. Curr Genet 49:69–84
132. Schwarze A, Kopczak MJ, Rogner M, Lenz O (2010) Requirements for construction of a functional hybrid complex of photosystem I and [NiFe]-hydrogenase. Appl Environ Microbiol 76:2641–2651
133. Scoma A, Krawietz D, Faraloni C, Giannelli L, Happe T, Torzillo G (2012) Sustained H_2 production in a *Chlamydomonas reinhardtii* D1 protein mutant. J Biotechnol 157: 613–619
134. Scott SA, Davey MP, Dennis JS, Horst I, Howe CJ, Lea-Smith DJ, Smith AG (2010) Biodiesel from algae: challenges and prospects. Curr Opin Biotechnol 21:277–286
135. Shanklin J, Cahoon EB (1998) Desaturation and related modifications of fatty acids. Annu Rev Plant Physiol Plant Mol Biol 49:611–641
136. Sheehan J, Dunahay T, Benemann J, Roessler PG A look back at the US Department of Energy's aquatic species program—biodiesel from algae, 1998, US Department of Energy's Office of Fuels Development: Golden, CO: National Renewable Energy Laboratory
137. Shen B, Allen WB, Zheng PZ, Li CJ, Glassman K, Ranch J, Nubel D, Tarczynski MC (2010) Expression of ZmLEC1 and ZmWRI1 increases seed oil production in maize. Plant Physiol 153:980–987
138. Shima S, Thauer RK (2007) A third type of hydrogenase catalyzing H_2 activation. Chem Rec 7:37–46
139. Shockey JM, Gidda SK, Chapital DC, Kuan JC, Dhanoa PK, Bland JM, Rothstein SJ, Mullen RT, Dyer JM (2006) Tung tree DGAT1 and DGAT2 have nonredundant functions in triacylglycerol biosynthesis and are localized to different subdomains of the endoplasmic reticulum. Plant Cell 18:2294–2313
140. Siaut M et al (2011) Oil accumulation in the model green alga *Chlamydomonas reinhardtii*: characterization, variability between common laboratory strains and relationship with starch reserves. BMC Biotechnol 11:7

141. Skjånes K, Knutsen G, Källqvist T, Lindblad P (2008) H_2 production from marine and freshwater species of green algae during sulfur deprivation and considerations for bioreactor design. Int J Hydrogen Energy 33:511–521
142. Stiebritz MT, Reiher M (2012) Hydrogenases and oxygen. Chem Sci 3:1739–1751
143. Stripp ST, Goldet G, Brandmayr C, Sanganas O, Vincent KA, Haumann M, Armstrong FA, Happe T (2009) How oxygen attacks [FeFe] hydrogenases from photosynthetic organisms. Proc Natl Acad Sci U S A 106:17331–17336
144. Stuart TS, Gaffron H (1972) The mechanism of hydrogen photoproduction by several algae. Planta 106:101–112
145. Suorsa M, Sirpio S, Aro EM (2009) Towards characterization of the chloroplast NAD(P)H dehydrogenase complex. Mol Plant 2:1127–1140
146. Surzycki R, Cournac L, Peltier G, Rochaix JD (2007) Potential for hydrogen production with inducible chloroplast gene expression in *Chlamydomonas*. Proc Natl Acad Sci U S A 104:17548–17553
147. Terashima M, Specht M, Naumann B, Hippler M (2010) Characterizing the anaerobic response of *Chlamydomonas reinhardtii* by quantitative proteomics. Mol Cell Proteomics 9:1514–1532
148. Terauchi AM et al (2009) Pattern of expression and substrate specificity of chloroplast ferredoxins from *Chlamydomonas reinhardtii*. J Biol Chem 284:25867–25878
149. Thannickal VJ (2009) Oxygen in the evolution of complex life and the price we pay. Am J Respir Cell Mol Biol 40:507–510
150. Timmins M, Thomas-Hall SR, Darling A, Zhang E, Hankamer B, Marx UC, Schenk PM (2009) Phylogenetic and molecular analysis of hydrogen-producing green algae. J Exp Bot 60:1691–1702
151. Timmins M et al (2009) The metabolome of *Chlamydomonas reinhardtii* following induction of anaerobic H_2 production by sulfur depletion. J Biol Chem 284:35996
152. Tolleter D et al (2011) Control of hydrogen photoproduction by the proton gradient generated by cyclic electron flow in *Chlamydomonas reinhardtii*. Plant Cell 23:2619–2630
153. Vieler A, Wilhelm C, Goss R, Sub R, Schiller J (2007) The lipid composition of the unicellular green alga *Chlamydomonas reinhardtii* and the diatom *Cyclotella meneghiniana* investigated by MALDI-TOF MS and TLC. Chem Phys Lipids 150:143–155
154. Vignais PM, Billoud B (2007) Occurrence, classification, and biological function of hydrogenases: an overview. Chem Rev 107:4206–4272
155. Vignais PM, Billoud B, Meyer J (2001) Classification and phylogeny of hydrogenases. FEMS Microbiol Rev 25:455–501
156. Vincent KA, Parkin A, Lenz O, Albracht SP, Fontecilla-Camps JC, Cammack R, Friedrich B, Armstrong FA (2005) Electrochemical definitions of O_2 sensitivity and oxidative inactivation in hydrogenases. J Am Chem Soc 127: 18179–18189
157. Vincent KA, Parkin A, Armstrong FA (2007) Investigating and exploiting the electrocatalytic properties of hydrogenases. Chem Rev 107:4366–4413
158. Voelker T, Worrell A, Anderson L, Bleibaum J, Fan C, Hawkins D, Radke S, Davies H (1992) Fatty acid biosynthesis redirected to medium chains in transgenic oilseed plants. Science 257:72–74
159. Wang ZT, Ullrich N, Joo S, Waffenschmidt S, Goodenough U (2009) Algal lipid bodies: stress induction, purification, and biochemical characterization in wild-type and starchless *Chlamydomonas reinhardtii*. Eukaryot Cell 8:1856–1868
160. Wijffels RH, Barbosa MJ (2010) An outlook on microalgal biofuels. Science 329:796–799
161. Winkler M, Heil B, Heil B, Happe T (2002) Isolation and molecular characterization of the [Fe]-hydrogenase from the unicellular green alga *Chlorella fusca*. Biochim Biophys Acta 1576:330–334
162. Winkler M, Kuhlgert S, Hippler M, Happe T (2009) Characterization of the key step for light-driven hydrogen evolution in green algae. J Biol Chem 284:36620–36627
163. Winkler M, Hemschemeier A, Jacobs J, Stripp S, Happe T (2010) Multiple ferredoxin isoforms in *Chlamydomonas reinhardtii* - their role under stress conditions and biotechnological implications. Eur J Cell Biol 89:998–1004

164. Winkler M, Kawelke S, Happe T (2011) Light driven hydrogen production in protein based semi-artificial systems. Bioresour Technol 102:8493–8500
165. Wykoff DD, Davies JP, Melis A, Grossman AR (1998) The regulation of photosynthetic electron transport during nutrient deprivation in *Chlamydomonas reinhardtii*. Plant Physiol 117:129–139
166. Yacoby I, Pochekailov S, Toporik H, Ghirardi ML, King PW, Zhang S (2011) Photosynthetic electron partitioning between [FeFe]-hydrogenase and ferredoxin:NADP$^+$-oxidoreductase (FNR) enzymes in vitro. Proc Natl Acad Sci U S A 108:9396–9401
167. Yohn C MM, Behnke C, Brand A (2011) Stress-induced lipid trigger. US Patent
168. Yoon K, Han D, Li Y, Sommerfeld M, Hu Q (2012) Phospholipid:diacylglycerol acyltransferase is a multifunctional enzyme involved in membrane lipid turnover and degradation while synthesizing triacylglycerol in the unicellular green microalga *Chlamydomonas reinhardtii*. Plant Cell 24:3708–3724
169. Zauner S, Jochum W, Bigorowski T, Benning C (2012) A cytochrome b_5-containing plastid-located fatty acid desaturase from *Chlamydomonas reinhardtii*. Eukaryot Cell 11:856–863
170. Zhang LP, Happe T, Melis A (2002) Biochemical and morphological characterization of sulfur-deprived and H_2-producing *Chlamydomonas reinhardtii* (green alga). Planta 214:552–561
171. Zhang M, Fan J, Taylor DC, Ohlrogge JB (2009) DGAT1 and PDAT1 acyltransferases have overlapping functions in *Arabidopsis* triacylglycerol biosynthesis and are essential for normal pollen and seed development. Plant Cell 21:3885–3901
172. Zheng P et al (2008) A phenylalanine in DGAT is a key determinant of oil content and composition in maize. Nat Genet 40:367–372

Index

S.M. Theg, F.-A. Wollman (eds.), *Plastid Biology*, Advances in Plant Biology 5, DOI 10.1007/978-1-4939-1136-3, © Springer Science+Business Media New York 2014

Printed by Printforce, the Netherlands